CONCEPTS IN PHYSICS

THIRD EDITION

CONCEPTS IN PHYSICS

THIRD EDITION

Franklin Miller, Jr.
Thomas J. Dillon
Malcolm K. Smith

Harcourt Brace Jovanovich, Publishers
Orlando New York Chicago Atlanta Dallas

THE AUTHORS

Franklin Miller, Jr.
Professor of Physics, Kenyon College, Gambier, Ohio
Dr. Miller is also the author of *College Physics* (now in its fourth edition) published by Harcourt Brace Jovanovich, Inc. He is the producer of various teaching films in physics, and served on the staff of Harvard Project Physics.

Thomas J. Dillon
Science Department, Concord-Carlisle High School, Concord, Massachusetts
Mr. Dillon has been a teacher of high school physics for more than 20 years. He has also been an active participant in all phases of the PSSC and IPS text, laboratory, and film development.

Malcolm K. Smith
Professor of Physics and Applied Physics, University of Lowell, Lowell, Massachusetts
Formerly with the Science Teaching Center at Massachusetts Institute of Technology, Professor Smith was executive editor of the PSSC high school physics text, and has taught high school physics and mathematics for several years.

EDITORIAL CONSULTANTS

Sigmund Abeles *Consultant, Science Education, Connecticut State Department of Education*, Hartford, Connecticut*

Carl F. Berger *Professor of Science Education, Associate Dean of School of Education, University of Michigan, Ann Arbor, Michigan*

Edward Shore *Instructor of Physics-Chemistry, The Putney School, Putney, Vermont*

*The work of Sigmund Abeles on the *Concepts in Physics* program was done in his private capacity, and no official endorsement by the Connecticut State Education Department is intended or should be inferred.

Printed in the United States of America ISBN 0-15-362352-7

Preface

As in earlier editions, our primary aim is to present the clearest possible treatment of the more significant concepts of physics. The conservation laws are the overriding theme of the book, and most of the topics are developed in terms of these laws. Moreover, every effort has been made to relate the ideas and techniques of physics to their historical context and to their meaning in today's world.

Much of current science and technology depends heavily on modern physics. By eliminating some traditional topics, the intimate relationship of classical to modern physics has been made clearer. For example, new sections expand the student's view of the relationship of solid state physics to classical concepts of electricity and magnetism.

In this edition, the metric system is used exclusively throughout. All references to other systems have been removed, in keeping with the national trend to join the world community in the standardization of measurement. Students become familiar with the metric system by repeated application to everyday situations rather than by memorizing a table of conversions from one system to another.

Apart from the change to metric system, the third edition retains the features of the earlier editions. The only mathematical prerequisite is a course in simple algebra; therefore, this book can be used to help all students become familiar with the basic ideas of physics. The scientific concepts that have been emphasized are a vital key to understanding important technological applications, as well as being of great importance in their own right. The text has many points of contact with everyday life: solved problems within chapters, discussion questions, and end-of-chapter problems of various levels of difficulty. Each of these has been chosen to "bring down to earth" some physical concept.

Sections labeled "For Further Study," with accompanying problems, follow many of the chapters. The more able students should be encouraged to dip into these sections on their own, or the sections can be assigned to an entire class. The text, however, is complete and self-sufficient without this extra material. References at the end of each unit will help students explore further the history of science and contemporary developments.

Electric current is discussed consistently in terms of positive current. Electric field and potential difference are thus defined in terms of a positive test charge and the right-hand rule. Unquestionably, there is a logical advantage in viewing current as a flow of positive charges, directed down from a region of high potential to one of low potential. The majority of students using this book will probably encounter this same development in later courses and in practical applications.

Once more we would like to acknowledge our indebtedness to the teachers, colleagues, and students who have influenced this work. Of particular value have been the contributions of Carl F. Berger, University of Michigan at Ann Arbor, and Edward Shore, The Putney School, Vermont, who have offered much helpful advice. The thoughtful comments of Dr. Eric Sheldon have been extremely useful. Thanks are also due to Mrs. Yvonne Smith for preparing the original manuscript for the publishers, and to Robert Mack, Margaret Frerking, and Bruce Holtje for their assistance in preparing problem solutions.

Franklin Miller, Jr.
Thomas J. Dillon
Malcolm K. Smith

Contents

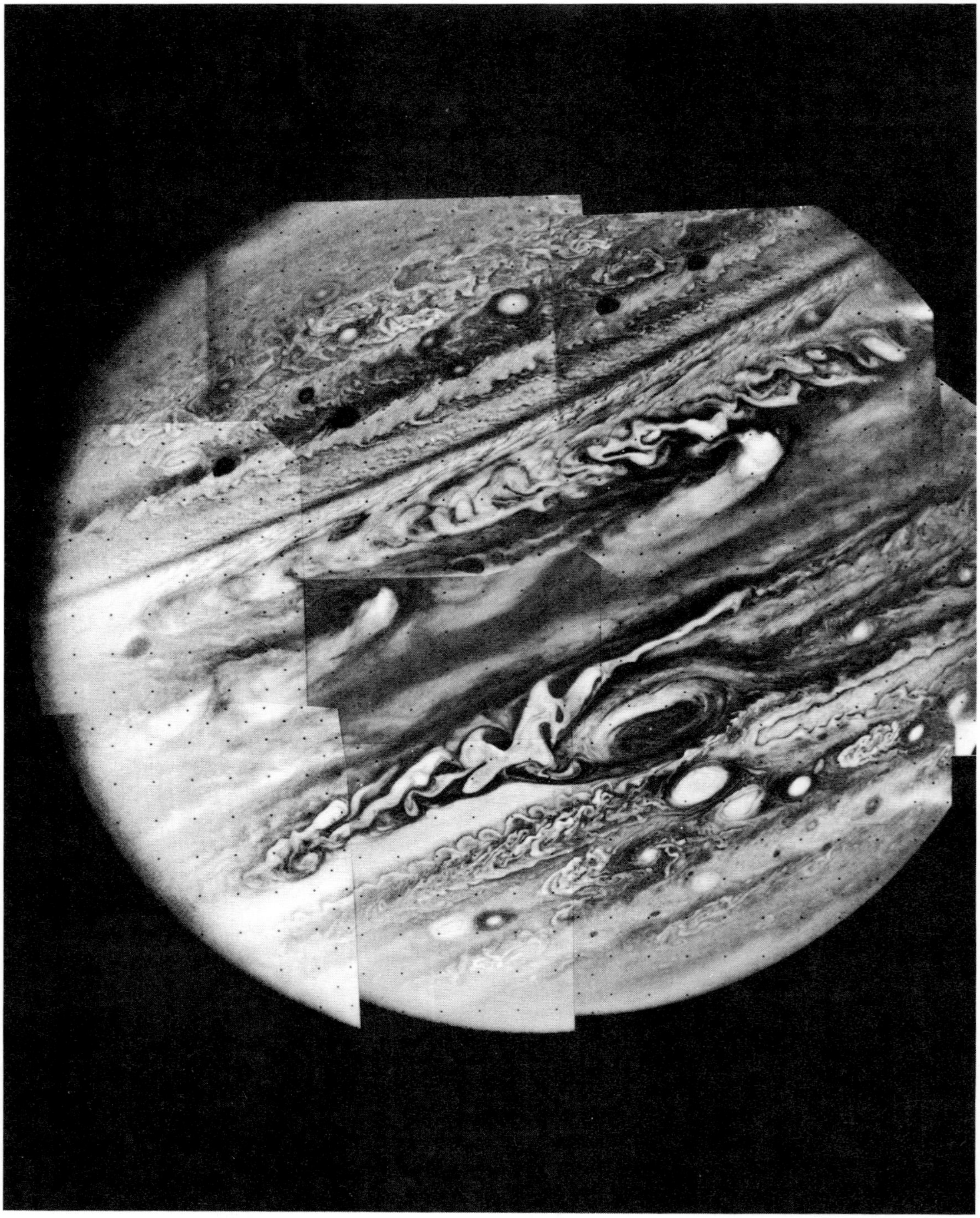

UNIT 1 INTRODUCTION TO PHYSICS

In your work in physics this year, you will engage in systematic investigations that will help you understand the universe. Like the great men and women of science before you, you will raise questions about the phenomena around you. You will carry out investigations and collect data. You will develop hypotheses. You will begin to recognize that innumerable phenomena can be explained by a surprisingly small collection of related concepts.

How can a rocket lift off from the earth? Why is the sky above us blue, while the sky on the moon is black? What is an atom? What is a star? How do pictures taken from spacecraft—such as this mosaic of Jupiter—reach us? You will be finding answers to many such questions in your study of physics.

The two chapters of this unit will serve as a preparation for your explorations. You will be introduced to some of the tools and techniques you will need. Let's begin.

1 The Science of Physics

People have always been curious about the world around them. In the modern world, science is more than curiosity. Physics and other sciences now form a major part of the background for decisions in government, business, and our individual lives.

In recent years, necessity has drawn our attention to new aspects of our world. For example, we have all been forced to think seriously about the concept of energy; the word has become familiar in everyday speech. We have come to realize that energy is an essential factor in our lives. We also recognize that the supply of energy is limited. But, in spite of the great need, we still don't know or understand all the principles that govern the production and use of this valuable quantity. We are facing difficult questions, such as: How do we most efficiently use our restricted supplies of energy? How do we go about developing new sources?

Finding the answers to these and other questions depends on careful study and research by men and women like yourselves. The answers will be found by learning more about the world around us. We say that we must discover alternate sources of energy; this can be accomplished only by people who thoroughly understand the principles of physics and other sciences.

We have already made tremendous strides in some areas, particularly in what is called technology—the application of scientific knowledge to useful devices.

A few years ago, slide rules were used in many calculations by scientists, engineers, and students. Today, slide rules have virtually disappeared, but pocket calculators are found in almost every pocket. The calculator has not just replaced the slide rule: the calculator is everywhere. People who never dreamed of owning a slide rule use a calculator to do their homework, balance their checkbooks, or total their shopping lists.

In your study of physics, the calculator will be a useful tool, making many calculations less time-consuming. But the calculator has even more significance in your study of physics. The heart of the calculator is the integrated circuit (commonly called IC). The IC is a direct application of research in the physics of solid materials.

Making ICs is a new technology that uses principles from many branches of science. Especially important are concepts of optics (the study of light) and the physics of solids. An IC is, in essence, an electronic device in which thousands of transistors and other components are packaged in a single very small "chip."

The transistor itself was invented by solid state physicists Bardeen, Brattain, and Shockley just over thirty years ago. The IC, containing many transistors, makes possible not only the calculator, but also the mini-computer, electronic market check-out, the digital clock, and many other devices. Clearly, physics is not something remote and other-worldly—it is a part of our lives.

A communications satellite is a dramatic example of space technology combined with integrated circuit electronics (Figure 1-1). Such satellites receive signals from earth stations, amplify them, and transmit them back to other earth stations at different points. Because the satellites are in orbit at great heights, the radio waves are not blocked off by the earth's curvature. Communications satellites have made international telephone calls and live television commonplace occurrences.

Intelsat V depends on many physical principles working together. The concepts of mechanics—of motions of all kinds—are used to plan the launch and determine the orbit. The concepts of electricity and magnetism are used to plan the electronic circuits. The concepts of waves determine how signals are transmitted to and from the satellite.

We have mentioned several important technological problems and applications that make use of physical principles. Your study of physics will be concerned with understanding these concepts and investigating how they are used.

Professional physicists, the men and women who work in the field, are continually trying to learn more about the world in which we live. We are still searching for answers to many questions. Some of the concepts and laws of physics that you will meet in this book have been known for centuries. But they are being revised even to this day as we make new discoveries and add new information to older ideas. Even though we have an impressive array of concepts and their applications, physics is far from "finished." We certainly do not know "all the answers." We continue to learn more and more about our world and the principles governing it.

1-1 PHYSICS AS A NATURAL SCIENCE

Aristotle, Archimedes, and even Galileo and Newton called themselves "natural philosophers." This name indicates that early scientists had all of nature as a field of study. But as scientific knowledge grew and grew, it became necessary for scientists to specialize. Even a genius could not hope to keep up with all of the discoveries and developments in astronomy, mathematics, biology, and many other fields. So it has been only in recent times that the special subject which we call physics became a distinct field of study.

Figure 1-1 Intelsat V. The solar cells of this communications satellite will provide 30 times more power than Intelsat I (Early Bird).

Physicists and chemists generally have been concerned with the more basic aspects of nonliving matter. Chemistry deals primarily with the rearrangements of the atoms that form molecules, and the analysis and synthesis of materials, whereas physics concerns itself with understanding and explaining the behavior of matter in general. The dividing line between chemistry and physics, however, can hardly be drawn in any clear fashion. A list of some scientific journals currently published in the United States serves to illustrate the overlapping areas of emphasis running from "pure physics" to "pure chemistry"—*Physical Review, Journal of Chemical Physics, Journal of Physical Chemistry,* and *Journal of the American Chemical Society.*

Like other sciences, physics has been developed through the combined efforts of dedicated workers in many countries (Figure 1-2).

Maria Goeppert-Mayer (1906-1972), Nobel laureate 1963, elaborated the shell model of the structure of the atomic nucleus.

Luis Walter Alvarez, Nobel laureate 1968, contributed to our understanding of short-lived subatomic particles.

John Bardeen, William Shockley, and Walter Brattain, Nobel laureates 1956, inventors of the transistor and pioneers in the field of solid state physics.

Chandrasekhara V. Raman (1888-1970), Nobel laureate 1930, discoverer of a light scattering effect that is a powerful tool in molecular structure and chemical analysis.

Hideki Yukawa, Nobel laureate 1949, author of the theory that explains the forces holding an atomic nucleus together in terms of a particle called the meson.

AT THE FRONTIERS OF SCIENCE

Chien-Shiung Wu, honored for her work in beta decay.

Hans Bethe, Nobel laureate 1967, theoretician, and developer of the revolutionary concept of how the stars produce energy.

Pavel A. Cerenkov, Nobel laureate 1958, discoverer of the Cerenkov effect, used in detecting and counting subatomic particles.

Research continues with the work of theoretical physicists such as Frederick Young, working in experimental laser research; and Shirley Ann Jackson, studying the physics of semi-conductors.

The great scientists of past and present always worked at the then existing frontier of science. We tend to accept Newton's laws of motion, which we use so often, as "handed down from on high," forgetting that Newton had to come to grips with nature in what was, in fact, "modern physics" in the late 1600's.

In a sense, it is true that modern physics is as old as physics itself. Yet today we give Galileo and Newton a prominent place in our study of science for more than historical reasons alone; Galileo and Newton also happened to be correct, and their methods are still of tremendous usefulness. The present-day physicist—the modern modern physicist, if you will—relies on Newtonian concepts in all work with large-scale (visible) objects moving at ordinary speeds (small compared with the speed of light). The orbits of space vehicles are exactly calculated from Newton's laws, which were formulated 300 years ago. The law of conservation of energy is still, after more than a century, a fundamental statement about nature and is used daily by research physicists and engineers. We would limit our understanding of physics if we neglected the contributions of those who went before us, and we would be unable to understand the concepts that have developed from some of the earlier laws.

1-2 DEVELOPING CONCEPTS IN PHYSICS

One objective of a physics course is to develop useful concepts. Students going on in engineering, research, or pure science will study a lot more physics after this course. Those going on in medicine, dentistry, agriculture, or geology may find only one more year of physics is necessary. However, perhaps half the students for whom this book is written do not intend to be scientists, and are studying physics for its "cultural" value. Our culture is one in which science plays an ever increasing role, and it is truly of cultural value to be able to understand a little of what goes on. Naturally, you will not be able to design a power plant as a result of diligent study of this book, but you *will* be able to appreciate some of the design problems. You will not be able to repair your TV set, but you *will* understand the basic laws of photoelectricity. Most important, you will come to appreciate the openness of scientific knowledge. What one group of men and women has discovered, another group, working independently of the first group, could have discovered just as easily. No country can have a monopoly on hydrogen bombs or TV sets. Whether you intend to be a writer, clergyman, politician, businessman, housewife, or musician, the concepts that you develop in your high school physics will help you feel at home in our present-day, science-oriented way of life.

You will be aided in your study of physics by your classroom discussions, problem-solving, laboratory program, and, of course, this book. The intent here is that all of these activities be interrelated. From the outset, you are strongly urged to *involve yourself.* A "passive" student, waiting to be taught, will miss much of the appreciation for, and excitement of, the ideas of physics.

Although you will soon notice that studying physics is a cumulative process, the basic laws of physics are remarkably few in number. This means that although this book may appear to contain many "formulas," the successful student soon realizes that most of them are simply restatements of a basic principle. For example, the "formulas" for centripetal force ($F = mv^2/r$) and rate of change of momentum ($F = mv/t$) are both restatements of the basic idea expressed by Newton's second law of motion ($F = ma$). Once you are able to regard physics in this manner (instead of listing a bunch of formulas to be memorized), you are well on your way to appreciation and understanding.

To be able to relate these concepts means that you should keep up to date in your assignments. Several study aids have been included in this book to help you. Use them. The questions and problems at the end of most chapters will help you review and consolidate your knowledge. The problems are graded in difficulty; all serve the purpose of aiding you to master *principles* through *practice.* Sections marked *For Further Study* also appear at the close of many of the chapters. These sections should give you a chance to "spread your wings" if you so desire. Don't overlook the references at the end of each unit. Many of the writings of the pioneers of physics, expressed in their own words, have a flavor and a lasting value. Their biographies are revealing and often inspiring. Through other references, you can make contact with the growing edge—the frontier—of contemporary physics.

And now let's get on with the task at hand, which is to learn the patterns, concepts, and processes of physics, a typical science. Many times during the next few months you will approach the frontier of physics, standing alongside pioneers of the past as they penetrate the barriers of their time. You will see where your contemporaries are striking forward. Let's hope physics will come alive for you as it has for them.

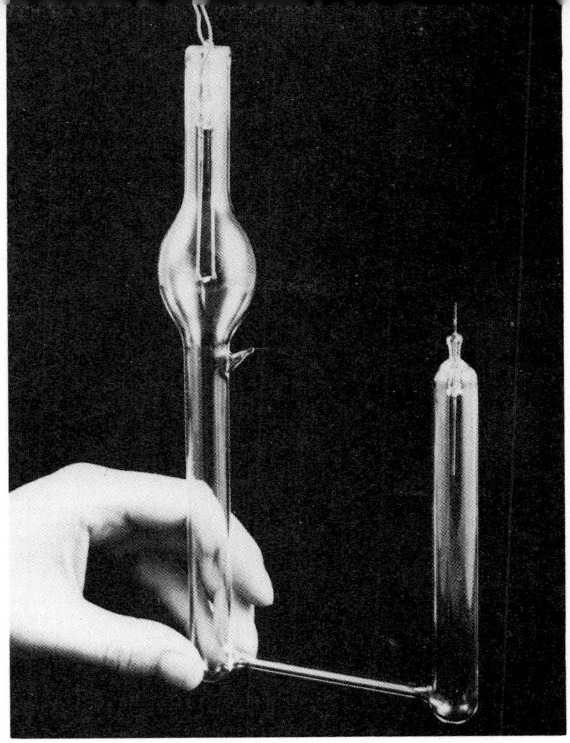

Figure 1-4 Krypton-86 lamp. A color photo of the spectrum emitted from the horizontal capillary tube of this lamp is opposite page 388.

1-3 MEASUREMENT AND UNITS IN SCIENCE

Precise measurements of physical quantities are essential to science. By the correlation of such measurements the basic laws are discovered. It is necessary, however, that measurements be made in a consistent set of units and that the basic units be well defined. For instance, if you wished to find the distance traveled by a car, you would not multiply its speed in kilometers per *hour* by the elapsed time in *minutes*. Common sense dictates that you must either convert the speed to kilometers per minute or convert the time to hours. Such quantities as speed, pressure, and area are *dimensional quantities.* A measured dimensional quantity has a numerical value dependent upon the system of units used. For example, the same area can be expressed as 9 square meters or 90,000 square centimeters.

Almost all countries of the world have now agreed to adopt a simplified form of the metric system—the International System of Units. This is abbreviated SI, to stand for the French words Système International. SI is quite similar to what was called the mks (meter-kilogram-second) system.

In SI the standard of length is the *meter* (m), which was defined in the late 1800's as the distance between two marks on a carefully protected bar of platinum-iridium alloy preserved at the International Bureau of Weights and Measures at Sèvres, France (Figure 1-3). In 1960 scientists developed a new standard. The meter is now defined as exactly 1,650,763.73 wavelengths of a certain orange light in the spectrum of a lamp containing a particular isotope of krypton (Figure 1-4).

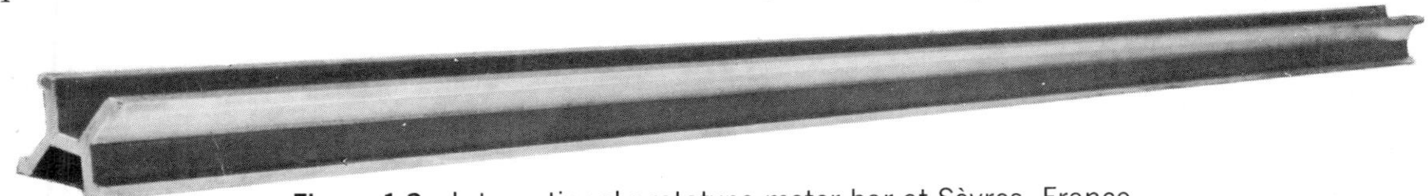

Figure 1-3 International prototype meter bar at Sèvres, France.

Figure 1-5 Prototype kilogram at Sèvres, France, showing apparatus used to protect it from atmospheric oxidation and corrosion.

For convenience, multiples and submultiples of the meter are used, such as the kilometer (1000 m), centimeter ($\frac{1}{100}$ m), and the millimeter ($\frac{1}{1000}$ m). (See Table 1-1.)

The SI unit of mass is the *kilogram,* which is now accepted as the mass of a certain block of platinum-iridium alloy known as the "international prototype kilogram" and preserved at Sèvres under controlled conditions, as shown in Figure 1-5.

When we speak now of the *mass* of a body we mean the property of matter that is compared on an equal-arm balance. Later on we shall make a distinction between *gravitational* mass and *inertial* mass. An unknown mass is measured by comparing it to a known mass (a standard) on an equal-arm balance. The two masses are balanced when the gravitational force is the same on both sides of the balance. If the balance is moved around from place to place—to a high mountain or even to the moon—the pull of the earth (or moon) will change, but the masses will still balance. From this it can be concluded that the mass of a body is a constant property, independent of where the measurement is made. The *weight* of a body, on the other hand, is a measure of the force of gravitation upon it, and is not constant. Weight can be measured with a spring scale. As the body is carried further away from the earth, the force of the earth's gravity upon it becomes less, and its weight also becomes less. Weight and mass are a common source of confusion in physics. For the present, it will suffice to remember the basic differences just mentioned. A more detailed discussion of mass and weight will come later.

Table 1-1

SI Units of Length

1 kilometer (km)	= 1000 meters (m) = 10^3 m
1 centimeter (cm)	= 1/100 m = 10^{-2} m
1 millimeter (mm)	= 1/1000 m = 10^{-3} m
1 micrometer (μm)	= 1/1,000,000 m = 10^{-6} m

A micrometer (μm) is also called a micron (μ). From this table it can be inferred that a kilogram (kg) = 1000 grams (g), a microsecond (μsec) = 10^{-6} second, etc. Exponents will be used often in this book to express both large and small numbers. (There is more about exponents in Section 2-2.)

The SI unit of time is the *second,* and indeed this unit is common to all systems of measurement and is familiar in daily life. The second was once defined as 1/86,400 of a mean solar day being based on the supposed constancy of the earth's rotation on its axis. In recent years, however, evidence has accumulated pointing to slight irregularities in the rotation of the earth on its axis; it has been found that the orbital motion around the sun is more nearly constant and therefore more suitable for defining a time unit. In 1955 the sec-

Figure 1-6 An atomic clock. A cesium beam is sent through the long tube and the vibration rate of cesium atoms is detected. This rate is used as a frequency standard to control time signals. With this type of clock the second is now defined as 9,192,631,770 vibrations of a cesium atom.

ond was redefined as exactly 1/31,556,925.9747 of a mean tropical year.* However, in 1967 this standard was replaced by one based on an "atomic clock" (Figure 1-6). These so-called clocks depend upon the vibrations of atoms. For intervals of up to a year or so, we can measure time about 100,000 times more accurately with atomic clocks than with astronomical means.

The mks system, or SI, is called an *absolute* system of units, since the meter, kilogram, and second are defined in a way that has nothing to do with local variations in the earth's gravitational field.

Occasional reference will be made to another system within the SI system: the *centimeter-gram-second* (cgs) *system* (which you have probably made use of in your chemistry classes). The cgs system differs from the mks system only in using the centimeter and the gram, which are submultiples of the meter and the kilogram. It, too, is an absolute system of units.

It is interesting to note the historical basis of the metric units. When the French authorities established the metric system in 1791, they intended that the meter should be exactly 1 ten-millionth of the distance from the equator to either pole; and that the gram should be the mass of one cubic centimeter of pure water at 4°C (a temperature at which the density of water is relatively independent of temperature change). These goals, however, were not quite realized, and so the prototype meter and kilogram do not have exactly the intended values.

The adoption of the metric system by the French revolutionaries was a manifestation of the Age of Reason. They wished to establish a "rational" basis for every aspect of life, and the decimal system fitted into such a scheme. In 1795 the *centime* was adopted as a monetary unit (it should properly have been called the "centi-franc"). The division of the right angle into 100 equal parts, instead of 90 degrees, was proposed, but never gained wide acceptance. The revolutionary calendar had months of 30 days, divided into three 10-day "decades" instead of 7-day weeks. This calendar was unpopular also, and was abolished by Napoleon in 1805.

* This is the year as used in the ordinary sense of the word, but since the length of the year is slowly changing at a known rate, the definition is made in terms of the earth's motion as of January 1, 1900.

2 Some Essential Mathematics

In your study of physics, you will apply the mathematics you already know to new situations. An advanced knowledge of mathematics is not required for this course, but, rather, a general knowledge of high school algebra and some aquaintance with similar triangles. Much of the material of this chapter should, then, be familiar to you. Emphasis will be not on what mathematics you have "had," but rather to what extent you are able to use it. For example, you probably know that if you plot y versus x, the equation $y = 3x$ will give you a straight line, while the equation $y = 3x^2$ gives some sort of a curve. You know this because you have "had" it in your mathematics course. What sort of a graph will you get if you plot y versus x^2 in the last equation? If you can see right away that this, too, must be a straight line, then you understand the main point of this discussion.

The material that follows will present some mathematical techniques that will make physics more understandable and more meaningful.

2-1 EQUATIONS AND FORMULAS OF PHYSICS

In Section 1-3, units and dimensions were discussed briefly, and it was shown that a dimensional quantity could have different numerical values in different systems of units. For example, the speeds for the cars in Figure 2-1—72 km/hr, 1200 m/min, and 2000 cm/sec—all represent the same value. They all have dimensions of speed (a distance divided by a time), but they are expressed in different units. In mathematics you learned not to add "dollars to doughnuts" if you wanted a meaningful answer. In physics, if two quantities are to be added or subtracted, they must also be expressed in the same units and, of course, quantities on both sides of an equals sign must have like dimensions. Consider the example below.

Example 2-1 A student derives an equation for the distance traveled by a falling body:

$$s = \tfrac{1}{2}v_f t^2 + v_i t$$

where the subscripts $_f$ and $_i$ are used to identify different values of the velocity v. In this equation, s is the distance fallen, v_i is the initial velocity of the body, t is the elapsed time, and v_f is the final velocity of the body just before it hits the ground. Is the equation possibly correct?

If we substitute mks units for distance, velocity, and time and ignore the $\frac{1}{2}$, which has no dimension, we have

$$\text{m} = \frac{\text{m}}{\text{sec}} \times (\text{sec})^2 + \frac{\text{m}}{\text{sec}} \times \text{sec}$$

Units are combined just like ordinary

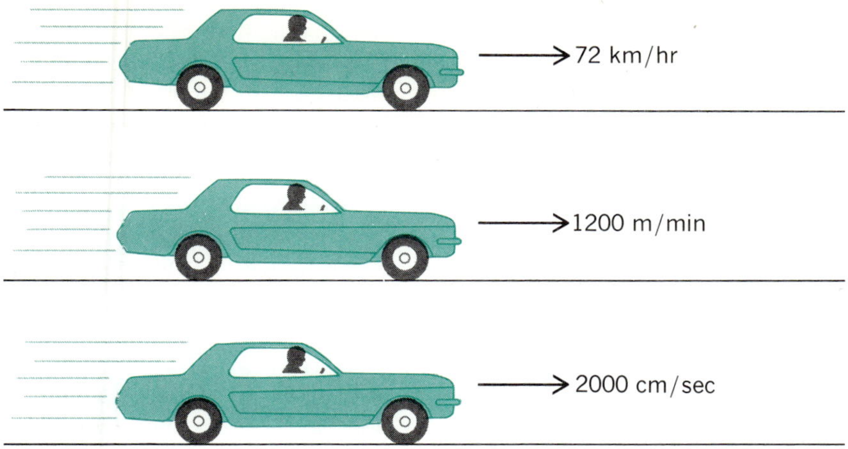

Figure 2-1 Can you determine which car is traveling at the fastest speed?

algebraic expressions. Canceling gives us

$$m = \frac{m}{\cancel{sec}} \times (sec \times \cancel{sec}) + \frac{m}{\cancel{sec}} \times \cancel{sec}$$

$$m = m \times sec + m$$

which is not true, so the equation must be wrong. How to make it correct is another matter, something we will take up later.

Cancellation of units can be a great help in testing an equation or formula for possible error. One should be careful, however. Just because an equation is *dimensionally* correct does not mean that the equation *is* correct. This can be illustrated in the following example.

Example 2-2 The time it takes an earth satellite to make one complete orbit depends on its distance from the earth's center. Check the following equation for possible correctness.

$$T = \sqrt{\frac{\pi^2 R^3}{4mG}}$$

where T is the period of the satellite (time for one complete orbit), in seconds; R is the distance, in meters; m is the mass of the earth, in kilograms; and G is a universal constant, in $m^3/kg \cdot sec^2$.*

Notice in this problem, as well as in the previous one, that it is not necessary to understand the meaning of the variables or how they are related in the situation being described. For the purpose of the problem, checking the dimensional consistency of an equation, it is only necessary to know the units in which each quantity denoted by a letter is measured.

Substituting the units for the various quantities, we find that the check is successful.

$$sec = \sqrt{\frac{\cancel{m^3}}{\cancel{kg} \times \cancel{m^3}/\cancel{kg} \cdot sec^2}}$$

$$= \sqrt{sec^2}$$

$$sec = sec$$

The equation is dimensionally correct. But this does not mean that the equation is correct. As a matter of fact, it is wrong because the numerical factor $\pi^2/4$ is incorrect. (As you will see later, the correct numerical factor for this expression is $4\pi^2$.) Cancellation of units never gives information about numerical constants. In spite of this, cancellation of units is a very powerful tool when properly used.

Let's use these two examples as an occasion for some remarks about how an equation should be written and how the information given in an equation can be used.

* Often, units will be presented in this form to save space. Thus $m^3/kg \cdot sec^2$ is understood to mean $\frac{m^3}{kg \cdot sec^2}$.

(a) The letters are *symbols* that stand for certain observable quantities. Usually, but not always, they are chosen to remind us of the name of the quantity, such as v for velocity, t for time, r for radius, etc. Unfortunately, the demand for symbols is so great that it is necessary to use capital letters as well as small letters (as with R in Example 2-2), subscripts (such as v_i and v_f in Example 2-1), and even Greek letters (as μ for micron in Table 1-1).

(b) Careful study of the symbols used in these examples shows two sorts of quantities: *variables,* such as v, t, and R, and *constants,* such as $\frac{1}{2}$, π^2, 4, m, and G. There is no automatic way of telling constants from variables other than knowing the exact meaning of the symbols. Often a subscript is used to single out some particular value of a variable (such as the small $_f$ after v, signifying the "final" value of v that is eventually reached in Example 2-1).

(c) Notice that m and G are constants with dimensions, while $\frac{1}{2}$, π^2, and 4 are numerical constants without dimensions. Indeed, G is properly called a constant of proportionality whose dimensions are chosen so as to balance the units in the equation. (You will find out more about this constant in Section 7-5.)

(d) With some justice, these relations may be called *equations* (since they both contain equals signs), or they may be called *formulas* (since the desired quantities, s or T, are all by themselves and can be calculated by substituting values for the other factors), or *laws* of physics. Perhaps the designation "formula" fits best in these cases, since the term "law" is generally reserved for statements of more general importance.

2-2 POWERS OF TEN, SCIENTIFIC NOTATION, AND ORDERS OF MAGNITUDE

Our galaxy is about 1,000,000,000,000,000,000,000 meters across, whereas nuclear particles have diameters of around 0.00000000000001 meter. Man is somewhere in between—about 2 meters tall. Considering the great differences among physical magnitudes, it is not surprising that we look for a more convenient way to express these very large and very small numbers.

Any number can be expressed as some power of 10 multiplied by another number between 1 and 10. Thus,

$$\begin{aligned} 4163.8 &= 4.1638 \times 1000 \\ &= 4.1638 \times 10^3 \end{aligned}$$

Moving the decimal point 3 places to the left decreases the number by a factor of 1000, or 10^3, so it is multiplied by 10^3 to keep the value the same. As another example,

$$\begin{aligned} 0.0000367 &= 3.67 \times \frac{1}{100{,}000} \\ &= 3.67 \times \frac{1}{10^5} \\ &= 3.67 \times 10^{-5} \end{aligned}$$

Here the number is increased by a factor of 10^5 by moving the decimal point 5 places to the right, so it becomes necessary to divide by 10^5 to keep the same value. The intermediate steps have been written down in these two examples, but in actual practice you should be able to write the final result directly. Thus,

$$\begin{aligned} 765{,}000 &= 7.65 \times 10^5 \\ 1320 &= 1.32 \times 10^3 \\ 0.034 &= 3.4 \times 10^{-2} \end{aligned}$$

Can you rewrite the galactic and nuclear diameters given at the beginning of this section in the form above?

This shorthand way of writing numbers is often called *scientific notation.* Aside from its convenience, scientific notation is often more informative than the ordinary method of writing numbers. When we say that a light-year is 9,460,000,000,000,000 meters, we immediately see that this length unit is a tremendous number of meters, but just how tremendous we don't immediately comprehend. Writing it as 9.46×10^{15} meters allows us to make immediate comparisons with other large numbers.

Multiplication and division. To multiply numbers expressed in scientific notation, we combine powers of 10 by addition of exponents and use ordinary multiplication for the numbers preceding the powers of 10. To divide such numbers, we subtract exponents and use division for the numbers preceding the powers of 10. These techniques are shown in the following examples:

Multiplication:

$$(2.5 \times 10^8)(3 \times 10^6) = (2.5 \times 3) \times 10^{8+6}$$
$$= 7.5 \times 10^{14}$$

Division:

$$\frac{42 \times 10^{-16}}{7 \times 10^{10}} = \frac{42}{7} \times 10^{-16-10}$$
$$= 6 \times 10^{-26}$$

$$\frac{(2 \times 10^6)(8 \times 10^{11})(3 \times 10^{-4})}{(4 \times 10^{-5})(10^3)} = \frac{(2)(8)(3)}{(4)(1)} \times \frac{10^{13}}{10^{-2}}$$
$$= 12 \times 10^{15}$$
$$= 1.2 \times 10^{16}$$

Sometimes it pays to revise a numerator or denominator to make the division easier.

$$\frac{1.8 \times 10^{16}}{9 \times 10^{-4}} = \frac{1.8 \times 10^{16}}{0.9 \times 10^{-3}}$$
$$= 2 \times 10^{19}$$

or

$$\frac{1.8 \times 10^{16}}{9 \times 10^{-4}} = \frac{18 \times 10^{15}}{9 \times 10^{-4}}$$
$$= 2 \times 10^{19}$$

Addition and subtraction. As in algebra, only *like* exponents can be added or subtracted. So if we wanted to add

$$\begin{array}{r} 2.30 \times 10^3 \\ +5.25 \times 10^4 \end{array}$$

2.30×10^3 must be changed to 0.23×10^4. We then have

$$\begin{array}{r} 0.23 \times 10^4 \\ +5.25 \times 10^4 \\ \hline 5.48 \times 10^4 \end{array}$$

Many times when dealing with large and small numbers we are only interested in the power of 10 and not its numerical coefficient. In such instances it is often sufficient to get just an approximate value of the number. For example, to get an idea of the mass of the proton, we would say that it is "on the order of" 10^{-27} kg. The added information that it is more nearly 1.7×10^{-27} kg in no way changes our comprehension. When we use numbers in this way we are speaking of the *order of magnitude* of the number. Thus, the order of magnitude of 1673 is 10^3; that is, 1.673×10^3 is closer to 10^3 than it is to 10^2 or 10^4. The order of magnitude of 0.009 is 10^{-2} because $0.009 = 9 \times 10^{-3}$, which is very close to $10 \times 10^{-3} = 10^{-2}$. An order-of-magnitude calculation can be a very helpful tool where only an approximate answer is desired.

2-3 SIGNIFICANT FIGURES

A number that represents the result of a physical measurement does more than merely signify the value of the measured quantity. Some indication of the *precision* of the measurement is given by the form in which the number is written. For example, if a car's speed is reported by observer A to be 45.3 km/hr, this might mean that the speed is somewhere between 45.25 and 45.35 km/hr; for if it were 45.36 km/hr, for instance, he would report it as 45.4. Another observer, *B*, might assign a greater uncertainty to the last figure. He might consider that 45.3 indicates a number between 45.1 and 45.5, or within some other range of a few tenths of a kilometer per hour on either side of the stated value of 45.3 km/hr.

A *significant figure* is one that is of *some* significance but does not necessarily denote a certainty. In the example just mentioned, the 3 of 45.3 is "significant," but the exact degree of significance depends on the experimental techniques, habits, and temperaments of observers *A* and *B*. It is difficult, therefore, to make any simple statement about significant figures other than to lay down the rule that *the last figure written down should be the first uncertain figure.*

When we read that the velocity of light is 2.9979×10^8 m/sec, we know that the last 9 is uncertain, but just how uncertain we cannot tell.

On the other hand, the 7 is certain. It would not be correct to write 2.99794×10^8, *knowing* that the last 9 is uncertain, for then the 4 would be entirely meaningless. This would be like trying to balance a checkbook to the nearest penny when you are uncertain about the dollars or dimes.

In your physics laboratory work you will usually strive for "three-figure accuracy." The density of iron, for example, is 7.86 g/cm^3, and writing it in this way indicates that the 6 is uncertain by some (unspecified) amount. If the apparatus is not capable of three-figure accuracy, then the answer must be rounded off as necessary, to 7.9 g/cm^3 or even to 8 g/cm^3.

The problem of significant figures can be illustrated by a hypothetical experiment in which two observers measure the time required for a stream of water to fill a bucket. Observer *A* gets 98 seconds; observer *B* gets 102 seconds. Each measures to the nearest second, so the last figure in each of the measurements is "significant." However, *A*'s result has two figures, and *B*'s has three. A simple criterion of "two significant figures" or "three significant figures" would be misleading here. A trained physicist, after taking several measurements, might report his result to be 98.1 sec ± 2.2 sec, meaning that 98.1 is his best estimate of the value and that the correct value is probably within 2.2 seconds of 98.1 seconds. This does not follow our rule that the last figure written down is the first uncertain figure, but by stating the probable error of 2.2 seconds the observer makes clear just *how certain* his value is, without any mention of significant figures at all.

Significant zeros. A special problem is posed by zeros, for zeros may be used either to indicate the location of the decimal point or to indicate a value. In the following examples, only the zeros that are underlined are significant; the other zeros are used to fill out the space to the decimal point and are not significant figures.

4.2$\underline{0}$1

4.2$\underline{0}$1$\underline{00}$ Here the final two zeros need not have been written; hence the fact that they *are* written means that they are significant. The uncertainty is in the fifth decimal place.

0.0042$\underline{0}$1$\underline{0}$

42.$\underline{0}$1

A number such as 86,400 grams is ambiguous, because it does not tell whether the two zeros merely locate the decimal point (in which case the error might be 100 or 200 grams) or whether they are really significant (placing the true value within several grams of being *exactly* 86,400 grams). There are several ways in which you could avoid this ambiguity. (a) Use scientific notation, and write 8.64×10^4 g *or* 8.6400×10^4 g to distinguish between the two possibilities. (b) Use a different unit, which will place the decimal point elsewhere. Write 86.4 kg *or* 86.400 kg to distinguish the two cases. (c) Use a probable error or a limit of error. Thus 86,400 ± 8 g is a definite statement, provided the precise meaning of the author's ± sign is known. In this book, methods (a) and (b) are preferred as useful compromises that are sufficiently precise for most purposes.

Arithmetic operations and significant figures. Consider the circumference of a circle whose diameter is 12.3 cm. This quantity can be determined by multiplying 12.3 by π, which is 3.1416. If the 3 of 12.3 is uncertain, then all numbers following the 3 are unknown and can be denoted by x. Carrying out the multiplication we get,

$$\begin{array}{r}
12.3x \\
3.1416 \\
\hline
738x \\
123x \\
492x \\
123x \\
369x \\
\hline
38.64168x
\end{array}$$

The product should be rounded off to 38.6, since the 4, 1, 6, and 8 are really *unknown* because an unknown x has been added to each of these digits. Even the 6 of 38.6 is uncertain because it depends on a 9 arising from the uncertain 3 of 12.3. Writing the answer as 38.6 cm expresses this correctly; the last figure written down is the first uncertain one.

Similar results are obtained for division.

Addition and subtraction behave somewhat differently. For example, an orange of mass 102.7 grams is added to a basket containing 512.83 grams of lemons and 1620.2 grams of grapefruit. The total mass is 2235.7 grams of citrus fruit:

$$\begin{array}{r} 102.7\,xx \\ 512.83x \\ \underline{1620.2\,xx} \\ 2235.7\,xx \end{array}$$

We can generalize as follows: In multiplication or division, the result has the same number of significant figures as the least accurate number entering into the multiplication or division. In addition or subtraction, the last significant figure of the result is in the same column relative to the decimal point as is the last significant figure of the least accurate number in the example.

These are approximate, rough-and-ready rules. The important thing to realize is that when you are working with numbers that are the result of a *physical measurement,* you should know the limitations of the measuring instruments. Measuring devices vary greatly in their degree of accuracy, as you can see by looking at those illustrated in Figure 2-2. It makes no sense at all to carry out the computation to five figures when your measuring instruments are only good to three.

Remember that numbers that are not the result of a physical measurement have unlimited accuracy. So if the mass of a metal cube was measured to be 0.495 kilogram, then the mass of 6 identical cubes would be 0.495 kg $\times$ 6 = 2.97 kilograms. In this case the 6 is the result of counting—it is an exact number—and, for the purposes of this problem, could be expressed as 6.00.

Significant figures in textbooks. Ideally, problems in a textbook should be formulated with strict attention to probable errors and significant figures. In the examples and problems in this text, the numbers are assumed to be *precise enough* to

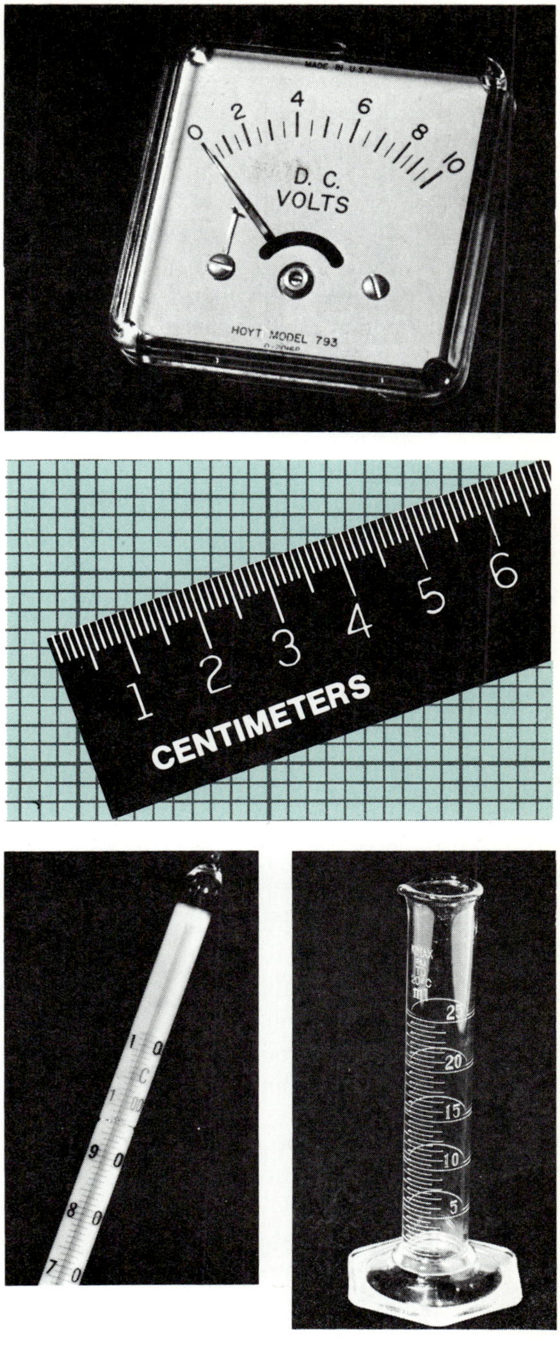

Figure 2-2 Some laboratory measuring devices. To how many significant figures can each of these be read?

yield a final answer having three significant figures, and final significant zeros are often omitted both from the statements of problems and their answers. Some readers may object, but this course of action can be justified on the following grounds:

(a) The rules about significant figures are themselves not rigid. Exceptions can easily be found, and some have already been illustrated. (Does 98 have the same number of significant figures as 102?) The three-figure rule used in this book is clearly stated and is definite enough to apply consistently.

(b) Examples and problems—hundreds of them—are scattered through the text to help you understand the principles of physics. Their purpose is *not* primarily to give a model for recording experimental data in the laboratory. Your laboratory manual will, rightly, be concerned with this aspect of the course, but in the text a working rule has been adopted for the sake of giving a simpler illustration of laws and principles.

(c) Where it serves a purpose, as in tables of certain properties of matter that have been observed experimentally, the data are given with the number of significant figures indicating (roughly) the precision of measurement.

(d) There is real value in simplifying the routine arithmetic associated with problem-solving. This is why "simple" numbers are often used. On the other hand, you probably want to check your work against something more definite than an answer that has been rounded off to one or two significant figures. When answers are given to three significant figures, it is assumed that the given data warrant such precision.

2-4 FUNCTIONS AND GRAPHS

The concept of "function" perhaps summarizes the spirit of scientific investigation. A function is a *relation*, or a *correlation*, between observables. In physics we seek to correlate cause and effect. Often we try to relate our observations of complex phenomena to our observations of factors that are under our control. One of the major tasks in studying physics is to state these relations in a concise and convenient form—because these statements become the laws of physics.

These relations can be expressed in several ways; the first, and perhaps the simplest, is called *direct proportion.* As an example, you will investigate the relation between the velocity a steel ball acquires when it is dropped from rest and the length of time it is falling. When this experiment is performed near the surface of the earth, you will find that after 1 second of fall the ball has a velocity of 9.8 m/sec; after 2 seconds the velocity is 19.6 m/sec; and at the end of 3 seconds the velocity is 29.4 m/sec. Continuing in this manner you will find that for each added second of fall the velocity increases by an equal amount—9.8 m/sec. You can conclude that the velocity is directly proportional to the time, and write this relation as

$$v \propto t$$

where the symbol $\propto$ can be read "is proportional to." Saying that the velocity is directly proportional to the time, or "varies as" the time, means that for every change in the velocity there is a corresponding change in the time: doubling the velocity doubles the time, tripling the velocity triples the time, etc. The statement $v \propto t$ can be made into an equation by replacing the proportionality symbol $\propto$ with an equals sign and some constant number that we shall call k. Thus the statement $v \propto t$ and the corresponding equation $v = kt$ are equivalent, and k is called the constant of proportionality.

Direct proportions and constants of proportionality are not new to you. You have used them often. When you write $C = \pi d$ to find the circumference of a circle, you are really saying that the circumference is directly proportional to the diameter, and in this case the proportionality constant has the value π. The expression for the area of a triangle, $A = \frac{1}{2}bh$, says that the area is directly proportional to both the base *and* the height with $\frac{1}{2}$ as the constant of proportionality. Sometimes the proportionality constant has the value 1, as can be seen in the formula for the area of a rectangle, $A = (1)(bh) = bh$. If we rewrite the equation for the falling ball, $v = kt$, in the form

$$v/t = k$$

the value of k is seen to be equal to the constant ratio v/t. By putting in any pair of observed values for v and t (for example, when $v = 19.6$ m/sec, $t = 2$ sec) we see that $k = 9.8$ m/sec².* Now that we know the value of the proportionality constant we can calculate the velocity for *any* time t. For example, when $t = 1.5$ sec,

$$\begin{aligned} v = kt &= 9.8 \text{ m/sec}^2 \times 1.5 \text{ sec} \\ &= 14.7 \text{ m/sec} \end{aligned}$$

The second way of expressing such relations is called *inverse proportion*. You will learn later that if the temperature of a gas is held constant, the pressure and volume of a given amount of the gas are related as follows,

$$P \propto \frac{1}{V}$$

where P is the pressure and V represents the volume. This statement tells us that the pressure is inversely proportional to the volume of the gas. When the pressure is increased, the volume decreases; when the pressure is decreased, the volume increases.

The equivalent statement $P = k/V$ tells you nothing more unless you know the value of the proportionality constant k. This can be found from experiment by measuring a pressure and a corresponding volume. Since both direct and inverse proportions are used many times throughout physics, it is important that you understand them.

* The actual quotient here is $\frac{\text{m/sec}}{\text{sec}}$, which is read "meters per second per second." This is sometimes written as m/sec/sec and also m/sec². These are all equivalent, but m/sec² appears to be the most convenient and is therefore used more often than the others.

Example 2-3 If a mass, starting from rest, is allowed to slide down a smooth inclined plane, the distance it slides is directly proportional to the square of the time it has been sliding. If it slides 0.3 m during the first 2 sec, how far will it slide during the first 4 sec?

The first part of the problem tells how the distance and the time are related. Specifically,

$$s \propto t^2$$

where s is the distance and t is the time. When this proportion is rewritten as

$$s = kt^2$$

k can be found by substituting the pair of values given.

$$0.3 \text{ m} = k \times (2 \text{ sec})^2$$

$$k = (0.3 \text{ m})/(2 \text{ sec})^2 = 0.075 \text{ m/sec}^2$$

This value of k numerically relates the distance and the time for *all* values of s and t.

$$s = 0.075 \text{ m/sec}^2 \times t^2$$

In particular, when $t = 4$ sec,

$$\begin{aligned} s &= 0.075 \text{ m/sec}^2 \times (4 \text{ sec})^2 \\ &= \boxed{1.2 \text{ m}} \end{aligned}$$

Example 2-4 The force that keeps a planet in its orbit around the sun is directly proportional to the mass of the planet and inversely proportional to the square of its distance from the sun. Jupiter is 320 times more massive than the earth and 5.2 times farther from the sun. Find the ratio of the force that holds Jupiter in orbit to the force that holds the earth in orbit.

The relation among the variables is

$$F = km/R^2$$

where F is the force, m is the mass of the planet, and R is the distance from the sun. There are no values given for F, m, and R, so we cannot find the numerical value of the constant of proportionality k. But it is not really needed because we are asked to

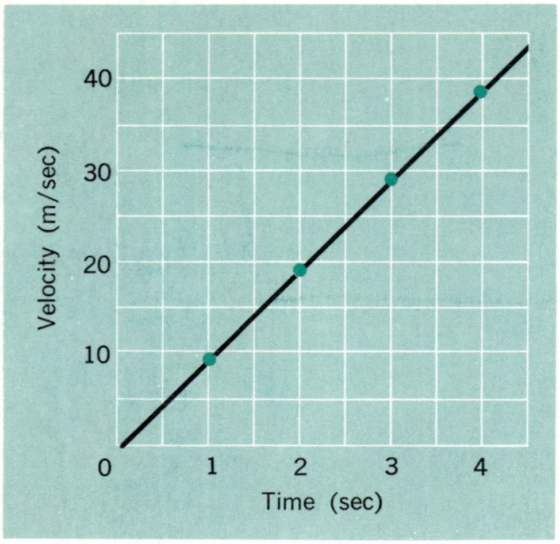

Figure 2-3 Velocity-time graph of a steel ball dropped from rest.

find a *ratio* of forces, not a numerical value for a particular force.

Referring to the preceding relation and noting the factors given in the problem for Jupiter, if the mass is increased by a factor of 320 and the distance increased by a factor of 5.2; then, for Jupiter, the right side of the expression is increased by a factor of $320/(5.2)^2$, so the left side (the force) must be increased by the same factor. Therefore, the force on Jupiter is $320/(5.2)^2 = 11.8$ times as much as the force on the earth. That is, the ratio of the forces is 11.8 to 1. You can see from this problem that knowledge of the constant of proportionality is not always necessary.

We have been examining ways of expressing functions in mathematical form. Another technique that is often used to display the behavior of functions is the *graph*, which is a straight line or curve connecting a series of plotted points. A good graph must be plotted carefully, but it is even more important to choose scales wisely and label the axes clearly. The table of values, at the lower left-hand column of this page, is taken from the observations made in our earlier experiment with the falling ball.

Time (sec)	*Velocity* (m/sec)
0	0
1	9.8
2	19.6
3	29.4
4	39.2

Note the features of the graph of these data shown in Figure 2-3.

1. A table of values is prepared showing the significant data for determining the locations of the points to be plotted.
2. Axes are well labeled, including units of measure.
3. On the velocity axis 2 squares = 10 m/sec. (Multiples such as 3 squares = 10 m/sec should be avoided. They make it hard to plot points and also hard to read values off the graph.)
4. Consistent with item 3, the graph is plotted to fill up as much of the allotted space as possible.
5. Plotted points are located by round dots shown in color.

The purpose of a graph is to give information about a relation. For instance, you can see from this graph that the velocity changes uniformly with time. This is called a *linear relation* (from the fact that the graph is a straight line). This is just another way of saying that the velocity is directly proportional to the time, $v = kt$.

As you can see, this is the same relation that was indicated earlier upon inspection of the data. You may argue that just by a quick look at the table of data above, it can be seen that velocity and time are proportional and that the graph will be a straight line. This may be true in the example we are discussing, but it is not always so. Graphing functions will prove to be one of the most useful tools you will have in your study of physics.

Even more information can be obtained from the graph in Figure 2-3. You have already noted that the velocity is changing uniformly with time. Is it possible to tell at what *rate* the velocity is

Figure 2-4 This is the same graph shown in Figure 2-3. The slope is found by dividing the increase in velocity, the "up" distance, by the corresponding increase in time, the "over" distance.

Figure 2-5 The time axis has been compressed so that the graph makes a steeper angle with the horizontal, but the slope of the line is the same as in Figure 2-4.

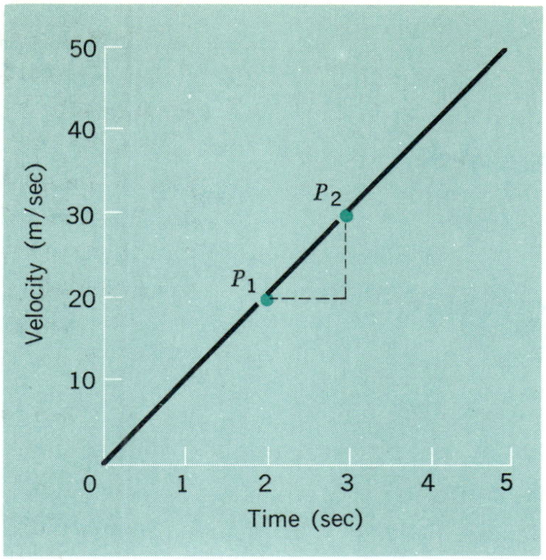

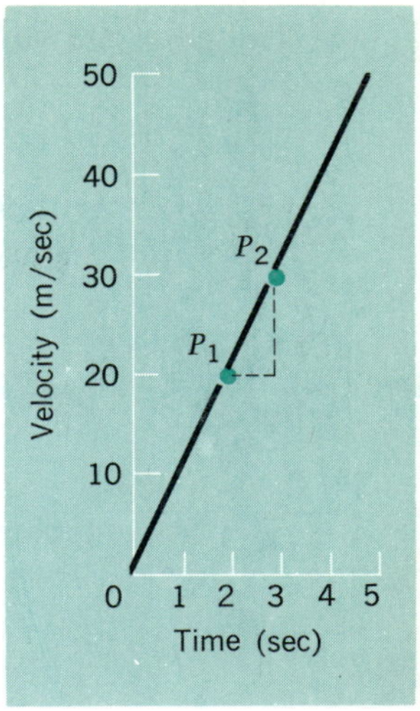

changing? The answer is yes. This information comes from the steepness of the graph. In Figure 2-4 the graph of Figure 2-3 has been reproduced with *any* two points P_1 and P_2 marked on the plotted line. The horizontal dashed line represents how much more "over" to the right P_2 is than P_1, and the vertical dashed line represents how much more "up" P_2 is than P_1. We now define the steepness of the graph as the ratio of the "up" distance to the "over" distance and call this ratio the *slope* of the line. If two different points had been chosen in Figure 2-3, the lengths of the "up" and "over" distances would have been different, but the *ratio* of the distances would always be the same. (Try it.) In other words, the slope of a straight line is the same everywhere along it.

As you can see from this graph, the "up" distance has a "length" of 9.8 m/sec and the "over" distance has a "length" of 1 second. The slope, therefore, is 9.8 m/sec divided by 1 second, or 9.8 m/sec^2. This, you will recognize, is the same value we got for k earlier, when we wrote $v = kt$ and substituted a pair of values for v and t. The slope of a straight line *is* the proportionality constant between the two factors that are being graphed. It is a measure of their constant ratio.

It is important to note that the angle the line makes with the horizontal axis has nothing to do with the slope as defined here. The angle is arbitrary and depends only on the scale to which the graph is plotted. A comparison of Figures 2-4 and 2-5 illustrates this. It should also be noted that we have been discussing the slope of a *straight* line only. In Chapter 3 we shall extend our definition of slope to include curved lines.

2-5 APPROXIMATE CALCULATIONS

Often a rough idea of the result of a calculation can give important insight into a physical problem. We can make a quick estimate when an exact solution is not needed. Quick approximate calculation also helps guard against large errors resulting from a misplaced decimal point.

"One-figure accuracy" or even worse is usually adequate for such rough calculations. Everything can be rounded off to one figure, the digits combined, and then the decimal point placed, combining whatever powers of ten are left over.

Example 2-5 How many seconds are there in a year?

$$(365.24)(24)(60)(60)$$
$$\approx (4 \times 10^2)(2 \times 10)(6 \times 10)(6 \times 10)$$
$$= (4)(2)(6)(6) \times 10^5$$
$$= (8)(36) \times 10^5$$
$$\approx (8)(40) \times 10^5$$
$$\approx 300 \times 10^5$$
$$= \boxed{3 \times 10^7}$$

The exact answer, to four significant figures, is 3.156×10^7 sec. Don't be alarmed at setting (8)(40) equal to 300. To be sure, 320 is an exact value for (8)(40), but we are only interested in a quick and easy estimate. ($\approx$ means approximately equals.)

Example 2-6 How many years are there in one 50-minute class period?

$$\frac{50 \text{ min}}{(365.24)(24)(60) \text{ min}} \approx \frac{\cancel{50}}{(4 \times 10^2)(20)(\cancel{60})}$$
$$\approx \frac{1}{80 \times 10^2}$$
$$= \frac{100 \times 10^{-2}}{80 \times 10^2}$$
$$= \boxed{1.2 \times 10^{-4}}$$

A more exact value is 0.951×10^{-4} year. Several points should be noted in this example.

(a) 50 and 60 were canceled just as if they had been equal. This is not right, but not far wrong either. Any short cut or approximation of this sort is permissible in a rough calculation.

(b) To divide by 80 with a minimum of pain, the 1 in the numerator was changed to 100×10^{-2} (really the same thing). Then 100/80 is easily seen to be just a bit more than 1—call it 1.2.

(c) The answer 1.2×10^{-4} is not very close to the correct value of 0.951×10^{-4}. However, if it were a question of deciding between 0.951×10^{-4} and 0.951×10^{-3}, it is easy to see from our approximate answer that the former is correct. Any approximate answer between, say, 0.7×10^{-4} and 2×10^{-4} would allow us to fix the decimal point, if an exact calculation gave an answer of the form $0.951 \times 10^{-?}$.

Example 2-7 What is the volume of a droplet of water in a fog if the radius of the droplet is 2.73×10^{-3} cm?

Using the formula for volume of a sphere, we have

$$V = (\tfrac{4}{3})\pi r^3$$
$$= (\tfrac{4}{3})\pi (2.73 \times 10^{-3})^3$$
$$\approx (\tfrac{\cancel{4}}{\cancel{3}})\cancel{\pi} (3 \times 10^{-3})^3$$
$$\approx 4(27 \times 10^{-9})$$
$$\approx 100 \times 10^{-9}$$
$$= \boxed{1 \times 10^{-7} \text{ cm}^3}$$

A more exact value for the volume is 0.85×10^{-7} cm^3.

Answers given by pocket calculators are usually misleadingly precise; they must be rounded off by considering the precision of the input data.

SUMMARY

Note: Key concept words and phrases appear in heavy print throughout the summary sections. These words should be noted carefully because they provide the foundation for your future growth in physics. Many of these words cannot be defined briefly; one's understanding of them broadens as one observes them in various contexts.

You have been looking at some ways in which mathematics can help in your study of physics, and have observed that it is necessary to be consistent in the choice of units when using equations and formulas. You also saw how the cancellation of units is often useful in checking equations for possible error.

Scientific notation was introduced as a convenient way of writing large and small numbers. Sometimes an order-of-magnitude estimate is all that is necessary when only an approximate result is desired.

The accuracy of numbers that are the result of a physical measurement is determined by the limitations of the measuring devices. No attempt has been made to give a rigorous treatment of **significant figures,** and it is enough to remember that calculations with such numbers are only significant to the extent of the significant figures in the data.

The graphing of experimental data often helps to point out the **relation** between the variables. The graph of a direct proportion is a straight line. The slope of the straight line is the **constant of proportionality** between the two variables which can be found by taking any two points on the graph and dividing the vertical increase between the two points by the horizontal increase.

QUESTIONS

1 List all the formulas you can think of from your mathematics courses that involve areas and volumes. Can you show, by substituting units for the symbols, that they are dimensionally correct?

For example, for the area of a trapezoid,

$$\begin{aligned} A &= \tfrac{1}{2}h\,(b_1 + b_2) \\ &= \text{cm}(\text{cm} + \text{cm}) \\ &= \text{cm}(\text{cm}) \\ &= \text{cm}^2 \end{aligned}$$

and square centimeters are a proper measure for area.

2 Try to estimate the area of the United States in square kilometers without using a pencil. (If you have to ask "Does this mean including Hawaii and the other islands?" then you miss the point.)

3 You will be using a standard ruler graduated in cm in your laboratory and for problem-solving. How many significant figures can you read from it when measuring the height of a page in this book?

4 Explain why the equation $y = 3x^2$ mentioned in the introduction to this chapter would be a straight line if y is plotted against x^2.

5 In Chapter 1 the meter was defined as exactly 1,650,763.73 wavelengths of the orange light in the spectrum of krypton 86. Discuss the accuracy of this measurement.

6 Re-write the number given in Question 5 to three significant figures and express it in scientific notation.

7 What would be the constant of proportionality in the equation that expresses the area of a rectangle in square meters, when the length and width are given in centimeters?

8 Estimate the number of cups of water needed to fill a bath tub.

PROBLEMS

GROUP A

1 The period of a pendulum (time for one complete vibration back and forth) is expressed by the formula

$$T = 2\pi\sqrt{\frac{l}{g}}$$

where T is the time, l the length of the pendulum, and g the acceleration due to gravity. In what units must g be expressed if T is in sec and l is in m?

2 Which of the following equations cannot be correct because of dimensional inconsistencies? In each case s is distance, v is velocity, a is acceleration (in units of length/time^2), and t is time.

(a) $s = v_1t + 3at^2$ (c) $s + \frac{1}{2}at^2 = vt$
(b) $v = v_1^2 - \frac{1}{2}as^2$ (d) $at^2 + v/t = 2s^2/t^3$

3 There are 31,556,925.9747 seconds in a year. Express this number in scientific notation.

4 Evaluate the following:

(a) $(1.6 \times 10^2)(3 \times 10^6)(2 \times 10^{11})$ (c) $\frac{(0.0022)(600)}{3 \times 10^{-10}}$

(b) $\frac{(1.6 \times 10^2)(3 \times 10^6)}{2 \times 10^{11}}$ (d) $\frac{1.08 \times 10^{10}}{(2 \times 10^6)(3 \times 10^{-2})}$

5 Express in scientific notation:

(a) 0.00336 (c) 0.0000014
(b) 4,673,000 (d) 7.14

6 Express in ordinary notation:

(a) 9.81×10^1 (c) 4.27×10^3
(b) 3.11×10^{-4} (d) 6.70×10^{-6}

7 Find the sum of 1.6×10^2, 2.7×10^3, and 4.8×10^4.

8 Subtract 2.83×10^{-4} from 6.29×10^{-3}.

9 During a laboratory experiment, a student measures a block of wood to be 2.76 cm long, 4.30 cm wide, and 1.75 cm high. What volume should she calculate for this block?

10 A circular piece of paper has a measured diameter of 7.4 cm. What is its area?

11 What is the height (in meters) of a person who measures up at 160.6 cm?

12 A tank contains 42.8 liters of water. How much will be left if 3.72 liters are removed?

13 What is the order of magnitude of the number of minutes in a year? (Do *not* make an exact calculation.)

14 Masses are hung on a suspended spring and the corresponding stretch is noted (measured from the original position of the lower end of the spring). The following results were obtained.

Mass (kg)	*Stretch* (m)
0	0.0
4	0.16
6	0.24
8	0.32
14	0.56
20	0.80

Make a graph of these data. (a) Does the graph have a constant slope? What is it? (b) Write the equation that describes the behavior of this spring. (c) What stretch would you expect if 9.3 kg were placed on the spring?

Assume the original unstretched spring was 1 m long. Can you sketch the graph you would then get if all the stretch measurements were made from the top of the spring? Can you write the equation for this graph?

15 Find the slopes of the following graphs (Figures 2-6, 2-7, and 2-8):

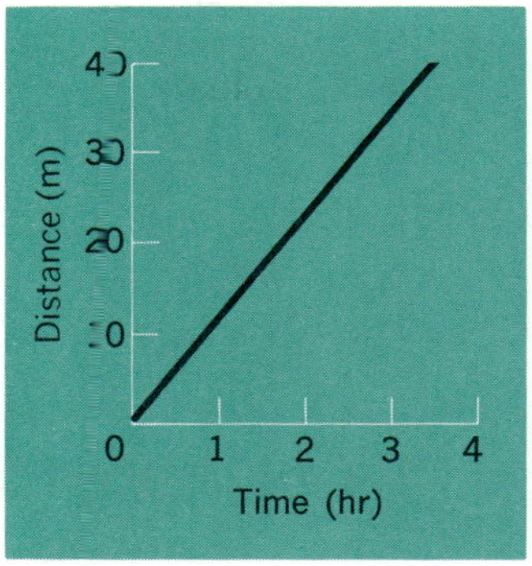

Figure 2-6

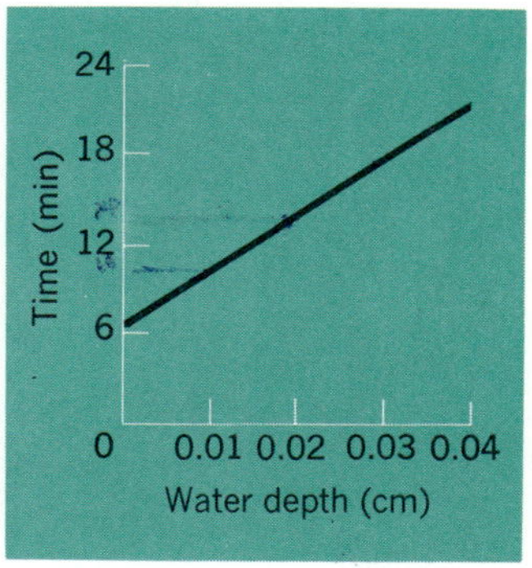

Figure 2-7

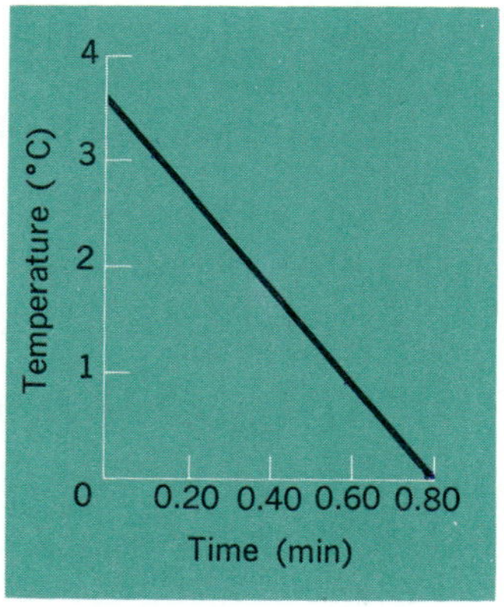

Figure 2-8

16 Make an approximate calculation of the fraction below, and then compare your result with the exact calculation.

$$\frac{(1590)(2\pi)}{41.7}$$

GROUP B

17 The radius of a circle inscribed in any triangle whose sides are a, b, c is given by

$$(1/s)\sqrt{s(s-a)(s-b)(s-c)}$$

where s is the abbreviation for $\frac{1}{2}(a + b + c)$. Check this formula for dimensional consistency.

18 The electric force between two charged objects is expressed as

$$F = K\frac{Q_1Q_2}{R^2}$$

where F is the force in newtons, Q_1 and Q_2 are the charges in coulombs, and R is their separation in meters. What units must K have to make this formula dimensionally correct?

19 Evaluate the following:

(a) $\dfrac{(2 \times 10^6)^3(0.04 \times 10^4)}{(3 \times 10^{11})(4.2 \times 10^7)}$ (b) $\dfrac{0.0000027}{10^{-11}}$

20 An atom of gold is about 2.5×10^{-10} m in "diameter." A thin square of gold foil, 1 cm on a side, is 1×10^{-6} m thick. Estimate (order-of-magnitude answer only) the number of gold atoms in the foil.

21 A baseball has a diameter on the order of 10^{-1} m. Estimate how many baseballs would fit in an empty swimming pool that measures 25 m long by 12 m wide.

22 The total force on the bottom of a tank of water is directly proportional to the depth of the water and the area of the bottom of the tank. When the water is 4 cm deep and the tank has a bottom area of 2 square meters, the force is 784 newtons. What is the total force when the depth is 2 m and the area of the bottom is 15 square meters?

23 The intensity of light that falls on a certain area from a given source is inversely proportional to the square of the distance from source to area. How far must a screen be placed from the source of light so as to have 9 times the intensity it had when it was 12 m from the source?

24 Estimate the value of the following:

(a) $\dfrac{(0.0332)(1.01 \times 10^4)^2}{\pi(18)^2}$ (b) $\dfrac{(1.6 \times 10^{-19})(6.02 \times 10^{23})}{84.5}$

25 A steel ball was allowed to roll down a long inclined plane and a student recorded the distance traveled at the end of each second as shown below.

Time (sec)	*Total distance* (cm)
0	0
1	4
2	16
3	36
4	64
5	100
6	144

(a) Use a full sheet of graph paper to graph these data. Put the time on the horizontal axis. If you have done this properly, you see that the graph is not straight, so it makes no sense to talk about a constant slope. However, the curve is smooth, which does suggest that there is some relationship between distance and time in this experiment.

(b) Add a third column to the table above and head it $(\text{time})^2$. Fill it in by squaring the times in the first column. Now try plotting distance vs. $(\text{time})^2$. Does this graph have a constant slope? What is it?

(c) Write the equation for this motion.

(d) How far does the ball roll in 2.5 sec?

UNIT 2

MOTION, FORCE AND CONSERVATION LAWS

Consider each of the following for a moment. There is the simple motion of a skier on a straight slope, and the complex motion of the stunt car pictured here. There is the great force needed to move an ocean liner, and the tiny force needed to remove an electron from its atom. There is the energy we get from a slice of bread, and the energy that we must provide to lift a rocket off the earth. There is the momentum of a gust of wind, of a charging rhino, and of the earth itself. Although this may seem like an array of unrelated phenomena, you will find that all these phenomena can be explained by a surprisingly few fundamental laws.

The next few chapters present some of the most important concepts in physics. It is impossible for a day of your life to go by without involving one or more of these concepts—motion, force, energy, and momentum. These concepts are basic to the understanding of almost all physical phenomena. It is in this unit, then, that we shall build the firm foundation for all that is to come.

3 Kinematics – The Description of Motion

We begin our study of physics with a description of the motions of physical bodies, ignoring for the time being the possible causes of motions. The study of motion, separate from its origin or cause, is called *kinematics* (from the Greek *kinein,* to move), and is concerned with the relations between speed, time, and distance. Fortunately, kinematics proves to be useful in its own right, since some interesting problems can be solved without considering the causes of motion. Even more important, the ability to think clearly about motion itself will be helpful to you when you approach *dynamics* (from the Greek *dynamis,* meaning power), the study of forces and the changes in motion that they cause.

The Italian scientist Galileo Galilei (1564–1642) became the most gifted experimental physicist of his time. Indeed, it was largely because of the special insight and example of Galileo that experimental methods became widely accepted as a part of physical science. He argued forcefully for the sun-centered theory of the solar system, and was persecuted for his teachings. Although Galileo did not invent the telescope, it was he who first used it to observe the moons of Jupiter and the phases of Venus, obtaining experimental evidence for the theory that the earth revolved around the sun. Galileo's experimental outlook, his reliance on observation as the final authority, and his skill in explaining his ideas to others make him one of the great men in the history of physics. His outlook pervades all present-day science.

Galileo realized that an adequate description of motion requires a quantitative, or mathematical, outlook. All the material in this chapter, and more, was known to Galileo, even though he was hampered by the primitive condition of mathematics in his day. With only the simple algebra that is taught in most high schools, we are better off than Galileo.

After you have learned to express the laws of kinematics in algebraic form, you will be able to proceed to the study of dynamics, with its numerous applications to basic science and to everyday life. *The study of kinematics and dynamics is a foundation for all that is to come* in physics, including space stations and cyclotrons, as well as older subjects such as pendulums and collisions between molecules.

3-1 TYPES OF MOTION

Consider these motions; two of which are illustrated with strobe photos (photos containing a series of exposures made at regular intervals):

(a) A coffee cup near the edge of a table slips off, falls to the floor, and breaks.

(b) A parachutist falls freely for 250 m, then

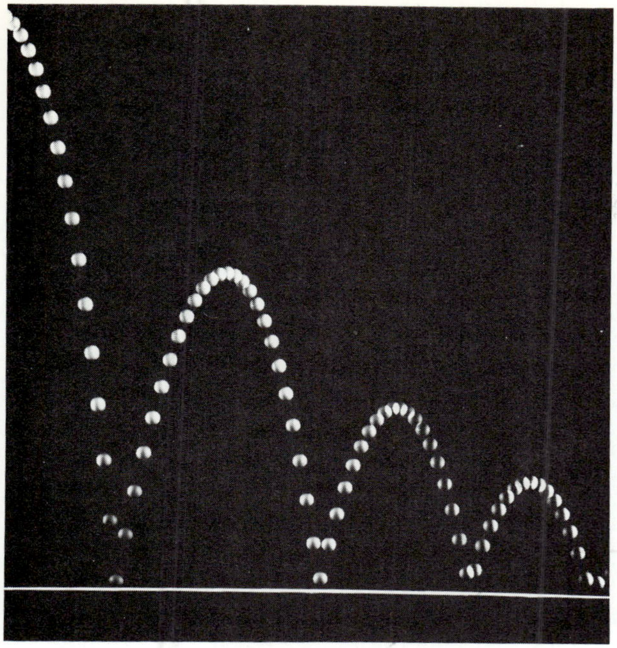

Figure 3-1

Figure 3-2

opens her chute and slows down to a steady speed of 40 km/hr; she continues to fall at this speed until she strikes the ground and comes to rest.

(c) A golf ball is dropped and bounces on the floor a number of times, always rising to successively lesser heights, and eventually comes to rest (Figure 3-1).

(d) A hockey puck glides across the ice, traveling in a straight line toward the goal without any appreciable loss of speed.

(e) The earth moves around the sun in an elliptical orbit at variable speed, moving a little faster when it is slightly closer to the sun (in January).

(f) A swimmer dives off a diving board, and the end of the board continues to vibrate up and down for a while (Figure 3-2).

(g) A truck driver avoids a collision by jamming on his brakes; he comes to a screeching halt without swerving to either side.

(h) An electron leaves the electron gun in the neck of a TV picture tube, passes immediately through the deflection coils, where it is deflected sideways, and then proceeds at constant speed in a straight line until it strikes the face of the tube and makes a spot of light.

(i) A child on a swing travels along a circular arc, picking up speed as he approaches the lowest point in his motion, and losing speed as he rises again.

We have only scratched the surface; countless additional types of motion easily suggest themselves for analysis.

Which of these motions is "simplest"? Most of the preceding descriptions involve not just one, but several distinct types of motion. For instance, in (b) there are four distinct periods: First, while the girl is freely falling; second, the short interval of time while the parachute has just opened and she is slowing down; third, the remainder of the trip in the air; and fourth, the small fraction of a second during which her speed is reduced to zero when she strikes the earth. Considering the *whole* motion in each of the examples, (d) is perhaps the simplest, since neither direction of motion nor speed changes. Motion (g) is probably the next most simple (assuming from the meager description that the truck loses speed at a steady rate). Motion such as (d) is called *uniform motion in a straight line,* and motion such as (g) is

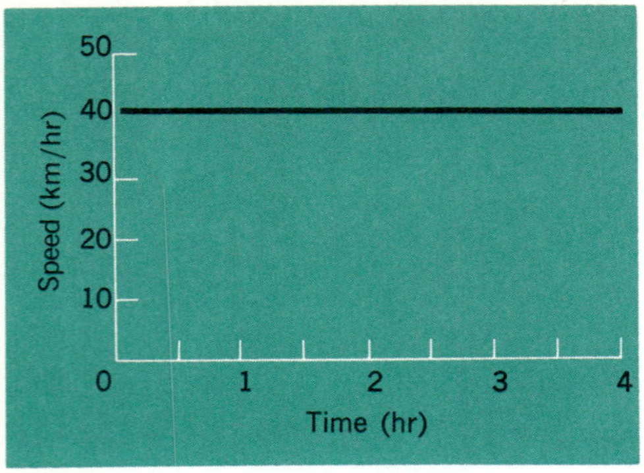

Figure 3-3 Speed-time graph of a car traveling at constant speed.

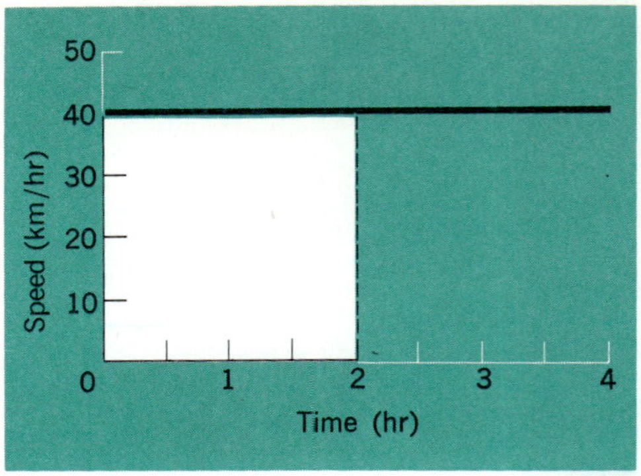

Figure 3-4 The distance traveled by the car in 2 hours is represented by the white area under the graph.

called *uniformly accelerated motion.* (In this case "deceleration" might seem a more descriptive word, but in physics the term "negative acceleration" is usually preferred.) Other motions such as (a), (c), and (h) are combinations of the two basic motions described above, but (e), (f), and (i) involve changing directions and are more complicated. Motion (b) is more complex to describe, since just after the parachute opens the girl loses speed, but not the same amount each second.

It is tempting to contemplate the causes of some of the motions described. For instance, in (b), is the effect of air resistance constant? Such questions will be studied in Chapter 4.

3-2 UNIFORM MOTION IN A STRAIGHT LINE

Let's first consider the simplest motion described above—constant speed in a straight line. The quantitative description of this motion is

$$s = vt \qquad \textbf{(Equation 3-1)}$$

where s is the distance traveled, v is the speed, and t is the time. You will recognize this as the familiar "distance equals rate times time" formula of earlier science and mathematics courses. The only difference is that the symbols are not the same as those with which you are perhaps familiar.

In the introduction to this chapter you were told that the laws of physics would be expressed in algebraic form. The above relation is one such example. You will find, however, that it is also beneficial to represent these motions by graphing them. In Figure 3-3 the speed versus the time is graphed for a car that is traveling at a constant speed of 40 km/hr. In this simple case the graph is a horizontal straight line which reflects the obvious statement that—for a car traveling at constant speed—there is no change in speed as time goes on. If we ask the question "How far does the car travel in 2 hours?" we see from the relation $s = vt$ that

$$s = 40 \text{ km/hr} \times 2 \text{ hr}$$
$$= 80 \text{ km}$$

In Figure 3-4 the area under the graph of Figure 3-3 from zero to the 2-hour mark has no color. This white square has an area equal to its base times its height. In this case the height is 40 km/hr and the base is 2 hr, so the area is 40 km/hr × 2 hr = 80 km. It may seem strange to you that "heights" are measured here in km/hr and "areas" measured in kilometers. A graph is nothing more than a scale drawing where, in this case, 10 km/hr is represented by a vertical length of 0.6 cm. Although the actual area of the rectangle can be expressed in cm^2 by measuring each dimension

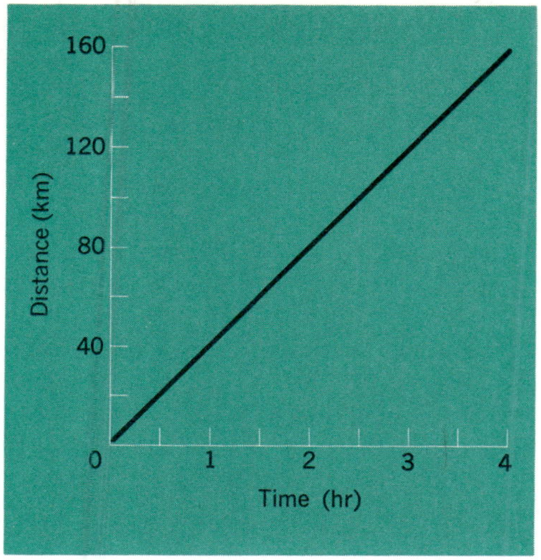

Figure 3-5 Distance-time graph of the car.

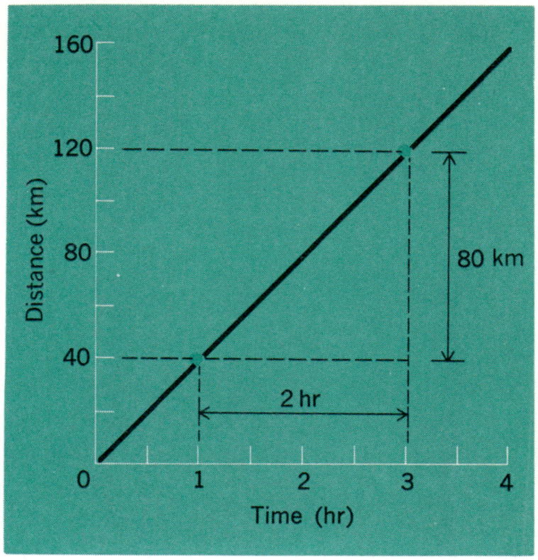

Figure 3-6 Determining the slope of the distance-time graph of Figure 3-5. The slope is the "up" distance divided by the "over" distance. For the two points chosen, 80 km/2 hr gives a velocity of 40 km/hr.

in cm, our "area" is expressed in kilometers because we have multiplied a height measured in km/hr by a base measured in hours.

We see from this example that the distance traveled by a car moving at constant speed is represented by the area under the speed-time graph. This should not be too surprising because the area under such a graph is a velocity (measured on the vertical or y-axis), multiplied by a time (measured on the horizontal or x-axis), and velocity times time is distance.

For motion at constant speed, the *displacement,* or distance traveled,* increases steadily as the time goes by. The table below is a record of the distance covered by this car for the first 4 hours of the trip.

Time (hr)	*Distance* (km)
1	40
2	80
3	120
4	160

* A more precise definition of *displacement* is given in the discussion of vectors (Section 3-6).

The distance-time graph of this motion is illustrated in Figure 3-5. The graph climbs steadily, indicating a uniform change in distance. We can confirm the relation between the speed-time graph and the distance-time graph for this motion quite readily. For example, in Figure 3-3, at the end of 2.5 hours the area under the graph represents a distance of 100 kilometers. In Figure 3-5, at the end of 2.5 hours we can read directly that the car has gone 100 kilometers.

There is another relation between these two graphs. In the preceding chapter we noted that the slope of a straight-line graph is the constant of proportionality between the two variables being plotted (page 19). So from Figure 3-5 we know that the distance s is directly proportional to the time t; $s = kt$, and k is the slope of the line. This is nothing more than a restatement of the basic relation $s = vt$, where v, the constant of proportionality between s and t, is the slope of the distance-time graph. By taking the slope of the graph in Figure 3-5, we see that it is a constant 40 km/hr. This is illustrated in Figure 3-6.

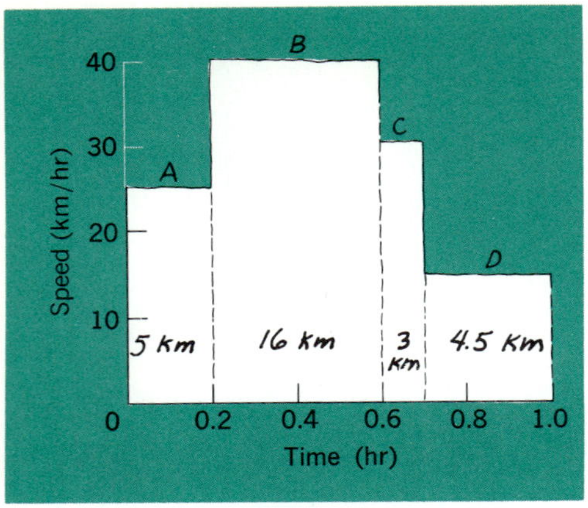

Figure 3-7

You are probably aware from past experience with the equation $s = vt$ that any questions asked about constant motion can be handled simply by using this relation. The really new concept here is the role that graphs can play in analyzing not only this simple motion, but, as you will see in the next section, more complicated motions as well. In these situations you may find that a graphical solution is much easier than the use of formulas.

Example 3-1 The table below is a record of a trip made by a car that travels at different speeds for various time intervals.

(a) During which interval did the car travel the farthest?

(b) During which interval was the car going the slowest?

(c) How far has the car gone after 0.35 hour?

Time interval	*Duration of interval* (hr)	*Speed during interval* (km/hr)
A	0.20	25
B	0.40	40
C	0.10	30
D	0.30	15

(d) What is the total distance traveled?

(e) How long did it take to travel the first 15 kilometers?

At least one of the questions can be answered by inspection of the table and the others by manipulation of the formula $s = vt$. For purposes of discussion, however, let's make a speed-time graph of the data and see what it tells us. Note that with data like these (the numbers are convenient multiples of 5 and 10) there is no need to make a carefully drawn graph as you would have to do when plotting experimental data. This takes extra time, which, in turn, defeats the purpose of our choice of a graphical solution. The graph in Figure 3-7 is a "sketch," but it is read with the accuracy of a carefully plotted graph.

A real car could not make the sudden jumps in speed shown in this graph, but we are keeping the problem simple. Each time interval has been set off with dotted lines and the areas of the rectangles formed represent the distance traveled during that interval. The areas have been computed in the sketch and are entered in each rectangle.

Inspection of the graph tells us that the answer to question (a) above is interval *B* because it has the largest area. The answer to question (b) is interval *D* because that is the lowest line on the graph. To answer question (c), we have to find the area under the graph from zero to 0.35 hour. This will be the sum of 5 kilometers and the area of a rectangle whose height is 40 km/hr and whose base is 0.15 hour, or 5 km + 6 km = 11 km. Question (d) is the total area under the graph, or 28.5 kilometers. Finally, for question (e), to find the time that corresponds to a distance of 15 kilometers, we see that it will fall somewhere in interval *B*. Specifically, we want the base of a rectangle whose area is 10 kilometers (the car has already gone 5 kilometers in interval *A*) and whose height is 40 km/hr. This would be a time of 0.25 hour. The total time is therefore 0.20 hr (interval A) + 0.25 hr = 0.45 hr.

It might prove interesting to make a distance-time graph for this problem. This is not necessary, of course, since we got all the information we wanted from the speed-time graph, but it will serve to test your understanding of this section. Note that the graph in Figure 3-8 is not sketched freehand, but carefully plotted and drawn with a ruler. Exact plotting is preferred here because the numbers involved (21 km and 24 km) do not fall conveniently on our choice of scale. You will also want to find slopes from this graph, and this becomes less troublesome when the lines are drawn with a ruler. With the practice that you will get from the end-of-chapter problems, you will soon be able to decide when a graph can or cannot be sketched.

From the graph, you can see that interval *B* represents the longest vertical rise and is therefore the answer to question (a). The slope of each straight-line portion of the graph represents the speed during that interval, and by inspection we see that interval *D* is the least "steep" and therefore corresponds to the slowest speed. The answers to questions (c), (d), and (e) can be read directly from the graph, and they agree with the answers we got from our analysis of the speed-time graph.

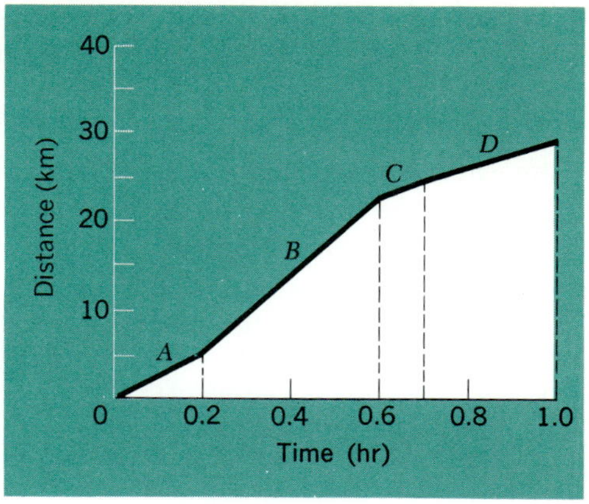

Figure 3-8

It is worth restating that the intent of this example is *not* to compare the relative merits of an algebraic versus a graphical solution, but rather to illustrate how graphs can be used to solve problems.

Let's look now at an example of uniform motion that does not lend itself to a graphical solution. We can use this example to illustrate some more of the good problem-solving techniques discussed in the preceding chapter.

Example 3-2 In a TV tube, an electron travels from the electron gun to the fluorescent screen, a distance of 45 cm, in 0.006 μsec. What is its velocity in m/sec? in km/hr?

$s = 45$ cm.
$t = 0.006\ \mu$sec
$v = ?$

($1 \mu\text{sec} = 10^{-6}$ sec)

$$s = vt$$

$$v = \frac{s}{t}$$

$$= \frac{(45\text{ cm})(1\text{ m}/100\text{ cm})}{(6 \times 10^{-3}\ \mu\text{sec})(10^{-6}\text{ sec}/1\ \mu\text{sec})}$$

$$= \boxed{7.50 \times 10^{7}\text{ m/sec}}$$

To change to km/hr, we make use of the fact that 36 km/hr is the same as 10 m/sec:

$$v = (7.5 \times 10^{7}\text{m/sec})\left(\frac{36\text{ km/hr}}{10\text{ m/sec}}\right)$$

$$= \boxed{2.70 \times 10^{8}\text{ km/hr}}$$

The method used in this simple problem helps to illustrate some of the "tricks of the trade" in problem-solving:

(a) First read the problem carefully and make a list of knowns and unknowns, setting them off in a box.

(b) Then search our memory (or the text) for an equation containing *these symbols and no others.* In this case the desired equation connecting *s*, *t*, and *v* is $s = vt$.

(c) Since you are asked not for *s* but for *v*, which is buried in the equation, try to solve *algebra-*

ically for the desired unknown, in this case obtaining $v = s/t$. This is usually done before substituting any numbers.

(d) Substitute into the revised equation the information from the box containing all the knowns and unknowns. The factor 1 m/100 cm in the numerator doesn't change the value of the answer, since 1 meter is the same as 100 cm, and hence 1 m/100 cm is just another way of writing 1. The solver thinks somewhat like this: "I have 45 cm in the numerator, but I want meters in the numerator. Therefore I write 45 cm $\times \dfrac{\text{m}}{\text{cm}}$ so that I can later cancel out the cm and have meters left. Finally, I fill in the numbers knowing that 1 meter equals 100 cm, getting 45 cm $\times \dfrac{1\ \text{m}}{100\ \text{cm}}$."* Note that, for the purposes of canceling out units, it is better to write $\dfrac{\text{m}}{\text{sec}}, \dfrac{\text{km}}{\text{hr}}$, etc., rather than use the slant line to indicate division (although for convenience of typesetting the slant line is often used in this text).

(e) In order that the final answer be clearly indicated, use some device such as a box or underlining.

Problem-solving is intended to help you master the meaning of physics; it should not become a battle of wits between the student and a collection of formulas pulled out of a magician's hat. *The most important single step is the first one—careful study of the problem to see what information is pertinent and just what is asked for.* The expert may carry these bits of information in his head until he hits upon the right equation; the beginner would do well to follow the example carefully and place the given data in a list, perhaps enclosed in a box.**

* In this particular example the 45 cm could have been changed to 0.45 m at the very start; the routine involving 1 m/100 cm is shown chiefly to encourage a habit that will prove useful later on in more complicated problems.

** Sample examples in this book or parts of examples will not always include the data box, but it is suggested that you construct them for yourself—at least until you have developed considerable skill in keeping the information in your mind.

In this section we got quite a bit of "mileage" out of the single relation $s = vt$. The ideas of this section and the discussion of Examples 3-1 and 3-2 merit close study. In the next section our discussion of motion will be extended to more complicated cases where the motion is not uniform but constantly changing.

3-3 UNIFORMLY ACCELERATED MOTION

The graph in Figure 3-9 shows the motion of a car that starts from rest and changes speed uniformly. We know the change in speed is uniform because the graph is a straight line. Is it possible to tell from this graph how far the car goes in 0.004 hour? In the last section we found that, for a constant speed, the area under the speed-time graph was the distance. This will also work for us here, but now it is necessary to find the area of a triangle. This should not surprise you because, for uniformly changing speed, it makes sense to multiply the *average* speed by the time to find the distance. In this case, the average speed is one half the final speed. This coincides with the formula for the area of a triangle, $\frac{1}{2}$(height)(base). So in 0.004 hour the car has gone a distance of

$$(\tfrac{1}{2})(20\ \text{km/hr})(0.004\ \text{hr}) = 0.04\ \text{km}$$

This is shown by the white area under the graph in Figure 3-10.

In general, the area under a speed-time graph represents the distance traveled. In Figure 3-11 the car is already moving at 10 km/hr when we start to time it. It speeds up uniformly for 0.03 hour and then travels at a constant speed of 30 km/hr for the next 0.03 hour. For the remainder of the trip it slows down uniformly until it comes to rest after a total elapsed time of 0.10 hour. The total distance traveled is represented by the total area under the graph. This can be found by dividing the area into rectangles, triangles, or trapezoids, and adding them all up. See if you can get 2.1 kilometers for this trip!

In common language, "to accelerate" means to speed up. In physics, however, our definition is more precise and you will find that "to speed up" is only one special case of acceleration. In order to define acceleration, we must first distinguish carefully between the terms *speed* and *velocity*. So far in this chapter, we have confined ourselves to the term *speed*, which simply describes how fast a body is moving. Velocity, on the other hand, describes both how fast and in what *direction* it is moving. This will be discussed more fully in Section 3-6. This chapter has dealt only with motion along a straight line. Such motion can only be in two directions, forward and backward (or up and down, etc.), so the direction can be indicated by the use of the algebraic signs plus (+) and minus (−). (See Example 3-4.) For reasons that will become apparent later on, we will define acceleration in terms of the change in velocity of a body. Specifically, *acceleration is the rate of change of velocity*, or the change in velocity divided by the time interval during which the change takes place.

What is the acceleration of the car in Figure 3-12? After 5 seconds, the velocity (or speed in the + direction) is 10 km/hr. (At the end of 10 seconds the velocity is 20 km/hr.) Evidently, the

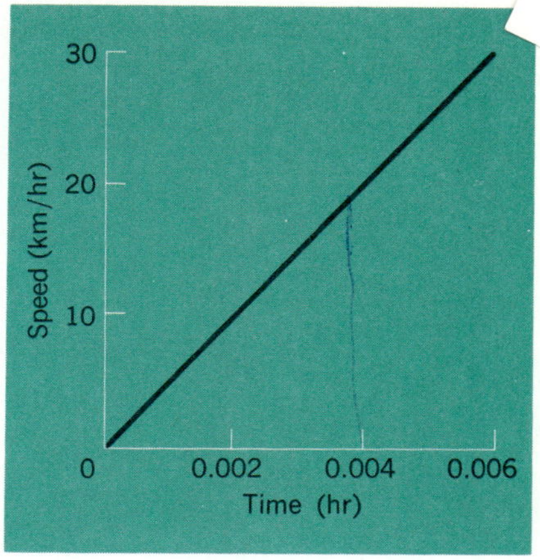

Figure 3-9

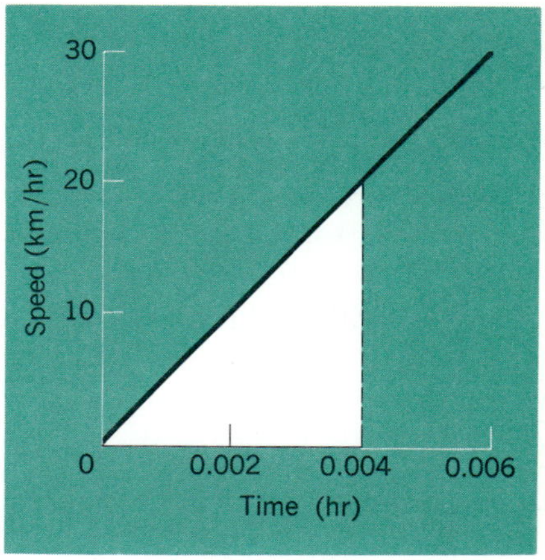

Figure 3-10

Figure 3-9 Speed-time graph of car that is changing speed uniformly. **Figure 3-10** Speed-time graph of the same car showing distance traveled by car in 0.004 hour—represented by white area under graph. **Figure 3-11** What is total distance traveled by car? **Figure 3-12** The velocity is changing by 10 km/hr each 5 sec.

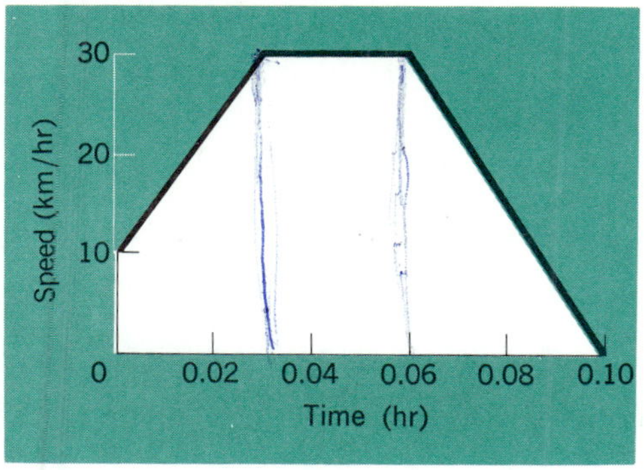

Figure 3-11

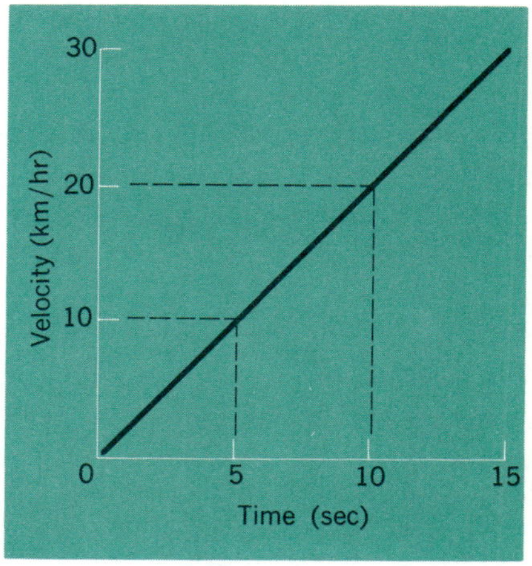

Figure 3-12

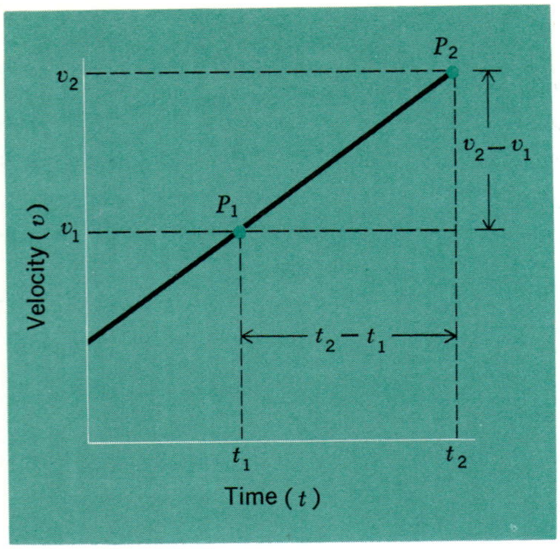

Figure 3-13 Slope of graph is $\frac{(v_2 - v_1)}{(t_2 - t_1)}$.

velocity is changing by 10 km/hr each 5 seconds, which, after simplifying, is read "2 kilometers per hour per second." (Accelerations are measured in m/sec^2, km/hr^2, km/hr·min, etc.—always a distance unit divided by two time units. See the footnote on page 17.)

From the above discussion you have probably become aware that the slope of the velocity-time graph is the acceleration. And from the slope of the graph in Figure 3-12 you will find that in this case it is 2 km/hr·sec.

In order to describe these ideas quantitatively, it is necessary to introduce some modifications of the symbols that have been used up to now. To distinguish between the many velocities a body may have during the course of a problem (like the accelerating car in Figure 3-12), we use subscripts. For example, we can write v_i and v_f for the initial velocity and the final velocity during the particular time interval being considered. The adjective "final" refers to the end of a particular time interval, though the motion may actually continue for an indefinite time. The more general form v_1, v_2, v_3, etc., can also refer to the different velocities of a body. Similar notation can be used for many other symbols, for example, t_1 and t_2 often refer to two different points in time.

These notations are used to express the slope of the graph in Figure 3-13. Here you see that at time t_1 the velocity has a value v_1 and at time t_2 the velocity is v_2. The change in the velocity is clearly $v_2 - v_1$ and the corresponding change in time is $t_2 - t_1$. But this is our definition of acceleration. Quantitatively, then,

$$\text{acceleration} = \frac{\text{change in velocity}}{\text{change in time}}$$

$$a = \frac{v_2 - v_1}{t_2 - t_1}$$

In general, when a body accelerates uniformly from some initial velocity v_i, to some final velocity v_f, during a time interval t (the time interval between the initial time t_i and the final time t_f),

$$a = \frac{v_f - v_i}{t} \qquad \textbf{(Equation 3-2)}$$

Example 3-3 A sports car traveling on a straight road has a speed of 22 km/hr which increases to a speed of 57 km/hr in 5 seconds. What is the car's acceleration?

Our box with the list of knowns and unknowns contains v_i, v_f, t, and a:

v_i = 22 km/hr
v_f = 57 km/hr
t = 5 sec
a = ?

$$a = \frac{v_f - v_i}{t}$$

$$= \frac{57 \text{ km/hr} - 22 \text{ km/hr}}{5 \text{ sec}}$$

$$= \frac{35 \text{ km/hr}}{5 \text{ sec}}$$

$$= \boxed{7 \text{ km/hr} \cdot \text{sec}}$$

If the car had not accelerated but continued at the initial velocity of 22 km/hr, the distance traveled could be found from $s = vt$. Because it did accelerate, we must use the *average velocity* over the time interval to find the displacement.

Thus

(displacement) = (average velocity)(time)

$$s = \frac{v_i + v_f}{2} t \qquad \text{(Equation 3-3)}$$

where $(v_i + v_f)/2$ is the average of the initial and final velocities. You may recognize this as the same reasoning we used to find the area under the graph in Figure 3-10.

Actually, the two preceding equations (Equation 3-2 and Equation 3-3) are sufficient for solving all problems concerning uniformly accelerated motion. However, for convenience, two additional equations are often used. They are obtained by algebraic manipulation of the basic equations and give us two new equations relating v_i, v_f, s, t, and a. By multiplying Equations 3-2 and 3-3 together, we get

$$as = \frac{(v_f - v_i)}{t} \frac{(v_f + v_i)}{2} t$$

which simplifies to

$$v_f^2 = v_i^2 + 2as \qquad \text{(Equation 3-4)}$$

If we solve Equation 3-2 for v_f, we get $v_f = v_i + at$. Substituting this expression for v_f in Equation 3-3 gives us

$$s = \frac{(v_i + v_i + at)}{2} t$$

which simplifies to

$$s = v_i t + \tfrac{1}{2} a t^2 \qquad \text{(Equation 3-5)}$$

Let's now collect these four equations together:

$$a = \frac{v_f - v_i}{t} \qquad \text{(Equation 3-2)}$$

$$s = \frac{v_i + v_f}{2} t \qquad \text{(Equation 3-3)}$$

$$v_f^2 = v_i^2 + 2as \qquad \text{(Equation 3-4)}$$

$$s = v_i t + \tfrac{1}{2} a t^2 \qquad \text{(Equation 3-5)}$$

Uniformly accelerated motion includes uniform motion as a special case. In uniform motion there is no acceleration, so by setting a equal to zero in Equation 3-5 we get $s = v_i t$, which is the equivalent of Equation 3-1.

There is a certain systematic property of these equations. Together they contain four different variables that are related to each other (v_i is not a variable; the initial velocity is a constant for any given problem). The variable s is missing from Equation 3-2, a from Equation 3-3, t from Equation 3-4, and v_f from Equation 3-5. When you use these equations (3-2 through 3-5), it will be to your advantage to select the most suitable for the problem you are solving and thereby avoid as much wasted effort as possible.

On some occasions it becomes necessary to refer to *changes* in certain quantities. A shorthand way of saying "change in," "increase in," etc., is to use the symbol Δ, which is the Greek capital *D*, and is read "delta." Therefore, the notation Δv, read "delta vee," means change in velocity. We sometimes replace $v_2 - v_1$, $v_f - v_i$, etc., with their equivalent, which is Δv in both cases. Our formula for acceleration would then be written as

$$a = \frac{\Delta v}{\Delta t}$$

3-4 ACCELERATION DUE TO GRAVITY

By far the most familiar example of uniformly accelerated motion is that of a freely falling body. If air resistance can be ignored,* a falling body picks up speed at the rate of about 9.8 m/sec each second. This constant is called the *acceleration due to gravity*, and is denoted by the symbol *g*. The value of this constant varies slightly from place to place on the globe; at 45° latitude and sea level, *g* is 9.80600 m/sec². (In Denver, Colorado, for example, *g* is 9.796 m/sec² and at the North Pole *g* is 9.832 m/sec². In problem work, *g* is taken to be exactly 9.80 m/sec² unless otherwise stated.

* Air resistance can be ignored if the falling body is heavy and has a relatively small surface area. Air resistance can certainly not be ignored for a falling raindrop, or for a man-parachute combination. But in these cases, the motion is not uniformly accelerated. The study of dynamics in later chapters will reveal the effects of friction, including air resistance, more clearly.

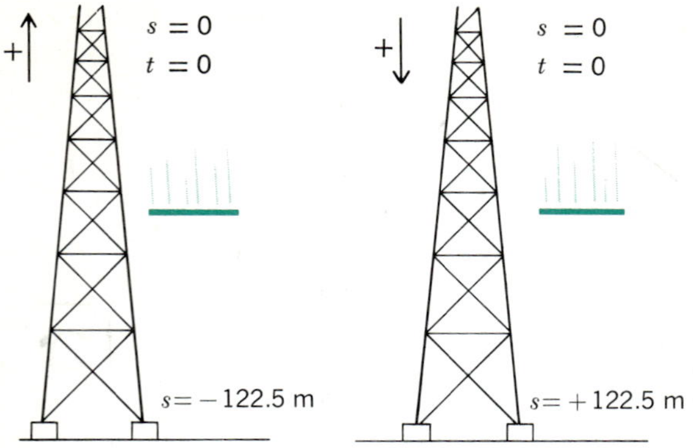

(a) **Figure 3-14** **(b)**

Example 3-4 On January 16, 1962, a steel TV tower was under construction. At 4:15 p.m., a girder broke loose and fell 122.5 m to the ground. How long did it take to reach the ground?

First we make a table of knowns and unknowns, assuming the positive direction to be *upward*. We also assume that $s = 0$ at the top of the tower, 122.5 m above the ground, since that is where the girder is when $t = 0$.

$s = -122.5$ m
$t = ?$
$a = -9.8$ m/sec^2
$v_i = 0$

It is well to indicate these facts in a diagram (Figure 3-14a). For this choice of positive direction the acceleration, which is downward, is *negative*. We therefore use -9.8 m/sec^2 for a.

Using Equation 3-5 and noting that $v_i t$ drops out because $v_i = 0$, we have

$$s = v_i t + \tfrac{1}{2}at^2$$

$$t = \sqrt{\frac{2s}{a}}$$

$$= \sqrt{\frac{2(-122.5\ \text{m})}{-9.8\ \text{m/sec}^2}}$$

$$= \sqrt{25\ \text{sec}^2}$$

$$= \boxed{5\ \text{sec}}$$

You need not worry about making the proper choice of signs here; just be careful that $s = 0$ at the location of the object at the start of the problem, when $t = 0$. Either direction may be taken as positive. For instance, the same problem can be solved by making the other direction positive (see Figure 3-14b).

$s = +122.5$ m
$t = ?$
$a = +9.8$ m/sec^2
$v_i = 0$

$$t = \sqrt{\frac{2s}{a}}$$

$$= \sqrt{\frac{2(+122.5\ \text{m})}{+9.8\ \text{m/sec}^2}}$$

$$= \sqrt{25\ \text{sec}^2}$$

$$= \boxed{5\ \text{sec}}$$

As long as you are consistent with the signs after a positive direction has been chosen, you should obtain the correct answer. You are then free to concentrate on the really important part of problem-solving—"setting up the problem." Try to grasp the *physical* essentials and ignore the irrelevant details (such as the time of day, the fact that the tower is made of steel, etc.). The next example, which introduces the concept of instantaneous velocity, shows the advantage of choosing the proper equation in order to make the algebraic work easier.

Example 3-5 In an acceleration test of a rocket sled, two markers were set up along a test track, 200 m apart. The sled passed the first marker with a speed of 10 m/sec, and when it reached the second marker its speed was 90 m/sec. What was the sled's average acceleration?

$v_i = +10$ m/sec
$v_f = +90$ m/sec
$s = +200$ m
$a = ?$

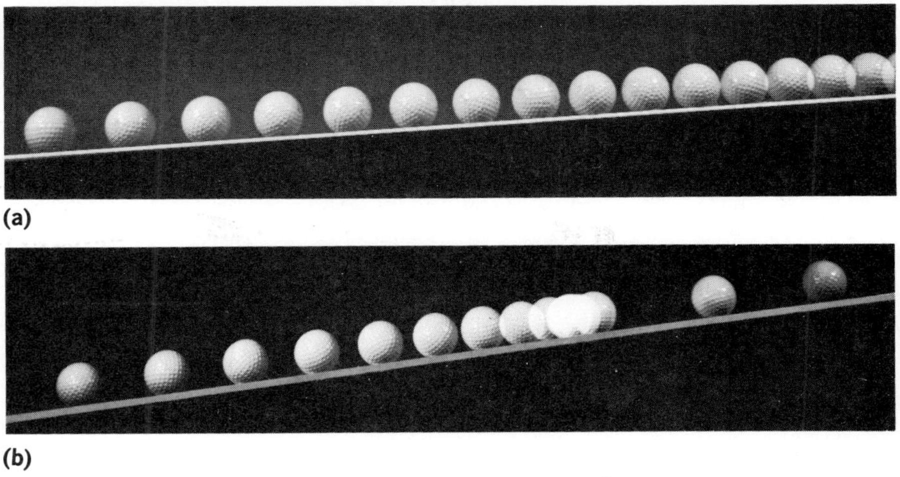

Figure 3-15 Two different kinds of motion. (a) Uniform acceleration of a golf ball rolling down an incline. (b) Non-uniform overall acceleration. Here the golf ball first accelerates uniformly down an incline and collides with an obstacle in the track. The ball bounces back and the obstacle is removed. The ball once more begins to accelerate down the incline.

$$v_f^2 = v_i^2 + 2as$$

$$a = \frac{v_f^2 - v_i^2}{2s}$$

$$= \frac{(90 \text{ m/sec})^2 - (10 \text{ m/sec})^2}{2(200 \text{ m})}$$

$$= \frac{(8100 - 100) \text{ m}^2/\text{sec}^2}{400 \text{ m}}$$

$$= \boxed{20 \text{ m/sec}^2}$$

This problem could also have been solved in two steps—using Equation 3-3 to find out that the test run lasted 4 sec, and then using Equation 3-2 to find the acceleration. Either method is correct. It is a matter of choice which you use.

3-5 AVERAGE VALUES AND INSTANTANEOUS VALUES

A subtle point appeared in the statement of Example 3-5. The sled's "average" acceleration was asked for—chiefly because there was not enough information given to allow a more detailed answer to be calculated. It is possible, for example, that the sled "eased up" in the middle of the test strip and came on with a rush at the end. Or it may have "coasted in" after a fast getaway. If we had known the speed of the sled as it passed every one of 200 markers spaced 1 meter apart, we could have computed the acceleration and velocity during each 1-meter interval. We would thus have been much closer to an exact description of the motion. However, our velocities computed in this way would still be average values, since the acceleration might not be uniform, even over a distance as short as 1 meter (Figure 3-15 shows a strobe-photo comparison of uniform and non-uniform acceleration over a given distance). Strictly speaking, *instantaneous* velocity (the velocity at one instant of time) can never be computed from actual measurements of displacement and time; but by using very short time intervals and correspondingly small displacements, we can arrive at an average velocity that closely approximates the instantaneous value. Also, actual measurements of velocity (even if they were possible) could lead only to average accelerations, not to instantaneous accelerations.

Lacking detailed information to the contrary, in Example 3-5 we are forced to assume that the acceleration is *uniform*. Our solution really gives us nothing more than the hypothetical constant acceleration which, if it were acting continuously, would produce the overall change in the sled's velocity.

These ideas can be given a graphical interpretation. You have already looked at the graphs for uniformly accelerated motion (Figure 3-13, for example). Here, the velocity increases steadily according to the equation $v_f = v_i + at$. However, displacement (the area under the graph) increases at a variable rate according to the equation $s = v_i t + \frac{1}{2}at^2$. These changes are shown

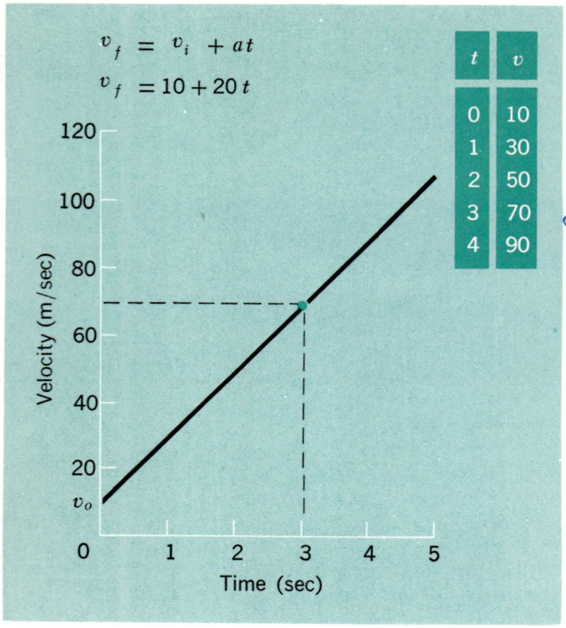

Figure 3-16 **(a)** Velocity-time graph for uniformly accelerated motion.

Figure 3-16 **(b)** Displacement-time graph for uniformly accelerated motion. The slope of each triangle represents the average velocity during a certain time interval.

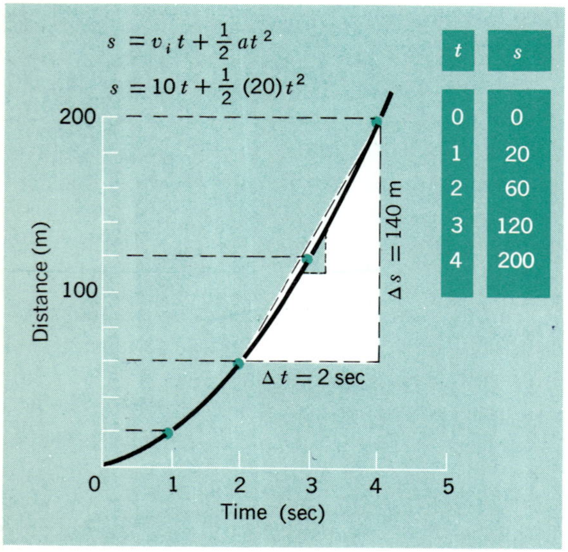

graphically in Figure 3-16, which applies to the sled of Example 3-5. The velocity graph is a straight line (Figure 3-16a), but the displacement graph is part of a parabola (Figure 3-16b). Recall that the slope of the distance-time graph gives the velocity (Figure 3-6). But in this problem there is no unique value for the velocity because the body is accelerating. In this case we have to talk about the *average velocity*. The average velocity during any interval is given by the slope of a chord drawn on the displacement graph. For instance, during the interval from $t = 2$ sec to $t = 4$ sec, the displacement increases by an amount Δs, which is 140 meters. The time interval Δt is 2 seconds, so the average velocity is $\Delta s/\Delta t = 140$ m/2 sec $=$ 70 m/sec. This is the slope of the dashed triangle in Figure 3-16b.

The *instantaneous velocity* is found by a process of using smaller and smaller time intervals. For instance, if the displacement is measured for a time interval extending from $t = 2.95$ sec to $t = 3.05$ sec (colored triangle), the displacement will be only 7 meters, and the ratio $\Delta s/\Delta t$ would be expected to be closer to the instantaneous velocity at $t = 3.00$ sec than the previous ratio, in which a larger time interval was used. By letting the time interval Δt become very small, the ratio $\Delta s/\Delta t$ will give the desired instantaneous velocity. This process is called "taking the limit of $\Delta s/\Delta t$ as Δt approaches zero." Such calculations are studied at length in calculus and will not concern us further in this book. An important point to note, however, is that as Δt approaches zero, the chord of the curve approaches the tangent to the curve at any given point of time. The instantaneous velocity, then, is the slope of the tangent line drawn on the displacement graph at the time in question. This is shown for the point $t = 3$ in Figure 3-17.

Similarly, since acceleration is the rate of change in velocity, we can say that instantaneous acceleration at any time is the slope of the velocity graph (Figure 3-16a) at the time in question.

As a matter of fact, it is not really necessary to take a limit in the problem illustrated in Figure 3-16. The average velocity for a 0.1-second interval centered around $t = 3$ sec is 7 m/0.1 sec = 70 m/sec, the same as the average value for a 2-second interval centered around the same time, $t = 3$ sec. It so happens that this is the same as the instantaneous velocity that can be read off from the velocity graph (Figure 3-16a), which shows $v = 70$ m/sec at $t = 3$ sec. This is true only because of the simple type of motion considered (*uniformly* accelerated motion). In such motion, the average velocity for any time interval equals the instantaneous velocity at the mid-time of the interval. However, for a more complex motion (motion not uniformly accelerated), this is not the case.

It was stated earlier that the study of kinematics and dynamics is the foundation for all that is to come in physics. In this chapter, graphs have been used to develop the kinematic equations. Although you will be able to solve *all* problems of uniformly accelerated motion by using Equations 3-2 through 3-5, it is important to keep in mind what they mean and where they came from. If you understand their origin—and the graphs give you that—then you are well on your way to using them successfully in your work.

3-6 VECTORS

Our study of the problems of kinematics would hardly be complete if we discussed only motion in a straight line. The motion of a bullet fired horizontally from a rifle; the motion of a particle of water rushing over a waterfall—these are examples of *projectile motion,* motion in which the body has two *simultaneous* velocities. The downward part of the motion is due to gravity, and the horizontal motion is practically constant if air resistance is small. Other types of motion also involve two simultaneous velocities—for example, a man walking from one side to the other on a moving train or swimming across the current in a fast-flowing river. To study the resulting motion, each part going on independently of the other, we need to use the concept of vectors and components of vectors.

A *vector* quantity has both magnitude *and* direction, whereas a *scalar* quantity has only magnitude. For example, an *upward* force, an *eastward* velocity, and a displacement *toward the northwest* are vector quantities. On the other hand, it is meaningless to think of "40 seconds toward the south" or "8 cubic meters downward"; these quantities are scalars, which are completely expressed by their magnitudes.

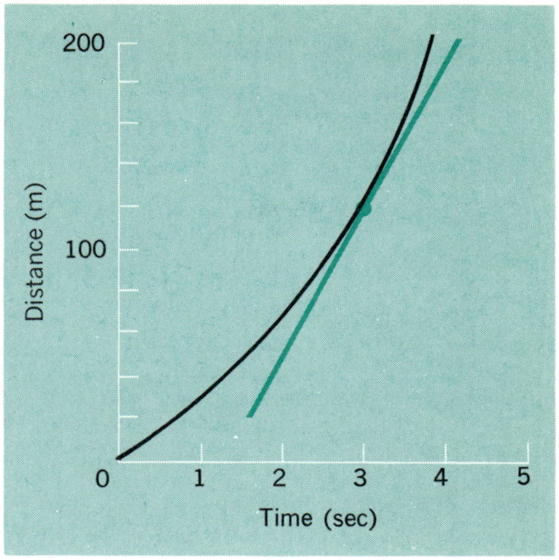

Figure 3-17 The instantaneous velocity at any time t is equal to the slope of the tangent line drawn at that point.

Table 3-1 compares various vector and scalar quantities. The first three vector quantities given are related to the scalar quantities in the second column as indicated by the double arrows; for instance, speed is the magnitude of the velocity. Such a correlation cannot be made for the remaining vectors.

Table 3-1
Vector and Scalar Quantities

Vector quantities		*Scalar quantities*
displacement	↔	distance
velocity	↔	speed
acceleration	↔	pick-up
force		time
momentum		volume
electric field strength		work
magnetic field strength		mass (inertia)

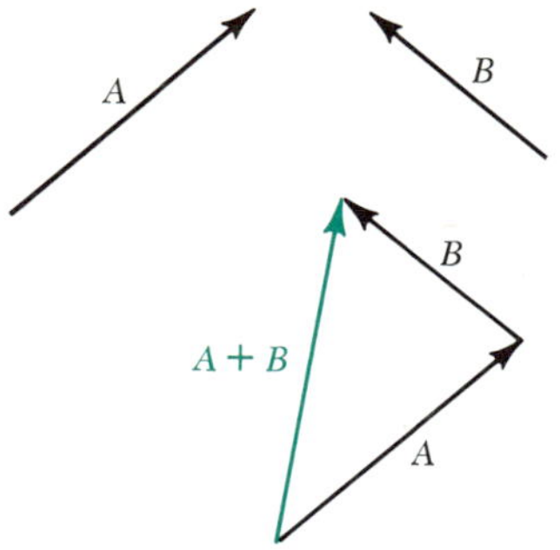

Figure 3-18 Vector addition: vector A + vector B = vector $A + B$.

A speedometer on a car is correctly named because it indicates only the *magnitude* of the velocity. You cannot tell by looking at the speedometer what the *direction* of the velocity is. Likewise, when a salesman extols the pick-up of a car, he doesn't care whether the acceleration is acting northward or eastward; he is interested only in its magnitude. Vectors are usually represented by arrows. The length of the arrow represents (to some scale) the magnitude of the vector, and the direction of the arrow indicates the direction of the vector.

The essential feature of vectors is that the combined effect, known as the *resultant,* or sum, of two or more vectors may be found by a geometrical rule known as *vector addition.* To add two vectors, place them head to tail (this may require moving one of them parallel to itself to maintain its direction) and the resultant is the vector drawn from the tail of one to the head of the other, as shown in Figure 3-18. If there are more than two vectors, they can be added in the same way and the resultant is the vector drawn from the tail of the first to the head of the final vector. Applied to displacement vectors, it works out like this:

Example 3-6 A girl delivering papers covers her route by traveling 3 blocks west, 4 blocks north, then 6 blocks east. What is her total displacement? What is the total distance she travels?

The displacements are laid off on a scale diagram (Figure 3-19) and, by measurement of the length of the resultant (or by simple geometry), the displacement is found to be of magnitude 5 blocks. (Can you see how to apply the Pythagorean theorem to this example?)

A protractor is used to find that the direction of the resultant is 53° north of east.

The distance traveled by the girl is a total of 13 blocks, found by adding up the various parts of her journey. This distance is a scalar quantity, and no particular direction can be associated with the figure of 13 blocks.

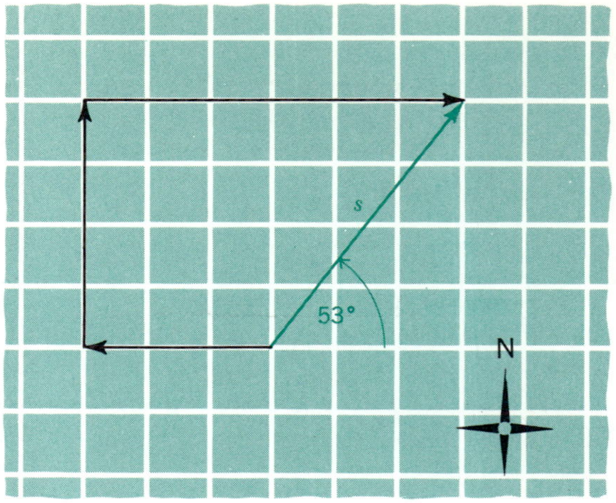

Figure 3-19 Vector addition of displacements.

The preceding example illustrates the distinction between displacement, a vector, and distance, a scalar. Another example: After a summer of vacation driving, the displacement is zero as the family car eases back into its garage, although the distance covered may be considerable. This is because a vector drawn from the beginning of the trip to the end of the trip has zero length.

Vector addition of velocities is necessary if a body has two or more velocities simultaneously, due perhaps to separate causes.

Example 3-7 A helicopter heads due northeast (an angle of 45° east of north) and has an air speed of 70 km/hr. Simultaneously, a wind of 30 km/hr blows from the north. What is the displacement of the helicopter 2 hr after leaving the heliport?

In 2 hr the helicopter would have gone 140 km due to its own air speed, and 60 km due to the wind. The two displacements take place simultaneously. Laying off the two displacements to scale (Figure 3-20), the resultant displacement is found (graphically) to be

$$\boxed{106 \text{ km, } 23.5^\circ \text{ N of E}}$$

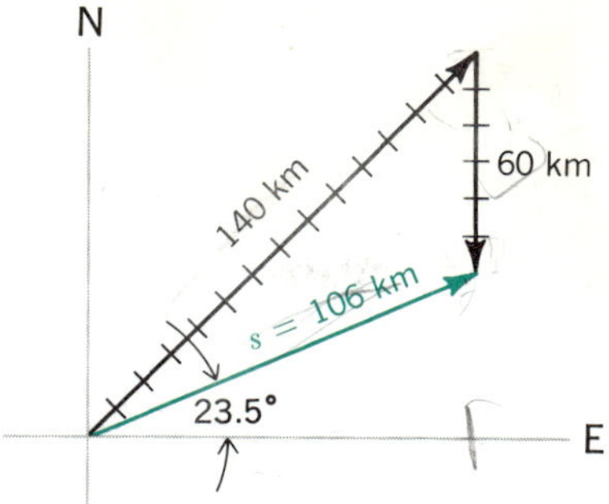

Figure 3-20

Example 3-8 Let's look at the same problem from a slightly different viewpoint. Instead of adding displacements, we can add (vectorially) the two velocities to get the magnitude and direction of the resultant velocity (Figure 3-21). In this case the "sum" of 30 km/hr and 70 km/hr is only 53 km/hr. The magnitude of the resultant displacement is found from

$$\begin{aligned} s &= v_i t \\ &= (53 \text{ km/hr})(2 \text{ hr}) \\ &= \boxed{106 \text{ km}} \end{aligned}$$

as before.

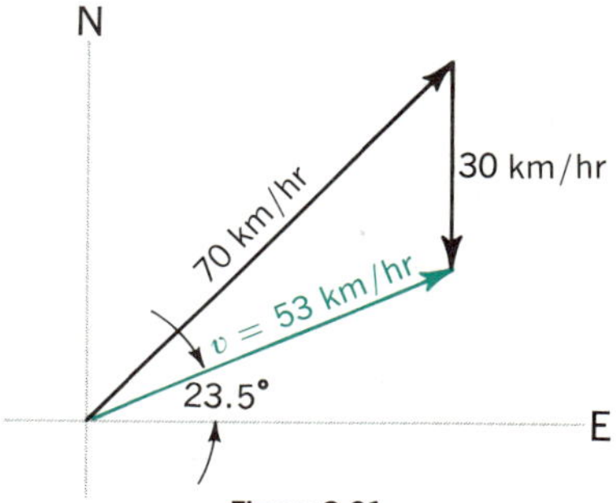

Figure 3-21

Any number of vectors can be added by the head-to-tail method. Thus, in Figure 3-22, R is the sum of vectors A, B, C, D, and E. The order of addition of these vectors is immaterial.

It is sometimes necessary to subtract two vectors. One method is to place the vectors tail to tail, then the *difference* is a vector drawn from the head of the *second* to the head of the first (Figure 3-23). This is the same thing as *adding* the *negative* of the second vector to the first (review Figure 3-18).

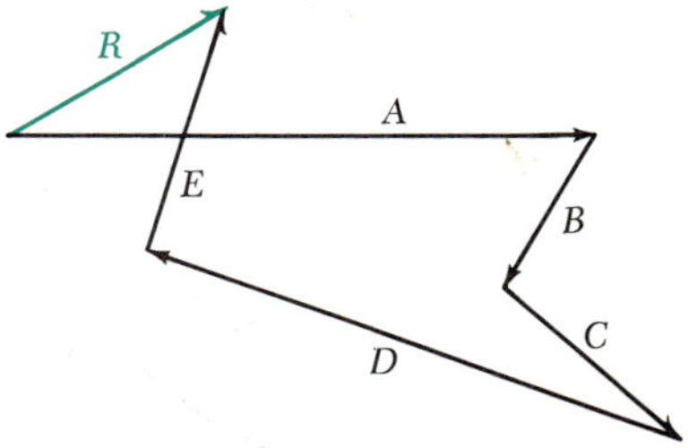

Figure 3-22

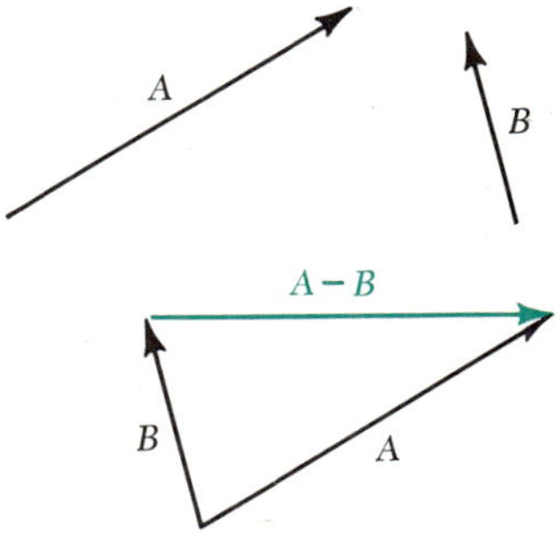

Figure 3-23

Figure 3-22 Head-to-tail method for addition of vectors.

Figure 3-23 Vector subtraction: vector A − vector B = vector $A - B$.

3-7 RELATIVE VELOCITY

Most of us are already familiar with the idea of relative velocity. A highway patrol car traveling at 125 km/hr overtakes a car traveling at 100 km/hr.

We say that the velocity of the patrol car relative to the car is 125 − 100, or 25 km/hr. Or if two cars are approaching each other, one going at 40 km/hr and the other going at 60 km/hr, the velocity of one relative to the other is 100 km/hr. Notice that in each of these examples the velocity vectors are parallel to each other (in the same or opposite directions). When vectors are parallel to each other they may be added or subtracted like scalars.

A ship is going southward at 30 km/hr, and a deck-walker moves northward at 2 km/hr relative to the deck. It is not hard to see that the velocity of the walker relative to the ocean is 28 km/hr southward. To avoid confusion in more complicated cases, let's put this simple problem, which can be solved by scalar subtraction, into a standard vector form (Figure 3-24) making north the positive direction.

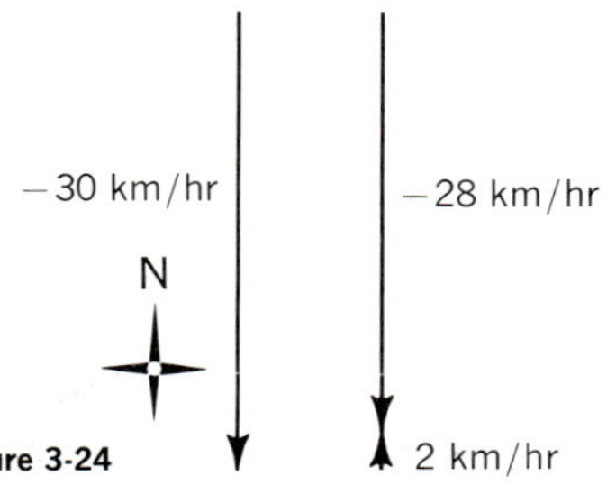

Figure 3-24

velocity of man relative to ocean

= (velocity of man relative to ship)

+(velocity of ship relative to ocean)

= (+ 2 km/hr) + (−30 km/hr)

= − 28 km/hr (i.e. southward)

For complete treatment of relative velocity, useful for situations in which the motions are not parallel to each other, additions should be carried out in a vector sense. If we use the symbol v_{MO} for the velocity of M relative to O, etc., we can write the vector sum as

$$v_{MO} = v_{MS} + v_{SO}$$

where it is understood that vector addition is to be performed. Note carefully the order of the subscripts. This scheme can be extended to include any number of bodies in relative motion, as you will see in our next example.

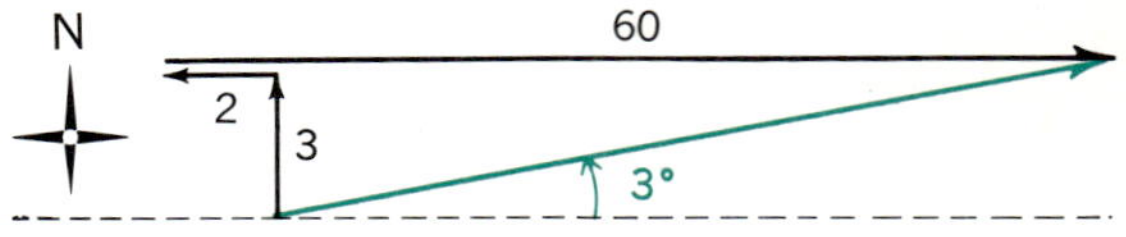

Figure 3-25

Example 3-9 A train T is moving eastward at 60 km/hr, a waiter W is walking toward the rear of the train at 2 km/hr, and a fly F is crawling north across the waiter's tray at 3 km/hr. What is the velocity of the fly relative to the earth E?

$$v_{FE} = v_{FW} + v_{WT} + v_{TE}$$

The vector diagram (Figure 3-25) expresses the vector equation written above. If a large (full-page), carefully constructed diagram is used, the velocity of the fly relative to the earth measures out to be slightly more than

58 km/hr, directed about 3° N of E

3-8 COMPONENTS OF A VECTOR

When one vector, labeled C in Figure 3-26, is the resultant of two other vectors, labeled A and B, then A and B are called the *components* of C. Any given vector can be resolved into components in infinitely many ways. But it is usually necessary to use components that are perpendicular to each other—as in (c). These are called rectangular components; if you add them together their sum is the original vector. For instance, if a ball is traveling through the air in a direction making an angle θ with the horizontal, we can show in a diagram (Figure 3-27a) the horizontal component of v (denoted by v_x) and the vertical component of v (denoted by v_y). Since the components are at right angles to each other, the Pythagorean theorem can be used for calculating v in place of a scale drawing. Therefore, $v = \sqrt{v_x^2 + v_y^2}$.

In finding rectangular components, it is often convenient to construct them along the perpen-

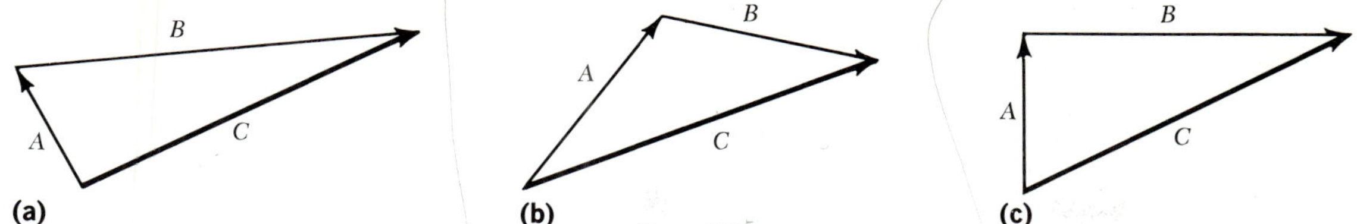

Figure 3-26

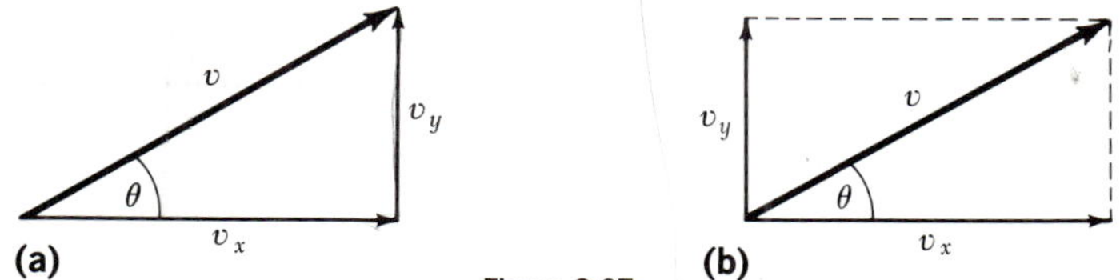

Figure 3-27

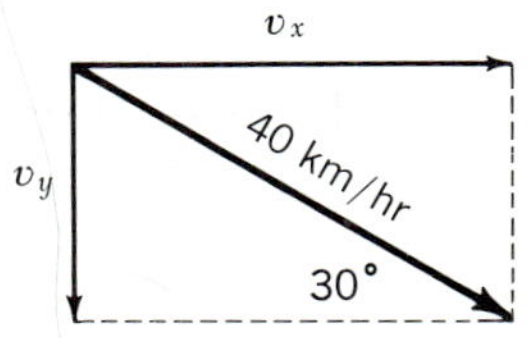

Figure 3-28

dicular directions asked for, as in Figure 3-27b, by completing the rectangle in which the resultant vector is the diagonal.

Example 3-10 A home-run ball moving at 40 km/hr is caught by a fan in the bleachers. Its path as it approaches the fan makes an angle of 30° with the horizontal (Figure 3-28). What are the horizontal and vertical components of the ball's velocity?

The first step in solving this problem is to draw the velocity to scale and construct a rectangle with horizontal and vertical sides. The horizontal and vertical components can then be measured directly from the scale drawing. Careful measurements give $v_x = 34.6$ km/hr and $v_y = 20$ km/hr. (Many of you may be able to solve this kind of problem using trigonometry or, in this particular case, using your knowledge of the 30–60–90 triangle. In most cases, graphical solutions will be used for the presentations in this book. In the section *For Further Study* at the end of this chapter, the same kind of problem is solved trigonometrically.) In this example, v_y would be called -20 km/hr or $+20$ km/hr, depending on which direction had been chosen as the positive direction.

Let's use this problem to clarify a source of confusion about components. What do we mean when we say that the ball in Example 3-10 has a velocity component downward? It is obvious that the direction of motion is along the path on which the ball is moving. The physical significance of the vertical component is that it tells us how fast the ball is dropping *while* it moves on its curved path. This is the subject of the next section.

3-9 PROJECTILES

The strobe photo in Figure 3-29 shows a ball that has been projected horizontally and photographed every $\frac{1}{30}$ second as it fell. Inspection of the vertical lines superimposed on the photograph shows that during each interval the horizontal component of the displacement is constant, indicating a constant horizontal velocity component. On the other hand, the vertical displacements are not constant, as indicated by the increased spacing of the horizontal lines; the vertical component of the velocity is increasing.

In projectile motion we really have two motions occuring at the same time. The only thing the two motions have in common is that they take place during the same time interval. The horizontal part of the motion is uniform motion in a straight line, since there is no horizontal resisting force. The vertical part of the motion is uniformly accelerated motion, since it is governed by the laws of falling bodies. Thus it

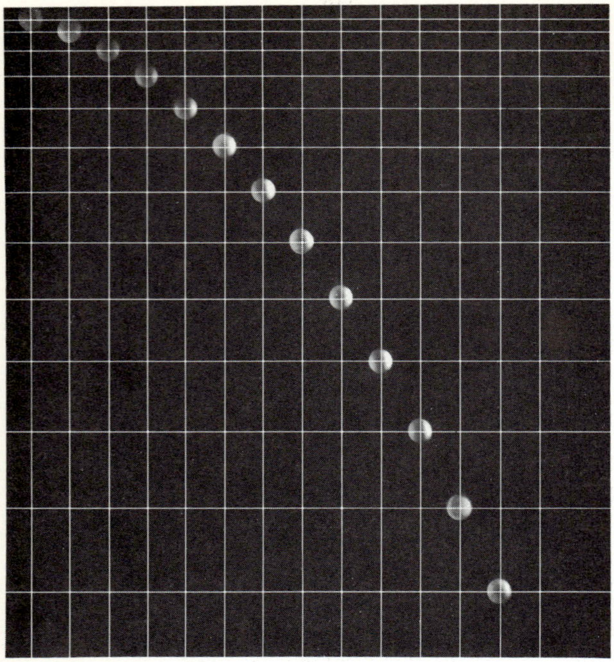

Figure 3-29 A multiple-flash photograph of a ball projected horizontally.

is necessary to resolve the velocity into its horizontal and vertical components at any instant, and treat each motion separately.

Example 3-11 A stone is thrown horizontally from a cliff 10 m high. The initial velocity is 5 m/sec. How far from the base of the cliff does the stone strike the ground (Figure 3-30)?

Consider first the *vertical* motion in order to find how long the stone is in the air. The vertical component of the initial velocity is zero, since the stone is thrown horizontally.

$v_i = 0$
$a = +9.8\ \text{m/sec}^2$
$s = +10\ \text{m}$
$t = ?$

$$s = v_i t + \tfrac{1}{2}at^2$$

$$10\ \text{m} = 0(t) + \tfrac{1}{2}(9.8\ \text{m/sec}^2)(t^2)$$

$$t = 1.43\ \text{sec}$$

Next consider the *horizontal* motion. Since $a = 0$ (there is no horizontal acceleration), $s = v_i t$ for this motion, and v_i is the full 5 m/sec because it is directed horizontally.

$$s = v_i t$$

$$= (5\ \text{m/sec})(1.43\ \text{sec})$$

$$= \boxed{7.15\ \text{m}}$$

Example 3-12 Calculate the velocity of the stone in Example 3-11 as it hits the ground.

Since velocity is a vector, both its magnitude and direction are needed for a complete answer.

Vertical motion:

$v_i = 0$
$a = +9.8\ \text{m/sec}^2$
$v = ?$
$s = +10\ \text{m}$

$$v_f^2 = v_i^2 + 2as$$

$$v_f^2 = 0^2 + 2(9.8\ \text{m/sec}^2)(10\ \text{m})$$

$$v_f = 14\ \text{m/sec}$$

$$= v_y$$

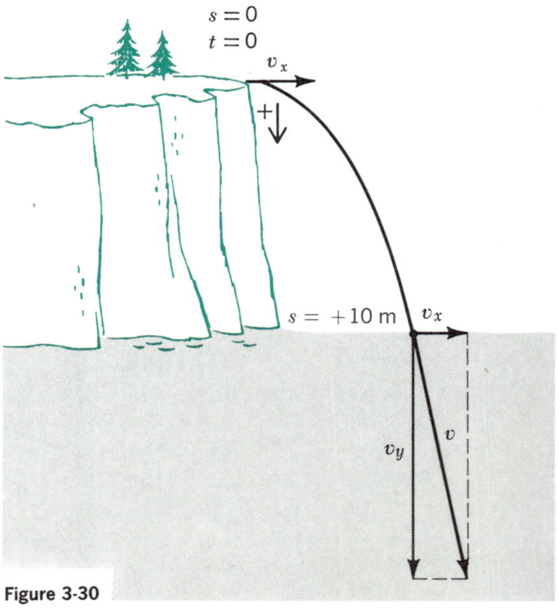

Figure 3-30

We interpret this answer as v_y, the vertical component of the final velocity.

Horizontal motion: The horizontal component of the final velocity is the original horizontal component (Example 3-11), which has remained unchanged (5 m/sec). The resultant final velocity is found by combining v_x and v_y:

$$\begin{aligned} v &= \sqrt{v_x^2 + v_y^2} \\ &= \sqrt{5^2 + 14^2} \\ &= 14.9 \text{ m/sec} \end{aligned}$$

The angle is found from the scale diagram (Figure 3-30) by measurement. Our answer is

$$v = \boxed{\begin{array}{l}14.9 \text{ m/sec at an angle} \\ 70^\circ \text{ below horizontal}\end{array}}$$

In Example 3-11 the same symbols were used for different quantities. Thus, in the box $v_i = 0$, whereas in the last line of the same problem $v_i =$ 5 m/sec. Similarly, in the box $s = 10$ m, while on the last line $s = 7.15$ m. These apparent contradictions will be cleared up if you remember that we are actually solving two separate problems when we solve a projectile problem. The only symbol that is necessarily the same for the two motions is t, the time interval, which is 1.43 sec for both parts of Example 3-11.

Example 3-13 A bomb is released from a jet plane flying horizontally with a speed of 900 km/hr at an altitude of 1000 m. Where will the bomb strike the ground?

In this case the bomb continues with the horizontal velocity of the plane. Just after it is released, the bomb is traveling forward at 900 km/hr. First we change 900 km/hr to m/sec.

$$\begin{aligned} 900 \text{ km/hr} &= 900 \text{ km/hr}\left(\frac{10 \text{ m/sec}}{36 \text{ km/hr}}\right) \\ &= 250 \text{ m/sec} \end{aligned}$$

Then we proceed to consider the two motions separately, as in all projectile problems. First, we find the time the bomb was falling.

Vertical motion:

$$\boxed{\begin{aligned} v_i &= 0 \\ s &= +1000 \text{ m} \\ a &= 9.8 \text{ m/sec}^2 \\ t &= ? \end{aligned}}$$

$$s = v_i t + \tfrac{1}{2}at^2$$

Since the original velocity has no vertical component, $v_i = 0$, and so

$$\begin{aligned} t &= \sqrt{\frac{2s}{a}} \\ &= \sqrt{\frac{2(1000 \text{ m})}{9.8 \text{ m/sec}^2}} \\ &= 14.3 \text{ sec} \end{aligned}$$

Horizontal motion: Since the horizontal acceleration is zero,

$$\begin{aligned} s &= v_i t \\ &= (250 \text{ m/sec})(14.3 \text{ sec}) \\ &= \boxed{3575 \text{ m}} \end{aligned}$$

The bomb strikes 3575 m ahead of the point directly below the point of release. Since the plane's velocity is 250 m/sec and the bomb's horizontal component of velocity is also 250 m/sec, the bomb always remains directly beneath the plane as it travels. We can say that the bomb always has the same *horizontal* position as the plane does, because its *horizontal* velocity relative to the plane is always zero.

In actual practice, the effects of air resistance are by no means negligible in projectile problems. The effect of air resistance is to lengthen the time in the air and also to cause the horizontal velocity to decrease somewhat from the original value. The exact study of projectile motion is called *ballistics,* and its mathematical complexities are such that electronic computers are often used.

SUMMARY

Kinematics is the description of motion, without regard to causes. **Dynamics** (to be studied in Chapter 4) is the study of the causes of motion. In **uniformly accelerated motion,** the **variables** are related to each other by the following equations:

$$a = \frac{v_f - v_i}{t} \qquad s = \frac{v_i + v_f}{2}t \qquad v_f^2 = v_i^2 + 2as \qquad s = v_it + \tfrac{1}{2}at^2$$

Motion with constant velocity is a special case of accelerated motion, in which $a = 0$.

These equations are perhaps best understood by their graphical interpretations. The area under the velocity-time graph represents the distance traveled and the slope of the velocity-time graph is the **acceleration.**

When the distance-time graph is a straight line, its slope gives the **velocity.** When the distance-time graph is a curve the average velocity equals the slope of a chord drawn between two points on the curve. The instantaneous velocity equals the slope of the tangent to this graph drawn through the point representing the time at which the velocity is desired. **Instantaneous velocity** is the limiting value of the average velocity as the time interval approaches zero.

Vector quantities have both magnitude and direction; **scalars** have magnitude only. The **resultant** of any number of vectors is found by the head-to-tail rule. Components of vectors are found by scale drawings.

If v_{AB} represents the velocity of A relative to B, then the velocity of A relative to another body Z can be found by vector addition: $v_{AZ} = v_{AB} + v_{BC} + v_{CD} + \cdots + v_{YZ}$

The motion of a projectile may be regarded as two motions that are independent and yet simultaneous. In the absence of air resistance, the vertical motion is uniformly accelerated, and the horizontal motion is one of constant velocity.

QUESTIONS

1 Although we have not talked about acceleration-time graphs, these can be constructed for any motion. What do you think is the significance of the area under the acceleration-time graph?

2 How would you measure the velocity of a jet airplane at the instant it passes by a control tower in which you are stationed?

3 Every automobile is equipped with an odometer on the dashboard. Does this measure a scalar or a vector quantity?

4 What is the description, in words, of the motion of the falling bomb in Example 3-13 as seen from the plane? What is the description, as seen from a speeding automobile moving 100 km/hr, parallel to the plane's motion, and in the same direction? The automobile was directly beneath the plane when the bomb was released.

5 A base runner sometimes takes a jump at the bag as he approaches first base. Is this more likely to result in an out than if he makes a level approach? Explain.

PROBLEMS

GROUP A

1 A driver travels the length of the Pennsylvania Turnpike (576 km) in 6 hr 40 min. What is her average speed (a) in km/hr? (b) in m/sec?

2 If, while you are driving along at 45 km/hr, your attention wanders for 0.5 sec, how far do you travel "blind" during that half second?

3 A snail traveling at a snail's pace (12 m/day) decides to slow down to only 5 m/day, and allows itself 2 min in which to make the change. (a) Express its initial velocity in km/hr; in cm/yr; in m/sec. (b) Compute the acceleration in m/day·min; in m/sec^2.

4 The distance-time graph of a car is shown in Figure 3-31. Make a velocity-time graph for this car.

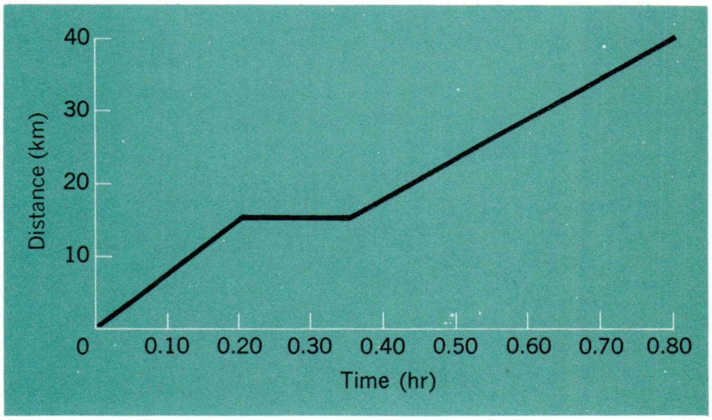

Figure 3-31

5 Two cars, side by side at the starting line, accelerated from rest at the same time. Their velocity-time graphs are shown in Figure 3-32. (a) At what time are the cars going at the same speed? (b) At what time does the "faster" car overtake the "slower" car? (c) How far from the start are they when the "slower" car catches up?

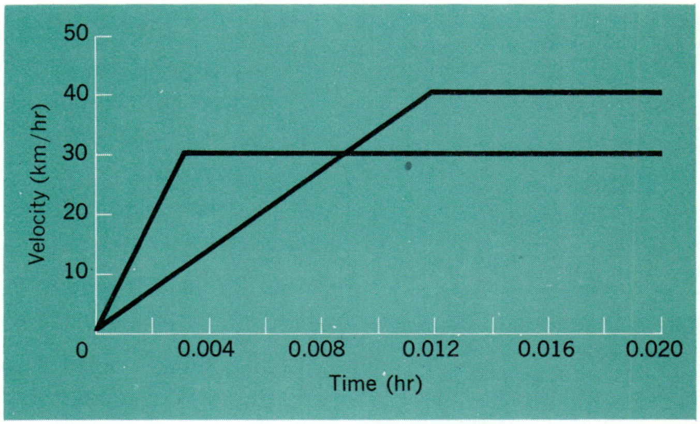

Figure 3-32

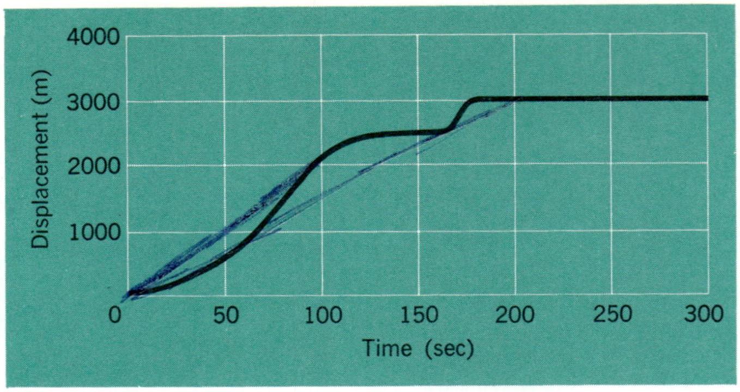

Figure 3-33 Displacement-time graph for motion of a subway train.

6 The displacement graph for a subway train traveling from one station to another is shown in Figure 3-33. (a) Estimate the instantaneous velocity 100 sec after leaving the station. (b) What is the instantaneous velocity 200 sec after leaving the station? (c) What is the average velocity during the first 200 sec?

7 (a) What is the instantaneous velocity of a falling body at the end of 2 sec of its fall? (b) What is the velocity at the end of 3 sec of fall? (c) What is the average velocity, averaged over the third second of the fall? (d) What is the distance traveled during the third second?

8 In a certain city the blocks are all square, the streets run east-west, and the avenues run north-south. A child starts from the corner of West Third Avenue and North Second Street, and walks south 5 blocks along West Third Avenue, crossing Main Street and reaching South Third Street. She then walks 4 blocks eastward, crossing Center Avenue to reach East First Avenue. Then she runs to the corner of North First Street and East First Avenue, and finally she goes along North First Street to West Fourth Avenue. (a) What is the distance traveled? (b) What is the displacement for the entire trip?

9 An airplane in a power dive is moving at 400 km/hr downward, making an angle of 30° with the vertical. If the sun is directly overhead, how fast is the plane's shadow moving along the ground?

10 All points on the equator are moving eastward at about 1660 km/hr due to the earth's rotation. (a) If a rocket is launched at the equator, aimed horizontally toward the east with a velocity of 8250 km/hr relative to the earth, what will its velocity be relative to the "fixed stars"? (b) What would it be if the rocket had been aimed toward the west?

11 A ball is thrown upward at 12 m/sec. Find the magnitude and direction of its velocity (a) 1 sec after being thrown; (b) 2 sec after being thrown.

12 A plane is traveling toward the northwest at 300 km/hr. What is the westward component of its velocity?

GROUP B

13 A bullet is fired at 660 m/sec and strikes a target that is 200 m from the gun. If the marksman hears the sound of the impact on the target 0.90 sec after he fires the gun, what is the speed of sound in this situation?

14 Nolan Ryan pitched a baseball that traveled from the pitcher's mound to home plate (a distance of 18.5 m) in 0.411 sec. (a) What was the speed of the ball in km/hr? (b) If the catcher allowed his mitt to recoil backward 0.075 m while catching the ball, what was the acceleration of the ball while it was being slowed down by the catcher?

15 A train traveling at 1 km/min is slowed down at a uniform rate of 1.1 m/sec^2. How far will it travel before its speed is half the original value?

16 A spacecraft increases its speed at the rate of 0.02 km/sec^2. How much time is required for the speed to increase from 7 km/sec to 8 km/sec? How far does it travel during this time?

17 A drop of water falls from a roof gutter, and 2 sec later another drop falls. Neglecting air resistance, how far apart are the two drops by the time the first one has reached a speed of 29.4 m/sec?

18 How much time does a driver "lose" on a long trip in which she must pass through 20 school zones, each 0.4 km long, if she slows down for each zone to 30 km/hr instead of maintaining a constant speed of 60 km/hr for the entire trip? Assume that the driver applies her brakes when she is 0.05 km from the start of each zone, and requires 0.10 km to regain full speed each time.

19 A ball is thrown vertically to a height of 19.6 m and allowed to strike the ground. If it loses one fourth its speed while in contact with the ground, how high does it rise on the rebound?

20 With what upward speed should a package be thrown in order to be caught easily by a person on a balcony 6 m above the ground?

21 A car moving at 108 km/hr makes a head-on collision with a stone wall. From what height would the car have to fall in order to make an equally hard collision with the ground?

22 A sailor drops his pocketknife from the top of a mast on a ship sailing eastward at 10 m/sec. The mast is 20 m high. Where does the knife hit the deck?

23 A glider has an air speed of 100 km/hr, and the pilot notices that, although she is headed due east, she is actually traveling northeast, and covers 70.7 km in 30 min. What is the magnitude and direction of the wind that is blowing her off course?

24 A swimmer can swim 4 km/hr in still water. (a) Starting from the west bank of a river that flows southward at 2 km/hr, in what direction should she head (somewhat upstream) so that she will travel directly across? (*Hint:* Velocity of the swimmer relative to the earth must be a vector directed toward the east.) (b) What will be the magnitude of the swimmer's resultant velocity? (c) How much time will be required for the trip if the river is 0.2 km wide?

25 If the swimmer in Problem 24 swims 0.2 km directly downstream and then swims back again to her starting point, how does her time for the round trip compare with what it would have been if she had swum to a point directly across from her starting point and back again, which is also a round trip of 0.4 km?

26 If the swimmer of Problem 24 heads straight across the river, so that she lands somewhat downstream, how long will it take her to cross?

27 A ball rolls with a speed of 0.6 m/sec toward the edge of a table that is 0.8 m above the floor. The ball rolls off the table. How long is it in the air? How far out from the edge of the table does the ball hit the floor?

28 Why does a hunter raise the barrel of his rifle when aiming at a distant target? If he aims directly at a target 115 m away, by how much will he miss the target if the muzzle velocity of the bullet is 330 m/sec?

29 A tail-mounted gunner fires a bullet horizontally at 500 m/sec (muzzle velocity) directly backward from a plane flying at an altitude of 100 m with a ground speed of 200 m/sec. How far behind the plane should a target on the ground be when the gun is fired?

30 A book is held 4.9 cm above a table top and released. How long is the book in the air?

31 A helicopter headed due north has an air speed of 40 km/hr, and the wind is 30 km/hr from the west. (a) What are the magnitude and direction of the plane's velocity relative to the ground? (b) How far does the helicopter travel in 12 min?

GROUP C

32 A late passenger, sprinting at 8 m/sec, is 30 m away from the rear end of a train when it starts out of the station with an acceleration of 1 m/sec^2. Can the passenger catch the train if the platform is long enough? (*Note:* This problem requires solution of a quadratic equation. Can you explain the significance of the two values you get for the time?)

33 A runner traveling at constant speed finds himself with 140 m still to go when 3 min 20 sec have elapsed in a 1500 m race. If for his home-stretch "kick" he accelerates at 0.20 m/sec^2 for 10.0 sec and then holds the new speed until the finish line, can he still run a 3 min 35 sec race?

34 A car and a truck are each traveling at 20 m/sec, and the car is 37.5 m behind the truck. The car driver decides to pass the truck, and he steps on the gas, producing an acceleration of 3 m/sec². (a) How long will it be before the car is alongside the truck? (b) How fast will the car be moving relative to the truck when they are side by side? (c) How far will the car travel while reaching the truck? (d) How far will the truck travel during the same time?

35 A flower pot topples from a penthouse window sill. A person on a lower floor observes that the pot passes a picture window 2 m tall in 0.1 sec. How far above the top of the picture window is the penthouse window sill?

Note: In the *For Further Study* sections at the ends of many chapters, you will have an opportunity to go beyond what is required of most students. These additional sections are of different types: some are proofs of statements assumed without proof earlier in the chapter; some are interesting sidelines rather than integral parts of the course; and some are extensions of material in the text. A knowledge of elementary trigonometry is needed for some of these sections.

FOR FURTHER STUDY

PROJECTILE MOTION AT AN ANGLE

So far, discussion of projectiles has been confined to cases in which the initial velocity is horizontal. If this is not the case, it is necessary to resolve the initial velocity into its horizontal and vertical components. The vertical component is used to find the time in the air, and the horizontal component of the initial velocity is used to find the horizontal distance that the projectile travels.

Example 3-14 A cat jumps off a table that is 1 m high. The initial velocity of the cat is 5 m/sec, at an angle 37° above the horizontal. How far out from the edge of the table does the cat strike the floor?

Resolving the initial velocity into its components (Figure 3-34), we

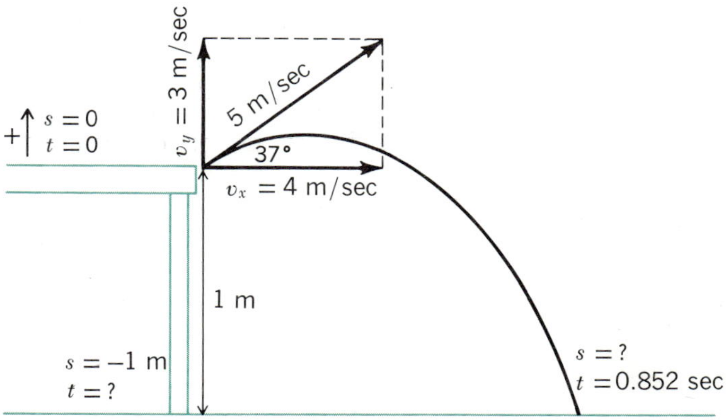

Figure 3-34

see that v_i in the y-direction can be found from $\sin 37° = v_i/5$, so $v_i = 5 \sin 37° = 3$ m/sec. (Note that 37° is the smaller angle in a "3–4–5 triangle," since $\sin 37° = 3/5 = 0.6$.) We make a table of knowns and unknowns for the vertical motion and solve for the time.

$v_i = +3$ m/sec
$s = -1$ m
$a = -9.8$ m/sec^2
$t = ?$

$$s = v_i t + \tfrac{1}{2}at^2$$
$$-1 = 3t - 4.9t^2$$
$$4.9t^2 - 3t - 1 = 0$$
$$t = 0.852 \text{ sec} \qquad t = -0.240 \text{ sec}$$

The quadratic equation has two roots. The positive answer is the only one that applies, since the position of the cat is desired when it strikes the floor *after* it jumps. The horizontal component of the initial velocity is $5 \cos 37° = 5 \times 0.8 = 4$ m/sec. The range R is the horizontal distance given by

$$s = v_i t$$
$$= (4 \text{ m/sec})(0.852 \text{ sec})$$
$$= \boxed{3.41 \text{ m}}$$

Suppose a gunner wishes to make a table showing how the horizontal range depends on the angle of elevation θ of his gun. We can derive an equation for this purpose, and check the dimensions on both sides of our equation when it is in its final form.

Although the mathematical solution of this problem is not difficult, let's first look at the physics of the situation. There are two extreme cases. (a) If $\theta = 90°$ (gun pointed straight up), the range R is obviously zero. (b) If the gun is pointed nearly horizontally ($\theta = 0$), the bullet is in the air for only a short time interval t before striking the ground, since the vertical component of the initial velocity is so small. Therefore the forward velocity component acts for only a very short time, and the range will be very small.

In some respects it is easier to work with symbols than with numbers. We can follow the same scheme as in Example 3-14, but using symbols;

Vertical motion:

$$v_y = v_i \sin \theta$$

and the total time that the projectile is in the air is

$$t = \frac{v_f - v_i}{a}$$

So for the *vertical motion* this becomes

$$t = \frac{v_i \sin \theta - (-v_i \sin \theta)}{a}$$

since it lands with the same vertical component of the velocity (except that the direction has changed) that it had originally. Thus,

$$t = \frac{2v_i \sin \theta}{a}$$

Horizontal motion:

$$v_x = v_i \cos \theta$$

and the range is given by

$$\begin{aligned} R &= v_x t \\ &= (v_i \cos \theta)\frac{2v_i \sin \theta}{a} \\ &= \frac{2v_i^2}{a} \cos \theta \sin \theta \end{aligned}$$

This is the equation we wished to derive. We now proceed to check our result in two ways. Any physical equation must have the same dimensions on both sides of the equals sign. In this case the check can be made as follows ($\cos \theta$ and $\sin \theta$ are ratios of lengths, and hence have no dimensions):

$$\text{m} = \frac{(\text{m/sec})^2}{\text{m/sec}^2}$$

$$\text{m} = \text{m}$$

The check has shown that our result is at least not obviously wrong, but we have only verified its dimensions, not the exact form. At any rate, the first hurdle has been passed. The equation is "not wrong" dimensionally, and is therefore more likely to be "right" than not.

A second check on our derivation is to see whether the answer is "reasonable." Does it fulfill the expected behavior listed under (a) and (b) at the beginning of our solution? If the angle is 90°, condition (a) is fullfilled, since $\cos 90° = 0$, and the factor $\cos \theta$ therefore makes R become zero. Likewise, condition (b) is fulfilled because the factor $\sin \theta$ becomes zero when θ is 0°. Thus, both of the preliminary expectations (a) and (b) are fulfilled, another indication that our derivation is probably correct.

Similar checks should be made for any physical derivation.

PROBLEMS

GROUP D

36 A diver takes off with a speed of 8.00 m/sec from a diving board 3 m high, at an angle of 30° above the horizontal. How much later does she strike the water?

37 A pilot cuts loose two of his fuel tanks in an effort to gain altitude. At the time of release, he was 120 m above the ground and traveling upward at an angle of 30° above the horizontal, with a speed of 84 m/sec. For how long were the tanks in the air?

38 A runner leaves third base at the instant the center fielder catches a fly ball, and 5.00 sec later crosses home plate. The fielder, who is 90 m from home plate, throws the ball 0.5 sec after catching it, and the runner is safe by an eyelash. (a) What was the horizontal component of the velocity of the thrown ball? (b) What was the vertical component of the thrown ball's velocity? (c) At what angle above horizontal, and with what speed, did the fielder throw the ball? (d) How high in the air did the ball go? (e) Explain how the fielder might have caught the runner if he had thrown the ball at a lower angle, allowing it to bounce once on the way to home plate.

39 A broad jumper takes off at an angle of 20° above the horizontal and jumps 0.60 m high. (a) What is her forward velocity? (b) How far does she jump?

40 On level ground, a ball is thrown forward and upward. The ball is in the air 2 sec, and strikes the ground 30 m from the thrower. With what speed and at what angle was the ball thrown?

4 Dynamics — The Causes of Motion

The study of certain simple motions in Chapter 3 was a beginning in kinematics, a science for which Galileo (1564–1642) was largely responsible. He placed emphasis upon experiment and upon the mathematical description of physical laws such as those concerned with the motions of falling bodies and projectiles. The methods of Galileo are, of course, characteristic of science as we know it today. Building upon the work of Galileo and others and making full use of the "new" outlook in his methods, Isaac Newton (1642–1727) created a system of mechanics that ranks as one of the great intellectual achievements of all time. This chapter presents the laws of motion as set forth by Newton—on these laws the foundations of the science of dynamics were laid.

Born in the year of Galileo's death, Isaac Newton was the most influential scientist of modern times. According to his own opinion, Newton was "in the prime of my age for invention" during the plague years 1665 and 1666, when he left Cambridge University and studied mathematics, mechanics, and optics at his farm home in Lincolnshire, England. When he returned to the university, his own teacher, Isaac Barrow, resigned his professorship so that the 26-year-old Newton could have it. By this time the young genius had made profound discoveries in many fields, but found himself so involved in controversy over his first published work, *Theory About Light and Colors,* that he came to care little about public recognition for his work. In 1684 he was persuaded by his friend, the astronomer Halley, to organize and publish his system of mechanics, including the theory of gravitation. Thus in 1686 Newton's *Principia* made its appearance—probably the most important book in physics ever written. Newton lived for forty years after writing the *Principia* and was a legend in his own time, but in later life he devoted himself mainly to theology and to the performance of his duties as Warden, and later Master, of the Mint. For a more complete story of Newton's life, consult the references at the end of this unit. A reproduction of the title page of Newton's *Principia* appears in Figure 4-1 on page 58.

4-1 FORCE AND MOTION

The role of force in everyday life is a familiar one. Pushes and pulls of all sorts—the pull of a magnet on a piece of iron, the push of the wind against a sail, the pushes and pulls exerted by muscles—are all easily tagged as forces. A horse pulls a wagon; a baby throws a rattle; a motor raises an elevator cage. These are common examples of force involving motion. However, motion or lack of

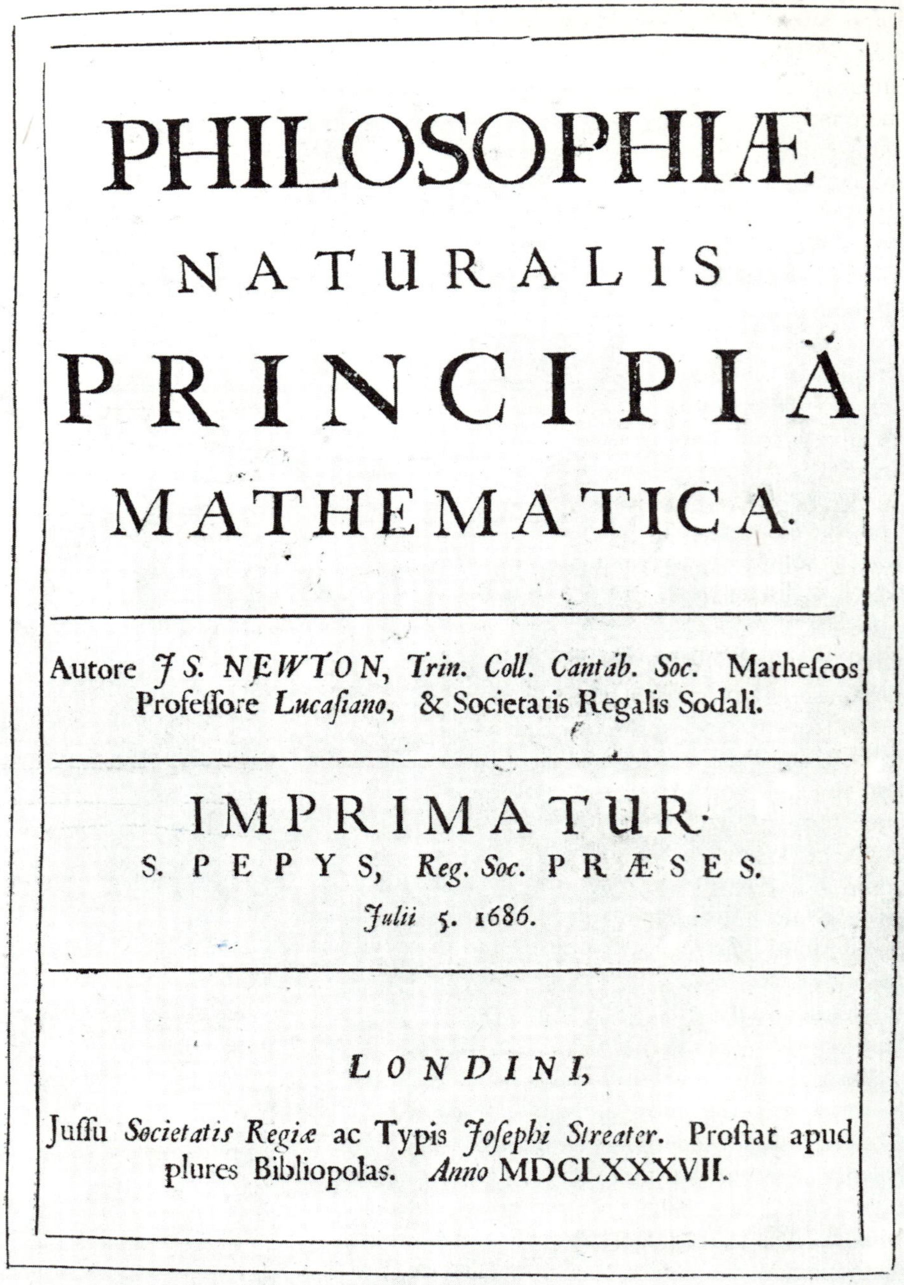

PHILOSOPHIÆ

NATURALIS

PRINCIPIA

MATHEMATICA.

Autore *J S. NEWTON, Trin. Coll. Cantab. Soc.* Matheſeos Profeſſore *Lucaſiano,* & Societatis Regalis Sodali.

IMPRIMATUR.

S. PEPYS, *Reg. Soc.* PRÆSES.

Julii 5. 1686.

LONDINI,

Juſſu *Societatis Regiæ* ac Typis *Joſephi Streater.* Proſtat apud plures Bibliopolas. *Anno* MDCLXXXVII.

Figure 4-1 Sir Isaac Newton (1642–1727) and the title page from his renowned *Principia,* in which the laws of motion that you will be studying here were first written. Like most scholarly works of the time, the *Principia* was published in Latin.

motion is really no indication of the existence of force. A horse can exert a force and yet not move a heavy block of stone; a baby can push on an immovable wall and produce no motion; and there must be a force (tension) exerted by an elevator cable to support the cage even when it is stopped at the sixth floor.

Obviously forces can cause motion, but they do not seem to be always correlated with motion. Where is the motion, if any, "caused" by the pull of gravity on a parked tractor-trailer? And where is the force, if any, causing the motion of the star Regulus as it drifts majestically through space? Questions such as these puzzled the natural philosophers of many centuries. Before Galileo, belief was widespread that a constant force was needed to produce a constant motion. This would seem to imply that in the absence of a force a moving object would come to rest. It is not difficult to see why this belief was held so strongly. After all, we must apply a constant push to keep a heavy box moving across the floor and, if we stop pushing, the box soon comes to rest.

Galileo was the first to deny this common misconception when he argued that a constant motion requires no force. He wrote, "any velocity once imparted to a body will be rigidly maintained as long as there are no causes of acceleration or retardation, a condition which is approached only on horizontal planes where the force of friction has been minimized." It was Newton, however, who clearly stated that *change* of motion is caused by forces and only by forces; unchanging motion, or "uniform" motion, requires no force.

4-2 NEWTON'S FIRST LAW

The motion of a body on which no net force acts is described by Newton's first law, often called the law of inertia:

Newton's first law of motion: A body remains at rest or, if in motion, it remains in uniform motion with constant speed in a straight line, unless it is acted on by an unbalanced external force.

The first law recognizes a property of matter called *inertia*—the tendency for all material bodies to stay at rest or, if in motion, to resist any change in their motion. The essential factor, here, is the absence of a *net* force—that is, an unbalanced force.

For example, if two equal and opposite forces act on a body, the net force is zero and the body's velocity remains constant. "Constant velocity" includes uniform motion in a straight line ($v =$ constant $\neq 0$) as well as a state of rest ($v =$ constant $= 0$). When a horse pulls a wagon along with constant velocity, the older view was that the force exerted by the horse is needed in order to *cause* the (uniform) motion. Newton's interpretation was that the forward force of the horse is balanced by a backward force of friction, the net force is zero, and hence the wagon continues to move forward because of its inertia. In the next section you will see how the inertia of a body is related to the *mass* of the body.

4-3 NEWTON'S SECOND LAW

In a way Newton's first law tells us what *doesn't* happen when *no* net force is applied to a body. To extend the laws of motion, we naturally seek a law describing what *does* happen when a net force *is* applied. We will investigate this situation by describing a series of experiments. These are experiments that can be performed in the laboratory.

Basically, we want to apply known forces to known masses and observe the motions produced. It will be necessary to minimize the effects of friction in these experiments so that the force we apply will be the net force. Although friction can never be removed entirely, let's consider some imaginary experiments where the force of friction is zero. The experiments that you probably will be doing in the laboratory, using modern apparatus that keeps friction at a minimum, should agree closely with the results that follow.

In the first experiment we apply a constant force to a convenient test body by pulling the body along horizontally with a spring balance to measure the pull. We continually adjust our muscular effort to keep the force constant (Figure 4-2). As we do this, our first observation is that the body accelerates, and, when careful measurements are made of the displacement and time, the acceleration is found to be uniform. In this experiment, a constant force produces a uniform acceleration. Can you describe the simultaneous motion of the hand that pulls the test body?

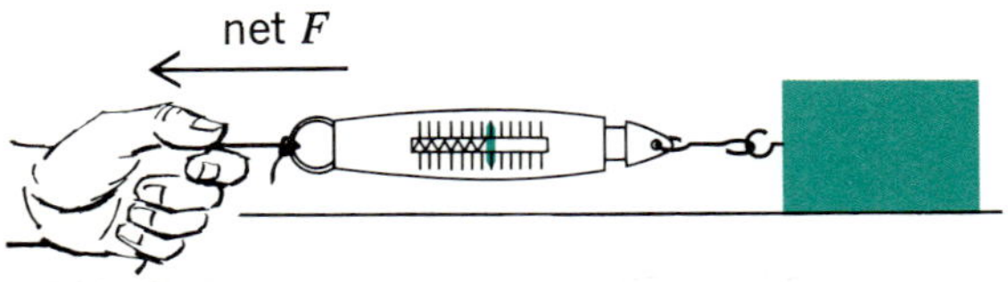

Figure 4-2

Let's repeat the experiment, but this time applying *twice* the force used in the first trial. Again a uniform acceleration results, but now the acceleration is *twice* that obtained with the initial force. When we repeat this experiment using three and four times the original force on the same body, we observe that the acceleration is three and four times the original acceleration.

A statement consistent with all these observations would be the following: *For a given body, the acceleration is directly proportional to the net force.* This can be expressed quantitatively as

$$\frac{\text{net } F}{a} = \text{constant} \qquad \textbf{(Equation 4-1)}$$

in which net F is the net force and a is the acceleration. This equation gives us a way to compare the magnitudes of two forces by letting them act, one after the other, on the same test body. That is, we can express the above relation as

$$\frac{F_1}{F_2} = \frac{a_1}{a_2}$$

We are, in effect, defining the magnitude of force by describing an experimental procedure for measuring it. This "operational" definition is a useful one, since experiment shows that it gives the same ratio for any given pair of forces F_1 and F_2, no matter what "test body" is used for the acceleration experiments.

Force is a vector, so it has direction as well as magnitude. In our experiments the acceleration was always in the direction of the net force. To complete the definition, then, the direction of a net force is said to be in the same direction as the acceleration it gives an object.

The next step is to determine quantitatively the concept of inertia—the tendency of an object to resist changes in motion. The constant in Equation 4-1 is surely somehow related to the inertia of the test body, for, if the body has a large inertia, a large net force is needed to produce a given acceleration, and so (net F)/a is large. Similarly, a smaller force would be needed to give the same acceleration to a body with less inertia. Let's, then, define the constant in this equation as the *inertial mass* (m_i) of the body. That is

$$\frac{\text{net } F}{a} = m_i \qquad \textbf{(Equation 4-2)}$$

By definition, the inertial masses of two bodies can be compared by acceleration experiments. For example, suppose we wish to find the inertial mass of an unknown object in relation to the inertial mass of some standard object. Our experiment consists of applying a given force (any force) to each body in turn, and observing the accelerations produced. If the force gives the standard object an acceleration of 2 m/sec^2, and later it is found that the same force gives the unknown object an acceleration of 3 m/sec^2, we find by using Equation 4-2 that the unknown body has less inertia than the standard body; its inertial mass is $\frac{2}{3}$ of that of the standard body. (Figure 4-3 shows such a ratio.)

We have now seen how to compare two forces with each other and how to compare two inertial masses with each other, but nothing has been said about the units in which force and inertial mass might be measured. Placing the emphasis upon acceleration, Equation 4-2 can be rewritten as

$$a \propto \frac{\text{net } F}{m_i} \qquad \textbf{(Equation 4-3)}$$

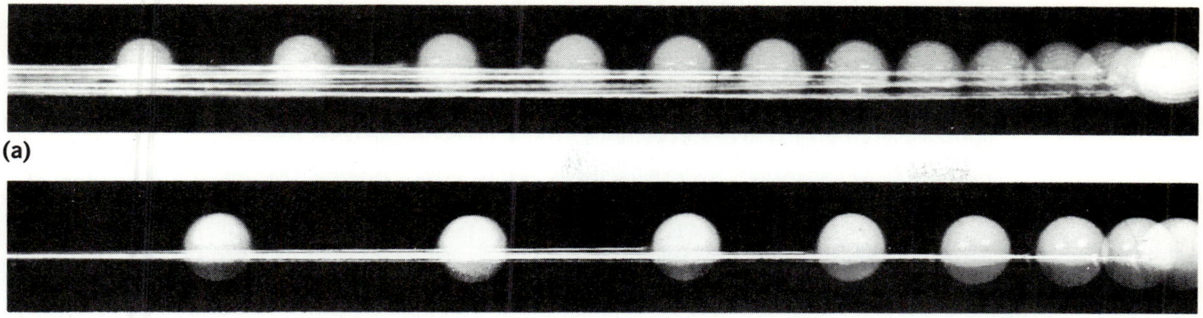

(a)

(b)

Figure 4-3 Strobe photos showing accelerations of different masses when a constant force is applied along the strings. Mass of the ball in **(b)** is about $\frac{2}{3}$ that of the golf ball in **(a).**

where a proportion is used rather than an equality, since units for F and m_i have not yet been established. In words we have

Newton's second law of motion: The acceleration produced by an unbalanced force acting on a body is proportional to the magnitude of the net force, in the same direction as the force, and inversely proportional to the inertial mass of the body.

In Equation 4-3, a and net F represent vectors that are parallel to each other, and m_i is a scalar quantity.

Why is it important to write the word "net" in front of the force in Newton's second law? Failure to write (or worse yet, to think) *net* force is the most common single source of error in problem-solving. For example, if a railroad engine pulls with a force of 20 tons in a forward direction and there is a retarding frictional force of 3 tons acting toward the rear of the train, the net force is a vector sum, or resultant, of four forces. The downward pull of gravity is balanced by the upward push of the tracks on the train, and the remaining two forces act in the same line. Therefore, ordinary algebraic addition can be used: $20 + (-3) = 17$ tons. The use of the word "net" will remind you that you must look carefully for all the forces on a body, not just the obvious one. Many simple problems apparently involving only one force are really highly simplified by neglecting friction. This is permissible, but we should know what is being neglected. Then too, in many problems the weight of a body is balanced out by an upward push of floor, table, or earth, as in the case just discussed. There is a tendency to forget that these forces are acting on the body, since they cancel out in so many cases.

Let's look for a moment at the relationship between Newton's first and second laws. In the second-law proportion, $a \propto (\text{net } F)/m_i$, if 0 is substituted for the net force, we get an acceleration equal to 0. Thus Newton's second law predicts no change in motion if no net force acts. But this is Newton's first law. It can be said, then, that the first law is just a special case of the second law; it is obtained from the second law by using a special value of the net force, namely zero.

4-4 INERTIAL MASS AND GRAVITATIONAL MASS

In Chapter 1, mass was introduced as that property of matter that is measured on an equal-arm balance. This is properly called *gravitational mass* because when measuring it in this way we are comparing the gravitational pulls on the objects. Perhaps you are wondering if the *inertial mass* defined in the preceding section is really a different property of matter or if it is simply the gravitational mass with which you are familiar.

From Equation 4-2, to measure the inertial mass of a body, we accelerate it with a known

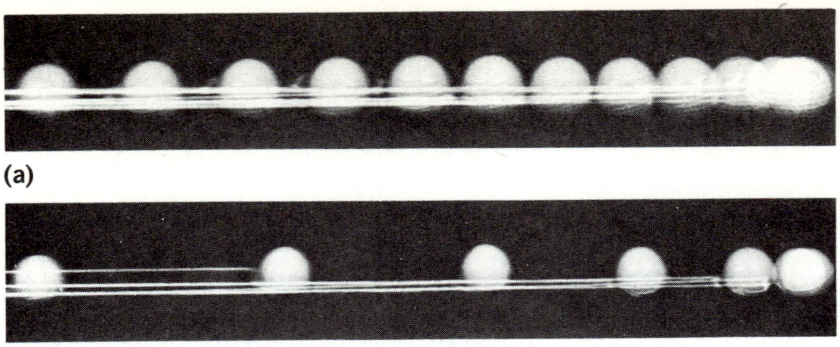

(a)

(b)

Figure 4-4 Acceleration caused by a constant force when the gravitational mass of the golf ball in **(a)** is twice that of the rubber ball in **(b).**

force and the ratio F/a is a measure of the inertial mass. This is a *dynamic* measurement. To measure the gravitational mass of a body, we compare it to a known standard on a balance. This is a *static* measurement. In the measurement of inertial mass, gravity plays no role whatsoever because in the horizontal plane it is canceled by the upward push of the table top. In the measurement of gravitational mass, gravity is the force necessary to make the balance operate. Gravitational mass and inertial mass are defined differently and they are measured differently. They are two distinctly *different* properties of matter, but they can be shown experimentally to be proportional.

If the gravitational mass (m_g) of body A is twice the gravitational mass of body B, we find that when each of these objects is accelerated by the same force, the acceleration of body B is *twice* the acceleration of body A (Figure 4-4). The inertial mass of a body is proportional to its gravitational mass. In symbols,

$$m_i \propto m_g$$

So far, we do not have a standard unit for inertial mass. We already have a unit for measuring gravitational mass—the kilogram. By letting the constant of proportionality in the above relation equal unity ($=1$), the proportion can be written as $m_i = m_g$. This allows us to use the same standard, the kilogram, as a unit for both. It does not imply that inertial mass and gravitational mass are the same thing. They are experimentally demonstrated to be *proportional,* but are set *equal* because of our wish to measure them both in the same units—kilograms.

To find the inertial mass of a body, it is not necessary to do a dynamic force-acceleration experiment. We simply measure the gravitational mass on a balance, and this gives us the inertial mass.

Since inertial mass and gravitational mass are both measured in kilograms, it is possible to drop the subscripts i and g and simply use the word *mass* in referring to them both.

4-5 NEWTON'S THIRD LAW—ACTION AND REACTION

Forces always occur in pairs. Everyone who has played tug of war knows from experience that he can exert a force only if the people on the other end of the rope are exerting a force on him. Replace the opposing team by a post fixed in the ground, and you have a one-man tug of war. The man exerts a force (on the post) and the post exerts a force (on the man). The rope serves to transmit the force from man to post and *vice versa.* Facts such as these are summarized in the following law:

Newton's third law of motion: Whenever one body exerts a force upon a second body, the second body exerts a force upon the first body; these forces are equal in magnitude, and opposite in direction.

Newton called these paired forces action and reaction. It is essential to realize that the two forces mentioned in Newton's third law always act on different bodies, and hence cannot add up to zero. Only if forces act on the same body can they be combined into a single net force.

Pairs of forces (action-reaction pairs) are all around us, but sometimes the reaction force is not obvious. Any action-reaction pair can be described in the form: "*A* acts on *B*, and *B* acts on *A*." When a book rests on a table, the book presses down *on the table,* and the table pushes up *on the book.* When a strong wind blows against a wall, molecules of air push against the wall, and the wall pushes back against the molecules of air. A more complicated example is that of a ball hanging by a rope from the ceiling. There are two complete action-reaction pairs involved here (four forces in all, each equal to the weight of the ball). In Figure 4-5, $A = -A'$, and $B = -B'$, in accordance with Newton's third law. But, in spite of the fact that *A* and *B* are equal and opposite, they do not constitute an action-reaction pair. Recall that action and reaction forces always act on different bodies, whereas *A* and *B* both act on the ball. Hence they must be equal and opposite for some other reason, not the third law. This reason is Newton's second law: the ball is in equilibrium, its acceleration is zero, and hence the vector sum of the forces acting *on the ball* must be zero. Only two forces, *A* and *B*, act on the ball, therefore $A + B = 0$, or $A = -B$. This has nothing whatever to do with the third law. If we now pull down slightly on the ball, the rope stretches a little and *B* (and also *B'*) increases to more than the weight of the ball. Releasing our pull on the ball, we find that for the moment it is no longer in equilibrium, since there is now a net force upward. Yet Newton's third law still holds. Forces *B* and *B'* are still of equal magnitude, but opposite in direction. So are forces *A* and *A'*. If we cut the rope, so that the ball falls freely, its weight is still *A* and it still exerts an upward force *A'*, equal to its weight on the earth.

It will help you identify pairs of action-reaction forces if you cast the description of each force in the form "________ acting on ________." Thus, in the example of the ball hanging from the ceiling, the force *A* is that of *earth acting on ball,* and

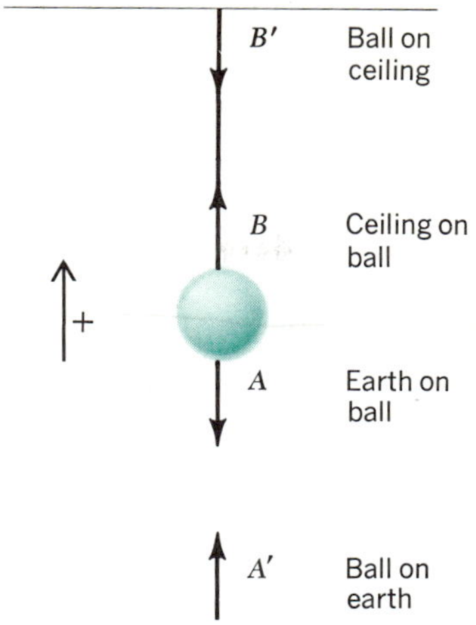

Figure 4-5 Action-reaction pairs.

its reaction force *A'* is the force of *ball acting on earth.* You can be sure that two equal and opposite forces are really an action-reaction pair if you are able to read the description of one force backward and arrive at the description of the other force.

Newton's third law is an important statement about the nature of force. In general terms, a force is an *interaction* between two bodies. Whatever the nature of the interaction, be it gravitational, electrostatic, magnetic, nuclear, or other, it is important to recognize that it takes two bodies to make a force. For example, during a tug of war, short-range electric forces act between neighboring atoms of the rope. Such forces, called elastic forces, occur as action-reaction pairs. Action and reaction are simply two aspects of the same force—two sides of the same coin, so to speak. It is impossible to have one without the other. The word "*inter*action" describes this situation nicely. Either force could equally well be called the action and the other force would then be the reaction.

4-6 UNITS FOR FORCE

Newton's second law is more useful when made into an equation. This can be done by first solving

Equation 4-3 for net force: net $F \propto ma$, then writing net $F = kma$, where k is a constant of proportionality. We already have units for mass and acceleration, but, as yet, we have not come across a unit of force. The simplest thing to do then, is to let $k = 1$. This allows us to write Newton's second law as

$$\text{net } F = ma \qquad \textbf{(Equation 4-4)}$$

and it also defines our new unit of force. Dimensionally, since mass is measured in kilograms and acceleration in m/sec^2, the right side of Equation 4-4 has units of kg·m/sec^2. This means that the left side, the force, must also be measured in units of kg·m/sec^2 in order to make the equation balance. Measuring force in units of kg·m/sec^2 may seem strange to you, but it is a perfectly good unit and it is the only one that will make the equation dimensionally consistent.

Because of its awkwardness, however, this collection of units has been given a name—the *newton* (N). 1 newton = 1 kg·m/sec^2. In particular, a newton is defined as that force which will give a mass of one kilogram an acceleration of one meter per second per second.

To gain a feeling for this unfamiliar force unit, the newton, let's imagine a 1-kilogram ball falling freely and acted on only by the force of gravity (its weight). The ball has a mass of 1 kilogram. Assuming air friction is negligible, the experiment shows that the acceleration of this freely falling body is 9.8 m/sec^2. Hence

$$\begin{aligned} \text{net } F &= ma \\ &= (1 \text{ kg})(9.8 \text{ m/sec}^2) \\ &= 9.8 \text{ kg·m/sec}^2 \\ &= 9.8 \text{ N} \end{aligned}$$

We see that the weight of the ball is 9.8 newtons, but we do not say that "1 kilogram equals 9.8 newtons" because the kilogram is a unit of mass, and the newton is a unit of force. However, it is correct to say that a body whose mass is 1 kilogram weighs 9.8 newtons on the earth. In the mks system, then, the newton is a unit of force, and since weight is gravitational force, the newton is also a unit of weight. It is analogous to the pound, which is the unit of force and weight in the English system. Weight and mass will be compared in detail in the next section.

You have just seen that the collection of units kg·m/sec^2 is equivalent to the single unit, the newton. Since a 1-kilogram ball weighs about 2.2 pounds, it can be shown that 9.8 newtons is about 2.2 pounds, or 1 newton is about 0.23 pounds. A quarter-pound stick of butter weighs about 1 newton.

The example that follows illustrates the use of mks units when solving a simple problem in dynamics with the help of Newton's second law.

Example 4-1 A car having a mass of 1000 kg coasts to a stop in 40 m. If the initial speed was 20 m/sec, what stopping force was supplied by the road acting on the car?

First, using the equations of kinematics (Chapter 3), we find the acceleration:

$$\boxed{\begin{aligned} v_i &= +20 \text{ m/sec} \\ v_f &= 0 \\ s &= +40 \text{ m} \\ a &= ? \end{aligned}}$$

$$\begin{aligned} v_f^2 &= v_i^2 + 2as \\ a &= \frac{v_f^2 - v_i^2}{2s} \\ &= \frac{0 - (20 \text{ m/sec})^2}{2(+40 \text{ m})} \\ &= -5 \text{ m/sec}^2 \end{aligned}$$

Now, using Newton's second law, we find the backward force of friction, f, exerted by the road on the car (Figure 4-6).

$$\begin{aligned} \text{net } F &= ma \\ f &= (1000 \text{ kg})(-5 \text{ m/sec}^2) \\ &= \boxed{-5000 \text{ N}} \end{aligned}$$

The answer appears as a negative force, which indicates that it acts in the negative direction, or backward. If the positive

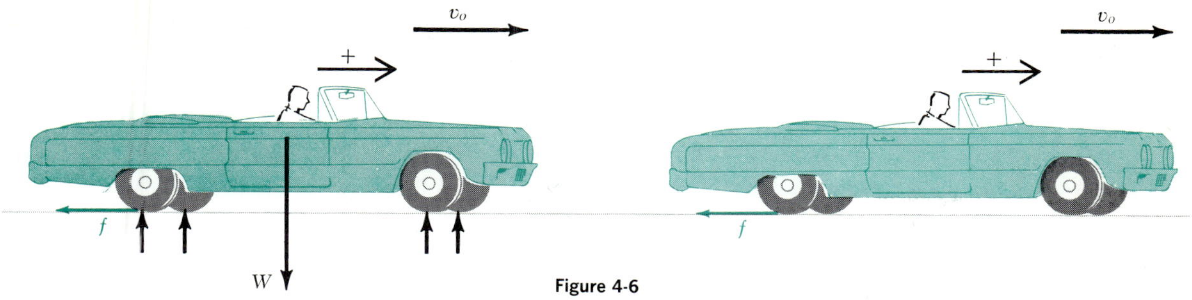

Figure 4-6

direction that was chosen had been toward the left, the force would have come out positive, but the acceleration would also have been positive. The *physical meaning* of the answer is the same regardless of which direction is chosen as positive.

Figure 4-6 includes two diagrams. The first diagram shows the downward force of gravity (the weight W of the car) and also shows the upward forces exerted on the four wheels of the car by the pavement. The sum of these four upward forces exerted by the pavement is equal to the downward force W, and so all these vertical forces have a vector sum equal to zero. The second diagram is simplified by omitting these vertical forces that cancel out, and shows only the force of friction f. In either diagram the net force is f, and it is f that is used in applying Newton's second law.

This next example illustrates that the net force is in the same direction as the acceleration it produces.

Example 4-2 A laborer pulls a 50 kg load along a frictionless road by a rope that makes an angle of 30° with the horizontal (Figure 4-7). If he exerts a constant force of 600 N, what acceleration does he give the load?

The direction of the acceleration is along the horizontal, and this is caused by the horizontal component of the applied force. This component can be found from a scale drawing to be 519 N (or using the geometry of 30-60-90 triangle, exactly $300\sqrt{3}$ N).

$$a = \text{net } F/m$$
$$= 519 \text{ N}/50 \text{ kg}$$
$$= \frac{519 \text{ kg}\cdot\text{m/sec}^2}{50 \text{ kg}}$$
$$= \boxed{10.4 \text{ m/sec}^2}$$

The cgs system is based on the centimeter, gram, and second, and the *dyne* is the unit of force. A dyne is that force which will give a mass of one gram an acceleration of one centimeter per second per second.

The dyne is a very small unit of force. A medium-sized pea has a mass of about 1 gram. If the pea is allowed to fall freely, acted on only by gravity, the only appreciable force acting will be the weight of the pea.

$$\text{net } F = ma$$
$$= (1 \text{ g})(980 \text{ cm/sec}^2)$$
$$= 980 \text{ g}\cdot\text{cm/sec}^2$$
$$= 980 \text{ dynes}$$

Thus a pea weighs almost a thousand dynes. In

Figure 4-7

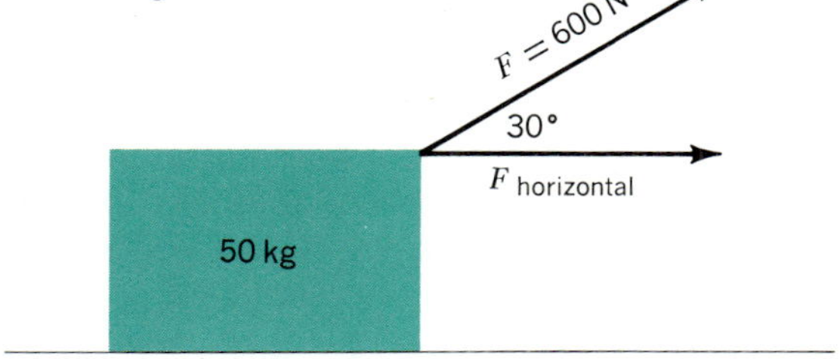

fact, a good-sized mosquito weighs about 1 dyne. Our discussion also shows that the collection of units g·cm/sec^2 is equivalent to the single unit, dyne.

The cgs system of units was formerly used for almost all work in physics involving the metric system, but it is gradually being replaced by the mks system. However, cgs units are often still used where very small forces are involved, as in our next example.

Example 4-3 Because of the pressure of a sound wave, the force on one side of a microphone diaphragm is increased to 1.0002×10^6 dynes. The force on the other side remains at its original value of 1.0000×10^6 dynes. What is the acceleration of the diaphragm, whose mass is 400 mg (0.4 g)?

$$\text{net } F = ma$$

$$(1.0002 \times 10^6 - 1.0000 \times 10^6) \text{ dynes} = (0.4 \text{ g})(a)$$

$$a = \frac{200 \text{ dynes}}{0.4 \text{ g}}$$

$$= \boxed{500 \text{ cm/sec}^2}$$

4-7 MASS AND WEIGHT

The weight of a body has already been defined as the force of gravity on it. It can be measured by suspending the object from a spring balance calibrated in newtons. For a 1-kilogram mass, this force is 9.8 newtons at the surface of the earth. The weight of a 10-kilogram mass is 98 newtons. In general, the weight of an object is seen to be proportional to its mass:

$$W = mg \qquad \textbf{(Equation 4-5)}$$

where g is the proportionality factor. Since weight and mass are measured in newtons and kilograms respectively, g has units of N/kg. In particular, g = 9.8 N/kg; it is a measure of the pull of gravity on each kilogram of mass at the earth's surface. If we think of the force of gravity extending out from the earth in all directions as a "force field," then we say the strength of the field is 9.8

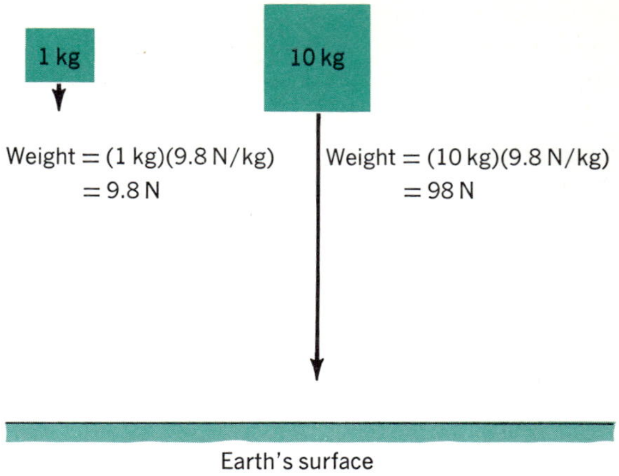

Figure 4-8

N/kg at the surface of the earth. (See **Figure 4-8**.)

You perhaps recognize Equation 4-5 as an alternate way of writing Newton's second law, where the weight W is the force of gravity and the factor g is the acceleration due to gravity. While it is true that g = 9.8 N/kg and g = 9.8 m/sec^2 are the same (check the units for yourself), it is best, when solving problems, to regard the former as a field strength and the latter as an acceleration due to gravity. This is illustrated in the following problem.

Example 4-4 A person whose weight is 980 N is on ice skates and is pushed horizontally by a net force of 90 N by the wind. What is the horizontal acceleration?

The acceleration is the net force divided by the mass. First we find the mass,

$$W = mg$$

$$m = \frac{W}{g}$$

$$= \frac{980 \text{ N}}{9.8 \text{ N/kg}}$$

$$= 100 \text{ kg}$$

Then from net $F = ma$

$$a = 90 \text{ N}/100 \text{ kg}$$

$$= \boxed{0.9 \text{ m/sec}^2}$$

Mass and weight are not the same. We can see from Equation 4-5 that they are proportional. However, people do say, incorrectly, "The apples weighed 2 kilograms." Physics students must guard against such looseness of expression. Correctly: "The applies had a mass of 2 kg, and a weight of (2 kg) (9.8 N/kg) = 19.6 N." Indeed, at the moon's surface, where gravitational field strength is only 1.6 N/kg (see Chapter 7), the apples still have a mass of 2 kg, but now weigh only (2 kg) (1.6 N/kg) = 3.2 N! A body's weight depends on the value of the gravitational field where it is weighed, while its mass is the same everywhere. In daily life, the general public does all its business in one place, of course. Hence to them, "the apples weigh 2 kilograms" is a perfectly acceptable statement.

4-8 ISOLATION OF BODIES IN PROBLEM-SOLVING

Newton's second law applies to any complete *system* or to any part of a system of bodies. (A system is a group of bodies that are interrelated in some way, such as the solar system, or the parts of a machine.) When solving a problem, it is essential that you decide clearly and definitely just *what* body or system of bodies is being considered. Obviously, if all forces on a system are considered, the total mass of the system must be used; if forces on only part of a system are considered, then the mass of only that part must be used. Let's illustrate this point by working a typical pulley problem.

Example 4-5 A cord connecting objects of mass 10 kg and 5 kg passes over a light, frictionless pulley (Figure 4-9). (a) What is the acceleration of the system? (b) What is the tension in the cord?

(a) First, we consider the system as a whole. The total mass is 15 kg. Two external forces act on the system. The force on the 10 kg mass is the force of gravity, its weight: $mg = +10\ \text{kg} \times 9.8\ \text{N/kg} = 98\ \text{N}$. The force on the 5 kg mass is: $-5\ \text{kg} \times 9.8\ \text{N/kg} = -49\ \text{N}$ where the signs are in accordance with the assumed positive direction shown in the diagram. The net force on the system is

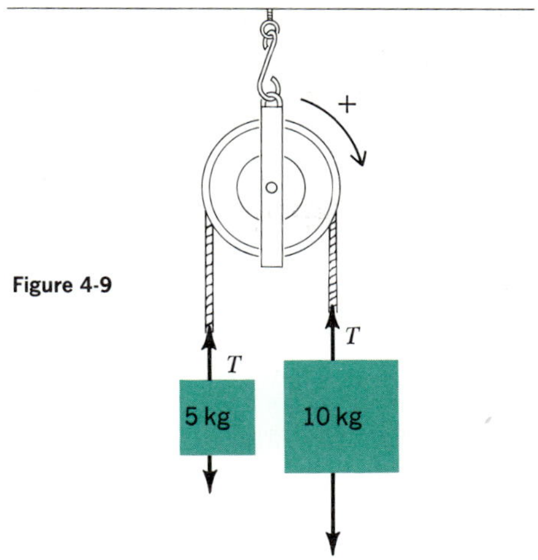

Figure 4-9

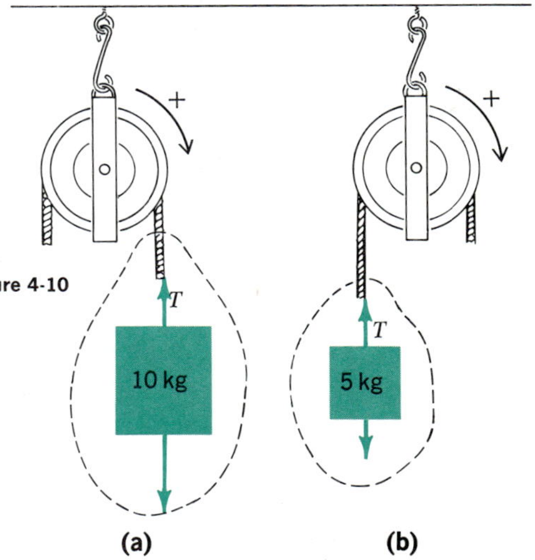

Figure 4-10

98 N − 49 N = 49 N. The tension in the cord is an internal force, acting between parts of the system, and does not enter into the calculation of the net force *on the system.*

$$\text{net } F = ma$$

$$98\ \text{N} - 49\ \text{N} = (15\ \text{kg})(a)$$

$$a = \frac{49\ \text{N}}{15\ \text{kg}}$$

$$= \boxed{3.27\ \text{m/sec}^2}$$

(b) To find the tension, we must consider either mass by itself. In Figure 4-10 dashed lines have been drawn around the mass to indicate this procedure. We apply Newton's second law to the 10 kg object (a), disregarding everything outside the dashed line. From this point of view, we do not know whether the tension is caused by a 5 kg object on the other side of the pulley, or by a spring attached to the cord, or perhaps, for all we know, by a little green man from Venus situated just outside the dashed line pulling upward on the 10 kg object. The essential point is that we don't care *what* causes the tension: we merely wish to compute its value, whatever the cause. Applying the second law to the 10 kg object, we have

$$\text{net } F = ma$$
$$98 \text{ N} - T = (10 \text{ kg})(3.27 \text{ m/sec}^2)$$
$$T = \boxed{65.3 \text{ N}}$$

As a check, the tension can be computed by isolating the 5 kg object (Figure 4-10b).

$$\text{net } F = ma$$
$$T - 49 \text{ N} = (5 \text{ kg})(3.27 \text{ m/sec}^2)$$
$$T = \boxed{65.3 \text{ N}}$$

It should not be surprising that the tension in the cord on the left side of the pulley is the same numerically as the tension on the right side of the pulley. According to Newton's third law, the force of the 49-newton object on the 98-newton object is equal and opposite to the force of the 98-newton object on the 49-newton object. The cord merely serves to transmit this force. Question: "If the tension is the same on both sides of the pulley, why does the system accelerate?"

Example 4-6 A tractor of mass 1000 kg is attached by means of a horizontal, "massless" chain to a log whose mass is 400 kg. The tension in the chain is 2000 N, and the backward force of friction exerted by the ground on the log is 800 N. How far will the log move in 2 sec, starting from rest?

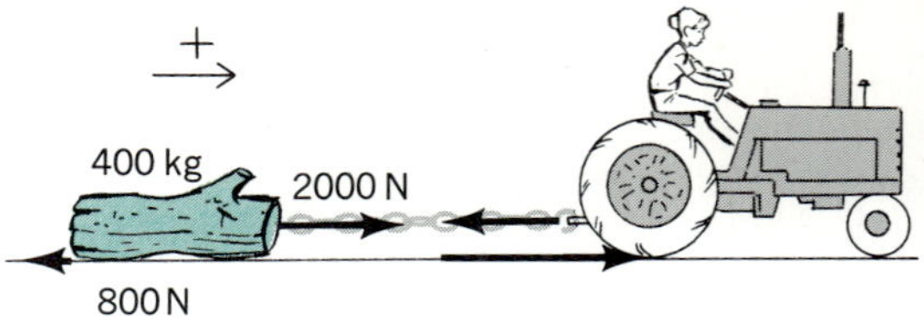

Figure 4-11

The problem is really two problems: First it is necessary to compute the acceleration, using Newton's second law; then we have a straight-forward problem in kinematics to find the displacement.

We begin by making a diagram (Figure 4-11), to show all the known forces and masses, as well as other forces whose values are not known. Then we isolate one of the bodies for consideration; the log is best for this purpose, since all the horizontal forces on it are known. We also choose a positive direction and indicate it on the diagram. Now Newton's second law can be applied to the log.

$$\text{net } F = ma$$
$$2000 \text{ N} - 800 \text{ N} = (400 \text{ kg})(a)$$
$$a = \frac{1200 \text{ N}}{400 \text{ kg}}$$
$$= +3 \text{ m/sec}^2$$

To find the displacement, we use Equation 3-5:

$s = ?$
$v_i = 0$
$t = 2$ sec
$a = +3$ m/sec^2

$$s = v_i t + \tfrac{1}{2}at^2$$
$$= 0 + \tfrac{1}{2}(3 \text{ m/sec}^2)(2 \text{ sec})^2$$
$$= \boxed{6 \text{ m}}$$

Note that the mass of the tractor is not used in this problem, since it is the *log's* motion that is being considered. Similarly, the forward force of the ground on the tractor is not used, because this force does not act on the log.

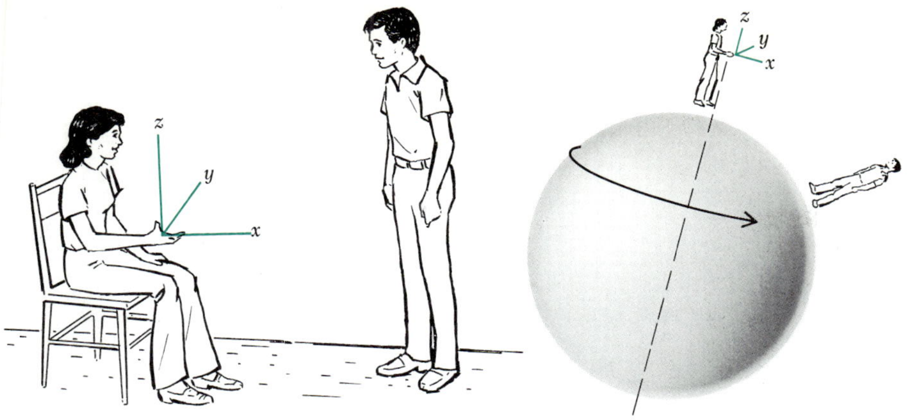

Figure 4-12 (a) (b)

4-9 FRAMES OF REFERENCE

What is your velocity right now? If you are standing still in the classroom, your neighbor would say it is zero because from his point of view you are not moving. But to someone situated just above the north pole, your velocity would seem to be about 1120 km/hr (if you are at the latitude of New York City) due to the earth's rotation. And to an observer on the sun, your velocity would be that of the earth in its annual orbit around the sun, about 107,000 km/hr!

Which of these observers is correct? Of course, they are all correct because each is describing your motion from a different point of view—in a different *frame of reference.* A frame of reference may be thought of as a set of three mutually perpendicular axes, as the x-, y-, and z-axes, relative to which all positions, velocities, and accelerations are measured. This is illustrated in Figure 4-12a. Here, your classroom neighbor is shown holding a model of a 3-dimensional coordinate system whose origin is at rest with respect to her. As she observes you, she notes that your position with respect to her origin does not change. She therefore concludes that you are at rest. (The fact that you both may realize there is a common motion about the sun for each of you is not considered in this description.) In her frame of reference, her coordinate system, there is no relative motion between the two of you, you are at rest relative to her. In Figure 4-12b your motion is now described by an observer suspended above the north pole. As the earth turns about her, she observes that you move in a circle with a certain tangential velocity as measured from her stationary (to her) coordinate system. We see, then, that no matter what frame of reference the observer places herself in, she regards this frame as being at rest (even though she may "know" that she is moving), and describes all other motions relative to that.

As long as we are content simply to *describe* a motion (this is kinematics), then any frame will do because they will all give a description correct for each particular observer. It is our natural tendency to place ourselves (mentally, if necessary) in the frame that gives the simplest description. In the above example, for instance, it is simplest for an observer to place himself on the surface of the earth—right next to you. On the other hand, when we want to identify the *causes* of motion (this is dynamics), we do not have this freedom of choice. Newton's laws of motion do not hold in all frames of reference. An *inertial* frame of reference, however, is one in which Newton's laws are valid. A frame of reference that is moving with constant velocity with respect to an inertial frame is also an inertial frame of reference. Not all frames are inertial; for instance, a person sitting in an automobile that is *accelerating* with respect to the earth is in a non-inertial, or accelerated, frame. For him, Newton's laws will not hold.

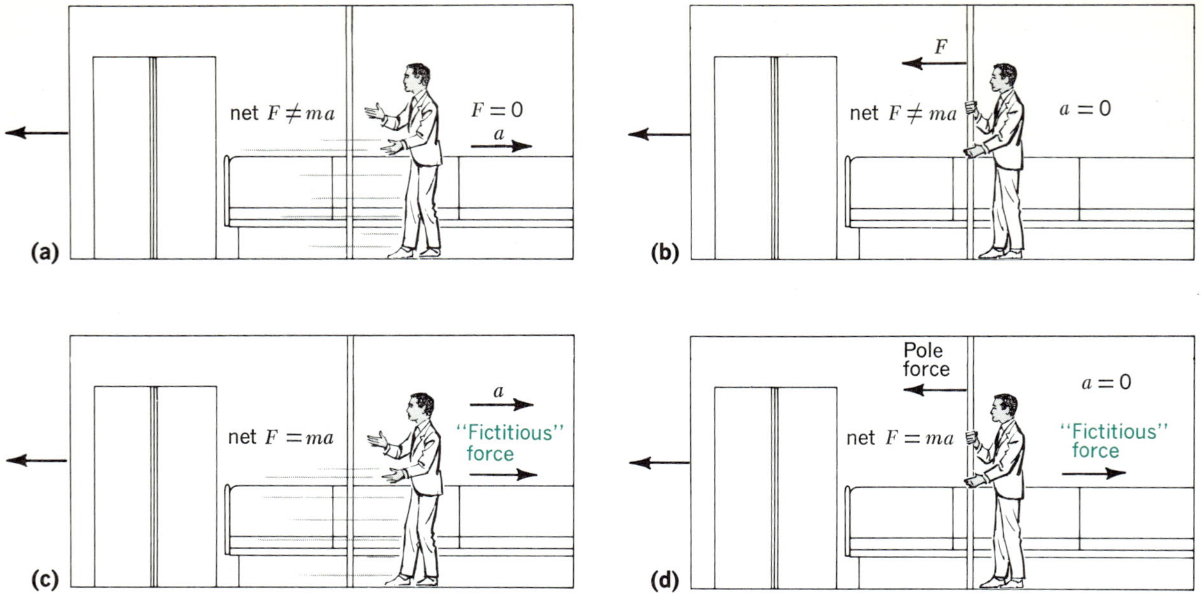

Figure 4-13 **(a)** As the train accelerates to the left, you think you are accelerating toward the rear of the car. **(b)** When you hold onto the pole you come to rest in your reference frame. **(c)** You introduce a fictitious force to explain your apparent acceleration. **(d)** This makes the net F on you apparently equal to zero.

What do we mean when we say that Newton's laws do not hold in a non-inertial reference frame? So that we may better understand our study of centrifugal force coming in Chapter 7, let's take the time now to look at an example in detail.

Imagine that you are standing in a subway train that is at rest in the station. There are no windows in this train and you can't see out, but you are at rest with respect to your surroundings (your frame of reference). To keep the problem simple, imagine further that you are standing on a perfectly frictionless floor. Now, unknown to you, the train starts to *accelerate* out of the station. It moves very slowly at first, with no noise, and so smoothly that you are not aware it is moving at all. Instead you feel that *you* are accelerating toward the back of the train! (Remember, you regard your frame of reference—the train—as being at rest, and you are standing on a frictionless floor.) You don't feel any horizontal forces acting on you, yet you are accelerating toward the back of the train. Newton's law says that a force is needed to produce an acceleration, but as far as you are concerned there are no forces present. For you, Newton's laws do not apply (Figure 4-13a).

Suppose now that you grab the pole that extends from the floor to the ceiling. Now you find that you come to rest; you always remain opposite the same point in your reference frame. You are at rest, but the pole is exerting a force on you (which you feel) to maintain your zero velocity (Figure 4-13b). Again it seems that Newton's laws do not hold because this "pole force" that is acting on you should produce an acceleration and from your reference frame it doesn't. No matter what experiment you perform in your accelerating subway-train frame of reference, you will find that Newton's laws do not apply. This is because your frame of reference is a non-inertial, that is, an accelerated, frame of reference.

There is a way to preserve your sanity, however, and that is to introduce a *fictitious* force in your accelerated reference frame so that Newton's laws *will* apply. This force is fictitious in that it has no origin—you simply make it up (or *postulate* it) to preserve Newton's laws. So when you find yourself accelerating toward the back of the train, you introduce the fiction that there is a force acting in that direction that produces this acceleration (Figure 4-13c). Or when you feel the pole force pulling on you while you maintain a zero velocity with reference to your surroundings, you introduce the *same* fictitious force acting toward the back of the train so that the net force on you is zero, which explains your lack of acceleration (Figure 4-13d). As long as you maintain this fictitious force, any experiment that you perform (in your frame of reference) will agree with Newton's laws.

It might be helpful to analyze your motion in the subway car as seen by an observer on the platform. (Assume he can see you in spite of the absence of windows.) As the train accelerates out of the station, he observes that you remain directly opposite him. (Remember that the floor is frictionless.) In his frame of reference your acceleration is zero and he has no need for the fictitious force that you introduced. Or when you grab the vertical pole, he watches you accelerate out of the station and he sees that the train motor is supplying the necessary force to cause this acceleration. Again, he has no need for the fictitious force (although in each case he can sympathize with you by recognizing your plight). This is because the platform is an inertial frame of reference in which Newton's laws of motion are valid. In summary, let's say that an inertial frame of reference is one in which Newton's laws of motion can always be applied. They can be applied in a noninertial frame only if the observer in that frame postulates a fictitious force.*

The surface of the earth is, strictly speaking, not an inertial frame of reference. Because of the daily rotation of the earth, any point on the surface has an acceleration toward the center of its circle of rotation. This will be explored further in Chapter 7. For now, it is enough to say that this acceleration is so small that for most purposes we can ignore this effect and treat our natural reference frame—the earth—as an inertial frame.

*These fictitious forces are also called inertial forces.

SUMMARY

Newton's first law serves as a qualitative (non-numerical) definition of the concepts of **force** and **inertia. Newton's second law,** net $F = ma$, tells us how much unbalanced, or "net," force is required to give acceleration to a mass. This kind of mass is called **inertial mass;** it is a quantitative measure of inertia—that property of matter which causes it to resist acceleration. Mass as measured on an equal arm balance is called **gravitational mass.** Although mass can be regarded in two distinct ways, inertial mass and gravitational mass can be experimentally demonstrated to be proportional. They can be set equal to each other by measuring them both in kilograms.

Force and mass have different dimensions, and can never have the same units. Consistent sets of metric units for use in Newton's second law may be summarized as follows:

$$\text{net } F = ma$$

$$\text{(mks)} \quad \text{N} = (\text{kg})(\text{m/sec}^2)$$

$$\text{(cgs)} \quad \text{dynes} = (\text{g})(\text{cm/sec}^2)$$

These are the units that should be used in Newton's second law, or in any equation derived from it.

Mass and **weight** are not the same. One is measured in mass units and the other in force units. They are proportional, however, as expressed by the relation $W = mg$.

An **action-reaction pair** can be identified by reading the description of one force backward, and arriving at the description of the other force. **Newton's third law** states that action and reaction forces are equal in magnitude and opposite in direction. Action and reaction always act on different bodies.

An inertial **frame of reference** is one in which Newton's laws of motion are valid. To apply Newton's laws in a non-inertial, or accelerated, reference frame, the observer must introduce a **fictitious force.** The surface of the earth is a non-inertial reference frame, but the acceleration is so small that it can usually be neglected.

QUESTIONS

1 Is the following statement true or false? "When a book is at rest, lying on a table, the downward force of gravity on the book is equal and opposite to the upward force of the table on the book. This is an example of Newton's third law."

2 Two sleds, A and B, on smooth ice, are connected by a light but stiff spring and then held far enough apart so that the spring is stretched somewhat. The sleds and their passengers, each of a different mass, are then released and the spring contracts, pulling them together. During the motion of A toward B and B toward A: (a) Are the forces on the sleds equal and opposite? If so, is this an example of Newton's third law? (b) Do the sleds have equal and opposite accelerations? (c) Do the sleds acquire equal speeds? (d) How can the mass ratio of the two sleds be found?

3 A horse is pulling on a heavy cart, and the cart and horse are both being accelerated. Is the force of the cart on the horse equal and opposite to that of the horse on the cart? If your answer is "yes," explain how it is that these equal and opposite forces give rise to the net force that is necessary to cause acceleration.

4 In Question 3, explain how the horse becomes accelerated. What is the origin of the net force upon him? In which direction is the net force on the horse?

5 If Galileo dropped two objects of unequal weight from the Leaning Tower of Pisa, why is it that the heavier one did not reach the ground first, since its greater weight would be expected to cause it to be accelerated more rapidly?

6 How would you go about determining whether two bodies have (a) the same weight? (b) the same inertial mass? (c) the same gravitational mass?

7 If a body of mass 10 kg were taken to the moon, where the acceleration due to gravity is $\frac{1}{6}$ as much as on the earth, what would be its weight? What would be its mass?

8 A freight train of 100 cars, each weighing 10 tons, is held together by couplings between the cars. Is the tension in the coupling between the third and the fourth cars the same as the tension in the coupling between the thirty-third and the thirty-fourth cars? Answer this question for two conditions: (a) if the train has constant velocity; (b) if the train is accelerating. (You can ignore the effects of friction in computing these answers.)

9 An iron ball is hanging by a string, and a similar string is attached to the lower side of the ball and hangs loosely. If the loose string is given a sharp jerk, it often breaks, but if it is given a slow and steady pull, the top string breaks. Explain.

10 How would you calculate the value of the fictitious force introduced in the discussion of Section 4-9? Is this value for the force unique for the situation described?

PROBLEMS

GROUP A

1 What is the net force acting on a spider whose weight is 1.0 dyne, if he is supported by a strand of silk that exerts a force of 0.8 dyne?

2 A force of 400 dynes acting on a body produces an acceleration of 80 cm/sec^2. What is the mass of the body?

3 What is the force, in newtons, required to give a ball of mass 2 kg an acceleration of 3 m/sec^2?

4 A body of mass 3 kg is acted upon by a net force of 10 N. What acceleration is produced?

5 During a satellite recovery, the satellite, having a mass of 100 kg, is partially supported by a parachute that supplies an upward force of 500 N. What is the net force on the satellite?

6 What net force is needed to give a mass of 50 g an acceleration of 40 cm/sec^2?

7 A force of 6 N is applied to a mass of 2 kg. The acceleration is measured to be 2 m/sec^2. What is the value of the frictional force in this experiment?

8 A child can exert a forward push of 50 N; what acceleration can he give to a loaded cart weighing 300 N?

9 A sled of mass 30 kg coasts over the ice with an acceleration of -0.5 m/sec^2. What is the retarding force of friction?

10 A 6-kg mass is moving at a constant velocity of 15 m/sec. (a) What force in newtons is needed to bring the mass to rest in 9 sec? (b) How far does the mass move while the force is acting?

11 A 2-kg mass is pulled on a frictionless table by a force of 8 N. If the mass starts from rest, what is the velocity 2 sec after the force starts acting?

GROUP B

12 A falling ball has a mass of 2 kg, and the upward force of air resistance is 11.6 N. What is the acceleration of the ball?

13 A golf ball of mass 60 g is struck by a golf club, and acquires a speed of 80 m/sec during the impact, which lasts for 2×10^{-4} sec. What force (assumed constant) is exerted on the ball?

14 A vertical rope is attached to a 35-kg cart. What tension in the rope is needed to cause the cart to acquire an upward velocity of 4 m/sec in 0.5 sec?

15 A 75-kg aviator in "free fall" acquires a velocity of 60 m/sec and then opens his parachute. After falling an additional 30 m, his velocity has been reduced to 8 m/sec. (a) What is the average acceleration of the aviator while his fall is being checked? (b) What is the average force exerted by the parachute?

16 A girl weighing 600 N wishes to reach ground from a treetop by sliding down a rope. The maximum upward force that the rope can exert without breaking is 300 N. (a) Can the girl slide down at constant speed? (b) What is the least acceleration with which the girl can slide down the rope?

17 A 60 kg sled is coasting with a constant velocity of 10 m/sec over smooth ice. It enters a rough stretch of ice 6 m long in which the force of friction is 120 N. With what speed does the sled emerge from the rough stretch?

18 A 10,000 kg rocket is acted upon by an upward thrust of 118,000 N. If the rocket is 16 m tall, how long is required for it to rise off the launching pad a distance equal to its own height?

19 An elevator of mass 1000 kg is supported by a cable that can sustain a force of 12,000 N. What is the greatest upward acceleration that can be given the elevator without breaking the cable?

20 A 3.0 N force and a 4.0 N force act on a 10 kg mass. The two forces act at an angle of 50° with each other. What is the magnitude and direction of the velocity of the mass 2 sec after the force starts acting?

21 A 100 kg rocket sled was moving at a constant speed v_i. A 5000 N force, acting for 20 sec opposite to the direction of its motion, slowed the sled down to $\frac{1}{3}v_i$. (a) How far did the sled move while it was slowing down? (b) What was the value of v_i?

22 A 1500 kg car is traveling at 20 m/sec and collides with a heavy truck. The

car moves 3 m forward while it is being brought to rest. What force (assumed constant) is exerted on the car by the truck?

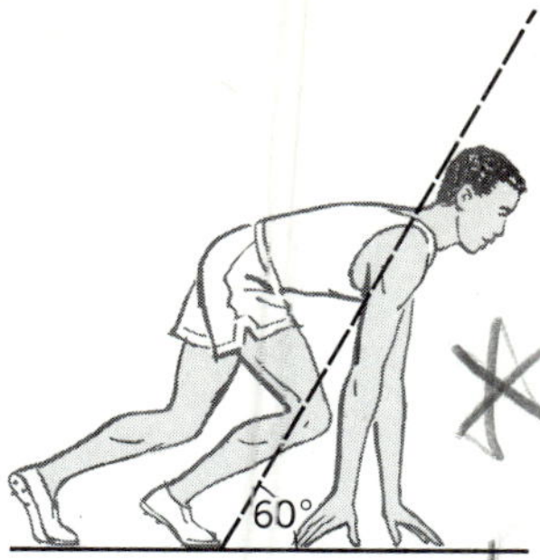

Figure 4-14

23 A 75 kg track star, at the start of a sprint, pushed on the ground with a measured force of 2000 N at an angle of 60°, as shown in Figure 4-14. What forward acceleration was produced?

24 Objects of mass 5 kg and 2 kg are connected by a light cord that passes over a horizontal frictionless rod. (a) What is the acceleration of the system? (b) What is the tension in the cord on the 5 kg side? (c) What is the tension in the cord on the 2 kg side?

GROUP C

25 The driver of a 600 kg sports car, heading directly for a railroad crossing 300 m away, applies the brakes in a panic stop. The car is moving at 40 m/sec, and the brakes can supply a force of 1500 N. (a) How fast will the car be moving when it reaches the crossing? (b) Will the driver escape collision with a freight train which, at the instant the brakes are applied, is still blocking the road and still requires 11.0 sec to clear the crossing?

26 Two boxes of mass 20 kg and 12 kg, respectively, are in contact on a horizontal frictionless surface. A horizontal force of 160 N is applied to the 20 kg box, pushing it against the other box. (a) Compute the acceleration of the system; (b) compute the force with which the two boxes push against each other.

27 A small paint bucket of mass 9 kg is attached to a larger paint bucket of mass 12 kg by a light rope 21 m long. The small bucket is at rest on the ground and the rope goes up and over a small frictionless pulley and is attached to the larger bucket, which is on a window sill 11 m above the ground. There is no slack in the rope. If the heavy bucket slips off the window sill, causing the smaller bucket to rise, and the two collide in midair, after what time will the big splash occur?

28 In Figure 4-15, the frictional force between the 30 kg block and the table is negligible. (a) If the peg is removed, what is the acceleration of the

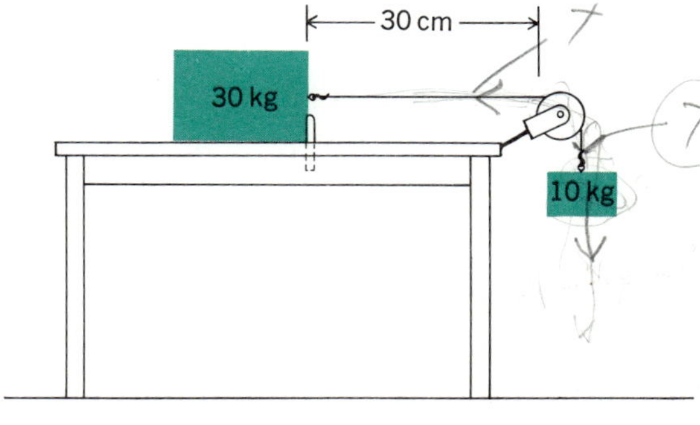

Figure 4-15

system? (b) How long will it take the block to hit the pulley? (c) What is the tension in the cord while the block is moving? (d) What is the tension in the cord after the block ceases to move?

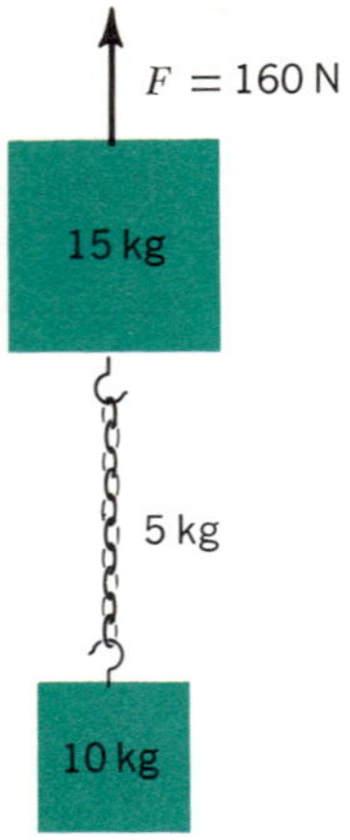

Figure 4-16

29 The two blocks in Figure 4-16 are connected by a heavy chain with a mass of 5 kg. An upward force of 160 N is applied to the top block. (a) How far and in what direction will the system move in 0.5 sec, starting from rest? (b) What is the tension at the top link in the chain?

30 The system shown in Figure 4-9 is held with the blocks opposite each other. In this position they are 3 m above the floor and 4 m below the pulley. The 5 kg block is then given a downward initial velocity sufficient to cause it to just reach the floor. How long after the initial push will the blocks again be opposite each other?

31 Bob and Joe, two construction workers on the roof of a building, are about to raise up a keg of nails from the ground by means of a rope passing over a pulley 16 m above the ground. Bob has a mass of 100 kg, Joe 80 kg, the keg 40 kg, and the nails 80 kg. They slip off the roof and the following unfortunate sequence of events takes place: Bob and Joe, hanging on the rope together, strike the ground just as the keg hits the pulley. Unnerved by his fall, Bob lets go of the rope, and the falling keg pulls Joe up to the roof where he cracks his head against the pulley but gamely hangs on. However, the bottom falls out of the keg of nails when it strikes the ground, and the empty keg rises as Joe returns to the ground. Finally, Joe has had enough and lets go of the rope and remains on the ground, only to be hit by the empty keg. Ignoring the possible mid-air collisions which merely add insult to injury, how long did it take this little drama to unfold? (To make the computations easier, use $g = 10\ m/sec^2$).

5 Conservation of Energy

Perhaps the most direct way of approaching conservation of energy would be by first defining the word "conservation," and then "energy." But unfortunately, this approach lies just outside our grasp. This is due mainly to the fact that energy is such a fundamental concept that a short, meaningful definition is impossible.

Wherever you turn and whatever you study, energy is involved. When gasoline is burned in an automobile, the energy stored in the gasoline is transferred to the up-and-down motion of the pistons. In turn, the pistons eventually transfer this energy to the forward motion of the car. If the moving car crashes into a stone wall, the energy is then transferred to the tumbling stones, to the sound of the crash, and even to the heating up of the wall and the car. In this example we can recognize several familiar forms of energy; the chemical energy stored in the gasoline, the mechanical energy of the moving pistons and car, and the heat and sound energy of the collision. It is also evident that energy is being changed from one form to another.

It seems that some of the major forms of energy within any system in the universe are what they are only as long as we do not tamper with them. Efforts to isolate chemical energy or different forms of nuclear energy, for example, only result in their changing to still other forms of energy. Preliminary inspection suggests that we no longer have the same system we had originally intended to study. Upon closer examination, however, and with sufficient attention paid to the tiny amounts of energy locked up in the atoms themselves, we are led to believe strongly that the total amount of energy in the system has remained constant. As in the car crash described above, the energy has simply been transformed from one form to another. In this chapter we shall discuss a method for measuring some of these energy changes. This will allow us to do the necessary bookkeeping in the energy ledger that will ultimately lead us to one of the great conservation laws in physics: *the law of conservation of energy,* which states that *the energy content of an isolated system (a system not acted upon by external forces) remains constant.*

5-1 WORK

After studying hard at your desk for half the night, you say to your friends the next day, "Boy, did I work hard on that assignment!" You take your watch to the jeweler and ask him to fix it because it doesn't work. Or, in the gym you work very hard when you hold a dumbbell over your head for 60 seconds. Whatever the general usages of the term may be, in physics we say that *work* is

Figure 5-1

done when a force causes a particle or other body to be moved, or displaced. A hockey player (Figure 5-1) does work when she strikes the puck, causing it to start to move; this work is done during the fraction of a second that the stick is in contact with the puck. You do work when you lift a book from a table top. In each case a force is being applied while a body is displaced. But it is only when the body is acted upon by a force having a component along the direction of motion that the work is done. Sometimes a force acts at an angle to the motion, such as when the downward force of gravity pulls on a thrown ball as it moves in its curved path. In such cases it is the *component* of the force along the direction of motion that enters into the work calculation. Quantitatively, work is defined as a product:

$$\text{work done} = \begin{pmatrix}\text{applied force}\\ \text{component}\end{pmatrix}\begin{pmatrix}\text{distance}\\ \text{moved}\end{pmatrix}$$

$$W = Fs \qquad \textbf{(Equation 5-1)}$$

where the distance moved (the displacement) is represented by s.

In any measuring system work units are the products of a force unit and a unit of distance; in the metric system certain work units have received special names. In the mks system, work and energy are measured in newton·meters and a newton·meter has been given the name *joule* (rhymes with "tool"). Thus if a 6-newton force moves a body 2 meters in the direction of the force, 12 joules (J) of work have been done. Similarly, in the cgs system, a dyne·centimeter is called an *erg*. Note that there is no new physics here. The joule and the erg are simply new words that are expressed below in terms of force and distance units already familiar to you:

$$1 \text{ joule} = (1 \text{ newton})(1 \text{ meter})$$

$$1 \text{ erg} = (1 \text{ dyne})(1 \text{ centimeter})$$

It is important to understand that no work is done if no motion is involved, no matter how much force is applied; and no work is done if there is no force component along the direction of motion, no matter how much motion there is. (This is the same as saying that, in Equation 5-1, $W = 0$ whenever F or s is zero.) Thus, from the viewpoint of the physicist, no work is done by the force required to hold a 7-newton book at arm's length (or the dumbbell over your head in the gym), since there is no motion. It also follows that no work is done by a force applied at right angles to a motion, since such a force has no component in the direction of motion.

Example 5-1 A man pulls a 50 kg stone a distance of 4 m at constant velocity along a rough road by applying a 250 N force along a rope that makes an angle of 60° with the horizontal (Figure 5-2a). He then stoops down and lifts the stone vertically to a height of 1 m and carries it at constant velocity a distance of 5 m (Figure 5-2b). How much work does he do on the stone?

In the first part of the job, the work done is the force applied in the direction of motion times the distance moved. The horizontal component of the 250 N force is found (from geometry or a scale drawing) to be 125 N (Figure 5-2a).

$$\begin{aligned} W &= Fs \\ &= (125 \text{ N})(4 \text{ m}) \\ &= 500 \text{ J} \end{aligned}$$

When he lifts the stone vertically, he

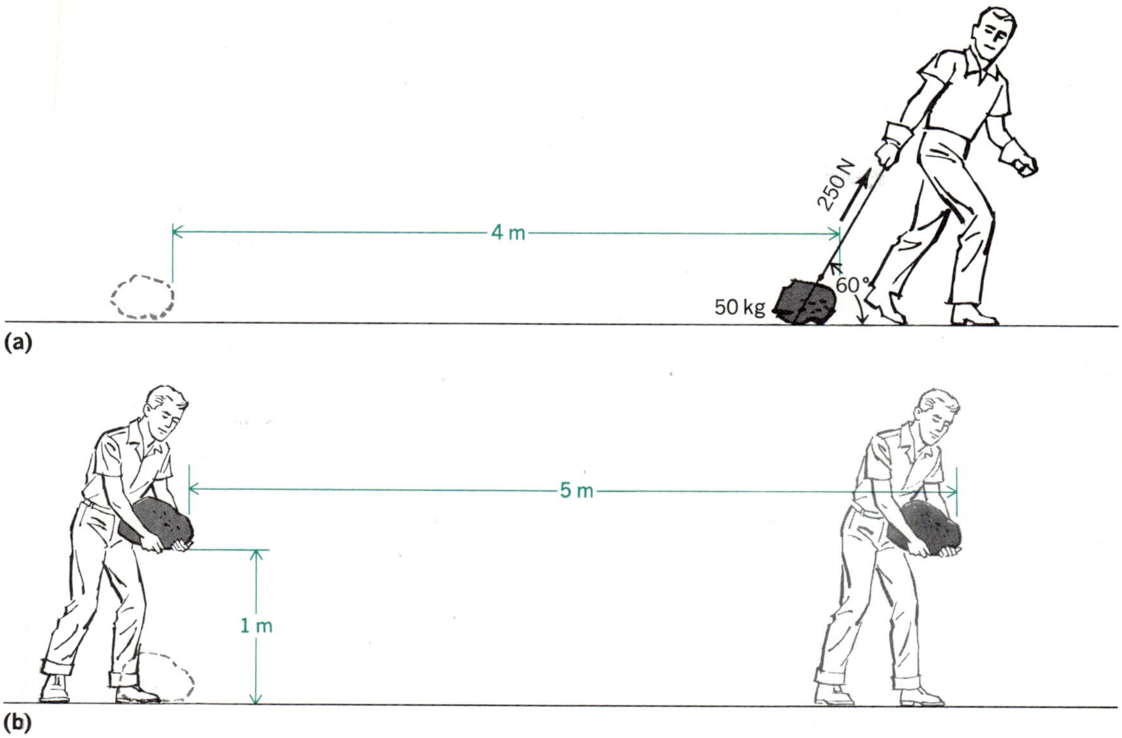

Figure 5-2

must apply a force equal to the weight of the stone. This force will be in the direction of the motion.

$$W = Fs = (mg)s$$

and, since $g = 9.8$ N/kg (Section 4-7)

$$W = (50 \text{ kg})(9.8 \text{ N/kg})(1 \text{ m})$$
$$= 490 \text{ J}$$

As he carries the stone along the horizontal, he must apply a *vertical* force to support it; this force has no component in the horizontal direction. Or saying it differently, there is no displacement in the direction of the force. In this case, then, there is *zero* work done as he carries it along. (We neglect any work done during the instant that the stone is accelerated from rest to a constant velocity. Here, a small horizontal force might be required to accelerate the stone if it had come to rest after being lifted. Of course, once the stone reaches constant velocity, it tends to keep moving in this manner because of its inertia. In mid-air there is very little force opposing its horizontal motion.)

The total work done on the stone is, therefore,

$$500 \text{ J} + 490 \text{ J} + 0 \text{ J} = \boxed{990 \text{ J}}$$

5-2 ENERGY

When work is done on a body or a system, energy is transferred to the system. The amount of work that is done is a *measure* of how much energy has been transferred. Thus, if seven J of work is done on a body, the body has gained seven J of energy. Two kinds of mechanical energy are involved here, and we can distinguish between them, depending on how the work was done on the body.

Figure 5-3 Surfer "caught" by the foam. Does the surfer's acceleration result in a slowing down of the wave?

Potential energy is the energy of a system of bodies due to the relative positions of the parts of the system. Often we say that potential energy (PE) is "due to position," but some examples will make it clear that it takes at least two bodies (a system) in order to have PE. When a bricklayer gives a brick PE by lifting it, he does work on a system, and two bodies are involved—the brick and the earth. In this raised position, the system of the brick and the earth has PE in the form of stored energy, as measured by the work done in separating them. We say the system has energy because when the parts are allowed to come back together again (by releasing the brick), some useful job can be done (like cracking a walnut). When you wind your watch, you change the relative positions of the various atoms of iron in the watch spring. The spring (a system of atoms) is distorted, and you have done work that appears as stored energy (PE) in the spring. We see, then, that when work is done to change the relative positions of the parts of a system—separating them or compressing them—energy (PE) is stored in the system by an amount equal to the work done. There is, however, an exception to the preceding statement. A system has PE only if the energy can be recovered, as when you do work against gravity or a spring. PE is not stored when you change relative positions by doing work against friction alone.

Kinetic energy is the energy of a body due to its motion. In order for its velocity to change, a body must be accelerated. The force required to do this is described by Newton's second law (Section 4-3). The work done by such a force is a measure of the kinetic energy (KE) transferred to the body. All bodies in motion possess KE and, because of their motion, they can change the velocity of other bodies. The energy of the winds and the waves is kinetic, as experienced by the surfer in Figure 5-3. The work done to speed up the surfer and her board is a measure of the KE that has been transferred to her from the wave she is riding. Again, we say that bodies in motion possess energy, because in coming to rest or slowing down, they can do some useful job as their energy is transferred, and this energy can be measured by the work done.

5-3 WORK AND ENERGY TRANSFER

This brief discussion of potential and kinetic energy points out the important role of work in all mechanical energy changes. KE can be transferred into a system or out of a system, as can PE. Indeed, PE can be converted to KE within the system. The falling brick loses PE and gains KE; or vice versa, as a ball is thrown straight up, it loses KE and gains PE. In each of these cases of energy transfer, however, work is done. Work is a measure of energy transfer, or more specifically, it is a measure of *mechanical* energy transfer.

To find out how work done on a body can appear as KE and PE, look again at Figure 5-2a. Let's assume that the stone is originally on a frictionless road not shown in the figure and that the man sets the stone in motion, pulling it 0.4 meter, by applying a horizontal force component of 125 newtons. The stone accelerates while it moves under the action of this force, and the work done by the man on the stone is (125 N)(0.4 m) = 50J. The stone therefore acquires 50 joules of KE. No work is done by or against gravity, since the force of gravity is downward, perpendicular to the motion. In fact, the weight of the stone has no bearing on the situation so far. Now suppose the stone comes to the stretch of road shown in the figure where the retarding force of friction is exactly 125 newtons, so that the man's forward force is balanced by the frictional force. The stone's already acquired velocity remains constant and, as it moves over the 4 meters the man, still exerting 125 newtons of force, does another 500 joules of work. In this case the work done is *not* a measure of the KE transferred to the stone because there is no change in the velocity of the stone (the net force on it is zero). The KE of the stone remains unchanged. Instead, the work done measures the change in the internal energy of the ground and stone as they both become slightly warmer. (This process is the subject of Chapter 11). In Figure 5-2b we see that the man does 490 joules of work on the stone when he lifts it 1 meter. In this raised position, the stone has no KE because it is at rest, but it does have PE due to the separation of the stone-earth system. As you will see in Section 5-5, it can do 490 joules of work if it is allowed to fall back to the ground.

5-4 FORMULAS FOR MECHANICAL ENERGY

A mechanical engineer must use a large number of formulas to cover the many applications of the laws of physics that pertain to his work. This is as it should be, but our use of formulas is quite different. We derive (and use) only a few formulas, chosen to demonstrate the essential factors upon which energy, for instance, depends. These formulas are basic, and are used over and over again.

If a body of mass m is raised vertically to a height h, the force of gravity being mg; then the work done is mgh. This work is a measure of the PE transferred to the body. Thus we have

$$\text{gravitational PE} = mgh \qquad \textbf{(Equation 5-2)}$$

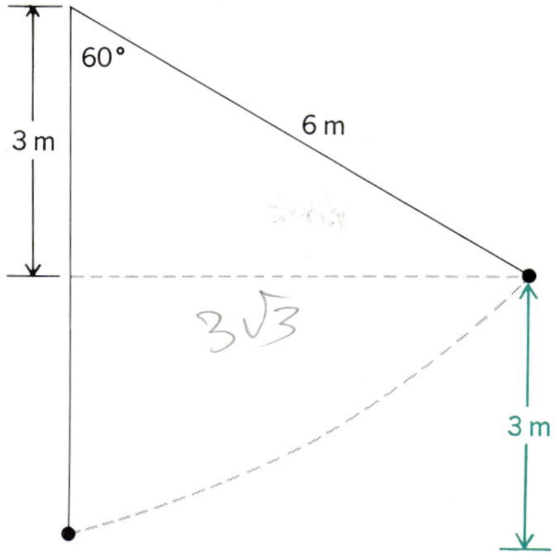

Figure 5-4

Whether or not the object is lifted straight up, the factor h in this formula refers to the vertical rise measured from some reference level where the gravitational PE is designated as zero. In problems using the above formula it is the change in PE that is required, so the reference level for zero PE can be quite arbitrary—sea level, the floor of a room, the top of a table.

Example 5-2 A 70 kg acrobat holds on to the end of a rope 6 m long and swings to a ledge where the rope makes an angle at 60° with the vertical. How much PE has she gained when she reaches the ledge?

A right triangle has been constructed in Figure 5-4, and it has been determined that the side adjacent to the 60° angle is 3 m, so the acrobat has been raised 6 m − 3 m = 3 m above her original position. The change in PE is determined as follows:

$$\begin{aligned}\Delta \text{PE} &= mg\,\Delta h \\ &= (70\text{ kg})(9.8\text{ N/kg})(3\text{ m}) \\ &= 2.06 \times 10^3\text{ N}\cdot\text{m} \\ &= \boxed{2.06 \times 10^3\text{ J}}\end{aligned}$$

Notice that in this case we did not need to define the lowest level as zero PE and then solve for the final PE. Instead, we found the change in PE directly.

Figure 5-5

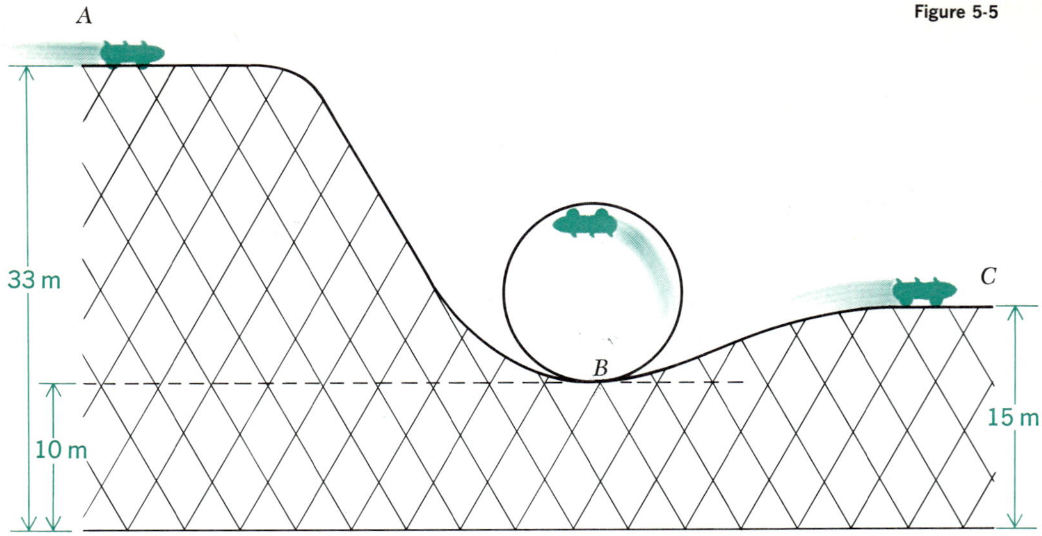

Example 5-3 A roller coaster of mass 3000 kg proceeds from point A (Figure 5-5) to point B and then to C. What is its PE at C? How much PE did the car lose in going from A to C?

We designate the ground as our reference level for 0 PE. At C the PE (relative to the ground level) is

$$\text{PE} = (3000\text{ kg})(9.8\text{ N/kg})(15\text{ m})$$
$$= 4.3 \times 10^5\text{ N}\cdot\text{m}$$
$$= \boxed{4.3 \times 10^5\text{ J}}$$

In going from A to C, the car falls 18 m.

$$\Delta\text{PE} = mg\,\Delta h$$
$$= (3000\text{ kg})(9.8\text{ N/kg})(18\text{ m})$$
$$= \boxed{5.3 \times 10^5\text{ J}}$$

The exact nature of the car's path (hills, loop, etc.) as it went from A to C has no bearing at all on the net change in PE during the over-all motion.

When a body is at rest, it has no kinetic energy. KE is given to the body when it speeds up under the action of a force F acting through the distance s. The KE gained is equal to the work done.

$$\text{KE} = \text{work} = Fs$$

and, since

$$F = ma$$
$$\text{KE} = mas$$

from Equation 3-4

$$v^2 = 2\,as$$

thus

$$\boxed{\text{KE} = \tfrac{1}{2}mv^2} \qquad \textbf{(Equation 5-3)}$$

Some readers may object to the above derivation of $\text{KE} = \frac{1}{2}mv^2$ on the grounds that it is limited to a uniformly accelerated body acted on by a constant force. Fortunately, it can be shown by advanced methods that the work done is given by $\frac{1}{2}mv^2$ regardless of the exact way in which the body reached its final velocity.

Note that the distance moved and the force applied do not appear in the formula for the kinetic energy. If a smaller force had been applied, the acceleration would have been less, and the body would have had to move farther in order to attain the same velocity v. The product Fs would have remained the same. Our two formulas for gravitational PE and KE (Equations 5-2 and 5-3) are easily remembered, and they serve to emphasize the significant physical quantities: weight (mg) and vertical rise for gravitational PE, mass and velocity for KE.

Example 5-4 A girl on a motorbike is scooting along at 20 m/sec. If she has a mass of 50 kg, what is her KE? By applying Equation 5-3 we have

$$\begin{aligned} KE &= \tfrac{1}{2}mv^2 \\ &= \tfrac{1}{2}(50\ \text{kg})(20\ \text{m/sec})^2 \\ &= 10{,}000\ \text{kg}\cdot\text{m}^2/\text{sec}^2 \\ &= \boxed{10{,}000\ \text{J}} \end{aligned}$$

Example 5-5 What is the KE of a 4000 kg truck moving along a turnpike at 30 m/sec? How does this compare with the PE stored in the truck if it were lifted 45 m straight up?

$$\begin{aligned} KE &= \tfrac{1}{2}mv^2 \\ &= \tfrac{1}{2}(4000\ \text{kg})(30\ \text{m/sec})^2 \\ &= \boxed{1{,}800{,}000\ \text{J}} \end{aligned}$$

For finding the PE we use Equation 5-2:

$$\begin{aligned} PE &= mgh \\ &= (4000\ \text{kg})(9.8\ \text{N/kg})(45\ \text{m}) \\ &= \boxed{1{,}760{,}000\ \text{J}} \end{aligned}$$

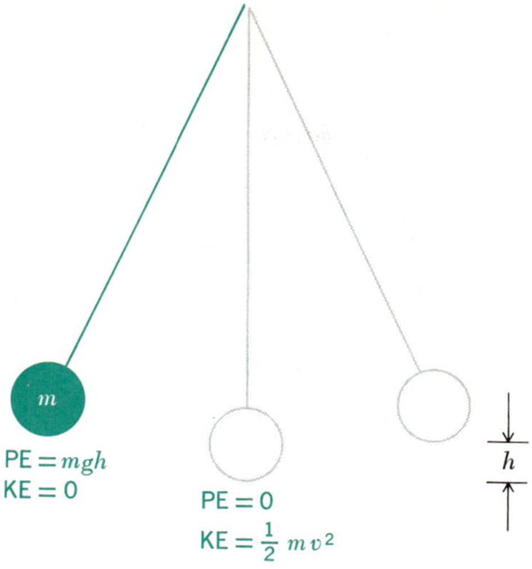

Figure 5-6 Conservation of Mechanical Energy: At any instant, the total mechanical energy (PE + KE) is constant.

5-5 THE ENERGY PRINCIPLE IN SOLUTION OF PROBLEMS

A swinging pendulum has maximum PE and zero KE at the beginning, or top, of each swing. Every time it reaches the lowest point of the swing, all of the PE has been converted to KE. In turn, this KE is sufficient to raise the bob again, converting all the energy back to PE (Figure 5-6). The mechanical energy of the system will always be conserved if we are able to neglect friction, and this energy principle provides us with a very convenient tool in solving mechanical problems. Remember our formula for kinetic energy, $KE = \frac{1}{2}mv^2$, came directly from Newton's second law. When we use this formula in a problem or a derivation, we are really only using Newton's second law in a "predigested" way that, as you can see from the way the formula was derived in Section 5-4, has already taken care of the kinematical part of the problem.

Example 5-6 In a region far removed from the earth's gravitational attraction, workers are assembling a space station. A section that has a mass of 3000 kg is floating through space at a speed of 1 m/sec toward a worker who is 3 m in front of a massive main bulkhead of the station that has already been assembled (Figure 5-7). If the worker can exert a force of 750 N, should he try to stop the oncoming section, or should he quickly move out of the way?

For comparison, we will first solve this problem by using the equations of Chapters 3 and 4, and then solve it by using the energy principle.

By Newton's second law,

$$\text{net } F = ma$$

and since the force is opposite to the motion,

$$\begin{aligned} -750\ \text{N} &= (3000\ \text{kg})(a) \\ a &= -0.25\ \text{m/sec}^2 \end{aligned}$$

This is the acceleration of the section while it is being brought to a stop. In

Figure 5-7

stopping, it travels a distance of (from Equation 3-4)

$$s = \frac{v_f^2 - v_i^2}{2a}$$

$$= \frac{0 - (1 \text{ m/sec})^2}{2(-0.25 \text{ m/sec}^2)}$$

$$= 2 \text{ m}$$

Since he is positioned 3 m from the bulkhead, the worker has no need to worry.

According to the energy principle, the worker is pushed back by the oncoming section, and work is done *on* him by the section.

$$\left(\begin{array}{c}\text{work done on} \\ \text{man by section}\end{array}\right) = \left(\begin{array}{c}\text{loss of KE of} \\ \text{moving section}\end{array}\right)$$

$$Fs = \tfrac{1}{2}mv^2$$

Therefore

$$(750 \text{ N})(s) = \tfrac{1}{2}(3000 \text{ kg})(1 \text{ m/sec})^2$$

$$s = 2 \text{ m}$$

As you can see, the energy method is much simpler.

Example 5-7 A bowling ball is returned to the bowler by an automatic pin-setter that transfers it to the return ramp 100 cm above the alley level (Figure 5-8). If the ball is given an initial speed of 0.6 m/sec at point *A*, what is its speed when it reaches the bowler on the short, level stretch *DE*? Ignore friction.

To solve the problem, we equate the initial mechanical energy to the final mechanical energy.

$$\text{KE}_1 + \text{PE}_1 = \text{KE}_2 + \text{PE}_2$$

$$\tfrac{1}{2}m(0.6 \text{ m/sec})^2 + (m)(9.8 \text{ N/kg})(1.0\text{m}) = \tfrac{1}{2}m(v^2) + (m)(9.8 \text{ N/kg})(0.4 \text{ m})$$

The mass cancels out,* and the solution of the equation gives

$$v = \boxed{3.48 \text{ m/sec}}$$

A direct solution of this problem would be more complicated. You would have to find the acceleration of the ball while on the slope *AB*, and the final velocity (at *B*) would become the initial velocity for a second problem to cover the interval *BC*. The energy method skips some of the steps, and gives an answer for the final velocity. It is important to remember, however, that the energy method cannot be used to find times or accelerations.

* You cannot count on the mass always canceling out; it does in Example 5-7, but not in Example 5-9. Hence, we keep the mass in each problem and cancel it out only when possible.

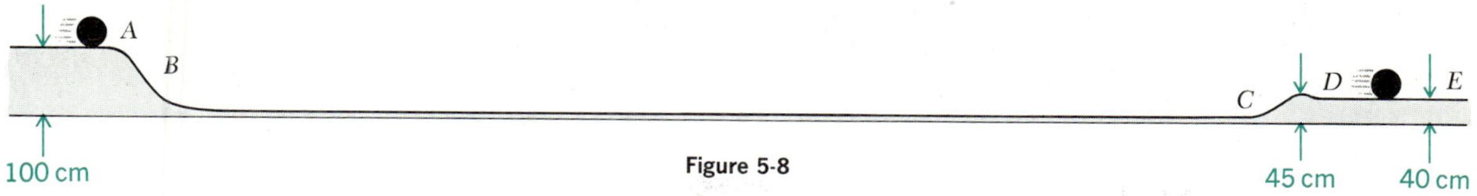

Figure 5-8

Example 5-8 A stone of mass 2 kg is thrown at 5 m/sec upward at a 30° angle from a cliff 20 m high (Figure 5-9). How fast will the stone be moving when it strikes ground? Ignore air resistance.

The direct method would make this into a projectile problem (Section 3-9), for which you would have to find the horizontal and vertical components of the velocity at *C*. Combining these components would give the magnitude of the velocity at *C*. The energy method skips the intermediate steps.

$$KE_1 + PE_1 = KE_2 + PE_2$$

$$\tfrac{1}{2}(2\text{ kg})(5\text{ m/sec})^2 + (2\text{ kg})(9.8\text{ N/kg})(20\text{ m}) = \tfrac{1}{2}(2\text{ kg})\,v^2 + 0$$

$$25\text{ J} + 392\text{ J} = (1\text{ kg})v^2 + 0$$

$$v = \sqrt{417\text{ J}/1\text{ kg}}$$

$$= \boxed{20.4\text{ m/sec}}$$

The angle of throw has nothing to do with this particular problem, although of course the angle does affect the maximum height of rise, time in the air, and distance out from the cliff. If the angle had been 20° instead of 30°, these latter quantities would have been different, but the final speed would still have been 20.4 m/sec, as is indicated by the fact that all the quantities in the energy equation would have been the same.

Figure 5-9

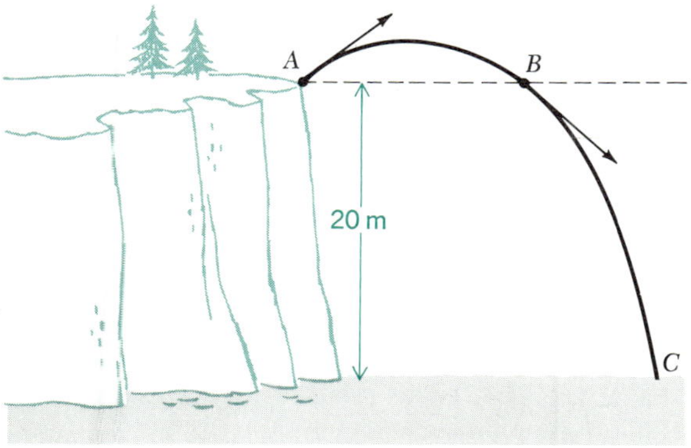

This final example illustrates a method of dealing with problems involving the PE in a compressed spring.

Example 5-9 A 4 kg mass is sliding along a frictionless table top with a velocity of 3 m/sec. It collides with a spring bumper and is momentarily brought to rest while the spring is compressed 0.2 m (Figure 5-10). If the spring has negligible mass, what is the average force exerted by the spring in bringing the mass to rest? Neglecting the friction in the spring, with what velocity will the mass recoil from the collision?

The mass, in coming to rest, does work on the spring by compressing it. This work

Figure 5-10

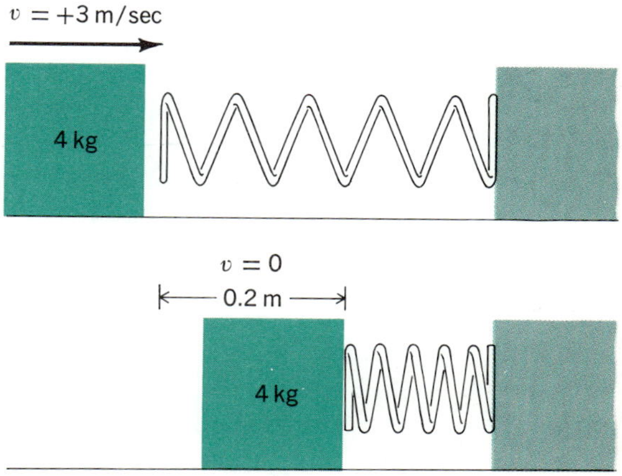

PE spring = KE block

comes from the KE of the mass, which is all converted into PE as the spring is compressed. The force exerted by the mass as it pushes on the spring is multiplied by the displacement to obtain the work done on the spring and, hence, its PE. In this problem the force is not constant, so the average force is used in calculating the work. Thus,

$$(\text{average } F)s = \tfrac{1}{2}mv^2$$

$$F = \frac{mv^2}{2s}$$

$$= \frac{(4 \text{ kg})(3 \text{ m/sec})^2}{2(0.2 \text{ m})}$$

$$= \boxed{90 \text{ N}}$$

As the spring expands, its PE is converted back into the KE of the mass. Since mechanical energy is conserved, the mass will leave the spring with the same KE and, hence, the same speed it had originally, but moving in the opposite direction. Hence,

$$v = \boxed{-3 \text{ m/sec}}$$

5-6 CONSERVATION OF ENERGY: RELATIVITY

The concept that energy can neither be created nor destroyed represents one of the great generalizations of the nineteenth century. Before the work of several nineteenth-century scientists, especially the English physicist James Joule (1818–1889), the possibility of converting various forms of energy to heat and vice versa was known, but the *quantitative* equivalence of heat, or thermal energy, and other forms of energy was yet to be discovered. Joule showed experimentally that when heat is converted to mechanical energy, or vice versa, the same amount of heat is always equivalent to the same amount of energy. This law of conservation of energy was then extended by Joule and others to cover "all" forms of energy and was first stated somewhat as follows: *The total amount of mechanical, thermal, chemical, electrical, and other energy in any isolated system remains constant.*

In 1905 Albert Einstein (Figure 5-11) broadened the law still further to include the equivalence of mass and energy. Einstein proposed, as part of the theory of relativity, that, a given amount of mass (1 kg) is always equivalent to the same amount of energy (9×10^{16} J), and vice versa. We in the present generation are familiar with the equivalence of mass and energy as shown in nuclear reactions. All the fragments, gases, ash, and so on, of a nuclear explosion would, if gathered together, have slightly less mass than the original uranium or hydrogen. The Einstein equation relating mass and energy is expressed simply as

$$E = mc^2$$

and can also be stated as

$$\Delta E = (\Delta m)c^2 \qquad \textbf{(Equation 5-4)}$$

Substituting the appropriate mks units shows that the energy is in joules:

$$\text{J} = (\text{kg})(\text{m/sec})^2$$

where c is a constant of nature, equal to the speed of light in a vacuum (3.00×10^8 m/sec).

Contrary to popular opinion, proof of the Einstein equation $\Delta E = (\Delta m)c^2$ is not highly mathematical, and can be derived by any student who has studied high school algebra. It took the genius of Einstein to investigate some of the *inconsistencies* of the then-understood laws of physics rather than to use high-powered mathematics to extend those laws. The Einstein equation stems from his proof that the mass of a body increases as its speed increases. The formula is

$$m = \frac{m_0}{\sqrt{1 - \frac{v^2}{c^2}}} \qquad \textbf{(Equation 5-5)}$$

where m is the mass of the body as judged by the observer, m_0 is the *rest mass*, which is *its mass when at rest relative to the observer*, and v is the velocity of the body relative to the observer. It follows that "mass" of a body is not a constant value. Two observers, equally competent, will measure the inertial mass (m in Equation 5-5) of a body differently, if their velocities relative to the body are different. Note that Equation 5-5 predicts that the mass becomes infinitely great as the relative velocity approaches c, the speed of light. Therefore, according to Newton's second law ($F = ma$, where m is approaching infinity), an ever-increasing force is required to accelerate the body, and a physical body can never reach the speed of light, since an infinite force is not possible.

Example 5-10 A hypothetical spacecraft travels at 2×10^8 m/sec away from the earth. A projectile is shot forward from the craft at 2×10^8 m/sec relative to the spacecraft. What is the velocity of the projectile relative to the earth?

We do not have to know the equation for relative velocity in order to look at certain features of the solution. Of one thing we can be sure—the velocity of the projectile relative to the earth must be less than 3×10^8 m/sec. Offhand, we would have expected it to be 4×10^8 m/sec, but the simple vector treatment of relative velocity (Section 3-7) is not valid at high speeds. The relativity result is 2.77×10^8 m/sec. We can reconcile these two apparently contradictory observations by noting that, relative to the people on earth, the projectile is moving faster than it is moving relative to the people in the spacecraft. As judged by the earthlings, the projectile therefore has more mass than as judged by the space travelers, and so the same force (reaction force of the fuel) produces less acceleration relative to the earth. The projectile hence receives less final velocity, as viewed from earth, than would have been expected if the mass had not changed.

Einstein reasoned that when a body is given kinetic energy, its mass is increased according to Equation 5-5. He concluded that, if *increase* of mass corresponds to *gain* in energy, the mass of a body even at rest must be equivalent to some energy. His conclusion (1905) was verified within less than 20 years, when Rutherford observed (1921) the conversion of mass into energy during the first experiments in artificial radioactivity. The increase in mass of a moving electron had actually been observed by Bucherer and Kaufmann before the theory of relativity (1901–04). No satisfactory explanation was made until the mass equation appeared as a sort of unexpected by-product of Einstein's re-examination of space, time, and relative motion.

Figure 5-11 Dr. Albert Einstein

There is no room for doubt that the theory of relativity correctly describes the equivalence of mass and energy. Nuclear physicists have verified the equation $\Delta E = (\Delta m)c^2$ in countless precise experiments. The law of conservation of energy must, therefore, be worded in such a way as to include the mass equivalent of energy. It has now been generalized in another way. An isolated system is one that does not have any interactions with objects outside the system, and the universe is, by definition, an isolated system. By extrapolation, the law of conservation of energy has been assumed to apply to the entire universe. Thus, finally, we state the law of conservation in its most general form:

The law of conservation of energy: The total amount of energy plus mass in the universe remains constant.

This generalization, evolved by many scientists over a period of several centuries, has proved to be of great significance for every branch of physical science. You will use it often in your future work.

The equivalence of mass and energy is now basic to our understanding of nature. Aside from the release of energy in a nuclear reactor and in uranium and hydrogen bombs, a less obvious but more important release of nuclear energy takes place all the time within the sun and other stars. As the end result of a chain reaction on the sun in 1 second 5×10^{13} metric tons of hydrogen are converted into nearly the same amount of helium. The mass balance between the hydrogen and helium is not exact, however, and the sun's mass is continually decreasing at the rate of 4 million metric tons per second.* This is the amount of mass that is radiated as energy. The sun has been producing radiant energy (and losing mass) at this rate for several billion years.

Yet it is reassuring to know that after all this time, because of the tremendous mass of the sun (1.97×10^{30} kg), more than 99.99% of the sun still remains.

As for the effects of relativity and the increase of mass with velocity, the next example shows that under ordinary circumstances, such increases are far too small to be noticed.

Example 5-11 A spacecraft having a rest mass of exactly 4 metric tons is moving at 3000 m/sec. What is its increase in mass? What is its KE?

Its new mass m can be obtained by using Equation 5-5 (see next column).

$$m = \frac{m_0}{\sqrt{1 - v^2/c^2}}$$

$$= \frac{4000.0000000 \text{ kg}}{\sqrt{1 - \frac{(3 \times 10^3 \text{ m/sec})^2}{(3 \times 10^8 \text{ m/sec})^2}}}$$

$$= \frac{4000.0000000 \text{ kg}}{\sqrt{1 - \frac{9 \times 10^6 \text{ m}^2/\text{sec}^2}{9 \times 10^{16} \text{ m}^2/\text{sec}^2}}}$$

$$= \frac{4000.0000000 \text{ kg}}{\sqrt{1 - 10^{-10}}}$$

$$= \frac{4000.0000000 \text{ kg}}{0.99999999995}$$

$$= 4000.0000002 \text{ kg}$$

$$\Delta m = 4000.0000002 \text{ kg} - 4000.0000000 \text{ kg}$$

$$= \boxed{0.2 \text{ mg}}$$

This increase is about equal to the mass of the ink used to write an address on a rocket-mail letter carried by the spacecraft.

The KE of the spacecraft is

$$\text{KE} = \tfrac{1}{2}mv^2$$

$$= \tfrac{1}{2}(4 \times 10^3 \text{ kg})(3 \times 10^3 \text{ m/sec})^2$$

$$= \boxed{18 \times 10^9 \text{ J}}$$

We may say that 18×10^9 J of KE is equivalent to 2×10^{-7} kg of mass. This checks with $\Delta E = (\Delta m)c^2$; each kg of mass increase corresponded to 9×10^{16} J $(=c^2)$ of energy increase, as it should.

This example clearly shows that the mass changes accompanying ordinary energy changes are tiny indeed. It is for this reason that we can apply the law of conservation of energy to the solutions of most of the problems that we encounter with a clear conscience, ignoring these small relativity effects. In other words, we usually need to consider only chemical, thermal, electrical, or mechanical energies.

The generalization that *all* forms of energy (including thermal energy, electrical energy, *mass energy,* and so forth) are interchangeable with mechanical energy is the substance of the law of conservation of energy.

* If four hydrogen nuclei with a combined atomic mass of 4.032 ultimately form one helium nucleus with an atomic mass of 4.003, then 4.032 − 4.003 = 0.029 atomic mass unit of matter is converted to energy.

5-7 CONSERVATION OF MASS IN RELATIVITY THEORY

Let's look critically at the popular statement "mass can be converted into energy." Recall from Equation 5-5 that the rest mass m_0 is the mass of a body when it is measured while at rest relative to the observer. The mass m is the *relativistic* mass of the body as measured by an observer when there is a relative velocity between mass and observer. The mass-equivalence of energy as given by the Einstein equation $\Delta m = \frac{\Delta E}{c^2}$ refers to the *change* in mass that results from the *change* in energy of a body. In this case, the relativistic mass is the rest mass *plus* the mass equivalence of energy. This last point will be illustrated in Example 5-12. With an understanding of the distinction between rest mass and relativistic mass, we shall find that it is the *rest mass* which can be converted into energy and it is the rest mass that measures what we ordinarily call *matter.* The total *relativistic mass* of the universe remains constant. It is in this sense that mass is conserved. Unless we use words carefully and consistently, we shall run into some difficulty in interpreting the law of conservation of mass and the law of conservation of energy.

The first difficulty is seen in Example 5-11 about the spacecraft. Surely this seems to be no example of the conversion of mass into energy, or vice versa, since both the spacecraft's mass *and also* its energy increased. This is like trying to keep account books with all credits and no debits. In fact, the theory of relativity considers that *any* energy increase ΔE, not just a KE increase, is accompanied by an increase of mass Δm. You are already aware that the Einstein equation says these changes are related by $\Delta m = (\Delta E)/c^2$. Considering the whole system, however, spacecraft plus fuel, we would find that the fuel had more mass before it was burned than afterward, because of its original PE.* In such problems, mass and energy are each conserved, if by "mass" we are referring to total (relativistic) mass. We might use the term "massergy" for the quantity that is conserved—mass plus the mass equivalent of energy. To see this more clearly, let's consider a simple example involving a falling body.

Example 5-12 A 100 kg safe is dropped from the top of a building that is 100 m high. Show the balance of matter and energy.

At the top, before the fall, the energy is all PE. We have:

$$\text{PE at top} = (100\ \text{kg})(9.8\ \text{N/kg})(100\ \text{m}) = 98{,}000\ \text{J}$$

Hence, by the law of conservation of energy,

$$\text{KE at bottom} = 98{,}000\ \text{J}$$

since the energy is all KE just before the safe hits the ground.

By the Einstein equation, the mass equivalent of 98,000 J of energy is

$$\Delta m = \frac{\Delta E}{c^2} = \frac{9.8 \times 10^4\ \text{J}}{9 \times 10^{16}\ \text{m}^2/\text{sec}^2} = 1.1 \times 10^{-12}\ \text{kg}$$

The equations that follow show how these various masses and energies of the safe are conserved.

SAFE AT TOP OF BUILDING		SAFE ABOUT TO STRIKE PAVEMENT

Conservation of mass:

$$\left\{\begin{array}{l} 100\ \text{kg} \\ +1.1 \times 10^{-12}\ \text{kg} \\ (\Delta m \text{ due to PE}) \end{array}\right\} = \left\{\begin{array}{l} 100\ \text{kg} \\ +1.1 \times 10^{-12}\ \text{kg} \\ (\Delta m \text{ due to KE}) \end{array}\right\}$$

Conservation of energy:

$$\left\{\begin{array}{l} \text{PE } 98{,}000\ \text{J} \\ \text{KE } 0 \end{array}\right\} = \left\{\begin{array}{l} \text{PE } 0 \\ \text{KE } 98{,}000\ \text{J} \end{array}\right\}$$

It can be seen that both mass and energy are conserved.

You may ask, "Where did the additional 1.1 ×

* In an exact treatment of the spacecraft problem, we would also have to consider the KE and the mass changes of the exhaust gases.

10^{-12} kg of mass come from in the first place?" Suppose the safe had been hoisted to the top by a pulley system activated by the muscular efforts of several workmen. The mass of the workmen must have *decreased* during the process, for they used up chemical PE, perhaps from a ham-and-eggs breakfast. In a sense, mass "flowed" from the men to the safe as it was raised. No matter how far back you carry the questioning, you will find that the eventual source of energy (and mass) was nuclear—more than likely from solar energy that was captured by plants, which certain hogs and hens then converted into ham-and-eggs-on-the-hoof, and so forth. But we know that the sun's energy comes to us as a result of nuclear changes. It would seem, therefore, that all or nearly all the energy changes on the earth ultimately come from mass changes in nuclear reactions, either in manmade nuclear reactors or in the gigantic natural reactor we know as the sun. What, then, is the difference, if any, between the changes taking place in the safe or in chemical reactions, and the changes taking place in a nuclear reactor? The difference is one of degree only.

We have seen, then, that mass is always conserved if we include the mass equivalent of KE or PE.

SUMMARY

Whenever a body is moved by a force that has a component along the direction of motion, **work** is done. The work done ($W = Fs$), is a measure of the **energy** transferred to or from the system. In the mks system, the unit for work and energy is the **joule** and in the cgs system it is the **erg. Potential energy** (PE) is due to the relative position of the parts of a system of bodies, and this includes gravitational, elastic, chemical, and some electrical energy, as well as nuclear energy. **Kinetic energy** (KE) is due to motion, and this includes much of what is known as thermal energy. In mechanical systems where friction is negligible and velocities are small compared to the velocity of light, mechanical energy is conserved, $\Delta PE = -\Delta KE$. Some formulas for energy are:

$$\text{gravitational PE} = mgh \qquad \text{KE} = \tfrac{1}{2}mv^2 \qquad \text{mass energy} = m_0c^2$$

The **mass** of a body depends upon its velocity relative to the observer. As the velocity increases, a body's mass approaches infinity according to the equation

$$m = m_0/\sqrt{1 - v^2/c^2}$$

When an isolated system undergoes change, the total (**relativistic**) mass of the system remains unaltered, and the total energy content of the system remains unaltered. These are the laws of **conservation of mass** and **conservation of energy.** The importance of the law of conservation of energy lies in the fact that it includes all forms of energy, such as thermal energy, electrical energy, and **mass energy,** as well as mechanical energy (KE and PE).

Rest mass is measured when at rest relative to the observer. It is rest mass that can be converted into a form of energy without violating the law of conservation of energy. No energy can be created or destroyed unless a corresponding amount of rest mass disappears or appears. The equivalence of mass and energy is governed by the Einstein equation,

$$\Delta E = (\Delta m)c^2$$

QUESTIONS

1 A car is coasting on a level stretch. If the driver turns on the headlights, where does the energy come from (a) if they are connected to a battery? (b) if they are connected directly to a generator geared to the wheels? Does the car slow down in either case?

2 Is there any source of energy available to us on earth that does not eventually trace back to the sun's energy? What about nuclear energy? What is the source of the sun's energy?

3 Compare the kinetic energies of a 30 g bullet traveling at 1200 m/sec and a 10,000 kg truck moving at 6 m/sec.

4 When an artificial satellite is moving in an orbit around the earth it has more energy than when it was on earth before launching. Where did this energy come from? Into what forms is the energy transformed while the satellite is circling the earth? What is the eventual fate of this energy as the satellite spirals toward the earth?

5 Is some of the sun's energy used up in causing the earth to move in an orbit (assumed circular for convenience)?

6 The tide in some parts of Norway rises some 12 m or more, and, when the tide goes out, the water dashes against the rocks and beach as it flows slowly out to sea, ready to be raised again by the moon in $12\frac{1}{2}$ hours as the earth turns. What becomes of the PE stored in the water each time the tide comes in? If the water were trapped in a dam, the potential energy could be used by man. Tidal energy stations have actually been constructed for this purpose. What is the ultimate source of this energy?

7 Is it true that a wound-up watch spring has more mass than the same spring when the watch has run down? Explain.

8 In Figure 5-6, derive a mathematical expression for the velocity of the ball at its lowest point. Compare this with the expression for the final velocity of a body dropped from the same height.

9 If you had to lift a weight a certain distance straight up, would you do less work if you used a pulley arrangement? Explain.

PROBLEMS

GROUP A

1 A child who weighs 250 N is sitting on a horizontal floor. (a) How much work is required to move him 1 m against an opposing force of friction equal to 30 N? (b) How much work is required to lift him 1 m straight up?

2 In Example 5-8, what is the velocity of the stone when it is 8 m above the ground?

3 In Example 5-9, what is the velocity of the mass when the spring is partially compressed with a PE of 10 J?

4 When a 50 kg girl walks a distance of 130 m from the bottom to the top of a hill, her PE increases by 9800 J. What is the vertical rise of the hill?

5 A wild goose of mass 4 kg is flying at a speed of 3 m/sec. What is the bird's KE?

6 What is the KE of an 8000 kg truck moving at 10 m/sec up a hill that makes an angle of 10° with the horizontal?

7 A fly takes off horizontally, from the edge of a table, with KE equal to 125 ergs. What force was exerted by the fly while it was accelerating through a distance of 2 mm?

8 The fly of Problem 7 weighs 24.5 dynes. What is its mass? What is its velocity just after takeoff?

9 Suppose an electron is moving at half the speed of light. Show that its mass is 1.15 times its rest mass.

GROUP B

10 Ten books, each 6 cm thick and weighing 20 N, are lying horizontally on a table. What work is required to construct from them a stack of books 60 cm high?

11 Repeat the derivation leading up to Equation 5-3, but do not assume that $v_i = 0$. Show that the work done against the reaction force of inertia is $\frac{1}{2}mv_f^2 - \frac{1}{2}mv_i^2$. Interpret your answer.

12 A .22 caliber bullet of mass 0.0026 kg is shot into a massive wooden target at 320 m/sec. It penetrates 0.08 m into the target. What is the average force exerted on the bullet by the wood? (Use the energy principle; do not find the acceleration).

13 A 75 kg boy, starting from rest, slides down a 30° hill that is 70 m long. He arrives at the bottom with a speed of 15 m/sec. How much thermal energy has been shared between the surface of the hill and the seat of his pants? (Use the energy principle; do not find the acceleration.)

14 A ping-pong ball of mass m rolls off the edge of a table that is 1 m high. When the ball strikes the floor its speed is 5 m/sec. How fast was it rolling when it left the table? (Use the energy principle.)

15 An egg of mass 50 g is dropped from a building 20 m high onto a target consisting of a thick block of foam rubber. (a) Find the KE of the egg just as it strikes the target. (b) The egg penetrates 20 cm into the rubber while stopping. What force is exerted on the egg?

16 A soccer player gives a ball of mass 1 kg an initial horizontal speed of 10 m/sec. The ball slides 15 m along the turf against an average force of 2.5 N. What is the final speed of the ball?

17 Use the energy principle to find with what speed the roller coaster of Figure 5-5 reaches point B. (Ignore friction.)

18 A stone of mass m is thrown horizontally at 10 m/sec from a cliff 10 m high. With what speed does the stone strike the ground below the cliff?

19 How fast would a rocket of mass 10^7 kg have to move in order for its mass to increase by 2 g?

20 How fast is a cosmic ray particle moving when its mass is double its rest mass?

21 A girl of mass m swings back and forth on the end of a rope 6 m long that is attached to the ceiling of a gymnasium. If she approaches within 2 m of the ceiling during each cycle, what is her speed as she passes through the lowest point of her swing?

22 If the ball in Figure 5-8 were at rest at point C, what speed must you give it so that it can just get over the hump at D? What speed would it have when it reaches point E?

23 A 2 kg stone is thrown at an angle with a velocity of 20 m/sec from a height of 10 m above the ground. It rises to a maximum height of 15 m as measured from the ground. (a) What is its KE at maximum altitude? (b) What is its speed when it is 5 m above the ground?

24 A pendulum bob, mass 0.5 kg, is displaced from the vertical until it is 0.25 m above its lowest point. Upon release, it rises to a height of 0.24 m on the other side. (a) How much energy was lost to friction? (b) Assuming that half the total frictional loss (due to air resistance and friction in the pivot) takes place during the downward swing, with what speed does the bob pass through the lowest point?

25 A heavy chain 40 cm long is stretched out on a frictionless table top, with one end of the chain a tiny bit over the edge. The chain is displaced and starts to pick up speed as it slides off the table. What is its speed when the last link of the chain is just leaving the table? (*Hint:* Assume the mass of the chain to be m.)

FOR FURTHER STUDY

MORE ABOUT RELATIVITY

While the equation $\Delta E = (\Delta m)c^2$ was one of the chief results of the Einstein theory of relativity (Section 5-6), the theory was enlightening in other ways that seem less spectacular and yet are equally important. Historically, the relativity theory originated in an effort to dispose of certain difficulties in the equations derived by James Clerk Maxwell (1831–1879), a Scottish physicist (shown in Figure 5-12). These equations described the "field theory," which links light, radio waves, and x rays together as various forms of electromagnetic radiation. It was recognized in the period 1890–1905 that certain inconsistencies existed

Figure 5-12 James Clerk Maxwell.

between the Maxwell equations and the concepts of space and time used by Galileo, Newton, and their successors. Einstein succeeded in showing that not Maxwell, but Newton, was at fault. It was not a question of some one particular equation being wrong; rather, Newton's underlying concepts about space and time were found to be inadequate in light of more recent knowledge.

This brief discussion about the origin of the relativity theory is given to show why c, the speed of light and all other electromagnetic waves, enters into what seem to be essentially mechanical problems.

The constant c happened to be first discovered by workers in the field of electricity, long before electromagnetic waves were known to exist. Maxwell discovered a *second* interpretation of this constant when he found it to be the speed of light and other electromagnetic waves. Einstein found a *third* interpretation of this constant, as $\sqrt{\Delta E/\Delta m}$. Further interpretations very likely await future investigators.

For these reasons, a detailed study of the relativity theory requires a thorough knowledge of electricity and magnetism beyond the scope of this book. However, it will be worth while at this time to write down the two postulates upon which Einstein based his theory:

1. If two systems are in relative motion with uniform velocity, an observer in one system can, by observing the second system, determine that there is *relative* velocity, but cannot determine anything more, such as whether one of the systems is "at rest."
2. Measurements of the speed of light always give the same numerical value, regardless of the relative velocity of observer and source of light.

The first postulate says, for instance, that if you were playing ping-pong on a train that was traveling in a straight line at uniform velocity relative to the ground, the bouncing ball would obey the same laws of physics as if you were outside, "at rest." You would not have to adjust your reflexes in order to play while on the train. It would be impossible to tell which system was "actually" in motion: the train moving forward at velocity v, with the earth stationary, or the ground moving backward at velocity v, with the train stationary. This is what is meant by the statement that the relativity theory gives up the possibility of measuring absolute space and time.

Figure 5-13

The second postulate is in agreement with experiment. A. A. Michelson and E. W. Morley in 1887 found that the velocity of a beam of light moving from east to west is the same as that of a beam of light moving from north to south (Section 23-2). It had been expected that the east-west velocity might be influenced by the velocity of the earth, but such was not the case. (Compare this with Problem 25, Chapter 3, in which we found that a swimmer going upstream and back took longer than he did swimming the same distance across the river and back.) The second postulate is perhaps easier to understand if we consider a "thought experiment" in which two searchlights are uncovered for a brief instant by two experimenters A (Figure 5-13a) and B (Figure 5-13b). Both are at the same point in space. B is in motion to the right relative to A with a relative velocity of 3×10^7 meters per second, one tenth the speed of light. After 1 second, a pulse of light has spread out from A's lamp a distance of 3×10^8 meters and has reached point X. However, after 1 second the light from B has spread out 3×10^8 meters from B. But B has moved to B' during the second and thus expects the light to reach Y', which is 3.3×10^8 from A. (Note that according to the second postulate, the velocity of the light away from B must be 3×10^8 m/sec, regardless of B's velocity.) From A's point of view, the light from B's searchlight should only have reached to Y at this same instant, for, if it had gone further its velocity relative to A would have been more than 3×10^8 m/sec.

How can the light be both at Y and Y' at the same time? Einstein's postulate seems to lead to contradictions. In the resolving of such contradictions Einstein was led to reconsider space-time relationships in general. He found that A and B are measuring their time intervals of 1 second by clocks that don't agree with each other, and the light was not "simultaneously" at Y and Y', because of this difference in the rates of ticking of their clocks. Eventually, a set of physical laws was found that was consistent with the two postulates. It is because of the second postulate that we are justified in speaking of "the" speed of light, equal to 3.00×10^8 m/sec, as a constant of the universe.

The formula for relative velocity is

$$v_x = \frac{v_x' + v}{1 + \frac{v}{c^2} v_x'} \qquad \text{(Equation 5-6)}$$

In this equation,* v_x is the velocity of a body as seen by an observer in one frame of reference (FR); v_x' is its velocity as seen by an observer in another frame of reference (FR'). The velocity of frame FR' relative to frame FR is v, and c is the speed of light. Let's see how this works out in a simple example where the velocities are small (see Section 3-7).

Example 5-13 A ship is moving southward at 30 km/hr, and a deck-walker moves northward with a velocity of 2 km/hr relative to the ship. What is the man's velocity relative to the water?

Here the water is one frame of reference FR, and the ship is the other frame FR'. Calling northward the positive direction, we have

frame FR = water frame FR' = ship

v_x = velocity of man relative to FR = ?

v_x' = velocity of man relative to FR' = +2 km/hr

v = velocity of FR' relative to FR = −30 km/hr

c = velocity of light

$= 3 \times 10^8$ m/sec

$= 1.08 \times 10^9$ km/hr

Placing these values into the formula we have

$$v_x = \frac{(+2 \text{ km/hr}) + (-30 \text{ km/hr})}{1 + \frac{(-30 \text{ km/hr})(+2 \text{ km/hr})}{(1.08 \times 10^9 \text{ km/hr})^2}}$$

$$= \boxed{-28.000000000000002 \text{ km/hr}}$$

In carrying through the calculation, it can be seen that the denominator is so close to 1 that it can be ignored; indeed we are not justified in carrying out the division to so many decimal places. Thus, on account of the smallness of v and v_x' in comparison with the velocity of light, the equation is essentially $v_x = v_x' + v$, the same as the *classical* result that would have been expected without the relativity theory. If either v_x' or v had been comparable in magnitude to c, then the result would have been quite different (Problem 26).

For the examples so far, the Einstein equation representing the relation between mass changes and energy changes has been expressed as

$$\Delta E = (\Delta m)c^2$$

* It is assumed that the relative velocity of the two frames is along the x-axis, and also that v_y' and v_z' are both zero, so that the motion is along the x-axis in each frame.

which shows that the increased energy of a body is accompanied by an increase in mass. We can also write $E = mc^2$ (Section 5-6), where E is the total energy (including the mass energy) and m is the total (relativistic) mass given by $m = m_0/\sqrt{1 - v^2/c^2}$. This equation places energy in a prominent place, without distinguishing too closely between the various forms that energy can take. The kinetic energy due to motion is then total energy minus rest energy. Thus

$$\text{KE} = mc^2 - m_0c^2 \qquad \textbf{(Equation 5-7)}$$

This is the relativistic definition of KE; we expect that at velocities which are small relative to c, Equation 5-7 will reduce to the familiar $\frac{1}{2}m_0v^2$ (see Problem 30). In fact, *all* relativity formulas must include the classical formulas of physics as special cases. We see that the KE of a body is not a definite quantity, since it depends upon m, which is not the same for all observers. This is nothing new; even in the classical relativity of Galileo and Newton the KE of a body depends on the observer. For example, consider an artificial satellite that is ejected in a forward direction from a launching rocket. The KE of the satellite is very much less as viewed from the rocket than as viewed from the earth (Problem 31), and, of course, is zero as viewed from the satellite itself.

PROBLEMS

GROUP C

26 In Example 5-10, prove that the velocity of the projectile relative to the earth is 2.77×10^8 m/sec.

27 The projectile of Example 5-10 emits a beam of light that can be seen both from the spacecraft and the earth. What is the velocity of this light as measured by an observer in the spacecraft? What is the velocity of the light as measured by an observer on earth?

28 Suppose that the projectile of Example 5-10 has a rest mass of 1000 kg. (a) What is its KE relative to the spacecraft? (*Hint:* Use Equation 5-7.) (b) What is its KE relative to the earth? (c) What is the value of the quantity $\frac{1}{2}m_0v^2$ where $v = 2.77 \times 10^8$ m/sec? (d) Does this latter quantity have any physical significance?

29 The answer to Problem 26 turned out to be less than c, as it should. Show that this will always be the case. That is, show that if v_x' and v are less than c, then the "sum" of v_x' and v, given by Equation 5-6, is also less than c. (*Hint:* Let $v_x' = ac$ and $v = bc$. The proof is then made by showing $(a + b)/(1 + ab) < 1$ where a and b are positive and less than 1.)

30 Combining Equations 5-7 and 5-5, show that

$$\text{KE} = m_0c^2(1 - v^2/c^2)^{-1/2} - m_0c^2$$

Simplify this expression, using the binomial theorem, and show that

$$\text{KE} = \tfrac{1}{2}m_0v^2 + (\quad) + \cdots$$

To prove that the correct expression for KE reduces to the classical formula as a special case, you will have to prove that the second and other terms in the infinite series become negligibly small compared with the first term as v/c approaches zero.

31 A satellite of mass 200 kg is ejected at 1000 m/sec in a forward direction from a launching rocket that is moving away from the earth at 7000 m/sec. Compute the KE of the satellite (a) as viewed from the rocket; (b) as viewed from the earth. (*Hint:* All velocities are much less than c, so classical mechanics is valid.)

6 Conservation of Momentum

In Chapter 5 you studied one of the basic principles of physics, the law of conservation of energy. Now we proceed to another unifying principle, the law of conservation of momentum. This law was not only known to Newton, but, as you will see, his statement of the second law of motion was actually in terms of change of momentum and not acceleration.

As a general principle, the law of conservation of momentum is of utmost value in physics. This law is still considered to be valid even in areas where Newton's original statement of the second law of motion has been replaced by a more inclusive theory.

6-1 DEFINITION OF MOMENTUM

You have already considered one measure of the motion of a moving body—its kinetic energy, $\frac{1}{2}mv^2$. Another measure of a body's motion is its momentum. Momentum is so fundamental that early writers such as Newton called it the "quantity of motion," or simply "motion." *The momentum of a body* is now defined as *the product of its mass and its velocity.*

$$\text{momentum} = (\text{mass})(\text{velocity})$$

Study

There are no special names for units of momentum; in the metric system, kilogram·meters per second and gram·centimeters per second are used. Momentum is always a mass unit multiplied by a unit of velocity. Since velocity is a vector quantity, momentum is also a vector quantity, having the same direction as the velocity of the body.

You will notice, however, that momentum, even though it is related to the velocity of a body, gives no indication of how fast a body is moving. As can be seen from the example below, two bodies having different masses and different velocities can have the same momentum.

Example 6-1 Compare the momentum of a tug boat of mass 10^6 kg moving along at 0.1 m/sec with that of a 2×10^3 kg truck whose speed is 50 m/sec.

For the tugboat:

$$\begin{aligned}\text{momentum} &= mv \\ &= (10^6 \text{ kg})(0.1 \text{ m/sec}) \\ &= \boxed{10^5 \text{ kg}\cdot\text{m/sec}}\end{aligned}$$

For the truck:

$$\begin{aligned}\text{momentum} &= (2 \times 10^3 \text{ kg})(50 \text{ m/sec}) \\ &= \boxed{10^5 \text{ kg}\cdot\text{m/sec}}\end{aligned}$$

6-2 NEWTON'S SECOND LAW IN TERMS OF MOMENTUM

Newton's second law (Section 4-3) can be restated in terms of change of momentum:

$$\text{net } F = ma$$
$$= m\left(\frac{v_f - v_i}{t}\right)$$
$$= \frac{mv_f - mv_i}{t}$$
$$= \frac{\Delta(mv)}{t} \qquad \textbf{(Equation 6-1)}$$

Thus, *net force equals rate of change of momentum.* Multiplying both sides by the time interval t, we obtain

$$(\text{net } F)t = \Delta(mv) \qquad \textbf{(Equation 6-2)}$$

It was in this form, rather than as net $F = ma$, that Newton originally stated his second law. This product of net force and the time during which it acts is defined as *impulse.* Equation 6-2 tells us that the impulse applied to a body equals the change of momentum of that body. Just as the change in the KE of a body is measured by the work (Fs) done on it, so is the change in momentum of a body measured by the impulse (Ft) given to it. To change the momentum of a body, a force must act for a certain time. As far as the total change in momentum is concerned, a large force acting for a short time can produce the same effect as a small force acting for a long time.

Example 6-2 A 70 kg water skier, moving at 20 m/sec, lets go of the towline. What impulse is needed to bring him to rest? If the impulse is provided by the water, which exerts an average force of −280 N on the skier, how long must the force act?

The initial momentum of the skier is

$$(70 \text{ kg})(20 \text{ m/sec}) = 1400 \text{ kg}\cdot\text{m/sec}$$

Since he is brought to rest, his final momentum is zero. The change in momentum is, therefore,

$$0 - 1400 \text{ kg}\cdot\text{m/sec} = -1400 \text{ kg}\cdot\text{m/sec}$$

To produce this change in momentum requires an impulse of

$$\boxed{-1400 \text{ kg}\cdot\text{m/sec}}$$

To find the time the force acts,

$$(\text{net } F)t = -1400 \text{ kg}\cdot\text{m/sec}$$
$$t = \frac{-1400 \text{ kg}\cdot\text{m/sec}}{-280 \text{ N}}$$

Since $1 \text{ kg}\cdot\text{m/sec}^2 = 1$ N (Section 4-6),

$$t = \boxed{5 \text{ sec}}$$

In this problem the force is negative because it acts opposite to the motion of the skier. For the same reason, the impulse is negative as is the *change* in momentum. This problem could also have been solved by finding the acceleration from net $F = ma$ and then the time from $a = \Delta v/t$. In general, however, momentum problems are handled more easily when the relation of Equation 6-2 is employed, rather than the kinematic equations.

If we consider momentum in combination with Newton's third law (action-reaction pairs), we are in a better position to understand several common phenomena. This is illustrated in the next example.

Example 6-3 A fire hose sends 20 kg of water each second on a burning building (Figure 6-1). The water leaves the nozzle at 60 m/sec and, for the sake of this problem, we will imagine that it *does not bounce back from the wall.* What is the force on the wall of the building?

Since the final velocity of the water is zero, in 1 second a certain amount of momentum of the water is canceled. To

Figure 6-1

do this, the wall exerts a force to the left.

$$\text{net } F = \frac{mv_f - mv_i}{t}$$

$$= \frac{(20 \text{ kg})(0) - (20 \text{ kg})(60 \text{ m/sec})}{1.00 \text{ sec}}$$

$$= -1200 \text{ N (to the left)}$$

This is the force of the wall on the water. By the third law, the force of the water on the wall is equal and opposite,

$$\boxed{1200 \text{ N (to the right)}}$$

Now suppose the water bounces back from the wall with half the speed it had originally, but in the opposite direction. The change in momentum of the water is now

$$\Delta mv = mv_f - mv_i$$

$$= (20 \text{ kg})(-30 \text{ m/sec}) - (20 \text{ kg})(60 \text{ m/sec})$$

$$= -600 \text{ kg}\cdot\text{m/sec} - 1200 \text{ kg}\cdot\text{m/sec}$$

$$= \boxed{-1800 \text{ kg}\cdot\text{m/sec}}$$

The force exerted by the wall, therefore, is

$$\text{net } F = \frac{\Delta mv}{t}$$

$$= \frac{-1800 \text{ kg}\cdot\text{m/sec}}{1.00 \text{ sec}}$$

$$= \boxed{-1800 \text{ N (to the left)}}$$

So the force of the water on the wall is

$$\boxed{1800 \text{ N (to the right)}}$$

From this example it can be seen that the water exerts a greater force if, instead of coming to rest, it bounces back with some velocity. This is because, in the second case, the water underwent a greater change in momentum.

6-3 LAW OF CONSERVATION OF MOMENTUM

In deriving the law of conservation of momentum, it is helpful to consider first an isolated system of two bodies at rest on a frictionless surface. They are placed together, separated by a compressed spring, one end of which is firmly attached to larger mass m_1 (Figure 6-2). When the thread that holds the spring compressed is cut the spring expands exerting equal but opposite forces on the two masses. The bodies receive different accelerations because their masses are different (Newton's second law). Suppose that the duration of the "explosion"* is t. Then, as was shown in the previous section, the impulse given to mass m_1, namely $-Ft$, is equal to the change in momentum of m_1; and the impulse given to m_2, that is $+Ft$, equals the change in momentum of m_2. According to Newton's third law, these changes are equal and opposite. In symbols,

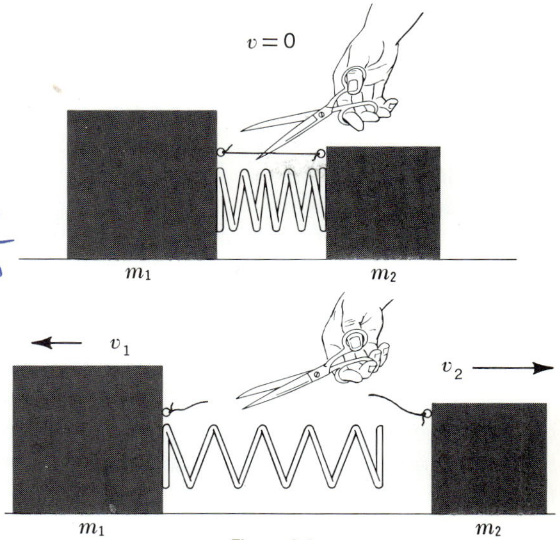

Figure 6-2

$$-Ft = \Delta(m_1 v_1)$$

$$+Ft = \Delta(m_2 v_2)$$

and these two equations say that

$$\Delta(m_1 v_1) = -\Delta(m_2 v_2)$$

Thus, the gain in momentum of one body ex-

*In the discussion of momentum, the term "explosion" will be used to describe the sudden, forceful breaking apart of two or more bodies, or of two or more parts of the same body.

actly equals the loss of momentum of the other body. The total momentum of the system has remained constant during the interaction. (In this case it is zero because both masses were at rest originally.)

A similar description can be used for bodies that are in motion before they interact. Suppose A and B, of mass m_1 and m_2, respectively, are moving toward the right at different speeds v_1 and v_2, with A overtaking and colliding with B (Figure 6-3). When they collide, A exerts a force $(+F)$ on B, and B exerts an equal and opposite force $(-F)$ on A. As before, these forces act for a common time t so that the impulses given to each body are equal and opposite, and therefore the changes in momentum are equal and opposite. To show that the total momentum of the system has not changed:

$$\text{total momentum before collision} = m_1v_1 + m_2v_2$$

total momentum after collision

$$= [m_1v_1 + \Delta(m_1v_1)] + [m_2v_2 + \Delta(m_2v_2)]$$

$$= m_1v_1 + m_2v_2$$

since $$\Delta(m_1v_1) = -\Delta(m_2v_2)$$

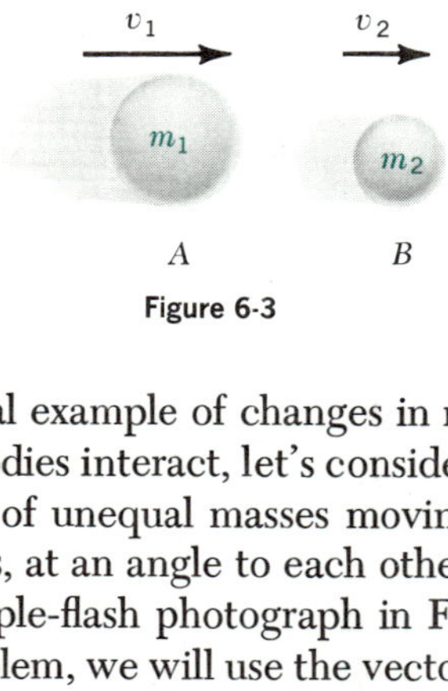

Figure 6-3

As our final example of changes in momentum when two bodies interact, let's consider the more general case of unequal masses moving at different velocities, at an angle to each other as shown by the multiple-flash photograph in Figure 6-4a. For this problem, we will use the vector nature of momentum in our analysis. Figure 6-4b shows the magnitude and direction of the velocity of each body before the collision as the vectors v_1 and v_2. After the collision takes place, the masses move off with velocities V_1 and V_2. In Figure 6-4b, then, the velocities have been drawn accurately from experimental data. In Figure 6-4c each velocity vector has been multiplied by the appropriate mass, thus producing momentum vectors that have been drawn to a different scale. (You can see from the relative magnitudes of these momentum vectors that m_2 is greater than m_1.)

To show that the change in momentum of one mass is equal and opposite to the change in momentum of the other, the initial momentum has been subtracted (vectorially) from the final momentum for each mass (Figure 6-4d). From the diagram it can be seen that the resultant vectors (representing changes in momentum) are equal in magnitude and oppositely directed. An equivalent statement would be that the total momentum has remained constant. That is, the total momentum before the collision is equal to the total momentum after the collision. This is shown by the addition of momentum vectors in Figure 6-4e.

These few examples serve to illustrate the fact that whenever two bodies interact, as long as there are no external forces acting on the system, the total momentum of the system remains constant. It is not hard to extend the proof to include a system containing any number of interacting bodies, for instance, a collection of 10^{25} molecules in a tank of compressed air. No part of a system can undergo a change of momentum unless another part also undergoes an equal and opposite change; the justification for this statement is Newton's third law. In general, the law of conservation of momentum can be stated as follows:

The law of conservation of momentum: The total momentum of an isolated system of bodies remains constant.

The importance of this law arises from its generality. It is true whether or not some mechanical energy has been transformed into thermal energy, sound energy, etc. It is true even if the colliding bodies are deformed permanently as in an automobile crash. It is true for any number of bodies making up a system. For instance, when a grenade explodes in mid-air, the fragments move in various directions with various speeds, but the

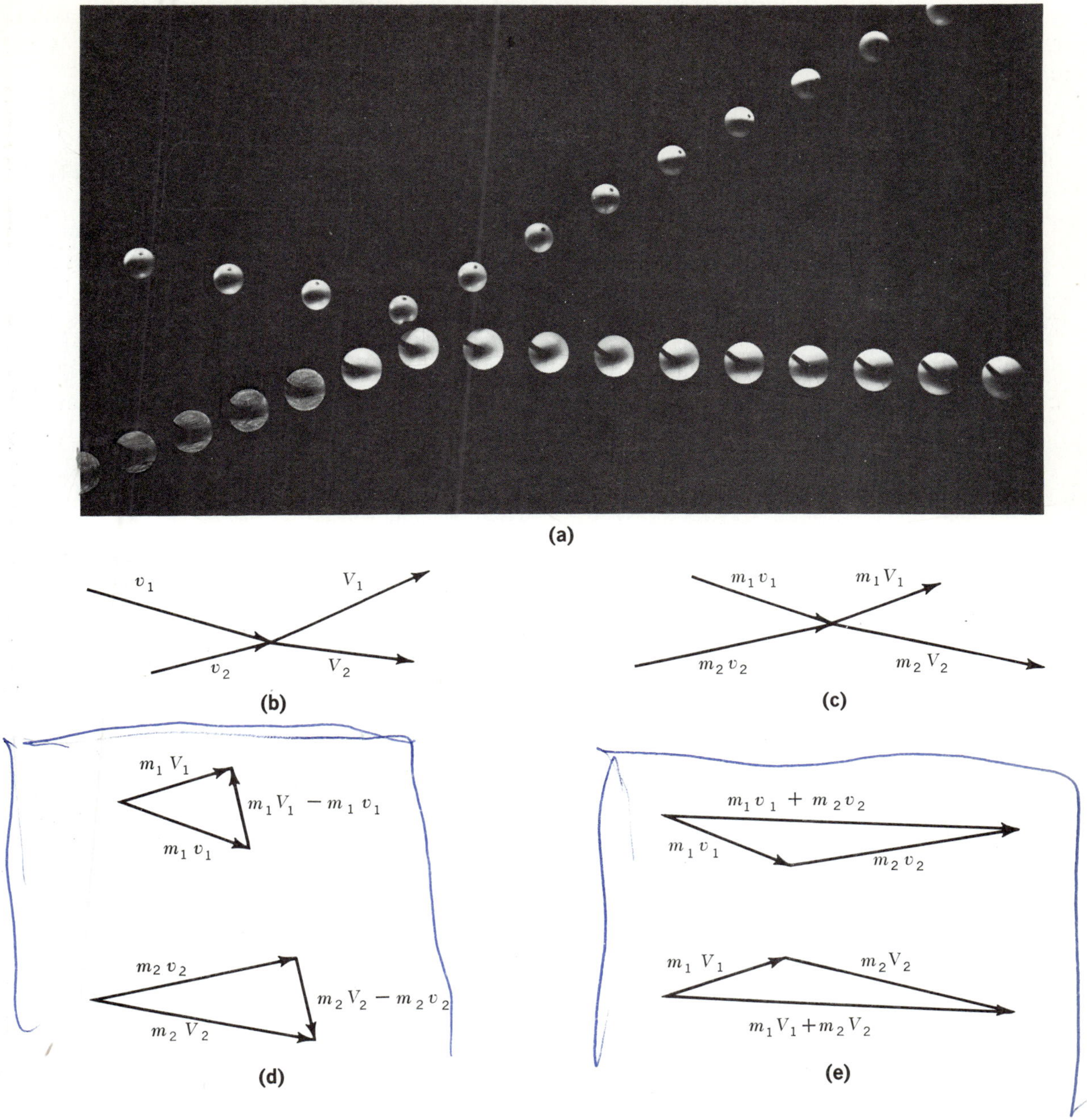

Figure 6-4 **(a)** Strobe photo showing collision of two masses, moving from left to right. **(b)** Velocity vectors representing magnitude and direction for two masses before and after the collision. **(c)** The momentum vectors are obtained by multiplying the velocity vectors by the masses. **(d)** The change in momentum of each mass is found by subtracting the initial momentum from the final momentum. The changes in momentum are equal and opposite. **(e)** The total momentum before the collision is equal to the total momentum after the collision.

vector sum of the momentum of all the fragments equals the momentum of the original body. This is true regardless of any energy released by the explosive in the grenade. The reason for the quite general validity of the law of conservation of momentum is that it follows from Newton's third law, which is always valid.

Example 6-4 A car of mass 600 kg traveling at 20 m/sec collides with a stationary truck of mass 1400 kg. The two vehicles adhere to each other after the collision. What is the magnitude and direction of their combined velocity (ignoring friction)?

After the collision, the car and truck form a single object of combined mass 2000 kg.

total momentum before impact = total momentum after impact

$$(mv)_C + (mv)_T = (mv)_{C+T}$$

$$(600\text{ kg})(20\text{ m/sec}) + (1400\text{ kg})(0) = (2000\text{ kg})(v_{C+T})$$

$$v_{C+T} = \boxed{\text{6 m/sec (in the same direction that the car was moving)}}$$

Example 6-5 A 2 kg gun fires a bullet of mass 0.05 kg at a velocity of 300 m/sec and, of course, recoils with each shot fired. What is the recoil velocity of the gun? Let v stand for velocities before firing; V for velocities after firing.

$$(mv)_g + (mv)_b = (mV)_g + (mV)_b$$

$$(2\text{ kg})(0) + (0.05\text{ kg})(0) = (2\text{ kg})(V_g) + (0.05\text{ kg})(300\text{ m/sec})$$

$$V_g = \boxed{-7.5\text{ m/sec}}$$

The law of conservation of momentum can be used as the basis of an experiment to measure mass. In a collision between two objects, let v_1 and v_2 be their respective velocities before collision, and let V_1 and V_2 be their velocities after collision.

$$m_1v_1 + m_2v_2 = m_1V_1 + m_2V_2$$

$$m_2(v_2 - V_2) = m_1(V_1 - v_1)$$

$$\boxed{\frac{m_2}{m_1} = \frac{V_1 - v_1}{v_2 - V_2}} \qquad \textbf{(Equation 6-3)}$$

The velocities v_1, v_2, V_1, and V_2 can be measured (for instance, with a movie camera); hence we have found one way of measuring the mass m_2 in terms of some "standard" mass m_1, such as the standard kilogram. In Section 4-3, you were told that to measure a property is to define it. The kind of mass that is being discussed here is inertial mass (Section 4-4), and we may say that Newton's laws of motion provide a way of defining inertial mass. Applying these ideas to the collision described in Example 6-4, we have $v_1 = 20$ m/sec, $v_2 = 0$, $V_1 = 6$ m/sec, $V_2 = 6$ m/sec. Therefore

$$\frac{m_2}{m_1} = \frac{6 - 20}{0 - 6} = \frac{14}{6}$$

If an observer had measured only the velocities, knowing nothing at the start about the masses, he would have been able to deduce the *ratio* of the two masses, and he would have found 14:6, as is actually the case (1400 kg:600 kg).

6-4 CENTER OF MASS

Imagine that you throw a hockey stick violently across the smooth ice of a pond. Once it leaves your hand there are no unbalanced forces acting on it, so it should continue to move in a straight line at constant speed. As you watch it sliding away from you, it is fairly obvious that the stick as a "whole" is moving in a straight line, but some parts of the stick have a very complex motion. These parts are not moving in straight lines but rotating in "wobbly" curves. Yet you feel that *something* is moving in a straight line. You are right, and this something is a special point called the *center of mass* (c.m.) of the body. It is defined as that point at which all the mass of a body can be considered to be concentrated when using

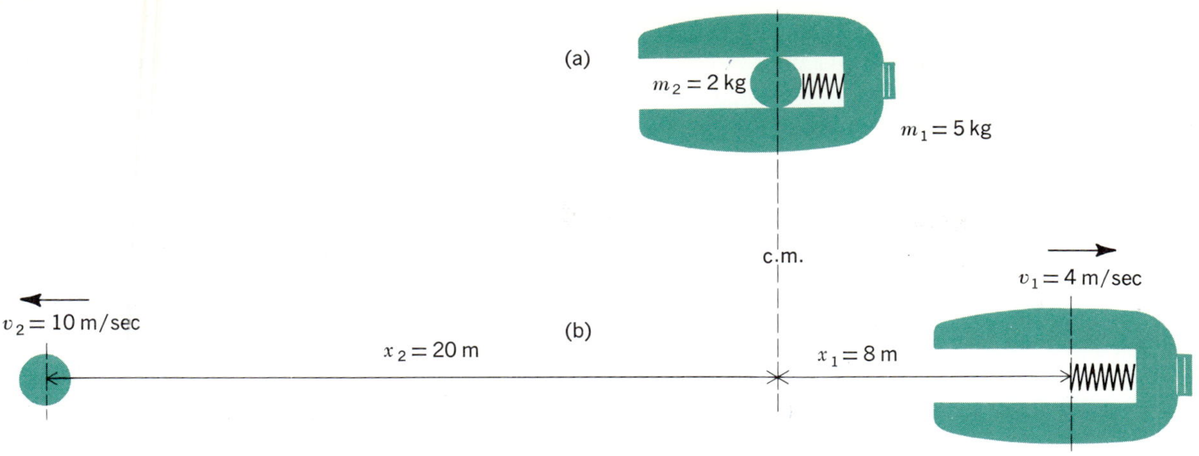

Figure 6-5

Newton's second law. You might imagine the whole body replaced by a particle of the same mass located at this point. The motion of the center of mass can be thought of as the average motion of all the particles that make up the body. The center of mass does not always lie "inside" the body. For example, the c.m. of an L-shaped piece of wood is an imaginary point outside the body.

This book often refers to extended bodies (trains, cars, balls, etc.) as if they were particles, that is, as points that have mass but no dimensions. In using Newton's laws, we have applied them as if all the mass of the body were concentrated at one point. Such treatment is valid and, although the proof of this assumption is not included, the concept of center of mass has been used in this way throughout the book.

How do we find the center of mass of a body or a system of bodies? For a symmetrical object like a sphere of uniform density, the center of mass is at the geometric center. For more complicated systems, such as the hockey stick or the several pieces into which a bomb explodes, the calculation is more difficult. We can illustrate the principle involved, however, by treating the simple case of two bodies.

Figure 6-5a shows a small cannon of mass 5 kg in which rests a projectile of mass 2 kg. The center of mass of the projectile originally coincides with that of the cannon, and the two masses are held together by a light but stiff spring. When the spring is released, the 2-kilogram mass moves off to the left at 10 m/sec with a momentum of 20 kg·m/sec. Because momentum is conserved, the 5-kilogram cannon has a recoil velocity of 4 m/sec to the right. What is the position of each mass 2 seconds after the "explosion"? In 2 seconds the larger mass moves 8 meters and the smaller mass moves 20 meters, as shown in Figure 6-5b. Since no external forces act on the system, the motion of the center of mass does not change, that is, the center of mass remains at rest. The center of mass of the two bodies 2 seconds after the explosion is, therefore, still located at the original position. The figure shows that the center of mass divides the distance between the masses in the ratio 20 to 8 (=5 to 2); and this is the inverse ratio of their masses. In symbols,

$$\frac{m_1}{m_2} = \frac{x_2}{x_1}$$

This result is general for any two masses. We have shown that the center of mass of two bodies is that point which divides the distance between them in an inverse ratio to their masses. For example, the earth is about 80 times as massive as the moon; the c.m. of the earth-moon system is about 4800 km from the earth's center, and about 384,000 km from the moon. This point is about 1600 km beneath the earth's surface.

In the following example, the law of conservation of momentum will be applied to a system whose c.m. is in uniform motion.

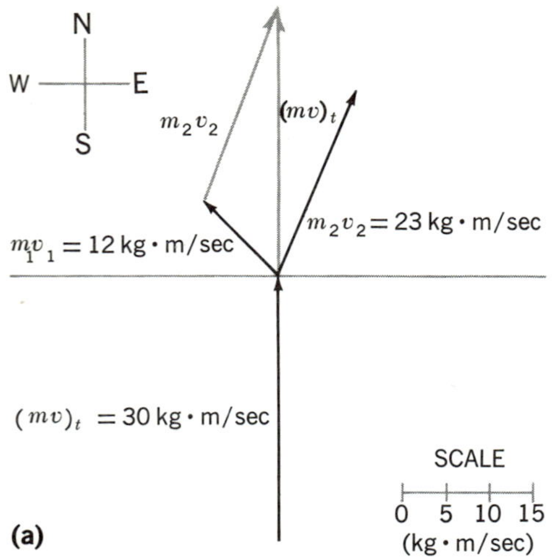

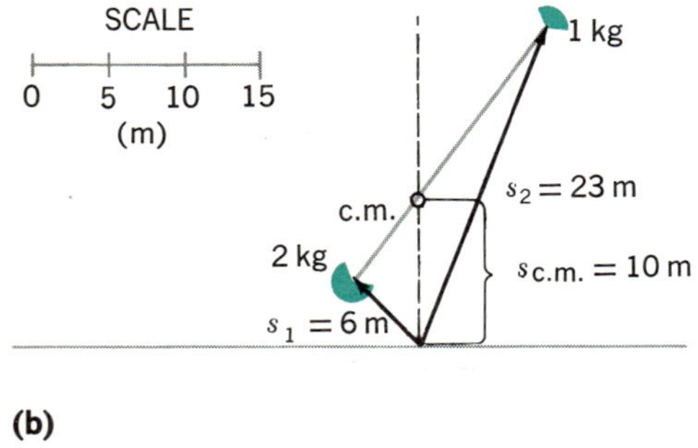

Figure 6-6

Example 6-6 Imagine a hypothetical situation wherein a 3 kg time bomb is sliding across the ice of a frozen lake due north at 10 m/sec when it explodes into two pieces. One piece of mass 2 kg, moves off 45° west of north at 6 m/sec. Find the magnitude and direction of the velocity of the other piece. Show that the motion of the center of mass of the system has been unaltered by the explosion.

Since no unbalanced external forces act on the bomb, we would expect its center of mass to proceed due north at 10 m/sec even after the explosion. To investigate this, let's use the law of conservation of momentum. To conserve momentum, the vector sum of the momentums of the two pieces must equal the original momentum of the bomb. In symbols,

$$m_1v_1 + m_2v_2 = (mv)_{\text{total}}$$

This is a vector equation, and its solution can be obtained from a scale drawing.

The 2 kg piece has a velocity of 6 m/sec and, therefore, a momentum of 12 kg·m/sec in the same direction as its velocity. This has been drawn to scale in Figure 6-6a. The momentum of the second piece can be found by subtracting the momentum of the first piece from the total, or original, momentum. This subtraction has been performed in the figure; and measurement of the momentum vector m_2v_2 gives 23 kg·m/sec at about 22° E of N. The velocity of this piece is its momentum divided by its mass,

$$v_2 = \frac{23\text{ kg·m/sec}}{1\text{ kg}}$$

$$= \boxed{23\text{ m/sec}}$$

and its direction is the same as the momentum,

$$\boxed{22°\text{ E of N}}$$

Now that we know the velocity of each piece, we can find out how far each one moves in a given time, say 1 second. In Figure 6-6b, the displacement vectors have been drawn for each piece 1 second after the explosion. From their velocities we know that the pieces will move 6 m and 23 m, respectively, in 1 second. The center of mass must lie on the line joining the two pieces and, since the masses are in the ratio 2:1, the center of mass must divide the line in the ratio 1:2. Measurement of the gray line in the figure shows this division to fall right at the intersection of the path the bomb would follow if it did not explode (dashed line). Finally, in 1 second the bomb, and also the center of mass, would have moved 10 m north of its starting place. Measurement of this dotted line shows it to be 10 m long. You can see, then, that the motion of the c.m. was not changed by the explosion.

6-5 ELASTIC AND INELASTIC COLLISIONS

Let's look closely at Example 6-4, which referred to the collision of a car and a truck. Before collision, all the KE of the system was in the car:

$$\tfrac{1}{2}mv^2 = \tfrac{1}{2}(600\text{ kg})(20\text{ m/sec})^2$$
$$= 120{,}000\text{ J}$$

since $1\text{ J} = 1\text{ kg}\cdot\text{m}^2/\text{sec}^2$ (Section 5-1). After collision, the system has less kinetic energy:

$$\tfrac{1}{2}mv^2 = \tfrac{1}{2}(2000\text{ kg})(6\text{ m/sec})^2$$
$$= 36{,}000\text{ J}$$

At first glance it seems that the law of conservation of energy is not valid here. There has been no obvious increase of gravitational PE, since all the motions are horizontal. Where, then, did the other 84,000 J of energy go?

We can account for the "missing" energy if we realize that *some* force must have been applied to the car in order to slow it down. At first, work was done against the elastic forces in the bumpers, etc. Then, as the elastic limits of the bumpers were reached, the metal crumpled and work was done against internal friction. The point is, regardless of the exact process, *some* force had to be exerted to slow the car down, and this force must have been exerted through *some* distance, however short. The car's velocity could not decrease instantaneously unless an infinite force acted on it. If no force had been exerted on the car, it would have sailed right through the truck and there would have been no collision! This hidden energy might properly be called PE if we could get it back again, but generally the "lost" KE in this type of collision is transformed—mostly into thermal energy as the colliding bodies are permanently deformed. A minute fraction of the KE may go into sound energy, light energy (if sparks fly), etc.

Whenever two bodies interact so that some KE is permanently lost, we call this an *inelastic impact*, or inelastic collision. Example 6-4 illustrates an inelastic collision because the KE after the collision was less than the KE before the collision. When a baseball is dropped to the floor, it bounces back only a small fraction of the height from which it was dropped. It has lost some of its KE during the collision. This is largely an inelastic impact. On the other hand, perfectly *elastic* collisions are those in which no KE is lost. A good steel ball bearing bouncing off a steel plate will return almost to its original height. This collision is very nearly elastic. None of the collisions that we experience in everyday life are ever perfectly elastic. Some of them, however, like the collision of two billiard balls, are nearly so, and by studying an idealized case we can learn a great deal about collisions.

To derive a general result, we will work with symbols rather than with specific numbers. Let v_1 be the initial velocity of a moving ball of mass m_1 heading straight for a stationary second ball of mass m_2. What will be the velocities V_1 and V_2 of the two balls after the collision? This is an elastic collision, so KE is conserved. In *any* collision free from external forces, momentum is conserved.

Study the equations below:

Conservation of energy:

$$\tfrac{1}{2}m_1v_1^2 = \tfrac{1}{2}m_1V_1^2 + \tfrac{1}{2}m_2V_2^2$$

Conservation of momentum:

$$m_1v_1 = m_1V_1 + m_2V_2$$

We have two equations in two unknowns (V_1 and V_2) and, since their solution involves nothing more than elementary algebra, it is omitted here. The results are given below. (You should work this out for yourself, however.)

$$V_1 = \frac{m_1 - m_2}{m_1 + m_2}v_1$$

$$V_2 = \frac{2m_1}{m_1 + m_2}v_1$$

Remember that these expressions for the velocity after collision only apply when (a) the collision is

Figure 6-7 Partially elastic impact of a bouncing ball, photographed at intervals of 1/60 sec. Relative velocity of separation is less than relative velocity of approach. Note the momentary storage of potential energy as the ball is deformed during impact.

elastic, (b) the motion is in a straight line, and (c) one body is initially at rest. One quick test of these equations gives a result well known to pool players. When the masses are equal (as is the case with billiard balls), the equations reduce to $V_1 = 0$ and $V_2 = v_1$. This means that the first ball stops dead and the second ball takes up the entire velocity v_1 of the first ball.

We can get one more interesting result from these equations. If one equation is subtracted from the other, the result is $V_2 - V_1 = v_1$. The left side of this equation is the relative velocity of separation of the two bodies (Section 3-7). Since $v_2 = 0$ (m_2 was at rest before the collision), the right side (v_1) is the relative velocity of approach of the two bodies. This provides an alternate definition of an elastic collision. A collision is elastic when the relative velocity of separation of the two bodies after impact equals their relative velocity of approach before impact.* In an inelastic collision, of course, the relative velocity of separation is less than the relative velocity of approach, as with the tennis ball in Figure 6-7.

Example 6-7 An alpha particle, mass 6.64×10^{-27} kg, is moving at 1.6×10^7 m/sec and collides elastically with a nitrogen atom, mass 2.32×10^{-26} kg, which is at rest. What fraction of the KE of the alpha particle is transferred to the nitrogen atom? With what velocity does the alpha particle recoil from the collision?

The KE of the alpha particle before the collision is

$$\tfrac{1}{2}mv_1^2 = \tfrac{1}{2}(6.64 \times 10^{-27}\text{ kg}) \times (1.6 \times 10^7\text{ m/sec})^2 = 8.50 \times 10^{-13}\text{ J}$$

To find the KE of the nitrogen atom after the collision, we first find its velocity from the equation developed for an elastic collision with one body at rest (page 107).

$$V_2 = \frac{2m_1}{m_1 + m_2} v_1 = \frac{2(6.64 \times 10^{-27}\text{ kg})(1.6 \times 10^7\text{ m/sec})}{6.64 \times 10^{-27}\text{ kg} + 2.32 \times 10^{-26}\text{ kg}} = \frac{2.12 \times 10^{-19}\text{ kg}\cdot\text{m/sec}}{2.98 \times 10^{-26}\text{ kg}} = \boxed{7.1 \times 10^6\text{ m/sec}}$$

The KE of the nitrogen atom is

$$\tfrac{1}{2}mv^2 = \tfrac{1}{2}(2.32 \times 10^{-26}\text{ kg})(7.1 \times 10^6\text{ m/sec})^2 = 5.85 \times 10^{-13}\text{ J}$$

*In the more general case, when both bodies are in motion, the equation is $V_2 - V_1 = v_1 - v_2$.

This represents a fraction of

$$\frac{5.85 \times 10^{-13}\ \text{J}}{8.50 \times 10^{-13}\ \text{J}} = \boxed{69\%}$$

that is transferred from the alpha particle.

For this elastic collision, the relative velocity of separation is equal to the relative velocity of approach. To find the recoil velocity of the alpha particle, we use

$$V_2 - V_1 = v_1$$

$$7.1 \times 10^6\ \text{m/sec} - V_1 = 1.6 \times 10^7\ \text{m/sec}$$

$$V_1 = \boxed{-8.9 \times 10^6\ \text{m/sec}}$$

6-6 THE SIGNIFICANCE OF THE CONSERVATION LAWS

You have already come across two important laws of physics, both of them expressed as conservation laws. Taken together, the laws of conservation of energy and conservation of momentum can give answers to most of the questions that might be asked about the mechanics of physical objects. These conservation laws are equivalent to Newton's laws; as a matter of fact, they are derived from Newton's laws of motion in the first place (Sections 5-4 and 6-3). It would be entirely possible to do it the other way around; assume the conservation laws and derive Newton's laws from them. Let's therefore ask the question, which is more fundamental—Newton's second law in the form net force = (mass)(acceleration) or Newton's second law in the form net force = (change of momentum)/(time)? So far, you have seen no reason to prefer one form of the laws to the other. Situations arise, however, in which the second way (involving change of momentum) is far better. What if the mass of a body changes while it is being acted on by a force? There could then be a change of momentum even if velocity were constant. Acceleration would be zero; and the equation net $F = ma$ would not apply.

Example 6-8 A train of empty coal cars is being pulled along a straight, level piece of track at 20 m/sec. If rain starts to fall, and the cars receive 300 kg of water per minute, how much additional force must the engine supply to keep the train moving at a constant speed of 20 m/sec?

Here the momentum changes because the mass changes. The amount of momentum that is added in each minute is (300 kg)(20 m/sec) = 6000 kg · m/sec. We now apply Newton's second law in the momentum form:

$$F = \frac{\Delta(mv)}{t} = \frac{6000\ \text{kg}\cdot\text{m/sec}}{60\ \text{sec}}$$

$$F = \boxed{100\ \text{N}}$$

A rather small force, to be sure, but the equation net $F = ma$ would have been inadequate to cope with the problem.

You will recall the change of mass of a moving body that is described by the relativity theory (Section 5-6). At high velocities, as the mass increases, the situation is so complex that in general the acceleration produced by a force is not constant, and is not even always in the same direction as the force. Nevertheless, the statement that net force = rate of change of momentum *is* true in relativity; force is defined in such a way as to *make* it true. Now force, mass, or any other concept can be defined in any way that appears useful. At ordinary velocities, the definitions of force in terms of acceleration and momentum change are equivalent, and can be used interchangeably. But if the velocities are high enough for relativistic changes, the definition of force as rate of change of momentum proves to be more fruitful. Evidently Newton's original formulation of his second law in terms of momentum rather than acceleration was an inspired guess that has proved valuable in ways not foreseen by Newton. It is for reasons such as we have just discussed that physicists consider the conservation of momentum and conservation of energy so fundamental. Indeed, physicists show a strong tendency to put other laws of nature into similar form wherever possible, as you will see in Chapter 14, when you study the law of conservation of electric charge.

SUMMARY

The **momentum** of a body is defined as the product of its mass and its velocity. Newton's second law may be restated in terms of change of momentum: The net force acting on a body equals the rate of change of its momentum.

Newton's third law may be used in deriving the **law of conservation of momentum:** The total momentum of an isolated system remains constant. When two bodies collide, they change momentum by equal and opposite amounts. When no unbalanced force acts on a body in motion, its **center of mass** moves at constant velocity in a straight line. During all collisions, momentum is conserved, but KE is conserved only in **elastic collisions.** In an **inelastic collision** the relative velocity of separation is less than the relative velocity of approach, and some KE is transformed to heat as the colliding bodies are permanently deformed.

The statement that net force equals rate of change of momentum serves as a general definition of force, which is useful in modern theories of physics where relativity effects are significant. The laws of conservation of momentum and conservation of energy are still the most effective expression of the laws of motion, applicable in modern as well as classical physics.

QUESTIONS

1 When an apple falls to the ground and strikes the earth without rebound what "becomes" of the momentum of the apple?

2 Can someone in a rowboat "create" momentum by his unaided efforts?

3 A prospector carrying a bag of valuable uranium ore is trapped on absolutely smooth ice on a frozen lake in northern Canada. Is there any way in which he can move across the ice to safety, or must he freeze to death?

4 When a jet plane lands on the deck of a carrier is this an elastic or an inelastic "collision"? What becomes of the momentum of the jet? What becomes of its KE?

5 Is it possible for a rocket to attain a speed greater than the velocity with which the exhaust gases leave it?

6 Compare the momentum of a 10,000 kg truck moving at 6 m/sec with that of a 30 g bullet moving at 1200 m/sec (see Question 3, Chapter 5).

7 An open freight car is coasting along on a smooth track when rain starts to fall thereby increasing the mass of the car. Does the velocity of the car remain constant?

8 A man stands on a wooden plank that in turn rests on a concrete floor. If he hammers one end of the plank with a heavy mallet, can he make the "bumpmobile" move (Figure 6-8)? If so, explain how.

9 A spaceship moving at constant velocity accelerates in a right-hand turn by ejecting gas from a jet on its side. Describe the motion of the center of mass of the system (ship plus ejected gas).

Figure 6-8

PROBLEMS

GROUP A

1 Compute the momentum of a golf ball that has a mass of 60 g and is moving with a velocity of 70 m/sec.

2 If in Problem 1 the impact between the golf club and the ball lasted for 2×10^{-4} sec, what was the rate of change of momentum?

3 A girl holds a 2 kg air rifle loosely and fires a bullet of mass 1 g. The muzzle velocity of the bullet is 150 m/sec. What is the recoil velocity of the gun?

4 If the girl holds the gun of Problem 3 tightly against her body, the recoil is less. Explain. Calculate the new recoil velocity, if the girl has a mass of 48 kg.

5 In a freight yard a train is being made up. An empty freight car, coasting at 10 m/sec, strikes a loaded car that is stationary, and the cars couple together. Each of the cars has a mass of 3000 kg when empty, and the loaded car contains 12,000 kg of bottled soft drinks. With what speed does the combined mass start to move?

6 A spaceman of mass 80 kg carries an empty oxygen tank of mass 10 kg. He throws the tank away from himself with a speed of 2 m/sec (measured relative to the so-called fixed stars). With what velocity does the spaceman start to move through space?

GROUP B

7 At 4:40 p.m. the momentum of a speedboat was measured as 20,000 kg·m/sec, and at 4:42 p.m. it was 50,000 kg·m/sec. Assuming that the boat was moving along a straight course and that its speed increased steadily (constant force), what forward force was exerted on the boat? By what was this force exerted?

8 What force, acting for 0.001 sec, will change the velocity of a 100 g baseball from 30 m/sec eastward to 40 m/sec westward? (Do not find the acceleration.)

9 In an elastic collision a 4 kg body at rest is hit head-on by a mass of 1 kg moving with a velocity of 0.5 m/sec. What is the final velocity of each body?

10 A ball of mass 3 kg, moving at 2 m/sec eastward, strikes head-on a ball of mass 1 kg that is moving at 2 m/sec westward. The balls stick together after the impact. (a) What is the magnitude and direction of the velocity of the combined mass after collision? (b) Compute the total KE of the system before and after impact, and find how many joules of KE were lost during collision.

11 (a) Solve Problem 10, this time assuming that the collision is perfectly elastic, that is, the balls do not stick together. (b) Compute the total KE of the system before and after collision, and show that no KE was lost in the elastic impact.

12 A neutron of mass 1 unit, having velocity v_0, makes a head-on elastic collision with a stationary oxygen nucleus of mass 16 units. (a) What forward velocity does the oxygen nucleus receive? (b) What fraction of the KE of the neutron is transferred to the oxygen nucleus?

13 One way of measuring the muzzle velocity of a bullet is to fire it horizontally into a massive block of wood placed on a cart. Assuming no friction, we then measure the velocity with which the wood (containing the bullet) starts to move. In one experiment the bullet had a mass of 50 g, and the wood and its cart had a mass of 20 kg. After the shot, a stopwatch was used to determine that the cart, wood, and bullet moved at constant velocity, traveling 8 m in 0.40 sec. What was the original speed of the bullet?

14 A life raft of mass 180 kg carries two swimmers of mass 50 kg and 80 kg, respectively. The raft is initially floating at rest; then the swimmers simultaneously dive off opposite ends of the raft, each with a horizontal velocity of 3 m/sec. With what velocity and in what direction does the raft start to move?

15 A basketball player jumps into the air with both arms extended vertically overhead. Just as she reaches maximum height she pulls one arm down so that her center of mass is moved to a point 4 cm closer to her toes. Do the fingertips of the other hand go higher or lower as a result of this maneuver? How much?

16 A 50 kg cart is moving across a frictionless floor at 2 m/sec. A 70 kg boy, riding in the cart, jumps off the cart so that he hits the floor with zero velocity. (a) What impulse did the boy give to the cart? (b) What is the velocity of the cart after the boy jumped?

GROUP C

17 Two objects of mass m_A and m_B are held together by a strong light thread, and are also acted on by a light spring that is compressed as shown in Figure 6-9. When the restraining thread is broken, the two objects fly apart with velocities $-v_A$ and $+v_B$. Use the law of conservation of momentum to solve for the ratio of the velocities, v_A/v_B. Then compute the ratio of the kinetic energies, KE_A/KE_B, and prove that the ratio of the KE's of the two objects is the reciprocal of the ratio of their masses.

18 The arrangement for measuring the speed of a bullet described in Problem 13 is not very practical since friction would prevent the cart from rolling at constant velocity, and the measurement of the short interval of time might prove difficult. A more precise arrangement is the ballistic pendulum (Figure 6-10). After the bullet lodges in the wooden block, the block swings over and the height of rise is measured. (a) In one experiment the bullet's mass was 6 g, that of the block of wood was 4 kg. The string was 5 m long and the rise was 7 cm. What was the forward velocity of the block when at its lowest position just after impact? (*Hint:* Use the energy principle.) (b) Using the velocity found in (a), apply the law of conservation of momentum to find the original velocity of the bullet.

19 An 80 kg man is standing on a uniform 400 kg plank that is lying on a smooth icy pond. The plank is pointing toward the shore, which is 10 m away from the point where the man is standing. If he walks along the plank toward the shore for a distance of 2 m as measured along the plank, how far is he now from the shore?

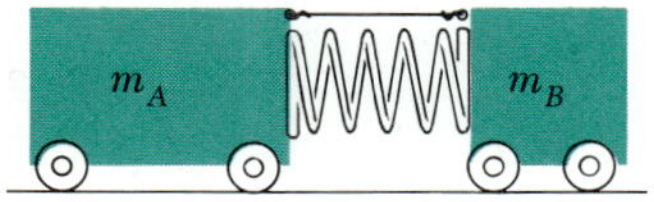

Figure 6-9

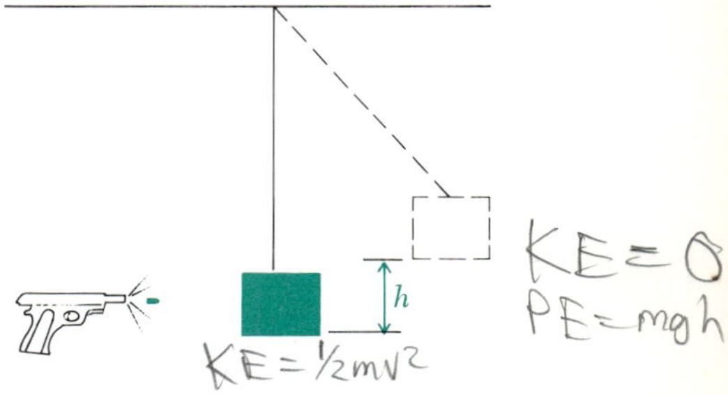

Figure 6-10

PE=0

7 Circular Motion, Gravitation, and Satellites

In Chapter 3 we discussed the kinematics of projectile motion near the surface of the earth. We saw that in the absence of air friction all bodies accelerate at 9.8 m/sec^2 under the action of the gravitational force. In this chapter Newton's laws of motion will be applied to a body in free fall, and the principles of circular motion will be developed. It was analyses such as these that brought Sir Isaac Newton, almost three hundred years ago, to the universal law of gravitation.

It was a triumph of scientific reasoning that Newton was able to arrive at this universal force law by applying his laws of motion to the paths of the moon and planets. These paths had been described already by the German astronomer Johannes Kepler (1571–1630) in his three laws of planetary motion.

7-1 FREE FALL

Whenever a body moves in the gravitational field of the earth so that the only force acting on it is the force of gravity, the body is said to be in free fall. Accordingly, a body that is thrown straight up, even when it is rising, is in free fall; once the object leaves the hand, if we neglect air resistance, the gravitational force is the only force acting on it.

When an airplane drops a bomb, the projectile moves in a parabolic* path. This, too, is free fall. It is not necessary for an object to be falling "straight down" to be in free fall; the only necessary condition is that gravity be the only force acting. The force of gravity on an object is called the object's weight, and is a vector pointing toward the center of the earth, with a magnitude mg (see Section 4-7).

To find out why a projectile in free fall moves in a curved path, we can apply Newton's second law to the motion. Figure 7-1 shows such a path with the object rising at point P_1 and falling at P_2. The gravitational force mg has a component along the path and also a component perpendicular to the path. At P_1 the component *along* the path (F_a) produces an acceleration along the path that changes the *magnitude* of the velocity. In this case the body is slowing down. The component *perpendicular* to the path ($F_\perp$) produces a perpendicular acceleration that changes the *direction* of the velocity, causing the body to move in a curved path. At P_2, the component of the force along the path speeds the body up, while the component perpendicular to the path deflects the body from its straight line path.

*The path is actually elliptical, but since the gravitational field lines near the earth's surface are essentially parallel, this ellipse can be approximated by a parabola.

These two components of the gravitational force act independently of each other. The component along the path changes the speed (magnitude of the velocity) but has no effect on the direction of the velocity. The component perpendicular to the path changes the direction of the velocity but has no effect on the speed. In the next section we consider the special case where the component along the path is zero, the body moves at constant speed with change in direction only.

7-2 UNIFORM CIRCULAR MOTION

Uniform motion in a circle, that is, motion at constant speed, is familiar to everyone. A rock being whirled on a string, the hands of a clock, a record turntable, all move at constant speed with change in direction only. In Section 3-3 acceleration was defined as the rate of change of velocity, $a = \Delta v/\Delta t$. In uniform circular motion, although there is no change in the *magnitude* of the velocity, there is an acceleration because the *direction* of the velocity is changing. Therefore, in spite of the apparent simplicity and constancy of the motion, the body is being accelerated because its velocity is changing. We can now derive a useful formula for the acceleration of a body that is moving uniformly in a circle.

Figure 7-2a shows a particle moving in a circle of radius R (a displacement vector) with velocity v. This could be, for example, a stone being whirled about on a string. Because the force producing the uniform circular motion always acts at right angles to the motion, there is no component of the force along the path to change the speed of the body; the force is a pure deflecting force that

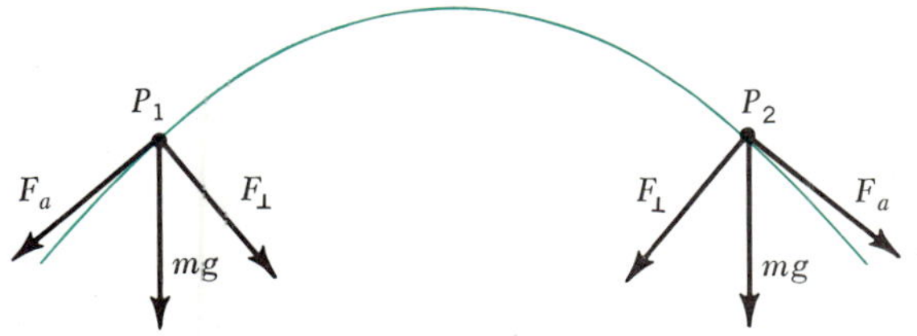

Figure 7-1

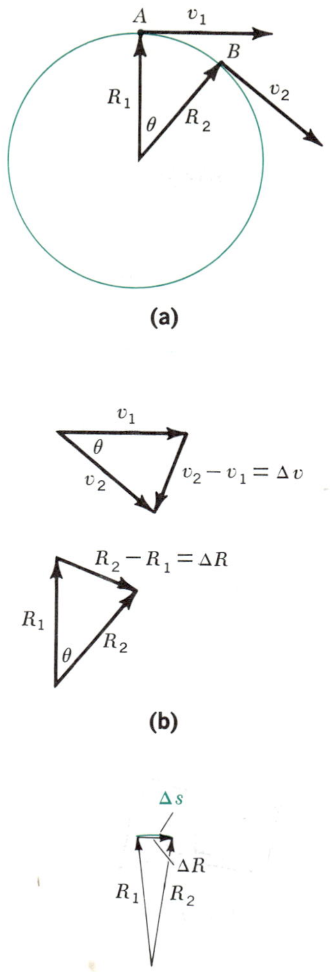

Figure 7-2

changes the direction of the velocity only. At a certain instant, the linear velocity is v_1. In a short time interval Δt, as the displacement vector rotates through the angle θ, the particle moves from A to B and the velocity changes to v_2 (numerically the same as v_1, but directed differently). In Figure 7-2b the velocity vectors and displacement vectors have been redrawn, and the vector subtractions give Δv and ΔR, respectively. You can observe that the angles in the "velocity" triangle (Figure 7-2b) and the angles in the "displacement" triangle are respectively equal. These

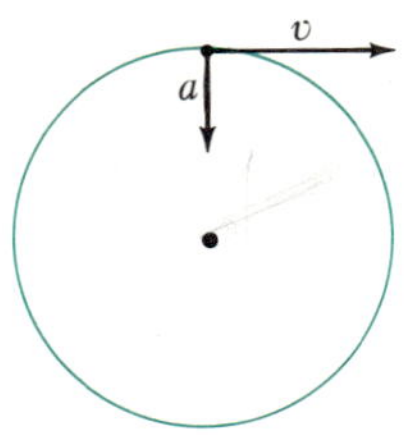

Figure 7-3

triangles, therefore, are similar and their corresponding sides are proportional. From the triangles, then,

$$\frac{\Delta v}{v} = \frac{\Delta R}{R}$$

or,

$$\Delta v = v\frac{\Delta R}{R}$$

where v and R represent either v_1 or v_2, R_1 or R_2. The acceleration, then, is

$$a = \frac{\Delta v}{\Delta t}$$

$$= \frac{v\Delta R/R}{\Delta t}$$

$$= \frac{v}{R}\cdot\frac{\Delta R}{\Delta t} = \frac{v}{R}\cdot v$$

$$a = \frac{v^2}{R} \qquad \text{(Equation 7-1)}$$

Substituting v for $\Delta R/\Delta t$ here is permissible if you remember that Δt is very small and in the *limit* (see Section 3-5) ΔR approaches Δs, the arc of the circle (see Figure 7-2c).

This expression for the magnitude of the acceleration in uniform circular motion is perfectly general and applies whenever there is a net force acting at right angles to the direction of motion (see Example 7-3). The *direction* of this acceleration vector will always be in the direction of the force that causes it. This follows, of course, from Newton's second law. We see, then, that when a body moves in a circle with uniform speed, there is an acceleration along the radius toward the center of the circle (Figure 7-3). For this reason we call it the *centripetal* ("center-seeking") *acceleration.*

It may seem strange to you that a body in uniform circular motion can have an acceleration, since it is not speeding up or moving nearer to the center of the circle. Remember that "speeding up" is an acceleration in the restricted sense of just a change in speed. Our broader definition of acceleration as a change in velocity includes a change in direction also. Actually, as a body moves in a circle it *does* fall in toward the center—but it never gets any closer. In the absence of the centripetal acceleration, the body would continue moving in a straight line, i.e., it would fly off on a tangent (Figure 7-4), and after a time t would arrive at some point such as C. The body actually arrives at B, and the distance CB represents a fall toward the center. In fact, it can be verified that the laws of falling bodies apply, and $CB = \frac{1}{2}at^2$, where a is the centripetal acceleration previously derived, equal to v^2/R.

Finally, we can develop an equivalent expression for the acceleration in terms of the radius of the circle and the period of rotation. Figure 7-2 shows that in one revolution the particle moves with velocity v a distance equal to the circumference of the circle ($2\pi R$). If this revolution takes a time T, called the period of the motion, then

$$2\pi R = vT$$

$$v = \frac{2\pi R}{T}$$

Substituting this into Equation 7-1 gives

$$a = \frac{v^2}{R}$$

$$= \left(\frac{2\pi R}{T}\right)^2\left(\frac{1}{R}\right)$$

$$a = \frac{4\pi^2 R}{T^2} \qquad \text{(Equation 7-2)}$$

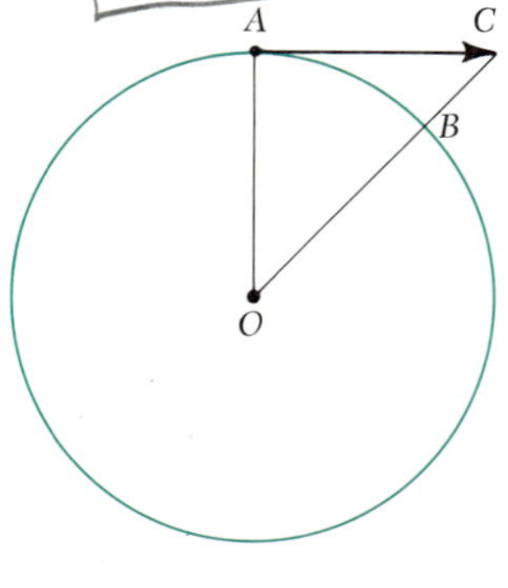

Figure 7-4

This relation, as expressed in Equations 7-1 and 7-2, has a wide range of application in physics. It is used to describe the motions of the planets, of earth satellites, and even electrons.

Example 7-1 What is the centripetal acceleration, due to the earth's rotation, of a body on the equator?

Assuming a spherical earth, the radius is 6.38×10^6 m. Its period of rotation is

$$24 \text{ hr} \times 3600 \frac{\text{sec}}{\text{hr}} = 8.64 \times 10^4 \text{ sec}$$

$$a = \frac{4\pi^2 R}{T^2}$$

$$= \frac{4\pi^2(6.38 \times 10^6 \text{ m})}{(8.64 \times 10^4 \text{ sec})^2}$$

$$= \boxed{0.034 \text{ m/sec}^2}$$

In view of this very small acceleration compared with g, it is clear why the earth's rotation can usually be neglected and why we can consider the earth as an inertial frame of reference (see Section 4-9).

7-3 CENTRIPETAL FORCE

The force that produces uniform circular motion is called *centripetal force*. This is the force that acts along the radius toward the center of the circle and produces the centripetal acceleration. Force and acceleration are related by Newton's second law, net $F = ma$. In the case of circular motion, $a = v^2/R$, therefore the centripetal force can be expressed as

$$\text{net } F = m\frac{v^2}{R} \qquad \text{(Equation 7-3)}$$

$$\text{net } F = m\frac{4\pi^2 R}{T^2} \qquad \text{(Equation 7-4)}$$

Although we set these equations off and give them each the special name centripetal force, they should be regarded as no more than a restatement of Newton's second law.

Any type of force can serve as a centripetal force. The following examples illustrate how stresses in materials (called elastic forces), gravitational forces, or magnetic forces can cause uniform circular motion of a body.

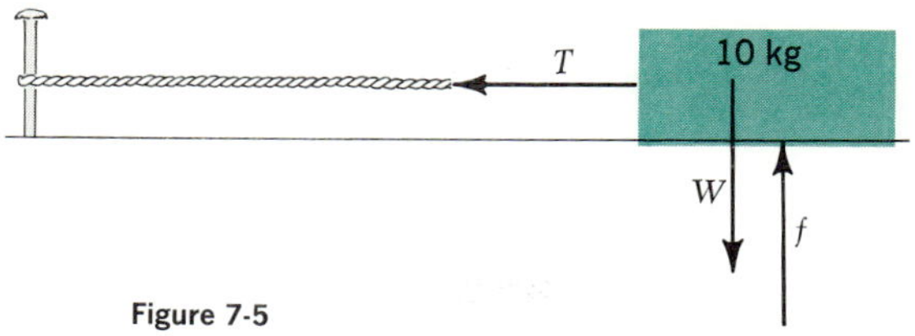

Figure 7-5

Example 7-2 A 10 kg block rests on a frictionless surface and is attached to a vertical peg by a rope (Figure 7-5). What is the tension T in the rope if the block is whirled in a circle of radius 2 m with a linear speed of 20 m/sec?

The block is acted on by three forces: its weight, the upward force of the surface, and the tension in the rope (an elastic force). The vertical forces balance and the net force that causes the block to move in a circle equals the tension in the rope.

$$T = \text{net } F$$

$$= m\frac{v^2}{R}$$

$$= (10 \text{ kg})\frac{(20 \text{ m/sec})^2}{(2 \text{ m})}$$

$$= \boxed{2000 \text{ N}}$$

To illustrate centripetal acceleration caused by gravitational force, let's consider a car going over the top of a hill.

Example 7-3 What is the force of a 600 kg sports car on the road surface as it goes at 10 m/sec over the crest of a hill having a radius of curvature of 30 m measured in the vertical plane (Figure 7-6)?

We are asked for the forces exerted by the car. However, Newton's second law must be applied to the motion of the car, and so we consider first the forces *on* the car: its weight mg, and the upward force of the road f.

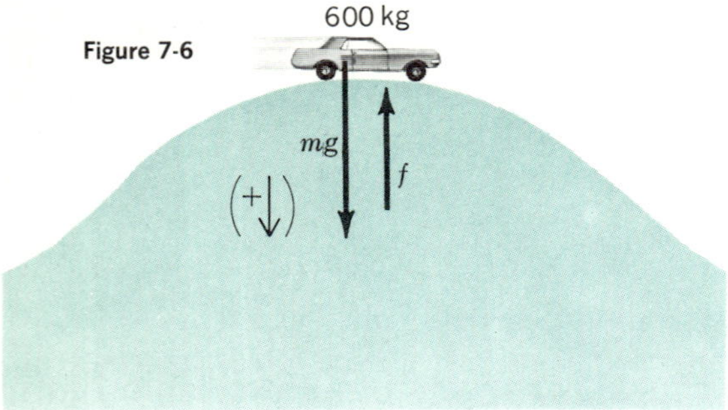

Figure 7-6

$$\text{net } F = m\frac{v^2}{R}$$

$$(600 \text{ kg})(9.8 \text{ N/kg}) - f = (600 \text{ kg})\frac{(10 \text{ m/sec})^2}{(30 \text{ m})}$$

$$5880 \text{ N} - f = 2000 \text{ N}$$

$$f = 3880 \text{ N (upward)}$$

This is the upward push of the road on the car. The third law tells us that the force of the car on the road is equal and opposite, that is,

$$\boxed{3880 \text{ N (downward)}}$$

As the car goes over the crest of the hill, it tends to "take off" and presses less against the road than if on a level stretch (where the downward force on the road would be 5880 N, the car's weight).

Magnetic force on a moving charge is responsible for the circular motion of a charged particle in an atom-smasher. Notice that in the next example it is not necessary to understand the nature of charged atomic particles or the workings of a cyclotron. As long as the motion is circular and the speeds involved are small compared to the velocity of light, we can apply the principles of circular motion with complete confidence.

Example 7-4 In a cyclotron a beam of protons (each of rest mass 1.67×10^{-27} kg) is moving in a circle of radius 80 cm. If an electromagnet supplies a force of 8.00×10^{-13} N on each proton directed toward the center of revolution, what is the magnitude of the velocity of the protons?

$$\text{net } F = m\frac{v^2}{r}$$

$$v = \sqrt{\frac{Fr}{m}} = \sqrt{\frac{(8.00 \times 10^{-13} \text{ N})(0.80 \text{ m})}{1.67 \times 10^{-27} \text{ kg}}}$$

$$= 1.96 \times 10^7 \text{ m/sec}$$

$$= \boxed{19{,}600 \text{ km/sec}}$$

7-4 CENTRIFUGAL FORCE

If there is a force *on* a body in circular motion directed toward the center of the circle, then, by Newton's third law, there must be a force directed *away* from the center exerted *by* the moving body. This force, exerted on whatever body supplies the inward force, is called the *centrifugal* ("fleeing the center") *force.* Centripetal force and centrifugal force are an action-reaction pair. Like all such pairs they never act on the same body and, although equal and opposite, they never cancel each other (see Section 4-5). For instance, in Example 7-2 the moving block pulls outward on the peg at the center with a centrifugal force of 2000 N. Perhaps a more descriptive term for this pull would be *reaction centrifugal force.*

Some of the popular statements about centrifugal force are confusing, if not actually incorrect. For example, it is often stated that in an automatic clothes washer "centrifugal force throws the water outward" as the drum containing wet clothes rotates rapidly. From the point of view of the housewife, an observer outside the drum, this statement is incorrect. A better way to describe what happens is as follows: When the drum begins to rotate, the molecular forces between water and fiber supply the necessary centripetal force to keep the drops of water moving in a circle within the drum. As the speed of rotation increases, a greater force is needed to hold the drops and sooner or later the molecular forces become insufficient. The water then flies off on a tangent, and moves away from the center of the drum because of lack of sufficient centripetal force. In other words, centrifugal force doesn't *cause* the water to fly outward. The water will do that anyway, since its natural tendency is to move in a straight line (Newton's first law) unless acted on by a centripetal (or other) force.

There is, however, a way in which an outward force *on* a rotating object might be postulated. To illustrate this we will perform an imaginary experiment and take the view of an "insider," one whose frame of reference is the rotating system itself. (See remarks on frames of reference in Section 4-9.) Imagine that you are sitting on a large rotating platform. The floor is frictionless, but you are "nailed down" so that as it rotates you remain fixed over the same spot on the floor. You can assume that the rotation is smooth and that you don't see yourself moving. Now you place a smooth block on the frictionless floor and attach a spring scale to it (Figure 7-7). The block starts to move away from you, extending the spring, and quickly comes to rest. What do you conclude? In your frame of reference the block is at rest, but this is only because you are exerting a force on it to keep it at rest.

The force that you are exerting is certainly real enough (you can read the value on the scale) and, if Newton's laws apply, this force *should* produce an acceleration of the block toward you, but it doesn't! To explain this, you assume that there is a force acting on the *block* opposite in direction to the force that you are applying. Thus your assumed force is trying to pull the block away from the center of the platform, but the net force on the block is now zero, and this explains the fact that the block remains at rest. The force that you assumed may be considered a *fictitious force*. It is fictitious because there is no agent causing it. But you need it to make Newton's laws of motion apply in your accelerated (non-inertial) frame of reference just as you needed a fictitious force in the example of the subway car in Section 4-9.

A person "outside" your system watching this little drama sees the block (and you) rotating in a circle. He observes that you are providing the necessary centripetal force on the block (through the spring) to produce the acceleration needed to

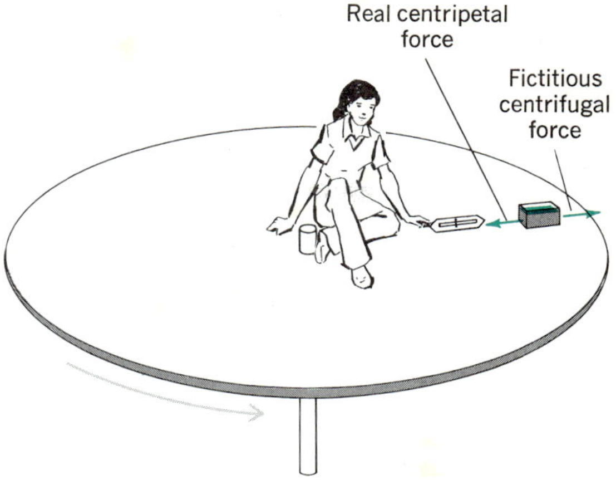

Figure 7-7

keep the block moving in a circle. He has no need for the *fictitious centrifugal force* that you postulated.

To summarize briefly, you should use the term centrifugal force, if at all, only when you understand the point of view from which you are describing a situation. When a mass is rotating about a fixed central pivot, the radial force outward exerted *by* the mass *on* the pivot is an example of *reaction centrifugal force*. On the other hand, the radial force outward *on* a rotating mass introduced to make the laws of motion valid in an accelerated frame of reference is an example of *fictitious centrifugal force*. In general, however, you will avoid much confusion if you take an outsider's view of things and place the emphasis on the *centripetal* force which acts on the rotating body to cause its acceleration toward the center.

7-5 GRAVITATION

So far, we have found that Newton's laws of motion can be applied to a wide range of phenomena. The discovery that the same laws also describe the motions of the planets, moons, and, indeed, all bodies in space, is one of the most fascinating stories in the history of science. Building upon the work of scientists such as Copernicus, Galileo, and Kepler, Newton brought the entire universe under one mechanical scheme.

Man's earliest concept of planetary motion was probably the geocentric (earth-centered) description set forth by the ancients, who pictured an earth at rest, with all planets, the sun, and the stars revolving about it. Acceptance of this theory hindered significant advances in planetary astronomy for nearly 2000 years. The motions of the planets as observed by early astronomers seemed tremendously complex and difficult to describe when considered from an earth-centered frame of reference. The first major breakthrough came with the heliocentric (sun-centered) model of the solar system as described by Nicolaus Copernicus, a Polish astronomer, about 1500 A.D. He demonstrated that the description of the motions of the moons and the planets was far simpler when viewed from an earth that rotated on its axis while revolving about the sun. His ideas, however, were not readily accepted, and it was not until a century later that Galileo's observations through his telescope of certain phenomena in space, namely the moons of Jupiter and the phases of Venus finally provided strong evidence for the Copernican system.

Perhaps the ultimate achievement of that time was Johannes Kepler's (a contemporary of Galileo) kinematic description of the solar system, which is best described in his three laws of planetary motion.

Kepler's Laws:

I The planets move in ellipses with the sun at one focus.

II The line joining any planet to the sun sweeps out equal areas in equal times.

III For any planet in the solar system, the cube of its average distance from the sun divided by the square of its period of revolution is a constant, $R^3/T^2 = K$.

Figure 7-8 illustrates laws I and II. Kepler's third law as it relates to each of the planets in the solar system is shown in Table 7-1.

This third law, or "law of periods" really applies to any system of satellites. Whenever a group of smaller bodies orbit about a much more massive object (e.g., the moons of Jupiter), R^3/T^2 is a constant for each of these systems. The value of this constant is different for each system, since, as you will see, it depends upon the mass of the central body. (See the color insert opposite page 132.)

Although the kinematics of planetary motion was fairly well established by Kepler and the others, the question of the *causes* of these motions still remained. What is the nature of the force that causes the planets to move in orbits as described by Kepler? This question was eventually answered by Newton. He showed that the force of gravity that causes bodies to fall to the earth also pulls

Table 7-1

Planet	*Mean radius of orbit (meters)*	*Period of revolution (seconds)*	R^3/T^2 (m^3/sec^2)
Mercury	5.79×10^{10}	7.60×10^6	3.35×10^{18}
Venus	1.08×10^{11}	1.94×10^7	3.35×10^{18}
Earth	1.49×10^{11}	3.16×10^7	3.35×10^{18}
Mars	2.28×10^{11}	5.94×10^7	3.35×10^{18}
Jupiter	7.78×10^{11}	3.74×10^8	3.35×10^{18}
Saturn	1.43×10^{12}	9.30×10^8	3.35×10^{18}
Uranus	2.87×10^{12}	2.66×10^9	3.35×10^{18}
Neptune	4.50×10^{12}	5.20×10^9	3.36×10^{18}
Pluto	5.9×10^{12}	7.82×10^9	3.34×10^{18}

These are modern values for the orbital radii and periods of the planets. In Kepler's day Uranus, Neptune, and Pluto had not been discovered.

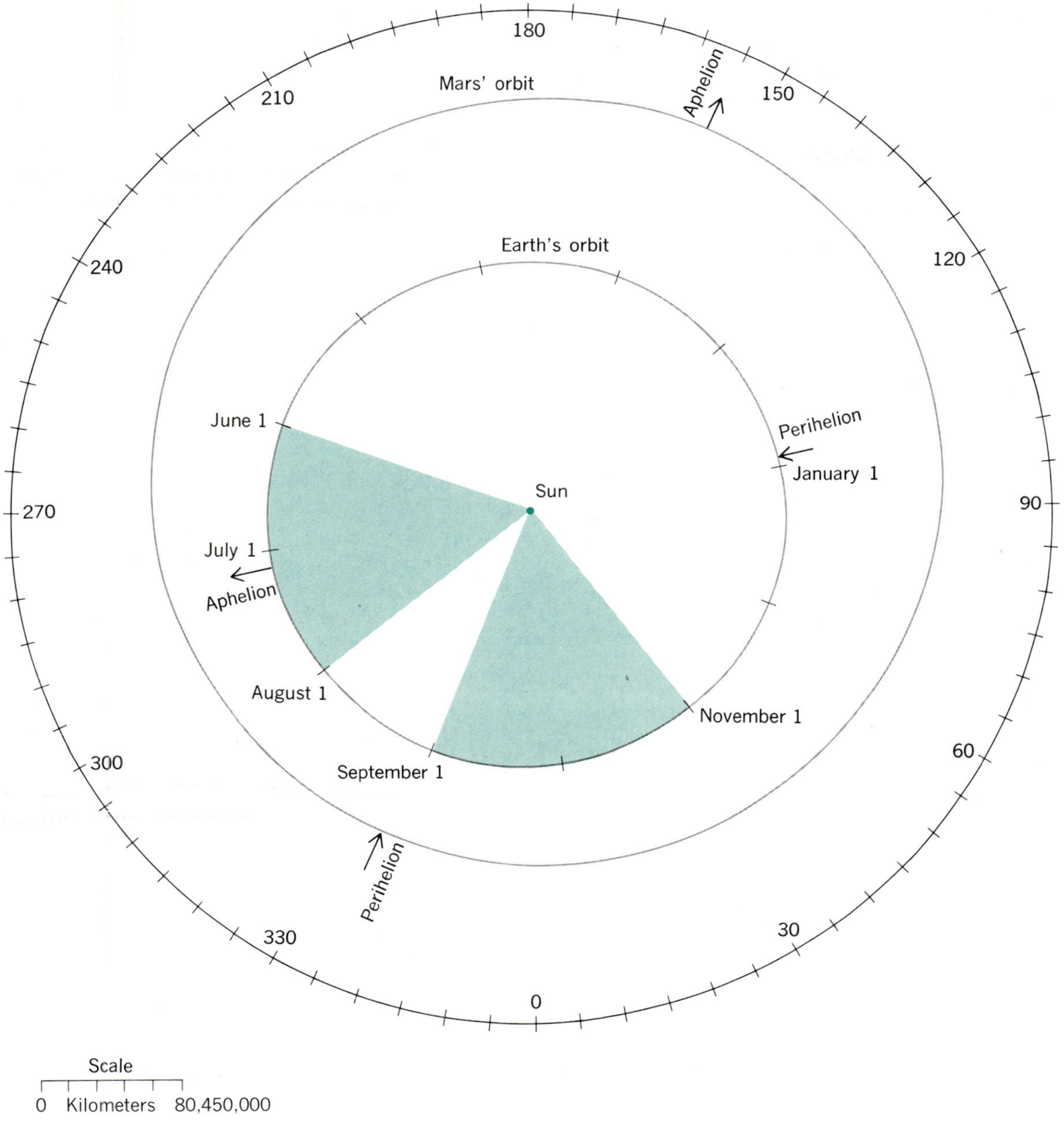

Figure 7-8 An accurate drawing of the orbits of the earth and Mars showing that they are slightly elliptical when compared to the outer reference circle. The colored areas represent areas swept off by a line connecting the earth and the sun during two equal periods of time (61 days each). In keeping with Kepler's second law, the areas are equal, even though the earth travels a greater distance during the period September 1–November 1 than June 1–August 1. You must measure the diagram carefully to verify the preceding statement.

on the moon. The moon is a falling body. It is the gravitational attraction between the sun and the planets that causes the planets to move in curved paths. Newton went on to show that gravitational attraction is a mutual force between all bodies. A falling stone is pulled by the earth, and the earth is pulled by the stone. The sun attracts the planets, and the planets attract the sun. Every piece of matter in the universe attracts every other piece of matter with a gravitational force.

Newton's ideas led to the development of this law:

Newton's law of universal gravitation: Every body in the universe attracts every other body with a force that is directly proportional to the product of the masses of the two bodies, and inversely proportional to the square of the distance between them.

This can be written in the form of an equation as follows:

$$F = G\frac{m_1 m_2}{R^2} \qquad \text{(Equation 7-5)}$$

where m_1 and m_2 are the masses, R the distance between the masses, and G the constant of proportionality.

It is interesting to speculate how Newton arrived at this law. Even though he left very accurate records of much of his scientific work, Newton has left the historians of science with some doubt as to the exact order in which his thoughts on universal gravitation had developed. It is possible for us, however, using the laws of motion we have already derived to reconstruct the line of reasoning Newton probably followed. By doing this we will also gain a closer knowledge of the problems involved in the analysis of satellite motion.

Although the planets travel in elliptical orbits about the sun, these orbits are very nearly circles, and we can treat them as circles with the sun at the center. This will simplify our calculations while introducing only negligible error.* For any planet circling the sun, the centripetal force on the planet is (from Equation 7-4)

$$F = m_p\frac{4\pi^2 R}{T^2}$$

where R is the radius of the orbit, T the period, and m_p the mass of the planet. We can eliminate T from this expression if we first multiply by R^2 as follows:

$$FR^2 = m_p\frac{4\pi^2 R^3}{T^2}$$

Now, notice the factor R^3/T^2 on the right side. From Kepler's third law and by observation, we know this to be a constant that can be replaced by the symbol K.

$$FR^2 = 4\pi^2 K m_p$$

$$F = 4\pi^2 K\frac{m_p}{R^2}$$

F is the force between the sun and *any* planet. It depends upon the mass of the planet and the planet's distance from the sun. It is surprising that the mass of the sun does not appear in this expression, since the sun *is* involved in every sun-planet interaction. The constant K also appears in the description of every sun-planet interaction, so perhaps the mass of the sun is contained in the constant K. That is,

$$K = K' m_s$$

where m_s is the mass of the sun and K' is some new constant. Putting this expression for K into the relation for the force we obtain

$$F = 4\pi^2(K' m_s)\frac{m_p}{R^2}$$

$$= 4\pi^2 K'\frac{m_s m_p}{R^2}$$

This relation says the force that holds a planet in its orbit depends upon the mass of the sun and planet, and the distance between them. Newton eventually assumed that this force of attraction exists between *any* two bodies in the universe. If we represent the constant factor $4\pi^2 K'$ by G (called the gravitational constant), we finally have

*Newton, of course, made no such simplifying assumption about the shapes of the orbits.

$$F = G\frac{m_1 m_2}{R^2}$$

where m_1 and m_2 are any two masses.

This result is based upon assumptions that are sometimes referred to as Newton's "famous guesses." To justify these assumptions and show that indeed this is the correct expression for the force of attraction between any two masses, it was necessary to put this force law to a test. Normally the force of gravitation is undetectable between bodies on the earth. It is masked by frictional forces of one kind or another, or by the tremendously larger force of gravitation due to the earth. Under special laboratory conditions, however, where the frictional forces can be almost eliminated, this gravitational force can be measured.

It was not until a century after Newton had proposed his law of gravitation that the English scientist Henry Cavendish (1731–1810) completed an experiment that demonstrated the gravitational force of attraction between small bodies a few centimeters apart. Two lead balls, each about 5 centimeters in diameter were attached to the ends of a lightweight rod suspended in a horizontal position by a long, fine wire about 100 centimeters in length (Figure 7-9). Two larger lead balls, each about 20 centimeters in diameter, were placed almost touching the small balls. The gravitational attraction between the balls caused the moving system to rotate to a new rest position determined by the stiffness of the wire. By carefully shielding the apparatus from air currents, Cavendish was able to measure the tiny force of gravitation which amounted to less than 5 ten-millionths of a newton!

The significance of Cavendish's experiment was not only that it proved there was a gravitational force between small objects, but also the determination of the constant G. Using his measured values of F, m_1, m_2, and R, he computed G. The modern value of this fundamental constant is

$$G = 6.67 \times 10^{-11}\ \text{N}\cdot\text{m}^2/\text{kg}^2$$

This unit for G is necessary to make Equation 7-5 dimensionally correct.

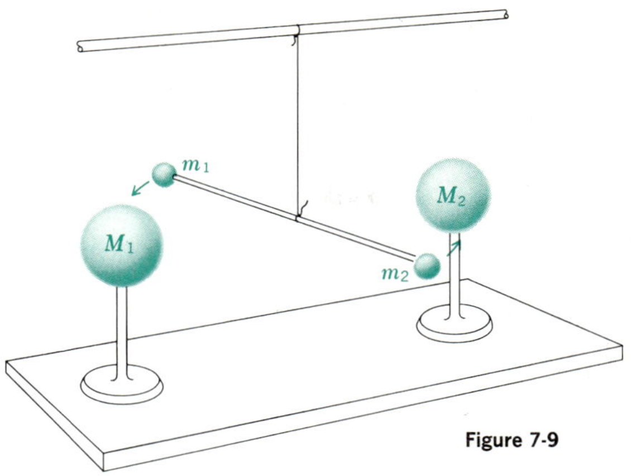

Figure 7-9

7-6 MORE ON THE EARTH'S GRAVITATIONAL FIELD

Consider a mass m at the surface of the earth. The gravitational force on it is

$$F = G\frac{m_e m}{R^2}$$

where m_e is the mass of the earth. The gravitational pull of the earth acts as if all its mass were concentrated at its center*, so R in this equation is the radius of the earth. The pull of the earth is also the weight of the body,

$$F = mg$$

so we set these two forces equal.

$$mg = G\frac{m_e m}{R^2}$$

$$g = G\frac{m_e}{R^2}$$

The gravitational field strength g at the surface of the earth is 9.8 N/kg. In Section 4-7 it was stated that this value decreases as one moves further away from the surface of the earth. We can see from the above relation just how the acceleration due to gravity changes. Because G and the mass of the earth are constant, this can be written as

*Newton delayed publication of his law of gravitation for more than a decade while he worked out a satisfactory proof of this statement.

$$g \propto \frac{1}{R^2}$$

In other words, the gravitational field of the earth decreases inversely as the square of the distance from its center. At a height of one earth radius above the surface of the earth (that is, 2 R_e from the center), g is one fourth of 9.8 N/kg, or 2.45 N/kg. The moon is about 60 earth radii from the earth. This would mean that the value of the earth's gravitational field at that distance is $(\frac{1}{60})^2 = \frac{1}{3600}$ of g at the earth's surface. Since the moon is a falling body, its acceleration is the same as the value of the earth's gravitational field at that distance. From the moon's period of revolution (about 27.3 days) and its orbital radius of about 60 earth radii, we can compute its centripetal acceleration.

Using the value 6.38×10^6 m for the radius of the earth, and Equation 7-2, we obtain

$$a_m = \frac{4\pi^2 R}{T^2}$$

$$= \frac{4\pi^2(60 \times 6.38 \times 10^6 \text{ m})}{\left(27.3 \text{ day} \times 24 \frac{\text{hr}}{\text{day}} \times 3600 \frac{\text{sec}}{\text{hr}}\right)^2}$$

$$= 2.7 \times 10^{-3} \text{ m/sec}^2$$

where a_m is the centripetal acceleration of the moon. This value for the moon's centripetal acceleration is in very good agreement with the gravitational field strength at the moon's altitude, $(\frac{1}{60})^2 g = \frac{1}{3600} \times 9.8$ N/kg $= 2.7 \times 10^{-3}$ N/kg as deduced from the inverse square law of force.

It was this calculation that suggested to Newton that the gravitational force that held the planets in their orbits was the same kind of force that held the moon in its orbit about the earth. He had deduced the inverse square law of force for the planets. By applying it to the moon, as we have done, he was able to show that the resultant acceleration was in agreement with the kinematic calculation of the moon's acceleration as found from its period and orbital radius. This is what eventually led him to assume that the force of gravitation is a mutual force that exists between any two bodies in the universe.

Example 7-5 Jupiter has a mass of 1.90×10^{27} kg and a radius of 7.18×10^7 m. What is the acceleration due to gravity on the surface of Jupiter?

Using the same procedure just outlined, we equate the weight of a mass m on Jupiter to the gravitational attraction between the mass and the planet.

$$mg = G\frac{m_J m}{R^2}$$

$$g = G\frac{m_J}{R^2}$$

$$= 6.67 \times 10^{-11} \frac{\text{N}\cdot\text{m}^2}{\text{kg}^2}\left[\frac{1.90 \times 10^{27} \text{ kg}}{(7.18 \times 10^7 \text{ m})^2}\right]$$

$$= \boxed{24.6 \text{ N/kg}}$$

This is about $2\frac{1}{2}$ times the value of g on the earth's surface. If you could walk about the surface of Jupiter, you would feel as if you were carrying $1\frac{1}{2}$ men on your back.

Example 7-6 Find the mass of the sun, using Kepler's constant from Table 7-1.

Although this seems like very little information to go on, the masses of bodies in the solar system can actually be determined by analyzing the motions of their satellites. The centripetal force that the sun exerts on a planet is

$$F = m_p \frac{4\pi^2 R}{T^2}$$

and the gravitational force between sun and planet is

$$F = G\frac{m_S m_p}{R^2}$$

In this case the centripetal force *is* the gravitational force, so

$$m_p \frac{4\pi^2 R}{T^2} = G\frac{m_S m_p}{R^2}$$

Solving for m_S, the mass of the sun, and noticing that the mass of the planet cancels out, we obtain

$$m_S = \frac{4\pi^2 R^3}{GT^2}$$

In this equation R and T are the orbital radius and period of *any* planet, but the ratio R^3/T^2 is the same for all planets. This, of course, is Kepler's constant, whose value is given in Table 7-1.

$$m_S = \frac{4\pi^2}{(6.67 \times 10^{-11}\ \text{N}\cdot\text{m}^2/\text{kg}^2)} \times (3.35 \times 10^{18}\ \text{m}^3/\text{sec}^2)$$

$$= \boxed{1.98 \times 10^{30}\ \text{kg}}$$

Example 7-7 Most of the several hundred earth satellites that have orbited the earth since 1957 take about $1\frac{1}{2}$ hours to complete a single revolution with average altitudes of a few hundred miles. A few of them, however, have been launched so that they travel at just the right speed to keep them in a "fixed" position over one point on the earth. At what height would such a satellite be if it remained over a certain point on the earth's equator as it moved in its circular orbit? The earth is assumed to be a sphere with radius $R_e = 6.4 \times 10^6$ m.

The satellite will remain over the same point on the equator if its period of revolution is the same as the earth's period of rotation, that is, 24 hours. As in Example 7-6, we could equate the gravitational force between earth and satellite to the centripetal force, and then solve for R. But in this example we will use another method, namely, comparing the satellite's motion to that of the moon by noting that Kepler's third law can be applied to any satellite system that follows an inverse square law of force.

$$\frac{R_m^3}{T_m^2} = \frac{R_s^3}{T_s^2}$$

where the subscripts m and s refer to the moon and the satellite. We are applying Kepler's third law to earth satellites without needing the actual value of the constant.

We want the altitude of the satellite, so the unknown in the equation is R_s, the distance of the satellite from the center of the earth.

$$R_s^3 = \frac{T_s^2}{T_m^2} R_m^3$$

$$R_s = \sqrt[3]{\frac{T_s^2}{T_m^2} R_m^3}$$

$$= R_m \sqrt[3]{\frac{T_s^2}{T_m^2}}$$

Instead of substituting numerical values in meters and seconds, we note that the moon is about 60 earth-radii away, that is, $R_m = 60\ R_e$, and its period is about 27 days.

$$R_s = 60\ R_e \sqrt[3]{\frac{(1\ \text{day})^2}{(27\ \text{days})^2}}$$

$$= 60\ R_e(\tfrac{1}{9})$$

$$= 6.7\ R_e$$

R_s is 6.7 R_e from the earth's center, so the satellite is 6.7 $R_e - R_e = 5.7\ R_e$ above the surface of the earth.

$$\text{altitude} = 5.7\ R_e$$

$$= 5.7(6.4 \times 10^6\ \text{m})$$

$$= \boxed{3.6 \times 10^7\ \text{m}}$$

This is 36,000 kilometers!

7-7 WEIGHTLESSNESS

In the last section you found that the strength of the earth's gravitational field at a distance of one earth radius from the surface of the earth is $\frac{1}{4} \times 9.8$ N/kg $= 2.45$ N/kg. At this distance a 100 kg man would have a weight (mg) of 100 kg $\times$ 2.45 N/kg $= 245$ N as compared to his weight of 980 N on the earth. It is clear from our definition of weight, $W = mg$, that because the mass remains constant, the weight depends upon the particular value of g at the place where the weighing is made. As a body moves further and further from the center of the earth, the gravitational field strength, and hence the weight, decreases. From $g \propto 1/R^2$ we can see that g is zero when R is infinite. In other words, a body has zero weight, it is weightless, only when it is infinitely far from the earth (and all other bodies in space).

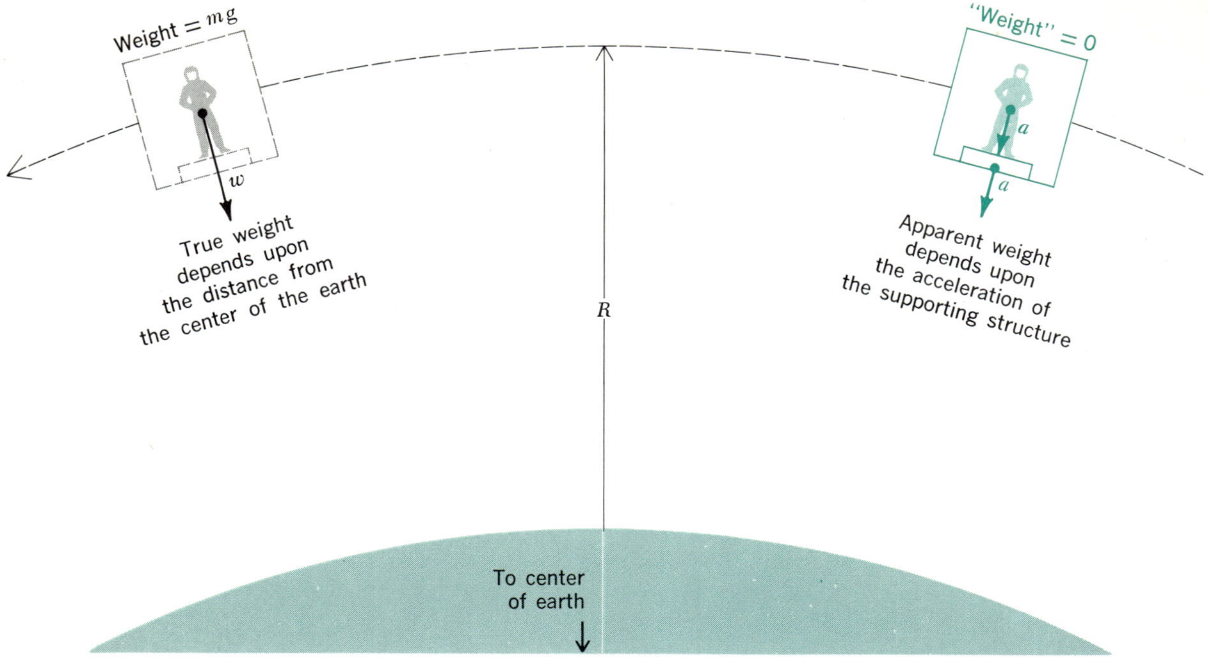

Figure 7-10 Everything in the orbiting spacecraft experiences the same centripetal acceleration.

From this we can conclude that a body is never truly weightless. Yet we often speak of astronauts in orbiting spaceships as being "weightless." The confusion in this use of the term arises not from our definition of weight, but from our method of measuring weight. That is, the force of the supporting structure must be considered if the term "weightlessness" is to make any sense. When you stand on a platform scale, the scale pushes up with a force equal to your weight. You feel this force and you can read its value on the scale. Figure 7-10 depicts an astronaut standing on a platform scale while orbiting the earth in a space capsule. The centripetal acceleration of the astronaut, the capsule, and the scale is v^2/R, directed inward along the radius of the orbit. Because everything has the same inward acceleration (they are all freely falling bodies), he senses no upward force on his feet. The scale will read zero and, in the absence of supporting forces, he feels "weightless." A better term to describe this situation is to say that his apparent weight is zero. His true weight is still mg; it depends upon his distance from the center of the earth, not his motion*.

As the capsule spirals in to touch down, it is no longer in free fall. The resistance of the earth's atmosphere and the retrorockets combine to exert a tremendous force on the capsule slowing it down. The capsule pushes on the astronaut, and he feels several times his true weight. Later, when the capsule's parachute opens, a tremendous upward force is exerted on the capsule and the astronaut. In this situation his apparent weight is said to be many "g's," that is, many times the weight he normally experiences at the earth's surface.

We see, then, that apparent weight is due to changes in the supporting force, while true weight depends upon position only. It is not necessary to be in a spaceship to experience weightlessness, as you will see by solving the next example.

* Weightlessness can also be studied by considering the fictitious forces arising in an accelerated frame of reference as in Section 4-9.

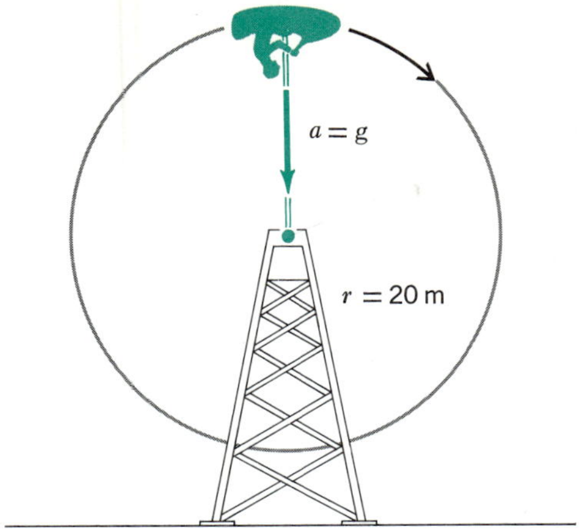

Figure 7-11

Example 7-8 The carnival ride pictured in Figure 7-11 moves in a circle of radius 20 m. How fast must it turn so that you feel weightlessness at the top of the rise?

To have zero apparent weight you must experience no forces from the supporting structures (seat, belt, etc.). This will only happen when you and your supports have the same acceleration. At the top of the rise, your acceleration due to gravity *alone* is 9.8 m/sec² downward. In order for the supports to exert no apparent force on you, they must have the same downward acceleration. This requires that the car move with an acceleration of 9.8 m/sec².

$$a = v^2/R$$

$$v = \sqrt{aR}$$

$$= \sqrt{(9.8\ \text{m/sec}^2)(20\ \text{m})}$$

$$= \boxed{14\ \text{m/sec}}$$

SUMMARY

Under the action of the gravitational force alone, a body will move in **free fall.** A projectile in free fall moves in a curved path whenever the force of gravity acting on it has a component perpendicular to the direction of motion, that deflects it from its straight-line path. A constant force acting at right angles to the motion of a body causes it to move in a circle at constant speed. Although the magnitude of the velocity in uniform circular motion is constant, the direction of the velocity changes. This gives rise to a **centripetal acceleration** that is constant in magnitude and directed inward along the radius of the circle.

The force that produces uniform circular motion is called a **centripetal force.** Any type of force can serve as the centripetal force. In the case of planetary motion, the centripetal force is the gravitational force exerted by the sun on the planets. **Newton's universal,** or **inverse square, law of gravitational force** describes the mutual force of attraction between any two bodies in the universe. For any group of satellites orbiting the same massive central body, R^3/T^2 is a constant. Gravitational field strength decreases inversely as the square of the distance from the central body.

The **actual weight** of a body depends upon its position in the gravitational field, while the changes in the **apparent weight** of a body are due to the acceleration of the supporting structures.

QUESTIONS

1 How is it possible for a body to be accelerated if its speed is constant?

2 Consider an atom of aluminum near the rim of a phonograph turntable turning at 45 rev/min. Is there any centripetal force acting on the atom? If so, what causes this force?

3 What is the source of the centripetal force on the pilot of a plane that is executing a vertical loop-the-loop when the plane is at the bottom of the loop curving upward?

4 "The earth doesn't fall into the sun because centrifugal force pushes it outward." Criticize this sentence.

5 In Example 7-4, should a relativistic formula have been used, taking account of the increase in mass of the proton due to its velocity? (*Hint:* How does the proton's velocity compare with the speed of light?)

6 In a game of marbles played on a perfectly smooth level surface is it possible for a marble to roll in an arc of a circle?

7 If you were in a freely falling elevator (and didn't know it), what explanation would you give for objects that you "dropped" remaining suspended in the air?

8 A man-made satellite takes 28 days to orbit the earth. What can you conclude about its mass and its altitude?

PROBLEMS

GROUP A

1 What is the apparent weight of the passenger in Figure 7-11 at the bottom of the circle?

2 What would be the apparent change in weight of a 100 kg man at the equator due to the rotation of the earth (see Example 7-1)?

3 What is the acceleration of the tip of the minute hand of a watch if the hand is 2 cm long?

4 A length of rope can just support a weight of 150 N without breaking. (a) At what speed can this same rope be used to whirl a 3 kg mass in a horizontal circle of radius 2 m? (b) What is the period of this rotation?

5 In aviation, a "standard turn" for level flight of a propeller-type plane is one in which the plane makes a complete circular turn in 2 min. If the speed of the plane is 170 m/sec, (a) what is the radius of the circle? (b) what is the centripetal acceleration of the plane?

6 A space explorer is 1 billion kilometers away from a certain star, and observes that the gravitational force of attraction exerted on the spacecraft by the star is 500 N. What will the force become when the craft has approached a position half a billion kilometers from the star?

7 From the expression developed in Section 7-6 for the gravitational field strength, find the mass of the earth.

8 According to the data in Table 7-1, which planet comes closest to the earth (assuming circular orbits)?

9 At what height above the earth's surface will an astronaut have a true weight that is one half his weight on the earth?

GROUP B

10 A fly of mass 2 g is sunning itself on a phonograph turntable at a location that is 4 cm from the axis. When the turntable is turned on and rotates at 45 rev/min, calculate the inward (centripetal) force needed to keep the fly from slipping.

11 A 35 kg boy is swinging on a rope 7 m long. He passes through the lowest position with a speed of 3 m/sec. What is the tension in the rope at that moment? (*Hint:* Use net $F = ma$. The net force is the resultant of the weight and the tension; see Example 7-3.)

12 A plane comes out of a power dive, turning upward in a curve whose center of curvature is 1300 m above the plane. The plane's speed is 260 m/sec. (a) Calculate the upward force of the seat cushion on the 100 kg pilot of the plane. (b) Calculate the upward force on a 90 g sample of blood in the pilot's brain.

13 Compute the gravitational force between a proton and an electron in the hydrogen atom, using the following data: mass of proton, 1.67×10^{-27} kg; mass of electron, 9.11×10^{-31} kg; radius of orbit of electron, 5.29×10^{-11} m.

14 (a) At what speed must a satellite of mass m be launched horizontally at the surface of the earth (radius $= 6.4 \times 10^6$ m) if its orbit is to be a circle just grazing the highest mountains? (Assume no air resistance.) (b) How much time would elapse between successive passes of this imaginary satellite?

15 The moon's mass is 7.35×10^{22} kg, and it moves around the earth approximately in a circle of radius 3.82×10^5 km. The time required for one revolution is 27.3 days. (a) Calculate the centripetal force that must act on the moon. (b) Verify numerically that gravitational attraction supplies the necessary force calculated in part (a), using the additional data that the mass of the earth is 5.98×10^{24} kg.

16 When a satellite revolves in an orbit around the earth, the centripetal force equals the gravitational attraction. (a) Derive a formula for the period T (time for one revolution) of a satellite in a circular orbit. (Your formula should include the radius R of the orbit, the mass M of the earth, and the gravitational constant G. Note that the mass of the satellite cancels out.) (b) Is your formula consistent with Kepler's third law? (c) Calculate the period of a satellite in a circular orbit of radius 7000 km.

17 With what speed does the earth orbit the sun? Mass of sun is 2×10^{30} kg, and the radius of the earth's orbit is 1.5×10^{11} m.

18 A planet of mass m has a moon revolving around it in a circular orbit of radius r and period t. What is the value of g at the moon's position?

19 A 60 kg girl is in an elevator that is accelerating downward at 1.4 m/sec^2. (a) What is the net force on the girl? (b) What is the apparent weight of the girl (i.e., the upward force of the floor upon her)?

GROUP C

20 If an Olympic athlete could put the shot 18.31 m in Helsinki (latitude 60.2° N), where g is 981.91 cm/sec^2, how much farther could he put it in Mexico City (latitude 19.5° N), where g is 979.69 cm/sec^2? (*Hint:* Consider the fractional change in g.)

21 Make a calculation to show that the gravitational force of the sun on the moon is always greater than the force of the earth on the moon, even at full moon. This being so, how is it that the moon goes around the earth and not around the sun? Or does it? (*Data:* Mass of sun, 2.0×10^{30} kg; mass of earth, 6.0×10^{24} kg; mass of moon, 7.4×10^{22} kg; average distance from earth to sun, 1.5×10^{11} m; average distance from moon to earth, 3.8×10^8 m.)

22 The communications satellite Echo I was orbited at an altitude of about 6.25×10^5 m. This 30 m aluminum-coated balloon was easily seen by the naked eye and looked like a bright star moving across the sky. Compute the period and velocity of Echo I. (*Data:* Mass of the earth, 6×10^{24} kg; radius of the earth, 6.4×10^6 m).

23 Show that the mass of a planet can be determined from the motion of its satellites by using the data of Midas II. This satellite had an altitude of 4.78×10^5 m at its closest distance (perigee) and 5.05×10^5 m at its furthest distance (apogee). It completed one revolution of the earth each 94.3 minutes. By finding its average altitude and using the value 6.38×10^6 m for the earth's radius, find the mass of the earth. How does your answer compare with the mass found in Problem 7?

FOR FURTHER STUDY

THE NATURE OF SCIENTIFIC THEORIES

You have seen how Kepler's laws led Newton to the law of gravitation. In the detailed working out of his theory, Newton did not obtain Kepler's third law in the exact form in which it was stated by Kepler. Newton obtained the following proportion (written for the case in which the two planets compared are Mars and the earth):

$$\frac{T^2_{\text{Mars}}\left(1 + \dfrac{m_{\text{Mars}}}{m_{\text{sun}}}\right)}{T^2_{\text{earth}}\left(1 + \dfrac{m_{\text{earth}}}{m_{\text{sun}}}\right)} = \frac{R^3_{\text{Mars}}}{R^3_{\text{earth}}} \qquad \text{(Equation 7-6)}$$

Kepler's form of the proportion (stated in words in Section 7-5) omits the factors in parentheses, involving the masses of the planets and the sun. Does this mean that Kepler was "wrong" and Newton "right"? Not at all. Kepler used existing data (collected during a lifetime of observations, chiefly of Mars, by his teacher, Tycho Brahe) to construct a theory that agreed with the experimental facts within his limit of observational error. Other theories might agree with observation equally well, and such theories would be considered just as "correct." It is to be expected that, at any time, our most firmly established laws of physics may suffer the same fate that Kepler's third law suffered. But note that the new law that replaces the old one must always contain the old law as a special case. For example, Newton's form of Kepler's third law (Equation 7-6) reduces to the Kepler form if the factors in brackets are set equal to 1; and this is very nearly true in view of the special circumstance that the sun is so much more massive than any of the planets. We say that Kepler's form is a first approximation to the law of planetary periods, and Newton's form is a second approximation. Are there third, fourth, . . . nth approximations? This is the way science progresses, each new step including all that went before.

As a matter of fact, Newton's law of gravitation has already received similar treatment at the hands of Einstein, and the Einstein law of gravitation (which you will encounter in a more advanced course than this one) includes Newton's form as a special case. Almost every prediction of the two laws of gravitation is the same, within experimental error. The predicted behavior of Mercury's orbit around the sun, however, is just noticeably different on the basis of the two laws, and the Einstein law agrees with the observation while the Newton law does not. This does not mean that we have "repealed" Newton's law of gravitation. It is a very useful and almost exact first approximation, valid for the situations in which we apply it. And revision has not ended with Einstein. Princeton scientist Robert H. Dicke recently uncovered evidence that the rotational speed of the sun is different from what Einstein had supposed. Since this new factor might be an indication that Einstein's explanation of the orbital behavior of Mercury is in error, an even newer and more accurate formulation may emerge.

Another example of the changing validity of a "law" of physics is supplied by Newton's second law. The "breakdown" of Newton's second law became evident in the early twentieth century as atomic phenomena became more clearly understood. The finishing touch was given in 1926 by Erwin Schrödinger, Max Born, Werner Heisenberg, and other physicists, who developed a new mechanics (quantum mechanics) to apply to a single atom. This does not detract in the least from the validity of Newton's laws as we use them in this text and as engineers use them for the most complicated calculations. We always deal with large-scale objects (marbles, cars, even dust particles) that consist of many billions of atoms grouped closely together. We do not expect a political poll to shed any light on a certain individual's voting preference, yet the poll does reflect the average behavior of a large group. Similarly, we should not expect a description of the average behavior of a group of atoms (Newton's laws) to predict the behavior of a single electron (quantum mechanics).

UNIT 2

References and Outside Reading

The purpose of the following list is to make you aware of additional reading that will extend and broaden most of the concepts you have met in this unit. These references should be easily available and are of sufficient variety to meet your individual interest.

The Birth of a New Physics, by I. Cohen, Science Study Series, Garden City, N.Y., Doubleday Anchor, 1960. An excellent treatment of the historical development of the dynamics of motion.

"The Conservation of Energy," by J. R. Mayer. Translated in *A Source Book in Physics* (see below).

"The Density of the Earth," by H. Cavendish. Reprinted in *A Source Book in Physics.*

"Galileo's Discovery of the Parabolic Trajectory," by S. Drake and J. MacLachlan, in *Scientific American,* March 1975. Shows that Galileo carried out careful experiments in projectile motion.

Gravity, by G. Gamow, Science Study Series, Garden City, N.Y., Doubleday Anchor, 1962. A brilliant discussion of one of the unsolved problems in physics.

Introduction to Special Relativity, by H. Smith, New York, W. A. Benjamin, Inc., 1965. An elementary treatment of a subject of concern to many, this book extends the treatment covered in this text and gives the reader an excellent preparation for more advanced treatments.

Knowledge and Wonder, by V. Weisskopf, Science Study Series, Garden City, N.Y., Doubleday Anchor, 1963. An award-winning book that summarizes what man has learned about himself and the world.

"Newton," by I. Cohen, in *Scientific American,* December, 1955. A biography by one of the leading authorities on the subject.

"On the Mechanical Value of Heat," by J. P. Joule. Excerpts in *A Source Book in Physics.*

One, Two, Three . . . Infinity, by G. Gamow, New York, Mentor Books, 1957. A must for everyone, young scientist or layman alike. The most interesting facts and theories of almost every branch of science are covered.

Kepler's Constant Throughout The Universe The following four pages attempt to illustrate our solar system and galaxy as an intricate machine that is repeating over and over again the motions given at the beginning of time. We can think of this whole scheme as a collection of individual satellite systems, each containing several small bodies revolving about a much more massive primary, as described by Kepler's laws of motion. On this page, the earth is shown with the Tiros weather satellite, a Mercury space capsule, and the moon. Each orbits the earth such that R^3/T^2 is a constant (Kepler's Third Law).

Here is the solar system seen from Tiros. It is not possible to preserve the relative sizes and distances in this drawing, so this is obviously not to scale. Of the nine known planets, all but two have satellites. Jupiter is the leader with twelve satellites. The largest, Ganymede, is bigger than the planet Mercury. The outer satellites of Jupiter are over 24 million

kilometers from the planet. These satellites, as well as the satellites of all the planets, orbit the primary body in accordance with Kepler's Third Law: R^3/T^2 for each of them is a constant, but a different constant for each satellite system.

Here we see what our galaxy, the Milky Way, would probably look like if viewed from the outside. It is a collection of over 100 billion stars rotating about the massive group of stars at the center. The galaxy is about 10^5 light-years in thickness and our sun is about half-way out from the center. But just as our own satellites orbit the earth, and the planets orbit the sun, so does the sun with its billions of stellar companions orbit the galactic nucleus. Our estimate that it takes about 200 million years for the sun to complete one circuit of the galaxy, and indeed our estimate for the mass of the galaxy and the number of stars it contains, are due in large part to the application of Kepler's Third Law.

Relativity and Common Sense, by H. Bondi, Science Study Series, Garden City, N.Y., Doubleday Anchor, 1963. A very easy-to-follow discussion of a subject of widespread interest.

"The Role of Music in Galileo's Experiments," by S. Drake, in *Scientific American,* June 1975. Describes how Galileo handled the problem of time measurement without having a reliable clock.

Sir Isaac Newton, by E. N. da C. Andrade, Science Study Series, Garden City, N.Y., Doubleday Anchor, 1965. A very popular biography of one of the giants in the history of science.

A Source Book in Physics, by W. F. Magie, New York, McGraw-Hill, 1935, also Cambridge, Mass., Harvard University Press, 1963. Selections from Galileo's *Two New Sciences* on acceleration and laws of falling bodies and projectile motion.

The Universe at Large, by H. Bondi, Science Study Series, Garden City, N.Y., Doubleday Anchor, 1960. The whole book is worth reading, but Chapters 9 and 10 offer a good discussion of gravitation and the motion of satellites.

Vectors—A Programmed Text for Introductory Physics, prepared by Basic Systems Inc., New York, Appleton-Century-Croft, 1962. A very good step-by-step exposition for those who want further material on vector operations.

The Watershed, by A. Koestler, Science Study Series, Garden City, N.Y., Doubleday Anchor, 1962. A revealing glimpse into the tortured life of one of the most remarkable men in the history of astronomy.

The World of Copernicus, by A. Armitage, New York, Mentor Books, 1954. A fascinating biography of the sixteenth-century astronomer, written in non-technical language.

"The Wright Brothers' Flight-Control System" by Frederick J. Hooven, in *Scientific American,* November 1978. The Wrights not only had to design and build a powered aircraft, but also had to make it controllable and develop the piloting skills to handle it.

UNIT 3 VIBRATION AND WAVES

The two conservation laws you have just explored have brought harmony out of seeming chaos. They allow us to predict the outcome of many complex events. Furthermore, whether the moving objects are baseballs or the sun and its planets, whether the motion is straight line, parabolic, or circular—you know that Newton's laws of motion can be applied with equal confidence.

You will now apply these laws to another type of motion—vibratory motion. You will find out how a vibration gives rise to a wave, and how a wave becomes a carrier of energy. The concept of waves can explain how sound reaches us from a plucked guitar nearby, and how radio signals reach us from a communications satellite. The concept of waves is the basis of the anechoic chamber pictured here. The chamber eliminates all radio waves except those from the communications satellite being tested.

The concept of waves will provide a key to the world of radiation and the microscopic world of the atom.

8 Vibration— Simple Harmonic Motion

It has often been said that nature is in unending motion. In Chapter 3 you studied the kinematics of linear motion and of parabolic motion near the surface of the earth. In Chapter 7 the discussion of uniform circular motion featured motions that were *periodic,* that is, in which an object repeatedly returns to its original position. Now you will encounter another class of periodic motion—vibratory motion. A *vibration* is a to-and-fro motion, generally along a straight line (linear vibration) or along an arc of a circle (angular vibration). A typical linear vibration is that of a load moving up and down on the end of a spring or the motion of a piston in the engine of an automobile (Figure 8-1a). A pendulum and the balance wheel of a watch illustrate angular vibratory motion (Figure 8-1b).

A brief look around us shows that there are many vibrations in nature. A raindrop strikes a leaf on a tree, and the leaf gyrates up and down; the tree itself sways in the breeze. The vocal cords of a bull moose vibrate many times per second, causing periodic compression of the air, which, in turn, causes vibrations of the hunter's eardrum. The ticking of a grandfather clock, the clang of a dropped piece of metal, the twang of a guitar string—these are some of the many sounds that result from countless types of vibratory motion. Because of the importance of vibrations in music, the term "harmonic motion" has come to mean the same as "vibratory motion." One particularly simple form of harmonic motion occurs frequently in both the large-scale world and at the atomic level, and it is with this form—called *simple harmonic motion* (SHM)—that we begin our study.

8-1 SIMPLE HARMONIC MOTION

First of all, it is obvious that a body must be acted on by a special kind of force in order to vibrate. The body, after passing through its equilibrium position (the position where no net force is acting upon it), must eventually stop and return; otherwise it would be lost forever and the motion could not repeat itself. Let's investigate the kind of force necessary to produce this type of periodic motion, by analyzing the back-and-forth motion of a mass (a load) on the end of a spring. In Figure 8-1a it is assumed that the spring has already been stretched (by the force of gravity) until the mass hangs motionless, in equilibrium at some point O. In this position the force F of the spring equals the weight W of the load. If the mass is moved to a position P slightly above O, the force in the spring decreases, F becomes less than W, and so the net F is downward. If the mass is released while above O, the acceleration, given by (net F)/m, is downward, and the mass returns to

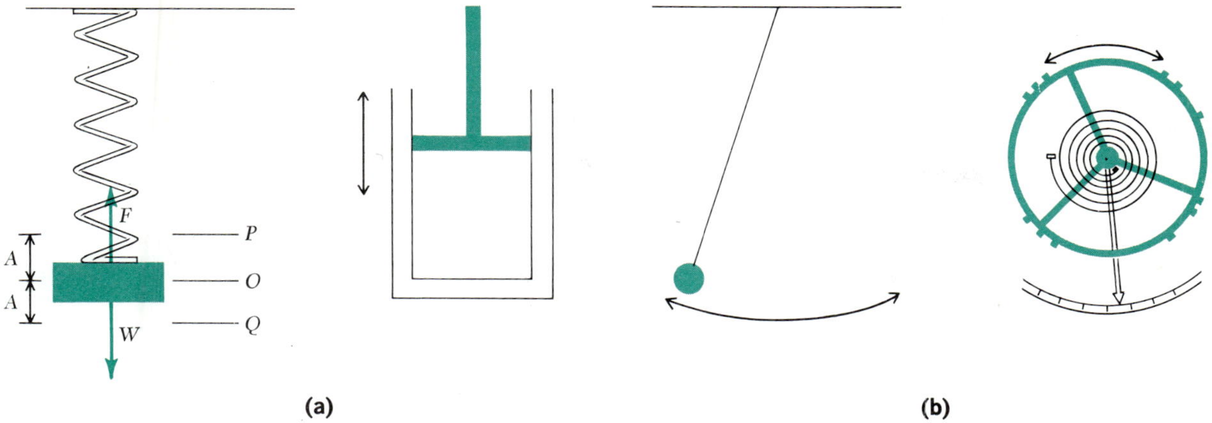

Figure 8-1 **(a)** Linear vibratory motion. **(b)** Angular vibratory motion.

O. Similarly, if the mass is pulled down below O to a position such as Q, the force in the spring increases, F becomes greater than W, and net F is upward. If the mass is released while below O, the acceleration is upward and once again the mass returns to O.

We can see, then, that the force and the acceleration are always directed toward the position of equilibrium. That is, when the mass is displaced upward from the equilibrium position, the force is directed downward, and when the mass is displaced below the equilibrium position, the force is directed upward. The force always acts to restore the mass to its equilibrium position. When this kind of experiment is repeated with different springs and different masses, one more thing can be noted; for small extensions of the spring, the force varies directly as the negative of the displacement. Such a force is called a *linear restoring force* and can be determined quantitatively by using the relation

$$F = -kx$$

where the minus sign indicates that force and displacement x* are opposite to each other. The displacement is the vector measured from the equilibrium point to the position of the mass at any time t, and the *amplitude* A is the maximum value of the displacement. The proportionality constant k is called the *force constant* of the spring. It is a measure of the spring's stiffness.

The restoring force, in the case of springs, for example, is the force with which the stretched spring pulls back on the mass that is distorting the spring. The mass, of course, exerts an equal and opposite force on the spring, and this force can be called a *distorting force*. The relation between force and stretch (or displacement), when written in terms of the distorting force, becomes

$$F = kx$$

This is known as *Hooke's Law*, after Robert Hooke (1635–1703), and is applicable only when the distorting force is not so great that it permanently distorts the spring.

In practice, if the mass is pulled away from its equilibrium position and released, it accelerates toward O and reaches O with some velocity. It coasts through the equilibrium position and immediately starts to lose speed, since the acceleration changes direction at O. Eventually, the mass momentarily comes to rest, then starts to move back toward the equilibrium position O. The vibration would continue indefinitely, if it were not for external forces such as friction due to air

* s is the general term for displacement in physics, while x is often used to denote displacement in SHM.

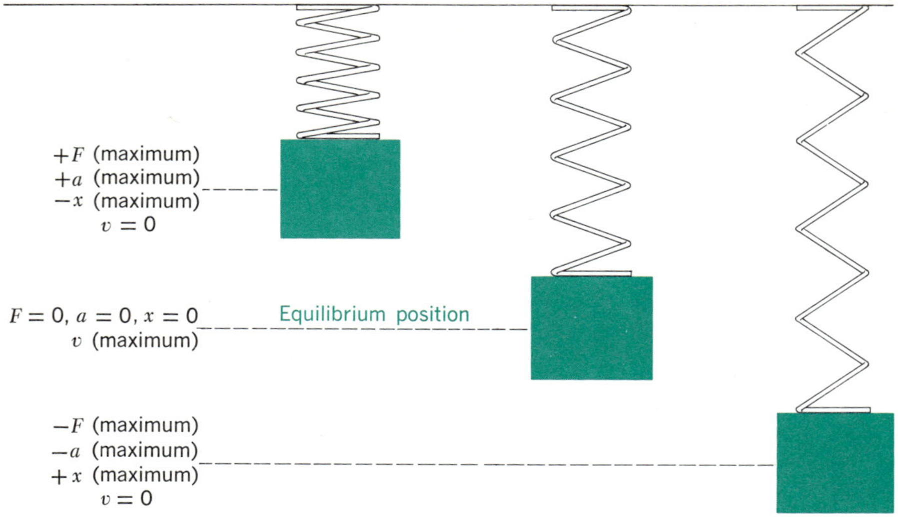

Figure 8-2 As the mass vibrates up and down, $F \propto -x$, and $a \propto -x$. These are the conditions that define SHM.

resistance and internal molecular friction of the spring.

Our observations of the vibrating mass are summarized in Figure 8-2. Here the downward direction has arbitrarily been chosen as positive. For SHM, the linear restoring force, F, and the acceleration are both proportional to the negative of the displacement. Whenever either one of these conditions exists, the motion is SHM. You should always remember that these relations hold only for small extensions of the spring. If the mass is too great, the spring will become distorted and no longer return to its equilibrium position.

8-2 PERIOD OF SHM: THE REFERENCE CIRCLE

The time for one complete to-and-fro vibration is called the *period* of the motion. It is not hard to guess which properties of the vibrating system determine the period. A stiffer spring will vibrate faster, whereas a larger mass will vibrate more slowly. To find the exact relation for the period of SHM we will first consider an experiment that relates SHM to uniform circular motion. This experiment (which you can reproduce yourself) is pictured in Figure 8-3. On a screen is cast the shadow of a ball attached to a rotating turntable, and alongside it is the shadow of another ball vibrating on the end of a spring. When the speed of the turntable is properly adjusted, the two shadows move side by side at all times. Our first step is to demonstrate that the projection of

Figure 8-3 The projection of uniform circular motion is simple harmonic motion.

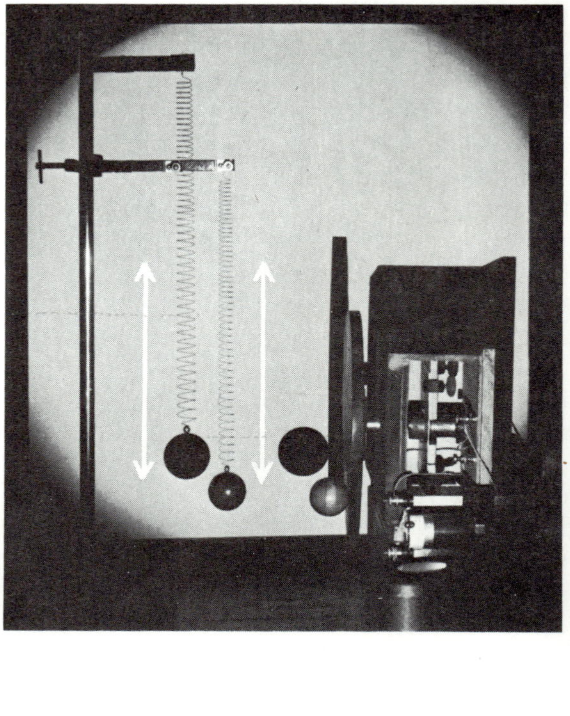

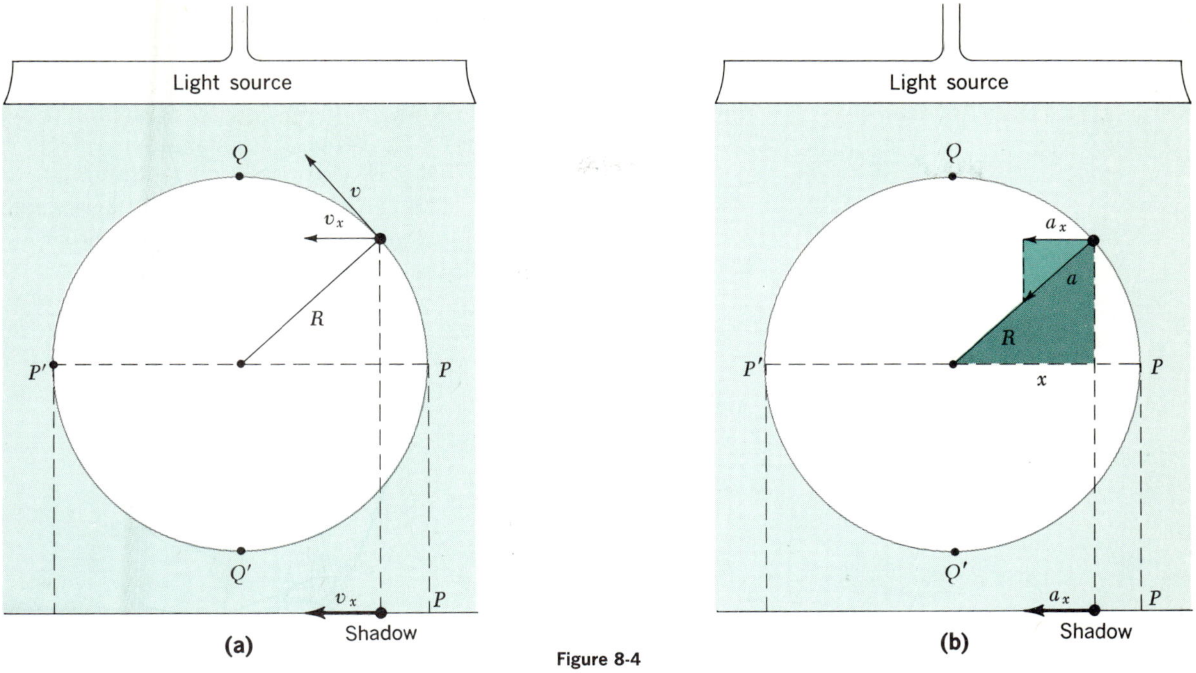

Figure 8-4

uniform circular motion (the shadow) is indeed SHM. To do this we make use of a *reference circle.*

Consider a particle in uniform circular motion with velocity v, and moving in a vertical circle of radius R (Figure 8-4a). If we imagine light shining vertically down on the particle, the shadow (or projection) of the particle moves back and forth along a line parallel to the diameter of the circle. Although the speed of the particle in uniform circular motion is constant, the speed of the shadow is continually changing. At any instant the velocity of the shadow v_x is equal to the *horizontal component* of the velocity of the reference particle. Thus, when the particle is at point P and its velocity is directed vertically upward, the horizontal component of the velocity is zero—the shadow is momentarily at rest. Again, when the particle is at point Q, momentarily moving in a horizontal direction, the horizontal component of the velocity is equal to the velocity itself, and the shadow has maximum velocity. You can see, then, that as the particle revolves in a circle, its shadow moves back and forth with a velocity that is zero at the end points of the motion and a maximum at the midpoint. So far our observations are in agreement with the fact that a vibrating mass in SHM is momentarily stationary as it changes direction at either end of its path and is moving at maximum speed as it passes the midpoint of the vibration (see Figure 8-2).

We now prove that the motion of the shadow is SHM. In Figure 8-4b the reference circle has been redrawn to show the acceleration of the particle. It has magnitude a and is directed inward along the radius. The acceleration of the shadow a_x is equal to the horizontal component of the acceleration of the reference particle. It should be clear that at the end points of the diameter, P and P', the horizontal component of the acceleration, and hence the acceleration of the shadow, is a maximum (its velocity v_x is momentarily 0 changing from $-$ to $+$, or vice versa). At the midpoint of the projected motion the acceleration is zero. (At points Q and Q' the horizontal component of the acceleration is zero.) These observations are also in agreement with our analysis of the acceleration of a vibrating mass in SHM.

Finally, in Figure 8-4b we see from the similar

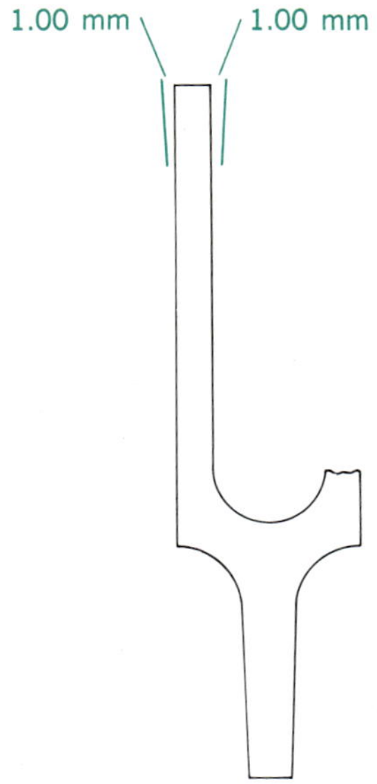

Figure 8-5

triangles (shaded in the figure) that $\frac{a_x}{x} = \frac{a}{R}$. But, since a and R are constant throughout the motion, $\frac{a_x}{x}$ = constant, or $a_x \propto x$. Acceleration is proportional to displacement—this is a general property of SHM. We can conclude, therefore, that the shadow—or, in general, the "projection" of uniform circular motion—moves in SHM.

Note also that R, the radius of the reference circle, is equal to the amplitude A of the SHM of the shadow. The time T for one complete vibration of the shadow is the same as the time for one complete revolution of the reference particle. Hence

$$T = \frac{\text{circumference of reference circle}}{\text{speed around the circle}}$$
$$= \frac{2\pi A}{v}$$
$$v = \frac{2\pi A}{T}$$

Use of this formula for v allows an easy calculation of the maximum velocity and maximum acceleration in any SHM.

Example 8-1 A prong of a tuning fork vibrates in SHM, with a period of $\frac{1}{500}$ sec. The prong moves through 2.00 mm, 1.00 mm on either side of center (Figure 8-5). What is its maximum velocity? What is its maximum acceleration?

The period is $\frac{1}{500}$ sec = 0.002 sec, and the amplitude is 1.00 mm = 1.00×10^{-3} m. At the midpoint of the motion,

$$v = \frac{2\pi A}{T}$$
$$= \frac{2\pi(1.00 \times 10^{-3}\ \text{m})}{2 \times 10^{-3}\ \text{sec}}$$
$$= \boxed{3.14\ \text{m/sec}}$$

At either end point,

$$a = \frac{v^2}{A}$$
$$= \frac{(3.14\ \text{m/sec})^2}{1.00 \times 10^{-3}\ \text{m}}$$
$$= \boxed{9870\ \text{m/sec}^2}$$

The result of this example illustrates the very large acceleration possible when the time interval between velocity changes is small.

We are now in a position to derive the formula for the period of a simple harmonic motion—the time for one complete back-and-forth vibration. It has been established that the projection of uniform circular motion is SHM, since the acceleration of the moving "shadow" is proportional to the displacement. Let's imagine a mass moving back and forth along a line, always coinciding with the shadow. Because $F = ma$, we know that the linear restoring force on the mass is proportional to the displacement (see Figure 8-2). For this motion, then,

$$F_x = -kx$$

where the subscript indicates that the force is along the horizontal direction. For uniform circular motion, the centripetal force is found by using the relation

$$F = -\frac{4\pi^2 m}{T^2} R$$

This is Equation 7-4, and the minus sign simply indicates that the inward force is opposite in direction to the radius vector R (the radius being a displacement from the center toward the outside of the circle). The horizontal component of this force (the force that "produces" the motion of the mass that coincides with the shadow) is

$$F_x = -\frac{4\pi^2 m}{T^2} x$$

Therefore,

$$-kx = -\frac{4\pi^2 m}{T^2} x$$

or, solving for the period T,

$$T = 2\pi \sqrt{\frac{m}{k}} \qquad \textbf{(Equation 8-1)}$$

This is the basic formula for the period of any SHM. It shows how the period is related to the essential properties of the vibrating system. This equation tells us that when k is large (a stiff spring), the period of vibration is small—a fact that can be easily observed. The equation also shows that a large mass vibrates more slowly, or when m is larger, T is larger.

Example 8-2 The springs of a 1000 kg car compress vertically 0.500 cm when a 100 kg man steps in. With the man in the car, how many vibrations per second does the body of the car make after being jarred while going over a bump?

First we find the force constant. (We can find this by considering the compression caused by the man.)

$$k = \frac{F}{x}$$

$$= \frac{(100 \text{ kg})(9.8 \text{ N/kg})}{5 \times 10^{-3} \text{ m}}$$

$$= 1.96 \times 10^5 \text{ N/m}$$

Next we find the time of a single vibration while going over the bump. (Note that now the total mass of the car and the man is involved.)

$$T = 2\pi \sqrt{\frac{m}{k}}$$

$$= 2\pi \sqrt{\frac{1100 \text{ kg}}{1.96 \times 10^5 \text{ N/m}}}$$

$$= 0.47 \text{ sec}$$

We see that 0.47 sec/vibration means 1/0.47 vibrations/sec. So the car bounces up and down 2.13 times per sec.

Actually, cars are built with shock absorbers that provide frictional *damping*, and the bouncing may last only one vibration or less.

In this example we were asked to find the number of vibrations per unit time and saw that this is the reciprocal of the period. The reciprocal of the period is called the *frequency:*

$$f = \frac{1}{T}$$

8-3 THE ENERGY PRINCIPLE IN SHM

As a body vibrates, there is a continual transformation of energy from KE to PE and back again. In particular, if friction losses are negligible, the *total* energy of the vibrating mass is the same at any point in the motion. There are two points at which the calculation of this energy is simple. First, at the end points of the motion, since the mass is motionless, the energy is all PE, stored in the system. Second, as the mass passes through the equilibrium position where the spring has normal extension and thus zero PE, the energy is all KE. Before we can make these calculations, it is necessary to define what is meant by the PE of a vibrating mass.

In Figure 8-6a (page 142) the body is in equilibrium and, since the mass is motionless and the spring is not compressed, the total energy of the system is zero. Now suppose the mass is moved by some outside agent from its equilibrium position through a distance x to some new point (Figure 8-6b), thus compressing the spring. The work done on the system is now stored in the spring as elastic PE. The work was done *against* a variable force ($-kx$, the linear restoring force exerted by

N= Kg·m/s² m/s²·1/m

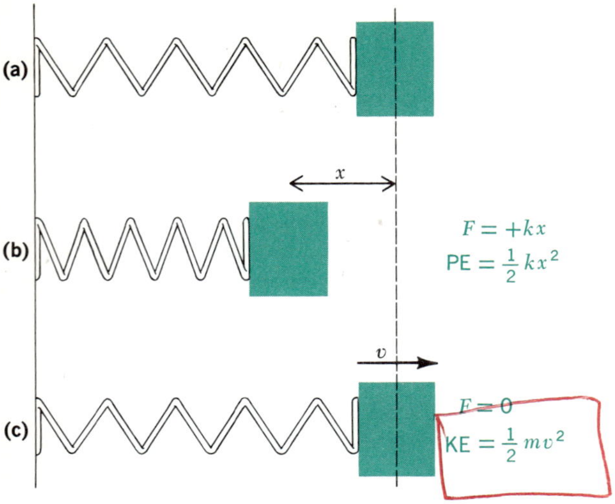

Figure 8-6

the spring on the mass). The variable force exerted by the outside agent was zero at the position in (a) and $+kx$ at the position in (b). The *average* force needed to compress the spring is, therefore, $(0 + kx)/2 = \frac{1}{2}kx$. The work done on the spring, and thus the PE stored in the spring, is this average force times the displacement x:

$$\text{PE} = (\tfrac{1}{2}kx)x = \tfrac{1}{2}kx^2$$

This expression gives the PE stored in a spring for any displacement x as it vibrates to and fro in SHM.

If the mass is now released, it moves back to its equilibrium position (Figure 8-6c) and, by the energy principle, the total energy at the position in (c) equals the total energy at (b). But at (c), the equilibrium position, there is no PE. All the energy appears as KE of the moving mass. Thus, at the equilibrium position, KE equals $\frac{1}{2}kA^2$, where A is the maximum value of the displacement x.

Example 8-3 A spring having a force constant of 250 N/m is loaded with a mass of 10 kg. (a) Find the period of vibration. (b) If the mass is displaced 30 cm and then released, what is the total energy of the vibrating mass? (c) Find the velocity with which it passes through the equilibrium position. (d) How much time is required for the mass to move the 30 cm from the end point of the motion to the equilibrium position?

(a)
$$T = 2\pi\sqrt{\frac{m}{k}} = 2\pi\sqrt{\frac{10\text{ kg}}{250\text{ N/m}}} = \boxed{1.256\text{ sec}}$$

(b) The total energy of the vibrating mass is the same at any point of the motion. At the end points x equals A, the amplitude of vibration, and the energy is all PE:

$$\text{PE} = \tfrac{1}{2}kx^2 = \tfrac{1}{2}kA^2 = \tfrac{1}{2}(250\text{ N/m})(0.30\text{ m})^2 = 11.3\text{ N}\cdot\text{m} = \boxed{11.3\text{ J}}$$

(c) At the equilibrium position the energy is all KE, and equal to the total energy of 11.3 J.

$$\tfrac{1}{2}mv^2 = 11.3\text{ J}$$
$$\tfrac{1}{2}(10\text{ kg})\,v^2 = 11.3\text{ J}$$
$$v = \boxed{1.50\text{ m/sec}}$$

As an alternative, we could have used the expression derived earlier for the maximum velocity in SHM.

$$v = \frac{2\pi A}{T} = \frac{2\pi(0.30\text{ m})}{1.256\text{ sec}} = \boxed{1.50\text{ m/sec}}$$

(d) The time required to move the 30 cm is $\frac{1}{4}$ of the complete period.

$$t = \tfrac{1}{4}T = \tfrac{1}{4}(1.256\text{ sec}) = \boxed{0.314\text{ sec}}$$

It would be incorrect to try to use the laws of uniformly accelerated motion (such as $s = \frac{1}{2}at^2$)

to find the time, since acceleration is *not* constant. The very idea of SHM is one of *variable* acceleration, according to the proportion $a \propto -x$. *The equations of uniformly accelerated motion cannot be used for SHM problems.*

8-4 THE SIMPLE PENDULUM

When a small, heavy pendulum bob is swinging at the end of a light string or fine wire, the pendulum is said to be "simple" because all moving parts that have mass are concentrated at one place, all moving parts have the same speed, and the restoring force acts at a single point. These assumptions are never quite true, since even the lightest string has some mass, even the smallest bob cannot be at one point but must be spread out over a finite volume, and all parts of the pendulum cannot be moving at the same speed. We will make these assumptions, however, for the idealized *simple pendulum.*

The bob of the pendulum moves from P to Q to P' and back again (Figure 8-7) with many of the characteristics of SHM: the speed is maximum at Q, and the acceleration is greatest at P and P' (when the velocity is momentarily zero, changing from + to −). To find out whether a pendulum does actually move in SHM, it is necessary to ask the usual question: Is the restoring force proportional to the displacement? In other words, *is* there a force constant?

What are the forces acting on a pendulum bob? Examination of Figure 8-7 shows that there are only two: the weight mg of the bob, and the tension T in the string. The weight vector has been resolved into two components: F_s, which is tangent to the path of swing, and $F_\perp$, which is at right angles to F_s. It should be clear that F_s is the only force capable of causing acceleration along the path of swing. In other words, F_s is the restoring force.

To see if this restoring force is proportional to the displacement, we note that the two angles marked θ in the figure are equal because their sides are respectively parallel. Hence, the two right triangles (shaded in the figure) are similar,

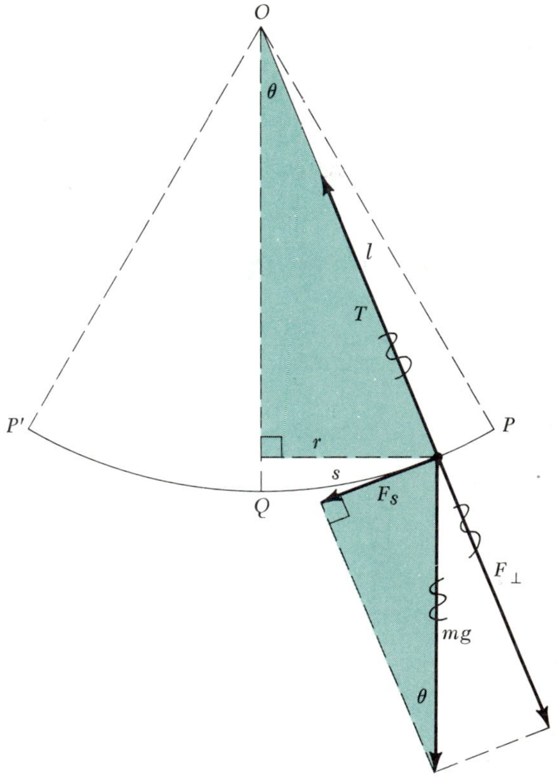

Figure 8-7

and we can write

$$\frac{F_s}{mg} = \frac{r}{l}$$

where r is the horizontal displacement of the bob from equilibrium, and l is the length of the string. Rewriting this proportion, we have

$$F_s = \frac{mg}{l} r$$

Since mg and l are constant for any given pendulum, this expression says that the restoring force is proportional to the *horizontal* displacement r of the bob. But to be SHM, the restoring force would have to be proportional to the displacement of the bob *along the arc,* that is, s. We see, then, that pendulum motion is *not* SHM. For very small angles of swing, however, r and s are very nearly equal in length and direction. If we

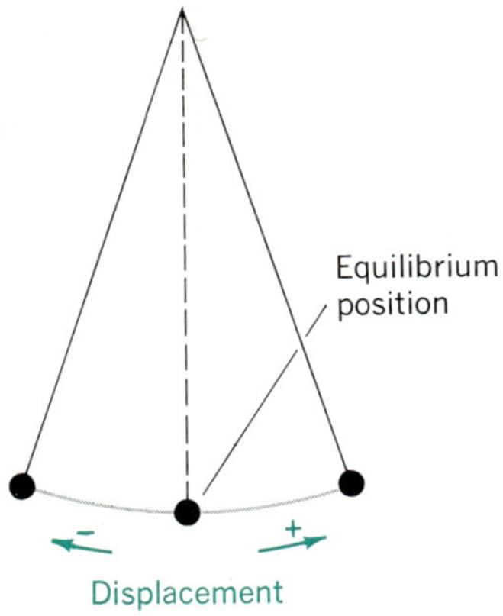

Figure 8-8

make this approximation, we then have

$$F_s \approx -\frac{mg}{l}s$$

where the minus sign indicates that the force and the displacement are oppositely directed, since the displacement is always away from the equilibrium position (Figure 8-8). But for any given pendulum, m and l are constant, therefore we see that the above expression has the form $F \approx -kx$ where

$$k = \frac{mg}{l}$$

For a simple pendulum, then, where the angular displacement is small, the motion is SHM.

We are now able to find the period of a simple pendulum. For any SHM, Equation 8-1 says

$$T = 2\pi\sqrt{\frac{m}{k}}$$

and for the simple pendulum, when the swing is small, $k = mg/l$. Substituting, we get

$$T = 2\pi\sqrt{\frac{m}{mg/l}}$$

$$= 2\pi\sqrt{\frac{l}{g}}$$

It is important to realize that this formula for the period of a simple pendulum is an approximation, but a very good one if the angle of swing is small.* It is also important to note that the symbol m canceled out because weight mg is proportional to mass m. This being so, the period of a pendulum is independent of the mass of the bob, depending only on the length of the string. On the other hand, since g appears in the formula, we have a means of determining the value of the acceleration due to gravity at any place.

Example 8-4 What is the acceleration due to gravity on a planet where a space explorer's simple pendulum 0.400 m long vibrates 100 times in 240 sec?

The period is (240 sec)/100 = 2.40 sec.

$$T = 2\pi\sqrt{\frac{l}{g}}$$

$$T^2 = \frac{4\pi^2 l}{g}$$

$$g = \frac{4\pi^2 l}{T^2}$$

$$= \frac{4\pi^2(0.400\text{ m})}{(2.40\text{ sec})^2}$$

$$= \boxed{2.74\text{ m/sec}^2}$$

Note that $T = 2\pi\sqrt{l/g}$ is a special formula, valid for a special case (simple pendulum vibrating with not too large an amplitude), while the similar-appearing formula $T = 2\pi\sqrt{m/k}$ is a general formula for the period of *any* SHM.

* The error is 0.1% if the angle POQ is 7°, and only 1% if it is as large as 23°.

SUMMARY

Simple harmonic motion (SHM) occurs whenever the **restoring force** on a body is proportional to the displacement of the body from its equilibrium position. The restoring force is described by $F = -kx$. Since the restoring force is proportional to the acceleration, we have SHM whenever the acceleration is proportional to the negative of the displacement, $a \propto -x$. Either of these conditions defines SHM. The **period** T of the SHM is the time for one complete vibration; the **frequency** f is given by $f = 1/T$.

The projection of uniform circular motion upon any straight line is SHM. The radius of the reference circle equals the **amplitude** A of the SHM. The speed v of a reference particle on the circumference of the circle equals the maximum velocity of a particle in SHM as it passes through the center: $v = 2\pi A/T$.

During a vibration KE is continually being transformed to PE (usually elastic), and vice versa. The motion eventually dies out as the (KE + PE) total is transformed into thermal energy through frictional losses.

The period of any SHM is given by $T = 2\pi\sqrt{m/k}$. As a special case, the motion of a simple pendulum turns out to be approximately SHM if the amplitude of swing is not too great. The simple pendulum's period is given by $T = 2\pi\sqrt{l/g}$, an equation that can be used to measure the acceleration due to gravity.

QUESTIONS

1 When a body is vibrating in SHM, is its acceleration zero at any point in the motion?

2 A pendulum clock keeps perfect time when in Los Angeles. Will it gain or lose when it is taken to a summer resort in the mountains near the city?

3 The presence of oil beneath the earth is often indicated by a "salt dome" over the oil-bearing rock. The salt dome has a smaller density than the surroundings. Explain how a geophysicist on the surface can use a delicate pendulum apparatus to determine the probable location of oil. (Review gravitation, Chapter 7.)

4 An airplane loops-the-loop in a vertical circle, picking up speed as it approaches the bottom of the loop and losing speed as it approaches the top of the loop. Does the shadow of the plane move across the ground in SHM?

5 A mass is vibrating on the end of a spring. Is it in equilibrium at any point of the motion? Answer the same question for the bob of a pendulum.

6 Give an example of SHM other than those mentioned in the text.

7 The velocity at the end points of a vibration is zero, but the acceleration is a maximum. Can you give another example (not SHM) where the acceleration of a body is not zero when the velocity is zero?

PROBLEMS

GROUP A

1 A piece cut from a bicycle inner tube is 0.7 m long when it carries a load of 25 N. The force constant k is 300 N/m. What will be the length of the piece of rubber when the load is 75 N?

2 A vibrating reed in a harmonica makes 200 vib/sec. What is the period?

3 What is the time required for 2400 vibrations of a loaded spring if the frequency is 4 vib/sec?

4 The piston of an engine moves a total distance of 0.2 m from one extreme point to the other. (a) What is the amplitude? (b) What is the displacement when the piston is 0.1 m from one end of its stroke?

5 The piston of Problem 4 makes 2 vib/sec. How fast is it moving when passing through the midpoint of its path?

6 Compute the time for one vibration of a spring whose force constant is 16 N/m, if the load has a mass of 1 kg.

7 Compute the length of a clock pendulum that ticks once each second. (*Hint:* It ticks twice during each complete vibration.)

8 What are the dimensions of (a) period? (b) frequency? (c) force constant?

9 What would be the period of a pendulum suspended from the top of a tall building on a light string 353 m long?

GROUP B

10 An object that weighs 25 N is moving with SHM of frequency 60 vib/min and amplitude 1.5 m. What is the restoring force when the object is at one of the end points of its motion?

11 A point on a horizontal vibrating piano string is moving up and down in SHM with a frequency of 200 vib/sec. The maximum displacement of the point below its normal position is 0.5 mm. (a) What is the maximum velocity of the particle, and where is this velocity reached? (b) What is the maximum acceleration of the particle?

12 A 30 g load hangs from a long, light spring. When pulled down 20 cm below its equilibrium position and released, it vibrates with a period of 2 sec. (a) What is its velocity as it passes through the equilibrium position? (b) By how much will the spring shorten if the load is removed?

13 A cart of mass 3 kg is held between two stretched springs, as shown in Figure 8-9. It is found that the cart moves to a position 0.1 m from its equilibrium position when a force of 60 N is applied. The cart is then released and allowed to vibrate. (a) What is the force constant for the system of springs? (b) How much work was required to move the cart the 0.1 m? (c) Calculate the frequency of vibration. (d) What is the velocity of the cart as it passes through its equilibrium position? (e) What is the maximum KE of the motion?

14 What energy must be given to a 20 kg mass suspended from a spring so that it will oscillate in SHM with a period of 0.5 sec and an amplitude of 2 cm?

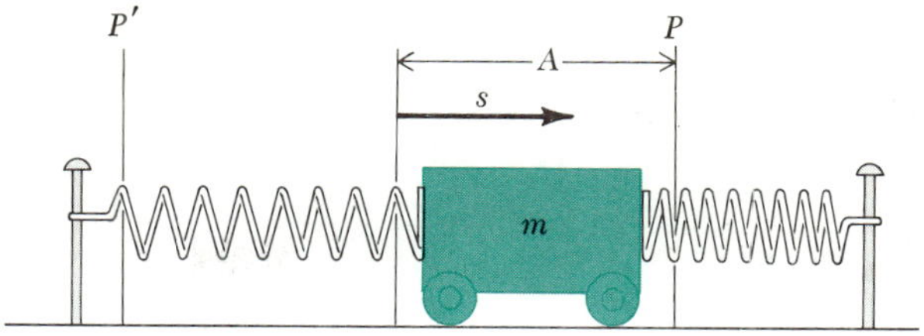

Figure 8-9

15 A raft of weight 4000 N is floating in a pond. When a 1000 N man climbs on board, the raft sinks 0.04 m deeper into the water to a new equilibrium position. If the man jumps off, how many vertical vibrations does the empty raft make in 5 sec?

16 A load of mass 400 g is hanging from a light spring whose force constant is 10 N/m. The load is pulled down 10 cm from its equilibrium position and released. (a) How long does it take for the load to reach the equilibrium position again? (b) What is its velocity then?

17 Derive a formula for the maximum acceleration of a particle in SHM, in terms of the amplitude A and the period T. Check your answer for dimensional consistency.

18 The value of the acceleration due to gravity on the moon is about $\frac{1}{6}$ that on the earth. What would be the period on the moon of a simple pendulum whose period on the earth is 2.0 sec?

GROUP C

19 Write an equation for the vertical motion of a body of mass 4 kg on a spring of force constant 36 N/m, if the motion has amplitude 2 m and the body passes upward through its equilibrium position at $t = 0$.

20 A particle vibrates in SHM with a period of 6 sec and an amplitude of 10 cm. (a) How long does it take the particle to move from one end of its path to a point 5 cm from the equilibrium position? (b) What is the particle's velocity at this point? (*Hint:* Refer to the reference circle described in Section 8-2.)

21 A light rubber band is hanging loosely from the ceiling. A load is attached to the free end of the rubber band, and it is observed that the load, when released, moves downward a distance of 40 cm before starting to rise again. What is the period of this SHM? (*Hint:* The mass is not given. Assign to it the value m, on the chance that it will cancel out of your final equation for the period.)

22 A small electronic computer placed on board a satellite must withstand a maximum acceleration of 10g. To test this, the computer is attached to a horizontal table which is driven from side to side in SHM at 20 vib/s. What must be the amplitude of vibration of this shake-test apparatus in order to give a maximum acceleration of 10g?

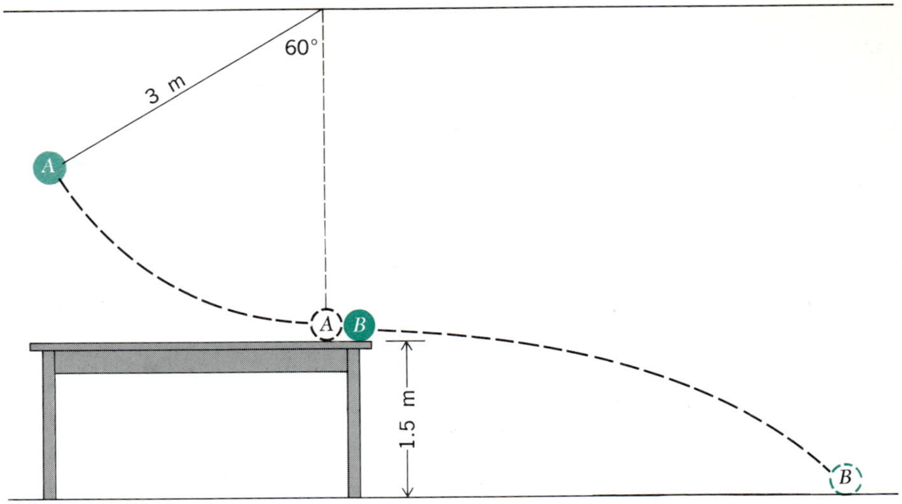

Figure 8-10

23 A steel ball A of mass 4 kg supported on a light cord 3 m long is held at a 60° angle from the vertical, as shown in Figure 8-10. A second steel ball B of mass 1 kg rests just at the edge of the table, which is 1.5 m high. Ball A is released and strikes ball B head-on in a perfectly elastic collision. (a) Approximately how long after A is released does B strike the floor? (b) How fast is A moving when it strikes B? (c) How far out from the base of the table does B strike?

9 Wave Motion — Disturbances and Their Propagation

You may wonder why the study of waves is so important that several chapters of this book are devoted to various kinds of waves. To most of us, "waves" tend to mean the waves on ponds, lakes, and oceans; but these are in fact just one special kind of wave. These waves on the surface of the water do illustrate one very important point, however. Imagine yourself standing on the edge of a lake while small waves are moving toward you. Wave after wave keeps coming in, but the water does not pile up, the water does not get higher and higher around you. Waves are moving in toward you, but the water as a whole is not moving appreciably. It is only the "lumpiness" called a wave that is moving. Figure 9-1 on the page that follows is an aerial photograph of ocean waves.

One major reason for wanting to learn about waves is that living organisms have always depended upon energy coming over great distances. There could be no life on the earth without the tremendous flow of energy from the sun; and one important way of describing this energy flow is by means of waves.

Furthermore, we all make use of a continual "feedback" of visible and audible information, transmitted as waves, that keeps us aware of what is happening in the world around us. As an example of the importance of this feedback, try a simple experiment: stand with your arms extended straight in front of you; bring the tips of your forefingers together with both eyes open; now try it with one eye closed; finally make the same attempt with both eyes closed. Many sense organs are involved in this simple operation, but the visible information reaching you in the form of light seems to be the most important of all.

From the point of view of physics, the mechanisms of information transfer (seeing and hearing, for example) and energy transfer (such as the flow of heat from the sun) are quite similar. For one thing, there just cannot be a transfer of information (a signal) without a transfer of energy. It is the energy of the light that affects our eyes when we see, it is the energy of a radio wave that affects the antenna of a TV set. Also, our most useful senses, seeing, hearing, and feeling, often involve the flow of energy over long distances. This flow is not instantaneous. Energy in the form of light, heat, or radio waves moves very fast, but it still takes a definite amount of time to move from one place to another. For example, it takes about $2\frac{1}{2}$ seconds for a radar pulse to go to the moon and back. Most of us have heard radiotelephone conversations with astronauts on the moon. But suppose that the signals took many minutes or longer to reach us. Can you imagine how much more difficult it would be to talk over such long distances?

We already know that energy can be carried by

Figure 9-1 Wave motion on the water's surface shows us much about the motion of waves that cannot be seen—such as light and sound waves.

a stream of bullets or by the blast of air molecules from a fan. This kind of energy transfer is called "convection." Now let's look at another way in which energy moves from place to place—by wave motion.

9-1 WAVES AND DISTURBANCES

There are many different types of waves—water waves, sound waves, earth waves. And there are electromagnetic waves which will be discussed in a later chapter: They include radio waves, the waves of a TV broadcast, heat radiation, visible light, and x rays.

All of us have had some first-hand experience with sound waves or water waves. Less familiar are the waves in the solid earth (commonly called earthquakes), or temperature waves (the so-called heat waves and cold waves that are mentioned on weather forecasts). Since this chapter is about waves in general, we want to find features or properties that are common to all waves. In other words, what distinctive features do all waves have? And what is more important, how is energy transmitted by a wave?

Water waves moving across a lake, waves in a flag on a windy day, and those made by repeatedly shaking a Slinky spring (Figure 9-2) are *periodic waves,* in which motions are repeated at regular intervals. Waves can also consist of only a single disturbance, never repeated at any given place. Such *shock waves,* or *pulses,* occur, for example, when a supersonic plane sets air molecules into rapid motion. An even more familiar example is the kind of pulse you can generate in a rope or a garden hose by sharply shaking one end (Figure 9-3).

The key word in discussing waves is "disturbance." There are many types of waves because there are many different ways of disturbing the physical state of a body or a substance. In water waves, the change of position of molecules is the disturbance; the water surface rises and falls. With the rope, when the end is swung from its normal position, the rope moves up and down as the pulse goes by. Changes of air pressure are the disturbance in sound waves; the pressure does not remain constant at any given point. In general, a wave may be thought of as the propagation, or spreading, of some kind of disturbance from one place to another. To understand this better, we need to know what kinds of disturbances commonly exist, and how these disturbances are "passed on" from point to point as the wave travels through a medium.

The idea of a disturbance can be illustrated even by the "wave of enthusiasm" that passes through a crowd waiting outside a theatre, hoping to get a glimpse of their favorite performers. When they are about to appear, the word reaches the first few rows of the crowd. Rumor starts the disturbance, then the enthusiasm (the disturbance) spreads at a more or less definite rate toward the people in the back rows. This is a nonphysical wave in which the disturbance is the state of unusual enthusiasm, rather than pressure, height, temperature, or some other physical state. The crowd is the "medium" through which the wave moves; the enthusiasm is passed on by sound and sight. As with most waves, there is (a) a disturbance, (b) a medium that can be disturbed, and (c) some mechanism by which the disturbance can be passed on through the medium.

Figure 9-3

Some waves, however, exist in a vacuum without the need of a material medium. Such waves depend upon disturbances in properties that do not require the presence of matter. Disturbances in gravitational, magnetic, or electric fields can exist in a vacuum, and therefore form waves that travel through a vacuum. Light and other electromagnetic waves are of this type.

Another very familiar type of wave is the starting wave when a traffic light turns green. What is the disturbance that is propagated along the line of cars? What is the medium that carries this wave? What is the connection between the cars?

9-2 SHOCK WAVES OF COMPRESSION

From the molecular standpoint, a compression wave is started whenever a solid object is struck. When a hammer strikes the head of a vertical iron nail, the molecules of iron at the top surface are acted on by a net force caused by the hammer.

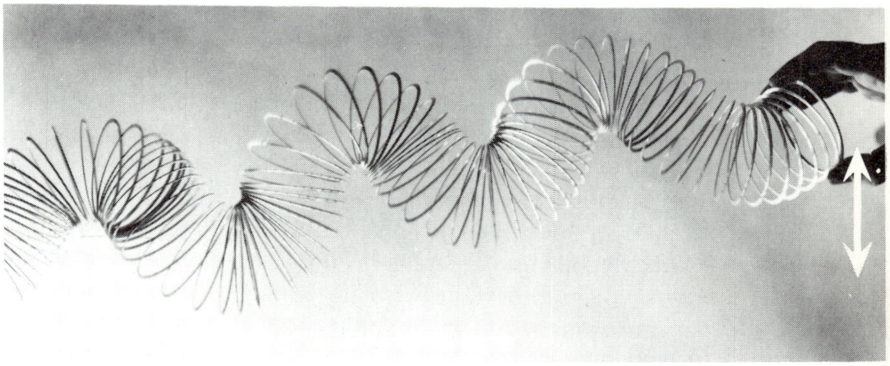

Figure 9-2 Transverse waves.

According to Newton's second law, these molecules then become accelerated and start to move downward. They, in turn, push on neighboring molecules, and eventually a pulse is transmitted to the tip of the nail. It is important to note that as the shock wave travels along the nail, a pressure is momentarily built up wherever the molecules are closer together than is normal. It is this squeezing together of the molecules—a disturbance—that gives rise to the force that pushes the next molecules along.

Figure 9-4

On a larger scale, a model of such compression shock waves can easily be observed when a stationary freight train backs up (Figure 9-4). Before the locomotive applies the initial force, the couplings between cars are loose, leaving perhaps just a bit of slack between each pair of cars. When the locomotive starts moving backward with a steady speed, one car after another is set into motion, and you can hear a "clank, clank, clank . . . " as the shock wave travels the length of the train. If in a 100-car train the pulse has reached as far as the thirtieth car, this means that the couplings between the first 30 cars are tight, and these cars, moving together as a rigid unit, are in the process of taking up the slack of the thirty-first coupling. The final 70 cars are still stationary at this instant. The velocity with which the pulse travels down the train is called the *wave velocity*. This is much faster than the speed of the locomotive; it will usually take only 5 or 10 seconds for such a shock wave to travel the length of a 100-car train. Such a shock wave is called a wave of compression. In similar fashion, if the locomotive starts forward, a shock wave of tension will be initiated.

It is worthwhile trying to make an "educated guess" as to the factors upon which the velocity of a compression wave should depend. Using the freight train model, each car must, in turn, be accelerated to a velocity equal to that of the locomotive. This requires time. The more heavily loaded the cars are, the greater their mass (inertia) will be and, by Newton's second law, the less the acceleration from a given force—and the longer the time required. The wave velocity, then, would be *less* for a loaded train. A similar argument holds for the model in Figure 9-5, which more closely resembles the structure of a solid body in which compression waves could travel. Here, the cars of our train model are replaced by small, heavy balls, and the couplings by stretched springs. The heavier the balls are, the harder they are to set in motion, and the slower the wave velocity. With stiffer springs, however, the wave velocity would be greater because the springs would cause the balls to accelerate more rapidly. In general, then, the speed of the waves increases when the spring stiffness increases, and decreases when the mass of the balls increases.

Let's now apply this reasoning to the speed of a wave in a medium such as air, water, or iron. Experiment shows that the wave speed is indeed larger in substances that are harder to compress and slower in a denser medium.

Figure 9-5

9-3 LONGITUDINAL AND TRANSVERSE WAVES

In some kinds of waves the disturbance is in the direction of motion of the wave. For example, in the freight train model each coupling is stretched in the direction of the train's motion and each car is displaced from its neighbors. The displacement is parallel to the direction of motion of the pulse. Such a wave is called *longitudinal*. The most familiar longitudinal waves are sound waves, in which the molecules are displaced back and forth in the direction the waves are moving. Figure 9-6

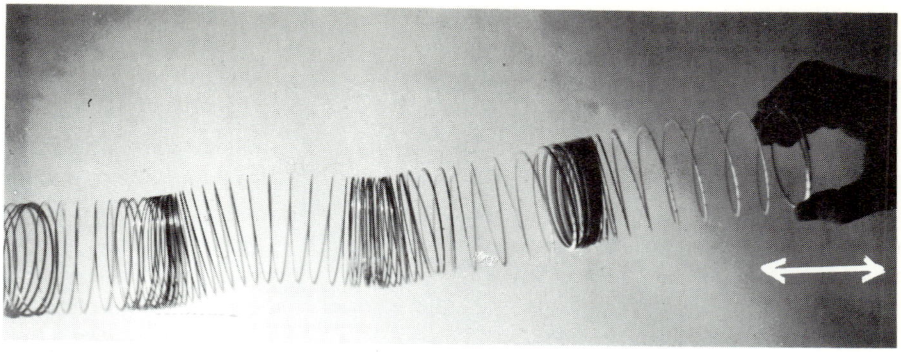

Figure 9-6 Longitudinal waves.

shows a model of longitudinal waves, made by moving one end of a Slinky back and forth in the direction of its length.

If the disturbance takes place at right angles to the direction of propagation of the wave, then the wave is called *transverse.* The Slinky in Figure 9-2 provides a model of transverse waves. The most familiar transverse waves are those on strings or ropes (Section 9-1). Light waves are also transverse, and will be discussed in greater detail later (Chapters 21 and 22). Water waves are approximately transverse, but the water molecules move somewhat back and forth as well as up and down. Small ripples on the surface of a liquid *are* very nearly transverse because the molecules do move almost in a vertical path, at right angles to the direction of motion of the wave. We will make frequent use of surface ripples on water as an example of transverse waves because they move slowly enough to make it easy to visualize what is happening to them.

If the displacements of the particles of the medium (or the disturbances in an electric field) are all in the same plane, then the wave is said to be *plane-polarized.* For example, suppose that as waves are moving along a rope, all the motion of the particles of the rope is in the vertical plane; such waves are *vertically polarized.* The polarization of waves on a rope is shown in Figure 9-7. In this model we determine the plane of vibration by observing the rope as it passes through a vertical and a horizontal slot.

Only transverse waves can be polarized. Longitudinal waves—sound waves, for example—would not be affected by a slot oriented at right angles to the direction of the wave motion. In longitudinal waves, all the motion is in the direction of propagation, and polarization has no meaning.

9-4 DESCRIBING WAVES AND VIBRATIONS

Waves are just one example of a physical phenomenon that appears quite complicated when we try to think about all aspects of it at once. A detailed photomicrograph of all the molecular motions involved in even a tiny part of a ripple would be tremendously involved and difficult to understand. Therefore, it is helpful for us to be able to draw diagrams that show only certain essential

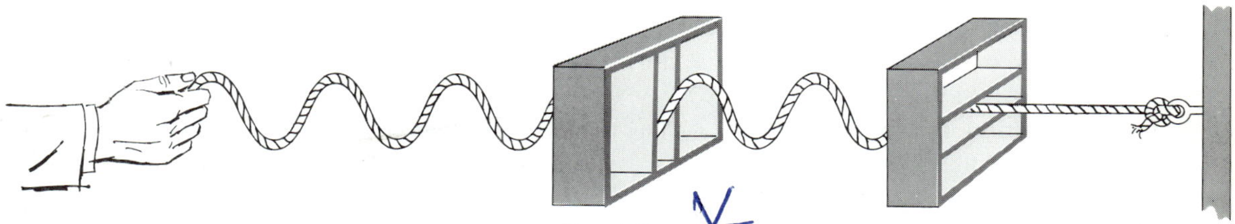

Figure 9-7

Figure 9-8

features, enabling us to focus more clearly on what is happening. Let's begin by drawing some graphs that show the movement of waves on the surface of a lake.

The wave shown in Figure 9-8 is a small water wave, and the disturbance is the displacement of the surface above or below the normal level of the water. It is important to note the difference between a moving *wave* and a *vibration*. To clarify this distinction, in the example that follows, several graphs show how the disturbance along the y-axis varies with distance from the source and with time.

Suppose a swimmer dives into a lake with a modest splash, and a short *wave train* of peculiar but definite shape spreads out from the spot where she enters the water. Let's consider a part of that wave train which moves toward someone fishing whose line enters the water at A (Figure 9-9). We start a stop watch when the swimmer enters the water; $t = 0$ at this instant. The wave train moves steadily (toward the left in the illustration) and finally moves past point A. As it does, the cork on the fishline at A starts to bob up and down, following the displacement of the water caused by the passing waves. We can now construct two related graphs: (a) displacement of the water surface as a function of distance from the swimmer at some particular time, and (b) displacement of the cork as a function of time at a given distance from the swimmer. That is, based on Figure 9-9, we plot y versus x for a particular t, and y versus t for a particular x.

Figure 9-10a is a graph of displacement y of the water as a function of position x, at the particular time $t = 11.5$ sec. In other words, we can think of Figure 9-10a as a flash photograph "freezing" the motion of the wave train at $t = 11.5$ sec. It may also be helpful to think of this graph as one frame of a movie film, while subsequent "frames" of the same film are shown in Figures 9-10b and 9-10c. Together they show how the wave moves with time. This is called a *wave-form* graph because it shows the shape, or form, of the wave. In Figure 9-10a, at the position of the cork A ($x = 275$ cm), the value of the displacement at $t = 11.5$ sec is $y = -3$ cm (below the normal water level). At other positions (corks at other values of x), the displacement may be quite different at the same time.

Figure 9-10b shows the position of the wave at $t = 11.8$ sec and Figure 9-10c shows the position at $t = 12.1$ sec. Note that the wave has preserved the same form, but has moved on to new positions.

If the person fishing is watching only the cork

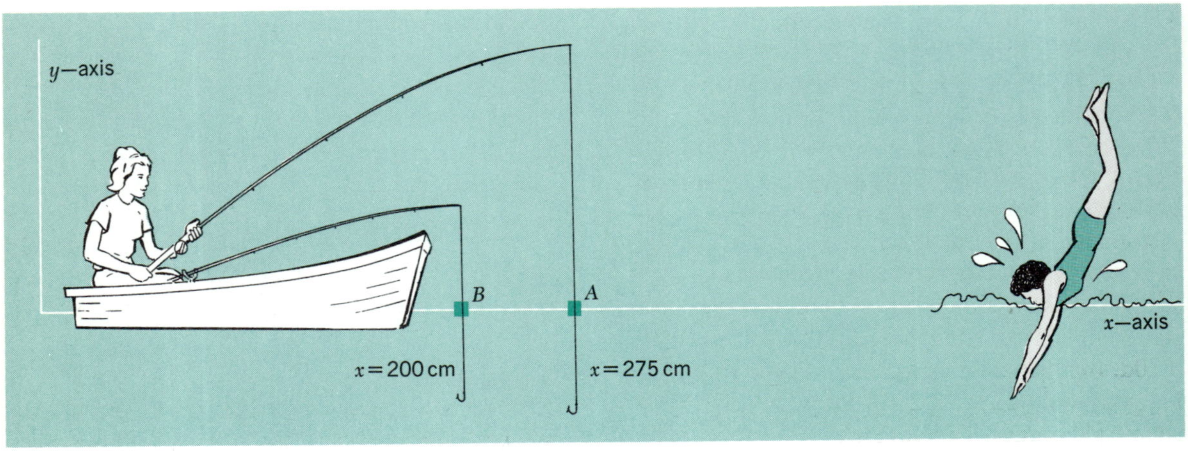

Figure 9-9

and does not know that the swimmer has dived, she merely notices that the cork is bobbing up and down. Figure 9-10d is a graph of what she sees, the displacement y as a function of the time t at the fixed position A, $x = 275$ cm. This kind of graph is called a *vibration graph* of the motion at a point. The cork remains stationary for 11.0 seconds until the leading edge (a in Figures 9-10 a, b, c) hits it. Then the cork rises up a bit, wiggles up and down twice, and remains at the high level for 0.1 second until the plateau cd has passed it. Next, the cork bobs down to a negative displacement as the trough e passes by, and finally, after a momentary positive value of y, the cork returns to zero level. The zero level is the "undisturbed" position, and as far as this graph is concerned cork A will remain at that level indefinitely.

For the cork B at $x = 200$ cm, the person fishing observes the same motion, but it starts at a later time. The graph of B's motion is shown in Figure 9-10e. This graph is the same as Figure 9-10d, except that the times are different.

Note that the *shape* of Figure 9-10a is exactly the same as the shape of Figure 9-10d, even though these graphs have different meanings. It is worth thinking a bit about this similarity: If the flash photograph reveals some characteristic kink in the wave form, that same kink will show up in the vibration graph at a given point. This is because the waves are moving at a steady speed and, therefore, a time interval can be thought of as equivalent to a definite distance.

It is important to note the distinction between a vibration and a wave. A *vibration* is a motion or disturbance of a single particle of some medium. On the other hand, a *wave* transmits a disturbance from place to place by coupling together the vibrations of many neighboring particles.

You studied vibrations in the previous chapter and found that some periodic vibrations, such as SHM, are simpler to describe than others, such as the complicated vibration of the cork in Figure 9-10d. In a general sense, even a single pulse that happens only once and is not periodic may be called a vibration. However, the term *wave* can be applied only to a large set of vibrations. It is

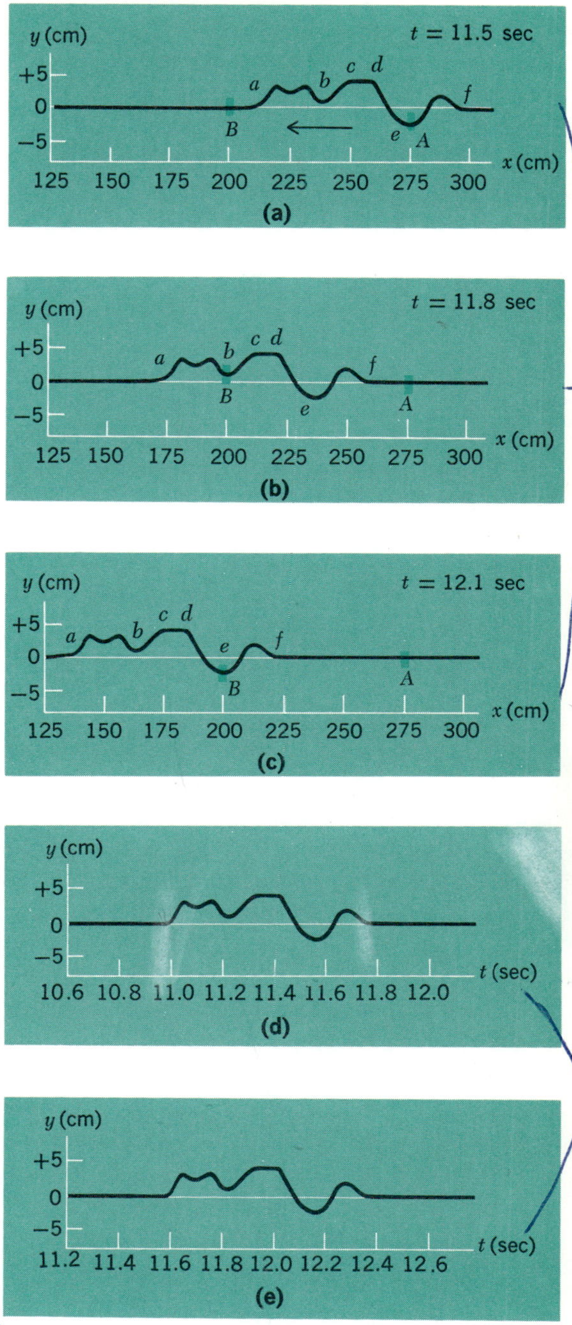

Figure 9-10 Wave form graphs at **(a)** $t = 11.5$ sec. **(b)** $t = 11.8$ sec. **(c)** $t = 12.1$ sec. **(d)** Vibration graph at A. **(e)** Vibration graph at B.

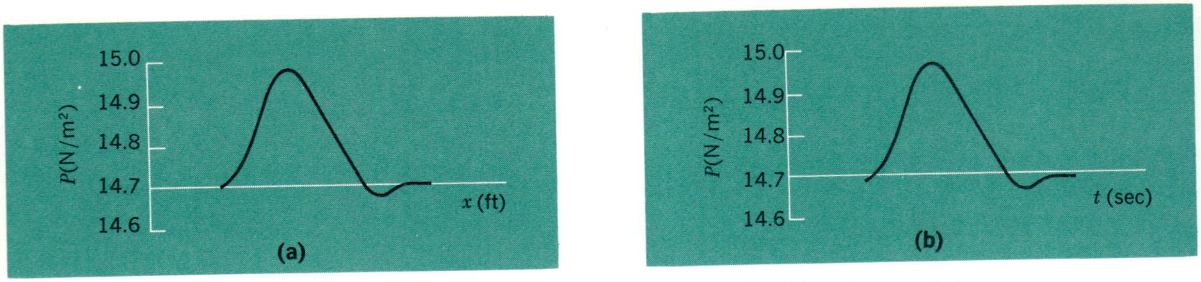

Figure 9-11 **(a)** Wave-form graph for a shock wave. **(b)** Vibration graph for same shock wave.

impossible to describe a wave with any single graph since it is necessary to have wave-form graphs at different positions (as in Figure 9-10a, b, c) or a series of motion picture frames. Or a number of recorders (perhaps corks with pens) making time records (vibration graphs), set up at a sequence of points spaced along the path of the wave, could provide a similar description.

The *wave velocity* is the rate at which any particular part of the wave progresses from place to place. For example, if we fix our attention upon the trough e of the wave form, we can calculate the velocity from the data given in the graphs. From Figure 9-10d, we see that the trough reached cork A at $t = 11.5$ sec and from Figure 9-10e, that it reached cork B at $t = 12.1$ sec. Since A and B are 75 centimeters apart (Figure 9-10a) and since it took the trough 0.6 seconds to go from A to B, the wave velocity is (75 cm)/(0.6 sec) = 125 cm/sec.

It is worthwhile to look briefly at the graphs of longitudinal waves (sound waves) in the same way that we studied transverse waves. The disturbance is the pressure P, rather than the displacement along the y-axis. The pressure is plotted first as a function of position x, and then as a function of time. Figure 9-11 (a and b) represents a "shock wave," or sudden sharp pulse. This could have resulted from a sudden compression, such as the explosion of a cap in a cap pistol, and it could be compared to the pulse made by the swimmer diving into the water. Here again, the vibration graph has the same shape as the wave-form graph. But there is an important distinction between the wave form in Figure 9-11a and that in Figure 9-10a. The water wave-form represented a disturbance that could actually be photographed. The graphs in Figure 9-11 represent the pressure variations that would have to be measured by some sensitive instrument. These pressure variations are associated with tiny displacements of the air molecules, which cannot be photographed in any simple way.

9-5 PERIODIC WAVES

We have not yet discussed waves that repeat themselves regularly, called *periodic*, because the general concept of waves applies equally well to single pulses, or shock waves, which are in some ways easier to think about. However, if the disturbance that forms the wave comes from a vibration that repeats itself regularly (it can be SHM, but does not have to be), the wave is periodic and can be described in new terms—namely *wavelength* and *frequency*.

Let's redraw Figures 9-10a and d, and assume that instead of a diver producing a single complicated splash, some device is periodically producing an up-and-down disturbance of the water.

The wave-form graph (Figure 9-12a) and the vibration graph (Figure 9-12b) are identical in appearance, which is true for any wave. In thinking about Figure 9-12a, it is helpful to view the medium as a set of coupled particles, all at different parts of their own SHM. For example, the particles, P, P', P'', and so forth, are all at the highest points of their motion at the same instant of time ($t = 3.0$ sec). Since they are at the extremes of their SHM, they are momentarily at rest.

Particles R, R', R'', etc., are all at their normal undisturbed positions at this instant. They are at the midpoints of their SHM, and are moving with their maximum speed, all in the same direction. Particles that have the same displacement and are moving in the same direction with the same velocity are said to have the same *phase* of vibration. For instance, in Figure 9-12a particles P and P' are "in phase" with each other, as are particles P and P'', R and R', R and R'', Q and Q', Q and Q''. In this case, P' and Q' are clearly out of phase because their displacements are in opposite directions. Points of zero displacement are out of phase if the particles are moving in opposite directions.

Note that when two vibrating particles are "out of phase," they differ in phase by one half cycle of their vibration. In the case of a wave, points separated by $\frac{1}{2}\lambda$ are out of phase (Figure 9-12a).

There are five quantities that are convenient to use in describing the various aspects of a periodic wave. Some of them have already been discussed in connection with vibration.

(a) The *period* T is the time for one complete vibration of any given particle of the medium.

(b) The *frequency* f is the number of vibrations per unit time for any particle, usually measured as so many vibrations per second. As with any SHM, the frequency is inversely related to the period, that is, $f = 1/T$.

(c) The *amplitude* A is the maximum displacement of any particle, measured from its equilibrium or undisturbed position.

(d) The *wavelength* λ (lambda) is the distance between any two successive particles that have the same phase. For example, in Figure 9-12 the distance from P to P' is λ; the distance from R' to R'' is also λ.

(e) The *wave velocity* v_w is the velocity with which any specific phase of the motion (crest, trough, or compression, etc.) is propagated through the medium.

There is an important relation that holds true for periodic waves of all kinds. Refer to Figure 9-12b and note that during the time of one period T, the point R' moves down and up and makes one complete vibration. In the same time this wave crest has traveled from P' to P'', and the crest that was at P has moved over to P'. That is, the crest moves one wavelength in an amount of time equal to the period of vibration of any individual particle. Thus,

$$v_w = \text{distance/time}$$
$$= \lambda/T$$

Recalling that $1/T = f$, we substitute in the above equation and get a frequently used relation that applies to any periodic wave,

$$v_w = f\lambda$$

wave velocity = frequency × wavelength

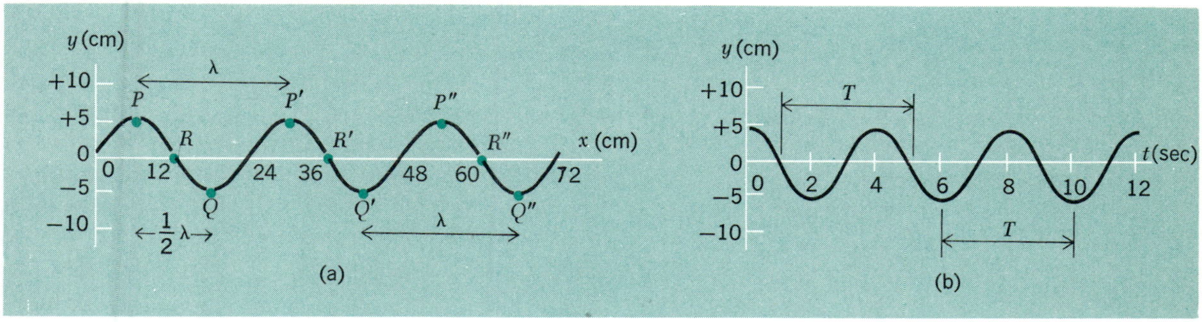

Figure 9-12 **(a)** Wave-form graph of a periodic water wave; snapshot at $t = 3.0$ sec. **(b)** Vibration graph at one point of same periodic water wave; record of the motion at the point R' at $x = 36$ cm.

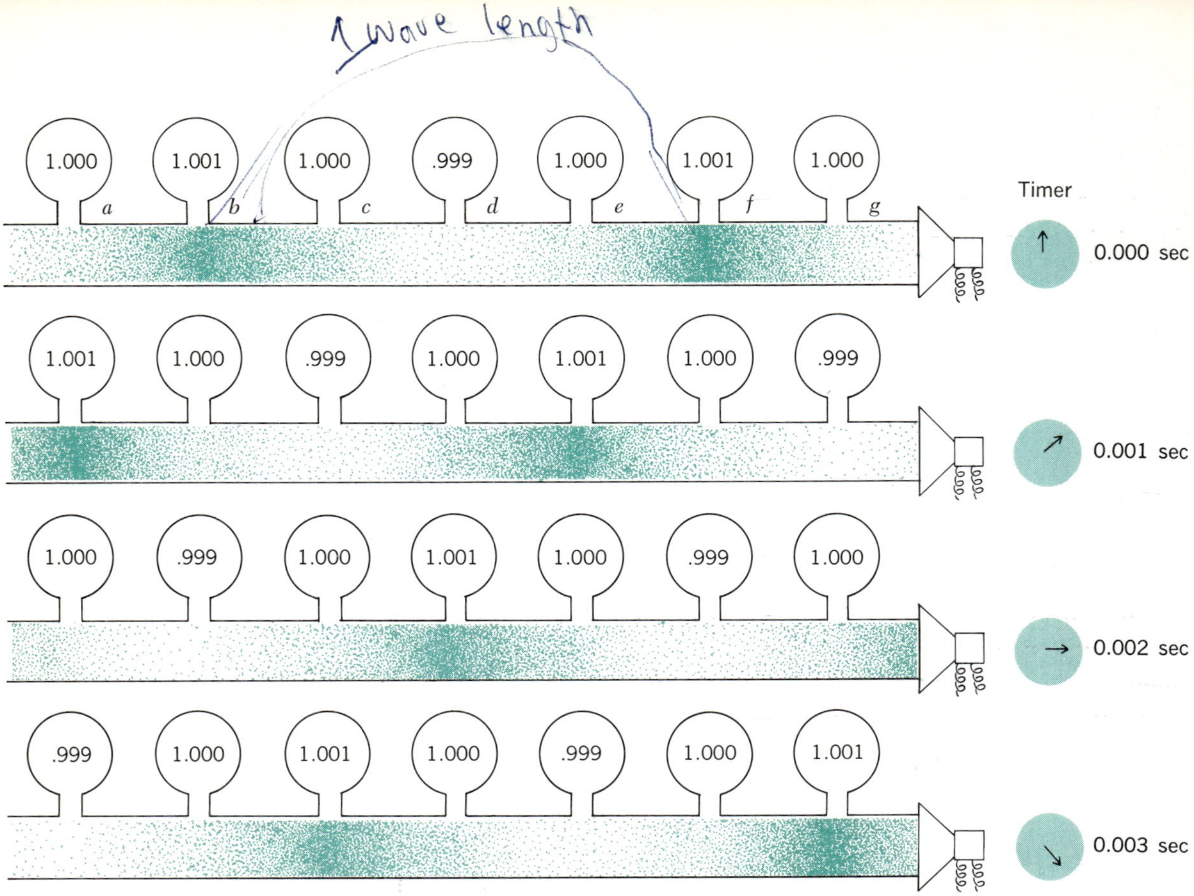

Figure 9-13 Periodic pressure waves.

Example 9-1 The wavelength of sound waves emitted by a whistle is 1.65 meters. If the velocity of sound in air is 330 m/sec, how many times per second does the pressure reach its maximum value (above atmospheric pressure) at a given point?

$$v_w = f\lambda$$

$$f = v_w/\lambda$$

$$= \frac{330 \text{ m/sec}}{1.65 \text{ m}} = 200 \left(\frac{1}{\text{sec}}\right)$$

$$= 200 \text{ sec}^{-1}$$

$$= \boxed{200 \text{ vib/sec}}$$

You can interpret 200 sec^{-1} as 200 "times" *per second,* or 200 vib/sec. "Vibration" is simply a count, it has no dimensions associated with it. One vibration is sometimes called a "cycle," and frequency can be expressed in *cycles per second.* Now it is customary to use the unit *hertz* (Hz) to mean 1 cycle per second. With this unit, a frequency of 1000 cycles/sec would be 1 kHz.

Example 9-2 What is the velocity of ocean waves when you observe that a crest passes your stationary ship every 10 seconds and you judge that the distance between crests is about 150 m? These figures are typical of ocean waves with an amplitude of about 7 m, so the ship is far from "stationary" vertically.

$$v_w = f\lambda$$

$$= \left(\frac{1}{10} \text{ sec}^{-1}\right)\left(150 \text{ m}\right)$$

$$= \boxed{15 \text{ m/sec}}$$

It is not easy to make a helpful picture of a longitudinal wave in terms of pressure differences. It is much more common to show such a wave as a series of regions of compression (in Figure 9-13, the dark areas, where many molecules are squeezed together) and "rarefaction" (lighter areas, which molecules have temporarily deserted). This model represents the notion that a longitudinal wave is transmitted by the motions of molecules bumping into one another as they move back and forth (vibrate) along the direction of propagation.

To view the wave as a series of pressure changes, we can imagine a series of quick-acting pressure gauges spaced at intervals along the path of the wave, perhaps along a pipe in which the wave is traveling. Such quick-acting gauges do exist, and are usually called microphones. A microphone, a familiar accessory with tape recorders, radio transmitters, etc., is a device that gives an electrical signal proportional to a varying pressure.

Suppose a loudspeaker creates periodic pressure variations at one end of a pipe in which the pressure is 1.000 atmospheres (atm) (Figure 9-13). As the wave progresses down the pipe, to the left the reading of each gauge goes through a sequence of values, ranging from say 1.001 atm in the condensation to .999 atm in the rarefaction.*

Gauges *a* and *e*, as you can see from the drawing, are in phase because their readings are always the same at any instant. The wavelength is the distance between any two adjacent *in-phase* pressure gauges, such as *a* and *e*, *b* and *f*, or *c* and *g*. In this case, the amplitude of the wave, expressed as a pressure change, is 0.001 atm, since this is the maximum deviation of the pressure from the normal value.

9-6 THE PRINCIPLE OF SUPERPOSITION

The principle of superposition is an extremely useful idea that tells us how to combine the effects of two or more waves under various conditions.

* As discussed later (page 202), pressure is measured in many different units. One atmosphere is the average, or normal value of atmospheric pressure at sea level.

Suppose a wave *A* creates a certain disturbance at a particular point; and another wave *B* creates a (perhaps different) disturbance at the same point. What is the net effect of the two waves acting together at that point? Perhaps it will not surprise you to learn that the net effect, or the total disturbance, is the algebraic sum of the two disturbances. This means, of course, that the plus and minus signs of the disturbances must be taken into account. The preceding statement summarizes the principle of superposition. But it is important to realize that disturbances combine in this straightforward, "linear" manner only when they are small. It is not an obvious phenomenon of nature that they should combine in this way. In fact, this simple principle does not always apply. For instance, it does not apply when the amplitudes are very large.

For an illustration of superposition, let's imagine waves of compression in an aluminum rod. Suppose a certain wave *A* is capable of producing some pressure P_A at a reference point and another wave *B* would also be able to produce a pressure (P_B) at that point. If both waves act together at the same time will the pressure be $P_A + P_B$? The answer is a qualified "yes," provided we don't go beyond a certain definite pressure. For compression waves having a very large pressure amplitude, the superposition of the two waves might actually deform the shape of the rod permanently or even cause it to break, even though each wave individually would not have this effect. In such a case the total disturbance is not equal to the sum of the separate disturbances.

Fortunately, the pressure amplitudes are very small for ordinary sound waves. It is remarkable that the human ear is such a sensitive instrument that it can detect a vibration of air molecules in which the displacement is a small fraction of the diameter of an atom. Under these circumstances we are indeed justified in using the principle of superposition. Waves on strings (say, a violin string being bowed) also combine this way, unless the amplitude of vibration is excessively large.

As another illustration of the use of the prin-

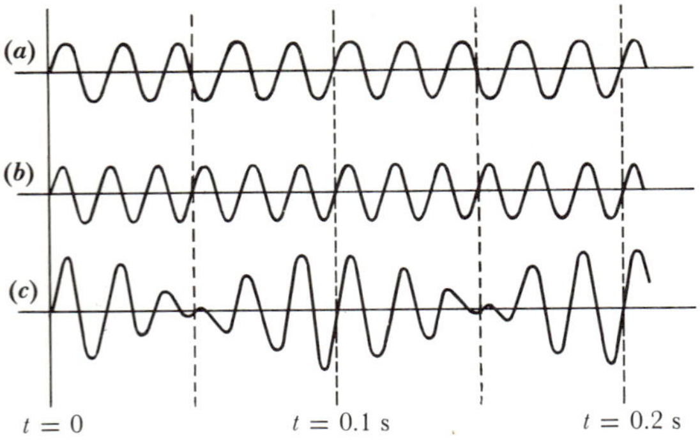

Figure 9-14 Beats between waves of different frequencies. **(a)** 50 vib/sec; **(b)** 60 vib/sec; **(c)** resultant vibration graph showing 10 beats/sec.

ciple of superposition, consider the effect of two periodic waves of different frequencies. We assume that both act on the same point, and in Figure 9-14 we plot the vibration graphs of the two waves and their superposition. The amplitude of the resultant wave rises and falls once during each 0.1 sec interval. These periodic changes in amplitude are called *beats*. The number of beats per second equals the difference between the frequencies of the two combining waves. In the example, waves of frequencies 50 and 60 vib/sec combine; they must be out of phase and cancel each other 10 times per second. The phenomenon of beats extends to all types of waves.

9-7 INTERFERENCE

We have studied wave motion in general, and have used sound waves and ripples on water to illustrate the propagation of a disturbance through a medium, but so far we have assumed that just one disturbance spreads out without meeting any obstacles or being absorbed. Actually, a number of waves will probably be passing through any particular region at one time. How do they interact with one another? Sound waves are reflected off a wall; water waves from several boats may reach a floating log at the same time; waves on a rope sooner or later reach the end of the rope and must be reflected or absorbed. What happens in these situations? The principle of superposition, discussed in the previous section, helps us find the answer.

The term *interference* is used to describe the superposition, or combination, of several waves traveling through the same region. In everyday speech, the word interference seems to imply that two things oppose or cancel one another. This is not the case with waves, for not only can there be destructive interference where the waves do tend to cancel, but we know already that there can also be constructive interference, in which the resultant amplitude is larger than that of either wave alone. Think of the aluminum rod described in Section 9-6.

Later on, when you study wave optics in Chapter 22, you will explore many different aspects of interference with reference to light waves, radio waves, and similar types of electromagnetic waves. For now, let's investigate one special case that is of particular interest as far as waves on strings and sound waves are concerned.

Suppose there are two identical waves—having the same wavelength, amplitude, and speed—that are traveling in exactly opposite directions. They might be sound waves in the air or transverse waves on a stretched string, for example. We can most easily analyze what happens in such a situation by using graphs. Let's consider two transverse waves moving along a string (Figure 9-15). The top drawing shows one wave alone on the string, moving to the right. Just below, in the same position, is the other wave alone, moving to the left. In the next drawing, both waves are shown together. The dotted line represents the wave moving toward the right; the dashed line represents the identical wave moving toward the left. At the instant shown in this graph, which has been called $t = 0$, the two waves are exactly out of phase. At the point P, for instance, the dotted-line wave tends to move the string down, and the dashed-line wave tends to move the same particle of string upward. Applying the principle of superposition, we can see that their algebraic sum equals zero, and the net effect is therefore zero

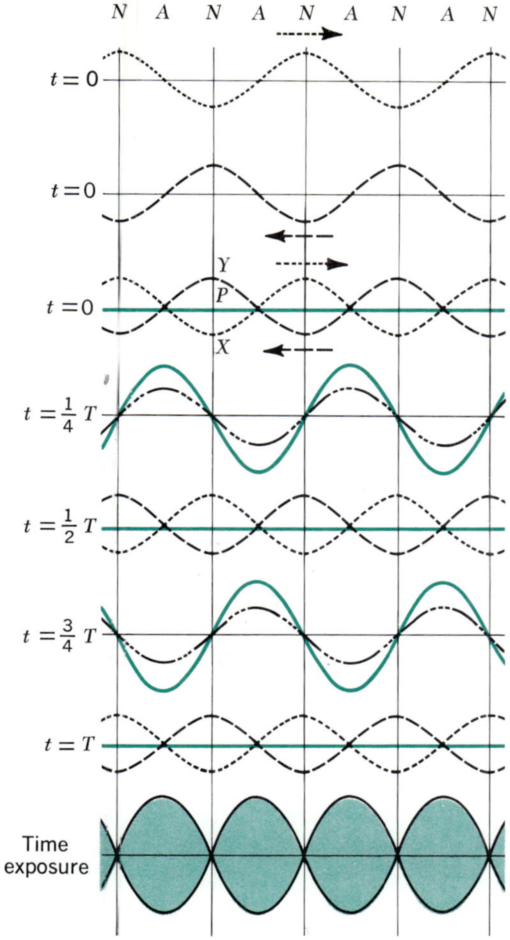

Figure 9-15 Stationary wave on a string, arising from interference of two identical waves traveling in opposite directions.

displacement. The same situation is true for any particle of the string, and therefore at $t = 0$ the entire string is straight—in its natural, undisturbed state. The heavy, colored line shows the superposition of the two waves, and you can see that here it has been drawn horizontally along the axis to indicate zero amplitude.

After a brief interval of time equal to $\frac{1}{4}$ period, each wave has moved $\frac{1}{4}$ wavelength. The dotted-line wave has taken the trough X to the right, and the dashed-line wave has taken the crest Y to the left. The two waves are now exactly in phase, and therefore cooperate to make a king-sized wave form, again shown by the colored line. Once more, it is evident that the superposition principle has been applied.

If this graphical procedure is continued, $\frac{1}{4}$ period at a time; a series of combined wave forms is obtained as shown by the colored lines in Figure 9-15. As you can see from the drawing, at certain times ($t = 0$, $t = \frac{1}{2}T$, $t = T$, etc.) the string is perfectly straight, that is, there is no net disturbance anywhere along its length. At other times ($t = \frac{1}{4}T$, $t = \frac{3}{4}T$, $t = \frac{5}{4}T$, etc.) the string is violently disturbed. Also note carefully that some points, marked N, never move from their normal positions. Such points of no net disturbance are called *nodes*. The points marked A, which have maximum disturbance, are called *antinodes*. It is easy to see from the graphs that nodes are half a wavelength apart and that antinodes are half a wavelength apart.

The whole effect that has just been described is called a *stationary wave*, or a *standing wave*. However, we have seen that this effect was produced by the interference of two waves, and that these waves were not at all stationary. It is the location of the nodes and antinodes that is stationary. This is shown in the final drawing, which is an imaginary time exposure. Here, we are not taking a flash photograph at a given instant, but are viewing the string as it looks over an extended time, such as several periods of the vibration. The waves are given a color tint here to emphasize the antinodes. The nodes are also clearly visible as fixed points along the string.

Although waves on a string were used for this example, periodic waves of any sort can combine to give stationary waves. One good place to look for them is at the edge of a body of water, where small ripples are reflected by a sharp obstacle such as a stone wall. Notice what happens when the incoming waves interfere with the reflected waves. The same effect takes place with waves of compression, light waves, and other periodic waves.

SUMMARY

The essential feature of all **waves** is that they are the propagation of some kind of disturbance, without the actual movement of matter from the point of origin to the place where the waves are received. Both energy and information can be carried by waves. There are many different kinds of disturbances; they may be mechanical, electrical, thermal, or even nonphysical, like the "wave of enthusiasm" mentioned in Section 9-1.

Compression waves in solids can be modeled by a series of small, heavy balls coupled by springs. The speed of a wave in such a medium increases with the stiffness of the springs and decreases with the mass of the balls.

If the disturbance is parallel to the direction of propagation, as in **sound waves** and other compressional waves, then the wave is said to be a **longitudinal wave.** In a **transverse wave** the disturbance is at right angles to the direction of propagation. Waves on strings and ropes as well as light waves are examples of transverse waves. Ripples on the surface of a liquid are approximately transverse. In the case of transverse waves, if all the disturbances take place parallel to each other, then the wave is said to be **plane-polarized.** Longitudinal waves cannot be polarized.

It is important to distinguish between the concepts of a **wave** and a **vibration.** A vibration refers to the displacement of a particle back and forth with time at a fixed position. A wave consists of the vibrations of many particles coupled together with a definite phase relation between neighboring particles. It is often helpful to show certain essential features of a vibration or a wave with graphs. Two graphs are required to represent a wave: (1) the vibration graph at any given point, to show how the disturbance changes with time; and (2) the waveform graph of the whole wave, to show how the disturbance changes with position. The two graphs for a given wave always have the same shape, even though they are plotted with different variables. The **wave velocity** refers to the rate at which some definite feature (such as a crest) of a wave moves from place to place.

Periodic waves are those that result from a regularly repeated series of disturbances, such as those caused by a motor driven "rippler" bobbing up and down on the surface of the water. In a wave, all those particles that have the same displacement and are moving in the same direction are said to be in **phase.** The **wavelength** is the shortest distance between two particles having the same phase of vibration. The **period** is the time required for one complete vibration of any particle. The **frequency,** the reciprocal of the period, is the number of vibrations per unit time, and is expressed as so many "cycles per second," or hertz. For all periodic waves, of whatever kind, wave velocity = frequency × wavelength. The **amplitude** of a wave is the maximum value of the disturbance.

The **principle of superposition** states that the combined effect of two waves is the algebraic sum of the effects of the two waves. This is so only if the amplitudes are not so great that the medium is deformed and unable to recover its regular shape. *Beats* result from the superposition of two periodic waves of different frequencies. When two identical waves, that is, waves of the same amplitude, frequency, wavelength, and speed travel in opposite directions through a medium, the superposition principle enables us to find the net disturbance at any point. The result is a stationary, or standing, wave. At certain points in the stationary wave, called **nodes,** the medium is always at rest. At other points, called **antinodes,** the medium undergoes vibration with amplitude equal to the sum of the amplitudes of the two waves. The distance from node to node is $\frac{1}{2}\lambda$; from antinode to antinode is $\frac{1}{2}\lambda$; and from node to antinode is $\frac{1}{4}\lambda$.

QUESTIONS

1 Give three examples of disturbances that can be propagated in a wave motion.

2 List three ways you could attract somebody's attention if he were in a second story room and you were on the ground outside.

3 Explain how the thermostat in your house illustrates the principle of feedback.

4 Does every wave have a wavelength? Does every wave have a wave velocity?

5 While a sound wave is moving past you in the air, is there any change in the density of the air?

6 A man is fishing from the end of a pier in a dense fog and all that he can see is the motion of the cork on his own fishline. The cork starts to bob up and down. Can the man be sure that a wave is moving past the cork's position?

7 In what sort of unit might you measure the amplitude of the wave of enthusiasm mentioned in Section 9-1? In what unit would you measure the wave velocity of this wave? How would you produce a periodic wave of enthusiasm? Could such a wave have a wavelength?

8 When a flag waves in the breeze, is this a real wave in the sense that was discussed in Section 9-4? If so, what is the disturbance? What is the medium? Approximately what is the period?

9 Two waves of equal frequency, wavelength, velocity, and amplitude are traveling in the same direction through a medium. A student says that the superposition principle predicts that the amplitude of the resultant wave is necessarily twice that of the individual waves. Is he correct? Explain.

10 Explain briefly the difference between a vibration and a wave.

11 What are the conditions necessary for the production of standing waves?

12 A glass of water is on a kitchen table, and the floor of the room vibrates because of a laundry dryer in action. For certain speeds of rotation of the dryer, stationary ripples are observed on the surface of the water. Explain.

13 A high-speed motion picture is taken of the water surface near a concrete wall at the shore of a lake. Would you expect any frame of the film to show the water surface absolutely flat?

PROBLEMS

GROUP A

1 What is the frequency of the compression (longitudinal) wave shown in Figure 9-13?

2 What is the wavelength of the compression waves in Figure 9-13 if the microphones are spaced at 2-meter intervals?

3 What is the distance between rarefactions in a sound wave whose velocity is 332 m/sec and whose frequency is 120 vib/sec?

4 How often does a compression pass an observer who is listening to a sound of wavelength 0.7 m and speed 338 m/sec?

5 What is the frequency, in vib/min, of the wave of Figure 9-12?

6 What is the shortest wavelength of sound that can be heard by a human being, if the frequencies of audible sound lie between 20 Hz and 20,000 Hz?

7 What is the wavelength of waves broadcast by a TV station whose frequency is 60 MHz? Express your answer in meters. (M = mega = 10^6)

8 At what time does cork A (Figure 9-10a) start moving? At what time does cork B start moving? In what direction (up or down) is cork A moving at $t = 11.4$ sec?

GROUP B

9 A 100-car train standing on a siding is started in motion by a locomotive. If there are 5 cm of slack in the couplings between cars, and the locomotive moves at a constant speed of 0.3 m/sec, how much time is required for the pulse to travel the length of the train?

10 The points R, R', and R'' are all in phase (Figure 9-12a). Are they moving upward or downward at $t = 3.0$ sec?

11 In which direction (to the right or the left) is the wave in Figure 9-12a traveling?

12 What is the wave velocity of the wave in Figure 9-12?

13 What is the velocity of the compression wave in Figure 9-13 if the microphones are spaced at 2-meter intervals?

14 A sound wave of velocity 600 m/sec has a wave length of 300 cm. At a certain instant one of the molecules of the medium is at its normal, undisturbed position. How long will it be before this same molecule is again at its normal position?

15 Many species of bats are known to find their prey (small insects) by a kind of sonar using short bursts of ultrasonic waves (above the limits of human

2. $\lambda = \frac{V}{f} =$

3. $V = 332\ m/s$
$f = 120\ vib/sec$

$\lambda = \frac{332\ m/s}{120\ vib/sec} =$ **2.76 m**

4. $\lambda = .67m$
$V = 338\ m/s$

$f = \frac{V}{\lambda} = \frac{338\ m/s}{.7m} = 482\ Hz$

*? 5. $\lambda = 24cm$
$V = 4$
$T = 4sec$
$f =$

$f = \frac{V_w}{\lambda}$

*? 6. $\lambda = \frac{V_s}{f} =$

16, $f = .33\ s^{-1}$

7. $W = \frac{V}{f} = \frac{3 \times 10^8 \text{ m/s}}{60 \text{ Hz}} = 5.0 \times 10^6 \text{ m}$

** ?9 $d = (5 \text{ cm})(100) = 5 \text{ m}$ $t = \frac{V}{d} = \frac{.3 \text{ m/s}}{.05 \text{ m}} =$ 6s $\lambda = .05 \text{ m}$

$V = .3 \text{ m/s}$ $V = .3 \text{ m/s}$

$V = \frac{\lambda}{T} \Rightarrow T = \frac{\lambda}{T} =$

?14. $f = \frac{V}{\lambda} = \frac{600 \text{ m/s}}{3 \text{ m}} =$ 2000 s^{-1}

$T = \frac{\lambda}{V} = \frac{3.0 \text{ m}}{600 \text{ m/s}} =$

$T = .0005 \text{ sec}$

15. $W = \frac{V}{f} = 3$ $\lambda = \frac{V}{f} = \frac{330 \text{ m/s}}{50000 \text{ Hz}} =$ .0066 m

Physics Problems

1. $\Delta P = .002$ $\quad V =$ $\quad F = \frac{V}{W}$

2. $V = WF$

   $t = .3s = 123\ cm/s$

3. $W = \frac{V}{f} = \underline{332m}$

4.

~~f. 1.~~ $V = f\lambda \Rightarrow f = \frac{V}{\lambda} =$

hearing). Find the wavelength if the frequency is 50 kHz and the speed of sound is about 330 m/sec.

16 Periodic SHM waves spread out over the surface of a lake where two men are fishing, 22 meters apart. Each man's boat bobs up and down 20 times per minute. At a time when F's boat is at its crest, G's boat is at its lowest point, and there is one additional crest somewhere between the two men. What is the speed of the water waves?

FOR FURTHER STUDY

THE DOPPLER EFFECT

When a car goes by fairly fast, blowing its horn, you have probably noticed that there is a peculiar change in the pitch of the horn. This effect is even more pronounced when a high-speed train goes by, sounding its horn continually. These are not "illusions" but a measurable phenomenon called the *Doppler effect.*

Named after the German physicist Christian Doppler (1803–1853), the Doppler effect occurs with all types of wave motion and has to do with the apparent change in wave frequency due to the relative motion of the source of the waves and the observer or "listener." Whether it is the source, the observer, or both that are moving, the effect of relative motion is that the apparent frequency becomes greater (producing a higher pitch in the case of sound) if the distance between the source and observer is decreasing. On the other hand, the apparent frequency is lowered if the source and the observer are moving away from each other. When a car sounding its horn moves quickly away from you the pitch is lower. But the exact relation for sound waves depends upon whether it is the source or the listener that moves. Instead of learning some general formulas, let's study the following solved problems which involve each of the cases just mentioned using the physics you already know. Finally, you will arrive at a simple formula that gives a good approximate answer for all cases—whether it is the source, the observer, or both, that are in motion.

Example 9-3 A train is moving away from a station platform at 33 m/sec, blowing a whistle whose frequency as measured on the train is 300 Hz. The speed of sound in air is about 330 m/sec. What is the apparent frequency of the whistle as measured by an observer on the platform?

Since it is the train that is moving, the sound, once emitted, travels through the air with the usual velocity of 330 m/sec. We can think of a sound wave as a series of pressure pulses in the air. With a frequency of 300 Hz, the whistle emits 300 pulses in each second. However, since the source is moving away from the observer, the wavelength is increased (Figure 9-16). During 1 second as the whistle emits 300 pulses, the train

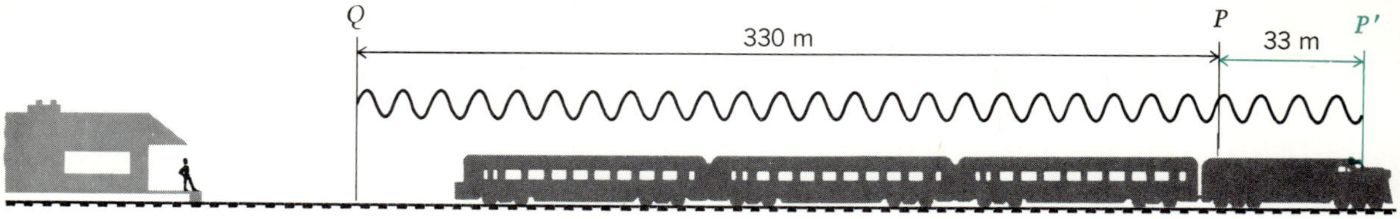

Figure 9-16 Doppler effect, source in motion.

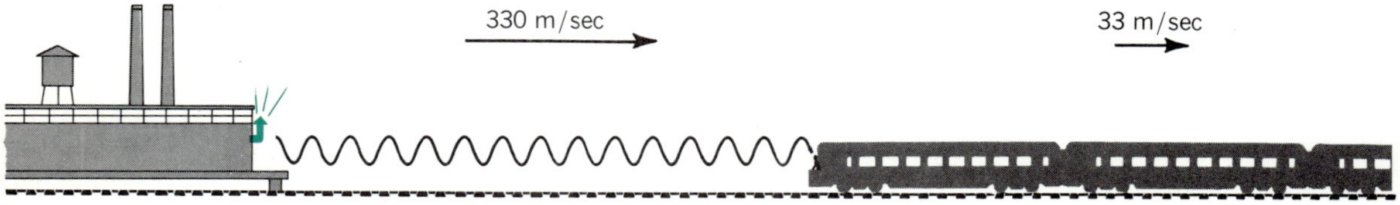

Figure 9-17 Doppler effect, observer in motion.

moves 33 m. The last pulse is emitted at P' the same instant that the first of the 300 pulses reaches the point Q. The wavelength is "stretched out"; there are 300 pulses in (330 + 33) m = 363 m, and so the new wavelength is

$$\lambda' = 363 \text{ m}/300$$
$$= 1.21 \text{ m}$$

The apparent frequency f' is

$$f' = v_w/\lambda'$$
$$= \frac{330 \text{ m/sec}}{1.21 \text{ m}}$$
$$= 272.7 \text{ sec}^{-1}$$
$$= \boxed{272.7 \text{ Hz}}$$

This example shows that, if the *source* is moving, the wavelength is altered but the wave velocity remains the same. On the other hand, if the observer moves, the wavelength remains the same, while the wave velocity is changed.

Example 9-4 Solve Example 9-3, assuming that the whistle is stationary and the observer is moving away from it at 33 m/sec.

Here the wavelength is unaltered because the source has no motion with respect to the medium. However, the observer is moving away from the source (Figure 9-17) and therefore the pressure pulses are not catching up to him as fast as if he were stationary. The wavelength is

$$\lambda = \frac{330 \text{ m/sec}}{300 \text{ vib/sec}}$$
$$= 1.1 \text{ m}$$

However, the apparent wave velocity v_w' is

$$v_w' = (330 - 33) \text{ m/sec}$$
$$= 297 \text{ m/sec}$$

Hence,

$$f' = v_w'/\lambda$$
$$= \frac{297 \text{ m/sec}}{1.1 \text{ m}}$$
$$= 270.0 \text{ sec}^{-1}$$
$$= \boxed{270.0 \text{ Hz}}$$

The Doppler effect can be summarized for sound waves in a simple way. The observed changes in frequency depend on v/v_w, the ratio of the relative velocity to the wave velocity. If the source moves, the *wavelength* changes by a fraction equal to v/v_w. If the observer moves, the *frequency* changes by a fraction equal to v/v_w. In both cases the observer will measure changes in the frequency with respect to some known "true" frequency. In many cases, if v/v_w is small compared with 1, it makes no difference which method is used. For instance, the answers to Examples 9-3 and 9-4 differ by only 1 percent. Figure 9-18 shows a photograph of water waves in a ripple tank in which there is relative motion between the source of the waves and the tank itself. Can you tell which, if either, is at rest?

According to Einstein's first postulate (Section 5-8), only relative motion can be observed. It is not possible to do any experiment that shows that one object is "really" in motion and another is "really" at rest. The examples you have just studied seem to violate the theory of relativity, for it seems as if we could determine which of two bodies (source or observer) is "actually" in motion. For instance, in comparing Example 9-3 with 9-4, suppose we know the relative velocity is 33 m/sec. If the apparent frequency is measured and found to be 272.7 Hz, we might conclude that the source is moving relative to

Figure 9-18

the air. If the frequency is found to be 270.0 Hz, we might conclude that the observer is moving relative to the air. Thus in this case, it seems possible for us to be able to decide which body is in motion.

We seem to have a paradox, or a conflict, between the theory and the apparently correct physical reasoning we just went through. To resolve the paradox, we should note that for sound waves there is a medium (air) that can serve as a reference with which to measure "real" velocities. For light waves there is no such medium. The Doppler effect for light waves is calculated in the relativity theory, and, as expected, only the relative velocity of source and observer enters into the equation.

Often it is convenient to use a simple approximate expression to calculate the frequency change. Exact expressions have been derived for two different situations, but the approximation is good enough when the frequency change is only a small fraction of the original frequency. This expression says that the fractional change in frequency $\Delta f/f$ is given by

$$\Delta f/f = v/v_w$$

where v is the *relative* velocity of source and observer, that is, the speed with which they are approaching or receding from one another. Note that this formula works if either source, observer, or both are in motion. The velocity v refers only to relative motion. The approximation gets better as v/v_w gets smaller. Let's use the approximation to solve a problem involving radio waves where the wave velocity v_w equals c the velocity of light.

Example 9-5 A jet plane approaches a control tower at 600 m/sec, emitting an electromagnetic wave of frequency 1000 megahertz. What is the apparent frequency as observed at the control tower?

The wave velocity is that of light and all other electromagnetic waves, 3.00×10^8 m/sec. This is so much greater than the relative velocity of 600 m/sec that we are quite safe in using the approximate equation for frequency shift. In other words, v/v_w is small compared with 1.

$$\begin{aligned}\Delta f/f &= v/v_w \\ &= 6 \times 10^2 \text{ m/sec} \big/ 3 \times 10^8 \text{ m/sec} \\ &= 2 \times 10^{-6} \\ \Delta f &= (2 \times 10^{-6})(10^3 \text{ MHz}) \\ &= 0.002 \text{ MHz}\end{aligned}$$

We know that the frequency is *increased*, since the source is approaching the observer. Therefore we *add* Δf to the true frequency to get the apparent frequency f'.

$$\begin{aligned}f' &= f + \Delta f \\ &= 1000 \text{ MHz} + 0.002 \text{ MHz} \\ &= \boxed{1000.002 \text{ MHz}}\end{aligned}$$

The so-called radar speed traps work on the principles outlined above. In such cases the policeman's radar set emits a wave of known frequency and the speed of the moving car is calculated from the change in frequency Δf observed in the "echo" wave reflected from the car.

PROBLEMS

GROUP C

17 A student walks at a speed of 1 m/sec. toward a stationary man. If the student is whistling a note of frequency 550 Hz, what is the apparent frequency heard by the man? (Assume the speed of sound to be 330 m/sec.)

18 (a) With what speed should an observer in a plane fly toward a stationary source of sound whose frequency is 1000 Hz so that the apparent frequency becomes 1500 Hz? (Assume speed of sound to be 330 m/sec.) (b) If the source of 1000 Hz is on the plane, with what speed should the plane fly toward a stationary listener so that the apparent frequency becomes 1500 Hz?

19 A listener and a source of sound waves are fixed relative to each other. If a strong wind blows toward the listener, (a) is the wavelength changed? (b) is the speed changed? (c) is the frequency changed?

UNIT 3

References and Outside Reading

The purpose of the following list is to make you aware of additional reading that will extend and broaden most of the concepts you have met in this unit. These references should be easily available and are of sufficient variety to meet your individual interest.

Analytical Experimental Physics, by Ference, Lemon, and Stephenson, The University of Chicago Press, Chicago, Ill., 1957. Chapter 33 is a fairly rigorous treatment of the science of musical sound. For those interested in music and the physics of music, this would be a good place to start.

"Architectural Acoustics," by V. O. Knudsen, *Scientific American,* November, 1963 (page 78).

Echoes of Bats and Men, by D. Griffin, Science Study Series, Garden City, N.Y., Doubleday Anchor, 1959. Applications of sound reflections in nature and industry.

Horns, Strings, and Harmony, by A. H. Benade, Science Study Series, Garden City, N.Y., Doubleday Anchor, 1960. Basic wave theory as it applies to musical instruments.

Moments of Discovery, by Christiaan Huygens, edited by Schwartz and Bishop, Basic Books Inc., New York, N.Y., 1958. Volume 1 of this series has the original account of Huygens' work on his wave theory.

"The Motion of the Ground in Earthquakes," by D. M. Boore, in *Scientific American,* December 1977. A discussion of seismic waves in the earth.

Perpetual Motion, by A. Stewart, Science Study Series, Garden City, N.Y., Doubleday Anchor, 1965. Chapters 6 and 7 offer a fairly rigorous treatment of atomic oscillations and vibrations.

The Physics of Music, Introduction by C. M. Hutchins, W. H. Freeman and Company, 1978. A collection of excellent articles on various aspects of sound and music.

"Shock Waves," by O. Laporte, *Scientific American,* November, 1949 (page 14).

Sound Waves and Light Waves, by W. E. Kock, Science Study Series, Garden City, N.Y., Doubleday Anchor, 1965. Approach to the fundamentals of wave motion.

A Source Book in Physics, by W. F. Magie, Cambridge, Mass., Harvard University Press, 1963. See Galileo's discussion of the pendulum.

"Tsunamis," by J. Bernstein in *Scientific American,* August, 1954 (page 60). A description of water waves in the Pacific Ocean having wavelengths of from 100 to 600 mi and periods of vibration from $\frac{1}{4}$ to 1 h. See also the letter by W. G. Van Dorn, *Scientific American,* January, 1955 (page 2). A related phenomenon, the exceptionally high tides in the Bay of Fundy, is discussed briefly in the *National Geographic Magazine* August, 1957 (page 156). This is a resonance phenomenon involving both a funnel-shaped ocean bottom and the production of stationary water waves whose antinodes are near the upper reaches of the Bay.

Waves and Beaches, by W. Bascom, Science Study Series, Garden City, N.Y., Doubleday Anchor, 1964. A fascinating account of the dynamics of the ocean surface.

Waves and The Ear, by van Bergeijk, Pierce, and David, Science Study Series, Garden City, N.Y., Doubleday Anchor, 1960. Chapter 3 pertains to the subject of this unit.

The next three references in this list apply to specific groups of instruments.

"The Physics of the Piano," by E. D. Blackman, *Scientific American,* December, 1965 (page 88).

"The Physics of Violins," by C. M. Hutchins, *Scientific American,* November, 1962 (page 79).

"The Physics of Wood Winds," by A. H. Benade, *Scientific American,* 1960 (page 145).

UNIT 4 HEAT AND ENERGY

Up to now, the law of conservation of energy has been restricted to mechanical systems where friction was assumed to be negligible.

Clearly, a world without friction is a make-believe world. So let's look now at the real world where friction is ever-present, a world where the atoms themselves contribute to the energy of the system by virtue of their motion.

The conservation of energy is really a matter of bookkeeping, and we now add one more page to the ledger by investigating the concept of thermal energy. A book slides across the table. Its kinetic energy has disappeared, but both book and table have increased their thermal energy. On the other hand, thermal energy can be changed into other kinds of energy—as in the novel structure shown, which may one day use the thermal energy of the ocean to supply electricity to a city of 10,000 residents.

We shall begin our study of thermal energy by investigating the concepts of temperature and heat.

10 Temperature and Expansion

A golf ball rolls across the green and comes to a stop just two centimeters short of the cup. In our study of mechanics we assumed, without much quantitative evidence, that such losses of mechanical energy due to friction become gains of thermal energy. Much of the next four chapters can be considered as an extension of the law of conservation of energy to include thermal energy. When the mechanical energy of a system disappears because of friction, the system warms up. We shall see that the energy of the system is simply being transferred from one form to another, and that the total energy of the system is conserved. In this unit we shall develop a *theory of heat,* which can be summarized by two important generalizations called the first and second laws of thermodynamics. To do this, it is first necessary to find out how temperature and thermal energy are measured.

10-1 TEMPERATURE

The sensations of "hot" and "cold" are among the first experiences of a newborn baby. Every known primitive language has words for hot, warm, cool, and cold. These four words apply to *temperature* —a term familiar to all of us. But if we are to work with temperature, we need more than just a general idea of what it means.

It is tempting to think of molecular motion as a fundamental property by which temperature can be defined, since most of the phenomena of heat are associated with the random motions of atoms or molecules. For instance, when a car remains in direct sunlight for a while, the iron atoms in the fender gain kinetic energy and move faster than before. The faster-moving atoms, acting against the forces of cohesion, will be able to vibrate in slightly larger paths, and so the iron expands. If an observer touches the hot metal, KE is transferred to his finger by a collision process, and the sensation of "hot" is produced in his brain. The pressure exerted by the air in a tire is due to the impact of molecules upon the walls of the tire; at higher temperatures the molecules move faster and the pressure is greater.

However, we do not define temperature in terms of molecular motion. The discussion of thermodynamics that comes later on (Section 13-6) makes it clear that mere absence of random molecular motion is no guarantee that a body possesses no heat or is at a temperature of absolute zero. There may be other forms of available energy (such as PE) remaining in the atoms, even though at ordinary temperatures these other forms of energy are negligible in comparison with the random KE of the molecules.

It is best not to attempt a complete definition of

temperature until Chapter 13, when a definition that will stand up later in your work will be presented. For the present, it is enough to say that many definitions are *possible,* if only because many types of thermometers have been invented. For ordinary work, common thermometers agree well enough with each other (and with the thermodynamic scale to be discussed later) that selecting one as a standard is unnecessary.

10-2 TEMPERATURE SCALES

The human senses are not precise enough to determine small changes in temperature. An anxious mother, feeling her child's brow, can distinguish between high fever and no fever at all, but such measurements are not quantitative. The science of medical diagnosis was greatly advanced by the invention of the "thermoscope" by Galileo in about 1593 (Figure 10-1). One of Galileo's students described as follows this first instrument for measuring temperature:

> Galilei took a glass vessel about the size of a hen's egg, fitted to a tube the width of a straw and about two spans long; he heated the glass bulb in his hands and turned the glass upside down so that the tube dipped in water contained in another vessel. As soon as the ball cooled down the water rose in the tube to the height of a span above the level in the vessel. This instrument he used to investigate the degrees of heat and cold.*

Galileo's instrument, crude by today's standards, was based upon the expansion of gases or liquids, and was followed by the development of thermometers that gave quantitative indication of temperature. (Early fever thermometers had no scales; they had to be individually calibrated before each use by making a mark at the point indicated when the bulb was placed in the mouth of a person believed to be in good health.)

Of the many temperature scales later developed at various times and places, only one is used predominantly throughout the world—the *Celsius scale.* Anders Celsius (1701–1744), a Swedish astronomer, suggested the 100-degree, or centigrade, interval between the melting point of ice and the boiling point of water. The Celsius scale is used in most countries and in scientific work the world over. It will be used in the United States when the metric system is in common use. On the Celsius scale, ice melts at 0° C and water boils at 100° C, making a range of 100 Celsius degrees between these points.

The term "centigrade" (originally used for this scale) refers merely to the division of a fixed interval into 100 degrees and could equally well be applied to a scale based upon the boiling point of sulfur as 0° and the melting point of silver as 100°, for instance. The use of water as a standard substance has prevailed, and the scale deserves to be called by the name of Celsius. How fortunate that the initial letter "C" can be used for both Celsius and centigrade!

Figure 10-1 Galileo's open-air thermometer, or thermoscope, as shown in a 19th century wood engraving.

* *A History of Science, Technology and Philosophy in the Sixteenth and Seventeenth Centuries*, by A. Wolf, London, Allen and Unwin, 1935.

Figure 10-2 A series of ceramic prongs of different compositions that have all had the same heat treatment. The degree of softening depends primarily on the kiln temperature.

Any property of a substance that depends regularly on temperature may be used as a temperature indicator. Such a property is called a *thermometric property* of the substance. For example, a yellow-hot piece of metal is hotter than a red-hot piece. Color is therefore a thermometric property, and thermometers can be made that depend on these color changes. The pressure of an enclosed gas also depends on temperature, and gas thermometers can be constructed in which change in temperature is indicated by change in pressure (Section 10-5). The electrical resistance of most metal conductors increases as temperature goes up, and so we have the resistance thermometer, consisting of a coil of platinum wire whose resistance indicates the temperature. An interesting thermometric property is used to determine (roughly) the temperature in a furnace. A series of ceramic prongs of different compositions is placed in the furnace; each one softens at a different temperature, and the furnace temperature is determined (to within a few degrees) by inspection of the row of prongs (Figure 10-2). It is easy to see that many different properties of substances can be used as thermometric properties; but thermometers based on the thermal expansion of solids and liquids are used most often —for reasons of convenience, ease of reading, and cost.

10-3 THERMAL EXPANSION OF SOLIDS

Almost all solids expand when heated. You certainly have noticed the "lines" that are placed every meter or so in a concrete sidewalk. These spaces allow for the expansion of the concrete on hot days so the sidewalk won't buckle. Steel rails on railroads are also placed with a small space between sections to allow for the expansion of the metal. Interlocking bridge joints serve the same purpose (Figure 10-3). Actually, when a body is heated, the expansion takes place independently in each of three directions. In this section we will consider only the *linear expansion* of a solid. Figure 10-4 represents the change in length of a certain brass rod whose length increases by a small amount for every small temperature increase. If the original length of the rod is represented by l, then the change in length will be Δl. Similarly, the change in temperature will be Δt. From the graph, it is easy to see that the change in length of the rod is directly proportional to the change in temperature, that is,

$$\Delta l \propto \Delta t$$

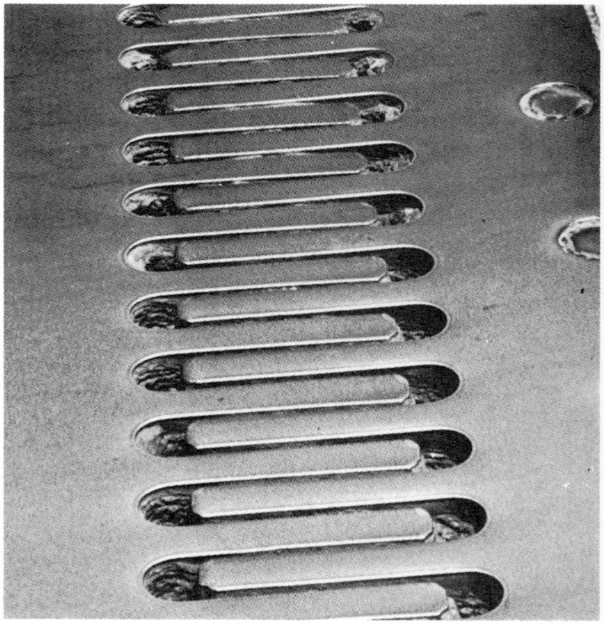

Figure 10-3 Interlocking bridge joint

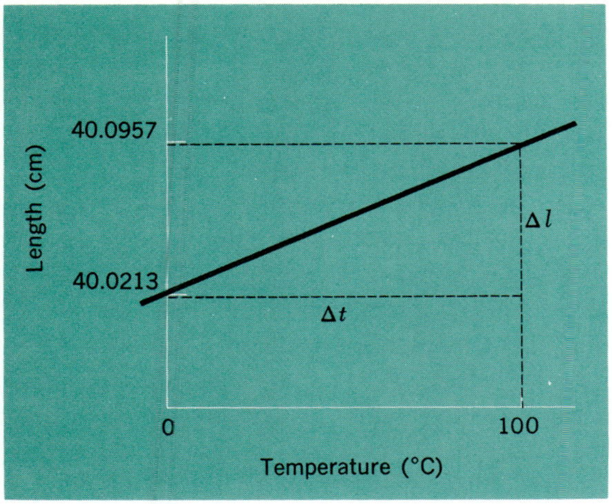

Figure 10-4 Uniform expansion of a certain brass rod.

However, we would expect that the change in length of a long brass rod for a given change in temperature would be more than that of a short rod. This is true because the longer rod contains a greater number of molecules, each of which moves slightly farther from neighboring molecules. Thus the change in length is also proportional to the original length,

$$\Delta l \propto l$$

Putting these observations together, we have

$$\Delta l \propto l\, \Delta t$$

We can now write the basic equation for the expansion of a solid as follows:

$$\Delta l = \alpha l\, \Delta t$$

where α (the Greek letter alpha) is the constant of proportionality, which is called the *coefficient of linear expansion* for the given substance. Different materials have different coefficients of linear expansion. Approximate values of α for various solids are shown in Table 10-1.

This expression gives the change in length of a solid when it experiences a temperature change. When the temperature is increased, the change in length Δl is added to the original length to get the new length. When the temperature is decreased, the rod shrinks by an amount Δl, and this change in length is subtracted from the original length to get the new length.

Table 10-1

Coefficients of Expansion

Substance	*Coefficient of linear expansion* α, $(C°)^{-1}$	*Coefficient of volume expansion* β, $(C°)^{-1}$
SOLIDS		
iron or steel	11×10^{-6}	33×10^{-6}
aluminum	26×10^{-6}	77×10^{-6}
brass	19×10^{-6}	56×10^{-6}
ordinary glass	8.5×10^{-6}	26×10^{-6}
Pyrex glass	3.3×10^{-6}	10×10^{-6}
fused quartz	0.4×10^{-6}	1×10^{-6}
platinum	9.0×10^{-6}	27×10^{-6}
concrete	12×10^{-6}	36×10^{-6}
lead	29×10^{-6}	87×10^{-6}
silver	19×10^{-6}	58×10^{-6}
LIQUIDS		
methyl alcohol		1134×10^{-6}
carbon tetrachloride		581×10^{-6}
glycerin		485×10^{-6}
mercury		182×10^{-6}
turpentine		900×10^{-6}
gasoline		960×10^{-6}
GASES*		
air		3670×10^{-6}
carbon dioxide		3740×10^{-6}
hydrogen		3660×10^{-6}
helium		3665×10^{-6}

* At constant pressure.

Example 10-1 Compute the coefficient of linear expansion for brass, using the data given in Figure 10-4. The original length of the rod was exactly 40 cm.

From the diagram,

$$\Delta l = 40.0957 \text{ cm} - 40.0213 \text{ cm}$$
$$= 0.0744 \text{ cm}$$
$$\Delta l = \alpha l\, \Delta t$$

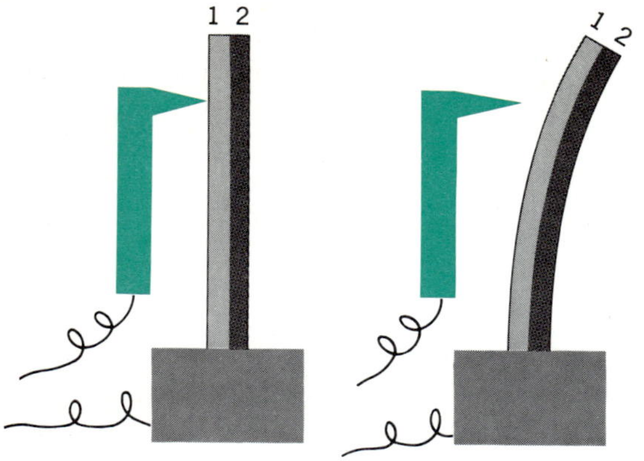

Figure 10-5 Thermostatic relay.

$$\alpha = \frac{\Delta l}{l\,\Delta t}$$
$$= \frac{0.0744\ \text{cm}}{(40\ \text{cm})(100\ \text{C}^\circ)}$$
$$= \boxed{18.6 \times 10^{-6}\,(\text{C}^\circ)^{-1}}$$

The unit for α is simply $1/\text{C}^\circ$ or $(\text{C}^\circ)^{-1}$, which represents the fractional change in length per Celsius degree. In words, a coefficient of $18.6 \times 10^{-6}\ (\text{C}^\circ)^{-1}$ means that any given length of the material changes by 18.6 parts per million of its original length for every Celsius degree change in temperature.

Example 10-2 A piece of steel bridge beam is exactly 50 m long in the summer when the temperature is 35° C. What is its length on a cold winter day when the temperature is −10° C?

Using the basic equation, we have

$$\Delta l = \alpha l\,\Delta t$$
$$= [11 \times 10^{-6}\,(\text{C}^\circ)^{-1}](50\ \text{m})(45\ \text{C}^\circ)$$
$$= 0.025\ \text{m}$$

Because the temperature has decreased, the beam has shrunk, so we subtract this amount from the original length. The new length is

$$50.000\ \text{m} - 0.025\ \text{m} = \boxed{49.975\ \text{m}}$$

The linear expansion of solids is demonstrated in the thermostats that operate furnaces, air conditioners, and the like. To gain sensitivity to small changes in room temperature, thin strips of different metals are welded or riveted together along their lengths, and clamped at one end. When the temperature changes, the combination (often called a bimetallic strip) must bend in an arc, as shown in the schematic diagram of a thermostat (Figure 10-5), where metal *1* has a greater coefficient of expansion than does metal 2. As the temperature rises, the electric contact is opened, turning off the furnace. In a thermometer of this type, a more sensitive indication is achieved by forming the bimetallic strip into a compact spiral that would be many centimeters long if unrolled.

The advantages of low coefficient of expansion are illustrated by the use of Pyrex brand glass for ovenware. Glass is a poor conductor of heat, and so, when placed in an oven, the inner and outer surfaces of a dish become hot before the glass in the interior has a chance to warm up. Because of thermal expansion, stresses are set up that may shatter the glass. One way to avoid this is to heat the glass dish very slowly to allow time for equalization of temperature throughout. Even rapid heating is possible if the glass has a low coefficient of expansion, as does Pyrex, for then the stresses will be small (see Table 10-1). Fused quartz has such a low coefficient of linear expansion that a white-hot piece of quartz at 1500° C may be removed from a flame and plunged into liquid air (−192° C) without shattering.

10-4 THERMAL EXPANSION OF LIQUIDS

When you fill a power lawn mower with gasoline and then run it for a few minutes, you have probably noticed some gasoline spilling out the "air-vent" hole in the gas-tank cap. The heat from the running motor has warmed up the gasoline in the tank and it expands. Like solids, most liquids expand when heated. But the atoms and molecules in a liquid are not tightly bound together as they are in a solid; it is easier for them to move around. Therefore the thermal coefficients of expansion are greater for a liquid than for a solid (see Table 10-1). Since liquids take the shape of their container, we are mostly interested in how the *volume* changes as the temperature is increased or decreased. The *coefficient of volume expansion* is defined as the fractional change in volume per degree change in temperature. Thus an equation similar to that for linear expansion can be used:

$$\Delta V = \beta V \, \Delta t$$

where ΔV is the change in volume, V the original volume, Δt the change in temperature, and β (the Greek letter beta) the coefficient of volume expansion. To find the final volume, the amount ΔV is added or subtracted, according to whether the temperature has increased or decreased.

Example 10-3 A quartz beaker is filled with 2000 cubic centimeters of turpentine at a room temperature of 20° C. If the beaker and its contents are put in a refrigerator so that the temperature becomes 5° C, what is the new volume of the turpentine?

The expansion of the quartz beaker can be ignored, since its coefficient is so small. In this problem Δt is 15 C°. From Table 10-1, β is 900×10^{-6} $(\text{C}°)^{-1}$. Hence,

$$\begin{aligned} \Delta V &= \beta V \, \Delta t \\ &= [900 \times 10^{-6} \, (\text{C}°)^{-1}](2000 \text{ cm}^3)(15 \text{ C}°) \\ &= 27 \text{ cm}^3 \end{aligned}$$

This amount should be subtracted from the original volume, since the temperature has decreased. Thus the final volume of the turpentine is

$$2000 \text{ cm}^3 - 27 \text{ cm}^3 = \boxed{1973 \text{ cm}^3}$$

The liquid-in-glass thermometer consists of a bulb and an indicating tube of small inside diameter ("bore"). Mercury is commonly used down to temperatures as low as −40° C, its freezing point. Alcohol or propane can be used down to temperatures below −100° C. If the bulb is of large volume and the indicating tube is of very

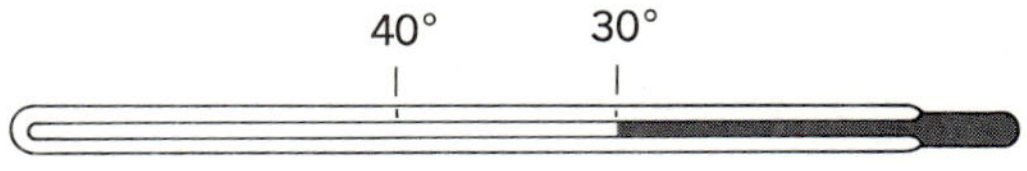

Figure 10-6

small bore, such a thermometer can be made sensitive enough to indicate changes of 0.001° C or less.

Example 10-4 A mercury thermometer (Figure 10-6) has a quartz bulb of volume 0.300 cm^3 and a stem of bore 0.0100 cm. How far does the indicating thread of mercury move when the temperature changes from 30.0° C to 40.0° C? Ignore the expansion of the quartz bulb.

From Table 10-1, the coefficient of volume expansion for mercury is 182×10^{-6} $(\text{C}°)^{-1}$. Hence

$$\begin{aligned} \Delta V &= \beta V \, \Delta t \\ &= [182 \times 10^{-6} \, (\text{C}°)^{-1}](0.300 \text{ cm}^3)(10.0 \text{ C}°) \\ &= 5.5 \times 10^{-4} \text{ cm}^3 \end{aligned}$$

This is the volume of a cylinder of mercury whose diameter is 0.0100 cm. The cross section of the tube is $\pi r^2 = \pi(0.00500 \text{ cm})^2 = 0.785 \times 10^{-4} \text{ cm}^2$. Since the volume of this cylinder of mercury is equal to the change in volume of the mercury (5.5×10^{-4} cm^3), and since volume = cross-sectional area × height, the height of rise is

$$\begin{aligned} h &= \frac{\text{volume}}{\text{area}} \\ &= \frac{5.5 \times 10^{-4} \text{ cm}^3}{0.785 \times 10^{-4} \text{ cm}^2} \\ &= \boxed{7.0 \text{ cm}} \end{aligned}$$

During its construction, the liquid-in-glass thermometer is sealed off at the end of the capillary tube to prevent loss of liquid by evaporation. The upper part is evacuated (Figure 10-7a) in order to prevent the building up of pressure as the mercury expands. Sometimes the space above the liquid is filled with gas, and a little cavity at the very top prevents excessive buildup of pressure

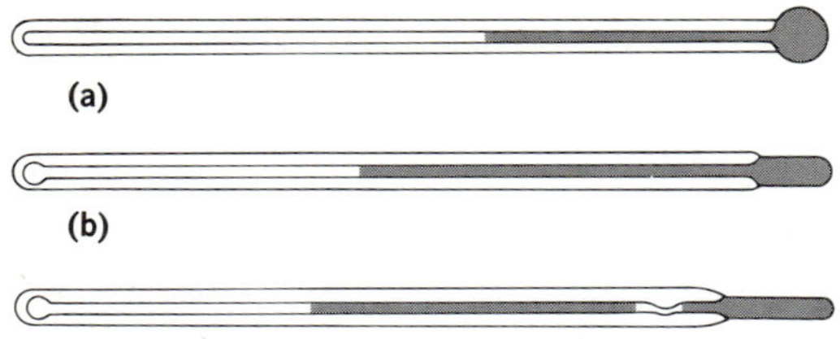

Figure 10-7 Various mercury-in-glass thermometers.

(Figure 10-7b). The clinical thermometer (Figure 10-7c) has a constriction just above the bulb so that the mercury thread is broken as the temperature falls. Surface tension keeps the mercury from flowing back through the narrow opening. In this way the patient's temperature can be read after the thermometer has been removed. Lowering the band of mercury requires hard shaking.

Most liquids expand more or less uniformly, as does mercury (Figure 10-8). However, the expansion of water shows a curious behavior; the volume of 1 gram of pure water varies with temperature as shown in Figure 10-9. The cause of this unusual behavior of water is not fully understood. In the region 0° to 4.0° C, water *contracts* as temperature increases, and the volume coefficient is negative. At the temperature 4.0° C, water has the least volume and maximum density, and the coefficient of expansion is zero. Above 4° C, it expands as the temperature increases and has the normal behavior of other liquids.

Biological life cycles are profoundly affected by the expansion characteristics of water. In a fresh-water lake, for example, as the surface water becomes chilled to 4° C, it becomes denser and sinks, thereby pushing up warmer water from below, which in turn is chilled at the surface. A continual mixing takes place, and the water deep in the lake remains at about 4° C until the colder (and less dense) water at the surface freezes. This is why lakes freeze from the surface down—a fortunate fact for the survival of living organisms in the water.

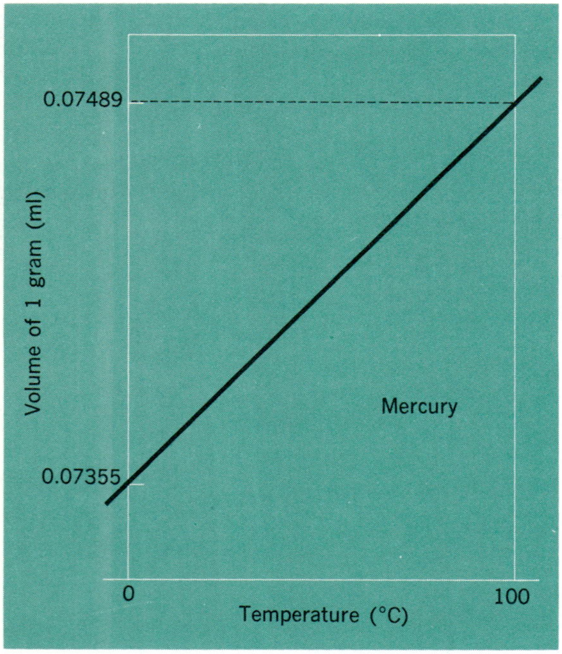

Figure 10-8 Uniform expansion of mercury.

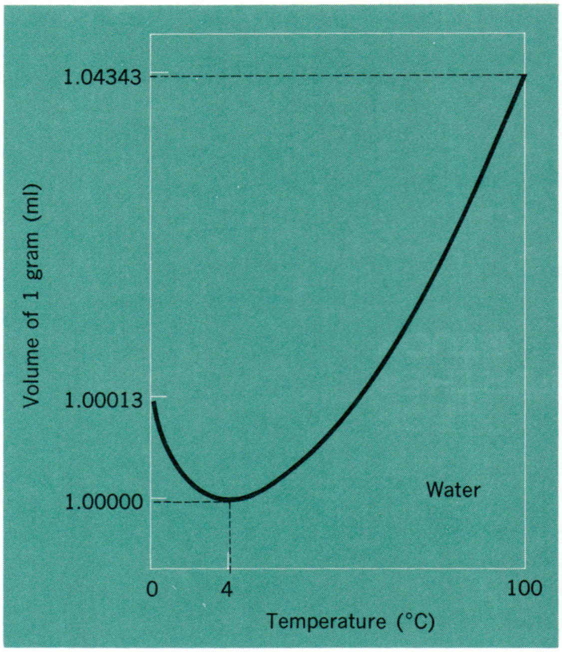

Figure 10-9 Non-uniform expansion of water.

(Scales of both graphs exaggerated for clearness.)

Figure 10-10

10-5 THERMAL EXPANSION OF GASES

You probably already know that an unconfined gas does not need a temperature increase to expand, it will do that anyway. It is a tendency of gas molecules to spread apart, or diffuse, rapidly. As you will see in Chapter 12, a gas will expand if the pressure is lowered, even if the temperature doesn't change. For the present, however, let's assume that the pressure on a confined mass of gas is held constant. This could happen, for example, in a tube of variable length, perhaps a cylinder with a piston and weight at the top that could move up or down as the gas expands or contracts, and thereby maintain a constant pressure on the gas regardless of its volume (Figure 10-10). As the temperature of the gas is increased the confined gas will expand, and lift the piston, making it possible for us to observe the volume change. Since the same weight is still pressing down on the gas, the pressure has remained constant.

A glance at Table 10-1 shows that the coefficients of volume expansion for gases are much larger than for liquids or solids. It is also evident that the coefficients for all gases are practically the same. This is to be expected, since the only differences between the various gases would be in the sizes of the molecules and in their cohesive forces (the forces of attraction between like molecules). Because the gas molecules are so widely separated and are so small, these effects are almost negligible.

In applying the standard expansion equation $\Delta V = \beta V \Delta t$ to gases, we must remember that the coefficient of volume expansion for a gas is measured at constant pressure. We must also be cautious in our choice of initial volume V. In contrast to solids or liquids, the change in volume of a gas may be a large fraction of the initial volume. The values of β listed in Table 10-1 are *based on the original volume at* 0° C. Thus hydrogen expands 0.00366 of its 0° volume for every degree rise in temperature, if the pressure remains constant.

A practical device for measuring such volume changes is the gas thermometer (Figure 10-11), which is capable of high precision and is a basic standard with which other types of thermometers may be compared and calibrated. The pressure of the gas is measured by the difference in the heights of the mercury columns B and C. If the gas container A is heated, the gas expands and pushes the mercury in arm B down and that in arm C upward. Thus the height h changes, and there is a pressure change as well as a volume change. The flexible connecting tubing allows arm C to be lowered, to bring the pressure back to the original value. In this way the change in volume is measured at a constant pressure, and the device

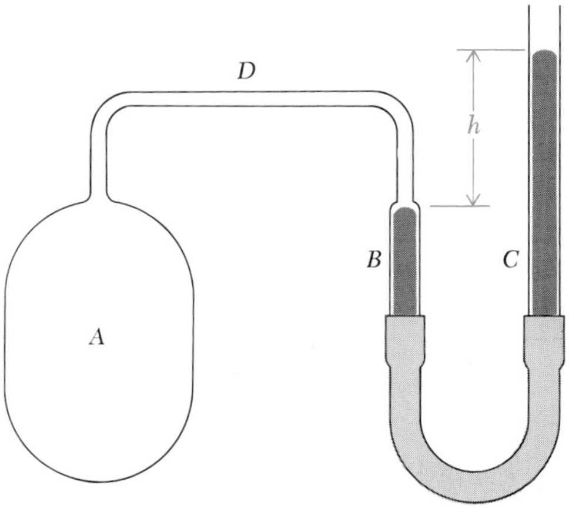

Figure 10-11 Gas thermometer. The flexible rubber tube allows adjustment of arm C so that either the pressure or the volume of the gas can be kept constant.

is often referred to as a *constant-pressure gas thermometer*. The connecting neck D is of small volume compared with the gas container A, but even so, correction must be made because this portion of the gas is not at the same temperature as that in A. The gas thermometer can produce accurate results over a wide range of temperatures, but cannot, of course, be used in extreme conditions, for example below the temperature at which the gas liquefies.

Example 10-5 The volume of the gas in a gas thermometer is 546 cm^3 at 0° C and 746 cm^3 at 100° C. What is the temperature when the volume is 628 cm^3?

The volume increased by 200 cm^3 for a 100 C° change in temperature. Each increase of 1 cm^3 corresponds to a temperature increase of 0.50 C°. The volume is 82 cm^3 greater than at 0° C.

$$t = (82\ \text{cm}^3)(0.50°\ \text{C}/1\ \text{cm}^3)$$

$$= \boxed{41°\ \text{C}}$$

SUMMARY

Temperature is a measure of how hot or cold a body is, as determined by some type of **thermometer.** The fixed points upon which the **Celsius scale** is based are the melting point of ice (0° C) and the boiling point of water (100° C). Various **thermometric properties** of matter may be used to define temperature scales. The most common thermometers make use of expansion of gases and liquids. At ordinary temperatures, such thermometers are in close, though not exact, agreement with each other.

The change in length of a solid is proportional to the original length and to the change in temperature, according to the equation $\Delta l = \alpha l\, \Delta t$. The factor α is the **coefficient of linear expansion.** Similarly, the change in volume of a solid, liquid, or gas is given by $\Delta V = \beta V\, \Delta t$, where β is the **coefficient of volume expansion.** All gases have approximately the same coefficient of volume expansion.

QUESTIONS

1 Name several thermometric properties of matter.

2 In the manufacture of the first light bulbs, the electrical leads were made of platinum. It happens that platinum has a coefficient of expansion nearly equal to that of ordinary glass (Table 10-1). Explain how this helps make a firm glass-metal joint.

3 If the stem of the thermometer described in Example 10-4 were engraved every 0.2° C, how far apart would the markings be?

4 Analyze the act of shaking down a fever thermometer, in terms of mass, forces, and acceleration.

5 Describe qualitatively the construction of a mercury-in-glass thermometer that indicates temperatures between 42° C and 46° C on a scale 20 cm long.

6 Why are the concrete slabs on a highway separated from each other by narrow cracks filled with tar?

7 In the equation $\Delta l = \alpha l \, \Delta t$, which length l should be used—the initial length or the final length? Does it really matter which length is used?

8 Why is it incorrect to say that hydrogen expands 0.00366 of its original volume for every °C rise in temperature?

9 A circular metal hoop is sawed through at one point, leaving a gap of 1.4 mm. If the hoop is then put in an oven and heated uniformly, will the air gap become larger or smaller?

10 Explain how you can loosen the metal cap on a bottle of catsup by holding it briefly in boiling water.

11 What is the approximate temperature of the lowest depths of a fresh-water lake covered with a layer of ice in winter time?

PROBLEMS

GROUP A

1 A metal rod, observed to be 102.302 cm long at 20° C, is 102.433 cm long at 70° C. Out of which of the metals listed in Table 10-1 might the rod be made?

2 A radiator in the attic of a house is connected to the furnace by a vertical iron pipe 12.0 m long. If the radiator rests snugly on the floor when the pipe is at 10° C, how high off the floor will it be lifted when the pipe contains steam at 100° C?

3 The steel main span of the Golden Gate bridge is 1260 m long. What allowance must be made for expansion if the temperature varies between 40° C and −15° C?

4 A motorist purchases 40 liters of gasoline on a cold day when the temperature is −20° C. What is the volume of this gasoline after it has warmed up to +20° C?

GROUP B

5 Several decades ago a "Homigrade" thermometer was introduced. On this scale, the melting point of ice is 0° H and the boiling point of water is 270° H. (a) What is the Celsius temperature corresponding to 60° H? (b) What is the Homigrade temperature corresponding to 20° C? (c) Express 1 Homigrade degree in terms of the size of the Celsius degree.

6 Newton used a thermometer filled with linseed oil on which the melting point of ice was 0° N, and body temperature was 12° N. What is the temperature on this scale of an ice-salt mixture that is at −15° C? Suggest one or more disadvantages of Newton's scale.

7 Starting at 0° C, through what temperature change would a given mass of hydrogen have to be heated in order to cause a 3.66% change in volume?

8 In assembling a steel bridge structure, a steel rivet 3.810 cm in diameter is

to be placed in a hole 3.807 cm in diameter. Would it be possible to do the job by first cooling the rivet in dry ice (−78° C)? Air temperature is 22° C.

9 An iron ball 5.000 cm in diameter is 0.001 cm too large to go through a hole in a brass plate when the ball and plate are at 20° C. At what temperature, the same for both ball and plate, will there be a snug fit?

10 An iron rim is to be placed around a wagon wheel whose circumference at 15° C is 270.562 cm. The rim's inside circumference at 15° C is 270.020 cm. To what temperature must the band be heated so that it will fit snugly on the wheel?

11 A 200 liter steel drum is just filled with turpentine on a cool morning when the temperature is 7° C. How much turpentine will overflow in the afternoon when the temperature is 26° C?

12 A lead rod and silver rod, each exactly 80 cm long at 0° C, are placed alongside each other, and clamped together at one end. Compute the separation of the free ends of the rods when the system is placed in a steam bath.

13 In precise surveying a steel tape is used that is calibrated at 18° C and is at a specified tension. If the tension is maintained constant but the temperature increases to 34° C: (a) What is the percent error in the reading of the tape? (b) Does the tape read too high or too low? (c) What would be the error, in centimeters, when measuring a lot 15 m wide?

14 A brass cannon has a bore 10.000 cm in diameter at 20° C. A steel liner, in the form of a thin cylinder of outside diameter 10.020 cm at 20° C, is to be inserted. To what temperature should the cannon be heated, so that the liner (still at 20° C) can be inserted?

15 A rod has original length l_o. When the temperature changes by Δt, its final length is l_t. The coefficient of linear expansion is α. Prove that $l_t = l_o\,(1 + \alpha\,\Delta t)$.

GROUP C

16 To what temperature should both the cannon and liner of Problem 14 above be heated so that the liner becomes loose and can be extracted?

17 A clock has a period of exactly $\frac{1}{2}$ sec at 20° C. The pendulum consists of a small, heavy bob on a thin brass rod. (a) What will the period be if the temperature rises to 30° C? (b) At the end of a 24-hour period at 30° C, how many seconds will the clock have gained or lost?

11 Heat and Thermal Energy

In the introduction to Chapter 10 it was stated that we would first have to find out how to measure temperature and thermal energy before we could develop a theory of heat. Although a definition of temperature has not been given yet, it has still been possible for us to measure it by using the thermal expansion properties of liquids, solids, and gases. Chapter 10 also stated that it is usually not necessary to choose a standard or a particular thermometer for temperature measurement, since most thermometers agree well enough with each other.

In this chapter you will be investigating the quantitative phenomena of heat and heat transfer. You will see that when material bodies interact and heat "flows" from one body to another body, there is a change in the thermal energy of each body; but the total energy of the system is conserved.

11-1 THE CALORIC THEORY AND ITS OVERTHROW

Until well into the late 1700's, people continued to think of heat as an invisible, weightless fluid. They called this fluid *caloric*. When a flame was put to a pot of water, for example, they said that caloric flowed from the flame to the pot, making the water hotter. This caloric was thought to reside in all heated bodies, and when a hot body was placed next to a cold one, the flow of caloric was thought to make the hot body colder and the cold body hotter. Some early measurements were made with fire in fireplaces serving as a source of caloric. It was assumed that a fire burning steadily produced in 40 minutes twice as much caloric as it did in 20 minutes. None of these ideas would contradict the notion that heat is a material fluid. As a matter of fact, we still speak of the "flow of heat," but no longer do we think that heat is actually a physical substance.

Perhaps one of the most colorful experimenters of his time, who put the caloric theory to rigid scientific tests, was Benjamin Thompson (1753–1814). Thompson was born and spent his youth in the New England town of North Woburn, Massachusetts. When a young adult, he left this country because of Tory sympathies during the Revolutionary War, and after a stay in England he became minister of war in Bavaria, where he received the title of Count Rumford. In 1797 he designed a simple but decisive experiment to test whether a fluid really did reside within a heated body. While supervising the boring of a cannon at the military arsenal in Munich, he noticed the great rise in temperature of the shavings that were left in the cannon by the boring

Figure 11-1 Count Rumford.

tool. This was in spite of the fact that water was continually poured into the bore of the cannon to minimize the heating during the boring process. Unlike many of his predecessors who must have observed this heat, Rumford (Figure 11-1) had some scientific training and an inquiring mind. He was interested in discovering basic scientific facts, even if they were not immediately useful. Rumford was finally led to press a *blunt* boring tool against the inside of a cannon barrel that was being turned by horses. The stationary tool exerted a considerable force—about 40,000 N—on the 50 kg cannon. After half an hour of turning, the cannon's temperature had risen to 54° C—strong evidence of the production of a large quantity of heat by friction. Only about 0.05 kilogram of gun metal had been scraped loose by the action of the borer, and Rumford reasoned that it was quite unlikely for so much caloric to have come from such a minute amount of metallic dust. (He had previously determined that metal powder or scrapings had the same thermal behavior as bulk metal.) Figure 11-2 is copied from Rumford's own drawing of one of his cannon drills.

In another experiment Rumford caused water to boil by the heat generated by friction. He relates, "It would be difficult to describe the surprise and astonishment expressed in the countenances of the bystanders, on seeing so large a quantity of cold water heated, and actually made to boil without any fire." And Rumford confessed to "a degree of childish pleasure" at the result of the experiment.

If the cannon were turned for a long enough time, an amount of caloric approaching infinity could be obtained from a finite amount of gun metal which itself suffered no apparent change. Or, stating it differently, caloric could be continually created from apparently nothing. Rumford rejected the idea of caloric for this reason, and concluded that heat is a form of motion.

The caloric theory was ingenious, and in its day explained all the then-known phenomena. It was "true" in the same sense that any physical theory is "true" (see page 131). The theory was overthrown when well-thought-out experiments gave new results that could not be explained by the existence of caloric.

11-2 HEAT—THE MODERN CONCEPT

Today we realize that in the case of Rumford's cannon boring, the mechanical energy of the drill was continually being transformed into the thermal energy of the cannon, the drill, and the scrapings. In such a case *heat flow* is the process by which the energy is changed from one form to another. But it is not necessary to dissipate mechanical energy to have a flow of heat. When a hot body is placed next to a cold body, the thermal energy of the cold body is increased as heat flows into it from the hot body. So heat flow can also mean the transfer of thermal energy from one place to another. You will recall from Section 5-2 that *work* was defined as a measure of mechanical energy transfer. Here you can regard "heat flow" as analogous to work; heat flow and work are both measures of the energy changes within a system. In this context the term "heat energy" would be as confusing as the term "work energy." We do

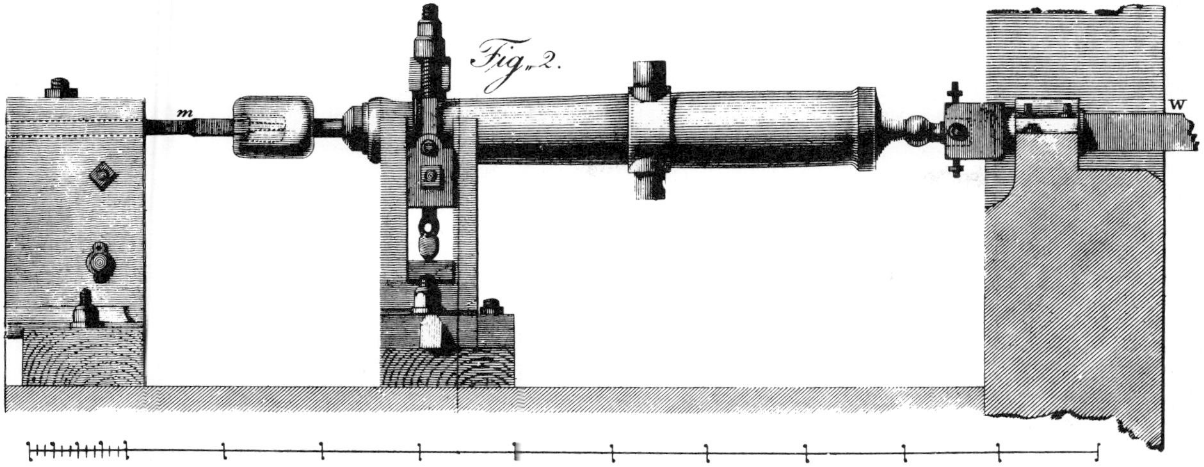

Figure 11-2 Count Rumford's crawing of his cannon boring apparatus. Shaft w (turned by horses) rotated the cannon barrel while the drill *m*, at left, remained stationary.

not say "work energy," but rather that work is done to change the energy of a system, and the amount of work done measures the changes in energy involved. Heat flow, like work, is a measure of energy transfer, but when heat flows from one body to another it is the *thermal energy* or *internal energy* of the bodies that is changed. More specifically then, the flow of heat is a measure of thermal energy transfer. And the term *heat* refers to thermal energy only when it is being transferred or changed. Once the flow of heat has stopped (when the bodies are the same temperature), we no longer have any need to speak of the heat in the body. Instead, as you will see in Chapter 13, the "thermal energy" of the body, which involves the motions of its molecules, is a more descriptive term.

11-3 THERMAL ENERGY—THE FIRST LAW OF THERMODYNAMICS

Heat flow is a measure of energy transfer, and the molecules of a body gain energy when heat flows into it. Similarly, the molecules of a body lose energy when heat flows out of it. It follows then that heat, like work, *can* be measured in the usual energy units, such as the joule. Indeed, there are times when we will be measuring heat in joules. However, before the modern concept of heat was developed, special units for "quantity of heat" were used and have persisted to this day. A *calorie* (cal) is the amount of heat needed to raise the temperature of 1 gram of water 1 Celsius degree.* Similarly, one calorie of heat is released when the temperature of 1 gram of water is lowered 1 Celsius degree.

Julius Mayer (1814–1878), a German physician, in 1842 published a paper in which he stated that "the warming of a given weight of water from 0° C to 1° C corresponds to the fall of an equal weight from the height of about 365 meters." We see from this that Mayer had a clear idea of the transformation of energy between mechanical and thermal forms.

What Mayer was saying is that the KE gained by the falling weight was numerically equal to the amount of energy needed to raise the temperature

* More precisely, from 14.5° C to 15.5° C. The Calorie, with a capital C, used by dieticians, is really a kilocalorie (kcal). It is the heat required to raise the temperature of one kilogram of water one Celsius degree. This Calorie is 1000 calories.

of an equal mass of water 1 C°. Specifically, using Mayer's figures, a 1-gram mass, say, falling from 365 meters would gain

$$\begin{aligned} \mathsf{KE} &= \tfrac{1}{2}mv^2 \\ &= \tfrac{1}{2}m(2as) \\ &= \tfrac{1}{2}(10^{-3}\ \mathrm{kg})(2)(9.8\ \mathrm{N/kg})(365\ \mathrm{m}) \\ &= 3.56\ \mathrm{J\ of\ energy} \end{aligned}$$

It takes 1 calorie of heat to raise the temperature of 1 gram of water 1 C°. Therefore Mayer was arguing that 1 cal = 3.56 J. Later investigators, notably James Joule (1818–1889), made more precise measurement of the amount of mechanical energy needed to heat water, and the modern value is *1 calorie is equivalent to 4.18 joules of energy.* This figure, 4.18 J/cal, is called the *mechanical equivalent of heat.* Figure 11-3 shows part of a page from Joule's original paper on the mechanical equivalent of heat published in 1849. The meticulously kept table of data, started at the bottom, was continued over three additional pages.

Much effort and ingenious experimentation lies behind this result. The implication is that whenever mechanical energy (KE or PE) is transformed into thermal energy, the rate of exchange is always 4.18 J for each calorie. In other words, the law of conservation of energy is valid if thermal energy is included as one form of energy. When so understood, the law of conservation of energy is also called the *first law of thermodynamics.*

in .. effect .. was obs...

1*st Series of Experiments.*—Friction of Water. Weight of the leaden weights along with as much of the string in connexion with them as served to increase the pressure, 203066 grs. and 203086 grs. Velocity of the weights in descending, 2·42 inches per second. Time occupied by each experiment, 35 minutes. Thermometer employed for ascertaining the temperature of the water, A. Thermometer for registering the temperature of the air, B.

TABLE I.

No. of experiment and cause of change of temperature.	Total fall of weights in inches.	Mean temperature of air.	Difference between mean of columns 5 and 6 and column 3.	Temperature of apparatus. Commencement of experiment.	Temperature of apparatus. Termination of experiment.	Gain or loss of heat during experiment.
1 Friction	1256·96	57°·698	2°·252—	55°·118	55°·774	0°·656 gain
1 Radiation ...	0	57·868	2·040—	55·774	55·882	0·108 gain
2 Friction	1255·16	58·085	1·875—	55·882	56·539	0·657 gain
2 Radiation ...	0	58·370	1·789—	56·539	56·624	0·085 gain
3 Friction	1253·66	60·788	1·596—	58·870	59·515	0·645 gain
3 Radiation ...	0	60·926	1·373—	59·515	59·592	0·077 gain
4 Friction	1252·74	61·001	1·110—	59·592	60·191	0·599 gain
4 Radiation ...	0	60·890	0·684—	60·191	60·222	0·031 gain
1	2	3	4	5	6	7

Figure 11-3 Reproduced from James P. Joule's original paper *The Mechanical Equivalent of Heat.*

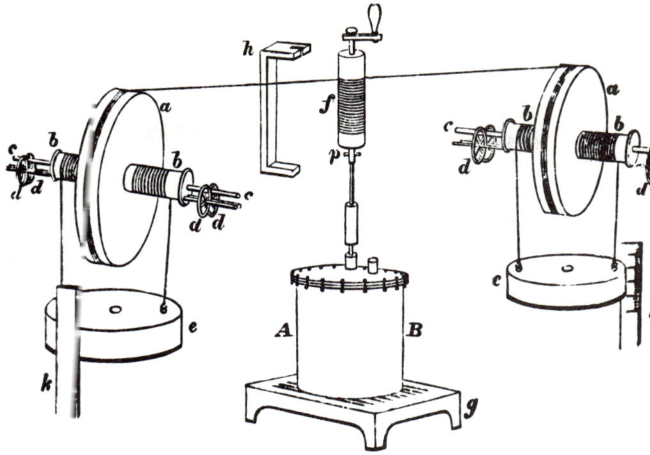

Figure 11-5 Early drawing of Joule's apparatus for determining the mechanical equivalent of heat, which led to the statement of the first law of thermodynamics.

The first law of thermodynamics did not spring into being all at once, but rather it evolved from the speculations and experiments of many workers. As long ago as 1797 Rumford had an inkling of the first law when he reflected on the production of heat in his cannon-boring experiment. He wrote, ". . . in case of necessity, the Heat thus produced might be used in cooking victuals. But no circumstances can be imagined in which this method of procuring Heat would not be disadvantageous; for, more Heat might be obtained by using the fodder necessary for the support of a horse, as fuel." Here he anticipated the energy relationships of metabolism, a biological field that he was unequipped to explore.

It was Joule, however, who made the first systematic measurements of the mechanical equivalent of heat. He churned water by a paddle system (Figure 11-4); he stirred mercury with iron paddles; he heated water by friction, passing it through a thick piston perforated with many small holes; he rubbed iron rings together under mercury; he measured the heat produced electrically in a wire through which electrons flowed. Using all these methods, Joule obtained results agreeing within 5 percent, truly a remarkable achievement. It is because of the variety of his methods that Joule's work is of such great significance. (Figure 11-5 shows an early drawing of Joule's complete "paddle" apparatus.) The German physiologist and physicist Hermann von Helmholtz (1821–1894) presented in 1847 a great paper that consolidated the work of others and gave the first systematic publication of the law of conservation of energy with convincing mathematical arguments.

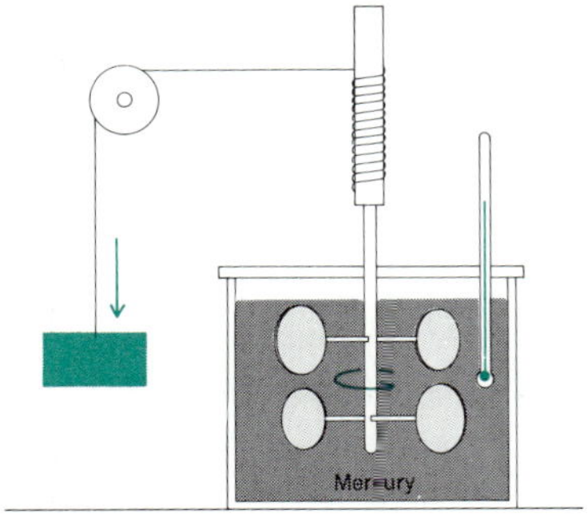

Figure 11-4 Cross-section of Joule's paddle device for measuring the mechanical equivalent of heat.

The decade 1840–1850 was a remarkably fruitful one, and it is perhaps pointless to try to assign "credit" to either Mayer, Joule, or Helmholtz for the first law of thermodynamics. This brief excursion into the history of science shows how a major discovery often develops gradually, with the "breakthrough" coming as a result of independent but simultaneous effort on the part of a number of workers. Present-day physics is in this tradition.

11-4 SPECIFIC HEAT

If a kilogram of lead and a kilogram of iron are placed side by side in front of a brisk fire that is burning in a fireplace, and if the blocks of metal are allowed to stand for, say, ten minutes the lead becomes warmer than the equal mass of iron. Apparently some property of the iron causes it to warm up more slowly than the lead. Similarly, suppose a pot of water and a solid block of metal of the same mass are each placed on a stove and brought to the same temperature. When they are removed from the source of heat, the metal block can be safely handled after a few minutes, but the water remains hot for a much longer time. In this case the water cools off more slowly than the metal block. From these examples we can see that substances differ markedly in the temperature change caused by the absorption or release of a given amount of heat.

This property of matter is described by the term *specific heat.* The specific heat of a substance is a measure of how much heat it can absorb per unit mass, per degree rise in temperature. According to this definition, the specific heat c of a substance equals the number of calories required to raise the temperature of 1 gram of the substance 1 C°. And since the definition of the calorie is in terms of water, it is evident that the specific heat of water is 1 calorie per gram per Celsius degree (1 cal/g·C°); that is, 1 calorie of heat is required to raise the temperature of 1 gram of water 1 C°. The specific heat of aluminum is 0.22 cal/g·C°, meaning that it requires only 0.22 calorie to raise the temperature of 1 gram of aluminum 1 C°. We see that because of the way the calorie was defined, water has become a sort of "standard substance" with which other substances may be compared easily. Table 11-1 shows that water has by far the largest specific heat of any common substance.

Now let's interpret the behavior of the substances mentioned at the beginning of this section. In the case of the kilogram of lead and the kilogram of iron, the table shows that iron has a higher specific heat than lead. This means that, for equal masses, the iron will warm up more slowly than the lead because the iron must absorb more heat. Similarly, water has a higher specific heat than any metal. The water will cool more slowly than the metal block, since it must release more heat per degree of temperature change.

Table 11-1

Specific Heats of Some Solids, Liquids, and Gases

Substance	*Temperature or temperature range* (°C)	*Specific heat,* (cal/g·C° or kcal/kg·C°)
SOLIDS		
iron and steel	20	0.11
copper	15–100	0.093
aluminum	15–100	0.22
glass	20–100	0.20
lead	20	0.0306
sodium chloride	20	0.21
ice	−3	0.50
wood	. . .	0.4
asbestos	20–98	0.195
silver	20	0.056
brass	15–100	0.094
mercury	0-100	0.033
LIQUIDS		
water	15	1.00
mercury	20	0.033
glycerin	50	0.60
methyl alcohol	20	0.60
benzene	20	0.41
GASES*		
steam	110	0.48
air	50	0.25
helium	20	1.24
neon	20	0.25

* At constant pressure.

How much heat is required to raise the temperature of 400 grams of aluminum from 40° C to 45° C? In the first place, if we had 400 grams of *water,* the heat required would be 400 calories for every degree, or (400)(5) = 2000 calories in all. But aluminum requires only 0.22 as much heat as does water, since the specific heat of aluminum is 0.22 cal/g·C° (Table 11-1). Therefore the required heat is (2000 cal)(0.22) = 440 cal. If 400 grams of aluminum were to cool down from 45° C to 40° C, 440 calories of heat would "flow out" of the aluminum.

We can say then, that in general, the heat

required to change the temperature of a substance depends upon the mass of the substance, its specific heat, and the temperature change. The equation

$$\Delta Q = mc\,\Delta t$$

describes this transfer. Here ΔQ is the amount of heat that flows in or out of the substance depending upon whether the temperature t increases or decreases, m is the mass, and c is the specific heat.

In many cases where a system of bodies is isolated and no mechanical work is done on or by the system, conservation of energy means conservation of *thermal* energy. Heat gained by one part of the system equals that lost by another part of the system, as illustrated below.

Example 11-1 An automobile radiator contains 20 kg of water. If 200,000 cal of heat is given to the cooling system by the motor, what is the rise in temperature?

Consider the isolated system water + cylinder block of motor.

Heat gained by water
= heat lost by cylinder block

$$mc\,\Delta t = 200{,}000 \text{ cal}$$

$$(20{,}000 \text{ g})(1 \text{ cal/g}\cdot\text{C}^\circ)(\Delta t) = 200{,}000 \text{ cal}$$

$$\Delta t = \frac{200{,}000 \text{ cal}}{20{,}000 \text{ cal/C}^\circ} = \boxed{10 \text{ C}^\circ}$$

As is the case in all problems of energy conservation, it is just a matter of doing a little "bookkeeping." When several substances, each at a different temperature, are mixed, thermal energy is transferred by heat flow until the whole batch reaches some equilibrium temperature It is usually not difficult to decide which parts of the system gain heat and which ones lose heat. From the first law of thermodynamics, we know that if the system is *thermally isolated,* the total energy of the system is conserved. This is illustrated in the next example where the *method of mixtures* is used to measure the specific heat of a substance.

Example 11-2 400 g of lead BB shot is heated to 100° C and dropped into a 100 g glass beaker containing 200 g of water at 20° C. The specific heat of glass is 0.20 cal/g·C°. The mixture is stirred until temperature equilibrium is reached, and the final temperature is 24.2° C. What is the specific heat of lead?

$$\begin{bmatrix}\text{heat gained}\\ \text{by water}\end{bmatrix} + \begin{bmatrix}\text{heat gained}\\ \text{by beaker}\end{bmatrix} = \begin{bmatrix}\text{heat lost}\\ \text{by lead}\end{bmatrix}$$

$$[(200 \text{ g})(1 \text{ cal/g}\cdot\text{C}^\circ)(24.2^\circ - 20.0^\circ)]$$
$$+ [(100 \text{ g})(0.20 \text{ cal/g}\cdot\text{C}^\circ)(24.2^\circ - 20.0^\circ)]$$
$$= [(400 \text{ g})(c)(100^\circ - 24.2^\circ)]$$

$$840 \text{ cal} + 84 \text{ cal} = (400 \text{ g})(c)(75.8 \text{ C}^\circ)$$

$$c = \frac{924 \text{ cal}}{(400 \text{ g})(75.8 \text{ C}^\circ)} = \boxed{0.0305 \text{ cal/g}\cdot\text{C}^\circ}$$

A final example, in which mechanical and thermal energy are both considered, will help clarify the energy principle as illustrated by the first law of thermodynamics. Many mechanical problems are highly simplified by neglecting friction. Let's now take friction into account.

Example 11-3 A 30 kg boy on a 5 kg sled takes a running start from the top of a slippery hill, moving at 4.00 m/sec. The hill is 60 m long and drops 2.00 m for every 10 m traversed (Figure 11-6, page 192). The boy and sled reach the bottom of the hill, moving at 10.0 m/sec. (a) How much mechanical energy is transformed into thermal energy? (b) If the steel runners of the sled have a mass of 1.500 kg and if we assume that half the heat generated remains in the runners, by how much does the temperature of the steel runners rise during the ride downhill?

(a) The top of the hill is 12 m above the bottom. The weight of the boy and sled together is 35 kg (9.8 N/kg) = 343 N. As the sled and boy slide down, some (but not all) of their initial PE is transformed into KE. The rest of the energy becomes ther-

Figure 11-6

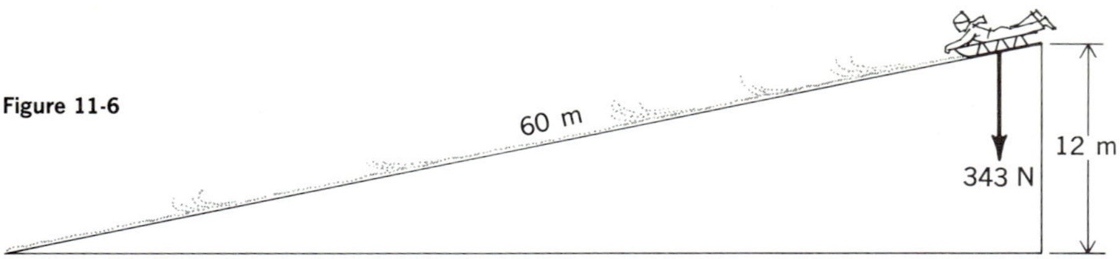

mal energy. By applying the first law of thermodynamics we get:

$$\text{PE}_1 + \text{KE}_1 = \text{PE}_2 + \text{KE}_2 + \Delta Q$$

where ΔQ is the thermal energy acquired by sled and ground.

$$(343 \text{ N})(12 \text{ m}) + \tfrac{1}{2}(35 \text{ kg})(4.00 \text{ m/sec})^2$$
$$= 0 + \tfrac{1}{2}(35 \text{ kg})(10.0 \text{ m/sec})^2 + \Delta Q$$
$$4120 \text{ J} + 280 \text{ J} = 0 + 1750 \text{ J} + \Delta Q$$
$$\Delta Q = \boxed{2650 \text{ J}}$$

This is the amount of energy that has been "dissipated" in heat. Of course it has not disappeared; it has merely been transformed into thermal energy, a less useful form. To express this thermal energy in calories rather than joules we use the fact that 1 cal = 4.18 J:

$$\Delta Q = (2650 \text{ J})(1 \text{ cal}/4.18 \text{ J})$$
$$= 634 \text{ cal}$$

(b) The example tells us that half the heat, or 317 cal, remains in the runners. To find the temperature rise of the steel runners is a simple problem involving the specific heat of steel (Table 11-1):

$$\Delta Q = mc\,\Delta t$$
$$317 \text{ cal} = (1500 \text{ g})(0.11 \text{ cal/g}\cdot\text{C°})\,\Delta t$$
$$\Delta t = \boxed{1.92 \text{ C°}}$$

Under some circumstances the rise in temperature due to friction can be considerable. A piece of sandpaper becomes noticeably warm while being used. A brake shoe on a train may become red-hot when it is pressed against a fast-turning train wheel in order to bring the train to a halt (see Figure 11-7). In other cases the temperature rise might be unobservably small. But it must be present according to the first law. The needle and the surface of a phonograph record become slightly warmer as the record spins. In some cases it is not so easy to see that "disappearing" energy is really being changed to thermal energy because of friction, as in the case of an object on the end of a spring. As the object vibrates up and down, there is a continual transformation from PE to KE and back to PE (Figure 8-5). However, the vibration dies out because there are "frictional" forces between atoms that oppose the motion of the spring. To move against these forces requires work, and there is no other source of energy for this work, unless the total mechanical energy of the system is to be gradually used up. The spring becomes warmer. As you will see in Chapter 13, the *random* KE, and therefore, the thermal energy of the atoms increases as the *ordered* KE (or vibration) of the whole body decreases and becomes zero. Energy is conserved.

11-5 CHANGE OF PHASE: LATENT HEATS

Matter exists in three states—solid, liquid, and gaseous. We live on a planet where the natural state of many common substances is the solid state. Except for the water in the rivers and oceans and the gases in our atmosphere, as we look around us the world is pretty much in solid form. But this does not imply that rocks, trees, and baseballs are always solids. Nor is water always liquid or the air always gaseous. On the surface of the sun, for example, all substances are in gaseous form, whereas on distant Pluto everything is probably frozen solid. It is evident, then, that the temperature of the surroundings determines the state in which a substance will exist. Not many of us have ever seen rock in its liquid or gaseous state, or air in its liquid or solid state. Let's review then, the behavior of a substance that is commonly found in all three of its states—H_2O.

One gram of water in the form of ice at −20° C is placed in a thin metal cup on a hotplate maintained at 130° C. According to Table 11-1 the specific heat of ice is 0.50 cal/g·C°, so 10 calories of heat are required to raise the ice to 0° C. Further heat is then absorbed, but it is found that, instead of the ice becoming warmer, the temperature remains 0° C while the ice melts. Experiments show that about 80 calories are required to melt 1 gram of ice, the temperature remaining constant at 0° C the entire time. Only when all the ice has become water at 0° C does the temperature again begin to rise. In the early days, before heat was known to be related to energy, it was expected that *any* absorption of heat would result in a temperature rise, so the behavior of a melting piece of ice was a puzzle. The 80 calories was called the *latent* heat of fusion*, since the effect of the heat was to cause melting (fusion) rather than even the tiniest rise in temperature.

As more heat is applied, the H_2O, in the liquid state now, warms up until 100° C is reached. Since the specific heat of water is 1 cal/g·C° the water absorbs 100 calories to get to this point. The temperature then remains at 100° C until *all* the water is converted to water vapor (steam).** The *latent heat of vaporization* of water at 100° C is considerably greater than its latent heat of fusion —about 540 cal/g is absorbed without any rise in temperature. Finally, if the water vapor is confined—say it is contained in a sealed box and placed in a furnace—the temperature again begins to rise. It can be heated as much as desired without any further latent heats, until the molecules are separated into hydrogen and oxygen

Figure 11-7 A railroad brake shoe.

atoms, and H_2O molecules cease to exist as such.

This whole process of changing 1 gram of ice to steam involves the *addition* of heat. The same gram of water can be taken through the reverse cycle, which would involve the *release* of heat. Starting with water vapor at 130° C, say, it is cooled down to 100° C. Since the specific heat of steam is 0.48 cal/g·C°, 1 gram of steam releases 14.4 calories of heat. As liquid drops of water form from the vapor, the temperature remains at 100° C until all the vapor is condensed. The latent heat, 540 cal/g, is released during this process. As the liquid H_2O cools down to 0° C, 100 calories of heat are given out; and then, as the H_2O changes from liquid to solid, 80 cal/g of heat are given off while the temperature remains at 0° C. Finally, the solid is cooled down to ice at −20° C as 10 more calories are released. This entire process is summarized in Figure 11-8. Here you can see that 744.4 calories are required to bring the ice from −20° C to steam at 130° C, and these same 744.4 calories are released during the reverse process.

* From the Latin word *latere*, to lie hidden or concealed. For a table of latent heats, see appendix page 480.

** Technically, "steam" is the same as "water vapor" and is invisible. The white visible cloud that is commonly miscalled steam is really tiny droplets of liquid water.

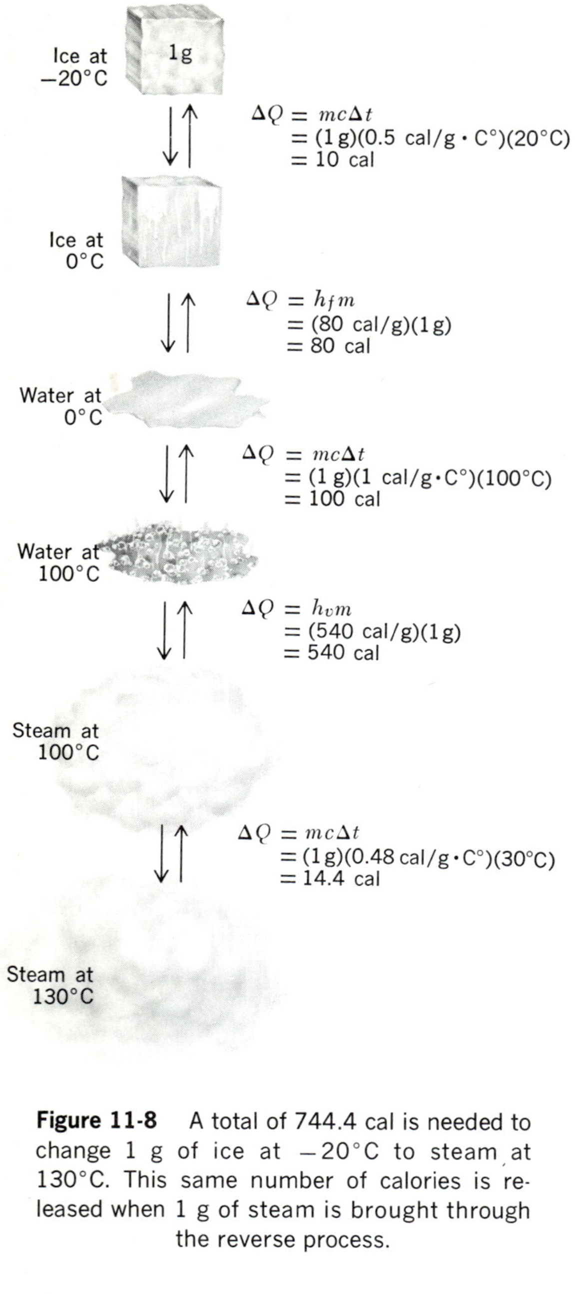

Figure 11-8 A total of 744.4 cal is needed to change 1 g of ice at −20°C to steam at 130°C. This same number of calories is released when 1 g of steam is brought through the reverse process.

It is important to note that the latent heats of fusion and of vaporization (denoted by h_f and h_v) involve no *temperature changes*, and are expressed in calories per gram rather than calories per gram per degree as are specific heats. Latent heats can be measured by the method of mixtures, as shown in the next example.

Example 11-4 A well-insulated copper cup of mass 100 g contains 400 g of water at 5° C. When 40 g of ice at −8° C is added, it is found that 23.6 g of the ice melts. What is the latent heat of fusion of ice?

Since not all the ice melts, we can assume that the final temperature is 0° C, with 16.4 g of ice floating in 423.6 g of water. For the specific heats of ice and copper we use the values from Table 11-1, which are 0.50 and 0.093 cal/g·C°, respectively.

$$\begin{bmatrix}\text{heat absorbed to}\\ \text{raise 40 g of ice to}\\ 0^\circ\text{ C}\end{bmatrix} + \begin{bmatrix}\text{latent heat ab-}\\ \text{sorbed to melt}\\ \text{23.6 g of ice}\end{bmatrix} = \begin{bmatrix}\text{heat given out}\\ \text{by water cooling}\\ \text{to }0^\circ\text{ C}\end{bmatrix} + \begin{bmatrix}\text{heat given out}\\ \text{by copper cup}\\ \text{cooling to }0^\circ\text{ C}\end{bmatrix}$$

$$(40\text{ g})(0.50\text{ cal/g}\cdot\text{C}^\circ)[0^\circ - (-8^\circ)] + (23.6\text{ g})(h_f) = (400\text{ g})(1\text{ cal/g}\cdot\text{C}^\circ)(5^\circ - 0^\circ) + (100\text{ g})(0.093\text{ cal/g}\cdot\text{C}^\circ)(5^\circ - 0^\circ)$$

$$160\text{ cal} + (23.6\text{ g})(h_f) = 2000\text{ cal} + 46\text{ cal}$$

$$h_f = \boxed{79.9\text{ cal/g}}$$

Example 11-5 What is the result of adding 10 g of steam at 100° C to 50 g of ice at 0° C?

We have to feel our way through this problem in order to find out whether all the ice melts. Then we can set up an equation to find the final temperature. During condensation, 10 g of steam *can* give up as much as

$$(540\text{ cal/g})(10\text{ g}) = 5400\text{ cal}$$

To melt all the ice requires (80 cal/g) (50 g) = 4000 cal. Hence all the ice *will* melt, and heat will be left over to warm up the water formed from the ice. The final temperature will be between 0° C and 100° C. We can call this final temperature *t*.

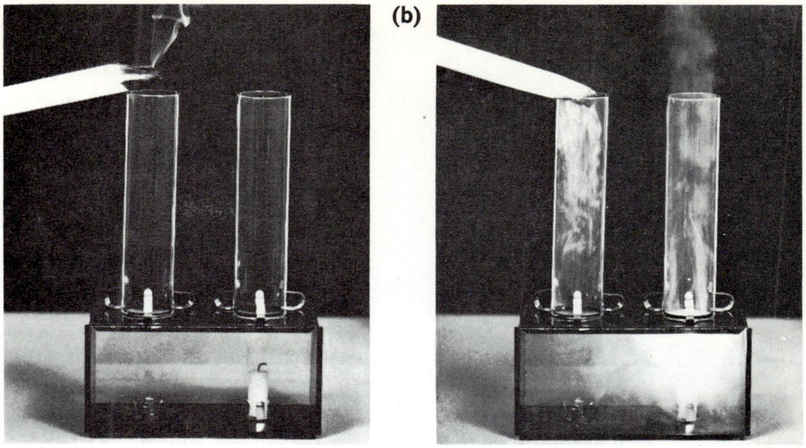

(a) (b)

Figure 11-9 A convection box. Why is smoke from the burning paper pulled into the box?

Heat given off = heat absorbed

$$\begin{bmatrix}\text{latent heat}\\ \text{of steam}\end{bmatrix} + \begin{bmatrix}\text{cooling of}\\ \text{hot water}\end{bmatrix} = \begin{bmatrix}\text{latent heat}\\ \text{of ice}\end{bmatrix} + \begin{bmatrix}\text{warming up of}\\ \text{cold water}\end{bmatrix}$$

$$(10\text{ g})(540\text{ cal/g}) + (10\text{ g})(1\text{ cal/g}\cdot\text{C}^\circ)(100^\circ - t) = (50\text{ g})(80\text{ cal/g}) + (50\text{ g})(1\text{ cal/g}\cdot\text{C}^\circ)(t - 0^\circ)$$

$$5400 + 1000 - 10t = 4000 + 50t$$

$$t = 40^\circ\text{ C}$$

The result is

60 g of water at 40° C

The latent heat given off when water freezes is put to good use on farms where produce is stored in an unheated shed. Large tubs of water are set in the shed, and as the outside temperature goes below 0° C, the water freezes. In so doing, it gives off its latent heat to the surroundings, keeping the shed and contents at about 0° C until all the water freezes. A similar action takes place when dew forms on the leaves of a plant. As water vapor in the air condenses and becomes liquid water, the latent heat of vaporization is given out, and this heat often helps to prevent plant damage by an early frost.

11-6 TRANSFER OF HEAT BY CONVECTION, CONDUCTION, AND RADIATION

We now regard heat as a form of energy rather than a material substance. With this understanding we can still talk about the flow of heat, which now means the transfer of thermal energy from one place to another. There are three fundamentally distinct ways in which heat may be transferred—by convection, conduction, and radiation.

The *convection* of heat is illustrated by the blower of a space heater in a large gymnasium. Hot air is moved by the fan from the point of origin to the spectators in the bleachers. The KE of the air molecules is transported by moving the molecules *en masse* to the desired point. Convection currents in gases and liquids are often caused by variations in density due to temperature changes. Hot air, which is less dense than cold air, rises from a point above a radiator and strikes the ceiling, spreading out through the room. Meanwhile, the cooler (and denser) air at floor level moves in to fill up the space vacated by the hot air, and a continuous flow takes place. The convection box in Figure 11-9 is a good model for showing this process. A similar phenomenon keeps the temperature of a lake at 4° C (the temperature of the densest water) before any of it cools to freezing (Section 10-4). Most of the changes in air temperature that appear on a weather map are due to convection currents, set up as masses of warm or cold air move over land and water.

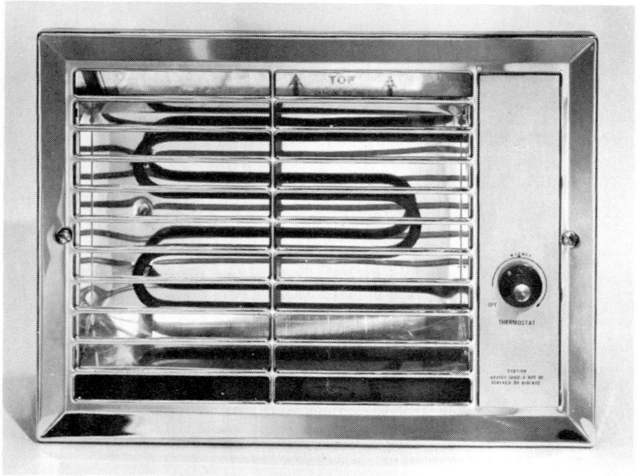

Figure 11-10 A radiant heater.

Heat *conduction* is a more intimate process, in which KE is passed on from one molecule to an adjacent molecule by a collision. When a silver spoon is placed in a cup of hot coffee, the vibrating silver molecules in the bowl of the spoon bump each other, but do not migrate away from their average positions. The slow-moving molecules in the handle eventually receive KE as it is passed along from one molecule to another, and so the handle becomes warm. As one would expect, heat flows by conduction only if there is a temperature difference, for only then (on the average) will fast molecules be striking slow molecules and be able to give up energy.

Substances differ markedly in their heat conductivity, and may be classified as insulators and conductors. Metals are especially good conductors of heat since a metal has free electrons, not attached to any particular nucleus. In addition to the transfer of KE by vibrating molecules, the KE can also be transferred by electrons as they collide with each other. These same free electrons are responsible for the conduction of electric currents too. It is no accident that silver and copper, which are the best conductors of heat, are also the best conductors of electricity.

The third process by which thermal energy is transferred from one place to another is called *radiation*. Unlike convection and conduction, the process of radiation does not require a physical medium. We say that heat comes to us from the sun by radiation. In the evening the earth radiates some of this heat back into the atmosphere. Our atmosphere, in turn, serves as a blanket to prevent complete radiation to outer space. In this way the night temperatures are kept from being unlivably cold as they would be, say, during night on the moon. When you place your hand next to a hot iron, you feel the heat "rays" coming from the iron. All heated bodies radiate. Heat radiation is an *electromagnetic wave* emitted by the heated body as a result of the vibrations of its electrically charged atoms. In Chapter 20 you will see that whenever an electric charge accelerates, electromagnetic waves are generated. These waves have several names, most of them familiar to you: infrared rays, ultraviolet rays, x rays, gamma rays, radio waves, and visible light. Although much more will be said about electromagnetic waves in later chapters, it is worth mentioning here that they all have one thing in common: They all travel with the same speed in a vacuum, 3×10^8 m/sec, the speed of light. It is not only true that radiation requires no physical medium, but also that electromagnetic waves reach their highest speed in space.

Not all surfaces are equally effective in emitting or absorbing radiation, even though their temperatures are the same. A white or shiny body is a poor absorber of radiation, since some of the incoming energy is reflected away from the interior. By the same token, a white or shiny body is a poor *emitter* of radiation, since radiant energy within the body is reflected inside. The surface is shiny from the inside as well as from the outside! Note the shiny reflector and the black emitter of the heater in Figure 11-10. A perfect absorber and emitter of radiations is called a *black body*. According to these ideas, a living-room "radiator" should be painted black for efficient radiation of heat. A light-colored paint job might be decorative, but is hardly in keeping with the laws of physics. However, most heat transfer from a "radiator" is by conduction, that is, by direct contact of air molecules with iron atoms or paint molecules.

SUMMARY

Heat flow is a measure of **thermal energy transfer** and, like work, can be measured in ordinary mechanical units such as joules. The **calorie** is an amount of energy (equal to 4.18 J) that will raise the temperature of 1 gram of water 1 C°.

The **specific heat** of a substance is the heat absorbed or given off per unit mass per unit change in temperature. In the cgs system, specific heat is measured in calories per gram per Celsius degree. When heat flows into or out of a body, the quantity of heat ΔQ that flows to cause a temperature change Δt is given by $\Delta Q = mc\,\Delta t$, *where* m is the mass in grams, c the specific heat, and ΔQ the heat that flows, in calories.

The law of conservation of energy, when modified to include heat as a form of energy, is called the **first law of thermodynamics.** Mechanical energy is not "lost" due to frictional processes; it is merely transformed into thermal, or internal, energy.

When heat flows into a body, the **internal energy** of the body increases in one or both of two ways. If the random KE of the molecules increases, the temperature increases. If the PE of the molecules increases, as in the separation of the molecules against cohesive forces during vaporization, the temperature does not increase, and the heat absorbed is called **latent heat.** Both processes are reversible; heat is given out when a body cools down (KE decreases) or when it condenses or solidifies (PE decreases).

Transfer of heat takes place in three ways. **Convection** is the motion of heated matter in bulk. **Conduction** is due to transfer of KE from molecule to neighboring molecule by collision. In metals, heat conduction and electrical conduction are both largely due to a cloud of free electrons within the metal. **Radiation** is the process by which energy is transferred by **electromagnetic waves.** Radiation includes all types of electromagnetic waves from radio waves to gamma rays. All electromagnetic waves travel with the same velocity (3×10^8 m/sec) through a vacuum. A good absorber is also a good emitter of radiation. A **black body** has a surface that absorbs and emits all radiations.

QUESTIONS

1 What is the difference between "heat" and "temperature"?

2 Correct the following statements: (a) "The heat of the sun's surface is 6000° C." (b) "Human body heat is 37° C." (c) "The heat that is needed to melt lead is 5.86 cal/g·C°." (d) "The specific heat of copper is 0.093 calories."

3 If two buckets of equal size are filled, one with water and one with glycerin, at 50° C, which will cool down to room temperature first? Why?

4 A book lying near the edge of a table is pushed over and allowed to fall to the floor. When the book strikes the floor, its KE disappears. What becomes of this KE?

5 An Eskimo places a stone near a fireplace, and then takes the warm stone to bed with him and rests his feet on the stone. Explain how convection, conduction, and radiation are involved in this native custom.

6 Why does a metal seem colder to the touch than wood, outdoors on a cold winter day?

7 Why are white clothes worn in the tropics?

PROBLEMS

GROUP A

1 How much heat is required to raise the temperature of 500 g of water from 40° C to 70° C? Express your answer in calories.

2 How much heat is wasted when a steamship engine pours 400 kg of water at 90° C into a lake that is at 5° C? Express your answer in calories.

3 In a laboratory experiment, 800 J of work was done against friction. How much heat, in calories, was generated?

4 It is found that 640 cal of heat is needed to heat 100 g of a substance from 6° C to 14° C. What is the specific heat of the substance?

5 How much heat, in calories, is needed to heat 100 kg of asbestos from 5° C to 40° C?

6 How much heat, in calories, is required to change 60 g of ice at 0° C to water at 20° C?

7 While condensing from a vapor into a liquid at 61.5° C, 20 g of chloroform liberated 1180 cal. What is the latent heat of vaporization of chloroform?

8 How much heat is needed to convert 1 kg of ice at 0° C to 1 kg of water vapor at 100° C?

9 How much heat, in calories, is given off by 200 g of water that cools from 100° C to 20° C?

10 How much heat, in calories, is needed to bring 500 kg of water from 15° C to within 12° C of the boiling temperature?

11 A metal block of mass 500 g is warmed from 20° C to 30° C when it absorbs 1200 cal of heat. Calculate the specific heat of the metal.

12 How much heat must be supplied to a 20 g ice cube, originally at 0° C, in order to convert it to 20 g of water at 30° C?

GROUP B

13 When 500 g of lead at 100° C is placed in an aluminum cup of mass 100 g that contains 200 g of olive oil at 20° C, the final temperature of the mixture is 29.3° C. What is the specific heat of olive oil?

14 In a laboratory experiment to determine the latent heat of fusion of ice,

46.9 g of ice at 0° C was placed in a cup that contained 265 g of warm water. The mass of the cup was 91.4 g and its specific heat is 0.22 cal/g·C°. All the ice melted, and the cup and contents were cooled from 32.4° C to 17.1° C. Calculate the latent heat of fusion of ice from these data. (*Note:* The lack of agreement between your answer and the accepted value is typical of the degree of precision commonly obtained in heat measurements in high school laboratories.)

15 An aluminum pan is placed on a stove at 4:05 P.M. The pan has a mass of 340 g and contains 300 g of water at 20° C. The flame supplies 100 cal of heat to the system each second. (a) When will the water begin to boil? (b) When will the temperature of the pan begin to rise above 100° C?

16 What is the result when 10 g of steam at 100° C is added to 20 g of ice at 0° C?

17 What is the result of adding 10 g of steam at 100° C to 60 g of ice at 0° C?

18 What is the result of adding 10 g of steam at 100° C to 90 g of ice at 0° C?

19 A craftsman hammers out a silver ashtray having a mass of 70 g. If the hammer head has a mass of 500 g and it strikes the tray with a speed of 10 m/sec, what is the rise in temperature of the tray after 30 strokes of the hammer? (*Note:* Silver is much softer than the iron hammer head. Hence it is reasonable to assume that the hammer rebounds with negligible speed. Assume also that 50% of the heat generated remains in the ashtray.)

20 An automobile of mass 1800 kg is traveling at 3 m/sec. How much heat, in calories, is developed when it is brought to rest?

21 Using the modern value of the mechanical equivalent of heat (4.18 J/cal), calculate the height of fall that Mayer stated to be "about 365 meters" (Section 11-3).

22 A lead bullet of mass 20 g is fired at 400 m/sec into a tree trunk, and it emerges on the other side with a speed of 300 m/sec. Assuming that 60% of the loss of mechanical energy is stored as thermal energy in the bullet, and that the specific heat is that given in Table 11-1, calculate its rise in temperature. Lead melts at 327° C. Will any of the bullet melt?

23 A 35 kg child slides down a playground slide made of steel, and arrives at the bottom with a speed of 1 m/sec. The slide is 3 m long and is inclined at 45°. How much heat, in calories, is shared by the child and the slide?

24 An ice skater of mass 100 kg starts to glide at 10 m/sec, and comes to rest after traveling 20 m. How much ice is melted? (Assume that half the heat generated by friction is absorbed by the ice, and also assume that the ice beneath the skates is just at the melting point.)

25 Steam is allowed to bubble through water in an aluminum cup. Some of the steam condenses into water, giving off its latent heat of vaporization, and then (as water) cools down until its temperature equals that of the surrounding water. The following data are taken: initial temperature, 15° C; final temperature, 40° C; mass of empty cup, 100 g; mass of cup plus cold water, 246 g; mass of cup plus warm water plus condensed steam, 253 g. Calculate the latent heat of vaporization of steam.

26 A lead ball of mass m is dropped onto a pavement from a height of 20 m. Assuming that 75% of the heat generated remains in the lead, calculate the temperature rise of the ball.

27 How much steam at 100° C would have to be added to 60 g of ice at 0° C for the entire system to become water at 30° C?

28 An ice cube of mass 20 g is in an aluminum cup of mass 40 g. The system is originally at −30° C. What will be the result of adding 6000 cal to the system?

29 With what velocity must a snow ball be thrown so that it completely melts upon striking a stone wall? (*Hint:* Write an equation involving the mass of the ball and hope that it cancels out in the calculation. You must assume that all the heat generated goes into melting the snow.)

GROUP C

30 A 2000 g copper ball is heated to 700° C and then placed quickly on a large block of ice that is at 0° C. How much ice will melt if you assume that all the heat given up by the copper goes into melting the ice?

31 A thermally insulated can contained 500 g of water and 500 g of ice. A 1 kg block of metal was heated to 400° C and dropped into the can. It was found that all of the ice just melted. If the mass of the metal had been twice as great, what would the final temperature of the mixture have been? (Neglect the heat loss to the can.)

32 A 60 g ice cube at −10° C is dropped into a beaker of water that is at 0° C. How much water will freeze onto the cube?

33 An 800 g copper flask contains 700 g of water and 90 g of ice. The flask is thrown over a cliff and its temperature after landing is 20° C. What was the speed of the flask just before it landed? (Assume that the change in thermal energy of the ground during the collision is negligible.)

12 Thermal Behavior of Gases

Now that you have become familiar with the concepts of temperature and heat flow, you are ready to proceed with an investigation of the thermal behavior of gases. In Section 10-5 it was shown that, for a constant pressure, the coefficients of volume expansion for all gases were practically the same. The present chapter examines this idea further by discussing "an ideal gas," which leads to the derivation of the general gas law—a crucial step in the development of the theory of heat.

12-1 IDEAL GASES

Although we are perhaps more familiar in daily life with the behavior of matter in the solid and liquid state, the gaseous state of matter is actually far simpler to understand. The reason for this is twofold. In a gas, the molecules are so far apart that the cohesive forces between them are usually very slight. Although the cohesive forces between the molecules of one gas may be different from those between the molecules of another gas, their effect on the behavior of either gas is so slight that it is negligible under ordinary circumstances. Again, it is natural to suppose that the behavior of two different gases would be affected somewhat by the different sizes of the individual gas molecules. Yet the volume actually occupied by the molecules of a gas is usually a tiny fraction of the total volume of the container in which the gas is confined. It has therefore been possible to develop the concept of an ideal gas. An *ideal* gas is one in which the effects of cohesive forces and molecular volume are negligible. Under these conditions, we could expect the laws describing the behavior of gases to be simple and universal, the same for all ideal gases, regardless of their composition. No real gas is, of course, an ideal gas, but under the usual conditions of temperature and pressure we shall not be very wrong if we consider all gases ideal.

12-2 PRESSURE

The term "pressure" has been used freely in this text because it is a concept familiar to almost everyone. Before we describe quantitative measurements, however, let's discuss briefly the physicist's definition of pressure and the units in which pressure is measured.

Pressure is defined as force per unit area:

$$P = \frac{F}{A}$$

Thus an object weighing very little can exert a tremendous pressure if the force it exerts acts only on a small surface area. On the other hand, the weight of 5.7×10^{16} metric tons of air creates only a relatively small pressure on the earth because the

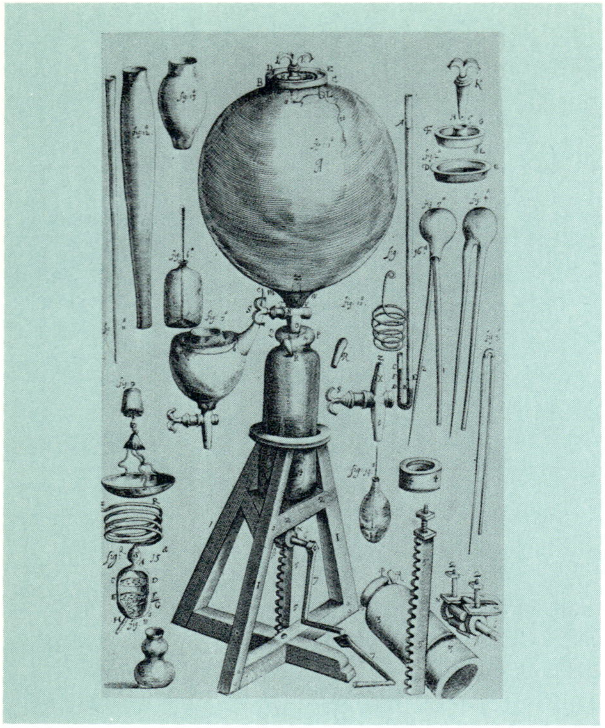

Figure 12–1 The improved air pump built by Robert Boyle with the help of his assistant, Robert Hooke. With this apparatus, Boyle confirmed the relationship of volume to pressure, now known as Boyle's law.

force is spread out over the entire surface of the globe. Pressure units are force units divided by area units, such as N/m², dynes/cm², etc. The SI unit of pressure is the pascal (Pa), equal to 1 N/m². For practical purposes, pressures are often expressed as so many atmospheres, where 1 *atmosphere* (atm) is normal atmospheric pressure*, approximately 1.01×10^5 N/m², or 1.01×10^6 dynes/cm². The *centimeter of mercury* (cm Hg) is used in many gas-law problems, and is defined as 1/76 atm. It is the pressure due to a column of mercury 1 centimeter high. The *millimeter of mercury* (mm Hg)** is, similarly, the pressure due to a column of mercury 1 millimeter high. These units can be readily manipulated with the aid of a few conversion factors:

$$\begin{aligned} 1 \text{ atm} &= 1.01 \times 10^6 \text{ dynes/cm}^2 \\ &= 1.01 \times 10^5 \text{ N/m}^2 \\ &= 101 \text{ kPa} \\ &= 760 \text{ mm Hg} \end{aligned}$$

12-3 BOYLE'S LAW

Four measurable quantities are of interest in describing the behavior of a gas. They are: the pressure, volume, mass, and temperature of a given sample of the gas. Together, these quantities determine the "state" of the sample. The first experimental results relating two of these were obtained by the English investigator Robert Boyle (1627–1691), who studied how the volume of a gas changed as the pressure was varied (Figure 12-1). In order to arrive at a physically significant law, Boyle simplified the problem by doing his experiments under controlled conditions. He kept the mass m of the gas constant (i.e., no leaks in the container), and he kept the temperature t constant. Under such circumstances, Boyle arrived at a simple relation. He found that when he increased the pressure on a confined gas the volume was reduced by the same factor. When the pressure was doubled, the volume was halved; with three times the pressure, the volume was one third as great (Figure 12-2). The pressure is inversely proportional to the volume of a gas. When mass and temperature are constant,

$$P \propto \frac{1}{V}$$

or

$$PV = \text{constant}$$

The graph of pressure vs. volume of 100 g of an ideal gas is shown in Figure 12-3. As the pressure increases, passing from state *1* to state *2* (shown in

* The normal atmosphere is defined as the pressure at the bottom of a column of mercury exactly 76 cm high, if the mercury is at such a temperature that its density is 13.5951 g/cm³ and it is located at a point where g = 980.665 cm/sec².

** The mm Hg is also known as the Torr, named for Evangelista Torricelli, who invented the barometer in 1643.

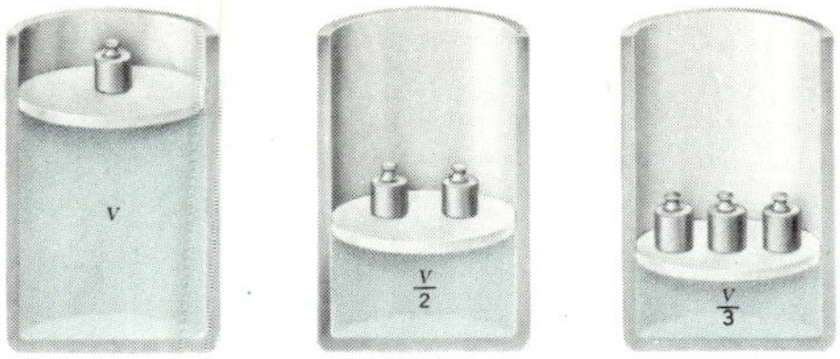

Figure 12-2 Pressure compared to volume of an ideal gas (neglecting weight of piston).

the figure), the product PV remains constant. Boyle's law is often written as

$$P_1V_1 = P_2V_2$$

assuming mass and temperature are constant. Here, the subscript 1 refers to the initial state of the gas, and the subscript 2 refers to the final state.

Example 12-1 An oxygen tank of volume 2.00 m^3 is filled to a pressure of 2.02×10^7 N/m^2. What was the original volume of the oxygen when it was at a pressure of 1 atm? (Assume no temperature change.)

In using Boyle's law, it is necessary to have similar units on each side of the equation. We know that 1 atm is 1.01×10^5 N/m^2 (Section 12-2), so the final pressure must be

$$P_2 = (2.02 \times 10^7\ \text{N/m}^2)\left(\frac{1\ \text{atm}}{1.01 \times 10^5\ \text{N/m}^2}\right)$$

$$= 200\ \text{atm}$$

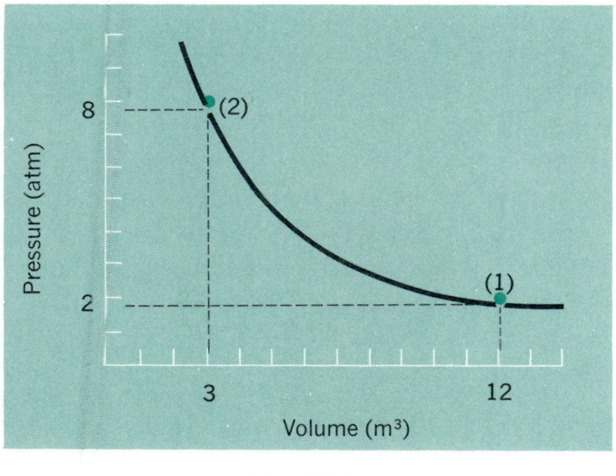

Figure 12-3

Applying Boyle's law, we have

$$P_1V_1 = P_2V_2$$

$$(1.00\ \text{atm})\ V_1 = (200\ \text{atm})(2.00\ \text{m}^3)$$

$$V_1 = \boxed{400\ \text{m}^3}$$

Solving the problem in another way will show that it is unimportant just which units are chosen for P or V, provided they are the same on both sides of the equation:

$$P_1V_1 = P_2V_2$$

$$(1.01 \times 10^5\text{N/m}^2)V_1 = (2.02 \times 10^7\text{N/m}^2)(2.00\ \text{m}^3)$$

$$V_1 = \boxed{400\ \text{m}^3}$$

Note that the mass of the oxygen has not changed. The same mass is involved both at 200 atm and at 1 atm.

12-4 THE EFFECT OF MASS—AVOGADRO'S NUMBER

The value of the "constant" in Boyle's law depends on several factors. First, let's imagine a sample of gas whose volume and temperature are held constant. If *more* gas is pumped into this fixed volume, that is, if the mass of the gas is increased, the pressure will increase. As the next chapter will show in detail, the pressure results from the impact of molecules as they strike the walls of the container. It is reasonable to expect that doubling the mass (that is, doubling the number of gas molecules) will cause the number of impacts per second to double, and hence the pressure will be doubled. Experiments show that the product PV is proportional to the mass of the gas. If m is the mass of gas, we can say that, at constant temperature,

$$PV = m \times \text{constant}$$

or

$$\frac{P_1V_1}{m_1} = \frac{P_2V_2}{m_2}$$

In other words, if m is increased, then so is PV, and PV/m remains the same.

Example 12-2 A compressed-air storage tank whose volume is 0.1 m^3 contains 3.00 kg of air at a pressure of 2.40×10^6 N/m^2. How much air would have to be forced into the tank to increase the pressure to 2.80×10^6 N/m^2, assuming no change in temperature?

$$\frac{P_1V_1}{m_1} = \frac{P_2V_2}{m_2}$$

$$\frac{(2.40 \times 10^6 \text{ N/m}^2)(0.1 \text{ m}^3)}{3.00 \text{ kg}} = \frac{(2.80 \times 10^6 \text{ N/m}^2)(0.1 \text{ m}^3)}{m_2}$$

$$m_2 = 3.50 \text{ kg}$$

The mass of air that is forced in is

$$3.50 \text{ kg} - 3.00 \text{ kg} = \boxed{0.50 \text{ kg}}$$

In general, when dealing with different kinds of gas, such as hydrogen (H_2) as compared with oxygen (O_2), it is necessary to use *equal numbers of molecules* rather than equal masses. But how can we determine the number of molecules in a given sample of gas? We can do this from our knowledge of the data of chemistry; specifically, by making use of the relative *atomic masses* of the elements and the *molecular masses* of the compounds. For example, when it is said that uranium has an atomic mass of 238 units and iron has an atomic mass of 56 units, the following ratio is implied:

$$\frac{\text{mass of 1 uranium atom}}{\text{mass of 1 iron atom}} = \frac{238}{56}$$

This knowledge of the *ratio* of atomic masses comes from careful experiments involving compounds containing uranium and iron. In this way the *ratios* of atomic masses were known many years ago. It is quite a different thing, however, to find out the mass of an individual atom.

Suppose we have 56 grams of iron in one lump, and 238 grams of uranium in another. These two chunks of matter are each said to contain one *mole* of metal. This term simply means that in each case the mass in grams is numerically equivalent to the atomic mass in atomic mass units (u).* The data from chemistry also tell us that *each chunk contains the same number of atoms.* To see that this statement is true, let's imagine two large sealed boxes *A* and *B*, with box *A* containing baseballs and box *B* containing marbles. The total mass of box *A* is measured to be 238 kilograms, and the total mass of box *B* is 56 kilograms. Suppose we reach in and remove one baseball from Box *A* and find it to have a mass of 0.238 kilograms. Then we pull out one marble from box *B* and find it has a mass of 0.056 kilograms. The ratio of the individual masses is

$$\frac{0.238 \text{ kilograms}}{0.056 \text{ kilograms}} = 238{:}56.$$

We conclude that there are as many baseballs in *A* as there are marbles in *B*, the greater total mass of *A* being exactly accounted for by the greater mass of the individual balls. Similarly, there must be some definite number of atoms in a mole of any element. This number is called *Avogadro's number*, N_A, named for Amadeo Avogadro, an Italian scientist (1776–1856) (Figure 12-4).

For a chemical compound, the word "mole" is used to refer to an amount whose mass in grams is equivalent to the *molecular* mass in u. A mole of molecules contains N_A molecules, just as a mole of atoms contains N_A atoms. For instance, the atomic mass of carbon is 12, and that of oxygen is 16. The molecular mass of CO_2 is $12 + 16 + 16 = 44$ u. Therefore 1 mole, or 44 grams, of CO_2 contains N_A molecules. There are N_A atoms of C and 2 N_A atoms of O, a total of 3 N_A atoms in the N_A molecules of CO_2.

Although the *existence* of Avogadro's number was known as soon as the existence of identical atoms was postulated, the *measurement* of the number is difficult, and Avogadro himself had no clear idea of its value. One early method of determining the number was to let a drop of oil spread

* This unit is denoted by u, to distinguish it from an earlier atomic mass unit (amu) defined in terms of ${}_8O^{16}$ instead of ${}_6C^{12}$.

out on the surface of water, forming a thin film. As the film spreads, it gets thinner until finally it becomes a monolayer, one molecule thick. If the original volume of the drop (found from its mass and density) and the area of the film are known, the thickness of the monolayer can be found. This, in turn, gives the approximate diameter of a single molecule, and hence the volume of the molecule can be estimated. In this way the mass of a single molecule also can be estimated, and the number of molecules in a known mass of the oil can be found. Modern measurements, which are far more precise but less direct than the method just described, have established the value of Avogadro's number as $N_A = 6.02 \times 10^{23}$ atoms per mole of element, or molecules per mole of compound.

By using Avogadro's number, then, it is possible to determine the number of molecules in a given sample of gas. For instance, there are actually more molecules in 8 grams of H_2 (molecular mass 2) than in 80 grams of O_2 (molecular mass 32). The 8 grams of H_2 represents (8 g)(1 mole/2 g) = 4 moles, and hence has $4 \times 6.02 \times 10^{23}$ molecules. Similarly, 80 grams of O_2 is only 80/32 = 2.5 moles, and contains only $2.5 \times 6.02 \times 10^{23}$ molecules. If temperature and volume are equal, we expect that the pressure exerted by the hydrogen sample would be greater because it contains a greater number of molecules, in the ratio of 4 to 2.5. Thus, even when dealing with different kinds of gas, we can say

Figure 12-4 Coin prepared by Academy of Science, Turin, Italy. Free translation: *Amadeo Avogadro. This commemorates the centennial of his molecular theory. September 24, 1911.*

that at constant temperature

$$\frac{PV}{n} = \text{constant}$$

where n is the number of moles of gas present.

Example 12-3 A certain tank contains 6.4 kg of oxygen at a pressure of 4.20 atm. What would the pressure be if the oxygen were pumped out and replaced by 6.6 kg of carbon dioxide (CO_2) at the same temperature?

The molecular mass of O_2 is 16 + 16 = 32; the molecular mass of CO_2 is 12 + 2(16) = 44.

$$6.4 \text{ kg of } O_2 = (6400 \text{ g})(1 \text{ mole}/32 \text{ g}) = 200 \text{ moles}$$

$$6.6 \text{ kg of } CO_2 = (6600 \text{ g})(1 \text{ mole}/44 \text{ g}) = 150 \text{ moles}$$

$$\frac{P_1V_1}{n_1} = \frac{P_2V_2}{n_2}$$

Here $V_1 = V_2$, since it is the same tank in both cases.

$$\frac{(4.20 \text{ atm})(\cancel{V_1})}{200 \text{ moles}} = \frac{(P_2)(\cancel{V_2})}{150 \text{ moles}}$$

$$P_2 = \boxed{3.15 \text{ atm}}$$

This really means that, at a given temperature and volume, the pressure exerted by any ideal gas does not depend on the type of molecules that make up the gas. This remarkable fact illustrates the essential simplicity of gases as compared with liquids or solids. You will encounter some of the reasons for this simplicity in Chapter 13, "The Theory of Heat."

12-5 THE EFFECT OF TEMPERATURE

Finally, in developing a general gas law it is important to know the effect of temperature change on the pressure exerted by a confined gas—volume and mass remaining constant. Experiment shows that the pressure of the gas increases

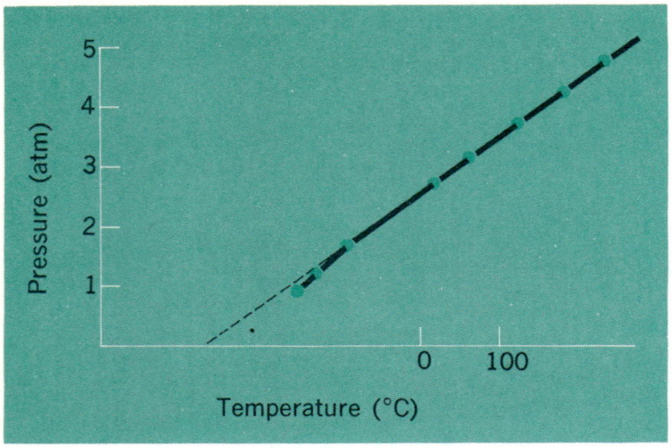

Figure 12-5 Pressure vs. temperature for a real gas.

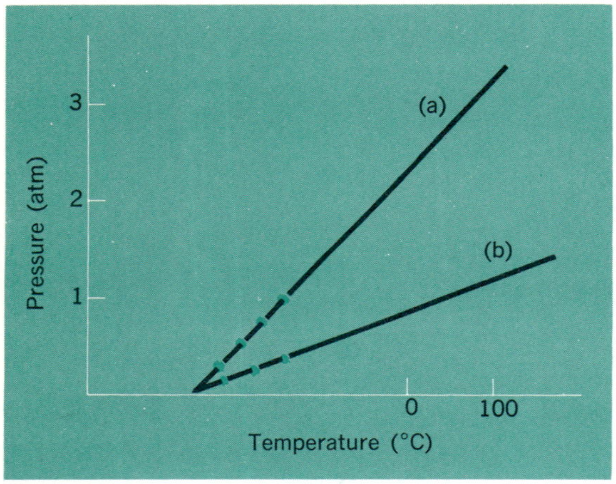

Figure 12-6 At a certain temperature, the pressure exerted by any ideal gas would be zero. **(a)** 4 kg of H_2 in 20 m^3; **(b)** 36 kg of O_2 in 30 m^3.

uniformly as the temperature increases (with certain exceptions to be noted later). The graph of pressure vs. temperature therefore should be a straight line. Note that in Figure 12-5 the graph of P vs. t is shown as a dashed line for all values below a certain temperature. This indicates that the pressure would become zero *if it continued to decrease at the same rate as it does around room temperature* (20° C). Any ideal gas is assumed to behave in this ideal way, but in actual practice, before the gas pressure becomes zero, the gas will liquefy at some temperature, and will no longer *be* a gas.

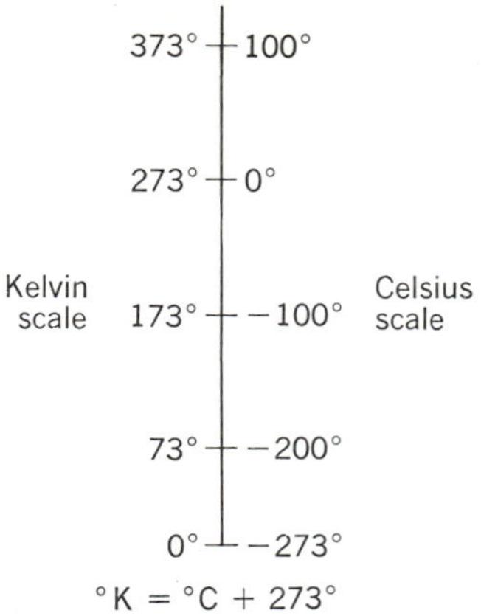

Figure 12-7

Figure 12-6 shows different masses of two different gases each enclosed in a container of a different volume. Again, when the graphs are extended, they meet at a common point, zero pressure at some certain temperature. From this and other experiments, we conclude that there is a certain temperature, *the same for all ideal gases*, at which the pressure would become zero if the gas remained an ideal gas all the way down. What is this temperature? By measuring the pressure of a gas at two temperatures (which may be 0° C and 100° C if desired) and extrapolating the graph down to zero pressure, it is possible to determine this temperature in the laboratory. Precise experiments have given −273.15° C as the value of this important constant, but it is usually rounded off to −273° C. This temperature is called *absolute zero.**

A new temperature scale based on this constant was devised by Lord Kelvin (1824–1907) and uses absolute zero as its zero point. The degrees on the absolute, or Kelvin, scale (°K) are the same size as the Celsius degrees. Therefore, since −273° C corresponds to 0° K, then 0° C corresponds to 273° K (Figure 12-7).

* Since the gaseous form does not exist at such a low temperature this definition of absolute zero is based on extrapolated values. Another definition will be given in Section 13-5.

Absolute temperature is thus defined as the Celsius temperature *t* plus 273°, and is denoted by *T*. Absolute temperature, measured in degrees Kelvin, represents the number of degrees a body is above absolute zero. The concept of absolute zero is of great significance especially in determining the efficiency of heat engines, as you will see in the next chapter. However, for the present, absolute zero will be used only to get a simpler equation for describing the effect of temperature on the pressure of an ideal gas.

We now come to the main point of this discussion. In Figure 12-8 the graph for a third gas has been added, but the general idea expressed in Figure 12-6 is repeated: The pressure of all ideal gases would become zero at $-273°$ C if they remained gases. In Figure 12-9 the pressure is plotted against the *absolute* temperature, with the origin shifted to absolute zero to emphasize the direct proportion that exists between *P* and *T*. You can see, then, that while the pressure *depends* upon the Celsius temperature *t* (Figure 12-8), it is actually *proportional* to the absolute temperature *T* (Figure 12-9). *This is the reason for using the absolute temperature in the gas laws*—it allows a simple proportion:

$$\frac{P_1}{T_1} = \frac{P_2}{T_2}$$

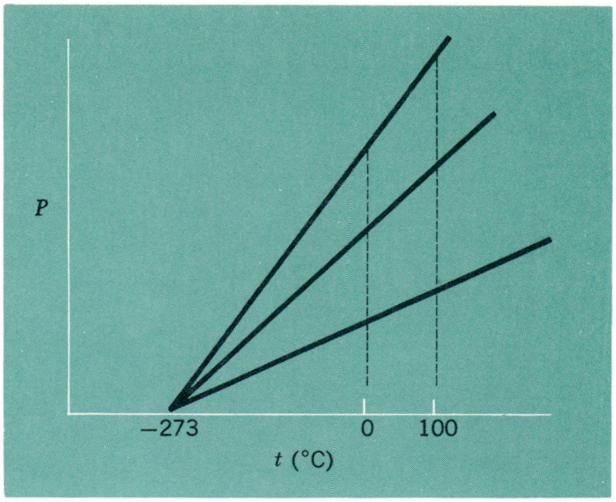

Figure 12-8

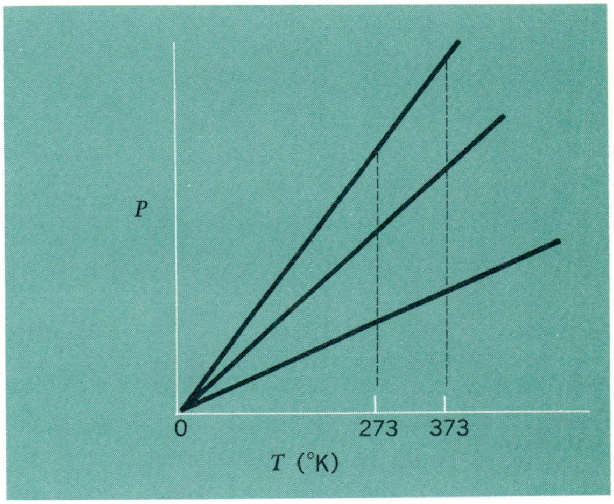

Figure 12-9

Example 12-4 A pressure gauge used on a truck tire reads 2 atm on a cold day when the thermometer stands at $-3°$ C. What will the pressure reading be after the tire has stood for a while in a heated garage at 27° C?

First, it is necessary to account for the atmospheric pressure, which is 1 atm. If the gauge read zero, the pressure of the air in the tire would be the same as the pressure outside, that is, it would still be at atmospheric pressure. At a gauge reading of 2 atm, the total (actual) pressure in the tire is 2 atm + 1 atm = 3 atm. Next, we convert both Celsius temperatures to absolute temperatures by adding 273°: $-3°$ C is 270° K, and 27° C is 300° K.

Since the mass and volume of the air remain constant, the pressure is proportional to the absolute temperature. Using the relation shown above, we have

$$\frac{P_2}{T_2} = \frac{P_1}{T_1}$$

$$\frac{P_2}{300°\text{ K}} = \frac{3\text{ atm}}{270°\text{ K}}$$

$$P_2 = 3.3\text{ atm}$$

This is the total pressure; the gauge will read

$$3.3\text{ atm} - 1\text{ atm} = \boxed{2.3\text{ atm}}$$

12-6 THE GENERAL GAS LAW

Our discussion so far has shown that the "state" of a sample of gas is determined by considering collectively the pressure, volume, mass, and temperature of the sample. We are now in a position to summarize the results of our treatment of these properties and, by so doing, arrive at the *general gas law* in which any or all of the properties just mentioned can be variable.

The first result to be entered in our summary is the statement in Section 12-3 that for an ideal gas, with mass and temperature constant,

$$PV = \text{constant}$$

and it was stated in Section 12-4, that at constant temperature

$$\frac{PV}{n} = \text{constant}$$

where n is the number of moles of gas. In the previous section we saw another important result: that the pressure is proportional to the absolute temperature, when mass and volume are constant. This gives still another relation:

$$\frac{P}{T} = \text{constant}$$

The value of the constants in the above equations are all different, and they depend upon the conditions in each particular problem. If we now combine these ideas, we can write a general gas law:

$$\frac{PV}{nT} = \text{a new constant}$$

and by giving the new constant the symbol R we have

$$\frac{PV}{nT} = R$$

where P represents the pressure, V is the volume, n stands for the number of moles, and T is the absolute temperature. Here at last is a constant that is really constant—not dependent on the mass of gas, its chemical formula, or the temperature. The value of this *universal gas constant, R*, depends on the units for P, V, and T. But for our purposes we will not actually have to use *any* number for R, if we remember one combination of P, n, T, and the corresponding experimental value of V. Such a set of values is based on the so-called *standard temperature and pressure* (STP) of 1 atmosphere pressure and 0° C (273° K): 1 mole of an ideal gas at a temperature of 273° K with a pressure of 1 atmosphere is found to occupy a volume of 22,400 cm³. You probably recall from your study of chemistry that 1000 cm³ is equal to 1 liter. Therefore, our experimental volume of 22,400 cm³ = 22.4 liters. Using the subscript $_0$ to denote standard conditions, the set of values becomes

$$P_0 = 1 \text{ atm}$$
$$n_0 = 1 \text{ mole}$$
$$T_0 = 273° \text{ K}$$
$$V_0 = 22.4 \text{ liters}$$

Several examples will make clear how to use the general gas law in various systems of units. In all cases, *absolute* temperatures are used.

Example 12-5 If 4 moles of gas exerts a pressure of P_1 N/m² when confined in a volume of 400 cm³ at 300° C, what would be the pressure of 200 moles of the same gas when placed in a 1000 cm³ tank at 600° C?

$$\frac{P_1V_1}{n_1T_1} = \frac{P_2V_2}{n_2T_2}$$

$$\frac{(P_1 \text{ N/m}^2)(400 \text{ cm}^3)}{(4 \text{ moles})(573° \text{ K})} = \frac{(P_2)(1000 \text{ cm}^3)}{(200 \text{ moles})(873° \text{ K})}$$

$$P_2 = \boxed{30.5 P_1 \text{ N/m}^2}$$

It is not necessary to worry about combining mks units (N/m²) and cgs units (cm³) as long as similar units appear on each side of the equation as in the above example.

Example 12-6 How much helium gas (molecular mass 4) is contained in a high-altitude balloon whose volume is 5000 liters, if the temperature is $-23°$ C and the pressure is 30 cm Hg?

Here we use the set of values for standard conditions to fill in one side of the gas law equation, recalling that 1 atm is 76 cm Hg pressure. The unknown is n_1, which represents the number of moles of helium.

$$\frac{P_0V_0}{n_0T_0} = \frac{P_1V_1}{n_1T_1}$$

$$\frac{(76 \text{ cm Hg})(22.4 \text{ liters})}{(1 \text{ mole})(273° \text{ K})} = \frac{(30 \text{ cm Hg})(5000 \text{ liters})}{(n_1)(250° \text{ K})}$$

$$n_1 = 96.2 \text{ moles}$$

The mass of helium is

$$(96.2 \text{ moles})(4.00 \text{ g}/1 \text{ mole}) = \boxed{385 \text{ g}}$$

Example 12-7 Solve the preceding example, but this time assume that the balloon is filled with hydrogen instead of helium. The molecular mass of hydrogen (H_2) is 2.02.

Just as before, the calculation shows that there are 96.2 moles of gas in the balloon. However, since the molecular mass is different, the mass of gas is also different.

The mass of hydrogen is

$$(96.2 \text{ moles})(2.02 \text{ g}/1 \text{ mole}) = \boxed{194 \text{ g}}$$

Example 12-8 A balloon containing 6 m^3 of hydrogen at 10° C and 1 atm pressure rises to a level where the temperature is $-30°$ C and the pressure is 0.2 atm. Assuming that the balloon bag is free to expand, what is its new volume?

$$\frac{P_1V_1}{n_1T_1} = \frac{P_2V_2}{n_2T_2}$$

$$\frac{(1 \text{ atm})(6 \text{ m}^3)}{(283° \text{ K})} = \frac{(0.2 \text{ atm})(V_2)}{(243° \text{ K})}$$

In this case the amount of gas is the same, so the value of n cancels out from both sides of the equation.

$$V_2 = \boxed{25.8 \text{ m}^3}$$

Example 12-9 What pressure is exerted by 64 g of oxygen (O_2), confined at 40° C in a flask of volume 5 liters?

If 32 g of O_2 (1 mole) occupied a volume of 22.4 liters at 0° C, the pressure would be 1.00 atm (expressed to 3 significant figures here).

$$\frac{(P)(5 \text{ liters})}{(64 \text{ g})(313° \text{ K})} = \frac{(1.00 \text{ atm})(22.4 \text{ liters})}{(32 \text{ g})(273° \text{ K})}$$

$$P = \boxed{10.3 \text{ atm}}$$

In none of the preceding examples was it necessary to use a numerical value for R, the gas constant in the general gas law. It was only necessary to remember the volume of an ideal gas at STP (22.4 liters for 1 mole of gas at 273° K and 1 atm pressure). It is not difficult to calculate the value of R, however, which is included here for reference: If P is in atmospheres, V in liters, and T in degrees Kelvin, then R has the value 0.0821 liter·atm/mole·K°; if P is in newtons per square meter, V in cubic meters, and T in degrees Kelvin, then R is 8.314 J/mole·K°. The latter value is the mks value for R.

SUMMARY

If the molecules of a gas exert no appreciable cohesive forces on one another, and if the molecules are small compared with the space between them, then the gas is an **ideal gas.** To a large extent these conditions are fulfilled by practically all gases at ordinary temperatures, with the result that all gases obey the same gas law regardless of chemical composition. The form of the gas law is simplified by using **absolute temperature,** which is equal to Celsius temperature plus 273°.

The pressure, volume, and temperature of an ideal gas are related by the **general gas law:** $PV/nT = R$, where the amount of gas is given by n, the number of moles, and R is the **general gas constant.** Boyle's law is a special case of the general gas law, with mass of gas and temperature held fixed.

QUESTIONS

1 Molecules of different gases are of different sizes. Explain why this fact is of little practical importance as far as the gas laws are concerned.

2 Starting with the general law, write a simpler law for the relation of volume and absolute temperature when mass and pressure remain constant. (This is called the Gay-Lussac law.)

3 How many moles of H_2O are there in a drinking glass that contains 360 g of water?

4 How many moles of H_2O are there in a piece of ice that has a mass of 360 g?

5 In Table 10-1, you can see that the coefficient of volume expansion for gases is about 0.00366 $(C°)^{-1}$. The fractional equivalent of this decimal is about $\frac{1}{273}$. Can you explain this fraction in terms of absolute zero?

Note: In all problems, assume atmospheric pressure to be 1×10^5 N/m^2 unless otherwise stated.

PROBLEMS

GROUP A

1 A large research balloon having a volume of 2000 m^3 is to be filled with helium at atmospheric pressure. The helium is stored in cylinders of volume 2 m^3 at a pressure of 2×10^6 N/m^2. How many cylinders are required?

2 Water is pumped into the bottom of a vertical storage tank whose volume is 200 m^3. When the pressure gauge shows that the air pressure in the tank is 3 atm above atmospheric pressure, what is the volume of water in the tank?

3 The cork of a popgun is inserted so tightly that a presssure of 3 atm is required to dislodge it. The air is admitted through a hole at A, which is 0.30 m from the cork at B (Figure 12-10). How far from A is the piston when the cork pops out, assuming no temperature change?

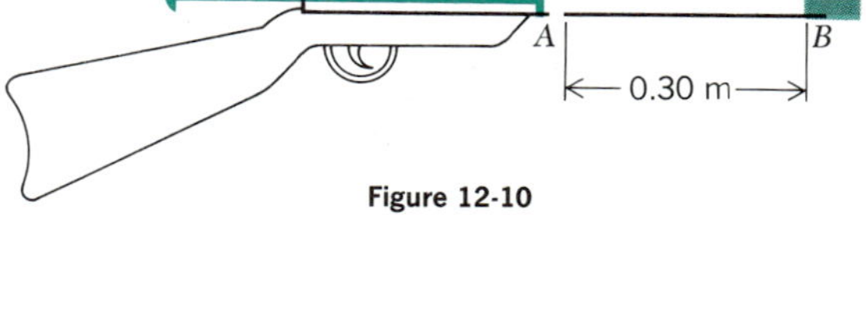

Figure 12-10

4 Convert to °K: 40° C; − 80° C.

5 Given that a certain mass of ideal gas occupies a volume of 30 liters at 300° K and 1 atm pressure, compute the volume of the same mass of the same gas at 1500° K and 3 atm pressure.

6 A sealed tank contains air at 27° C and a pressure of 6×10^5 N/m^2. What will be the pressure if the temperature goes up to 127° C?

7 A tank containing 0.1 m^3 of nitrogen at room temperature and at 8×10^6 N/m^2 is connected through a valve to an empty tank of volume 0.4 m^3. The valve is opened and the nitrogen allowed to expand. After the system returns to room temperature, what is the pressure in the tank?

8 The volume of a truck tire is 30 cm^3 and it contains air at 10° C. A gauge reads 1.6×10^6 dynes/cm^2 above atmospheric pressure. What will the gauge read if the temperature rises to 30° C?

9 A certain amount of ideal gas occupies 1 m^3 at 200° K and 2 atm. What pressure will the same mass of gas exert if confined in 0.25 m^3 at 500° K?

GROUP B

10 Writing Boyle's law in the form $PV = k$, show by dimensional analysis that the constant k has the same dimensions as does work.

11 A vertical cylindrical tank is 42 cm tall, 2.00 cm in radius, and is open at the top. Atmospheric pressure is 1.013×10^5 N/m^2. A close-fitting cylindrical plug of mass 5.00 kg is inserted at the top, and is let fall inside. If the temperature of the trapped air does not change, how far from the top of the cylinder is the base of the plug when it comes to rest?

12 (a) How many moles of ammonia gas (molecular mass 17) are contained in a tank of volume 50 liters if the temperature is 40° C and the pressure is 10 atm? (b) What is the mass of this gas? (c) How many molecules of ammonia are in the tank?

13 A scientist stores 12 g of hydrocyanic acid (a highly poisonous gas whose molecular mass is 27) at a pressure of 900 mm Hg. Overnight the container develops a slight leak and the pressure drops to 840 mm Hg. What mass of the gas has escaped?

14 What pressure is exerted by 800 g of sulfur dioxide (molecular mass 64) that is confined in a tank of volume 60 liters at 50° C?

15 A soccer ball of constant volume 2.24 liters is pumped up with air to a gauge pressure 0.86 atm at 20° C. (a) How many moles of air are in the ball? (b) What mass of air is in the ball? During the game, the temperature rises to 30° C. (c) What mass of air must be allowed to escape to bring the gauge pressure back to its original value? (The molecular mass of air may be taken to be approximately 29.)

16 A tank contains 5 liters of oxygen under a pressure of 1104 cm Hg. It is put into connection with a second tank, cubical in shape and 40 cm on an edge, containing hydrogen at 69 cm Hg pressure. Assuming that the temperature does not change, calculate the resultant pressure, which will be the same in each tank.

17 From the data given in Example 12-5, calculate the value of the pressure P_1.

18 What is the weight, in newtons, of the oxygen in the tank of Example 12-1, if the temperature is 40° C? (This is as much an exercise in conversion of units as it is an exercise in physics!)

13 The Theory of Heat

The concepts of heat, temperature, and molecular motion have already been discussed, but now it is time to look at certain aspects of heat more closely. In particular, we will now form a unified theory tying together various fundamental phenomena of heat. Some of this theory can be developed by building upon Newton's second law. Also, since some processes in nature seem to proceed in one direction only (but not in reverse), we will be able to formulate some general statements relating heat and molecular motions. The theory of heat has been a stepping stone for the advance of knowledge in many areas of physics.

13-1 HEAT AND TEMPERATURE

What is the difference between heat and temperature? In Section 11-1 you found that when mechanical energy is transformed into heat, the thermal (internal) energy of the system is increased. In fact, the amount of mechanical energy lost is equal to the increase of the internal energy of the system. In the overall process, the total energy of the system is conserved. Mechanical energy and thermal energy, then, are simply the different forms that the energy can assume. You also found that whenever a hot body is placed next to a cold body, there is a thermal energy transfer as heat flows from the hot body to the cold body. The thermal energy of the cold body increases, but this is at the expense of a corresponding decrease in the thermal energy of the hot body. From the molecular point of view, then, heat is a measure of thermal energy, which is *the sum of the KE and PE of the molecules of a body.* (The fact that the body as a whole may have KE and PE due to its motion or position is not considered in measuring its thermal energy.) For a gram of ice to melt at 0° C, 80 calories of heat must be added to the ice. This increases the PE of the ice molecules as they are pulled apart against the cohesive forces binding them together. Heat flowing into a body may also appear as increased KE of the molecules as they are speeded up. In either case, the total thermal energy in the body is equal to the *sum* of the KE and PE of its molecules. On the other hand, *temperature* depends upon the *average* KE per molecule. As an illustration, consider a swimming pool full of cold water at 5° C and a teacup full of hot water at 90° C. Although the teacup has a higher temperature (it is "hotter"), there is much more thermal, or internal, energy in the pool. This is because, even though the average KE per molecule in the pool is less than the average per molecule in the cup, there are many more molecules in the pool. Thermal energy, or internal energy, therefore, is simply the *total* mechanical energy of the molecules

Figure 13-1

of a body; temperature is a measure of the *average* KE per molecule; and heat flow is a thermal energy transfer caused by a difference in temperature.

To say that temperature depends upon the average KE per molecule is really over-simplifying the situation. A more precise definition (which will be demonstrated in detail in Section 13-2) would be the following: *absolute temperature of a gas or liquid is proportional to the average translational* KE per molecule due to random motions.*

Each word in the above statement has definite significance: "proportional," since the factor relating absolute temperature to average KE depends on the choice of temperature scale; "average" because not all the molecules have the same KE, that is, not all are moving at the same speed, and, if they were, collisions would soon destroy equality; "translational," since rotational or vibrational types of KE are without effect in causing collisions in a gas and therefore cannot cause thermal energy to be transferred (a basic effect of temperature difference); "kinetic" energy because increases of PE, such as a latent heat, takes place without rise in temperature; "per molecule," since we are distinguishing between temperature and thermal energy; due to "random" motions because the "ordered" motions of all the molecules of a body are considered to be mechanical energy. The molecules of a baseball are given ordered motions when it is thrown. These "organized" motions of translation and rotation are not in themselves indicative of a temperature rise, but are classed as KE of the ball as a whole. When the ball strikes a catcher's mitt and is stopped, the KE of the ball is transformed into increase in random KE of the molecules, and the temperature rises (see Figure 13-1).

13-2 THE KINETIC THEORY OF GASES

To see why physicists believe that absolute temperature is proportional to the average translational KE per molecule, we will investigate what is called the kinetic theory of gases. In particular, let's try to study the motion of gas molecules as they bounce around at random, colliding with one another and the sides of their container. In developing the kinetic theory of gases, we can make use of Newton's laws of motion and the general gas law described in Chapter 12. Our objective is to substantiate the interpretation of temperature that was given in the last section, that is, to relate the absolute temperature of a gas to its thermal energy.

The first step in the analysis is to use Newton's second law to derive a formula for the pressure P exerted by the molecules in a closed box of volume V. Let's assume that there are N molecules, each of mass m, in the box, and to keep the discussion simple, that the molecules all have equal speeds.

* In translational motion all particles of a body are moving parallel to each other in the same general direction without rotating about an axis.

(A derivation that considers the different speeds of the molecules is given in *For Further Study* at the end of this chapter.) Also, for the moment, let's imagine that the molecules are all moving in the same direction. In Figure 13-2 this is shown as the x-direction, that is, all the molecules at that particular instant are moving toward the face bc of the box. We would like to find the force on the face bc due to impacts of molecules as they strike and bounce back. This problem is similar to the problem of the fire hose sending water against a burning building (Example 6-3). Now, as then, it is convenient to use Newton's second law in the form: Net force = rate of change of momentum. Each gas molecule in the box has a momentum $+mv$ before collision with the wall, and a momentum $-mv$ after collision, assuming that it bounces back without loss of energy. The change of momentum for each molecule is

$$(+mv) - (-mv) = 2mv$$

at each collision with the face bc. In a time t, a molecule moves a distance equal to (v m/sec) $\times$ (t sec) $= vt$ meters. Also, the molecule has to travel twice the length of the box, a distance of $2a$ meters, between successive collisions at the same surface. The number of collisions for each molecule in time t is, therefore

$$\frac{\text{distance traveled in time } t}{\text{distance traveled between collisions}} = \frac{(vt \text{ meters})}{(2a \text{ meters/collision})} = \frac{vt}{2a} \text{ collisions}$$

Now we apply the second law:

$$\text{net force} = \frac{\begin{pmatrix}\text{change in}\\ \text{momentum}\\ \text{per collision}\end{pmatrix}\begin{pmatrix}\text{total}\\ \text{number of}\\ \text{collisions}\end{pmatrix}}{\text{time}} = \frac{(2mv)(vt/2a)}{t} = \frac{mv^2}{a}$$

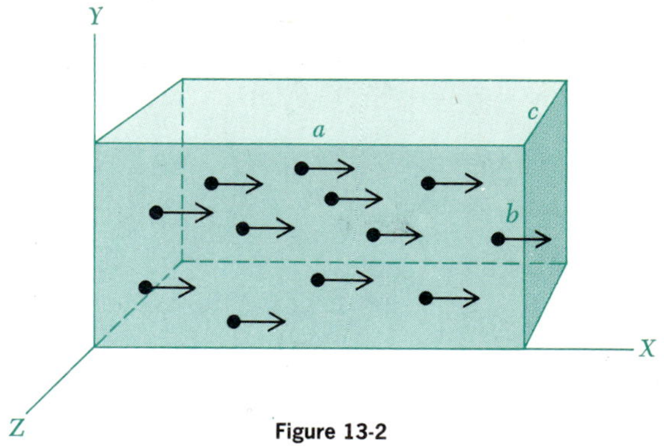

Figure 13-2

Thus the force exerted by each molecule is $(mv^2)/a$. There are N molecules, so the total force is $(Nmv^2)/a$. The pressure on the face bc is force/area:

$$P = \frac{Nmv^2}{(a)(bc)} = \frac{Nmv^2}{V}$$

where $V = abc$ = volume of the box. To make our derivation more realistic, we should take account of the fact that the molecules are moving in random directions, not just parallel to the x-axis. This introduces a factor of $\frac{1}{3}$, since on the average a molecule has a component of velocity along any one of three axes. Only $\frac{1}{3}$ of the momentum changes are "effective" in any one direction*. The result is, therefore,

$$P = \frac{1}{3}\frac{Nmv^2}{V} \qquad \textbf{(Equation 13-1)}$$

or

$$PV = \tfrac{1}{3}Nmv^2 \qquad \textbf{(Equation 13-2)}$$

Recall that our objective was to relate the temperature of a gas to its thermal energy. Before continuing, let's stop a moment in order to observe how this equation can be used to compute the speeds of molecules in gases.

*This is proved in the *For Further Study* section at the end of this chapter.

Example 13-1 What is the speed of the molecules in a flask containing 1 mole of oxygen (O_2) at 0° C and 1 atm pressure?

If there is one mole of oxygen present, the volume is 22.4 liters = 22,400 cm^3 = 2.24×10^{-2} m^3 (see Section 12-6). The pressure of 1 atm is equal to 1.01×10^5 N/m^2 (see Section 12-2). First, we solve Equation 13-2 for the velocity:

$$3PV = Nmv^2$$

$$v = \sqrt{\frac{3PV}{Nm}}$$

Since N is the number of molecules and m is the mass of each molecule, the product Nm is the total mass of gas. In this case the mass of a mole is 32 g = 3.2×10^{-2} kg since the molecular mass of O_2 is 32.

$$v = \sqrt{\frac{3(1.01 \times 10^5\ \text{N/m}^2)(2.24 \times 10^{-2}\ \text{m}^3)}{3.2 \times 10^{-2}\ \text{kg}}}$$

$$= \sqrt{21.2 \times 10^4\ \text{m}^2/\text{sec}^2}$$

$$= 4.60 \times 10^2\ \text{m/sec}$$

$$= \boxed{460\ \text{m/sec}}$$

This is an average speed. Individual O_2 molecules may have speeds much greater or smaller than this value.

Returning to Equation 13-2, $PV = \frac{1}{3}Nmv^2$, let's look at it more closely. Since KE $= \frac{1}{2}mv^2$, we recognize mv^2 as just twice the translational KE per molecule. Therefore, replacing mv^2 by 2KE, we get

$$PV = (\tfrac{1}{3}N)(2\text{KE per molecule})$$

or

$$PV = (\tfrac{2}{3}N)(\text{KE per molecule}) \quad \textbf{(Equation 13-3)}$$

This, then, is the result of applying Newton's second law to calculate the pressure due to moving molecules.

Recall that the general gas law based upon experiments with gases at various temperatures and pressures also makes a statement about the product PV:

$$PV = (nR)(T) \quad \textbf{(Equation 13-4)}$$

Since n, the number of moles, is proportional to N, the number of molecules (each mole has 6.02×10^{23} molecules), we can write Equations 13-3 and 13-4 in words as follows:

13-3: pressure times volume is proportional to number of molecules times KE per molecule

13-4: pressure times volume is proportional to number of molecules times absolute temperature

The conclusion is immediate and far reaching: *Absolute temperature is proportional to KE per molecule.* This is what we set out to prove. In so doing, we have not only strengthened our understanding of temperature; we have also strengthened our belief in the adequacy of Newton's laws. The derivation has been made for an ideal gas. The fact that most real gases deviate slightly from the general gas law can be accounted for by a detailed analysis including factors that we have neglected, such as the slight cohesive forces between molecules and the volume of the molecules themselves.

Although our derivation was made for gases, the conclusion is also true for liquids and solids. Imagine an aluminum cup full of water, sitting on a table at 20° C (Figure 13-3). The water and aluminum molecules are being continually bombarded by air molecules*, and there would be a tendency for an exchange of energy (heat flow)

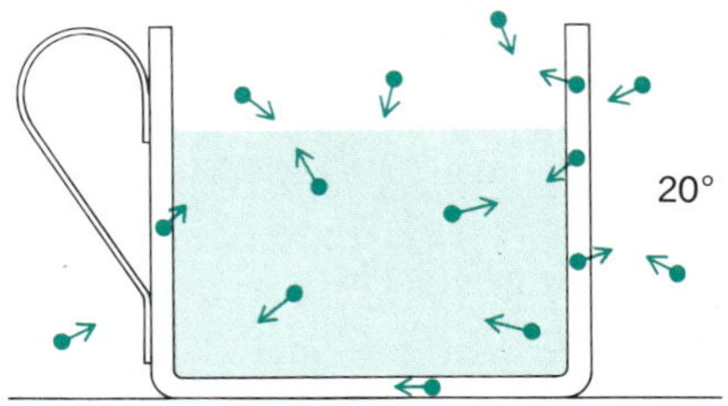

Figure 13-3

* Strictly speaking there are no *air* molecules. A better expression would be molecules of the gases that make up the air.

unless the average KE's were the same for all the molecules. We know that if the temperatures of air and water are the same, there is no heat flow between them. Hence the average KE's are the same in air and water at the same temperature. Similarly, if no heat flows through the cup the average KE of the vibrating aluminum molecules is the same as that of the air and the water molecules.

Here is an example that will show the usefulness of the kinetic theory of gases in verifying an earlier hypothesis. In 1811 Avogadro suggested that, at any temperature and pressure, the number of molecules per unit volume is the same for all gases (Section 12-4). The kinetic theory, which came later, gives a direct proof of Avogadro's brilliant hypothesis. Solving Equation 13-1 for the number of molecules divided by the volume, we get for the number per unit volume $N/V = 3P/mv^2$. Now the product mv^2 is proportional to the absolute temperature T. Thus at any given temperature and pressure, both P and mv^2 are fixed. Therefore N/V, the number of molecules per unit volume, is completely determined by the temperature and the pressure, and this ratio is independent of other things, such as the identity of the gas. We now know that if the temperature is 0° C and the pressure is 1 atm, the ratio N/V equals 6.02×10^{23} molecules/22.4 liters for any ideal gas.

13-3 THE FIRST LAW OF THERMODYNAMICS

The science of thermodynamics is a powerful method for analyzing the relations among temperature, quantity of heat, and mechanical work. The first law of thermodynamics has already been referred to as simply a restatement of the law of conservation of energy to include thermal (internal) energy (Section 11-3). This can now be elaborated on with the following quantitative example. Whenever mechanical energy (KE or PE) is transformed into thermal energy, the amount of work done in the transformation is equal to the increase in thermal energy of the system. This is illustrated in Figure 13-4. The falling weight has 100 J of energy just before impact with the table. In coming to rest, 100 J of work is done in transforming the KE of the weight into an increase in thermal energy of the weight and table. The quantity of heat developed in this process is 100 J(1 cal/4.18 J) = 23.9 cal. In the overall process, the total energy of the system is conserved.

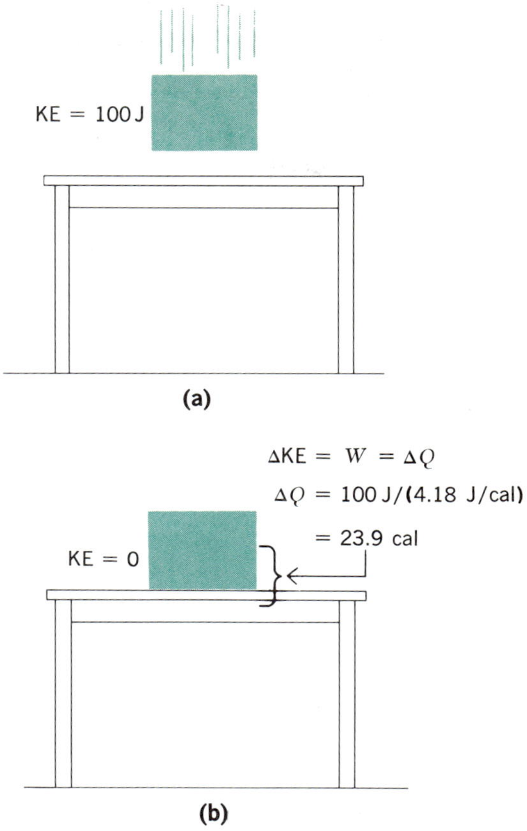

Figure 13-4

You already know that it is not necessary for work to be done to change the thermal energy of a system (Section 11-2). If we allow heat to flow from one place to another, we accomplish the same thing. It is in this respect that work and heat flow are analogous. They are both means for energy changes. In Figure 13-4b, for example, the weight and table could just as well have been placed next to a hot furnace (Figure 13-5). The in-

Figure 13-5

ternal energy of the system would increase as heat flowed from the furnace to the system of weight and table. There would be no way of telling whether the temperature rise of the system (weight and table) was due to the mechanical energy of the falling weight being transformed into thermal energy or whether the increase in thermal energy was due to the transfer of heat from the furnace. In keeping with these ideas, let's look again at the first law of thermodynamics restated now in a more rigorous manner.

The first law of thermodynamics: All the heat added to a closed system can be accounted for as mechanical work, an increase in thermal energy, or both.

Example 13-2 A 200 g hailstone at 0° C is falling at 20 m/sec when it strikes the ground. If $\frac{1}{5}$ of the heat generated remains in the ground and $\frac{4}{5}$ remains in the hailstone, how much ice is melted?

The closed system here is earth + hailstone. The original KE of the hailstone is

$$\begin{aligned} \text{KE} &= \tfrac{1}{2}mv^2 \\ &= \tfrac{1}{2}(0.200 \text{ kg})(20 \text{ m/sec})^2 \\ &= 40 \text{ J} \end{aligned}$$

Of this energy, $\frac{1}{5}$, or 8 J, remains in the ground, and 32 J is available to melt some ice. The latent heat of fusion of ice is 80 cal/g, and hence the amount melted is

$$\begin{aligned} m &= \frac{(32 \text{ J})(1 \text{ cal}/4.18 \text{ J})}{80 \text{ cal/g}} \\ &= \boxed{0.0957 \text{ g}} \end{aligned}$$

The investigations of Joule and others showed that the rate of exchange, 4.18 J = 1 cal, is the same whether the heat is generated by friction between solid bodies, by internal molecular friction (when water or mercury is stirred), by electrical heating in wires, or in some other way.

13-4 THE SECOND LAW OF THERMODYNAMICS

It is more difficult to give a brief and all-inclusive statement of the second law of thermodynamics, partly because there are so many ways of stating it, all of them equivalent. The second law expresses a *trend* in nature. To put it very simply, heat tends to flow from a hot place to a cold place. If we mix 1 kilogram of boiling water at 100° C with 1 kilogram of ice-cold water at 0° C, the result is 2 kilograms of water at 50° C; such calculations were made in Chapter 11. Now the total energy of the system has remained constant—as much heat was lost by the hot water as was gained by the cold water (first law of thermodynamics). However, nothing in the first law says that the reverse process could not take place: Heat *might* leave the cold water (thereby freezing it) and flow into the hot water (causing it to boil). For that matter, the first law would still hold even if the molecules of the 2 kilograms of 50° water were to sort themselves out into two groups, with the faster ones in one part of the bucket and the slower ones in another part, setting up a temperature difference again. All these unlikely happenings are contrary to the second law, for in each case heat would be flowing spontaneously from cold to hot. The second law says that it is highly unlikely that heat will flow from cold to hot—so unlikely that we can safely consider it an impossibility in everyday life.

On a very small scale, say a dozen molecules in a small region of a container of gas, random collisions might act together to make little "pockets" of hot spots (faster molecules) and cold spots (slower molecules). Averaged out, there is no lasting observable effect. Ordinarily we deal not with a dozen, but with 10^{23} or so molecules. Under

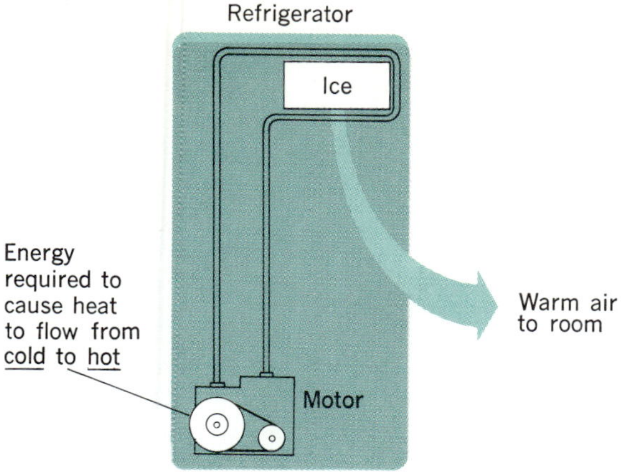

Figure 13-6 Energy must be expended to remove heat from a refrigerator.

such circumstances, highly improbable means practically impossible. Such considerations of probability are at the very heart of the second law.

It *is* possible to get heat to flow "uphill," but it requires expending some energy and doing some work. The refrigerator is a good example: heat is caused to flow from a cold place (the ice cubes) to a warm place (the kitchen). This is done, however, only because a motor is turning, expending electrical energy as the price to be paid for the backward transfer of heat (Figure 13-6).

The unavailability of heat is a key problem. There is a tremendous amount of thermal energy in the oceans, for instance, but it is not available for doing work unless there is a region of lower temperature to which the heat can flow. Energy is "available" only when it can be converted into mechanical or "useful" work.

The science of thermodynamics had its origin in the studies of N. L. Sadi Carnot (1796–1832), a young French engineer and physicist who tried to improve the efficiency of steam engines. His conclusions were general and apply to steam engines, diesel engines, gas turbines, and all other devices that convert heat into work. A *heat engine* is a device for getting useful mechanical work out of the heat supplied from a burning fuel. One way of representing a heat engine is shown by the diagram in Figure 13-7. There is a source of thermal energy A from which heat flows to the engine. The engine transforms some of this heat into useful work and exhausts the "unused" heat to a reservoir B. In a steam engine, for example, A would be the boiler in which the superheated (heated above 100° C) steam is generated, and B would be the condensing chamber that receives the cooled-down steam after it has passed through the engine (cylinder and piston). According to the second law of thermodynamics, A must be at a higher temperature than B for heat to flow from A to B. The heat Q_2 rejected by the engine to the reservoir is less than the heat input Q_1 received by the engine from the source. The remainder of the heat is converted into useful work W. By the first law, the heat added to the engine must equal the work done by the engine plus the heat rejected:

$$Q_1 = Q_2 + W$$

Hence

$$W = Q_1 - Q_2$$

How efficient is such an engine? In general, the efficiency of the engine can be expressed by the

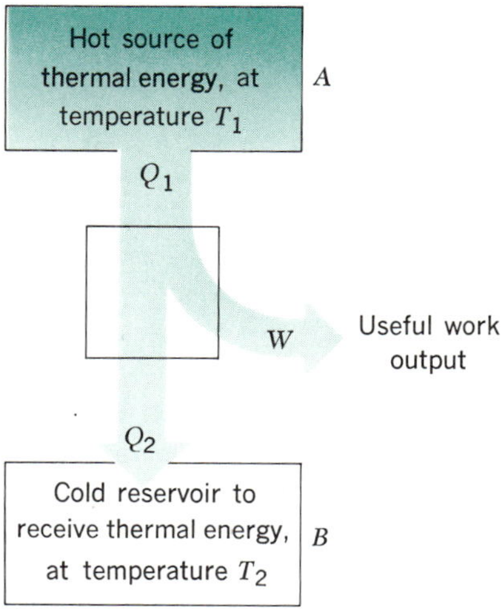

Figure 13-7 Division of energy in a heat engine.

ratio of useful work output to heat input:

$$\text{efficiency} = \frac{W}{Q_1} = \frac{Q_1 - Q_2}{Q_1}$$

Carnot proved that for an ideal engine (one in which there are no frictional losses or losses due to heat conduction) the efficiency $(Q_1 - Q_2)/Q_1$ would have a maximum value given by

$$\frac{Q_1 - Q_2}{Q_1} = \frac{T_1 - T_2}{T_1}$$

where these temperatures are absolute temperatures of source and reservoir. Even if we neglected frictional losses and other losses and imagined an engine that had this theoretical Carnot efficiency, it would still not be 100 percent efficient. To achieve 100 percent efficiency the cold reservoir would have to be at absolute zero. Then, if $T_2 = 0$, the efficiency would be T_1/T_1, or 100 percent. This would mean that $Q_2 = 0$, the engine would reject no heat, and thus *all* of Q_1, the heat input, would be converted into useful work.

Since it is impossible to have a cold reservoir that is at absolute zero, a heat engine in which a given amount of heat is converted completely into mechanical work cannot be constructed. This is another way of stating the second law of thermodynamics.

Actual heat engines, then, are less than 100 percent efficient, even in the absence of frictional or other losses. An engine with a maximum possible (Carnot) efficiency, that is, where $(Q_1 - Q_2)/Q_1$ has its maximum value, is called a *Carnot engine*. As you know, most real engines such as a steam engine or a gas turbine, cannot be designed to have the Carnot efficiency, even in theory. Frictional losses and conduction of heat through the metal parts of the engine lower the efficiency still more. To illustrate these ideas: In a simple type of steam engine, the heat input might be at 200° C. The steam is allowed to expand, pushing a piston, and it condenses into water at 100° C. According to Carnot's theorem, the greatest possible efficiency for this operation is

$$\frac{T_1 - T_2}{T_1} = (473°\text{ K} - 373°\text{ K})/473°\text{ K} = 21\%$$

In other words, at least 79 percent of the heat input is unavailable for doing work, and this 79 percent is carried away in the hot water that is the end result of the cycle. In practice, frictional and other losses cut down the overall efficiency of a piston-type steam engine from 21 percent to about 10 or 12 percent. The efficiency can be improved by operating at higher temperatures and by improved mechanical design.

Example 13-3 The most efficient engine of the piston type yet constructed is a turbocharged, dual fuel engine developed by the Cooper Industries (Figure 13-8). The working substance, a mixture of fuel and air, operates between temperatures of 1870° C in the firing chamber and 430° C in the exhaust. The overall efficiency, including all frictional losses, is such that for every 6.85×10^9 cal of heat used, 1.20×10^{10} J of useful mechanical energy is produced. Calculate (a) the Carnot efficiency, (b) the actual efficiency.

(a) For this engine, with $T_1 = 2143°$ K and $T_2 = 703°$ K, the Carnot efficiency is

$$(2143°\text{ K} - 703°\text{ K})/2143°\text{ K} = \boxed{67\%}$$

(b) The efficiency of the engine as actually constructed is

$$\frac{\text{work output}}{\text{work input}} = \frac{1.20 \times 10^{10}\text{ J}}{(6.85 \times 10^9\text{ cal})(4.18\text{ J/cal})} = \boxed{42\%}$$

The term *entropy* is used to describe the *unavailability of energy*. Entropy is a measurable property of a body, just as is its temperature or its internal energy. In the example of mixing 1 kilogram of hot water and 1 kilogram of cold water (page 218), *some* energy was available at the start of the operation, since a heat engine could (conceivably) have been operated to make use of

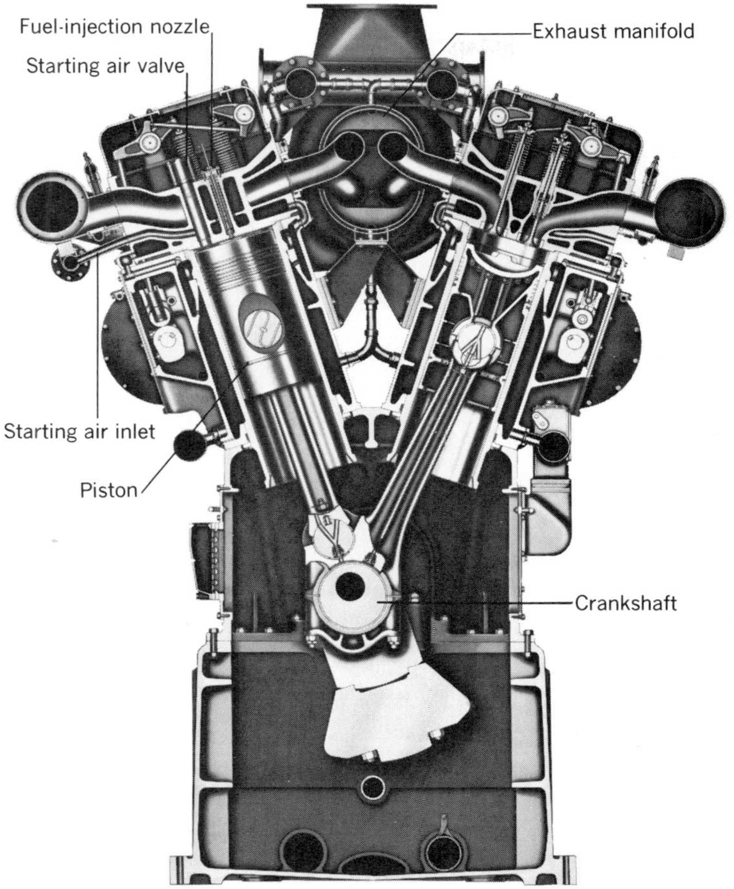

Figure 13-8 Cross-section of the Cooper-Bessemer 16-cylinder gas diesel turbocharged engine, which can be adapted for use either as a conventional diesel or spark-ignited gas engine.

the temperature difference between the two bodies of water. However, after the mixing this energy is no longer available because there is no temperature difference. The energy of the system has remained constant (first law), but the entropy of the system has increased. Now more of the energy is unavailable than at the start of the process.

The entropy of a system can be thought of as a measure of the "disorder" of its molecular motions. In the mixing of the hot and cold water, as we have said, heat *might* leave the cold water and flow into the hot water. This would be a *decrease* of molecular disorder, or a *decrease* in entropy, as shown in Figure 13-9a on page 222. But such a heat flow from cold to hot does not occur, according to the second law of thermodynamics. The reverse process, however, the tendency for molecules to move toward a disordered grouping as the system reaches equilibrium temperature, is the most likely thing to happen (see Figure 13-9b). The entropy of the system increases as the molecular grouping becomes more disordered or mixed up. Entropy is such a fundamental property that

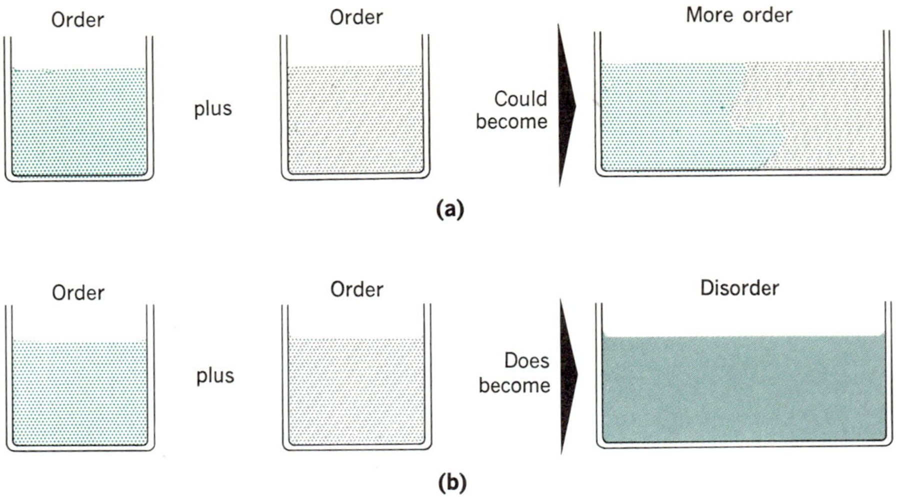

Figure 13-9

the second law may be stated in terms of it.

The second law of thermodynamics: The entropy of the universe never decreases; during any process, the entropy either remains constant or else increases.

Eventually, when the entropy of the universe has reached its maximum value, when its molecules have attained maximum disorder, everything will be at the same temperature (estimated to be a few degrees above absolute zero), and there will be no way to convert the thermal energy of the universe into useful mechanical work. This "heat death" is far in the future.

13-5 ABSOLUTE ZERO

So far, absolute zero has been used simply as a tool for describing the behavior of ideal gases. The kinetic theory of gases showed that for an ideal gas the KE of the molecules is zero at absolute zero. At temperatures of 5 or 10 degrees above absolute zero, however, most gases are far from ideal, if indeed they have not yet condensed into the liquid or solid state. It is questionable, therefore, whether our definition of absolute zero has much meaning, since the gas laws used to define absolute zero no longer work well near absolute zero. There is a better way of defining absolute zero, based on Carnot's work with heat engines.

We found that the efficiency of a Carnot engine would be 100 percent if the temperature of the cold reservoir were at absolute zero (Section 13-4). For 100 percent efficiency, *all* the heat Q_1 would be converted into useful work, so that no heat would be rejected to the reservoir. This leads to Kelvin's definition of absolute zero:

Absolute zero is the temperature of a reservoir to which a Carnot engine would reject no heat.

And this is the only definition that will stand up under close scrutiny.

Absolute zero is sometimes defined (erroneously) as the temperature at which all random molecular motion stops. If there *is* random molecular motion that can be utilized, then surely we are not at absolute zero. But this is not the whole story. Even if the molecules were devoid of random translational, rotational, or vibrational KE, there might still be sources of available energy, such as magnetic energy, electric energy, or some as yet unknown form of internal energy. Such molecules (in theory, at least) could run an engine and so would *not* be at absolute zero, since there would then be a lower temperature to serve as a reservoir.

Of course, the magnetic and other energies are usually of no significance. At ordinary temperatures, the molecular KE so far overshadows these smaller energies that it may well be said that absolute temperature is proportional to the KE of the molecular motion. At very low temperatures, below 1° K, magnetic and other energies are the major source of available energy since molecular KE is, by comparison, already zero.

The quest for ever lower temperatures is one of the active fields of research in physics. A temperature of 10^{-6}° K has been attained. In some respects the properties of matter (its electrical conductivity for example) are radically different at very low temperatures, and low-temperature research is bringing us a better understanding of the properties of matter and the basic laws of physics.

SUMMARY

Absolute temperature of an ideal gas is proportional to the average random translational KE per molecule. **Thermal energy** is total internal energy, including PE and KE. The pressure exerted by the molecules of a gas can be computed from Newton's second law, and this allows determination of molecular speeds from observed data on temperature, pressure, and volume. At any given temperature, molecular speeds are less for molecules of greater mass, since the average KE's of molecules of all gases are the same at any given temperature.

The **first law of thermodynamics** is an extension of the law of conservation of energy, including thermal energy along with mechanical energy. The **second law of thermodynamics** may be stated in several ways: (a) heat tends to flow from a hot place to a cold place; (b) energy tends to become more and more unavailable for conversion from thermal energy to mechanical energy; (c) the **entropy** of the universe tends toward a maximum. **Heat engines** are devices for extracting useful work from a source of heat. It is impossible to convert all the heat from a source with 100 percent efficiency unless the cold reservoir is at **absolute zero.** A heat engine operated in reverse is a refrigerator in which heat flows from a cold place to a hot place. An external source of mechanical work is required to perform this feat.

At absolute zero, random molecular motions would cease. In addition, all other forms of available energy would have to be removed from the molecules.

QUESTIONS

1 What is the difference between "heat" and "temperature"?

2 Give two conditions, having to do with the molecules of a gas, under which the general gas law might fail to represent the behavior of that gas.

3 Suppose the molecules of a gas, rather than having a finite (though small) diameter, were tiny points. Would they make more or fewer collisions per second against the walls of the box in Figure 13-2? (Assume the same temperature, that is, the same speed.) Would the pressure be greater for molecules as tiny points or with a finite diameter?

4 In deriving Equation 13-1, we ignored the collisions that must be taking place between molecules in the box. Explain why this is justified. (*Hint:* See Section 6-3.)

5 The plunger of a bicycle pump is suddenly pushed down, compressing the air within the cylinder. Explain why the air becomes warm.

6 In a real gas there are small cohesive forces between molecules. If the valve of a tank of compressed air is opened and the gas allowed to expand suddenly, the metal parts of the valve become cold. Explain, on the basis of the kinetic theory of gases.

7 In some parts of the Indian Ocean the surface water is 23 C° warmer than the water 100 m below the surface. Would it be possible to operate a heat engine continuously, using this temperature difference?

8 Would it be possible to cool a room by placing an electric refrigerator with its door open in the room and allowing a fan to blow cool air from the ice cube compartment out into the room?

9 Construct a flow diagram, similar to Figure 13-7, that will apply to a a refrigerator. Show the work input, the heat absorbed, and the heat rejected for each. How would you define a "coefficient of performance" of a refrigerator (analogous to the efficiency of an engine)? (*Note:* the coefficient of performance of a typical household refrigerator is about 5.)

PROBLEMS

GROUP A

1 The total random translational KE of the water molecules in a 20 liter can is 4.4×10^6 J; the total random translational KE of the water molecules in a 400,000 liter swimming pool is 8×10^{10} J. Which body of water—can or pool—is the higher temperature?

2 A molecule traveling between two walls 0.8 m apart has an average speed of 320 m/sec. How many collisions does it make with one wall in 25 seconds?

3 An oxygen molecule (O_2) bounces back and forth between two walls of a cubical container 20 cm on an edge, and makes 500 round trips each second. What is its speed?

4 A heat engine absorbs 10 cal from the source of heat, and does 4.18 J of mechanical work. What is its efficiency?

5 What is the maximum efficiency of an engine that operates between fixed temperatures of 500° K and 400° K?

6 A heat engine performs 1000 J of work, and at the same time rejects 4000 J of thermal energy to the cold reservoir. What is the efficiency of the engine?

7 An engine that has 10% efficiency does 120 J of work. How much thermal energy does it take in from the hot reservoir?

8 An engine whose maximum (Carnot) efficiency is 40% obtains energy from a hot reservoir at 500° K. What is the temperature of the cold reservoir?

GROUP B

9 A machine gun fires 20 bullets per second. Each bullet has a mass of 1 g, is moving at 250 m/sec, and bounces back from the target with its speed unaltered. What is the force on the target?

10 Calculate the average speed of the molecules in a tank of H_2 gas (molecular weight 2.02) if the temperature is 819° C and the pressure is 3 atm.

11 A total mass of 150 kg of a certain gas is confined in a tank of volume 8 m^3, and the pressure is 5×10^5 N/m^2. What is the average speed of the molecules?

12 Suppose, in Example 13-1, that the pressure were 3 atm instead of 1 atm, the temperature still being 0° C. Show that Equation 13-2 will give the same velocity. In other words, show that the molecular speed for a given gas depends only on temperature, and not at all on pressure.

13 Make a dimensional check for Equation 13-2.

14 The best modern vacuum pumps are able to reduce the pressure in a container to about 10^{-12} mm Hg. (a) At such a pressure, how many molecules are in 1 cm^3 of gas at 0° C? (b) Assuming each molecule to be at the corner of a cube x cm on an edge, calculate x, which is the approximate average distance, in cm, between molecules in the gas.

15 At 0° C the average speed of certain molecules is v. What will the average speed of these same molecules become if the temperature rises to 819° C?

16 An engine, working at maximum efficiency between temperatures of 300° K and 500° K, rejects 1500 cal of heat during a certain time. How much work is performed during this period?

17 A Carnot engine takes in 50 kcal of heat from a reservoir at 900° K and performs 3×10^4 J of work. What is the temperature of the cold reservoir?

GROUP C

18 Starting with Equation 13-2, prove that the average speed of molecules of a gas of molecular mass W at a temperature T° K is given by the equation $v = \sqrt{3RT/W}$—where W is the molecular mass (commonly called molecular weight).

19 Prove that Equation 13-2 is also equivalent to $P = \frac{1}{3}dv^2$, where d is the density of the gas.

20 Make dimensional checks for equations derived in Problems 18 and 19.

21 A household refrigerator can be considered to be a heat engine operating in reverse, with the hot source at room temperature (say 27° C) and the cold reservoir at the temperature of the freezing compartment (−21° C). (a) What is the maximum efficiency of such a device, if operating as a heat engine? (b) Operating as a maximum-efficiency engine, how much work output would be obtained if enough heat flowed into the cold reservoir to transform 1 kg of ice cubes at −21° C into water at 27° C? (c) If this maximum-efficiency engine were reversed to act as a refrigerator, how much work input would be needed to move enough heat from the ice-cube compartment to the hot source in order to change 1 kg of water at 27° C to ice at −21° C.

FOR FURTHER STUDY

The Distribution of Molecular Speeds

In Section 13-2 we derived an expression for the pressure exerted by molecules moving in a box: $P = \frac{1}{3}Nmv^2/V$. The derivation is marred by certain weaknesses. For one thing, it was assumed that all molecules have the same speed v. Now let's proceed with a more detailed study of molecular motions, leading to the same result derived earlier with the aid of some simplifying assumptions.

We assume identical molecules, each of mass m, but suppose that the molecules have different speeds $v_1, v_2, \ldots, v_N$. Suppose also that the molecules are moving through the box in various directions, not all parallel to the x-axis, as in Figure 13-10. Molecule *1* has velocity v_1 (a vector), and the components of its velocity along the x-, y-, and z-axes have magnitude denoted by v_{1x}, v_{1y}, and v_{1z}. Molecule *2* has components v_{2x}, v_{2y}, and v_{2z} (for simplicity only v_{2x} is shown), and so forth for molecules *3*, *4*, . . . , *N*. Note that molecule *3* is moving along the y-axis, so v_{3x} and v_{3z} are both zero.

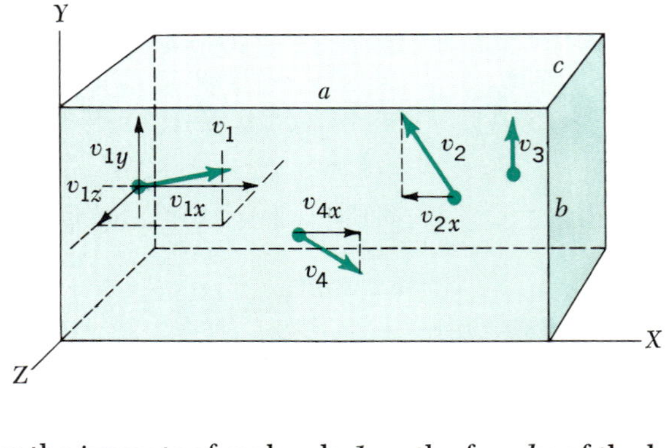

Figure 13-10

Let's consider the impacts of molecule *1* on the face bc of the box. Molecule *1* approaches this face with a velocity v_{1x}, and the x-component of the change of momentum at each collision is $2\,mv_{1x}$. Molecule *1* makes $v_{1x}/2a$ collisions per second with face bc. By Newton's second law, the force on bc due to repeated impacts of molecule *1* is the rate of change of momentum:

$$F_1 = (2mv_{1x})\frac{v_{1x}}{2a}$$

$$= \frac{mv_{1x}^2}{a}$$

This is similar to the result obtained in Section 13-2, where the molecule itself was considered to be moving in the x-direction. (The components of velocity v_{1y} and v_{1z} do not contribute at all to the force on bc.) Similarly, molecule 2 causes a force on bc equal to

$$F_2 = \frac{mv_{2x}^2}{a}$$

The total force of all N molecules on face bc is the sum of the forces caused by the separate molecules:

$$F = \frac{mv_{1x}^2}{a} + \frac{mv_{2x}^2}{a} + \cdots + \frac{mv_{Nx}^2}{a}$$

$$F = \frac{m}{a}(v_{1x}^2 + v_{2x}^2 + \cdots + v_{Nx}^2)$$

$$F = \frac{m}{a}(\Sigma v_x^2) \qquad \text{(Equation 13-5)}$$

The Greek letter Σ (sigma) is used here to denote a sum. To evaluate the sum, we recall that the magnitude of a vector is found by applying the pythagorean theorem to the perpendicular components. This theorem is general and applies to vectors in three dimensions as well as two. Thus $v_1^2 = v_{1x}^2 + v_{1y}^2 + v_{1z}^2$. And the same holds true for v_2, v_3, etc. Let's make a table and form the sum of each column:

$$\begin{aligned}
v_1^2 &= v_{1x}^2 + v_{1y}^2 + v_{1z}^2 \\
v_2^2 &= v_{2x}^2 + v_{2y}^2 + v_{2z}^2 \\
v_3^2 &= v_{3x}^2 + v_{3y}^2 + v_{3z}^2 \\
&\vdots \\
v_N^2 &= v_{Nx}^2 + v_{Ny}^2 + v_{Nz}^2 \\
\hline
\Sigma v^2 &= \Sigma v_x^2 + \Sigma v_y^2 + \Sigma v_z^2
\end{aligned}$$

As an example, suppose molecule *1* has a large velocity, more or less diagonally through the box, with components (3, 2, 3) in some system of units. Suppose molecule *2* is moving upward and to the left, with components (−1, 2, 0). If molecule *3* is moving upward, its components would be (0, 3, 0). The table of values would look like this:

$$\begin{aligned}
v_1^2 &= 9 + 4 + 9 \\
v_2^2 &= 1 + 4 + 0 \\
v_3^2 &= 0 + 9 + 0 \\
\hline
\Sigma v^2 &= 10 + 17 + 9
\end{aligned}$$

We now make the assumption that the molecules are moving in *random directions.* That is, there is no breeze blowing through the box, and there is no preferred direction of motion for the molecules. In the example with only 3 molecules, the y-direction is distinctly favored; but in the case of a measurable quantity of gas,* it is surely safe to say that if the molecules are in random motion, then

$$\Sigma v_x^2 = \Sigma v_y^2 = \Sigma v_z^2$$

Therefore, the equation following Equation 13-5 becomes

$$\Sigma v^2 = \Sigma v_x^2 + \Sigma v_x^2 + \Sigma v_x^2$$

or

$$\Sigma v_x^2 = \tfrac{1}{3}\Sigma v^2$$

Substituting into Equation 13-5, we obtain

$$F = \frac{m}{a}\left(\frac{1}{3}\Sigma v^2\right) \qquad \text{(Equation 13-6)}$$

Since there is no speed common to all the molecules, the next step is to introduce some sort of average speed. Guided by Equation 13-6, the mean-square speed $\overline{v^2}$ can be defined as

$$\overline{v^2} = \frac{v_1^2 + v_2^2 + \cdots + v_N^2}{N}$$

$$= \frac{\Sigma v^2}{N}$$

It is the average, or "mean," of the squares of the speeds of the N molecules. Multiplying by N, we see that

$$\Sigma v^2 = N\overline{v^2}$$

and hence

$$F = \frac{1}{3}\frac{m}{a}N\overline{v^2}$$

The rest of the proof is similar to that in Section 13-2:

$$P = \frac{\text{force}}{\text{area}}$$

$$= \frac{1}{3}\frac{m}{abc}N\overline{v^2}$$

$$PV = \frac{1}{3}Nm\overline{v^2} \qquad \text{(Equation 13-7)}$$

This means that PV is proportional to the average of the squares of all the speeds of the molecules, rather than to the square of any one speed as in Equation 13-2.

* Even a millionth of a mole of gas contains 6×10^{17} molecules.

In Example 13-1 we found that $v = \sqrt{3PV/Nm}$. How shall we interpret speed computed in this way? From Equation 13-7 it can be seen that $\overline{v^2} = 3PV/Nm$, so $\sqrt{\overline{v^2}} = \sqrt{3\ PV/Nm}$. Thus we are calculating the *root-mean-square* (rms) *speed*—the square root of the mean of the squares of the speeds. Without further specification, we mean the rms speed ($\sqrt{\overline{v^2}}$ or v_{rms}) when we speak of the "speed" of the molecules in a gas. There are several other ways of describing the distribution of molecular speeds. For instance, we could simply average the speeds. This would result in the *mean speed* (v_m). Another index of molecular speeds which is sometimes useful is the *most probable speed* (v_p), that speed at which the greatest number of molecules move.

PROBLEMS

GROUP D

22 Ten molecules have speeds as follows, in arbitrary units: 1, 3, 0, 2, 2, 4, 3, 6, 1, 2. Calculate (a) the most probable speed; (b) the mean speed; (c) the rms speed.

23 Prove that the total translational KE of N molecules moving in random directions is given by $\frac{1}{2}$(total mass)($v_{\text{rms}}{}^2$), and hence, insofar as KE and absolute temperature are concerned, the molecules behave as if all had the same speed, equal to the rms value.

UNIT 4

References and Outside Reading

The purpose of the following list is to make you aware of additional reading that will extend and broaden most of the concepts you have met in this unit. These references are easily available and are of sufficient variety to meet your individual interest.

"Negative Absolute Temperatures," by Warren G. Proctor, in *Scientific American,* August 1978. The strange concept of negative absolute temperatures: ". . . not colder than zero but . . . hotter than infinity."

"Amateur Scientist" section, in *Scientific American,* May, 1956, page 149, reports on two unusual simple rubber-band heat engines. (Rubber under tension contracts when heat is applied.)

Count Rumford, by S. C. Brown, Science Study Series, Garden City, N.Y., Doubleday Anchor, 1962. The remarkable biography of a man with a dual personality.

"Count Rumford," by M. Wilson, in *Scientific American,* October, 1960.

Cryogenics, by M. McClintock, New York, Reinhold Books, 1964. An interesting and descriptive account of the important subject of extreme low-temperature research and some of its applications.

The Early Development of the Concepts of Temperature and Heat: The Rise of the Caloric Theory (Harvard Case Histories in Experimental Science No. 3) by D. Roller, Cambridge, Mass., Harvard University Press, 1950. See especially Rumford's investigation on weight ascribed to heat, his experiments on heat produced by friction, and Davy's experiments. *Caution:* The early writers sometimes used the words "heat" or "degree of heat" where we use "temperature" today. The context will tell whether temperature or quantity of heat is meant.

Heat Engines, by F. Sandfort, Science Study Series, Garden City, N.Y., Doubleday Anchor, 1962. Thermodynamics in theory and practice. See especially Chapters 3, 4, 5, and 6.

Knowledge and Wonder, by V. Weisskopf, Science Study Series, Garden City, N. Y., Doubleday Anchor, 1963. See Chapter 4. (On Unit 2 reference list.)

"Alternative Automobile Engines," by David G. Wilson, in *Scientific American,* July 1978. Examines the advantages and limitations of engines other than the familiar gasoline type.

One, Two, Three . . . Infinity, by G. Gamow, New York, Mentor Books, 1957. See Chapter 8. (On Unit 2 reference list.)

The Quest for Absolute Zero: the Meaning of Low Temperature Physics, second edition, 1978, by K. Mendelssohn, New York, Halsted. Describes the men involved in research in low-temperature physics, and some of the important and startling phenomena that arise.

Robert Boyle's Experiments in Pneumatics (Harvard Case Histories in Experimental Science No. 1), by J. B. Conant, Cambridge, Mass., Harvard University Press, 1950. See especially the Forward, the Introduction, and Boyle's law.

A Source Book in Physics, by W. F. Magie, New York, McGraw-Hill, 1935, also, Cambridge, Mass., Harvard University Press, 1963. Selections from the writings of Newton, Amontons, Fahrenheit, Taylor, Black, Rumford, Davy, Gay-Lussac and others. See especially the work of Black on latent heat, Rumford on heat produced by friction, and Davy's experiments on heat produced by friction.

Water: The Mirror of Science, by Davis and Day, Science Study Series, Garden City, N.Y., Doubleday Anchor, 1961. Chapters 1 and 5. More on the internal energy of water.

"What is Heat?," by F. J. Dyson, in *Scientific American,* September, 1954. Further development of the concepts of heat, temperature, and thermal energy.

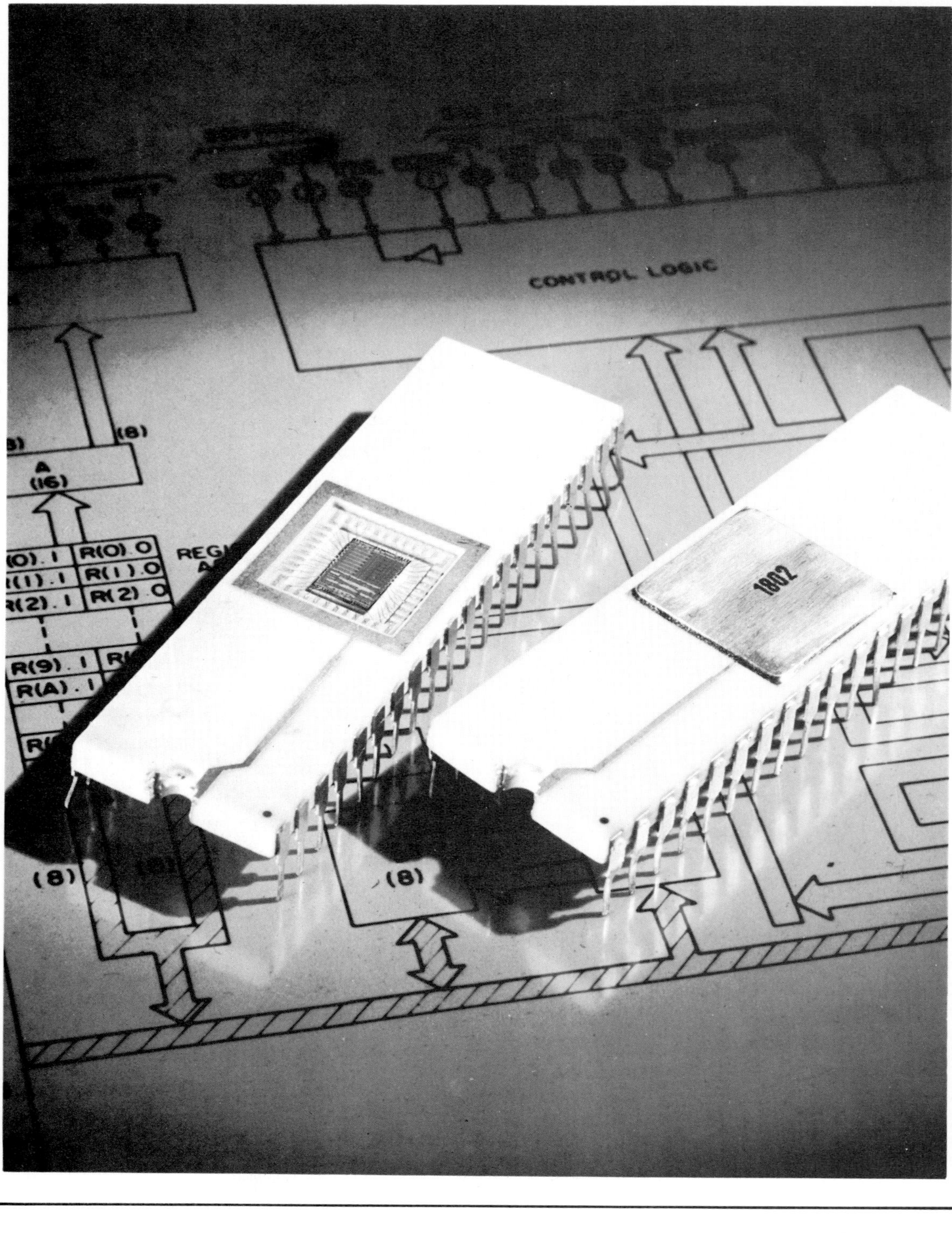
CONTROL LOGIC
(8)
A
(16)
R(O).1
R(O).O
R(1).1
R(1).O
R(2).1
R(2).O
R(9).1
R(A).1
(8)
(8)
1802

UNIT 5 ELECTRICITY AND ELECTROMAGNETIC RADIATION

You have seen how Newton's laws of motion and the concepts of energy and momentum are applied to the motions of great numbers of very small particles, atoms and molecules. The concepts made clear that thermal energy is just another form of energy, the random mechanical energy of these particles. You have also investigated the forces of interaction that arise when bodies collide.

You will now meet a new kind of force—the electric, or Coulomb, force. This force is associated with electric charge. Newton's laws do apply to charged particles, but the motions of vast numbers of tiny charged particles would be virtually impossible to analyze by these laws. The energy concept, however, allows us to make calculations about their overall behavior. It also helps us understand some of the recent achievements in science and technology—such as the microprocessor shown here against a block diagram of its internal structure.

14 Electric Forces – Charges at Rest

From electric light to television, from simple ammeter to complex computer, it is hardly necessary to point out that most of our modern devices depend upon the use of electricity. The study of electricity has a long history. Some electrical phenomena were certainly familiar to the ancient Greeks, though no clear concepts were developed until the eighteenth century. Then came many profound changes in our understanding and application of electrical effects in the nineteenth and twentieth centuries.

This chapter and the six that follow will be devoted to the study of electricity. We shall begin with certain simple experimental facts, and for this we will use some terms that have not been clearly defined yet. You will gain an idea of the meaning of these terms by working with them.

14-1 ELECTRIC AND MAGNETIC FORCES

The picture of an atom as a dense central nucleus surrounded by a cloud of electrons is probably quite familiar to you (Figure 14-1). The nucleus is said to have a "positive electric charge," and the electrons are said to have a "negative electric charge." You will find out more about the terms positive and negative as we investigate electrical effects experimentally. The experiments described in the text will introduce the concept of a "charged body;" closely related is the concept of "electric charge." A charged body can be any object that possesses an electric charge. It may be of ordinary size (a balloon or a cloud) or very small (an electron or an atom). Sometimes you will consider the motion of charged bodies, and sometimes just the motion of the charge itself.

In terms of the atomic model, a charged body is usually one that has gained or lost some electrons in the outer clouds of its atoms. If a body has lost electrons, it has a deficiency of negative charge, and thus has a net positive charge. If a body has gained electrons, it has a net negative charge.

The fundamental electrical effects are really forces of two general kinds. The kind of force found between charged bodies both at rest and in motion, and which therefore does not depend upon motion, is called the *Coulomb force.* This will be discussed in Section 14-4. In addition to the Coulomb forces that are *always* present, there are forces that arise only when two charged bodies are in motion relative to each other. These forces due to the *motion* of charged bodies are called *magnetic forces* and are discussed in detail in Chapter 18. Magnetism is thus a subdivision of electricity, dealing with the effects of charges that are in motion.

Another way to look at these two kinds of force is to make a distinction between electric charge and electric current. When electric charge is moving from place to place, such a flow of charge is called an *electric current.* Using this terminology, we can say that Coulomb forces are caused by electric charges, as such, while magnetic forces are caused by electric currents.

14-2 ELECTRIFICATION OF BODIES

It has long been known that small bits of paper, fluff, and so on, can be "picked up" by a rod of glass, hard rubber, amber, or plastic that has been rubbed with fur, silk, or cloth. The phenomenon was observed by the Greeks, whose word for amber was *elektron.* In the sixteenth century, following the suggestion of William Gilbert, the word "electrified" was used to describe a piece of amber, glass, or other material that could exert a force on a small test body such as a scrap of paper on a silk thread. These long-range electric forces could be observed at distances up to several centimeters. Not until the early nineteenth century did investigators come to the conclusion that there are two, and only two, kinds of electricity; two kinds of charge. For instance, let body *A* (Figure 14-2), be a glass rod that has been rubbed with silk, and let *B* be a rubber rod that has been rubbed with fur. We observe a force of *attraction* between *A* and *B;* if *C* happens to be a glass rod rubbed with silk (that is, if *C* is identical to *A*), then we find that *A* is *repelled* by *C.*

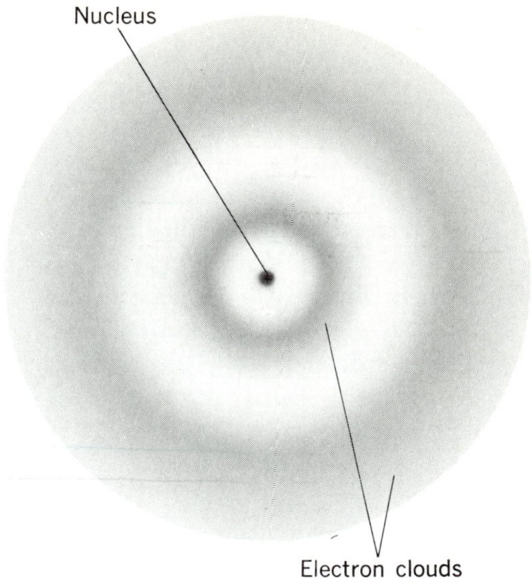

Figure 14-1

Body *A* and body *B*, as just described, are oppositely charged and these are the only two kinds of charges known. No one has ever succeeded in finding a third kind of charge that will be attracted by both *A* and *B*, or repelled by both *A* and *B*.

Historically, there has been a great deal of needless confusion over the naming of these two kinds of charge. The kind of electricity usually found on rubbed glass was called vitreous, which simply means "derived from glass." We now call this kind of charge positive (+). The kind usually

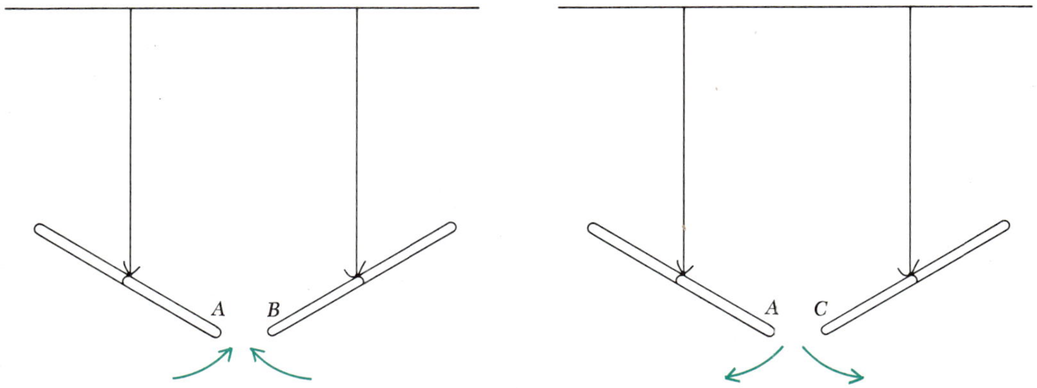

Figure 14-2

found on rubbed amber was called resinous because amber is a fossilized resin. This, of course, is now called negative (−) charge. Charles du Fay (1698–1739) advocated a "two-fluid" concept of electricity, in which vitreous electricity and resinous electricity were placed on equal footing. Benjamin Franklin (1706–1790) proposed a "one-fluid" theory in which vitreous electricity was an excess of positive electricity, and resinous electricity was a deficiency of positive electricity. Franklin therefore called resinous electricity "negative." We now know there *are* two kinds of electric charge, and so the two-fluid theory is actually "correct." The one-fluid theory has merit, however, since in many large-scale applications (such as the familiar electric currents in metal wires) only one type of electricity is free to "flow." Franklin's choice of the terms positive and negative has caused some confusion, since the free electrons that move through wires and light our lamps and turn our motors are of the negative sort. This negative electricity actually exists, and is not merely a deficiency of positive electricity.

In a solid body, the nuclei are fixed in position except for the slight to-and-fro vibrations that are the manifestation of thermal energy. At higher temperatures the nuclei vibrate with greater amplitude. However, unless the solid melts, nuclei so rarely leave their places in the crystal structure that they take little part in the electrical phenomena that are observed in solids. On the other hand, in certain substances some of the orbital electrons are rather easily detached from the cloud surrounding a nucleus, and become the "free electrons" which play a large part in electrical effects.

Electrification by rubbing really means electrification by contact. If a glass rod is touched momentarily to a piece of silk, electrons have a tendency to leave the glass and attach themselves to the silk. This leaves an excess positive charge on the rod at the point of contact. Rubbing the glass with the silk is a more effective way of electrifying it because in this way more parts of the surface are brought into contact. The primary mechanism is still one of contact, and surface layers of excess charge are created. There is a negative layer on the silk (excess electrons) and a positive layer on the glass (deficiency of electrons). This result is shown in Figure 14-3, where the silk and the rubbed glass have opposite charges and attract each other. The process of electrification by friction is, as yet, very imperfectly understood.

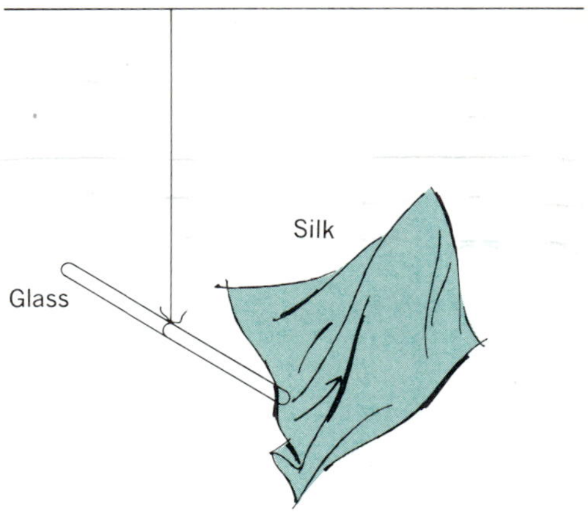

Figure 14-3

14-3 ELECTROSTATIC EXPERIMENTS—CONSERVATION OF CHARGE

We will first investigate electrical effects *qualitatively*, that is, we will make observations but *not* numerical, or *quantitative*, measurements. A very simple instrument will enable us to observe the behavior of electric forces and the motions of electric charges. The *electroscope* (Figure 14-4) has a

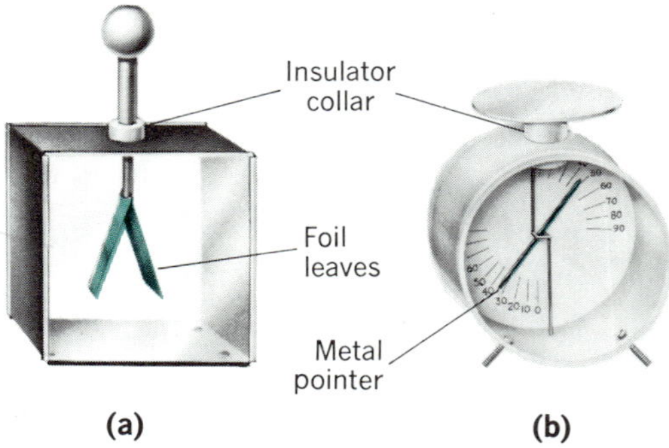

Figure 14-4 Types of electroscopes.

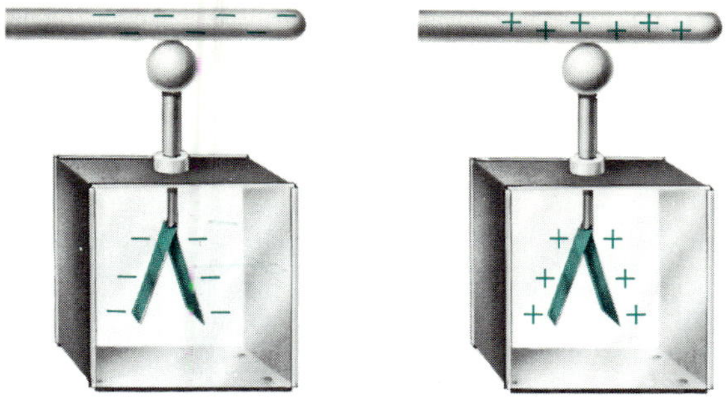

Figure 14-5

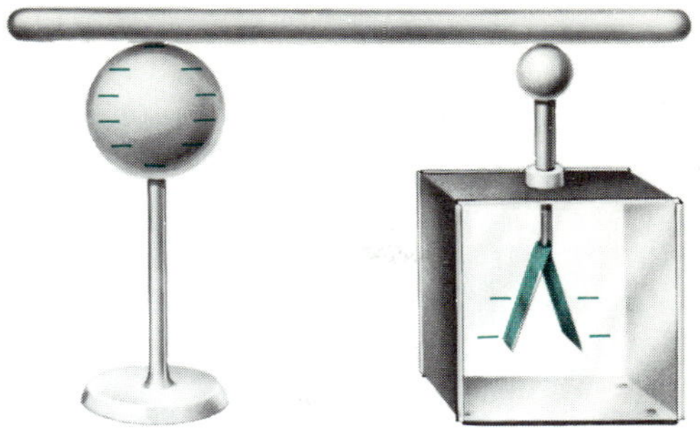

Figure 14-7

central rod with a flexible leaf of gold or aluminum foil or a very light metal pointer pivoted on the rod. The leaf or the pointer must be free to move. The outer case may be connected to the earth by a metal wire although this is not necessary for all experiments. The central rod is electrically isolated from the case by a collar of plastic or some other insulating material. Sometimes an electroscope is constructed with two movable leaves as shown. Below are several experiments that can be performed with the aid of an electroscope and give important information about charges and charge distribution.

(a) Touch a charged rod of either sign to the knob or plate at the top of an electroscope. The leaves diverge (Figure 14-5). What is the explanation? One interpretation is that some of the surface charge on the rod is shared among the knob, the rod, and the leaves. If the rod is removed, the leaves remain spread apart, indicating that some charge has remained on the leaves. We conclude that the repulsion of like charges spreads the leaves apart until equilibrium is reached with the gravitational force. This suggests the first experimental fact: *like charges repel each other.*

(b) Now suppose that an electroscope is charged with a certain sign, say negative. The leaves are diverged due to the repulsion of the negative charges (Figure 14-6). If a negatively charged rod is now brought near the knob, not touching it, the leaves diverge still further. We interpret this as a repulsion of negative charges driving additional negative charge to the leaves. If a positively charged rod is brought up, however, the leaves collapse a little and this can be interpreted as an attraction of unlike charges. The rod has attracted some of the excess negative charges that were on the leaves. Here we see two distinct types of behavior that can be summarized as follows: *electric charges are of two kinds; unlike charges attract each other.*

(c) The experiments described in item (b) also illustrate that *electric forces act at a distance.* The mathematical relation between electric force and distance will be discussed in Section 14-4.

(d) In Figure 14-7 an uncharged rod is placed

Figure 14-6

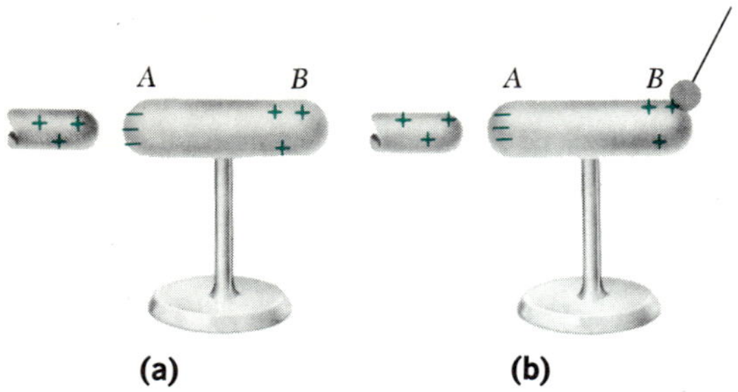

Figure 14-8

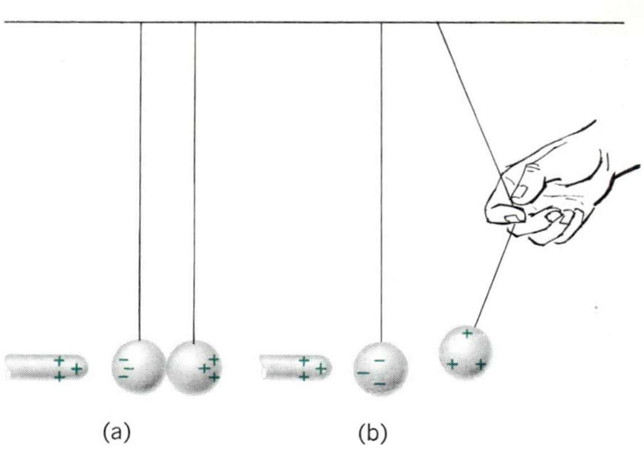

Figure 14-10 Charging a metallic body by induction.

in contact with a charged sphere and an electroscope that was originally uncharged. The leaves begin to diverge, but at a rate that depends on the rod. If the rod is made of copper, iron, aluminum, or some other metal, the leaves diverge instantly. If, however, the rod is made of glass or plastic, the leaves diverge slowly if at all. We interpret these results by saying that the charge flows easily through the metal, but only with difficulty through the glass or plastic. Thus we say that the metal is a good *conductor,* but the plastic is a poor conductor, or an *insulator*. This kind of experiment enables us to distinguish between conductors and insulators and to get a rough idea of their relative conductivities. We can draw the general conclusion that *substances differ markedly in their electrical conductivity.*

The electroscope itself is designed with these differences in mind. The leaves must be good conductors so that charges can flow. Therefore they are made of metal foil. The supporting collar at the top of the case must be a good insulator so that charge applied to the knob will not "leak" to the surrounding case.

(e) There are several kinds of experiments that illustrate the effect called *charging by induction.* If a positively charged rod is brought near an uncharged metal body mounted on an insulating stand (Figure 14-8), free electrons in the metal are attracted toward the rod, leaving an excess positive charge at the end *B* of the body. A charge has therefore been induced at each end of the metal body. No charge is created, but the charges that were already there have been separated into two equal and opposite amounts. This interpretation can be tested qualitatively with a "proof plane"—a small metal disk or ball on an insulating handle that enables us to carry a small sample of a charge. When the proof plane is touched to either end of the metal body, it receives a sample of the charge to be tested. The sign of the charge on the proof plane can then be tested with the aid of an electroscope, as in experiment (b).

Another induction experiment consists of "grounding" one end of the metal body as shown in Figure 14-9. The "ground" is a connection by a wire to the earth or to a metal water pipe. If the

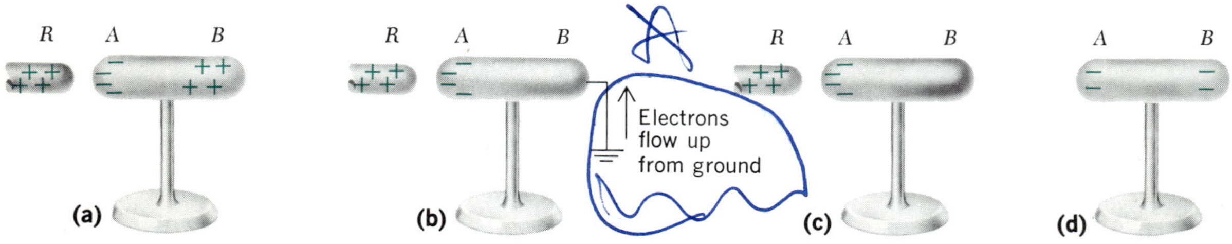

Figure 14-9 Separation of charge by induction.

charge on the rod in Figure 14-9a is positive, then a negative charge is held bound at the left end of the body. When the body is grounded (Figure 14-9b) electrons flow from the ground to end B and neutralize the positive charge there. Upon removing the ground connection (c), the negative charge is still bound at end A. Finally, when the charged rod R is removed, the negative charge spreads out and the body is left with a net excess of negative charge (d).

Figure 14-10a shows two metal spheres (both originally neutral) hanging on insulating threads, when a positively charged body is brought nearby. If the charged body is removed, the two spheres will become neutral again, but if the spheres are separated while the charged body remains nearby (Figure 14-10b), the induced charges are held in place by the electric force, and the spheres retain their charges. A considerable amount of positive and negative charge can be isolated in this way.

Induction takes place even in good insulators, but there is no separation of charge since the charges cannot move. Nevertheless the effect is noticeable when a plastic pen is rubbed on clothing and held near a small piece of paper. The molecules of the paper are forced out of their normal shape (Figure 14-11) and this produces the effect of a layer of positive charges at the surface. (If the pen had been positively charged, the induced surface layer would have been negative. In either case, the force of attraction between unlike charges would cause the pen to "pick up" the paper.)

(f) A small piece of fur glued to an insulating handle is stroked against a piece of hard rubber. It is easy to show qualitatively that the rubber gains a net negative charge and the fur gains a net positive charge. Simply check the signs of the charges with a charged electroscope. For a quantitative test of the equality of these charges, the following induction method can be used: A metal cup on an insulating stand is connected by a wire to an electroscope (Figure 14-12). Originally the cup and electroscope are uncharged. If the negatively charged piece of rubber is held in the cup, not touching the metal, a separation of charge takes place by induction, and the outer surface of the cup becomes negatively charged. The electroscope leaves share this charge, and they diverge a certain amount.

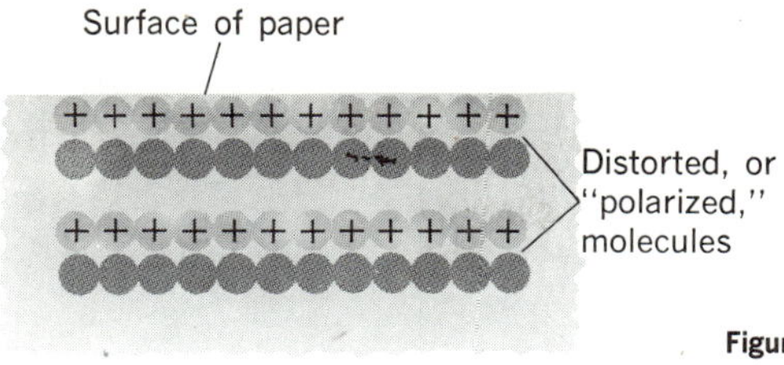

Figure 14-11

If the rubber rod is removed, and the fur is inserted, the leaves diverge the same distance as before, but this time the electroscope is positively charged. As a final test of the equality of the charges, the rubber and the fur are both inserted, not touching each other or the metal cup. No induction takes place, and the electroscope leaves do not diverge. This kind of evidence indicates that *when charges are separated by contact, equal amounts of positive and negative charge are pro-*

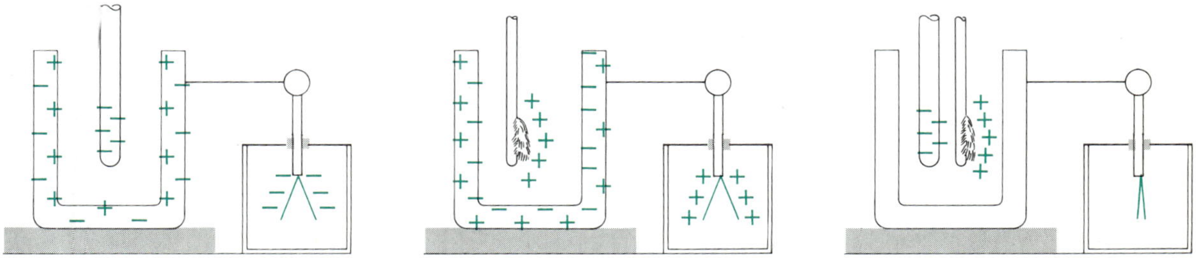

Figure 14-12 Positive and negative charge is produced in equal amounts.

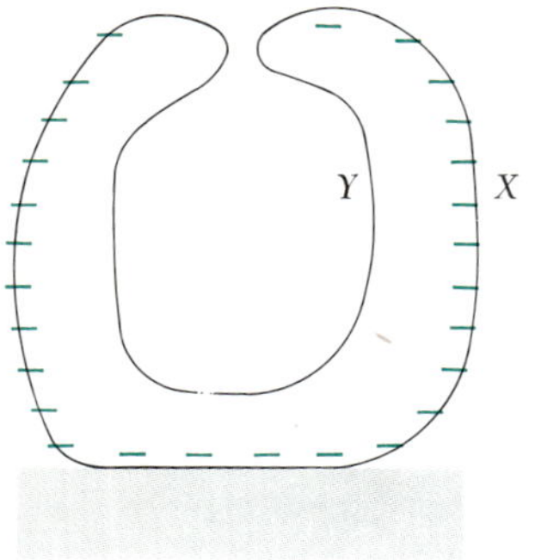

Figure 14-13 Any excess charge is distributed over the *outside* surface of a hollow conductor.

duced. Similar experiments led Michael Faraday (1791–1867) to the idea that electric charge is indestructible.

(g) A hollow metal conductor, such as that in Figure 14-13, is given an excess charge (say it is negative). Testing with a proof plane and an electroscope shows that there is charge at an outside point such as X, but no charge at an interior point Y. A large number of similar tests leads to the conclusion that *the charge on a hollow conducting body lies entirely on the outer surface of the body.* Although we cannot show it experimentally with simple equipment, it is also true that the charge lies entirely on the surface of a *solid* conductor.

We can express the results of these and other experiments in the form of a law of conservation of electric charge.

Law of conservation of electric charge: The algebraic sum of the electric charge in any closed system remains constant.

In other words, charge can be neither created nor destroyed, though it can be moved from place to place. This is a fundamental conservation law, with the same validity as the other conservation laws dealing with mass, energy, and momentum. The facts of relativity have required us to reexamine the conservation of energy and mass, and we should be cautious about saying that any conservation law is absolute and final. The conservation of charge has been tested on the atomic scale and the law is now believed to be true with a high degree of certainty. Perhaps some modification of the law will eventually have to be made, but any new law will include the familiar conservation law as a special case. This is the way scientific advances come about.

14-4 COULOMB'S LAW

How does the electric force between two charged bodies depend upon the distance between them and the magnitudes of the charges? This question

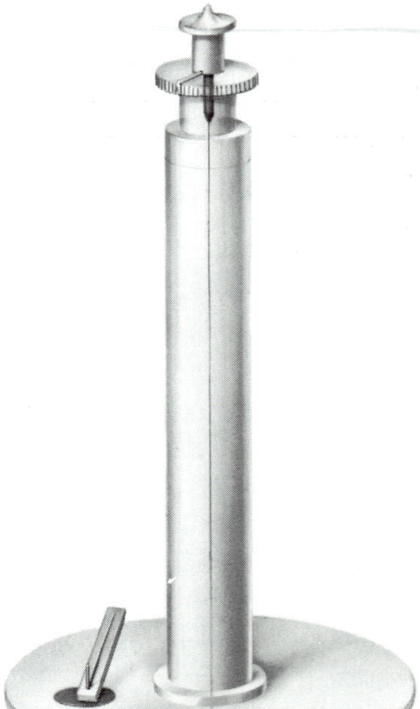

Figure 14-14 Coulomb's device for measuring the force between electric charges.

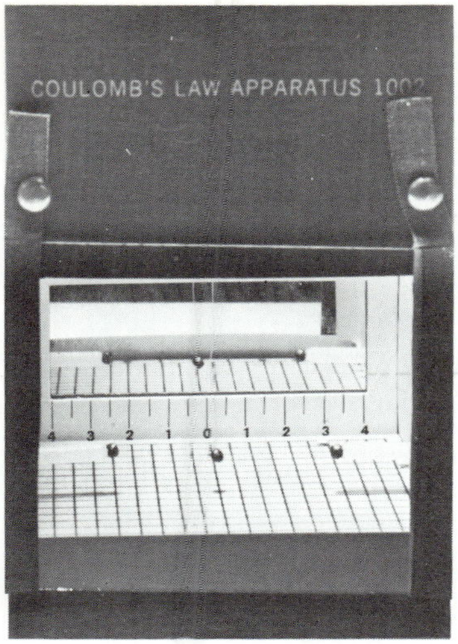

Figure 14-15 Modern laboratory apparatus for demonstrating Coulomb's law.

Table 14-1

Relative Force F	*Relative Separation r*
1.00	10.0
1.23	9.0
1.56	8.0
2.04	7.0
2.77	6.0
4.00	5.0
6.24	4.0
11.1	3.0
25.0	2.0
100.0	1.0

was investigated by the French military engineer and scientist Charles-Augustin Coulomb (1736–1806), who also made important studies of friction and the twisting of wires. To measure the small forces involved, Coulomb in 1788 used a torsion balance similar in principle to the one used ten years later by Cavendish in his measurement of the force of gravitation described in Section 7-5 (see Figure 14-14). A small, charged metal sphere was placed a known distance from another similarly charged sphere, suspended by a fine cord, that was free to move. The force was calculated from the observed twist of the cord.

A simple piece of apparatus now enables us to repeat Coulomb's experiment easily in the laboratory (Figure 14-15). One tiny conducting object is free to deflect when forces act on it. If the deflection is very small, its magnitude is a direct measure of the force, and the actual force can be calculated from the angle of swing and the weight of the object. A second small conductor can be placed at various distances from the suspended sphere. Both spheres are charged, and the deflection of the suspended sphere is then measured. Table 14-1 gives a set of data that could be obtained from a careful experiment of this type.

Now we are faced with the question that continually confronts the experimenter: How can one best interpret the data? What mathematical law describes the relation between force and separation? As you know, plotting a graph can be one of the most useful techniques. Figure 14-16a shows force F plotted vs. separation r. The shape of the graph suggests that there is some kind of inverse relationship between force and separation, therefore in 14-16b force is plotted against the reciprocal of the separation, $1/r$. We have still not found the linear relation that we are looking for. In Figure 14-16c force is plotted against $1/r^2$, and the straight line indicates that there is a direct proportionality between force and $1/r^2$. In more usual terms, the electric force is inversely proportional to the square of the separation.

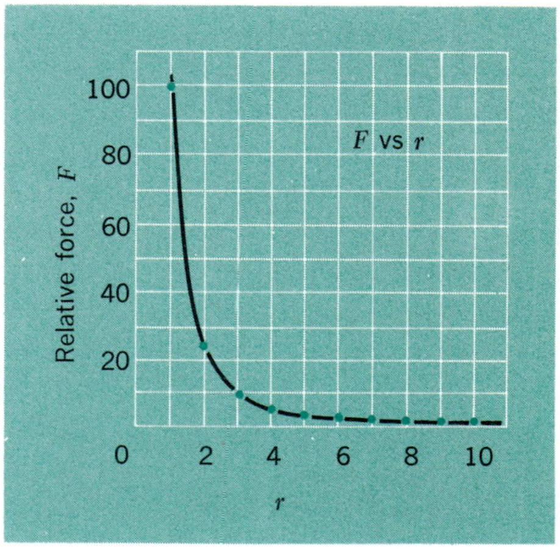

(a)

Figure 14-16

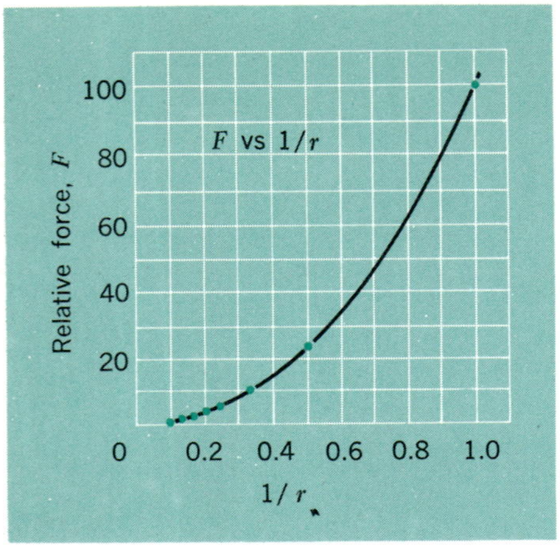

(b)

Figure 14-16

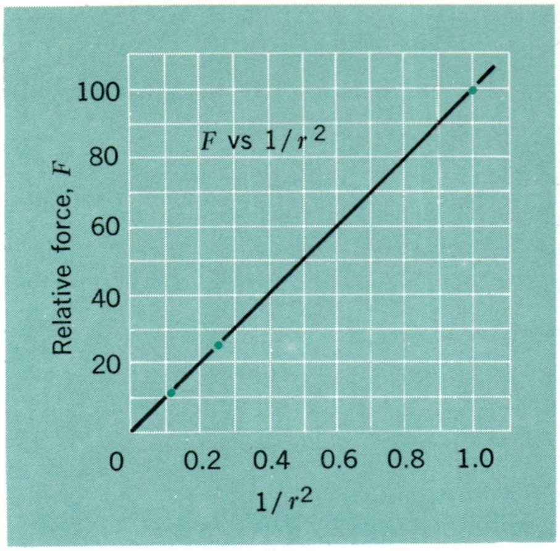

(c)

Figure 14-16

Using this same apparatus it is possible, though not very easy, to show that the electric force also depends upon the magnitude of the charges. All of these experimental facts have been summed up in Coulomb's law.

Coulomb's law: The force between two charges at rest is directly proportional to the product of the magnitudes of the charges, and inversely proportional to the square of the distance between them.

This can be compactly expressed in symbols as

$$F = k\frac{QQ'}{r^2}$$

where F is the electric force, Q and Q' are the magnitudes of the two charges involved, and r is the distance between them. The value of the constant k will have to be determined by experiment.

Coulomb's law provides a way to define magnitude of charge quantitatively. Suppose that Q' in the law is some fixed "test charge" and that in a given experiment the separation r is constant. Then the magnitude of any charge Q is measured by the force it exerts on the test charge. The ratio of the magnitudes of two charges is equal to the ratio of the forces they exert on any test charge (see Figure 14-17). In symbols,

$$Q_1/Q_2 = F_1/F_2$$

This definition agrees with the experimental fact that the total force exerted by n equal charges all at one point is n times the force exerted by any one of them placed at the same point. This is an operational definition of electric charge that defines the quantity by telling how to measure it. It is similar to the definition of gravitational mass, $m_1/m_2 = F_1/F_2$, where the forces are the attraction of the earth. Force has been previously defined in terms of a_1/a_2, the ratio of the accelerations of a test body, and inertial mass has been defined as F_1/F_2, the ratio of the forces necessary to cause given accelerations (see Section 4-3).

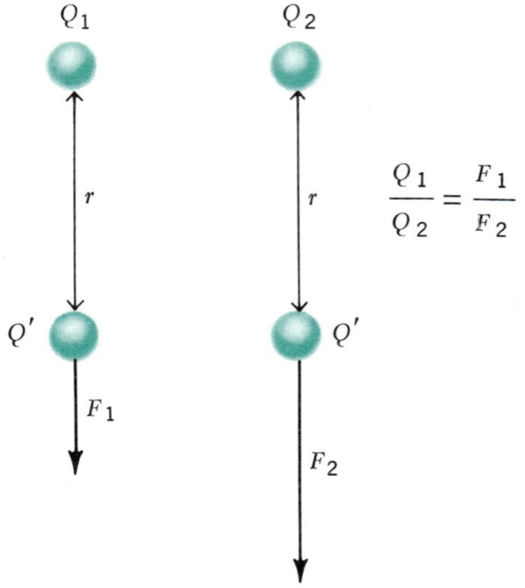

Figure 14-17

The similarity is impressive between Coulomb's inverse-square law for electric forces and Newton's inverse-square law for gravitational forces, $F = Gmm'/r^2$. Some scientists believe that this similarity is more than accidental, but the precise connection, if any, has not yet been found by even the best theoretical physicists. The most remarkable fact is the relative strength of the two forces, which is particularly dramatic in the case of charged subatomic particles where the electric force is enormously greater than the gravitational force. We must for the present, then, assume that Coulomb's law is an independent law of physics, based upon experimental evidence.

The constant k in Coulomb's law can be determined as soon as we adopt a system of units for charge, distance, and force. In this book mks units are always used for electrical quantities. The mks unit of charge is the *coulomb* (C), and it would seem logical to define the coulomb in terms of a Coulomb force experiment, that is, by the proportion

$$Q_1/Q_2 = F_1/F_2$$

that we have just discussed. Some earlier systems of units did just this. However, scientists are continually striving to define the units of measurement so they can be reproduced with the greatest possible accuracy. A Coulomb force experiment is difficult to do accurately, but measurements with electric current can be done with great precision. Therefore the coulomb of charge is now defined in terms of the mks unit of current, the *ampere* (A). The ampere will be defined precisely in Section 18-4, but for the present we need only say that the current I measures the *rate of flow of charge.* A coulomb is the quantity of charge carried past a given point in 1 second by a current of 1 ampere. Symbolically,

$$Q = It$$

We could write 1 A = 1 C/sec and this would be a true and sometimes useful statement of the relation between current and charge. However, it is not a definition, and it is important to realize that the ampere of current is the basic electrical unit. The coulomb is a derived unit in the same sense that the joule is a derived unit in mechanics.

Example 14-1 How much charge flows through the bulb in a desk lamp in 1 hr if the current through the bulb is 0.5 A?

$$\begin{aligned} Q &= It \\ &= (0.5\ \text{C/sec})(3600\ \text{sec}) \\ &= \boxed{1800\ \text{C}} \end{aligned}$$

A problem in language arises; strictly speaking it is charge that flows, and current exists in a wire. Current means *rate of flow of charge.* In this book such inexact statements as "current flows through a wire" will be avoided.

One of the principal building blocks of the atom is the electron, and the charge of the electron suggests itself as a natural unit of electric charge. This charge $-e$ is the most fundamental of all atomic constants and has been the subject of many investigations since about 1897. It is now known that all electrons are identical and that they all have a negative charge equal to 1.6020×10^{-19} coulomb. In Chapter 24 the work of R. A. Millikan is described. Millikan not only measured the charge e with great precision, but also showed that it is indeed a natural unit. Electric charge comes in packages that are multiples of e, and has never been found in any packages smaller than e. Accurate to three significant figures, the magnitude of the charge of the electron is

$$e = 1.60 \times 10^{-19}\ \text{C/electron}$$

It is also possible to define the coulomb in terms of the charge of the electron; the reciprocal of e is

$$\frac{1}{1.6020 \times 10^{-19}\ \text{C/electron}} = 6.2422 \times 10^{18}\ \text{electrons/C}$$

In other words, a coulomb might be defined as the charge carried by this number of electrons. Such a definition would be a permanent one, based on atomic standards, but its precision would fall far short of the precision of atomic standards of length and time mentioned in Section 1-3.

Example 14-2 How many electrons flow through a flashlight bulb in 0.001 sec if the current is 50 milliamperes?

$$Q = It$$
$$= (50 \times 10^{-3}\text{ A})(10^{-3}\text{ sec})$$
$$= (50 \times 10^{-6}\text{ C})$$

Number of electrons

$$= (50 \times 10^{-6}\text{ C})\left(\frac{1\text{ electron}}{1.60 \times 10^{-19}\text{ C}}\right)$$
$$= \boxed{3.12 \times 10^{14}\text{ electrons}}$$

Now that we have defined the coulomb as a unit of charge, and already have mks units for force and distance, we can measure the value of the constant k in Coulomb's law. In principle, k can be found by experiments in which known charges are separated by a measured distance, but in practice a better value can be obtained by indirect experiments. The value of k is

$$\boxed{k = 9.0 \times 10^9\text{ N}\cdot\text{m}^2/\text{C}^2}$$

Example 14-3 What is the electrical force of attraction between a positive charge of 4 microcoulombs (μC) and a negative charge of 5 μC, separated by 30 cm?

$$F = k\frac{QQ'}{r^2}$$
$$= \frac{\left(9\times10^9\,\frac{\text{N}\cdot\text{m}^2}{\text{C}^2}\right)(4\times10^{-6}\text{ C})(5\times10^{-6}\text{ C})}{(0.30\text{ m})^2}$$
$$= \boxed{2.00\text{ N}}$$

14-5 ELECTROLYSIS

Many liquids such as oil, alcohol, or pure water are poor conductors of electricity. However, water solutions of common salt and of other chemical compounds, including various types of "salts," are good conductors. Such solutions, known as *electrolytes*, differ from non-conducting solutions (such as sugar in water) in that the dissolved substance is broken up into freely moving charged fragments called *ions*. In a typical case, a solution of silver nitrate ($AgNO_3$) consists of positively charged silver ions (Ag^+) and negatively charged nitrate ions (NO_3^-).

If a solution of silver nitrate is placed in a cup with two metallic electrodes A and B, charge can be caused to flow by connecting the electrodes to sources of opposite charge, as shown in Figure 14-18. Within the electrolyte the flow of charge is brought about by a motion of ions. The positive Ag^+ ions are attracted to the negative electrode A, and, when they touch it, each Ag^+ ion receives an electron from the electrode. The silver ions become silver metal and stick to the electrode. In this way the electrode becomes silver plated and gains mass. At electrode B electrons leave the solution and become free electrons in the metal.

In the solution, the carriers of charge are positive and negative ions moving in opposite directions; in the metal, the carriers of charge are free electrons. The reactions at the electrodes cannot

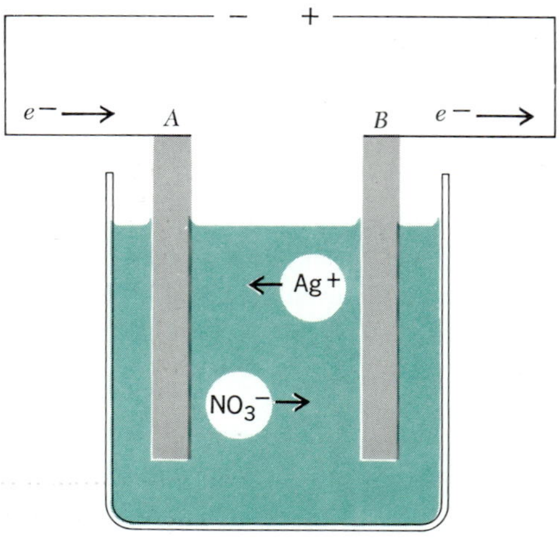

Figure 14-18

always be easily predicted, but it is usually safe to say that the positive ions will accept free electrons from a metal electrode and will become neutralized and "plate out" at the negative electrode.

It is an easy matter to calculate the mass of substance that is deposited during electrolysis. We need to know the atomic mass of the ion, its excess charge, and the total quantity of charge that flows. Many ions can exist in several states of charge. For instance, copper has atomic number 29, which means that the neutral atom has an electron cloud of 29 electrons. The Cu^+ ion has 28 electrons in the cloud (it is lacking 1 electron), and the Cu^{++} ion has only 27 electrons in the cloud (it is lacking 2). To plate out each Cu^{++} ion, 2 electrons must flow through the circuit in order to neutralize the doubly charged ion.

Example 14-4 A current of 8 A is maintained for 10 min through an electrolytic cell containing Cu^{++} ions. The mass of a single copper atom is 1.06×10^{-22} g. What mass of copper is deposited?

$$\begin{aligned} Q &= It \\ &= (8 \text{ A})(600 \text{ sec}) \\ &= 4800 \text{ C} \end{aligned}$$

The number of electrons that flows through the cell is

$$\begin{aligned} n &= (4800 \text{ C})(1 \text{ electron}/1.60 \times 10^{-19} \text{ C}) \\ &= 3.00 \times 10^{22} \text{ electrons} \end{aligned}$$

Since each atom requires 2 electrons for neutralization, the number of atoms of Cu is

$$\frac{3.00 \times 10^{22} \text{ electrons}}{2 \text{ electrons/atom}} = 1.50 \times 10^{22} \text{ atoms}$$

The mass m of Cu deposited is, then,

$$\begin{aligned} m &= (1.50 \times 10^{22} \text{ atoms})(1.06 \times 10^{-22} \text{ g}) \\ &= \boxed{1.58 \text{ g}} \end{aligned}$$

The study of electrolysis brings out the close connection between the particles of matter and particles of electricity. Several decades before the discovery of the electron, scientists realized that if matter were made of particles, the facts of electrolysis required the existence of "particles of electricity" having a charge of the order of magnitude of 10^{-19} C. Using the results of experiments in electrolysis and Avogadro's number, it was possible to calculate the charge of the "fundamental particle of electricity" encountered in chemistry (1.60×10^{-19}C/electron). And from the work of Millikan we know directly the value of charge of the electron (1.60×10^{-19} C/electron). These two values prove to be identical and show us that we are dealing with the same particle in both cases—the electron, the natural unit of electricity.

SUMMARY

Electric forces between charges that do not depend on relative motion are called **Coulomb forces,** and electric forces between charges that do depend on relative motion are called **magnetic forces.**

When two dissimilar substances are placed in close contact, **electrons** tend to leave one substance and go to the other. If the substances are **insulators,** the electrons are bound to the nuclei, and transfer of **electric charge** takes place only at the points of contact. Electrification of insulators by friction is effective because contact is made at many points. In metallic **conductors,** free electrons can flow through the metal and take part in the conduction process.

There are two kinds of charge, positive and negative, and the **law of conservation of charge** states that the algebraic sum of the electric charge in any closed system remains constant. A neutral, or uncharged, body has equal amounts of positive and negative charge. Like charges repel each other, and unlike charges attract each other.

The law of force between stationary electric charges is **Coulomb's law,** which states that the force is directly proportional to the magnitudes of the charges and inversely proportional to the square of the distance between them. The mks unit of charge is the **coulomb,** which is the charge carried past a given point in 1 second due to an **electric current** of 1 **ampere.**

In **electrolytes,** current is carried by free-moving **ions** usually in a water solution. An ion has an excess or deficiency of electrons, and the net charge of an ion is a small multiple of the electron charge. When an ion reaches a metal electrode, its charge may be neutralized by the flow of free electrons to or from the electrode.

QUESTIONS

1 Why is magnetism considered a subdivision of electricity?

2 In what respects is the modern electrical theory a two-fluid theory? In what respects is it a one-fluid theory?

3 Do protons contribute to the flow of electricity through a metal? No

4 Why is a plastic pen electrified more strongly when it is rubbed vigorously on a coat sleeve than when it is merely touched to the cloth?

The whole surface of the pen is charged and not just one spot

5 Explain the action of a rubbed comb that picks up bits of paper. Why do the pieces eventually fly off the comb with an initial velocity? Why is it difficult to pick up pieces of paper on a damp day?

6 Suggest a method by which Coulomb might have placed exactly half as much charge on one of two identical metal balls as on the other.

7 Correct the following statement: "The charge on the raindrop was 463 μA."

8 Why should the body of an airplane be connected to a "ground wire" before a refueling operation begins?

PROBLEMS

GROUP A

1 In 5 minutes 3000 C of free electrons enters one end of a conductor, and 3000 C moves out the other end. What is the current through the conductor?

2 A service station charges a battery for 10 hr, using a current of 20 A. How much charge passes through the battery?

3 How many electrons are needed to carry a picocoulomb ($= 10^{-12}$ coulomb) of charge?

4 What force does a charge of 10^{-10} C exert on a charge of 4×10^{-8} C that is 2 m away?

5 Two charged bodies exert a force on each other of 24 millinewtons. What will be the force between the same two bodies if the distance between them is doubled?

GROUP B

6 Assuming that the electron in a hydrogen atom is 5×10^{-11} m from the nucleus (which consists of a proton), (a) compute the force of electrical attraction between the electron and the proton. (b) Compute the gravitational attraction between the electron and proton. (c) What is the ratio of these two forces? (Refer to Table 1 on page 476.)

7 What would be the force of repulsion between a gram of electrons on the moon and another gram of electrons on the earth 3.70×10^5 km away? The mass of the electron is given in Table 1 on page 476.

8 Two helium nuclei (alpha particles), each of charge $+2e$, approach each other to within a distance of 10^{-10} m in a vacuum. What is the force of repulsion? (e refers to the magnitude of the charge of the electron; it may be either $+$ or $-$.)

9 A charge $Q_1 = +10\ \mu C$ is on the x-axis at $x = 0$, and a second charge $Q_2 = -2\ \mu C$ is on the x-axis at $x = 5$ m. A third charge $Q_3 = -10\ \mu C$ is placed on the x-axis at $x = 15$ m. What is the magnitude and direction of the force on Q_2?

10 Iron articles are sometimes plated with cadmium to increase their resistance to rusting. Calculate the total mass of cadmium deposited on a group of iron objects by a current of 50 A during an 8 hr working day, if the cadmium is present in the form of Cd^{++} ions. The mass of a cadmium atom is 1.87×10^{-22} g.

11 A student performs a laboratory experiment in which she plates copper from a solution containing copper ions (which may be Cu^{+} or Cu^{++}). The initial mass of the electrode was 22.043 g. After a current of 0.400 A was passed through the solution for 25.0 min, the plate was removed, dried, and found to weigh 22.441 g. What can you conclude about the charge of the copper ions in the solution?

GROUP C

12 Two ping-pong balls, each of mass 10 g, are painted with aluminum paint to make their surfaces good conductors of electric charge, and are suspended from the same point by threads 60 cm long. When equal charges are given to the two balls, they come to rest in an equilibrium position in which their centers are 60 cm apart. Calculate the amount of charge on each ball's surface.

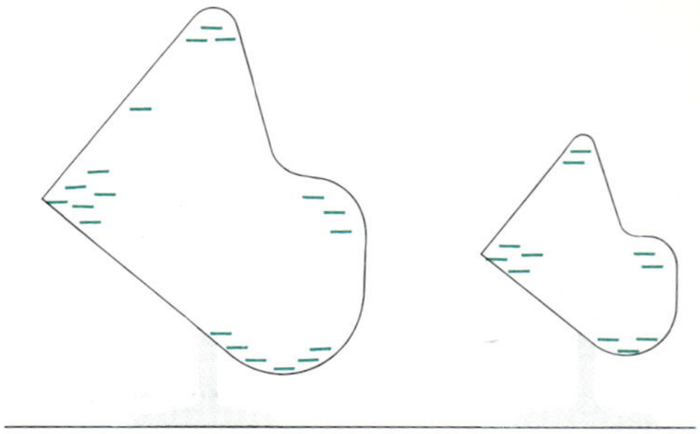

Figure 14-19

FOR FURTHER STUDY

SCIENTIFIC PRIORITY AND COULOMB'S LAW

Over the years, there have unfortunately been arguments and controversy between scientists, each claiming to have first discovered some physical law or invented a certain device. Robert Hooke (1635–1703) was a versatile English scientist who took the matter of receiving credit for his discoveries very seriously. In 1676 he announced his law of elasticity in the form of an anagram "ceiiinosssttuu." His intent was to protect his priority, that is, to make the announcement prior to anyone else. He used the anagram form to keep secret the nature of his discovery. He unraveled the anagram two years later as *Ut tensio sic uis,** which can be translated "As the extension, so the force." Or, in everyday usage, we might say "the extension that results when stretching a wire or spring is proportional to the force applied" (see Section 8-1). Christian Huygens (1629–1695) used a similar Latin anagram to "pin down" his priority in the matter of the true shape of Saturn's rings.

It is sometimes difficult to determine who should receive credit for a major discovery. As an example of this kind of problem, let's look closely at the unfolding of the inverse-square law for electric forces, known as Coulomb's law. The action of a force at a distance had been introduced to physics through Newton's law of universal gravitation, published in 1687. When it became clear that electric forces decrease rapidly as distance (r) increases, a natural assumption was made that $F \propto 1/r^n$, where n is some exponent. This assumption that the force depends on some *power* of r was justified by experiment. It was generally known by 1760 that the charge on the surface of an irregularly shaped body (Figure 14-19) is distributed irregularly, but for similarly shaped bodies the charge distribution is the same. This observation plus some geometrical reasoning is sufficient to show that the force law must be of the form

$$F \propto 1/r^n$$

and not, for instance, of the form

$$F \propto a^{-kr} \qquad \text{or} \qquad F \propto 1/r^n + 1/r^m$$

* In the twentieth century a Latin phrase might still seem to many to be a "hidden message," but in 1678 Latin was the accepted language for communicating learned discoveries.

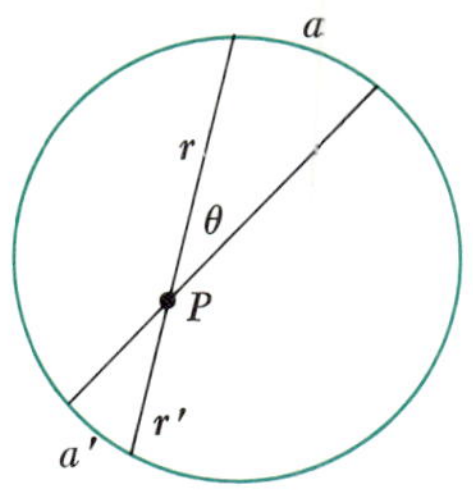

Figure 14-20

The problem, then, was to determine the value of the exponent. In 1760 Daniel Bernoulli guessed that $n = 2$, but had no proof for his guess.

Joseph Priestley (1733–1804), the discoverer of oxygen, was informed in 1766 by his friend Benjamin Franklin that cork balls inside a metal cup were totally unaffected by the electrification of the cup. From Newton's work it was known that if the earth were a hollow spherical shell, a body anywhere within the shell would experience no force of gravitation. This is a consequence of the fact that gravitation follows an inverse-*square* law. Hence Priestley drew the correct conclusion that, since the cork balls experienced no force, the law of electric force must also be an inverse-square law.

To prove this statement, consider a point P somewhere inside a uniformly charged hollow sphere (Figure 14-20) and assume that the inverse-square law is true. Construct a double cone through P, as shown, intersecting the surface in areas a and a'. By simple geometry, it can be shown that

$$\frac{\text{area } a}{\text{area } a'} = \frac{r^2}{r'^2}$$

Since the shell is uniformly charged, this means that the charges on the areas a and a' are in this same ratio. Thus $Q/Q' = r^2/r'^2$, which can be rearranged as

$$\frac{Q}{r^2} = \frac{Q'}{r'^2}$$

If, now, the inverse-square law is true, a test charge Q_0 placed at P will experience two opposing forces kQQ_0/r^2 and $kQ'Q_0/r'^2$, and these forces are equal by the equation just derived. The net force on the test charge due to both charged areas is zero at any arbitrary point P because the greater charge on area a has been exactly compensated for by the greater distance r. The entire sphere can be divided up in this way by cones passing through P, and so the net force on a charge anywhere inside the spherical shell is zero. To complete the proof, one must also show that no exponent other than 2 will give this result. Priestley apparently did not take account of the fact that the electrified cup was not a sphere.*

Priestley published his conclusion in London in 1767. For some reason the paper seems to have escaped the notice of other scientists of his day, or perhaps he himself failed to present his case forcefully enough. At any rate, in 1769 John Robison of Edinburgh attempted a direct experimental proof of the law, measuring the forces between small charged bodies. He arrived at the exponent 2.06—satisfactory enough in view of the difficulty of the experiment.

A few years later Henry Cavendish (1731–1810), who was to become known for his work on Newton's law of gravitation a decade later, turned his attention to the problem and chose to expand upon Priestley's indirect method. He placed a metal sphere of diameter 30.7 cm within two close-fitting hemispheres 33.8 cm in diameter. The inner sphere was completely enclosed but nowhere touched the outer hemisphere except by means of a removable wire (Fig-

* Analysis by Karl Friedrich Gauss (1777–1855) extended the proof to include any *closed* metallic cup of whatever shape.

Figure 14-21

ure 14-21). Cavendish now electrified the system as strongly as possible, allowing every opportunity for charge to flow to the inner sphere by means of the connecting wire. He then withdrew the wire, isolating the inner sphere, and finally separated and removed the hemispheres. He tested the inner sphere by touching it with a pith ball, and found it to be electrically neutral. His conclusion was similar to Priestley's, that the absence of any flow of charge to the inner sphere showed that electric forces obey an inverse-square law.

Cavendish went further; he tested the sensitivity of his apparatus by deliberately charging the inner sphere to the least detectable amount. In this way he placed limits of accuracy upon his conclusion, and stated that the exponent must lie between 1.96 and 2.04. This important experiment of Cavendish's, as well as his gravitational experiment (Section 7-5), deserves to be studied carefully by students interested in the development of scientific methods. Cavendish's results were not published during his lifetime.

The next investigator of the law of electric force was Coulomb, who published his researches in a series of papers during the years 1788 and 1789. Coulomb no doubt expected to get an inverse-square law, as did all the others. He had already invented the torsion balance, independently of the Rev. John Michel (1724–1793) (whose torsion balance was applied by Cavendish to the gravitational problem), and he had studied the torsion of wires. It was a natural step for Coulomb to apply the torsion balance to the measurement of the exceedingly small forces of electrical repulsion and attraction.

Following Coulomb, other workers repeated Cavendish's experiment with more sensitive detectors of charge. In 1838 Faraday actually put himself inside a 4-meter wire cage and had the cage charged to an extent that artificial lightning would jump to the walls of the room. He felt no effect upon himself, and a sensitive electroscope inside the cage showed not the slightest indication of charge inside the cage. A century after Cavendish, Maxwell in 1870 used a delicate electrometer and found no effect, leading to a value of the exponent between 1.9999 and 2.0001. The present "record" in the inverse-square race is held by S. J. Plimpton and W. E. Lawton, who showed in 1936 that n lies between 1.999999999 and 2.000000001.

In view of all this activity, why then has the law become known as Coulomb's law when Coulomb's experiment was neither the first nor the most precise? Scientific priority is based upon *publication* of results in a form suitable for the use of others. For this reason, Cavendish lost out because he never published his method or his results, even though he kept an excellent record of his experiments in his private notebooks. It was Maxwell who brought Cavendish's notes to light and edited and published them for the first time in 1879, a century too late. It is hard to explain Cavendish's eccentricity in this respect, since he did publish some of his researches in the usual channels, including the gravitational experiment for which he is famous. He was best known in his lifetime as a chemist, and is one of those to whom the discovery of the composition of water is attributed. He inherited great wealth and retired from much of public life and from society to devote himself to scientific work. As Maxwell said,

Cavendish cared more for investigation than for publication. He would undertake the most laborious researches in order to clear up a difficulty which no one but himself could appreciate, or was even aware of, and we cannot doubt that the result of his enquiries, when successful, gave him a certain degree of satisfaction. But it did not excite in him that desire to communicate the discovery to others which, in the case of ordinary men of science, generally ensures publication of their results. How completely these researches of Cavendish remained unknown . . . is shown by the external history of electricity.

Coulomb, on the other hand, contributed to his subject through a series of well-organized and comprehensive papers. He described his apparatus in detail, discussed the possible sources of error, and, most important, checked his result by an entirely different experiment. His first method used the torsion balance to measure repulsive forces between two small, similarly charged spheres. His second method involved attraction, instead of repulsion, and was dynamic instead of static (he timed oscillations of a small charged body placed near a charged sphere several feet in diameter). Having obtained an inverse-square law by two different methods, he felt able to announce the law as proved.

The subsequent experiments of Faraday, Maxwell, and others contribute nothing new but merely extend the precision of the value of the exponent. Such experiments have to be made, of course, for if they fail to come out exactly 2 (within experimental error) the consequences would be disastrous for much related electrical theory. For its scientific importance, which includes publication and comprehensiveness, Coulomb's work stands out, and we justifiably speak of the law as Coulomb's law.

PROBLEM

GROUP D

13 Construct a proof of the relation $\frac{\text{area } a}{\text{area } a'} = \frac{r^2}{r'^2}$, on page 249, assuming that the angle θ is small. In general, what is the approximate shape of the areas a and a'?

15 Electric Energy—PD, EMF, and Resistance

Our survey of electric charge in Chapter 14 centered on Coulomb's law, which describes how the electric force between charges at rest depends on the magnitudes of the charges and their separation. One might expect that Coulomb's law and Newton's laws could be used to predict the motions of charged particles, just as the law of universal gravitation and Newton's laws can be used to predict the motion of a falling body or a satellite. This is indeed possible in simple cases. However, electric forces present a much more complex problem when many charges are involved, as inside a wire in which an electric current exists.

15-1 THE ENERGY METHOD IN ELECTRICITY

Earlier in this book, in working with mechanical problems, you used the law of conservation of energy to obtain a relatively easy solution to a problem, when the use of Newton's laws would have been too difficult. The examples in Section 5-5 illustrate the use of the energy principle in the solution of problems.

Energy considerations give us much useful information about the behavior of electric charge. We know that energy is transferred to a body whenever work is done on it, whether the work is done against gravitational, electric, or magnetic forces, or if the body is accelerated and gains KE (Section 5-2). When a charge is moved against an electric force exerted on it by other charges, work is done and the charge gains electric PE. Conversely, when a charge moves in the direction of an electric force, it loses electric PE and gains KE.

We speak of *electric* PE only to remind us how the work is done and energy is transferred. Electric PE is in no way different from any other PE except in its origin. There is no need to worry about how the force on a charge changes as it is moved from point to point; when using the energy method in the solution of problems, we can pass over the details and deal only with the total work done.

It is important to realize what we have already done and what we are proposing to do here. We have extended the law of conservation of energy to include electric energy, as well as mechanical energy, thermal energy, and mass energy. The right to make such an extension is by no means obvious, and it has been based upon an experimental foundation. During his historic experiments, Joule measured the heat produced by electric energy. Thus he measured the "electrical equivalent of heat" as well as the "mechanical equivalent of heat" (Section 11-3). When taken all together, Joule's experiments show the equivalence of electric, thermal, and mechanical energies—a truly remarkable fact of nature.

15-2 ANALOGIES—BENEFITS AND DANGERS

Models and analogies can be of great value in physics if they are used with care and discrimination (see Section 9-2). It is important, for example, to guard against the danger of believing that a model or analogy is an exact representation of some physical system. One should always regard a model critically and remember that an analogy means no more than: under certain special conditions, the physical system being studied behaves as if . . . . With these restrictions always in mind, we now turn briefly from electricity to certain aspects of fluid mechanics, which we will study as an analogy to help us gain a better understanding of electricity.

15-3 PRESSURE OF FLUIDS

The general term *fluid* includes any substance that has no rigidity; thus both liquids and gases are fluids. One important quantity used to describe the state of a fluid is its *pressure*, which you first met in studying Boyle's law and the kinetic theory of gases (Section 12-2). Recall that pressure was defined then as force per unit area,

$$P = F/A$$

Pressure units are force units divided by area units such as N/m^2.

The pressure at a certain depth in a container of liquid is easy to compute; it is just the weight of the liquid standing above the area, divided by the area. The volume of liquid is found from the base area A and the height h (Figure 15-1). The mass of the liquid is its volume times its density.* In the

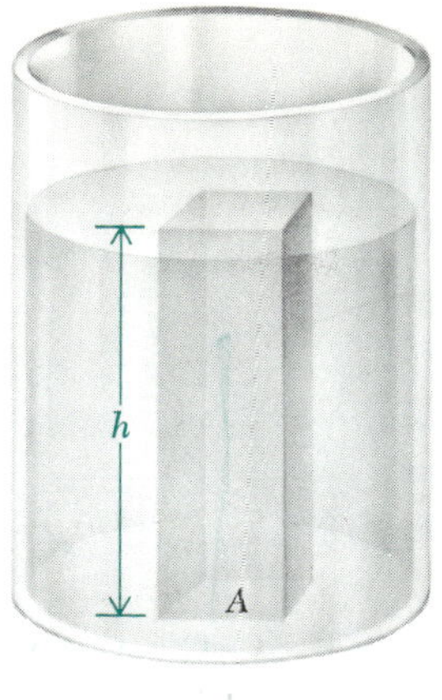

Figure 15-1

case of the volume marked off in Figure 15-1,

$$m = (Ah)(d)$$

The weight of the volume of liquid is

$$mg = Ahdg$$

The pressure is force per unit area, or $Ahdg/A$. Therefore,

$$P = hdg$$

Thus the pressure at any point in a uniformly dense liquid at rest depends only on the height of the liquid above the point, the density, and the value of g.

Example 15-1 Water stands at a height of 35 cm in a glass tube. What is the pressure at the bottom of the tube?

Water has a nearly constant density of 1.0×10^3 kg/m³, or 1.0 g/cm³

$$P = hdg$$
$$= 0.35 \text{ m} \times 10^3 \text{ kg/m}^3 \times 9.8 \text{ m/sec}^2$$
$$= 3.4 \times 10^3 \text{ kg}\cdot\text{m}^2/\text{m}^3\cdot\text{sec}^2$$

and since $1 \text{ N} = 1 \text{ kg}\cdot\text{m/sec}^2$ (Section 4-6),

$$P = \boxed{3.4 \times 10^3 \text{ N/m}^2}$$

There are many instruments for measuring

* The *density d* of a substance is a measure of how closely its constituent matter is packed, and is defined as *the ratio of the mass m of a given quantity of matter to its volume.* Thus density d is mass per unit volume, $d = m/V$. The density of water is 1 g/cm³, or 10^3 kg/m³.

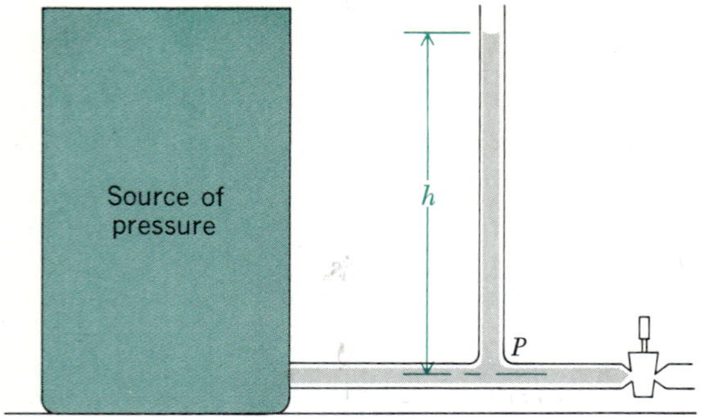

Figure 15-2 The pressure P is indicated by the height h of the liquid in the tube.

pressure, but we will make use of the simplest one —the water-filled manometer—which consists of a transparent vertical tube connected to the liquid whose pressure we want to measure (Figure 15-2). The pressure at the bottom of the tube is measured by the height of the liquid in the tube (times the appropriate conversion factor). For example, when the height is in cm, then

$$P = h(\text{cm}) \times \frac{1\ \text{m}}{100\ \text{cm}} \times 10^3\,\frac{\text{kg}}{\text{m}^3} \times 9.8\,\frac{\text{m}}{\text{sec}^2}$$

$$P\left(\frac{\text{N}}{\text{m}^2}\right) = h(\text{cm}) \times 98\,\frac{\text{N}}{\text{m}^2\cdot\text{cm}}$$

Stated in words, to obtain the pressure in N/m^2, multiply the height of the water column in a manometer (in cm) by the conversion factor 98. (For many cases, 100 is accurate enough.)

15-4 FLUIDS IN MOTION—BERNOULLI'S EQUATION

Figure 15-3 shows a combination of a pipe, a valve, and several manometers forming a simple fluid circuit. The valve is tightly closed, and no water is flowing. The pipe is horizontal and we will use its center line as a reference level. The water in the container (or "standpipe") is at a height h above the reference level, as is the water in all the manometers (except M_5) indicating that the pressure is the same everywhere between the standpipe and the valve, and is equal to hdg.

If we open the valve somewhat and allow a moderate flow of water, we find that the levels of the water in the manometers fall slightly, but they still remain at about the same height, except for M_5, which is close to zero.

However, if we open the valve wide and allow the water to flow rapidly (Figure 15-4), we see that while the pressure at M_1 is nearly the same as before, the pressure at M_2 has become less, at M_3

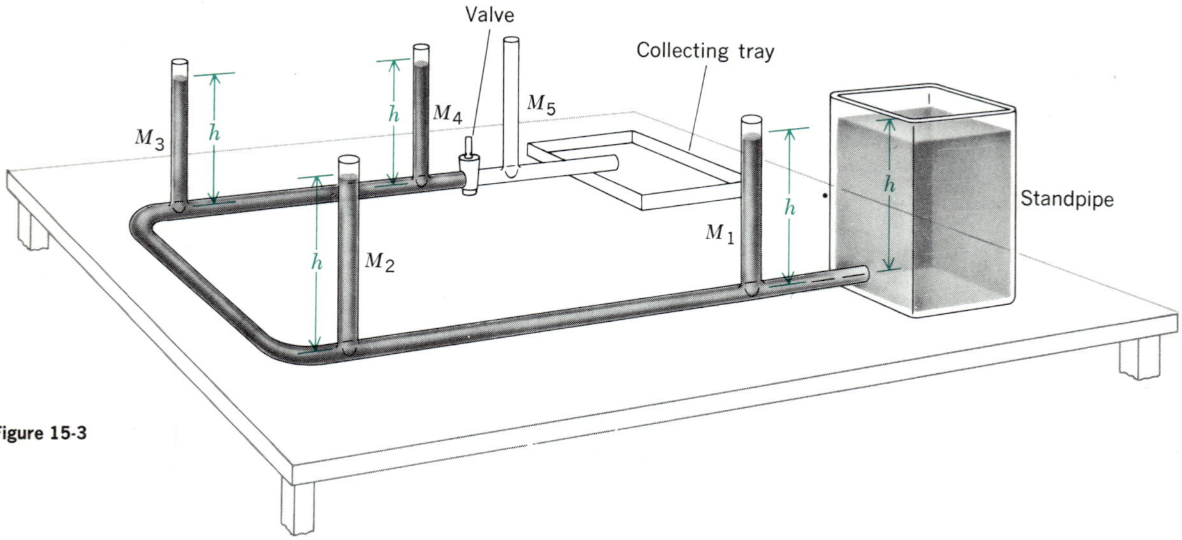

Figure 15-3

still less, and finally almost no pressure is indicated at M_4 and M_5. How do we interpret this observation?

Many aspects of the behavior of fluids in motion can be described by an equation credited to Daniel Bernoulli (1700–1782). Bernoulli's principle is usually stated in the form of an equation:

$$\frac{1}{2}dv_1^2 + P_1 = \frac{1}{2}dv_2^2 + P_2$$

where d is the density of the fluid, v is the velocity of the fluid at a point, and P is the pressure at the same point. This means that the quantity $\frac{1}{2}dv^2 + P$ is constant at all points in an ideal fluid system where all points are at the same level. In the equation, the subscript 1 is used to refer to one point in the system and the subscript 2 can refer to any other point.

Bernoulli's principle is, in fact, a statement of how the conservation of mass and the conservation of energy apply in the ideal case of a fluid of constant density, with no turbulence, and with no energy losses. With these restrictions, it is clear that we are dealing only with ideal conditions, but there are many cases in which the equation can be a useful one.

Let's now try to understand the terms of Bernoulli's equation by seeing how they apply to the situations described previously.

15-5 APPLICATION OF BERNOULLI'S EQUATION

We look first at the case where the valve is closed and the water is at rest (Figure 15-3). This is a special case of Bernoulli's equation in which the velocity v of the water is zero. The terms $\frac{1}{2}dv^2$ equal zero and therefore P_1 is equal to P_2, which is just another way of saying that the pressure is the same everywhere at a given level in a liquid at rest.

Now, how do we explain what happens when the valve is opened slightly, allowing a moderate flow of water? Recall that the pressure fell, as was indicated by a decreased height of liquid in the manometers.

Let's look closely at the term $\frac{1}{2}dv^2$ in Bernoulli's equation and find a meaning for it in terms of familiar mechanics. First of all, it is similar to the expression $\frac{1}{2}mv^2$ for the KE of a moving mass (Section 5-4), so we might assume that $\frac{1}{2}dv^2$ expresses a KE. But what KE does it refer to? Remember the definition of density, $d = m/V$. By substituting this into $\frac{1}{2}dv^2$, we have

$$\frac{1}{2}dv^2 = \frac{1}{2}mv^2/V$$
$$= \text{KE}/V$$

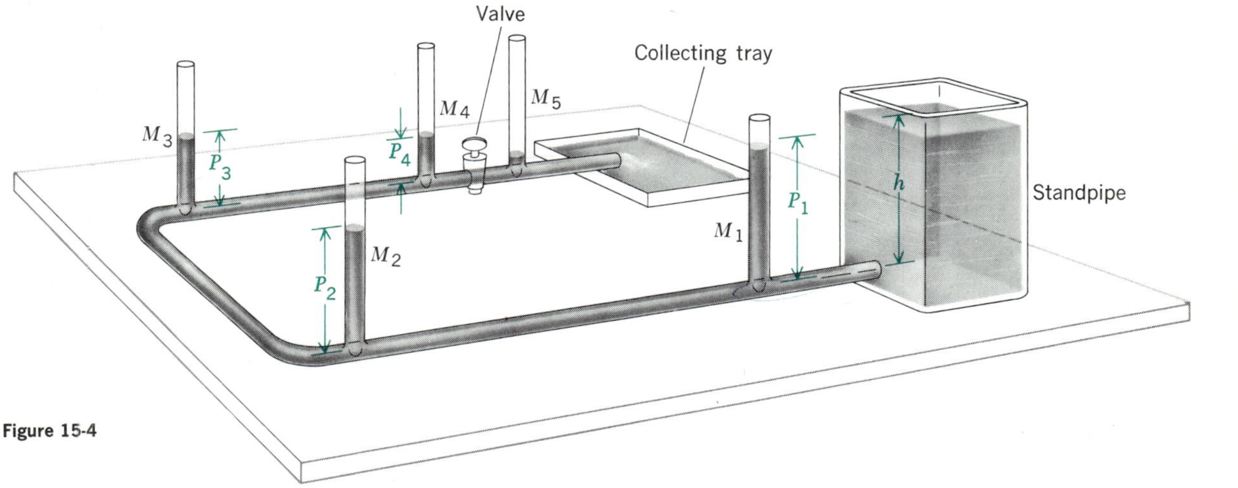

Figure 15-4

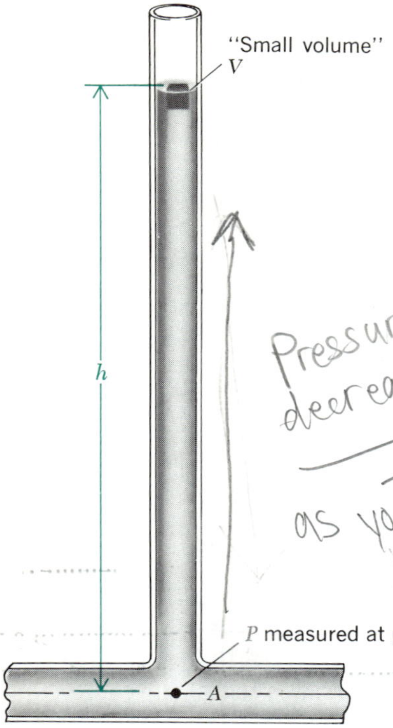

Figure 15-5

This says that $\frac{1}{2}dv^2$ is the KE *per unit volume* of the liquid, or, the KE/V of a small volume located at a point in the liquid. The "small volume" means only that the dimensions of this volume are so small that we make no significant error by thinking of it as being located at just one point.

The pressure P can be interpreted in very much the same way since the units of P are also energy per unit volume. By writing the pressure in terms of its units, and then multiplying by meter/meter (= 1), we have

$$P = \frac{\text{newton}}{(\text{meter})^2} \times \frac{\text{meter}}{\text{meter}}$$
$$= \frac{\text{newton} \cdot \text{meter}}{(\text{meter})^3}$$
$$= \text{joules}/(\text{meter})^3$$

since 1 joule = 1 newton·meter (Section 5-1). This check of units suggests that pressure is related to energy per unit volume.

Pressure actually *equals* PE per unit volume; to see this look at point A in Figure 15-5. This point is at the reference level in a pipe; the pressure is P, and there is a manometer directly above it. We know that the pressure is measured by the height of the water column in the manometer. Bear in mind that the water in the manometer was not always at the height h. It was lifted there, and work was done against the force of gravity by the pressure of the liquid. The source of this energy is the standpipe. Note that the manometer samples the work that can be done by the pressure in lifting a small volume to the height h. For this reason a manometer should be a small-bore tube, so as not to sample away too much liquid and too much energy.

Let's calculate the work required to lift a small volume of water to the top level of the liquid in the manometer, which is the weight of the small volume times the height (Section 5-4), or

$$\text{work} = mgh$$
$$= (Vd)(g)(h)$$

Rearranging, and dividing by the volume V, we obtain

$$\text{work}/V = \frac{Vhdg}{V}$$
$$= hdg$$

Thus we see that hdg, the expression for pressure, also equals the work per unit volume that can be done by that pressure. In other words, pressure measures the PE/V stored at a given point in a liquid system. It is important to keep in mind that we are speaking of a very small volume—so small that removing this volume of liquid for sampling purposes would not affect h. Therefore, for constant h, Bernoulli's equation can be stated in words as follows:

kinetic energy per unit volume at a given point	+	energy stored per unit volume as pressure at a given point	=	constant
$\frac{1}{2}dv^2$	+	P	=	constant

Next we must understand the constant in the Bernoulli equation. It is natural to think that it refers to the total energy per unit volume (E/V) at any point in the system, whereas the individual terms of the Bernoulli equation refer to the KE and PE at the point. This total energy per unit volume must have come from an outside source. In this case, energy is supplied by means of the standpipe (Figure 15-6). The standpipe itself is not a *source* of energy. It is a device in which the energy of our muscles is stored (by lifting water to fill the standpipe) and then transformed into PE as the water in the horizontal pipe is compressed, and also into KE if the water flows. The total E/V at any point divides among $\frac{1}{2}dv^2$ and P in such a way that the total is constant for points at any level, such as the reference level represented by the line in the horizontal pipe. If v increases (due to a constriction in the pipe) then P decreases. Some devices, such as an aspirator, can produce a fairly good vacuum (less than 0.05 atm) in this way.

Finally, we must explore the situation in which water is flowing rapidly and the pressure drops steadily all around the circuit. The system is not losing or gaining any water, and we are continually lifting water from the tray to the standpipe. Also, the piping has essentially the same cross-section area all the way around. Therefore the water moves at the same velocity everywhere *in the pipe*, and $\frac{1}{2}dv^2$ has the same value everywhere. Nonetheless, we observe a steady drop in pressure (Figure 15-4) that must be explained. It is certainly clear that we no longer have "ideal" conditions, and Bernoulli's equation no longer applies in its simple form.

In what situations does Bernoulli's equation fail to describe a fluid's behavior? First, the equation applies only in what is called streamline flow where the molecules of fluid move from point to point without any rotational motion or turbulence. In streamline flow, the effects of fluid resistance, or "viscosity," are small (directly proportional to the speed). Yet even in streamline flow, Bernoulli's equation applies only if v is small enough so that fluid resistance is negligible. Real fluids moving at high speeds always exhibit a resistance to motion, called the *drag force*, resulting from turbulence and viscosity. The drag force increases as the velocity increases, but the exact relation is not simple. In general, it can be stated with regard to drag that

$$R_v \approx kv^n$$

which says that the drag force R_v is approximately proportional to some power of the velocity.

At this increased speed, the drag force R_v has become a significant factor, and energy is being dissipated. R_v times the length of pipe over which the drag force acts is equal to the work that transfers mechanical energy out of the system to heat the pipe and the water. A sufficiently sensitive thermometer would show that the water was warmer when it reached the end of the circuit than when it was at the beginning. The important fact is that the drag force causes a definite dissipation of energy per unit length, which is measured by the drop in pressure along the length of a pipe. This dissipated energy is not "lost," but, as usual, appears as thermal energy.

For a further example of the behavior of a fluid circuit, let's attach, between manometers 2 and 3, a simple motor consisting of a very light paddle wheel to be turned by the flowing water. A thread is wound around the paddle-wheel shaft and can be used to lift weights. After this apparatus is set

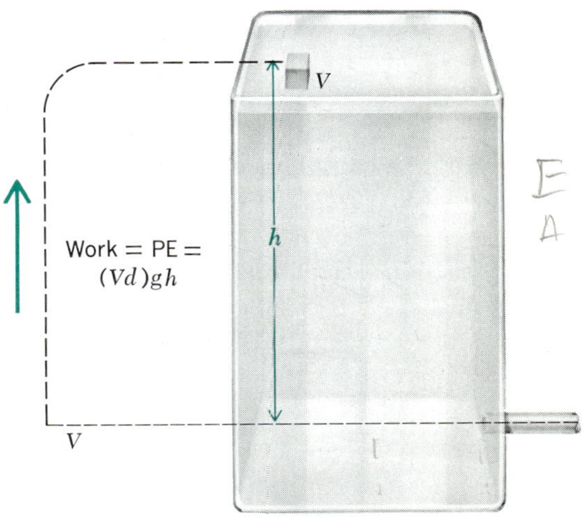

Figure 15-6

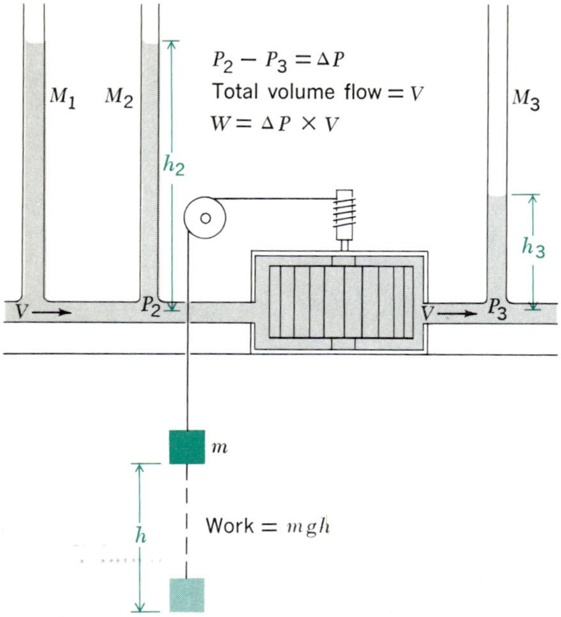

Figure 15-7

in motion, manometers M_1 and M_2 will stand at about the same level, but there will be a distinct pressure difference ΔP between the readings of M_2 and M_3 (see Figure 15-7).

By measuring the total flow of water V through the motor, and the pressure difference ΔP, we can calculate the total amount of energy used (work done) to lift the weight:

$$W = \Delta P \times V$$

remembering that pressure represents energy per unit volume.

The energy involved in lifting the weight is measured by the work *mgh* required to lift a body of mass *m* to a height *h*. In the ideal case, $\Delta P \times V$ and *mgh* should be equal, but in the real world energy will always be dissipated by heat flow due to the viscous drag and other frictional forces. However, careful bookkeeping of pressures, volumes, heights, masses, and temperatures would show that all the energy can be accounted for. This is simply a statement of the law of conservation of energy as it applies to these circumstances when we take all forms of energy into account.

15-6 ELECTRIC POTENTIAL DIFFERENCE

In our study of electric energy, the first fundamental quantity we need to understand is *potential difference* (PD)—and we can gain some insight into the meaning of this term by referring once more to the hydraulic model. The water standpipe is shown again in Figure 15-8. We have already discussed the concept of pressure and pressure difference between two points in the system but now let's consider two different levels —one near the top of the standpipe just below the water's surface, and the other a lower level—separated by a distance Δh.

A very small volume of water at the lower level has potential energy PE_1. The same small volume of water near the top of the container has potential energy PE_2. The crucial point about this difference in potential energy, $PE_2 - PE_1$, is that it is directly equivalent to the work required to move the small volume of water from the lower level to the level near the top of the container. We have already shown that this work is

$$W = mgh$$
$$= (Vd)g(\Delta h)$$

or the work per unit volume is

$$W/V = dg\,\Delta h$$

This work per unit volume does not depend on the exact value of either height, but only on the

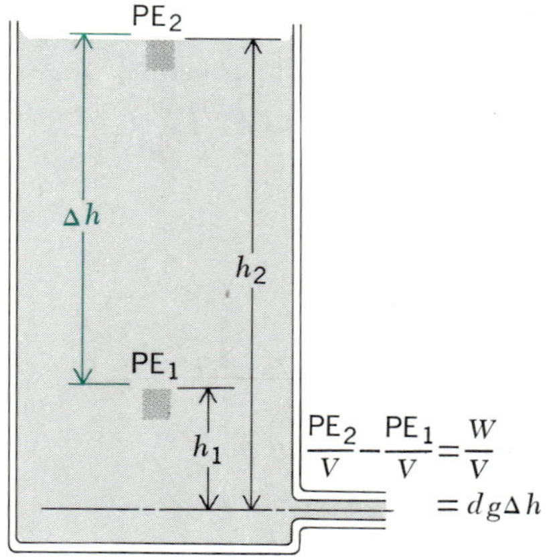

Figure 15-8

difference Δh. Our equation indicates that the difference in pressure ($dg\ \Delta h$) between two points in a fluid is exactly equivalent to the work required to move a unit volume from one point to the other. Such a difference in pressure is directly analogous to a difference in electric potential. To explore the behavior of charged bodies rather than liquids, we shall use a small *positive test charge* (rather than a small volume) that we imagine to be moved from point to point. The test charge must be small enough so that it does not exert appreciable electric force on the surrounding bodies. That is, it must not cause any changes in the original distribution of charges. Remember that this is a hypothetical test charge and not any specific charged particle.

If electric forces act on it, the test charge gains or loses PE as it moves from point to point. In the same way, a small volume of liquid gains or loses PE as it is lifted or falls under the action of gravity. The PE *per unit charge* at a particular point is called the electric potential and is analogous to the fluid pressure (energy per unit volume). The mks unit for electric potential is the *volt* (V).* Two points differ in electric potential by 1 volt if 1 joule of work is required to move 1 coulomb of charge from one point to the other. Calling the two points A and B, in symbols we have

electricity	*water*
$V_{AB} = W_{B\to A}/Q$	$P_{AB} = W_{B\to A}/V$
$\mathrm{V} = \dfrac{\mathrm{J}}{\mathrm{C}}$	$\dfrac{\mathrm{N}}{\mathrm{m}^2} = \dfrac{\mathrm{J}}{\mathrm{m}^3}$

where V_{AB} is the difference in potential (and P_{AB} is the difference in pressure) between points A and B. $W_{B\to A}$ is the work done against electric (or gravitational) forces to move a test charge Q (or volume of water V) from B to A.

Note that in electricity the potential difference is the meaningful quantity while in the hydraulic example it is differences of pressure, or energy per unit volume, that are of interest. Examples of potential difference are not hard to find. Let's discuss a few representative situations in detail.

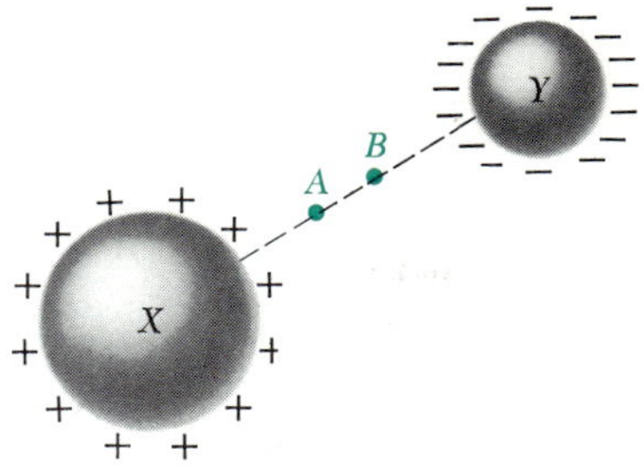

Figure 15-9 Potential difference; A is a higher potential than B.

(a) Two metal spheres are charged, with X having an excess of positive charge and Y an excess of negative charge (Figure 15-9). Since the test charge is defined as a positive one, work must be done on it to bring the positive test charge from B to A. Therefore the positive charge has greater PE at A than at B.

(b) Water tends to flow downhill. This is often explained by saying that the force of gravity pulls the water down the hill. In many cases it is more helpful, as we have seen, to look at fluid flow from the point of view of energy changes. When water flows downhill its gravitational PE is less in the lower position. A heavy particle, or small volume of liquid, always seeks the lowest attainable level, where its gravitational PE is least.

Similarly, *positive* electric charge tends to flow from points of high potential to points of low potential, "downhill" so to speak. We don't have to worry about the details of electric forces to know that their net result is to push or pull charges into a position of least PE. In Figure 15-10 the pear-shaped object, the connecting wire, and the leaves of the electroscope are all at the same potential. If this were not the case, the charges in the metal would readjust their distribution, flowing away from regions where they had a higher PE into regions in which they had a lower PE. This is comparable to the original static case of the fluid

* V (not italicized) is the abbreviation for volts. Do not confuse it with V (italics), which is used to represent volume and also PD.

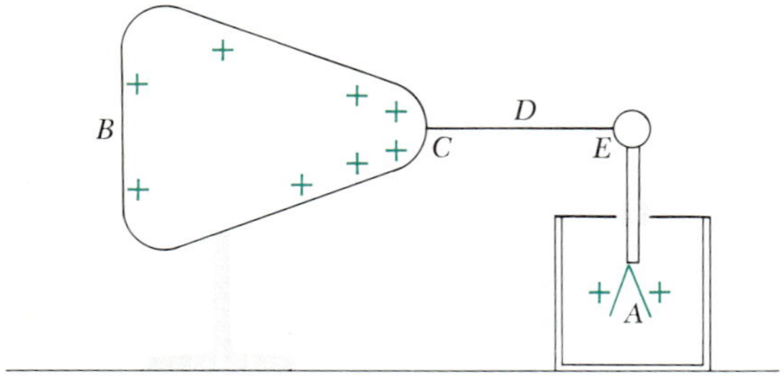

Figure 15-10

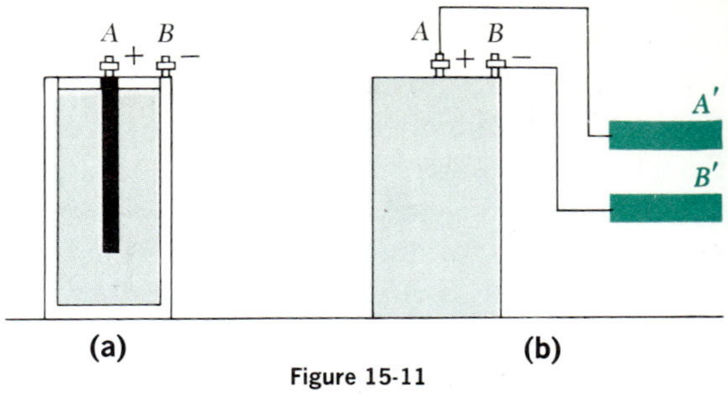

Figure 15-11

circuit, where all parts of the circuit at the same level were at the same pressure so long as water was not flowing. If there *were* a difference of pressure, water would flow to equalize this difference.

(c) An ordinary dry cell (Figure 15-11a) is a chemical device which, when no circuit is connected to it, provides a slight excess of positive charge on the positive terminal and a slight excess of negative charge on the negative terminal. We can find by experiment that 1.5 joules is required to move a charge of +1 coulomb from B to A through the air. Therefore the PD is 1.5 J/C, or 1.5 volts. If the terminals are connected by wires to a pair of plates (Figure 15-11b), the plates become charged (this process takes a small fraction of a second), and then the PD between A' and B' is also 1.5 volts.

Example 15-2 What is the PD between two points if 200 J of energy is required to move 40 C from one point to the other?

$$V_{AB} = \frac{W_{B \to A}}{Q} = \frac{200\ \text{J}}{40\ \text{C}} = 5\,\frac{\text{J}}{\text{C}} = \boxed{5\ \text{V}}$$

Example 15-3 How much energy is required to move an electron between the two terminals of a large x-ray tube whose PD is 4 million volts?

$$W_{B \to A} = V_{AB}Q = (4 \times 10^6\ \text{V})(1.60 \times 10^{-19}\ \text{C}) = \boxed{6.40 \times 10^{-13}\ \text{J}}$$

Example 15-4 An electron leaves the heated cathode of a radio or TV vacuum tube (Figure 15-12) with negligible initial velocity, and is accelerated through an applied PD of 500 V. What is the velocity of the electron as it approaches the accelerating electrode A?

While the electron is being repelled by the cathode K and attracted by the anode A, it loses PE, just as a baseball does when it falls toward the earth. The change in PE equals the PD times the charge. Now we apply the energy principle:

$$\text{loss of PE} = \text{gain of KE}$$

$$QV = \tfrac{1}{2}mv^2$$

$$(1.60 \times 10^{-19}\ \text{C})(500\ \text{V}) = \tfrac{1}{2}(9.11 \times 10^{-31}\ \text{kg})(v^2)$$

$$v = \sqrt{\frac{2(1.60 \times 10^{-19}\ \text{C})(500\ \text{V})}{9.11 \times 10^{-31}\ \text{kg}}} = \sqrt{1.77 \times 10^{14}\ \text{C}\cdot\text{V/kg}} = \boxed{1.33 \times 10^7\ \text{m/sec}}$$

Figure 15-12

A note on units: In this example we arrived at velocity = $\sqrt{C \cdot V/kg}$. This fine collection of units can be shown to be equal to m/sec, as follows:

$$\sqrt{\frac{C \cdot V}{kg}} = \sqrt{\frac{C \cdot (J/C)}{kg}}$$
$$= \sqrt{\frac{J}{kg}}$$
$$= \sqrt{\frac{N \cdot m}{kg}}$$
$$= \sqrt{\frac{(kg \cdot m/sec^2)m}{kg}}$$
$$= \sqrt{\frac{m^2}{sec^2}}$$
$$= m/sec$$

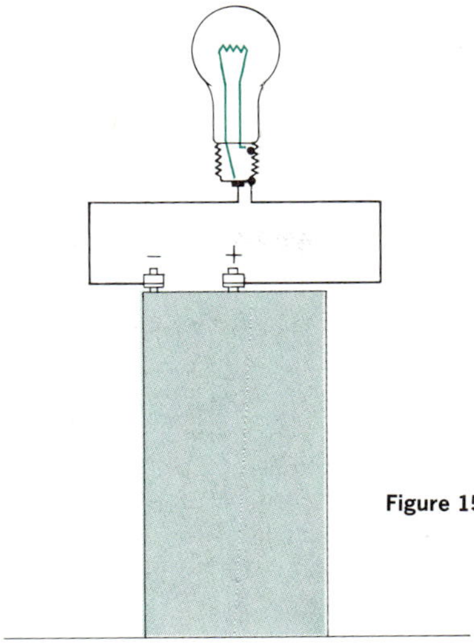

Figure 15-13

Whenever a large velocity turns up in a problem, as in Example 15-4, we must consider the possibility of relativity effects. The answer for Example 15-4 is about $\frac{1}{20}$ the speed of light ($\frac{1}{20}c$), and so we realize that a small error has been made in using the simple expression $\frac{1}{2}mv^2$ for KE. However, relativity effects depend on v^2/c^2 (Section 5-6), which in this problem is

$$\frac{(1.33 \times 10^7 \text{ m/sec})^2}{(3.00 \times 10^8 \text{ m/sec})^2} = 0.0020$$

Since in this case v^2/c^2 is small compared to 1, the error in our "classical," or "nonrelativistic," treatment is not a serious one.

15-7 CURRENT AND EMF

Much of the usefulness of electricity in our daily lives arises from the possibility of continuous flow of charge through an electric circuit. A simple circuit consisting of a dry cell connected to a bulb is illustrated in Figure 15-13. The bulb becomes warm and lights up, and we know that some agency must be supplying the energy radiated as heat and light. We believe that this agency is the cell and that in some way it is forcing charges to move in the circuit.

How does the cell operate? A full answer would involve a great deal of complicated chemistry and is beyond the scope of this book. We know that a PD of 1.5 volts exists between the two terminals. How did the charges become separated to produce this PD? We are dealing here with chemical forces as well as electric forces. Chemical forces in the cell impel charges to move to the terminals against the electric forces that build up as the charges accumulate at the terminals (like charges repel each other). This movement of charge continues until just enough charge is separated so that electrostatic forces balance the chemical forces. The statement that a PD of 1.5 volts exists between the terminals means that for every coulomb of positive charge that is moved through the battery, chemical forces do 1.5 joules of work against electric forces.

If, now, the dry cell's terminals are connected by a wire and a small lamp, there will be a current, and the lamp will light. It is natural to inquire about the nature of this electric current. What does go on inside the metal wire? Why does the lamp light? Here again a full answer would be extremely complicated, but we can make a useful model (this time quite different from the liquid model) to help our thinking. The characteristic feature of metals that makes them good conductors is the ease with which electrons can be detached from the metal atoms. Within a metal there is always a large number of free electrons

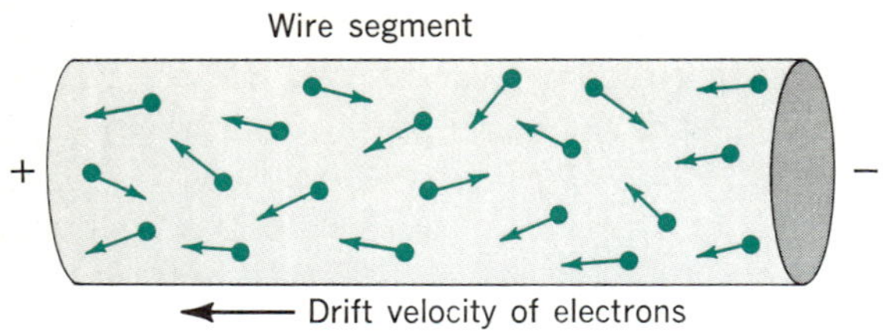

Figure 15-14

rushing around in all directions among the atoms. The electrons form a kind of "gas" within the metal, in the sense that they are constantly carrying out rapid random motions much like gas molecules in the kinetic theory. So long as their motion is random, there is no net flow of charge in any direction.

Yet, when the ends of a wire are subjected to a PD, the electric forces set up a "drift velocity" in the direction of these forces. This small drift velocity is superimposed upon the very much larger random velocities of the electrons as they rush around within the metal. If we look at any cross section of a wire, we see virtual chaos; but on the average there will be more electrons moving in one direction than another (Figure 15-14).

The electric force causes the electrons to accelerate, and to collide more frequently with ions of the metal, transferring their excess energy to the metal. In this view, the electrons first have PE at the terminals of a battery. In the wire this PE is transformed into KE, and then it is rapidly further transformed into thermal energy by the collisions. However, not all the KE is transformed into thermal energy. Some of the energy may do work if the current is passing through a motor. It may be converted to sound energy if it passes through the loudspeaker of a TV set; or if the current is in the antenna of a TV transmitter, the energy may be radiated. Sometimes all the energy is dissipated as thermal energy, but in any case, whatever part is dissipated heats the wire. In a lamp, this heating makes the filament give off light.

This again is analogous to the liquid case where the water in the standpipe initially has PE that is converted into KE as it begins to move. The drag forces, however, steadily convert a fraction of the water's KE into thermal energy. But not all of this energy has to be dissipated, since some of it could be used to turn a paddle wheel or other device.

It is important to note that the drift velocity of the electrons is extremely slow, only a few centimeters per second. An electrical disturbance, however, such as a changing electric force, or a changing PD, propagates as a wave at the speed of light.

Physicists generally look upon an electric current in a metal wire as the *net* flow of positive electric charge arising from the drift velocity of the electrons and *not* as the motion of electrons themselves. You will recall that in Section 15-6 PD was discussed in terms of the movement of a positive test charge. It has become a widely accepted convention to think of the direction of the current from the positive terminal to the negative terminal, even though, in the special case of metal wire, it is only the negatively charged electrons that move. However, electric circuits do not consist solely of metal wires. There are electric currents in liquid electrolytes, gases, and plasmas; there are currents through solid "semiconductors" that, as you will see later, make up the transistors and diodes of modern electronics. In all of these examples, there are various kinds of posi-

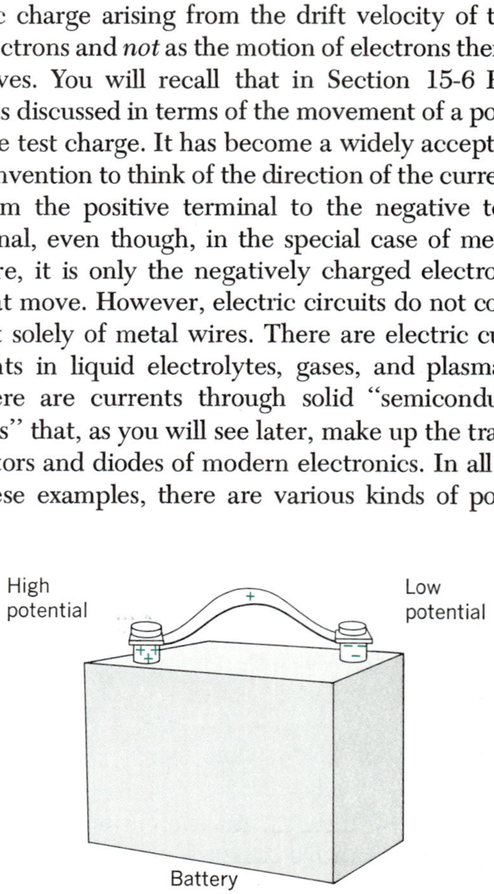

Figure 15-15 The PD is the work required to move a positive test charge from the negative to the positive terminal of the battery. The direction of the current is the direction this positive charge moves when it is free to move, from + to −.

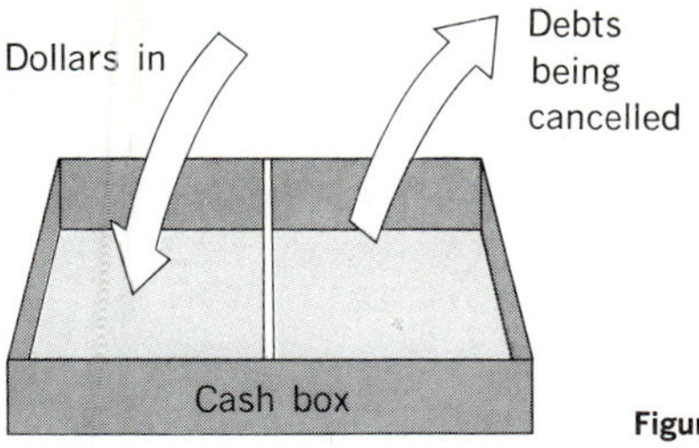

Figure 15-16

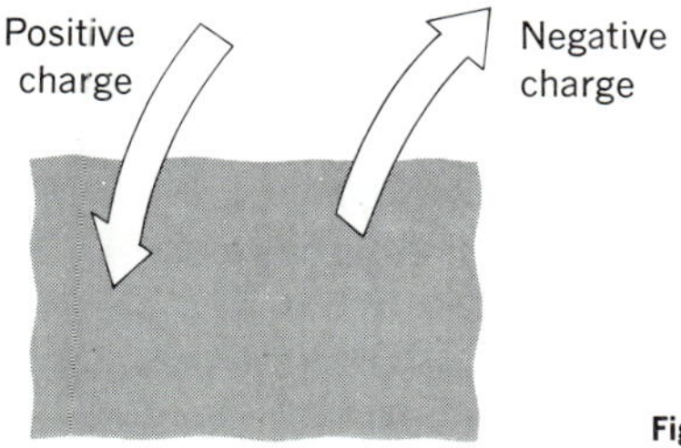

Figure 15-17

tively charged ions and positive "holes" that move and form all or part of the electric current. There is nothing particularly favored about thinking of an electric current as a movement of negative electrons, except in the case of a metal wire. Therefore, it has long been a matter of agreement among physicists, engineers, and electricians that the current should be referred to as a flow of positive charge from the positive terminal to the negative terminal. This does not represent any law of nature, but is just a matter of convenience. We therefore speak of an accumulation of positive charge (having a positive potential) as being a point of "high potential" and of an accumulation of negative charge as a point of "low potential" (Figure 15-15). This agrees with our definition of potential in terms of a positive test charge. It takes work to bring a positive test charge to a point of high potential against the electric force of the positive charges there.

In general, then, an electric current is said to be a flow of positive charge from a point of high potential (accumulation of positive charge) to a point of low potential (accumulation of negative charge). What is actually moving? In the special case of a metal wire, it is a drift of electrons in response to electric forces. Yet a drift of electrons to the right can be considered exactly equivalent to a drift of positive charge to the left. As a crude analogy, think of your own payments and receipts of money. Receiving "positive money" has exactly the same effect as canceling "negative money," a debt (Figure 15-16). In the same way, negative charges moving away from a point have just the same effect as positive charges moving toward the point (Figure 15-17). The net positive charge is increased; in the money example, your net assets have increased.

In the case of liquid electrolytes, gases, and plasmas, both positive ions (atoms that have lost 1 or more electrons) and negative ions (either free electrons or atoms that have picked up an extra electron or so) are able to move and form a current. If you could stand at a point in an electrolyte and watch the motion of charges, you would see positive ions going in one direction and negative ions in the other (Figure 15-18). The current would be in the direction that the positive ions are moving, and numerically it would be proportional to the sum of the number of positive ions moving one way each second and the number of negative ions going the other way each second.

Charges move because there is a PD, and PD results from an accumulation of positive charge in one place and negative charge in another. These charges cause electric forces that push the moving charges through the conductor. This is like the hydraulic example, where the accumulation of water in the standpipe sets up forces that cause

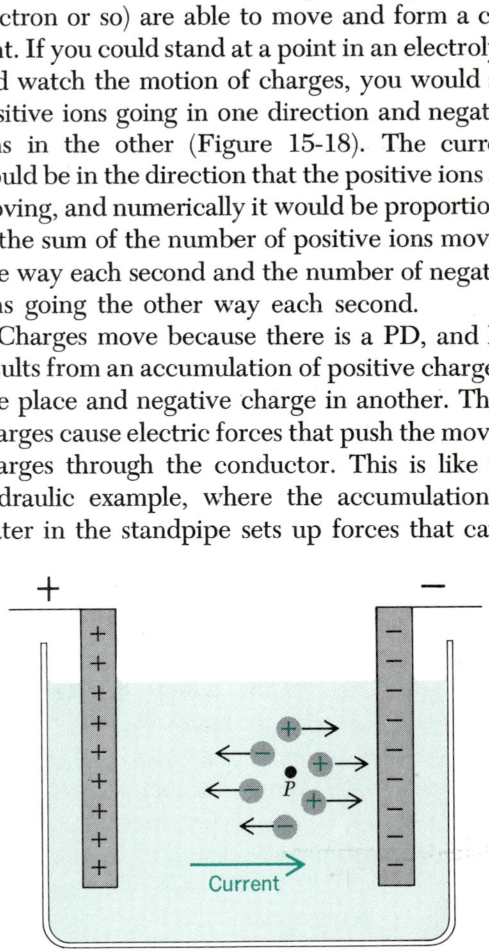

Figure 15-18 An observer is at point P in an electrolyte watching the motion of ions. A PD is set up by the charges on two electrodes.

the water to move in the pipe. In the simple circuit of the battery, the wire, and the lamp (Figure 15-13), the battery builds up the necessary PD.

The essential action of the battery is to transform energy from chemical energy to electric PE, and in this sense it continually supplies energy to the circuit. Our discussion of the cell can be paralleled by other sources of electric energy. In a generator, mechanical energy is transformed into electric energy. In a thermocouple, thermal energy is transformed into electric energy. All such devices are called *seats* (*or sources*) *of emf** because they transform nonelectric energy into electric energy; that is, they supply an *emf* to the circuit**.

The numerical value of an emf is denoted by $\mathcal{E}$ and is measured in joules per coulomb, or volts. Thus, if a coulomb of charge passing through the battery gains 6 joules of electric energy, the battery's emf is 6 volts.

The concepts of PD and emf are easily confused because both are measured in terms of energy per unit charge. The distinguishing feature of emf is the continuous transformation of nonelectric energy into electric energy. Referring back to the hydraulic model, a pump continually supplying energy to the water would be a close analog of a device that has an emf.

On the other hand, we can illustrate PD by imagining a wire connected between two charged bodies, as in Figure 15-9. There is a PD between the bodies, but no emf, since there is no transformation of nonelectric energy into electric energy. When the connection is made, charge will flow momentarily, but there will be no continuous current because the charges will soon be equally distributed between the two bodies and there will no longer be a PD. This is much like the situation where a flow can be set up briefly from the energy of water stored in a small tank, but this energy will be quickly exhausted as the liquid level falls. It is always true that there must be a continuous transformation of nonelectric energy into electric energy in a circuit to obtain a continuous flow of electric charge.

15-8 THE ELECTRIC CIRCUIT—JOULE'S LAW

There must be emf in any circuit in which a continuous current exists. The electric energy supplied by the source of emf can, of course, later be converted into mechanical or chemical energy. In addition, it has been stated already that a certain amount of the electric energy is transformed into heat.

There is a fundamental difference in the way electric energy is utilized in a motor and in a lamp bulb. The motor is (electrically speaking) reversible, and, if it were turned by a crank powered by a muscle or a waterfall, it would act as a generator to send electric energy back into the circuit. It becomes a source of emf. The paddle wheel shown in Figure 15-7 is closely analogous to an electric motor, in that it is a device through which energy can be added to or taken out of the liquid circuit and transformed into other forms. The lamp bulb, on the other hand, is not reversible. It is impossible to generate electric current in a wire by building a fire under a lamp bulb connected to the two ends of the wire. This is why a lamp bulb is said to "dissipate" electric energy—it is a one-way street for electric energy. It is not, and cannot be, a source of emf. Such a conductor, in which electric energy is dissipated is called a *resistor.* In fluid terms, a pipe in which the drag is dissipating energy is analogous to a resistor. Every conductor (except a "superconductor")† is a resistor, in the sense that the flow of charge is impeded to a greater or lesser extent, and some energy is dissipated as it is transformed into thermal energy.

* The letters emf stand for a now largely obsolete term, electromotive force. Since the quantity is *not* actually a force, we will use only the initials.

** This process is reversible. Electric energy becomes chemical energy when an auto battery is being charged.

† Superconductors are substances that, when placed in temperatures approaching absolute zero, possess no electrical resistance at all. Mercury and lead are examples.

During his investigations of the mechanical and electrical equivalent of heat, Joule studied the production of heat in resistors (now often called *Joule heat*). He found that the rate of production of thermal energy in a metallic resistor is proportional to the square of the current. The rate of production, or use of energy, or the rate of doing work, is called power P. Thus power $P = \frac{\text{energy}}{\text{time}}$, measured in the mks system as joules per second or *watts*. Joule's experimental results, now referred to as Joule's law, are expressed as

$$P = \text{constant} \times I^2$$

for a given resistor at a given temperature. The constant in Joule's law represents the resistance R of a resistor:

$$R = \frac{P}{I^2}$$

The mks unit of resistance is the *ohm*. A resistor has a resistance of 1 ohm if it dissipates energy at the rate of 1 watt when the current is 1 ampere. Rearranging the above gives another form of Joule's law:

$$P = RI^2$$

$$\text{watts} = (\text{ohms})(\text{amperes})^2$$

The symbol Ω (the Greek capital letter *omega*) is often used to represent ohms.

Example 15-5 What is the resistance of a 40 W lamp bulb that is lit at full brilliance by a current of $\frac{1}{3}$ A?

$$R = \frac{P}{I^2} = \frac{40\ \text{W}}{(\frac{1}{3}\ \text{A})^2} = \boxed{360\ \Omega}$$

Example 15-6 An electric iron has a resistance of 10 Ω. How much thermal energy is produced in 10 minutes if the current is 10 A?

$$P = RI^2 = (10\ \Omega)(10\ \text{A})^2 = 1000\ \text{W}$$

Thermal energy is being produced at the rate of 1000 J/sec.

$$\text{energy} = Pt = (1000\ \text{J/sec})(600\ \text{sec}) = \boxed{6 \times 10^5\ \text{J}}$$

Converting to heat units (Section 11-3), we obtain

$$(6 \times 10^5\ \text{J})\left(\frac{1\ \text{cal}}{4.18\ \text{J}}\right) = \boxed{1.44 \times 10^5\ \text{cal}}$$

Joule's law is based on experiment, and it tells us something about the behavior of matter; yet it is not of the same logical importance as Newton's laws of motion or Coulomb's law. Newton's laws are very general statements about the relation between any force and the resulting motion. Coulomb's law describes the behavior of electrical forces in a wide range of circumstances. Joule's law, however, merely describes the proportionality between electric current and energy dissipation in certain types of material such as metals. Nevertheless, Joule's law is an important generalization that is the basis for much circuit theory, including Ohm's law, which will be discussed next.

15-9 OHM'S LAW

In many practical cases we are interested in how a resistor affects the current, rather than the rate at which heat is being developed. For such calculations another relation, known as *Ohm's law*, is useful. This law can be derived directly from Joule's law, but we must bear in mind that it is quite limited in its application, and can be used only under certain conditions that will be discussed later on.

In 1827, several decades before Joule's investigations, Georg Simon Ohm, a German physicist, discovered experimentally the law that bears his name.

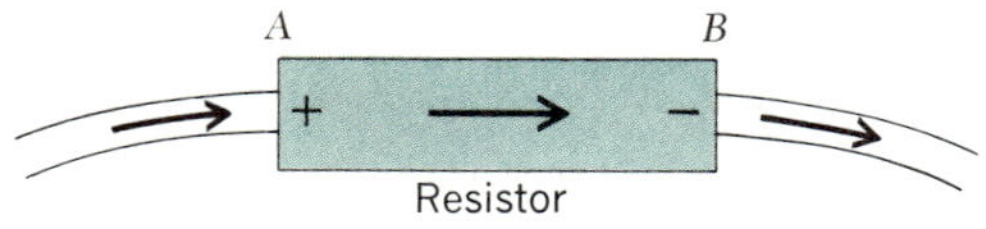

Figure 15-19

Consider a resistor AB (Figure 15-19) whose resistance is R, and through which the current is I. We must assume that the resistor contains no sources of energy. Since there is a flow of charge through the resistor, and heat is being produced, these charges must be losing electric energy. In other words, there is a PD between the terminals of the resistor; A has a positive potential relative to B. To compute the PD requires only a little algebra and the use of some basic definitions.

$$\text{PD} = \text{work per unit charge: PD} = \frac{W}{Q}$$

$$\text{charge} = (\text{current})(\text{time}) = It\text{, then PD} = \frac{W}{It}$$

$$\text{power} = \text{work per unit time} = \frac{W}{t}\text{, then PD} = \frac{P}{I}$$

$$\text{Joule's law: power} = I^2R\text{: PD} = \frac{I^2R}{I}$$

Hence,

$$\text{PD} = IR$$

or, since V is often used to represent PD in an equation,

$$V = IR$$

This is Ohm's law for part of a circuit. In words:

Ohm's law: The PD between any two points in a circuit equals the product of the current and the resistance.

Ohm's law is valid only for metallic resistors at a definite temperature and carrying steady currents. The law is approximately true for other types of resistors, but fails completely for the conduction of electricity by gases.

There are two equations for resistance: $R = P/I^2$ (Joule's law) and $R = V/I$ (Ohm's law). The first equation is a definition of R and is always true. It reflects the fact that in any conductor (except a superconductor) energy is always being dissipated as heat, and the resistance is the constant factor relating the current and the power dissipated. Ohm's law is a statement about the behavior of metals and can be derived from fundamental considerations about the drift of free electrons in metals.

In Chapter 17 we will study electric circuits in detail; we will derive formulas for resistors and emf's in different circuit combinations. For the present, let's consider a very simple circuit in which a device D is connected to the two terminals of a "black box" B (Figure 15-20). The "black box" is often a useful concept in the study of electricity (and other branches of science); it simply means a device whose *behavior* we will investigate without worrying about the details of its internal construction. To some extent, all aspects of nature are "black boxes." Without knowing anything about the nature of B and D, we can say that, if there is a current of I amperes out of the positive terminal of B, and if the PD across the terminals of B (and D) is V volts, then energy is being supplied by B, and absorbed by D at the rate of P watts, where

$$P = VI$$

This follows directly from the definition of the volt as 1 joule per coulomb (Section 15-6):

$$\text{watts} = (\text{joules/coulomb})(\text{coulombs/second})$$

If B contains an emf $\mathcal{E}$, and the current through the source of emf is I, the rate at which nonelectric energy is being converted into electric energy

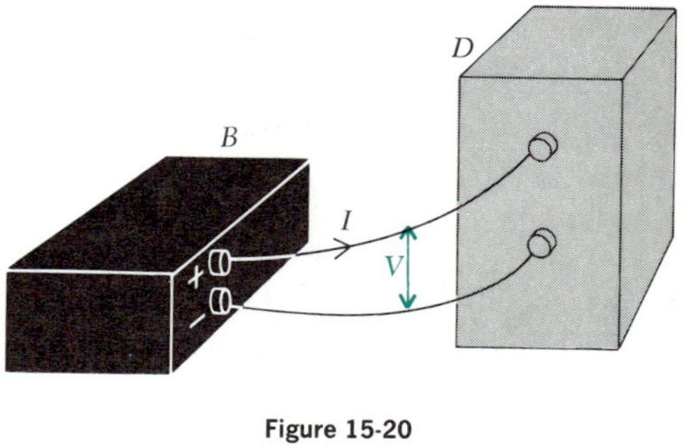

Figure 15-20

at the source of emf is

$$P = \mathcal{E} I$$

These power equations apply to any conductor, even if it does not have a constant resistance. Note that $\mathcal{E}$ is used above because we are concerned here with the transformation of nonelectric energy into electric energy at a source of emf.

Example 15-7 Suppose that in Figure 15-20, the device D consists of a motor that uses 400 W, and a lamp bulb. If $V = 100$ V and $I = 5$ A, how much electric power is received by the bulb?

The power input to D is

$$VI = (100 \text{ V})(5 \text{ amp}) = 500 \text{ W}$$

Therefore power to the bulb is

$$500 \text{ W} - 400 \text{ W} = \boxed{100 \text{ W}}$$

Example 15-8 If the black box B of the preceding problem dissipates energy at the rate of 40 W, calculate the value of the emf inside the box.

The box is delivering electric energy to device D at the rate of 500 W, and also electric energy is being transformed into thermal energy inside the box at the rate of 40 W. Hence the total rate of production of electric energy from nonelectric forms must be 540 W.

$$P = \mathcal{E}I$$

$$\mathcal{E} = \frac{P}{I} = \frac{540 \text{ W}}{5 \text{ A}} = \boxed{108 \text{ V}}$$

Example 15-8 may help us to clarify the difference between emf and PD. Both are measured in volts, and refer to changes in energy (work) per unit charge—that is, joules per coulomb. Emf refers to the rate at which electric energy is made available to a circuit at a seat, or source, of emf (such as B in Figure 15-20).

PD, however, refers to the difference of electric PE per unit charge between two points in a circuit, such as the terminals of D in Figure 15-20. We may think of the PD as the energy, or work, required to move a unit charge from one point to another in the presence of other charges.

Electric energy may be measured in joules (= watt·seconds). Another unit, commonly used by power companies, is the kilowatt-hour (kWh), which is the energy produced, transformed, or dissipated in 1 hour if the power is 1000 watts. One kilowatt-hour equals (1000 J/sec)(3600 sec) $= 3.6 \times 10^6$ J. Of course, the unit kWh is not reserved for electric energy; it can equally well be used for thermal, mechanical, or other forms of energy.

15-10 RESISTIVITY

To compare the resistances of various substances, we need to define a property that does not depend on the size or shape of the conductor. We know that resistance is the ratio of the power dissipated to the square of the current in a conductor. Other things being equal, then, we would expect resistance to increase with the length of the conductor. Experiment has confirmed this belief, as well as the belief that the current will increase if the conductor has a larger cross section. Thus we can then say that resistance R depends on conductor length L and area A in the following way:

$$R = \frac{\rho L}{A}$$

The constant ρ (Greek letter *rho*) is called the *resistivity*, and in mks units it is the resistance of a cube of material 1 meter on a side.

Table 15-1 on page 268 shows that silver has the smallest resistivity of any metal, with copper not much greater. Fused quartz, an excellent insulator, has a resistivity more than 10^{25} times that of silver.

Example 15-9 In a power station, a "bus bar" designed to carry many amperes of current is in the form of a slab of copper 2 m long and 1 cm × 10 cm in cross section. (a) What is the resistance of the bar at 0° C? (b) What PD is needed to cause a current of 5000 A through the bar?

(a) The cross section is $(10^{-1}\ \text{m}) \times (10^{-2}\ \text{m}) = 10^{-3}\ \text{m}^2$. To find the resistance, we look up the resistivity of copper in the table and substitute into the formula:

$$R = \frac{\rho L}{A}$$

$$= \frac{(1.54 \times 10^{-8}\ \Omega\cdot\text{m})(2\ \text{m})}{10^{-3}\ \text{m}^2}$$

$$= \boxed{3.08 \times 10^{-5}\ \Omega}$$

(b) To find the necessary PD, we use Ohm's law:

$$V = IR$$

$$= (5 \times 10^3\ \text{A})(3.08 \times 10^{-5}\ \Omega)$$

$$= 15.4 \times 10^{-2}\ \text{V}$$

$$= \boxed{0.154\ \text{V}}$$

Such a small PD would be negligible in a power station, where the PD between the terminals of the generator might be several hundred volts.

Example 15-10 What will the resistance of the bar of Example 15-9 become if it is stretched out to form a long wire that is 1 mm × 1 mm in cross section?

The length of the wire equals the volume divided by the area of the cross section:

$$L = \frac{(2\ \text{m})(10^{-1}\ \text{m})(10^{-2}\ \text{m})}{(10^{-3}\ \text{m})(10^{-3}\ \text{m})}$$

$$= 2 \times 10^3\ \text{m}$$

The wire has the same amount of copper as the bar, but it is now in the form of a wire 2000 m long (2 kilometers long), and its cross section is now only $10^{-6}\ \text{m}^2$.

$$R = \frac{\rho L}{A}$$

$$= \frac{(1.54 \times 10^{-8}\ \Omega\cdot\text{m})(2 \times 10^3\ \text{m})}{10^{-6}\ \text{m}^2}$$

$$= \boxed{30.8\ \Omega}$$

Table 15-1

Approximate* Resistivities ρ, all at 0°C

Substance	ρ (Ω·m)
silver	1.52×10^{-8}
copper	1.54×10^{-8}
gold	2.27×10^{-8}
aluminum	2.63×10^{-8}
tungsten (wolfram)	5.0×10^{-8}
iron	11.0×10^{-8}
platinum	11.0×10^{-8}
constantan (60 Cu, 40 Ni)	49×10^{-8}
mercury	94×10^{-8}
Nichrome (60 Ni, 24 Fe, 16 Cr)	100×10^{-8}
carbon	3.5×10^{-5}
germanium	2×10^{0}
silicon	3×10^{4}
boron	1×10^{6}
wood (maple)	3×10^{8}
Celluloid	4×10^{12}
glass	$10^{11} - 10^{13}$
amber	5×10^{14}
sulfur	1×10^{15}
mica (colorless)	2×10^{15}
fused quartz	5×10^{17}

* Exact values depend on purity, heat treatment, etc.

SUMMARY

The energy principle was first used in this chapter to explain a simple fluid system as an analog of an electric circuit. Although a fluid circuit is far from identical to an electric one, it illustrates certain important principles, such as the transformation of energy into various forms, the necessity of supplying energy to maintain a continuous flow, and the dissipation of energy because of resistance.

Liquid **pressure** P measures PE per unit volume of a liquid, and the difference in pressure ΔP between two points measures the gain or loss of energy per unit volume between those two points. In the fluid model there had to be some way of transforming "outside" energy (from human effort, perhaps) into energy of the liquid, and this was accomplished by lifting water to the top of the standpipe that served as a device where the system was continually supplied with energy.

Then in electricity a similar situation was described, one in which a unit positive test charge gains or loses energy. **Electric potential** (analogous to liquid pressure) is the electric potential energy per unit charge. The **potential difference** (PD) between two points is measured in joules/coulomb, or volts. The PD is 1 volt if 1 coulomb of charge gains or loses 1 joule of electrical energy when moved from one point to the other.

A device that can transform mechanical, chemical, thermal, or other nonelectric energy into electric energy is called a **seat of emf,** or **source of emf.** Like PD, emf's are also measured in joules/coulomb, or volts, but emf is distinguished from PD because emf refers to the rate at which electric energy is supplied to the circuit from some outside source. PD refers to the difference in energy per unit charge between any two points in a system. Common devices that have emf's are electric generators, cells, and thermocouples. The liquid standpipe is analogous to a device with an emf, and ΔP is the liquid analog of PD.

Joule's law states that the rate of production of thermal energy, or **Joule heat,** in a conductor is proportional to the square of the current. The proportionality constant is called the **resistance** of the conductor; it equals the power dissipated in heat flow, divided by the square of the current.

Ohm's law states that for a metallic conductor at any given temperature, the ratio of PD to current is a constant, and this constant is the resistance as defined by Joule's law. A conductor's resistance is 1 ohm (Ω) if a current of 1 ampere (A) generates heat at the rate of 1 watt (W), or (an equivalent statement) if there is a current of 1 A when the PD is 1 volt (V).

As you go on with the study of electricity, the liquid analogy will continue to be helpful. For example, in a later chapter you will find that flow through the pipes will be a help to you in understanding the flow of electric charges through resistors.

QUESTIONS

1 How is fluid pressure related to potential energy?

2 What is the difference between electric potential and electric PE?

3 Use the energy method to discuss the statement "water seeks its own level."

4 Use Bernoulli's equation to predict what will happen to the pressure in a moving liquid as it moves through a pipe of decreasing cross-sectional area.

5 When a dry cell is furnishing current through an external circuit, electrons inside the cell are moving toward the negative terminal. Explain how they are able to do this.

6 Explain why the standpipe might be regarded by some people as a misleading analogy for an electric battery.

7 A single object cannot have PE in isolation, but must be part of a system involving at least two bodies and an interaction force. What interaction force affects the water in a standpipe and what is the other body involved?

8 Both PD and emf are measured in volts. What is the difference between these concepts?

9 Suppose that a device were built for converting relativistic mass-energy directly into electric energy and vice versa. Could such a device be considered a source of emf? Explain your answer.

10 Does a resistor that obeys Ohm's law necessarily obey Joule's law? Does a resistor that obeys Joule's law necessarily obey Ohm's law?

11 Could you construct two wires of the same length, one of copper and one of iron, that would have the same resistance at the same temperature?

PROBLEMS

GROUP A

1 In a system like that shown in Figure 15-4, the water level in the standpipe is 15 cm above the reference level. What is the pressure in the standpipe at the reference level? What is the pressure P_4 when the valve is closed?

2 What do the pressures calculated above mean in terms of energy? Give the numerical values in proper units.

what if there was an initial speed.

3 Water is flowing in a system like that of Figure 15-4. The water level in the standpipe is kept at a constant 20 cm by repeatedly lifting 50 cm^3 cups of water from the reference level in the collecting tray to the top of the standpipe. (a) What is the pressure P_s at the reference level in the standpipe? *Hint:* The speed is nearly zero at the reference level. (b) Assume that the viscous drag is negligible, and calculate the speed of the water when the water level in M_1 is 19 cm.

4 The pipe in Problem 3 has a cross-sectional area of 0.50 cm^2. What is the volume flow rate?

5 Refer again to the system of Figure 15-4 and assume that the valve is opened wider and water is flowing at the rate of 3 liters/min. What is the speed of the water if the cross-sectional area of the pipe is again 0.50 cm^2?

6 In the circumstances described in Problem 5, drag is not negligible, the pressure P_1 is 14.9 cm water and P_2 is 12.9 cm water. What is $\Delta P = P_1 - P_2$ in N/m^2? What is this ΔP in terms of energy?

7 What is the PD between two points if 2 J of energy is required to move 20 C from one point to the other?

8 How much energy is needed to move a charge of +5 C from *A* to *B*, if the potential of *A* relative to a certain point is +100 V and the potential of *B* relative to the same point is +112 V?

9 What is the resistance of a 1000 watt heater in which the current is 10 A?

10 What is the current through a 25 Ω resistor if the PD between its terminals is 100 V?

11 What PD is needed to cause a current of 5 milliamperes through a resistance of 2 megohms?

12 What is the resistance of the filament of a radio tube if the current is 0.06 A when the PD is 6 V?

13 Calculate the energy expended in lighting a 100-watt lamp bulb for 30 min. Express your answer in kWh.

GROUP B

14 A certain wire has a resistance of 20 Ω. What is the resistance of another wire, made of the same material, that is 4 times as long and has twice the diameter?

15 Starting from rest, a charge of -3×10^{-8} C is moved from *A* to *B* in Figure 15-9 by a muscle that does 11×10^{-6} J of work. The test particle arrives at *B* with 5×10^{-6} J of KE. What is the PD between *A* and *B*?

16 What speed will an electron acquire if it is accelerated through a PD of 45 V in a small vacuum tube?

17 A current of 4 A through a battery is maintained for 20 sec, and in this time 480 J of chemical energy is transformed into electric energy. What is the emf of the battery? What power is available for heating and other uses?

18 A bird whose feet are 5 cm apart perches on a power line that carries 1000 A. If the resistance of the wire is 50×10^{-6} Ω/m, what is the PD between the bird's feet?

16 The Concept of Electric Field and Capacitance

In Chapter 14 you found that electric forces exist and that they act at a distance. But this is almost no theory at all, for while it is essential to describe what is happening, it is equally important to explain a phenomenon in terms of familiar laws and facts of physics. In other words, we need a model for electric forces, but we must keep in mind the limitations of any model, and remember that a model is not the same thing as reality (see Section 15-2).

One model that has been proposed for electric forces is the concept of electric field, introduced by the English scientist Michael Faraday (1791–1867). Faraday was a remarkable and largely self-taught man who (in spite of little formal education) progressed from laboratory assistant to Director of the Laboratory of the Royal Institution in just twelve years. His accomplishments in many fields of physics and chemistry were of tremendous importance. The unit of capacitance (described later) is named the *farad* in honor of Faraday's work.

The electric field concept, as you will soon see, gives us a way of thinking about and visualizing what happens in the region surrounding an electric charge. In particular, Faraday's notion of lines of force enables us to make calculations and predictions involving electric forces that would be very difficult to do otherwise.

16-1 MEASURING AN ELECTRIC FIELD

To understand how the field concept is applied, suppose we have a positively charged sphere, as in Figure 16-1. We explore the electric force in the region around the sphere by moving a small, positive test charge to various locations, such as *c*, *m*, *b*, *n*, *a*, and measuring the force on the charge at these points. The forces are represented as F_1, F_2, etc. The force F_3 is larger than F_2, because point *m* is closer to the sphere than point *n*. We can make a measurement and draw a force vector at any point surrounding the sphere. Such a collection of vectors is called a *vector field*, or simply a *field*.

The direction of the field is defined as the direction of the force on a *positive* test charge. The strength, or intensity, *E* of the electric field is defined as the force on a test charge divided by the magnitude of the test charge. This can be expressed as

$$E = F/Q$$

Electric field strength equals force per unit charge, and *E* is measured in newtons per coulomb.*

* This is similar to our definition of gravitational field strength (Section 4-7), which is measured in newtons per kilogram.

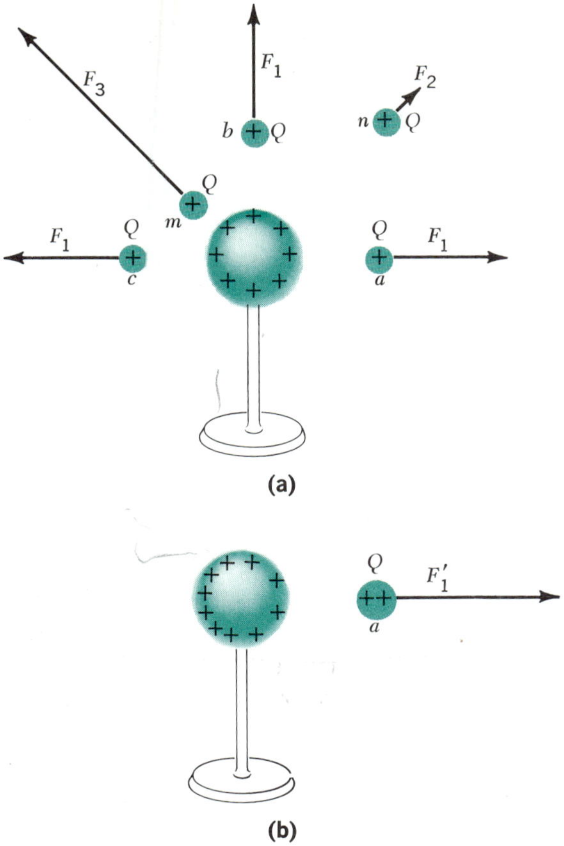

Figure 16-1

The test charge itself is assumed to be so small that it has almost no effect on the distribution of the charges that cause the field. Suppose, for instance, that a very small test charge Q is placed at a, b, or c near a uniformly charged metal sphere (Figure 16-1a). The force vectors labeled F_1 will all be equal in magnitude if a, b, and c are equidistant from the sphere's surface. But if the test charge is larger it will exert appreciable electric forces on the charges on the sphere. They will become redistributed, as in Figure 16-1b, and the force on the test charge at a will be F_1'. The ratio F_1'/Q will be equal to the field strength at point a, to be sure, but this ratio is the field strength due to the redistributed charge and is not the same as the original field strength we wished to measure.

It is not unusual in physics for a measuring device (the test charge, in this case) to have an effect on what is being measured. (Remember, for example, that it is important to have a manometer sample only a small portion of the energy of liquid pressure.) To overcome this difficulty we imagine using a test charge so small that making it still smaller would cause no appreciable change in the ratio F/Q. In other words, the ratio F/Q approaches a limit as Q (and F) approach zero, and this limit is E.

Example 16-1 Calculate the magnitude and direction of the electric field at a point 50 cm directly above a charge $Q' = -2 \times 10^{-6}$ C.

Since by definition the field strength equals the force on a unit *positive* charge, and a positive charge would be attracted by the given negative charge, we see at once that the direction of the field at the point in question is downward (Figure 16-2). The magnitude of the force on a test charge Q is found from Coulomb's law:

$$F = \frac{kQQ'}{r^2}$$

and since $E = \dfrac{\text{force}}{\text{charge}}$, we have

$$E = \frac{kQQ'/r^2}{Q}$$

$$= \frac{kQ'}{r^2}$$

$$= \frac{(9 \times 10^9\ \text{N}\cdot\text{m}^2/\text{C}^2)(2 \times 10^{-6}\ \text{C})}{(0.50\ \text{m})^2}$$

$$= \boxed{7.2 \times 10^4\ \text{N/C}}$$

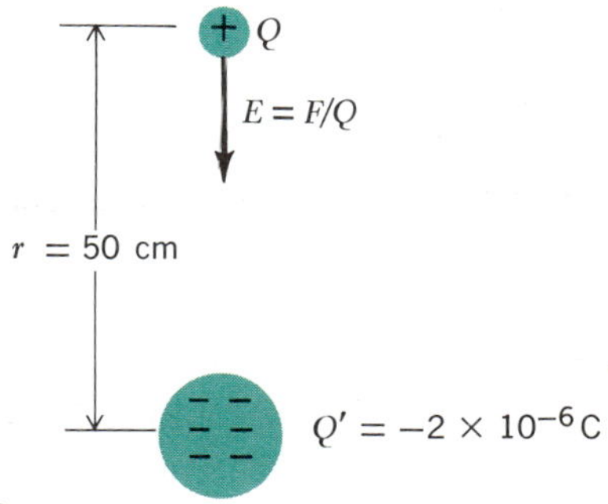

Figure 16-2

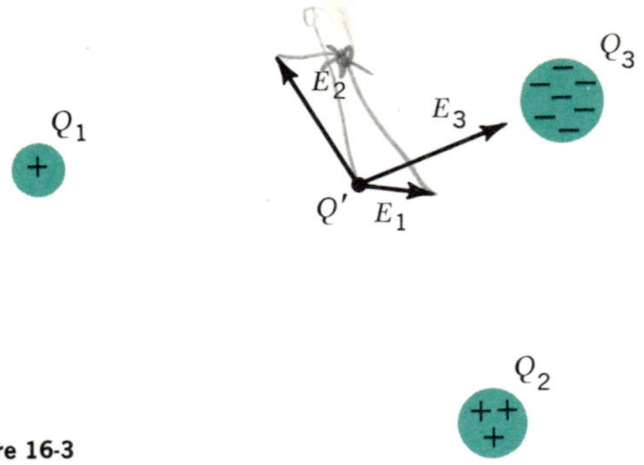

Figure 16-3

Figure 16-4 Smoke tunnel test of jet-wing airfoil showing slip stream, made visible by smoke particles, passing over the wing.

The magnitude and direction of the electric-field vector E are not always as easily calculated as in the example just given. If there are many charges, perhaps on several conductors, it might be a difficult task to compute the field. Vector addition would be necessary, with one vector contributed by each of the charges (Figure 16-3). In this case the net electric field E is the vector sum of E_1, E_2, and E_3. It is important to note that, regardless of the origin of a field, the field can be measured by use of a positive test charge. Here, as elsewhere in physics, a concept (electric field) is defined by describing a measurement procedure. We have arrived at an *operational definition** by setting up a procedure (measuring the force per unit charge) that *defines* what is meant by electric field.

The concept of field is not limited to force fields such as we have been discussing. For an example of a vector field that is not a force field, consider a river of varying depth, flowing in a curved path, perhaps having here and there some swirling eddies or stagnant places. Or consider the flow of air past an airfoil in a wind tunnel as in Figure 16-4. The flow pattern can be defined as a velocity field, for at every point a velocity could be measured both in magnitude and direction. Also, there are scalar fields; as an example, we could measure the density of the sea at various latitudes, longitudes, and depths. Since density is a scalar quantity, this collection of density values, one for every point in space, is a scalar field.

16-2 LINES OF FORCE

Faraday was not inclined to think in mathematical terms and therefore used what he called *lines of force* to help him visualize the electric fields. Maxwell and others who came after him also made extensive use of these lines.

A line of force is a line drawn at all points in a region in such a way that it represents the direction of the force field at these points. Figure 16-5 shows some of the lines of force surrounding a positive point charge. These lines represent the *directions* of the electric forces that would be felt by a positive test charge as it moved around the central charge. The lines are directed away from the positive charge, since like charges repel each other and the test charge is also positive.

Figure 16-5

* For other examples of this "operational" viewpoint, see Section 4-3 and 10-2, where operational definitions of inertial mass and temperature, respectively, are discussed. Also, in Chapter 13 the definition of absolute zero was an operational one made in terms of a procedure involving Carnot engines.

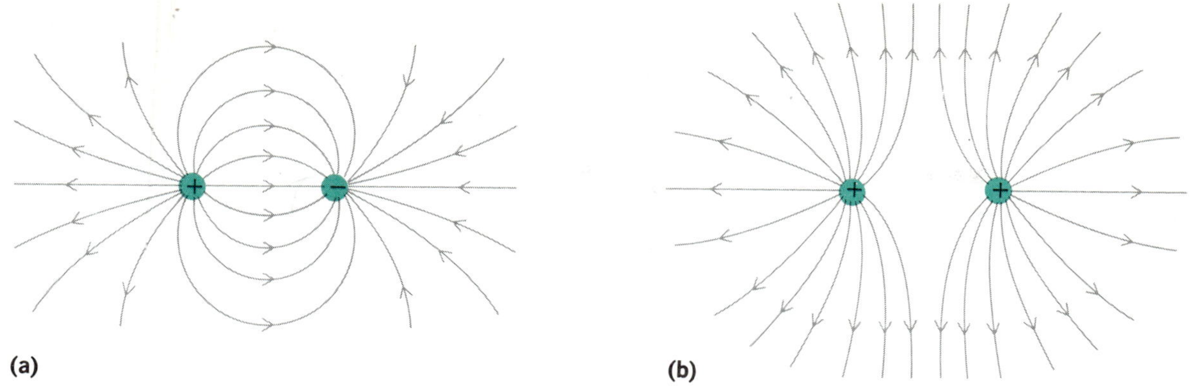

Figure 16-6 (a) Electric field in the neighborhood of a pair of unlike charges of equal magnitude; (b) electric field in the neighborhood of a pair of like charges of equal magnitude.

In principle, the lines should extend all the way to infinity, since there would be *some* Coulomb force on a test charge no matter how far away it was. At some point the force would become indistinguishable to a given measuring instrument, but there is no point where the force can be said to be zero.

Only a representative number of lines are drawn here; lines could just as well be drawn through other points, such as *P*. Note also that the field is in fact a three-dimensional pattern, but this would be extremely difficult to draw, and fields are therefore usually represented in just two dimensions. The field near a pair of equal but opposite charges is shown in Figure 16-6a, and the field near a pair of equal and like charges is shown in Figure 16-6b. Figure 16-7 contains photographs of these typical field patterns.

Lines of force originate on positive charges and terminate on negative charges; no line starts in mid-air, so to speak. Figure 16-6b may seem to violate this statement. Since there are only positive charges in the drawing, where do the lines terminate? We must imagine that there are nega-

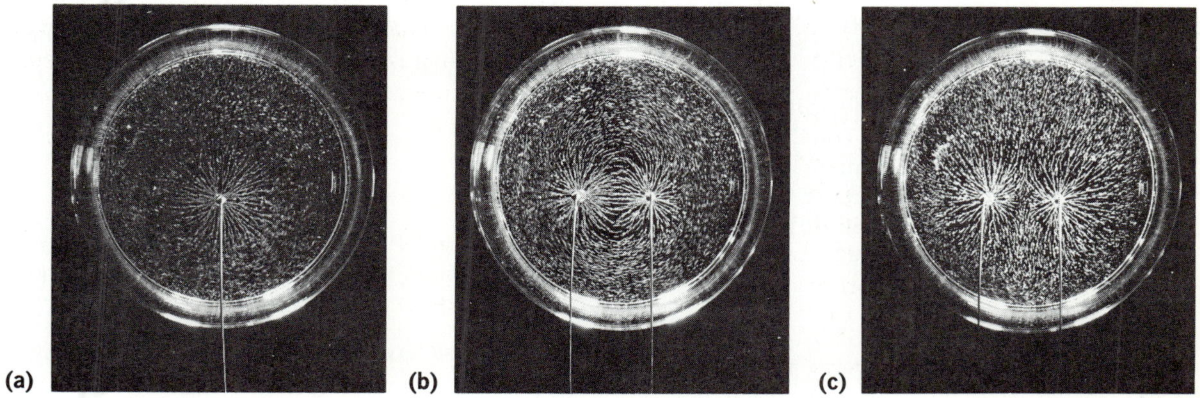

Figure 16-7 Photographs of electric force field patterns formed by grass seeds in an insulating liquid. (a) A single charged rod. (b) Two rods with equal and opposite charges. (c) Two rods with the same charge.

tive charges off somewhere at a great distance.

Another useful feature of the lines of force idea is that the "density" of the lines can be used to indicate the strength of the field. The field is strongest where the lines are closest together.

16-3 ELECTRIC FIELD AND ELECTRIC POTENTIAL

There is a close relation between field and potential, which leads to a new interpretation of electric field strength. Since

$$\text{work} = \text{(force)(distance moved)}$$

when the distance moved, or displacement, of a unit charge is parallel to the force, we can write

$$\frac{\text{work per unit charge}}{\text{distance moved}} = \text{force per unit charge}$$

or, since potential difference (PD) is equivalent to work per unit charge (see Section 15-6),

$$\frac{\text{potential difference}}{\text{distance moved}} = \text{field strength}$$

$$\boxed{\Delta V/\Delta s = E}$$

The change in potential per unit distance* is called the *potential gradient.* We see, then, that the magnitude of the field strength E equals the potential gradient. Potential gradients are measured in volts per meter.

Example 16-2 A fine iron wire 25 cm long is connected between the terminals of a battery whose terminal voltage is 6 V (see Figure 16-8). What are the magnitude and direction of the electric field inside the wire?

$$E = \Delta V/\Delta s$$
$$= 6 \text{ V}/0.25 \text{ m}$$
$$= \boxed{24 \text{ volts/meter}}$$

* The distance moved must be measured along the direction of the lines of force.

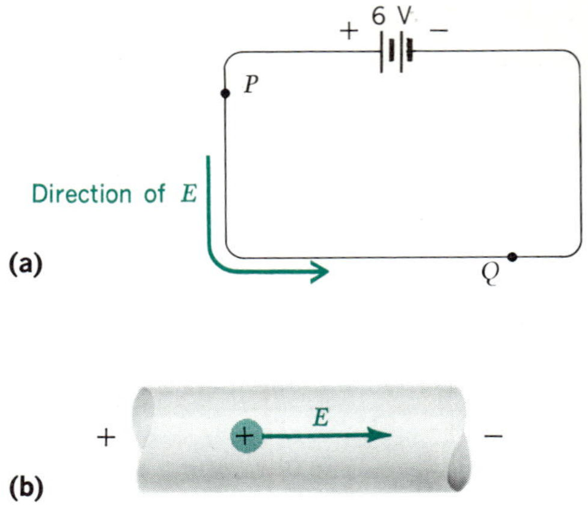

Figure 16-8 **(a)** The direction of the field is along the wire from the positive terminal toward the negative. **(b)** A greatly magnified view shows that the field is in the direction of the force on a positive test charge within the wire.

The direction of the field is toward the negative terminal of the battery, since a positive test charge inside the wire would move in this direction. The direction of field is always in the direction of *decreasing* potential**.

Example 16-3 In Example 16-2, what is the PD between point P of the wire, which is 5 cm from the + terminal of the battery, and point Q, which is 15 cm from the + terminal?

The field inside the wire is uniform, equal to 24 V/m. Hence

$$\Delta V = E\,\Delta s$$
$$= (24 \text{ V/m})(0.10 \text{ m})$$
$$= \boxed{2.4 \text{ V}}$$

Point P is 2.4 V higher in potential than is point Q, since P is closer to the + terminal.

We now have two mks units for field strength: newtons per coulomb (force per unit charge) and

** Note the complete analogy between electric and gravitational fields. Near the earth, a body falls downward in the direction of decreasing gravitational potential.

volts per meter (potential gradient). It is easy to show that these units are equivalent, remembering the definition of the volt:

$$E = \frac{\mathrm{V}}{\mathrm{m}} = \frac{\mathrm{J/C}}{\mathrm{m}} = \frac{\mathrm{J}}{\mathrm{C \cdot m}}$$

$$= \frac{\mathrm{N \cdot m}}{\mathrm{C \cdot m}} = \frac{\mathrm{N}}{\mathrm{C}}$$

In most engineering applications, the volt per meter or the volt per centimeter is used to measure electric field. For instance, the field strength required to cause a spark in dry air is about 10^6 V/m.

16-4 CAPACITANCE

When two conductors are placed close to each other but do not touch each other directly, the combination is called a *capacitor*. The capacitor is an important device in electronics, and later, when you come to discuss this and other applications of electricity, you will see in some detail how capacitors are used in electrical circuits. For now, it is enough to say that there are two main uses, both of which depend upon the capacitor's ability to store charge, and thus store energy. One use is the direct application of energy storage. In the photographer's electronic flash gun, electric energy from a small battery is slowly stored up in a capacitor, and then released in a brief but intense pulse when the flash gun is fired. Another application is found in radio and TV, where capacitors are part of circuits used to select the desired station or channel.

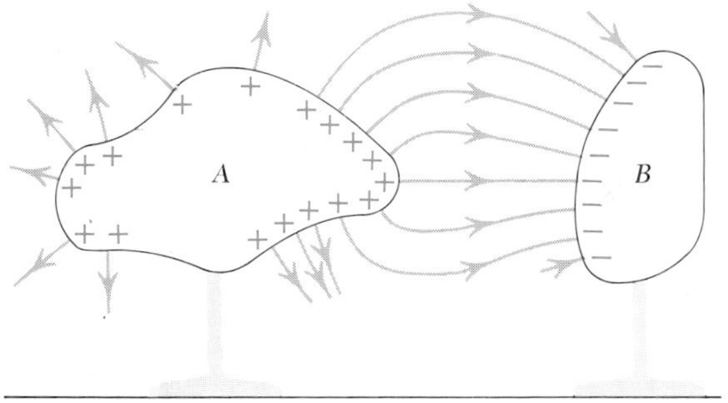

Figure 16-9

Usually, the conductors in a capacitor are in the form of parallel plates, but this is not necessary. Figure 16-9 shows two arbitrarily shaped conductors, with some positive charge on one plate and some negative charge on the other plate. If the charges on the surface of a conductor are at rest, there can be no component of electric field parallel to the surface, for if there were, these charges would move in response to the electric force. Also, if the electrons inside a metal are at rest, on the average (except for random thermal motion), we can deduce that there is no electric field at any point within the metal; if there were any electric field, the free electrons in the metal would drift in the direction of the electric force and give rise to a current. The absence of electric field in conductor *A* means that no work would be required to move a test charge around; therefore, all points of the conductor are at the same potential. We can conclude, then, that if the charges are at rest, all points on the surface of a conductor are at the same potential. Such a surface is called an *equipotential* surface. The charges automatically adjust themselves so that all points on the surface of *A* are at the same potential. The surface of *B* is also equipotential, but at a potential different from *A*. Therefore there is a single PD between body *A* and body *B*. In Figure 16-9, *A* is at a higher potential than *B* because it would require work to move a positive test charge from *B* to *A*.

The PD between the plates of any capacitor is directly proportional to the charge on either plate. Assuming equal and opposite charges on the plates, if the charge is increased, the field between the plates is increased in proportion to the charge, since electric force is proportional to the magnitude of the charge (from Coulomb's law). This is illustrated in Figure 16-10, where the same capacitor is charged with a small charge Q_1 in (a) and with a large charge Q_2 in (b).

In Chapter 15 PD was defined as work done, per unit test charge; and if a test charge is moved from the bottom plate to the top plate along any of the lines of force, the work done in (b) would be greater, because the force is greater while the displacement is the same. Therefore the

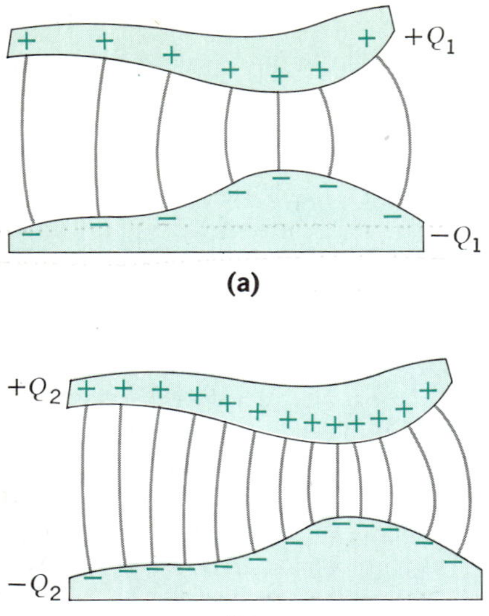

Figure 16-10 A capacitor.

potential difference V is proportional to the charge.

For any given pair of conductors, the ratio Q/V would be a constant, and this constant is called the *capacitance* C of the capacitor.

$$\text{capacitance} = \frac{\text{charge on either plate}}{\text{PD between the plates}}$$

$$C = \frac{Q}{V}$$

$$\text{farads} = \frac{\text{coulombs}}{\text{volts}}$$

the *farad* (F) being the mks unit for capacitance. A capacitor has a capacitance of 1 farad if 1 coulomb of charge causes a PD of 1 volt between the plates. The farad is an inconveniently large unit. For most applications, the microfarad ($1 \mu F = 10^{-6}$ F) or the picofarad ($1 \text{ pF} = 10^{-12}$ F) is used.

The symbols for a capacitor are

the last one representing a variable capacitor.

For the special case of a pair of parallel plates close together and separated by a vacuum, it can be shown that

$$C = \frac{A}{4\pi kd}$$

where A is the area of each plate, d is their separation, and k is the constant ($9.00 \times 10^9 \text{ N} \cdot \text{m}^2/\text{C}^2$) that enters into Coulomb's law (Section 14-4). Putting in the numerical value of k, we get

$$C = 8.85 \times 10^{-12} \frac{A}{d}$$

C is in farads if A is in square meters and d is in meters, as you can see from the following check of the units of this equation.

$$C = \frac{A}{4\pi kd} = \frac{\text{m}^2}{(\text{N} \cdot \text{m}^2/\text{C}^2)(\text{m})} = \frac{\text{C}^2}{\text{N} \cdot \text{m}}$$

$$= \frac{\text{C}^2}{\text{J}} = \frac{\text{C}}{\text{J/C}} = \frac{\text{C}}{\text{V}} = \text{F}$$

One way of charging a capacitor is to connect the two plates to a source of PD, such as a battery. In Figure 16-11 the battery (represented by the symbol —|ı|ı—) causes a momentary current from the bottom plate, through the battery, to the top plate. The PD between A and B is 45 V. Just enough charge (given by $Q = CV$) flows so that the top plate is at a potential of +45 V with respect to the bottom plate (which is then at a potential of 0 V). For the special case of a parallel-plate capacitor with plates close together, the lines of force are parallel, and the field between the plates is *uniform* both in magnitude and direction.

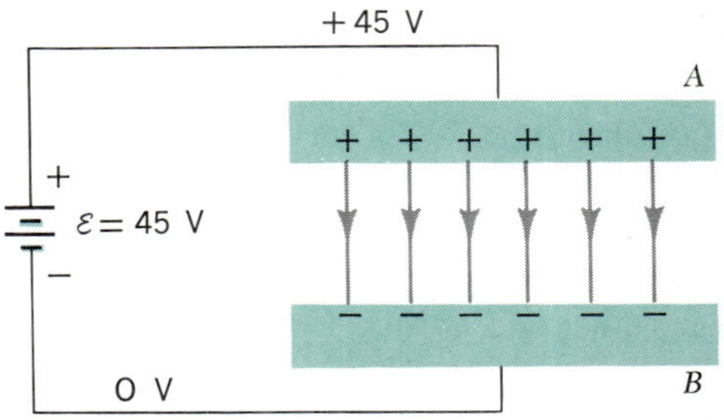

Figure 16-11

If the two plates of a capacitor are not separated by a vacuum, the measured capacitance is increased by a factor that depends on the electrical nature of the insulating medium between the plates. Such a medium is called a *dielectric*. The *dielectric constant* K of a medium is the ratio of capacitances: if C_{med} is the capacitance of a capacitor whose plates are separated by the medium, and C_{vac} is the capacitance of the same pair of plates separated by a vacuum, then

$$K = \frac{C_{med}}{C_{vac}}$$

This equation is an operational definition of the dielectric constant, since it specifies a way of measuring K. Dielectric constants for some substances are given in Table 16-1.

For any parallel-plate capacitor, the definition of K requires that the equation for capacitance be modified to read

$$C = \frac{KA}{4\pi kd}$$

16-5 THE USE OF CAPACITORS IN CIRCUITS

Even though capacitors in electric circuits are used to store charge and energy, somewhat as a bucket can be used to store a quantity of water, you should not assume that the capacitance of a capacitor is the "amount of charge it can hold." Since $Q = CV$, the charge Q can be increased to any desired value, with a corresponding increase in the potential difference V. The limit is reached only when the insulation between the plates breaks down and a spark jumps across the gap. The water analogy is fairly close if we think of a cylindrical storage tank (like the standpipe of Section 15-4) of very great height. As water is pumped in, the pressure rises, and the "capacity" of the tank is limited only by the ability of the seams and joints at the bottom of the tank to withstand the pressure of the water. At any given time, the amount of water stored in the tank is proportional to the pressure P, just as the charge stored in a capacitor is proportional to the potential difference V.

Example 16-4 A capacitor whose capacitance is 20 μF is momentarily connected to a dry cell of emf 1.5 V, and then disconnected. How much charge is stored in the capacitor?

The resistance of the connecting wires and the internal resistance of the cell are of no consequence except to determine how rapidly the capacitor becomes charged. In a typical case, after 10^{-6} sec or 10^{-8} sec or some similar very short time, the capacitor is fully charged and the current has ceased. The plates have reached a PD equal to the emf of the cell. Using the definition of capacitance $C = Q/V$, we have

$$\begin{aligned} Q &= CV \\ &= (20 \times 10^{-6}\,\text{F})(1.5\ \text{V}) \\ &= 30 \times 10^{-6}\ \text{C} \\ &= \boxed{30\ \mu\text{C}} \end{aligned}$$

Example 16-5 In a certain experiment a uniform electric field is obtained by connecting a pair of parallel circular plates to a source of PD such as the battery (represented by the symbol −|ı|+ in Figure 16-11). If the plates are 10 cm in radius and

Table 16-1

Dielectric Constants*

vacuum	1
dry air at 1 atm	1.0006
water	80
carbon tetrachloride	2.24
benzene	2.28
castor oil	4.67
methyl alcohol	33.1
glass	4–7
amber	2.65
wax	2.25
mica	2.5–7

* All at 20°C.

are separated by 3 cm, what charge on the plates will give a field of 400 V/cm in the space between the plates?

If the field is to be 400 V/cm, we find that the PD between the plates must be

$$V = (400\ \text{V/cm})(3\ \text{cm})$$
$$= 1200\ \text{V}$$

We next compute the capacitance of the capacitor formed by the plates. They are separated by air, whose dielectric constant is very nearly 1.

$$C = 8.85 \times 10^{-12} \frac{KA}{d}$$
$$= (8.85 \times 10^{-12}\ \text{C}^2/\text{N}\cdot\text{m}^2) \times \left(\frac{(1)(\pi)(1 \times 10^{-1}\ \text{m})^2}{3 \times 10^{-12}\ \text{m}}\right)$$
$$= 9.27 \times 10^{-12}\ \text{F}$$

Now we can compute the charge on the plates:

$$Q = CV$$
$$= (9.27 \times 10^{-12}\ \text{F})(1.20 \times 10^3\ \text{V})$$
$$= \boxed{1.11 \times 10^{-8}\ \text{C}}$$

The physical construction of capacitors depends upon the desired capacitance and the PD that must be sustained without breaking down the dielectric. In the formula

$$C = KA/4\pi kd$$

either K, A, or d can be varied. For low values of C (such as those required in the tuning stage of a radio receiver), parallel plates separated by air are used (Figure 16-12a). The capacitance is changed by rotating one set of plates to give a variable effective area of overlapping. Small "trimmer" capacitors (b) have a mica dielectric, and C is changed by changing the separation d. Capacitors of fixed value are often constructed of two rolled-up aluminum foils separated by waxed paper (c). When complete, they often resemble (d). Compromise must be made between high capacitance (which requires small separation d) and the ability to withstand high voltages without rupture (which requires a large d). Capacitors are therefore rated by "working voltage" as well as capacitance. For about a dollar, you can buy capacitors of 1000 μF rated 6 V, 40 μF rated 450 V, or 0.0005 μF rated at 10,000 V. Some larger oil-filled capacitors are designed to store a small charge at a PD of a million volts or more, for use in artificial lightning experiments.

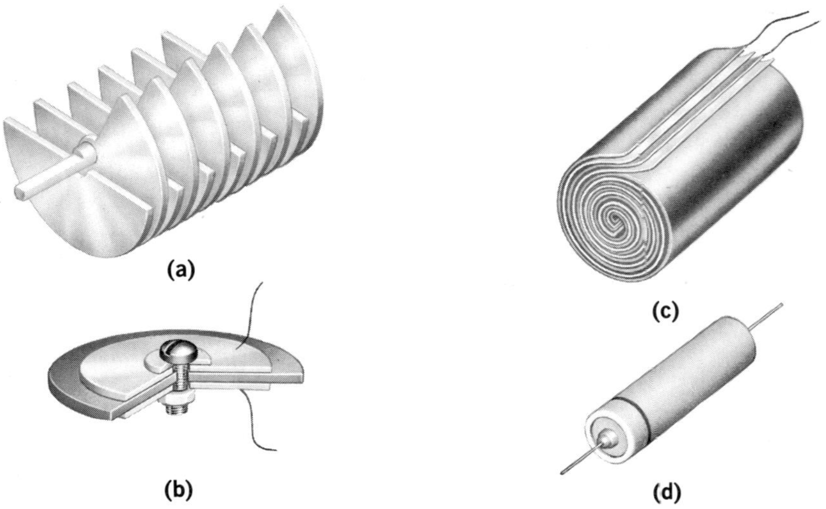

Figure 16-12 Typical capacitors: **(a)** Variable 30–300 pF, rated at 300 V. **(b)** 5–40 pF, rated at 100 V. **(c)** Tubular fixed capacitor, 0.05 μF, rated at 600 V. **(d)** Tubular capacitor in cover-case.

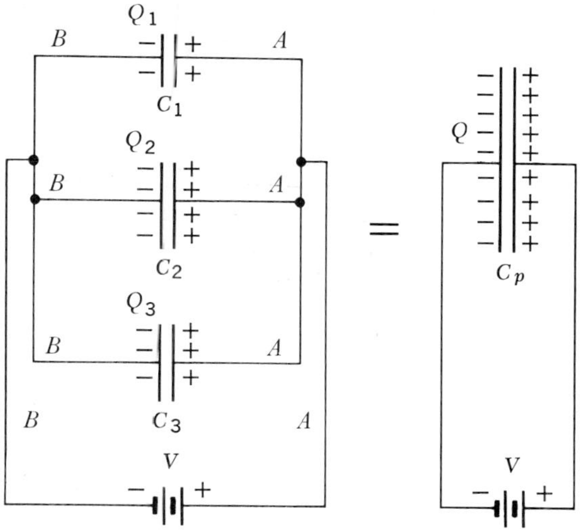

Figure 16-13 Capacitors connected in parallel.

As parts of electric circuits, or *circuit elements*, capacitors are notable for the fact that they do not allow a steady current to exist. Instead, a charge builds up, and the dielectric material or vacuum between the plates is subject to an electric field.

When capacitors are used in circuits, two or more capacitors are often used in combination. What is the effective capacitance of such a combination? In Figure 16-13, three capacitors are shown connected in what is called a *parallel* combination. After the charges reach their equilibrium positions, all conductors marked A are part of an equipotential surface. Likewise, all conductors B are part of a different equipotential surface. Therefore the PD is V across each capacitor. The charge leaving the battery divides, and each of the capacitors receives a different quantity of charge. The charge on each capacitor is given by the relation $Q = CV$. The total charge that leaves the battery is

$$Q = Q_1 + Q_2 + Q_3 + \cdots$$

Hence, if C_p is the equivalent capacitance of the parallel combination,

$$C_pV = C_1V + C_2V + C_3V + \cdots$$

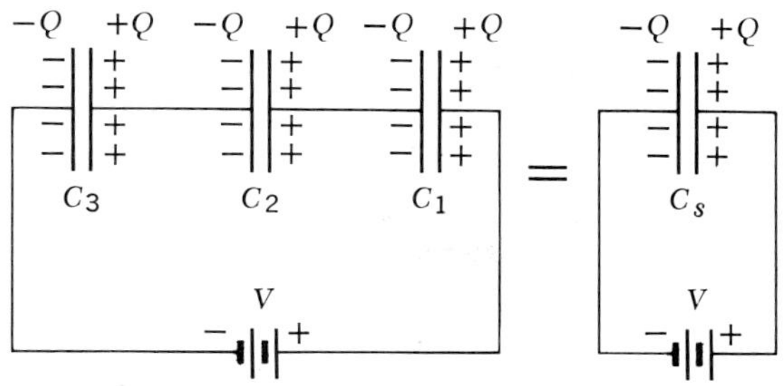

Figure 16-14 Capacitors connected in series.

Canceling V, we get

$$C_p = C_1 + C_2 + C_3 + \cdots$$

In other words, when capacitors are connected in parallel, the charge is shared among the capacitors and the equivalent capacitance is the sum of the individual capacitances.

Capacitors may also be connected in *series*, that is, one after the other in a line as shown in Figure 16-14. Here the PD is not the same across each of the capacitors shown. A certain charge $+Q$ leaves the battery and flows onto one plate of C_1. This repels an equal charge $+Q$ to the right-hand plate of C_2, leaving a charge $-Q$ on the left-hand plate of C_1. Similarly, the plates of C_3 acquire charges $+Q$ and $-Q$. The total PD is the sum of the PD's of the individual capacitors, but the charges on all the capacitors are the same, equal to the charge that flows from the battery.

$$V = V_1 + V_2 + V_3 + \cdots$$

$$\frac{Q}{C_s} = \frac{Q}{C_1} + \frac{Q}{C_2} + \frac{Q}{C_3} + \cdots$$

Here C_s refers to the equivalent capacitance of the series combination. Canceling Q, which is the same for each capacitor, gives

$$\frac{1}{C_s} = \frac{1}{C_1} + \frac{1}{C_2} + \frac{1}{C_3} + \cdots$$

In words, the reciprocal of the equivalent capacitance of a number of capacitors connected in series is equal to the sum of the reciprocals of the individual capacitances.

Series and parallel combinations of resistors will be investigated in Section 17-2.

SUMMARY

An **electric field** exists in any region where a small test charge would experience an electric force. The magnitude, or intensity, of the field equals the force per unit charge, and the direction of the field is defined as the direction of the force on a small positive test charge. Another example of a force field is a gravitational field, whose magnitude is defined as force per unit mass. In general, a **field** may be thought of as a collection of vectors or scalars that have measurable values throughout a region of space. Lines of force are used to represent the direction of an electric field.

If the free charges in a conductor are at rest, they reside on the surface and are distributed in such a way that all points of the conductor are at the same potential. Lines of force "originate" on positive charges and "terminate" on equal negative charges. Electric field strength can also be measured in terms of **potential gradient,** which is the rate of change of potential with respect to distance, measured along a line of force.

The **capacitance** of a **capacitor** is defined as the stored charge per unit PD between the plates. For a parallel-plate capacitor, the capacitance depends on the area and separation of the plates, and on the material between the plates. The **dielectric constant** of a medium is the factor by which the capacitance of a capacitor is increased when immersed in the medium. Formulas exist for the combined capacitance of capacitors in series and in parallel.

QUESTIONS

1 Is electric field strength E a vector quantity or a scalar quantity? Is electric potential V a vector quantity or a scalar quantity?

2 Why must a *small* test charge be used when measuring an electric field?

3 Why are the lines of force in Figure 16-9 perpendicular to the surfaces where the lines and surfaces meet?

4 Is the gravitational field at the bottom of a deep mine greater or less than at the surface of the earth? What is the gravitational field at the center of the earth?

5 On the weather map in your local newspaper, wind velocities and temperatures are shown at various points. Explain how such a map illustrates the following concepts: (a) vector field; (b) scalar field; (c) temperature gradient. Where is the temperature gradient the greatest?

6 Two plates of metal that have fixed charges $+Q$ and $-Q$ are immersed in a tank of oil. If the oil is pumped out, does the electric field at a point midway between the plates increase, decrease, or remain the same?

7 Ten equal capacitors are connected (a) in series and (b) in parallel. Is the resultant capacitance in each case greater than, less than, or equal to the capacitance of each capacitor?

a) less
b) greater

PROBLEMS

GROUP A

1 The force on a charge of -3×10^{-7} C is measured and found to be 0.24 N in a downward direction. What are the magnitude and direction of the electric field at this point?

2 What electric force is exerted on a charge of 6×10^{-8} C that is placed in a field whose strength is 5×10^4 N/C?

3 Three points *A*, *B*, and *C* are on a straight line, at intervals of 20 cm. The potentials of the points are +30 V, +80 V, and +130 V, respectively. What is the potential gradient?

4 A charge of +4000 μC is on one plate of a capacitor, and −4000 μC is on the other plate. The PD between the plates is 200 V. What is the capacitance of the capacitor?

5 What charge flows from a battery of emf 12 V when a 100 μF capacitor is connected to it?

GROUP B

6 Two charges of value $Q_1 = +40\ \mu$C and $Q_2 = +100\ \mu$C are 7 m apart in air. At a point *P*, which is on the line between the charges, 2 m from Q_1, calculate (a) the field due to Q_1; (b) the field due to Q_2; and (c) the resultant field at point *P*.

7 The moon has, in round numbers, $\frac{1}{81}$ as much mass as the earth. Show that, on a journey from moon to earth, the resultant gravitational field is zero when a spacecraft is $\frac{1}{10}$ of the way to the earth. (*Hint:* Let M = mass of earth, x = distance from earth, and y = distance from moon.)

8 A parallel-plate capacitor has square plates 10 cm × 10 cm, separated by 1 mm of glass. The dielectric constant of the glass is 5. Calculate the capacitance of the capacitor.

9 Two horizontal metal plates, each 2 m × 3 m, are separated by 5 mm of air. The top plate has a charge of -4×10^{-6} C, and the bottom plate has a charge of $+4 \times 10^{-6}$ C. (a) Calculate the capacitance of the plates. (b) Calculate the PD between the plates. (c) Calculate the magnitude and direction of the field at a point somewhere in the air gap between the plates.

10 A parallel-plate capacitor has a slab of glass (of dielectric constant 4) separating the plates. The capacitor is first connected to a 6 V battery, and then disconnected from the battery. The PD between the plates is observed to be 6 V. The slab of glass is now pulled out. What is the new PD between the plates?

11 A parallel-plate capacitor of capacitance 4×10^{-8} F has a charge of 120

μC. The plates are separated by 1 mm of air. What is the magnitude of the electric field at a point between the plates?

12 An electron is located at a point at which the electric field is 6000 V/m. What is the force on the electron? What is the acceleration of the electron?

13 It is found that 0.3 J of work is needed to move a charge of 0.75 millicoulomb from one insulated conductor to another insulated conductor that is 20 cm away. What is the average field strength in the region between the conductors?

14 Three capacitors having capacitances 10 μF, 20 μF, and 60 μF are available. What is the maximum value of the capacitance that can be formed from these three? What is the minimum value? Show the connections for each case.

GROUP C

15 If the breakdown field strength for dry air is 3×10^6 V/m, what is the maximum charge that can be stored in a parallel-plate air capacitor whose plates are each 40 cm × 50 cm? (*Hint:* The spacing of the plates is not given. Call it *d*, carry *d* through your calculation, and trust that it will cancel out of your final answer.)

Why is distance negligible

16 A capacitor is designed to have a capacitance of 0.1 μF and is to operate safely at a PD of 1000 V. The dielectric is mica, for which the dielectric constant is 3 and the breakdown field is 100 kilovolts/mm. What must be the area of each plate, allowing a safety factor of 4?

4×PD = 4V

FOR FURTHER STUDY

Energy Storage in Capacitors

The act of forcing charge into a capacitor is analogous to the stretching of a spring. At the start, when the capacitor is "empty," it is easy to move a few electrons from one plate to the other through an external circuit. However, as the capacitor becomes charged, an electron approaching the negative plate is repelled by the charges already on that plate; and an electron leaving the positive plate is attracted back by the excess of positive charge on that plate. Thus, the more charge on a capacitor, the more work is required to place additional charge on the plates. Similarly, when filling a storage tank with water to a height *h*, the higher the water level, the harder it is to force more water in at the bottom of the tank. The work done to increase potential energy is $mg\,(\frac{1}{2}h)$ or $\frac{1}{2}wh$, where $\frac{1}{2}h$ is the average height to which the water is raised. In a similar fashion, the more a spring is already stretched, the more work is required to stretch it an additional small amount. You saw in Section 8-3 how to calculate the PE stored in a stretched spring, using the average force instead of the final force. It was shown that, if Hooke's law is applied, the average force is just half the final force. The PE for a stretch of A was found to be $\frac{1}{2}(kA)(A)$, or $\frac{1}{2}kA^2$.

The same reasoning is used to find the work done in charging a capacitor. To move a charge Q coulombs through a PD of V volts requires QV joules. However, the PD varies from 0 to the final value V; the average PD is $\frac{1}{2}V$, and the work done is $Q(\frac{1}{2}V)$. Therefore, the PE of a charged capacitor is given by

$$\text{PE} = \tfrac{1}{2}QV$$

Since $Q = CV$, this equation can also be written in the equivalent forms

$$\text{PE} = \tfrac{1}{2}CV^2$$

or

$$\text{PE} = \frac{1}{2}\frac{Q^2}{C}$$

Let's note the close analogy with the formula for the PE of a spring.
PE of a charged capacitor:

$$\text{PE} = \frac{1}{2}\frac{1}{C}Q^2$$

PE of a stretched spring:

$$\text{PE} = \tfrac{1}{2}(k)A^2$$

The reciprocal of the capacitance is a measure of the "stiffness" of a capacitor, just as the force constant k is a measure of the stiffness of a spring. If $\frac{1}{C}$ is very large, the capacitor absorbs only a small charge with a large rise in its PD, and it might be said to be very stiff. Mechanical analogies of this sort are useful in discussing electrical oscillations, as you will see in Section 20-2.

PROBLEMS

GROUP D

17 A 40-millifarad capacitor is charged to a PD of 30 V. How much electric PE is stored in this capacitor?

18 A photographer's flash-bulb outfit consists of a capacitor that is slowly charged by five 90 V batteries in series, and then discharged quickly through a xenon-filled glow tube. The tube requires 100 J of energy for a successful discharge. What must be the capacitance of the capacitor?

19 The electric field surrounding the earth is about 100 V/m in the air near the surface. How much electric energy is stored in a layer of air 2 m thick over a football field that measures 91.4 × 48.8 m? *Hint:* Imagine a conducting sheet above the field 2 m above the ground forming a capacitor with the ground. Find the capacitance of this capacitor, and find the PD from the potential gradient.

17 Electric Circuits and Their Components

So far we have discussed several of the elements, or parts, of a circuit such as sources of energy (seats of emf), resistors, and capacitors. A surprisingly large number of simple circuits involving only emf's and resistors are in common use and can be investigated with the help of Ohm's law. These include many instruments that are used routinely in engineering, in the physics laboratory, in the chemistry laboratory, and in medical research. Several kinds of circuits will be discussed in this chapter, but no new theories will be introduced. Your understanding of PD, emf, and Ohm's law will be sufficient for analyzing each of the circuits.

In drawings of electric circuits, certain symbols are standard. A resistor is represented by a zigzag line . A variable (adjustable) resistor is drawn as  or . A battery, or cell, is represented by 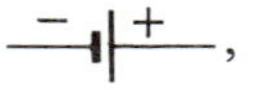, where the long line is the positive terminal and the short line is the negative terminal. Two or more cells in series are shown as . Unless specifically stated otherwise, all connecting wires in diagrams are assumed to have zero resistance. The junction between two connected wires is indicated by a thick spot. In this book current direction is represented by colored arrowheads as in Figure 17-1.

In discussing electric circuits, we will make use of two fundamental conservation laws—the law of conservation of charge and the law of conservation of energy. The law of conservation of charge tells us that charges do not appear or disappear anywhere in a circuit, and the law of conservation of energy tells us that the total work done *against electric forces* in taking a test charge around any complete circuit is equal to zero. Thus no net change in potential can exist after a charge goes completely around a circuit. This means that if we pick some point as the zero of potential, the test charge will again have zero potential after traveling around the circuit and returning to the starting point. As the charge moves through a battery (or other source of emf), it will gain energy; then, as it moves through a resistor, a certain amount of this energy will be dissipated as heat, and as it goes through a motor, energy will be converted to mechanical work. When the charge returns to the starting point, the gains and losses of energy will be equal. There will be no net change.

Changes of potential in a circuit can arise in several ways. One way is when charges flow through a resistor; the PD across the resistor is given by Ohm's law, stated as $V = IR$. The product IR is called the "IR drop" in the resistor. Another way in which potential may change is in a cell or battery. In a cell there is a PD equal to the emf. Also, if the cell has an *internal resistance r*, there is a PD equal to Ir inside the cell, which we will now investigate in some detail. Let's begin by finding out how to apply the conservation laws to several simple circuits.

17-1 TERMINAL VOLTAGE OF A CELL

A cell or battery is no exception to the rule that "irreversible" Joule heat (given by Joule's law $P = I^2R$) is produced by a current. This means that the cell has internal resistance, which for a fresh "dry" cell is about 0.05 ohm; but, as the moist electrolyte inside the cell dries out, this internal resistance may increase to as much as 100 ohms or more. Also, hydrogen gas is released by electrolysis during the passage of charge through the cell and collects around the carbon rod. This increases the resistance because a gas is a poor conductor of electricity. The cell is said to become *polarized* during use.

From the information given we might expect the internal resistance of a cell to vary according to its age and past history as well as the amount of current through it. Yet in problem-solving we can usually consider the internal resistance to be a constant, at least as a first approximation.

Taking the internal resistance into account, we find that the PD between the terminals of a cell, the *terminal voltage*, is not always equal to the cell's emf. Consider a 6.08-volt storage battery that is delivering 4 amperes to a headlamp of resistance 1.5 ohms, as shown in Figure 17-1. The internal resistance of the battery is represented as a resistor r drawn next to the battery symbol. The battery as a whole is represented by a box that includes $\mathcal{E}$ and r, with terminals A and B. We must not forget that the junction point, such as X in Figure 17-1, is never available to us, and we could never connect a voltmeter between A and X to measure the emf $\mathcal{E}$ directly.

Example 17-1 In Figure 17-1, calculate (a) the PD across the headlamp terminals PQ, (b) the terminal voltage of the battery (AB), and (c) the internal resistance of the battery.

(a) By Ohm's law, across the headlamp

$$V = IR$$
$$= (4.00\ \text{A})(1.50\ \Omega)$$
$$= \boxed{6.00\ \text{V}}$$

(b) By terminal voltage of the battery we mean the PD between its terminals, which we can determine by using the law of conservation of energy. The 6.00 V IR drop across PQ means that 6.00 J of energy is dissipated for each coulomb of charge that flows through the resistor. Where does this energy come from? Since the connecting wires are assumed to have no resistance and contain no sources of energy, they may be safely ignored. They neither supply nor dissipate energy. The energy can come only from the battery. We have seen that 6.00 J/C is dissipated in the "outside" circuit PQ. Therefore the battery must supply a net energy of 6.00 J/C and the PD across the terminals must be the same as the PD across the headlamp, that is

$$\boxed{6.00\ \text{V}}$$

(c) The emf of the battery ($\mathcal{E} = 6.08$ V) indicates that 6.08 J/C is transformed from chemical to electric energy. The terminal voltage is only 6.00 V. Hence, the IR drop across the internal resistance r is

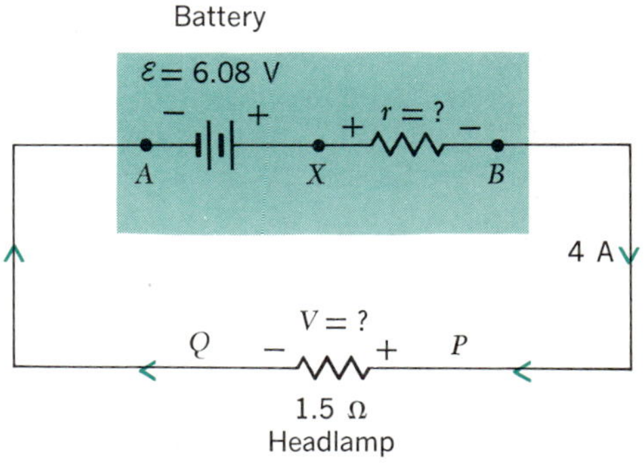

Figure 17-1

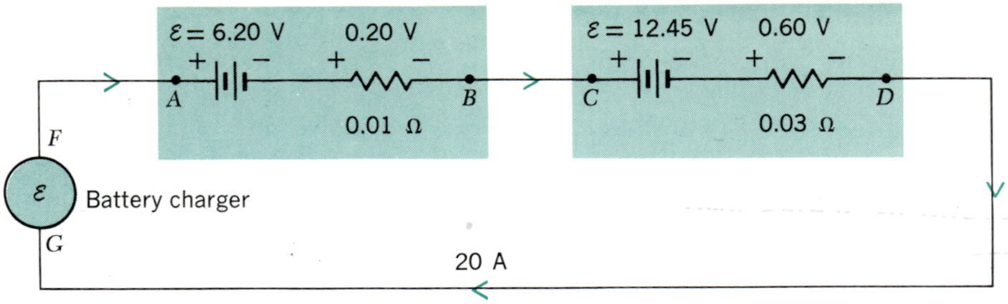

Figure 17-2

$$Ir = \mathcal{E} - \text{terminal PD}$$
$$= 6.08\text{ V} - 6.00\text{ V}$$
$$= 0.08\text{ V}$$
$$r = 0.08\text{ V}/4\text{ A}$$
$$= \boxed{0.02\ \Omega}$$

In general, we will call the terminal voltage V, and it is clear that V will be less than the emf whenever the battery is supplying energy. In the case just discussed

$$V = \mathcal{E} - Ir$$

where $\mathcal{E}$ is the emf and r is the internal resistance of the battery. It is also possible for the terminal voltage of a battery to be greater than its emf, if charge is being forced through a battery opposite to the battery's normal direction of charge flow. This is what happens to storage batteries when they are being charged.

Example 17-2 Two batteries, one of emf 6.20 V and the other of 12.45 V, are being charged by a battery charger that sends 20 A through the cells (Figure 17-2). The internal resistances of the batteries are 0.01 Ω and 0.03 Ω, respectively. What is the terminal voltage of each battery?

Ohm's law can be used to calculate the Ir drop inside each battery. In the 6.20 V battery, $Ir = (20\text{ A})(0.01\ \Omega) = 0.20\text{ V}$. In the 12.45 V battery, $Ir = (20\text{ A})(0.03\ \Omega) = 0.60\text{ V}$. These internal Ir drops are labelled + to − as shown, since charge is being forced through the internal resistances of the batteries, and the positive end of a resistor is always the one at which the conventional current enters. The PD from A to B, then, is the algebraic sum of the emf and the internal Ir drop:

$$V = 6.20\text{ V} + 0.20\text{ V}$$
$$= \boxed{6.40\text{ V}}$$

Similarly, for the 12.45 V battery, the PD from C to D is

$$V = 12.45\text{ V} + 0.60\text{ V}$$
$$= \boxed{13.05\text{ V}}$$

These terminal voltages are *greater* than the emf's of the batteries. In this case $V = \mathcal{E} + Ir$.

In general, then, the terminal voltage of a battery or other source of energy is given by

$$V = \mathcal{E} \pm Ir$$

Whether to use the + or the − sign depends upon the direction of the current through the battery, that is whether the battery is supplying energy to the outside circuit or receiving energy from it when being charged. When the current through a cell, battery, or other source of emf is zero, then $V = \mathcal{E}$ and the terminal voltage equals the emf. This points out one way to measure emf—by measuring the terminal voltage under circumstances where there is no current through the battery. The potentiometer circuit (Section 17-6) is designed to do just this. An ordinary voltmeter is ruled out for this purpose, since there must be *some* current through the voltmeter to produce a deflection of the needle.

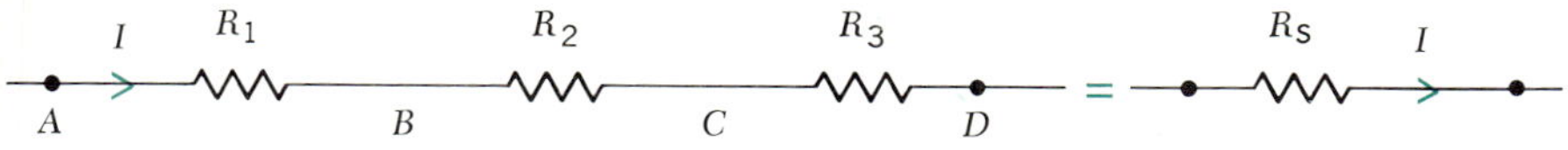

Figure 17-3 Resistors connected in series.

17-2 RESISTORS IN SERIES AND IN PARALLEL

It is possible to simplify many (but not all) complex circuits by mentally replacing a group of resistors by a single resistor equivalent to the group (recall that this was done for capacitors in Section 16-5).

The resistors R_1, R_2, and R_3 in Figure 17-3 are said to be connected *in series* because charges move through the resistors one after the other, making no short cuts. In terms of the fluid analogy, resistors in series are like pipes connected end to end; the same amount of water must flow through all of them, for it is assumed that there are no leaks and no fluid is being added. With the resistors, we know that charge is conserved—that no charge is added or lost as the charges move through the resistors in turn.

Let's call the resistance of the series combination R_s, the current I, and the PD from A to D V_s. Then by Ohm's law we know that $R_s = V_s/I$, even though we do not yet know what V_s and R_s are.

We can arrive at the PD of the combination V_s by making use of the law of conservation of energy and by remembering that the PD across a resistor measures the energy per coulomb dissipated as charge flows through the resistor (Section 15-6). V_s represents the total energy dissipated by the combination when a coulomb of charge flows through. V_1 is the energy dissipated per unit charge in R_1, V_2 is the energy dissipated per unit charge in R_2, and so on through the resistors. By the law of conservation of energy, the total energy dissipated must equal the sum of the energies dissipated in the individual resistors. The following proof is general and can apply for any number of resistors in series, although for simplicity only three resistors have been shown in the diagram above.

$$V_s = V_1 + V_2 + V_3 + \cdots$$

$$R_s = \frac{V_s}{I}$$

$$= \frac{V_1 + V_2 + V_3 + \cdots}{I}$$

$$= \frac{IR_1 + IR_2 + IR_3 + \cdots}{I}$$

Hence, for resistors in series,

$$R_s = R_1 + R_2 + R_3 + \cdots$$

The essential features of our proof are (a) that the PD's are added for resistors in series and (b) that the current is the same through each resistor in a series group.

The resistors in Figure 17-4 are connected *in parallel.* In a parallel connection, the incoming current divides as you saw in Section 16-5. In the case of resistors, some of the current goes through each resistor. In fluid terms, imagine three branching pipes connected to a single pipe at point E of Figure 17-4, and rejoining at point F. The moving fluid divides at the junction, and some goes through to each of the branches. Again, since we do not add water or allow leaks, the flow out of point F equals the flow into point E. In terms of an electric current the law of conservation of charge tells us that the total current leaving point F equals that coming into point E. The PD's across each resistor are the same, and each is equal to the PD measured between points E and F. The PD between E and F is the amount of

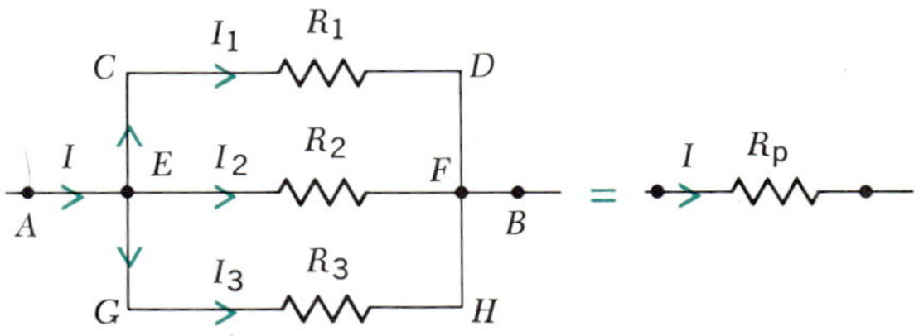

Figure 17-4 Resistors connected in parallel.

energy dissipated per unit charge flowing from E to F, and this amount is the same whichever path is taken.

The currents can be added together as follows:

$$I = I_1 + I_2 + I_3 \cdots$$

By Ohm's law, the equivalent resistance R_p of the parallel combination is

$$R_p = \frac{V}{I}$$

and therefore

$$\frac{1}{R_P} = \frac{I}{V}$$

$$= \frac{I_1 + I_2 + I_3 + \cdots}{V}$$

$$= \frac{I_1}{V} + \frac{I_2}{V} + \frac{I_3}{V} + \cdots$$

So, for resistors in parallel,

$$\frac{1}{R_P} = \frac{1}{R_1} + \frac{1}{R_2} + \frac{1}{R_3} + \cdots$$

The essential features of this proof are (a) that the total current equals the sum of the currents through the parallel resistors, and (b) that the same PD exists across each resistor in a parallel group.

Example 17-3 Resistors of 6 Ω, 12 Ω, and 4 Ω are connected in parallel. What is the equivalent resistance of the combination?

$$\frac{1}{R} = \frac{1}{6} + \frac{1}{12} + \frac{1}{4}$$

$$= \frac{2}{12} + \frac{1}{12} + \frac{3}{12}$$

$$= \frac{6}{12}$$

$$R_p = \frac{12}{6}$$

$$= \boxed{2\ \Omega}$$

It will be helpful to remember that the combined resistance of any number of resistances in parallel is always less than the smallest resistance in the combination. This can be illustrated by an analogy from highway engineering. The opening of a parallel bypass, no matter how narrow, will always allow a greater flow of cars; hence the "resistance" of the combination is less than that of either the road or the bypass taken separately. Perhaps the easiest way to remember how to combine resistors is to bear in mind that the quantity used above, $1/R$, or the reciprocal of the resistance, is called the "conductance." It is a measure of how easily charge can flow through a resistor (or any conductor or set of conductors). Therefore, for resistors in parallel, add the conductances; and for resistors in series, add the resistances. You may have noticed that these rules for resistors are just the opposite of the rules for combinations of capacitors given in Section 16-5.

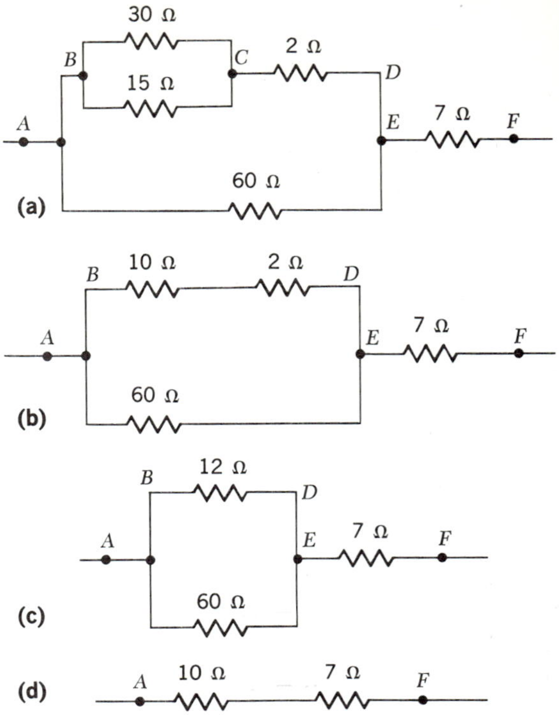

Figure 17-5 Successive stages in the simplification of a circuit diagram.

Example 17-4 What is the net resistance between A and F in Figure 17-5a? (All resistances are in Ω.)

This circuit is a combination of series

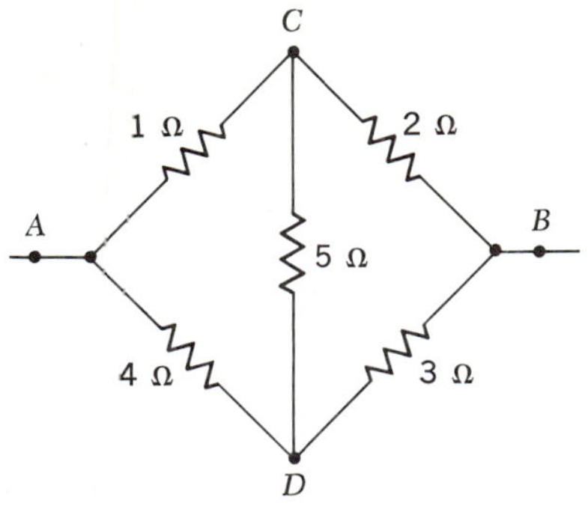

Figure 17-6 A nonsimple circuit.

and parallel units. We start from the "inside" and use our two basic formulas.

Step 1: Replace the parallel combination *BC* by a single resistor.

$$\begin{aligned} \frac{1}{R} &= \frac{1}{30} + \frac{1}{15} \\ &= \frac{1}{30} + \frac{2}{30} \\ &= \frac{3}{30} \\ R &= \frac{30}{3} \\ &= 10\ \Omega \end{aligned}$$

Substituting 10 Ω for the combination, we can redraw the circuit as in Figure 17-5b.

Step 2: Replace the series combination *BD* by a single resistor.

$$\begin{aligned} R &= 10 + 2 \\ &= 12\ \Omega \end{aligned}$$

We are now able to redraw our circuit as in Figure 17-5c.

Step 3: Replace the parallel combination *AE* by a single resistor.

$$\begin{aligned} \frac{1}{R} &= \frac{1}{12} + \frac{1}{60} \\ &= \frac{6}{60} \\ R &= \frac{60}{6} \\ &= 10\ \Omega \end{aligned}$$

This allows us to redraw the circuit as in Figure 17-5d.

Step 4: Finally, using the series formula, it can be seen that the resistance from *A* to *F* is

$$\begin{aligned} R &= 10\ \Omega + 7\ \Omega \\ &= \boxed{17\ \Omega} \end{aligned}$$

Not all circuits can be simplified by this technique. The circuit of Figure 17-6 is actually a "nonsimple" circuit that cannot be further simplified. It is impossible to use the method outlined above to find the resistance from *A* to *B*. The difficulty is that the 5-ohm resistor is neither in series nor in parallel with any other resistor. Currents in nonsimple circuits can be found by the systematic use of Ohm's law, but we will not go into such methods here.

17-3 SOURCES OF EMF IN SERIES AND IN PARALLEL

When several sources of emf are connected in series, they can be replaced by a single emf equal to the algebraic sum of the individual emf's. This follows from the energy principle—the total energy gained by a unit charge in going through a series of emf's equals the sum of the energies gained in the individual sources of energy. Three batteries are shown in Figure 17-7 and the direction of current, or positive charge movement, is indicated by the arrow. In moving through battery *1*, each coulomb gains 4 joules of energy, and in moving through battery 2, each coulomb gains 13 joules. Battery 3, however, is connected "backwards," in the reverse direction from the others, and so each coulomb *loses* 10 joules when moving through this battery. (Of course, this energy is not

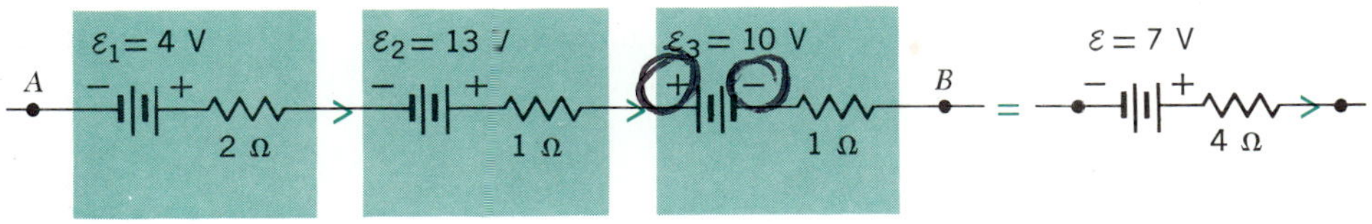

Figure 17-7

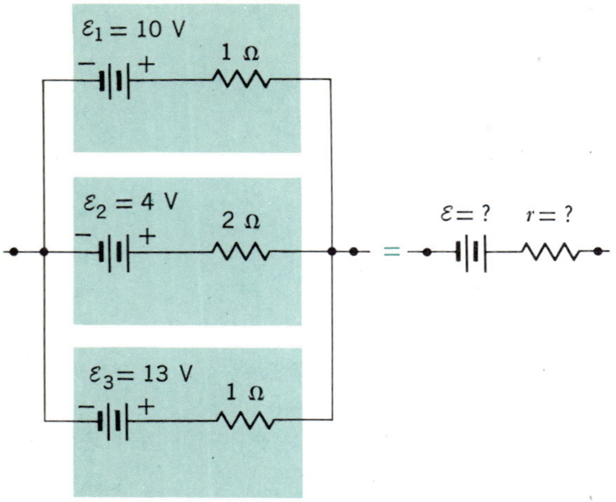

Figure 17-8 Batteries connected in parallel; equivalent $\mathcal{E}$ and r cannot be found in any simple way.

really lost. The battery may become charged as some of the electric energy is converted into chemical energy, and the balance of electric energy appears as heat in the battery.)

Therefore the net emf between A and B is

$$\mathcal{E} = +4 \text{ V} + 13 \text{ V} - 10 \text{ V} = +7 \text{ V}$$

Since the cells are in series, their internal resistances can be added up, to give 4 ohms.

The situation is more complicated when batteries are connected in parallel. In Figure 17-8 it is impossible to make an easy computation of the effective $\mathcal{E}$ and r of the combination. Certain currents exist within the batteries, and all we know is that the terminal voltages of all three batteries will be equal because they are connected in parallel. The 4-volt battery in this combination is probably being charged by current from the other batteries, but how much current is going through it we cannot say with ease. There is only one case in which we can make a simple treatment of cells in parallel. If they are identical in emf *and in* internal resistance, then the net emf equals the emf of each cell, and the net internal resistance equals the parallel resistance of all the cells (and hence is $1/n$ times the internal resistance of any one cell).

17-4 OHM'S LAW FOR A COMPLETE CIRCUIT

We have now discussed the conditions under which a circuit can be replaced by the simplified equivalent of an emf connected to a single resistor. Ohm's law takes a simple form in such a situation: *The current through the circuit equals the net emf divided by the net resistance of the circuit.* Thus we find that there are two rather different aspects of Ohm's law.

Ohm's law when applied only to a part of a circuit:

$$I = \frac{V}{R}$$

and *Ohm's law* when applied to a complete simple circuit:

$$I = \frac{\text{net } \mathcal{E}}{\text{net } R}$$

It is important to realize that these equations are to be applied in different situations, even though both may be expressed as "amperes = volts/ohms." The first form is always true (for any metallic resistor carrying a steady current). The second applies only to certain "simple" circuits in which there *is* a net emf and a net resistance (Figure 17-9b). In the solution of problems, both forms of Ohm's law are used to calculate unknown currents and PD's.

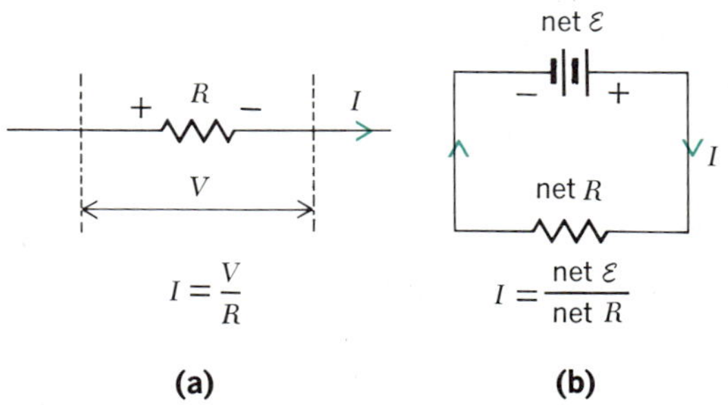

Figure 17-9 Two aspects of Ohm's law: **(a)** shows Ohm's law for a part of a circuit; **(b)** shows Ohm's law for a complete simple circuit.

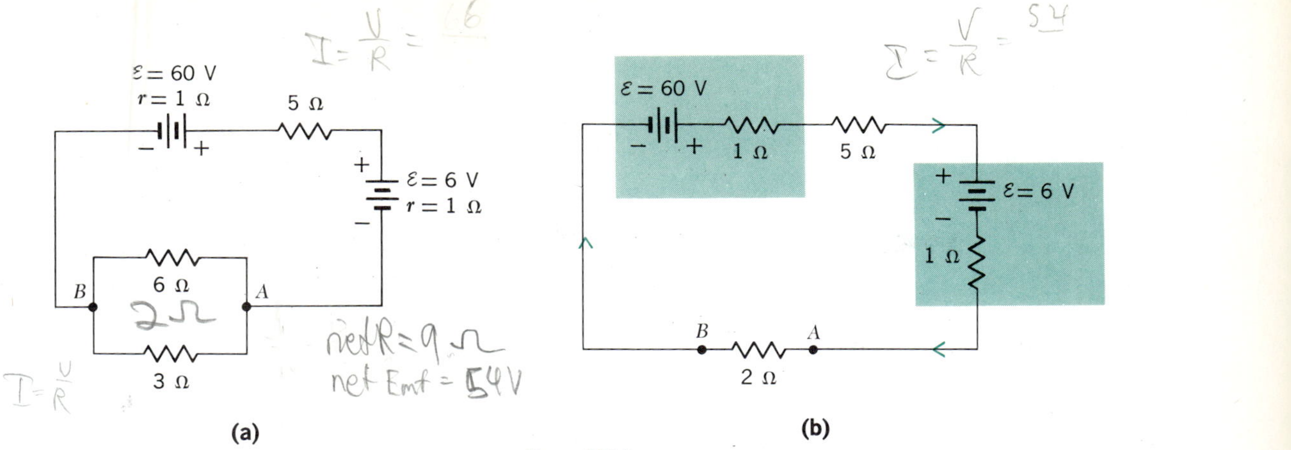

Figure 17-10

Example 17-5 What is the current in the 5 Ω resistor of Figure 17-10a?

First we simplify the circuit and find the total current. For the 6 Ω − 3 Ω combination, $1/R = \frac{1}{6} + \frac{1}{3}$, whence $R = 2\ \Omega$. We now redraw the original circuit as in Figure 17-10b, remembering to put in the resistance of the batteries. It is now a series combination, and the net resistance is given by $R = 1\ \Omega + 5\ \Omega + 1\ \Omega + 2\ \Omega = 9\ \Omega$. To find the net emf, we observe that the two batteries tend to send current around the circuit in opposite directions, so that the net emf is 60 V − 6 V = 54 V. Ohm's law can now be applied to the entire circuit:

$$I = \frac{\text{net } \mathcal{E}}{\text{net } R} = \frac{54\ \text{V}}{9\ \Omega} = \boxed{6\ \text{A}}$$

This is the current through each battery and through the 5 Ω resistor.

Example 17-6 What are the currents through the 3 Ω and 6 Ω resistors in Figure 17-10a?

Here we apply Ohm's law to a portion of the circuit. The section AB is redrawn in Figure 17-10b. Since the current entering point A is 6 A and the net resistance from A to B is 2 Ω, Ohm's law (for a part of a circuit) tells us that $V = IR = (6\text{A})(2\ \Omega) = 12$ V. This is the PD across each of the two resistors. The current through each resistor is found by further application of this form of Ohm's law:

$$I_1 = \frac{12\ \text{V}}{6\ \Omega} = \boxed{2\ \text{A}}$$

$$I_2 = \frac{12\ \text{V}}{3\ \Omega} = \boxed{4\ \text{A}}$$

As a check, we find that I_1 and I_2 add up to 6 A, which is the total current entering point A. This illustrates the law of conservation of charge.

Example 17-7 Calculate the terminal voltage of each battery in Figure 17-10b.

To find the terminal voltages, we use the previous solution, in which the current through each battery was found to be 6 A. In Figure 17-11 we fill in the Ir drops in the internal resistances of the batteries, paying close attention to signs. (Remember that the direction of the current is from + to − through a resistor.)

The terminal voltage of the 60 V battery is found by going from C to D through the battery. A rise of 60 V is followed by a drop of 6 V, so the net PD is +60 V − 6 V = +54 V, with D positive relative to C.

Similarly, in going through the other

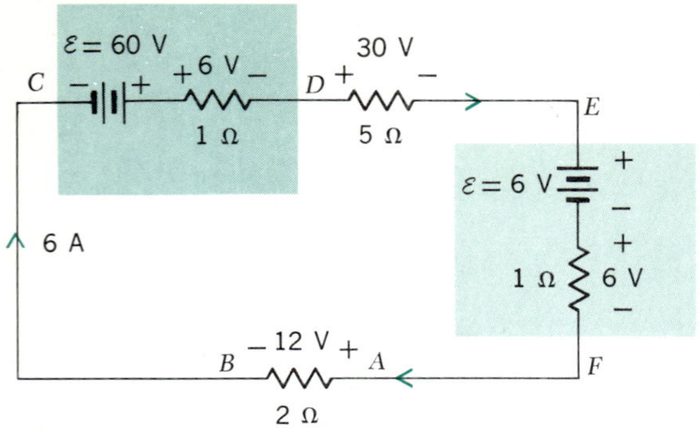

Figure 17-11

battery from E to F, we find a drop of 6 V followed by another drop of 6 V. The net PD is $-6\text{ V} - 6\text{ V} = -12\text{ V}$, and the terminal voltage of this battery is 12 V, with E positive relative to F.

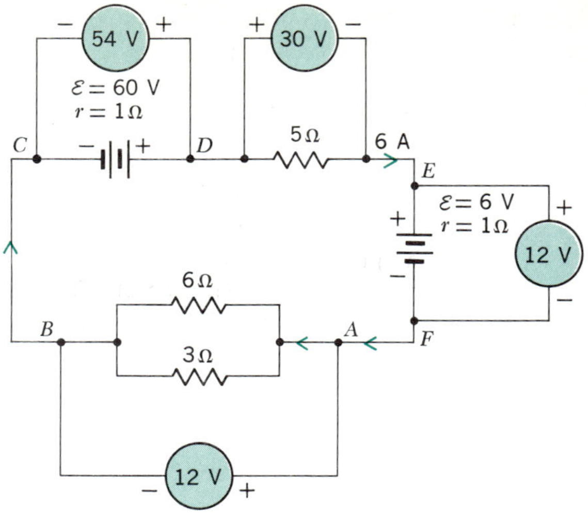

Figure 17-12

Suppose some ideal voltmeters* are connected in parallel with the various parts of our circuit. In Figure 17-12 the two terminals of each of the voltmeters (shown as circles with PD's inside) are connected to the points between which the PD is to be measured in each case. According to the calculations of the preceding examples, the voltmeters would read as shown in this figure. Energy conservation is illustrated here, and the net change of potential around the circuit is zero, as it should be. All of the changes around the path *ABCDEFA* add up to zero:

$$-12\text{ V} + 54\text{ V} - 30\text{ V} - 12\text{ V} = 0$$

We can interpret this by saying that the total energy per unit charge taken out of the circuit by dissipation into heat and by the charging of cells must equal the nonelectric energy converted into electric energy per unit charge in the sources of emf.

We can learn still more from our study of the circuit of Figure 17-10. Let's make an energy check, or, since power is "energy flow per unit time," let's make what amounts to the same thing —a power check. In any resistor (including the internal resistances of a battery), the Joule heating is at the rate of $P = I^2R$, where the power P is measured in watts if I is in amperes and R in ohms. The table below shows that heat is produced at a total rate of 324 watts. We can account for this by the action of the two batteries. The 60-volt battery is delivering current in the "normal" direction, and chemical PE is being transformed into electric energy at the rate of $P = \mathcal{E}I = (60\text{ V})(6\text{ A}) = 360\text{ W}$. The 6-volt battery is being charged as the charge is forced "backward" through it. Here electric energy is being transformed into chemical PE at the rate of $P = \mathcal{E}I = (6\text{ V})(6\text{ A}) = 36\text{ W}$. The net rate of production of

Rate of Joule Heating

R (*ohms*)	I (*amperes*)	V (*volts*)	P (*watts*)
3	4	12	48
6	2	12	24
1	6	6	36
5	6	30	180
1	6	6	36
			324

* That is, voltmeters of very high resistance, which would not divert any appreciable current from the circuit (more information on voltmeters will follow, in Section 17-5).

electric energy is 360 W − 36 W = 324 W; and the table shows that this electric energy is transformed into heat in the various resistors and in the batteries themselves at a total rate of 324 W.

Example 17-8 A portable generator is delivering 80 A to a string of lamps at a carnival (Figure 17-13). The terminal voltage of the generator is 100 V, and its internal resistance is 0.05 Ω. (a) What is the emf of the generator? (b) Ignoring friction, what mechanical power must the engine that turns the generator supply?

We represent the generator schematically as an emf in series with a resistance of 0.05 Ω.

(a) The Ir drop inside the generator is (80 A)(0.05 Ω) = 4 V. We know the emf of the generator is:

$$\mathcal{E} = V + Ir$$

$$\mathcal{E} = 100\text{ V} + 4\text{ V} = \boxed{104\text{ V}}$$

(b) The rate of transformation of mechanical energy to electric energy* is

$$P = \mathcal{E}I$$
$$= (104\text{ V})(80\text{ A})$$
$$= \boxed{8320\text{ W}}$$

In the commonly used unit of horsepower (hp), 1 horsepower equals 746 W. Thus:

$$P = (8320\text{ W})\frac{(1\text{ hp})}{(746\text{ W})}$$
$$= 11.2\text{ hp}$$

This is the rate at which the gasoline engine does work to supply the electric energy.

Example 17-9 Suppose the load in the preceding example consists of a number of 50 W lamps connected in parallel. How many lamps are used? (The four drawn in Figure 17-13 merely represent an unknown amount.)

In each lamp, $P = VI$. Hence the current in each lamp is

* $\mathcal{E}I$ rather than VI is used here, since $\mathcal{E}$, an emf, represents the reversible transformation of energy.

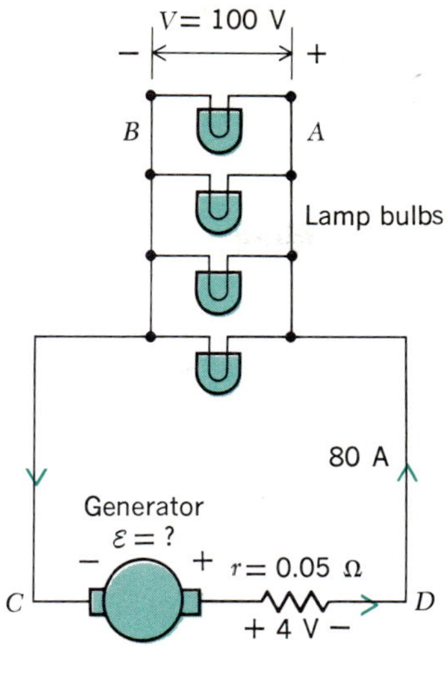

Figure 17-13

$$I = \frac{P}{V}$$
$$= \frac{50\text{ W}}{100\text{ V}}$$
$$= 0.5\text{ A}$$

Since the total current is 80 A, the number of lamps is

$$\frac{80\text{ A}}{0.5\text{ A/lamp}} = \boxed{160\text{ lamps}}$$

Check: The resistance of each lamp bulb is 100 V/0.5 A = 200 Ω. Connecting 160 of these lamps in parallel gives a total resistance R_p:

$$\frac{1}{R_p} = \frac{1}{200} + \frac{1}{200} + \cdots + \frac{1}{200}$$
$$= \frac{160}{200}$$
$$R_p = \frac{200}{160}$$
$$= 1.25\ \Omega$$

The total current through the system is

$$\frac{V}{\text{net } R} = \frac{100\text{ V}}{1.25\ \Omega}$$
$$= 80\text{ A}$$

which checks with the given data.

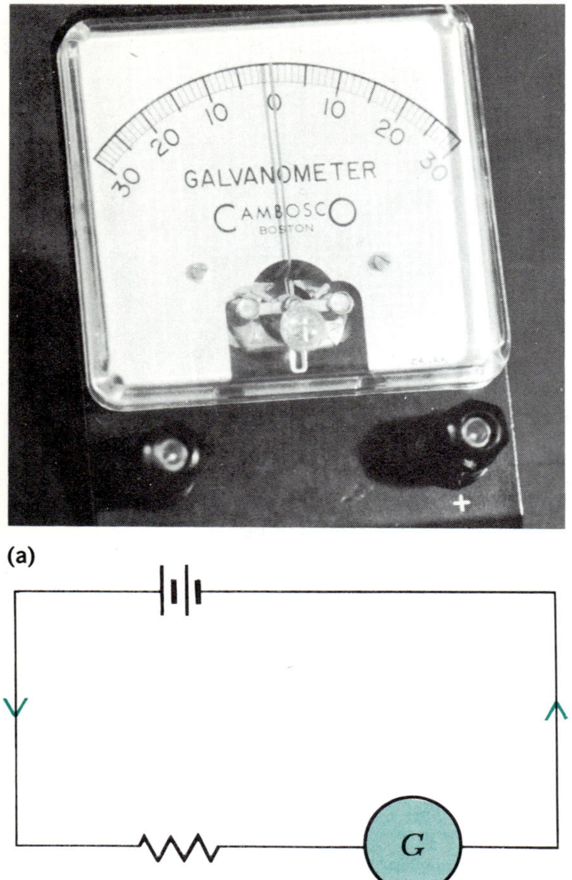

(a)

(b)

Figure 17-14

17-5 THE AMMETER AND THE VOLTMETER

As their names indicate, the ammeter measures current expressed in some multiple of the ampere and the voltmeter measures potential difference expressed in some multiple of the volt. The most commonly used variety of both these instruments is based on the galvanometer, which is fundamentally a current-indicating device depending on magnetic forces. We will discuss the magnetic force on a wire carrying current in Chapter 18, and we shall study the galvanometer in detail in Chapter 19. At this time let's just consider the galvanometer as a device whose pointer needle deflects in proportion to the current that passes through it. Figure 17-14a shows a typical laboratory galvanometer. Our immediate problem is to learn how to use it in practical measurements of current and potential difference.

Small currents can be measured by a galvanometer without any modification. Figure 17-14b shows a galvanometer G connected in series with other circuit elements to measure a small current. Note that all the current passes through G.

A typical galvanometer movement might have a resistance of 100 ohms and it might require 200×10^{-6} ampere (200 microamperes, or 200 μA) for a full-scale deflection. By "full-scale" we mean that the indicator needle swings to the limit of the scale, for example, to a 30 mark in Figure 17-14a. A current of 160×10^{-6} ampere would give a deflection 80 percent of full-scale, and so forth. Usually, however, we need to measure larger currents—for instance, the current of 2 amperes through a lamp bulb or a small motor. A galvanometer is designed to respond to very small currents and therefore is far too sensitive for such a measurement. Furthermore, there is the practical difficulty that too large a current would damage a galvanometer by "burning out" the coil.

To make a galvanometer less sensitive and thereby able to measure larger currents, a resistor (called a *shunt*) is connected in parallel with it (Figure 17-15). This resistor then carries most of the current, but leaves a definite small fraction passing through the galvanometer. We can calculate just what this fraction is and can mark off a new pointer scale accordingly.

Let's compute the value of the shunt resistor needed to convert our galvanometer into an *ammeter* that will deflect to full-scale for a current of 2 amperes. We can do this by first calculating the PD across the galvanometer at full-scale deflection, and then determining how much "extra" current must be carried by the shunt. The "extra" current is the difference between the maximum current we desire to measure and the maximum current that can pass safely through the galvanometer coil. This extra current and PD can be used in Ohm's law to calculate the shunt resistance.

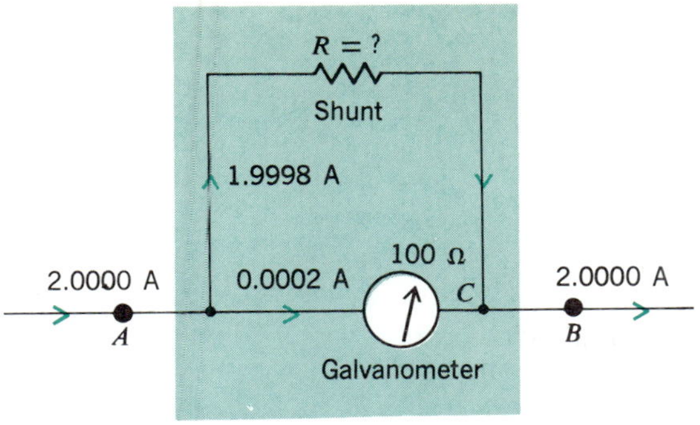

Figure 17-15 Conversion of a galvanometer into an ammeter, using a low-resistance shunt.

In Figure 17-15 we see that the PD from A to B is $(100\ \Omega)(2 \times 10^{-4}\ \text{A}) = 2 \times 10^{-2}\ \text{V}$ when the galvanometer needle deflects full-scale. The extra current through the shunt would be 1.9998 amperes when the needle is deflected full-scale by the current of 0.0002 amperes through the galvanometer. By Ohm's law, then,

$$R_{\text{shunt}} = \frac{2 \times 10^{-2}\ \text{V}}{1.9998\ \text{A}}$$
$$= 0.01\ \Omega$$

The shunt has a resistance of 0.01 ohm, a very small resistance indeed.

Another way of looking at the action of the shunt is to consider how the current divides. In Figure 17-15, the shunt has a resistance that is about 1/10,000 that of the galvanometer. The wire at A is carrying 2 amperes, and the current splits in the ratio of 10,000 to 1; that is, $1/10{,}001 \times 2$ ampere through the galvanometer, and $10{,}000/10{,}001 \times 2$ ampere through the shunt. To three-figure accuracy, with $1/10{,}001 \times 2 = 2.00 \times 10^{-4}$ A, the instrument reads full-scale as desired. After adjustment of the shunt, the final step in the conversion of our galvanometer into an ammeter is to mark off the scale with values ranging from 0 to 2 amperes, so that it will be direct-reading. The shunt is usually mounted inside the case, out of sight.

A galvanometer is modified in a somewhat different way when a *voltmeter* is desired. The original instrument could be used to measure a small PD. Full-scale deflection requires 200×10^{-6} ampere through a resistance of 100 ohms, as we saw before in our calculations for the ammeter shunt, and the required PD is given by

$$V = IR$$
$$= (200 \times 10^{-6}\ \text{A})(100\ \Omega)$$
$$= 0.020\ \text{V}$$

The meter reads 20 millivolts full-scale.

To measure a dry cell's terminal voltage, however, would require a larger range. This is made possible by a high-resistance resistor called a *multiplier*, connected in series with the galvanometer coil. Figure 17-16 shows the construction of a voltmeter that reads 2 V full-scale, starting with the same galvanometer used for the ammeter of Figure 17-15. The galvanometer coil and the multiplier are in series, and at full-scale deflection each carries the current of 2×10^{-4} ampere. Our procedure is to find the R that will allow the full scale I to pass when the required maximum PD is applied. By Ohm's law, the resistance of the circuit from A to B is

$$R_G + R_{\text{multiplier}} = \frac{2\ \text{V}}{2 \times 10^{-4}\ \text{A}}$$
$$= 10{,}000\ \Omega$$

Hence,

$$100\ \Omega + R_{\text{multiplier}} = 10{,}000\ \Omega$$
$$R_{\text{multiplier}} = 9900\ \Omega$$

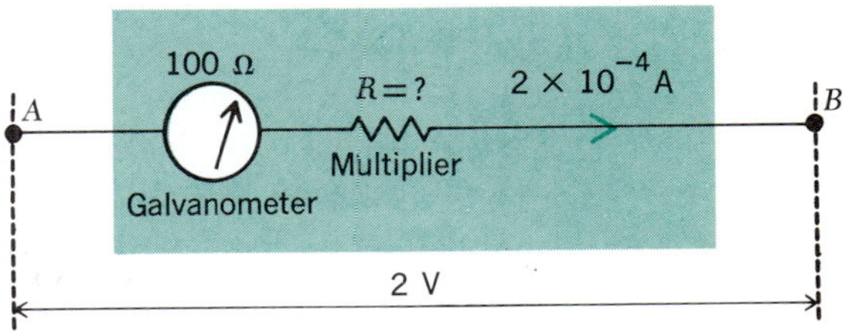

Figure 17-16 Conversion of a galvanometer into a voltmeter, using a high-resistance multiplier.

It is necessary, then, to connect a resistor of value 9900 Ω in series with the coil, in order to obtain a voltmeter of range 0 to 2 volts. If the applied PD is reduced from 2 volts to 1.6 volts, the current through both coil and multiplier is reduced, and the deflection of the needle is 80 percent of full-scale. The scale should be marked off with values ranging from 0 to 2 volts, so that the voltmeter is direct-reading. The multiplier is usually mounted inside the case, out of sight.

The voltmeter just described has a high resistance (10,000 Ω total), and it diverts no more than 0.0002 ampere from the device to which it is connected. An *ideal voltmeter* is one whose resistance is infinite, and which therefore draws no current from the circuit whose PD is being measured. Voltmeters used in most high school physics laboratories have resistances such that they require about 0.005 to 0.01 ampere for full-scale deflection. Voltmeters used for research may need as little as 10^{-5} ampere. If even less disturbance of the circuit is required, electronic voltmeters are available that use currents of 10^{-9} amperes or less.

Ammeters and voltmeters, then, are really modifications of the same basic instrument—the galvanometer. Ammeters have very low resistance, and are connected in series with the circuit element for which the current is to be measured. All the current passes through them. Voltmeters have high resistance, are connected in parallel with the circuit element for which the PD is to be measured, and draw as little current as possible.

17-6 THE POTENTIOMETER

You already know that an ordinary voltmeter cannot be used to measure the emf of a battery or cell. Such a voltmeter measures the terminal voltage correctly enough, but, since the cell must supply some slight current through the voltmeter, the reading of the voltmeter scale is $V = \mathcal{E} - Ir$, which is less than the emf $\mathcal{E}$.

Example 17-10 A voltmeter of resistance 1000 Ω is connected to an "old" dry cell whose emf is 1.500 V and whose internal resistance is 100 Ω (Figure 17-17). What is the reading of the voltmeter?

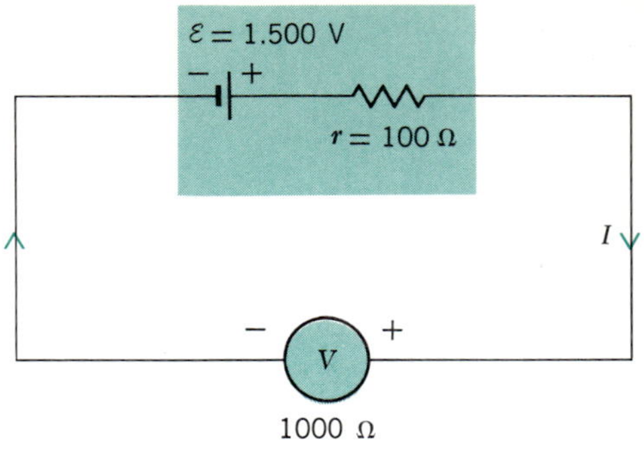

Figure 17-17 The voltmeter reads less than the emf of the cell.

The net resistance for the entire circuit is

$$1000\ \Omega + 100\ \Omega = 1100\ \Omega$$

Hence,

$$I = \frac{\text{net } \mathcal{E}}{\text{net } R} = \frac{1.500\ \text{V}}{1100\ \Omega} = 0.001364\ \text{A}$$

Now, applying Ohm's law to the voltmeter alone, we have

$$V = IR = (0.001364\ \text{A})(1000\ \Omega) = \boxed{1.364\ \text{V}}$$

Compared with the emf of 1.500 V, this is a considerable error, due entirely to the current through the voltmeter.

The potentiometer circuit* measures an emf by balancing one PD against another in such a way that no current is drawn. In the basic potentiometer circuit of Figure 17-18, a "working battery" sets up a steady current through the closed loop *ABCDEA*. The purpose of the working battery is

* Invented by the German physicist Johann Poggendorff (1796–1877).

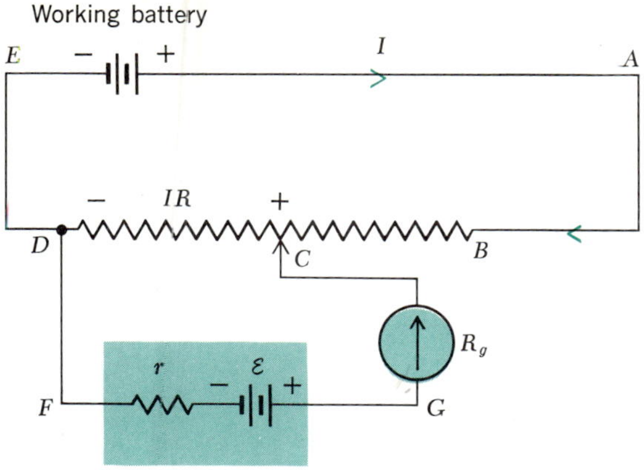

Figure 17-18 Potentiometer circuit.

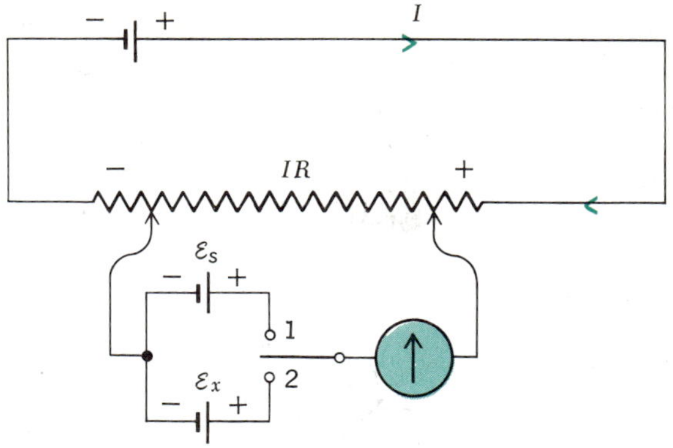

Figure 17-19 Comparison of two emfs, using a slide-wire potentiometer.

to set up a PD between two points of the resistor *BD*. This PD results from the *IR* drop in the resistor due to the steady current *I* provided by the working battery. The portion of resistor labeled *CD* is made variable by means of a sliding contact at point *C*. This enables us to make *CD* any fraction that we wish of *BCD*.

The sliding contact is adjusted so that there is no current through the galvanometer (labeled R_g). When there is no current through *G*, then the points *G* and *C* must be at the same potential, and it is clear that *D* and *F* are at the same potential. Therefore the terminal voltage of the test battery (between *F* and *G*) equals the *IR* drop in the resistor *CD*.

How does this circuit allow the emf of a test battery to be measured? Suppose the test battery has emf $\mathcal{E}$ and internal resistance *r*, and the galvanometer's resistance is R_g. There is no current through *r* or R_g when the potentiometer is in balance. Therefore there are no *IR* drops in *r* and R_g when the potentiometer is in balance, and no little + and − signs are written down for them. This means that the terminal voltage of the cell equals its emf; in the equation $V = \mathcal{E} - Ir$ we have $I = 0$, and thus $V = \mathcal{E}$. Therefore, when the potentiometer is in balance, the *IR* drop of *CD* equals the emf of the test battery. The internal resistance of the cell is of no consequence because there is no current through it.

Figure 17-19 shows how the unknown, or test, cell $\mathcal{E}_x$ is compared with a *standard cell* whose emf E_s is known. The potentiometer is first "standardized" by throwing the switch to position 1 and adjusting the sliding contact for balance—zero current through the galvanometer. This gives a length l_1 proportional to a resistance that we shall call R_1, but whose value we do not have to know. The length l_1 represents the emf of the standard cell. Then the potentiometer is again balanced with the switch thrown to position 2, giving a length l_2 proportional to another resistance R_2. We do not have to know the value R_2, as the length l_2 represents the emf of the unknown cell.

$$\frac{\mathcal{E}_x}{\mathcal{E}_s} = \frac{IR_2}{IR_1} = \frac{R_2}{R_1} = \frac{l_2}{l_1}$$

$$\mathcal{E}_x = \mathcal{E}_s\left(\frac{l_2}{l_1}\right)$$

Several forms of potentiometer have been developed for special purposes; many are made direct-reading (in volts) by proper adjustment of the working current *I*. One common type of standard cell is the Weston cell, which uses as its negative electrode a cadmium-mercury amalgam instead of the more common zinc. Such a cell has an emf that is reproducible and is relatively independent of temperature.

The potentiometer is a most valuable tool for the chemist and biologist. Since the emf of a cell depends upon its electrodes and on the nature and concentration of the electrolyte, measurement of emf gives valuable information to the chemist. Also, numerous potentials exist in living tissue. For instance, there is a steady emf of about 50 millivolts between the inside and outside of the giant axon nerve cell of the squid *Loligo forbesi*, when no nerve impulse is being transmitted. This emf rises sharply to about 95 or 100 millivolts during nerve action. These potentials can be measured best with a potentiometer, since the potentiometer requires no current from the source and the high internal resistance of biological tissue has no effect on the measurements.

SUMMARY

The **law of conservation of charge** asserts that there are no gains or losses of charge in a circuit, and the law of conservation of energy asserts that the algebraic sum of the changes in potential around any closed path is zero. These laws offer a systematic way of applying **Ohm's law** to a circuit.

The **internal resistance** of a cell depends on its physical construction and its past history. When there is current through a cell the terminal voltage of a cell differs from its emf because of internal resistance. If the cell is discharging, $V = \mathcal{E} - Ir$, and, if the cell is being charged, $V = \mathcal{E} + Ir$.

Formulas for the equivalent resistance of resistors in series and in parallel can be used to reduce many circuits to a simple circuit. Not all circuits can be studied by such elementary means, however. Emf's in series can be added up, but there is no simple way to obtain the equivalent emf to represent a number of emf's connected in parallel, unless they are identical in emf and internal resistance. For a simple circuit for which a net emf and net resistance exist, Ohm's law states that $I =$ net $\mathcal{E}$/net R.

The deflection of a **galvanometer** is proportional to the current through its coil. An **ammeter** is a galvanometer with a low-resistance **shunt** in parallel with the coil. A **voltmeter** is a galvanometer with a high-resistance **multiplier** in series with the coil. A voltmeter will usually have some effect on the circuit to which it is connected unless the resistance of the voltmeter is sufficiently large.

The **potentiometer** is used for measuring the value of an emf without drawing current from the cell. The disturbing effect of a cell's internal resistance is thus eliminated.

QUESTIONS

1 In Figure 17-20, tell where + and − signs should be placed to indicate the polarity of the PD across resistors *A*, *B*, *C*, *D*, and *E*.

2 Why is the internal resistance of a cell not a constant?

3 Under what circumstances is the terminal voltage of a battery greater than its emf?

4 Under what conditions could the terminal voltage of a battery be zero?

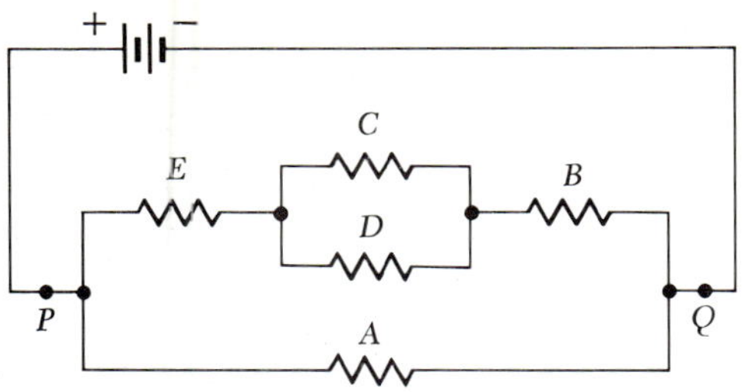

Figure 17-20

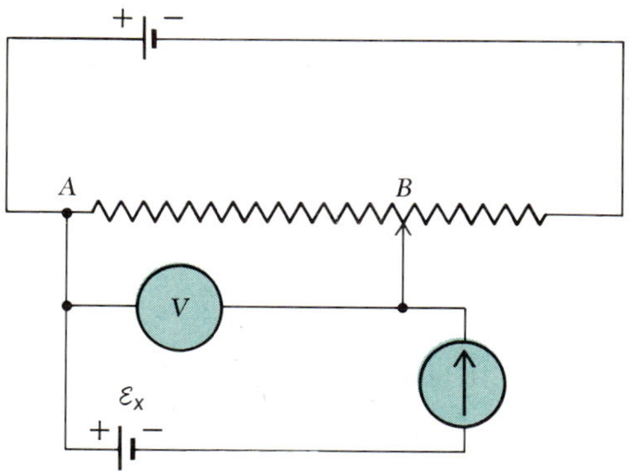

Figure 17-21

5 Under what conditions does the terminal voltage of a cell equal its emf?

6 Is the mechanical construction of a voltmeter essentially different from that of an ammeter? Explain.

7 In what way is a potentiometer superior to a voltmeter for measuring a PD? In what way is a voltmeter superior?

8 The circuit of Figure 17-21 is a potentiometer that requires no standard cell. Show that, if the sliding contact B is adjusted so that the galvanometer reads zero, the voltmeter reading equals the emf of the unknown cell E_x.

PROBLEMS

GROUP A

1 A test for a fresh dry cell is to connect the terminals directly to the terminals of a heavy-duty ammeter. (*Warning:* Don't try this with an ammeter of insufficient range. Why not?) If the ammeter reads at least 30 A the cell is considered fresh. What is the internal resistance of a fresh dry cell?

2 In Figure 17-22, what is the PD across each of the three resistors?

3 Five 10 Ω resistors are connected (a) in series and (b) in parallel. Compute the net resistance in each case.

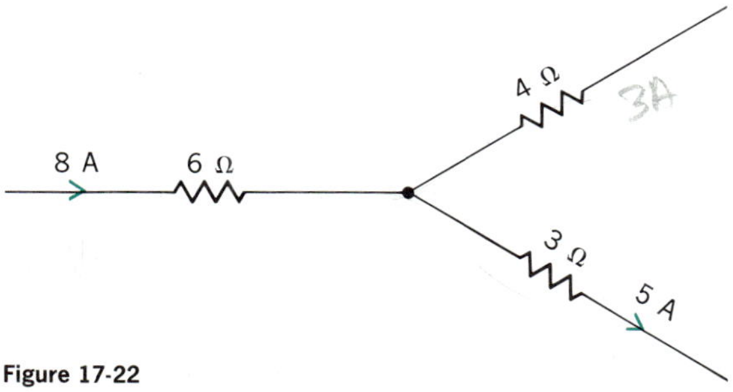

Figure 17-22

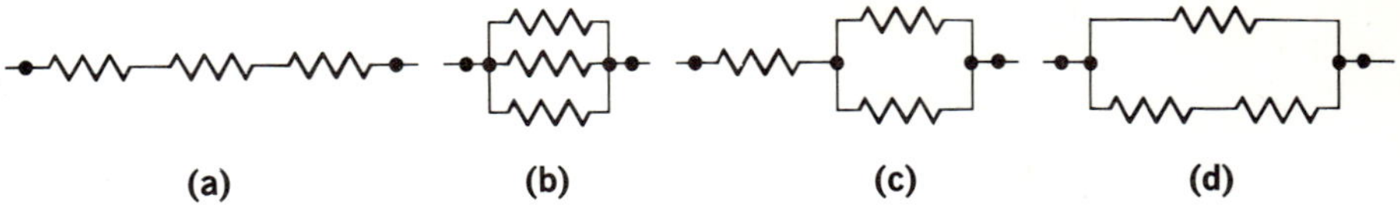

Figure 17-23

4 Three equal resistors, each of resistance 6 Ω, can be connected in four different ways (Figure 17-23). What is the resistance of each combination?

5 A string of Christmas-tree lights consists of 8 bulbs in series connected to a 120 V source of PD. (a) What is the PD across each bulb? (b) If one bulb is removed from its socket while the string is plugged in what is the resulting PD across the socket's terminals? (c) While this bulb is removed, what is the PD across each of the remaining seven bulbs?

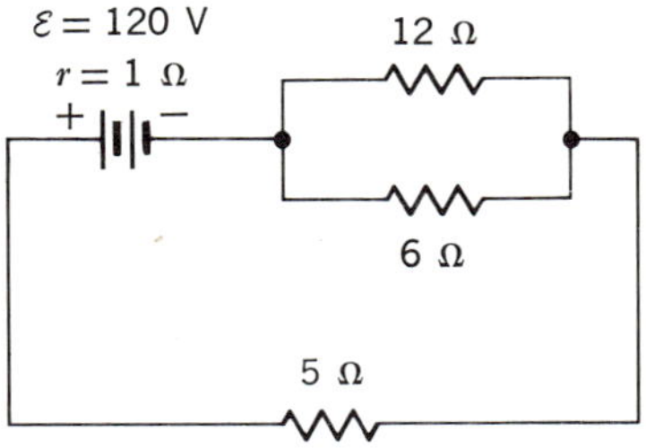

Figure 17-24

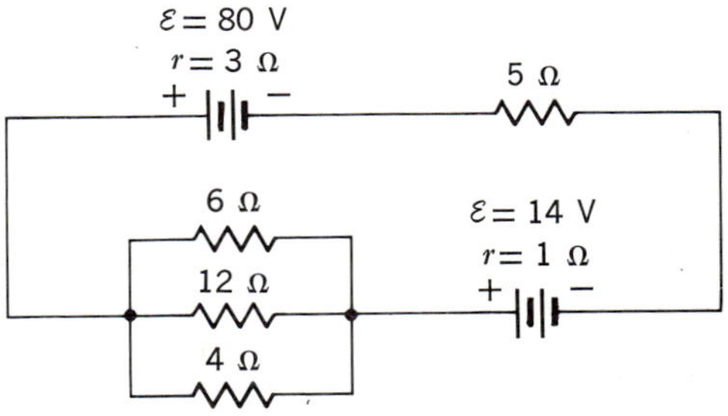

Figure 17-25

GROUP B

6 What is the resistance between P and Q in Figure 17-20, if $A = 15\ \Omega$, $B = 20\ \Omega$, $C = 10\ \Omega$, $D = 10\ \Omega$, and $E = 5\ \Omega$?

7 (a) In Figure 17-24, calculate the current through each resistor. (b) At what rate, in watts, is chemical energy being transformed into electric energy in the battery? (c) Make a table of power dissipated as heat in the various circuit elements and show that the sum of these powers agrees with your answer to (b).

8 In Figure 17-25 (a) what power is dissipated in the 5 Ω resistor? (b) What is the current in the 6 Ω resistor? (c) What is the terminal voltage of each battery?

9 In Figure 17-26, AB represents a radio "B battery" of emf 45 V and negligible internal resistance. Calculate the potential of each terminal of the battery with respect to G.

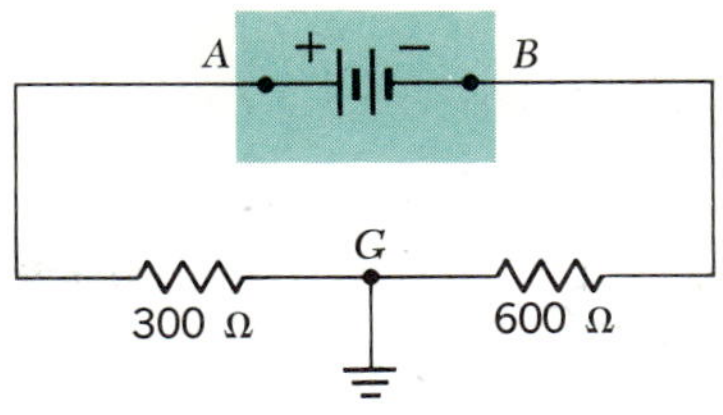

Figure 17-26

10 Two cells are connected in parallel, + to + and − to −. Cell 1 has $\mathcal{E} = 1.5$ V and $r = 1\ \Omega$; cell 2 has $\mathcal{E} = 2.1$ V and $r = 2\ \Omega$. What is the common terminal voltage of the cells when connected together in this way, with no external load?

11 A milliammeter reads 1 milliampere full-scale and has a resistance of 50 Ω. How would you convert this to a voltmeter reading 10 V full-scale, by use of a single additional resistor? Draw the circuit and compute the value of the resistor that is needed.

12 What is the value of a single resistor that would convert the milliammeter of the preceding problem into an ammeter reading 5 A full-scale? Draw the circuit and compute the value of the resistor that is needed.

13 In Figure 17-2, (a) what is the PD between points F and G? (b) The battery charger has an emf of 20 V. What is its internal resistance?

18 Principles of Electromagnetism

The ancients knew that certain naturally occuring iron ores called magnetite could be used as navigational aids, for a piece of magnetite (often called a lodestone) would point in a north-south direction when suspended by a thread. According to legend, these lodestones were first found near the ancient city of Magnesia in Asia Minor, and thus became known as *magnets*. As early as A.D. 121 the Chinese knew that a piece of iron could be "magnetized" by bringing it near a lodestone, and navigation with the aid of the magnetic compass dates back ten centuries or more.

Without knowing it, the early natural philosophers who studied magnetism were really dealing with electric forces—but of a kind different from the electrostatic (Coulomb) forces exerted by a rubbed piece of amber that is at rest. The early investigators knew them just as *magnetic* forces, and you will see that, unlike electrostatic forces, they are exerted only by charges in motion. An ordinary magnet fits into this scheme because a magnetic force exists between certain moving electrons in the iron of the magnet and other moving electrons in the iron that it attracts. When a compass needle points north, it does so because of a magnetic force between some moving electrons in the iron atoms of the needle and (presumably) moving charges deep within the earth.

18-1 MAGNETIC FIELD

The basic ideas of a magnetic field are easy to understand when we consider an isolated charged particle moving in a vacuum, perhaps an electron on its way to the front surface of a TV picture tube, or a proton moving in a particle accelerator. The magnetic forces are extra forces due to motion, over and above any Coulomb forces that would be acting even if the charge were stationary.

In order to examine magnetic forces in detail, it is necessary to gain understanding of the concept of magnetic field. In Chapter 16 you studied the concepts of electric field, and gravitational field. In each case, you found that the field is explored by measuring the force on a suitable test object. The magnitude of a gravitational field is force per unit mass while the magnitude of an electric field is force per unit charge. In the same way, it can be said that a magnetic field exists in any region of space where magnetic force would be exerted on a suitable test object.

At this point, however, before giving further consideration to the forces just mentioned, it is important to clarify what is meant by the term *electromagnetic forces*. In Chapter 14 and again in Chapter 16 we discussed the Coulomb force—the force that arises between charges simply because they are charged—without reference to their motion. Since this force is the only force observed when charges are at rest, it is sometimes called the electrostatic force. The term Coulomb force is better, however, because this force is found not only with static charges but also continues to act when the charges are in motion. Magnetic force is a "velocity-dependent" force, since it arises from the relative motion of two charges. Electromagnetic forces are the sum total of both the Coulomb forces and the magnetic forces.

As a test object for investigating a magnetic field we can use a moving charge. This time, because it is familiar, we will use a negatively charged electron moving down the axis of a TV picture tube or the cathode-ray demonstration tube as shown in Figure 18-1. Actually, the cathode-ray tube shown is somewhat simpler than a modern picture tube, but the general arrangement is the same. There is a source of the electron beam, a cathode, at one end, and at the other end a surface that the beam strikes (Figure 18-1a). By varying the voltages on the electrodes of the tube, we can change the velocity of the electrons. It would also be possible to do this with various other charged particles, such as doubly charged positive helium ions, etc. It would be necessary to modify the tube for such experiments, but it is quite possible.

A magnetic field is set up in the tube by using a large permanent magnet. In a particular field, the magnetic force on a test charge depends both on the magnitude of the charge Q and its velocity v. Thus $F \propto Qv$, and we can use the ratio F/Qv $\left(\text{force per unit charge-velocity product—for example, } \frac{\text{N}}{\text{C}\cdot\text{m/sec}}\right)$ as a measure of the strength of the magnetic field at the point in question.

When there is no magnetic field, the electron beam will travel straight through the tube and make a glowing line across the center of the fluorescent screen (Figure 18-1a). When a field is present, the magnetic force will deflect the beam and the line will be bent in a different direc-

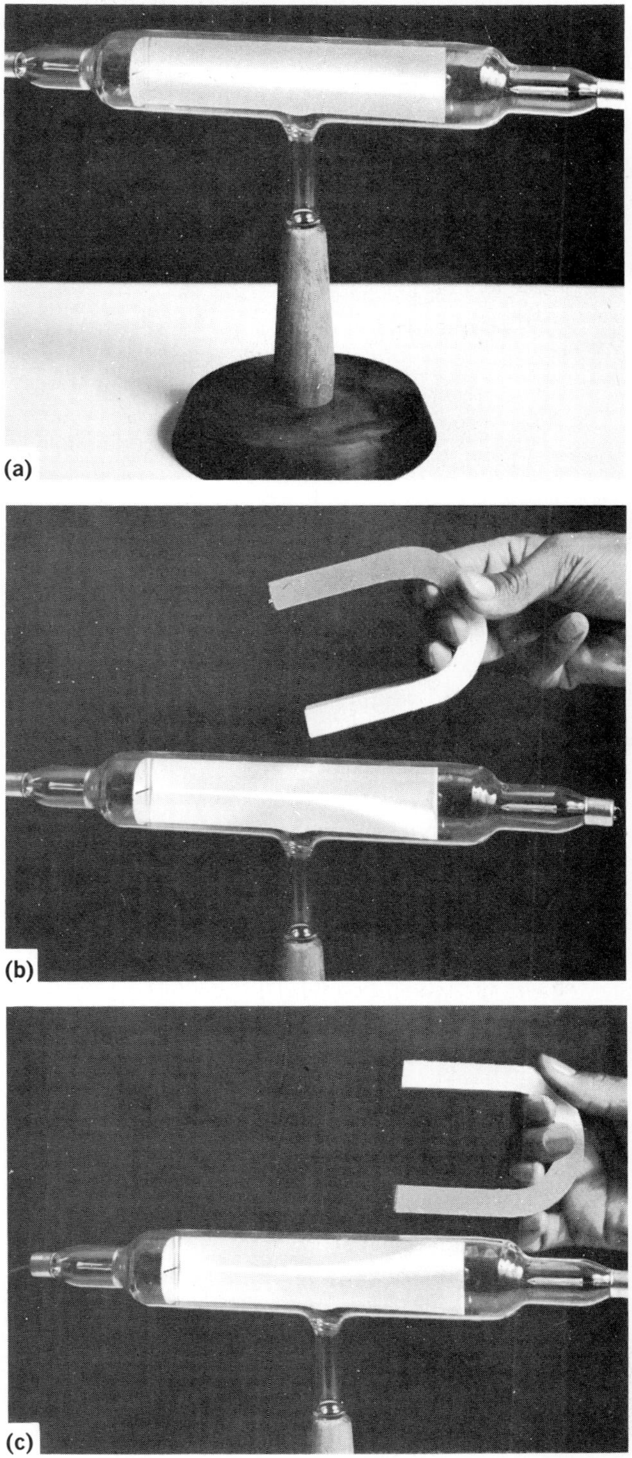

(a) (b) (c)

Figure 18-1 Cathode-ray (Crookes') tube. What causes the beam first to be deflected one way and then the other?

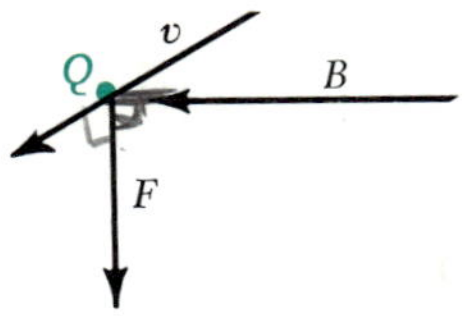

Figure 18-2 The force is maximum when v is perpendicular to B. F is perpendicular to both v and B.

tion (Figures 18-1b, c). There is no need to go into the details of the calculation now, but it is relatively simple to determine the magnetic force from a measurement of this deflection. We have defined the magnitude of a magnetic field by describing how to measure it, but we have not yet considered whether a magnetic field has a direction associated with it (whether it is a vector field.) A magnetic field does have a direction, and it is important to know how to measure that direction. A little imaginary experimentation with the electron-beam tube gives useful clues.

Imagine that we have a large magnet and that we point the tube and the electron stream in various directions with respect to the magnet. When we do this we find that the electrons are deflected (showing the presence of a force) when the tube is pointed in practically any direction. The deflected beam shows that the force is greater when electrons move in some directions than in others and that directions in a particular plane give us maximum deflection. The fact that stands out most clearly is that when the tube is held in a certain direction there is no deflection and no force on the moving charges in the electron beam. Of all the directions we have tried, this is the one that is unique—the direction of zero force. We say that the direction of a magnetic field is the direction of motion of a moving charge when it experiences no force in that field. On the other hand, more careful experiments like this one would show that the force on a moving charge is maximum when the direction of motion is perpendicular to the direction of the field.

On the basis of these definitions, a magnetic field can be described at any point by a vector B, called the magnetic field strength. The magnitude of B is

$$B = F/Qv$$

F is the magnetic force on a test charge Q. The velocity v must be in the direction to give maximum force, that is, at right angles to B. The direction of the vector B, however, is the direction in which the charge moves when experiencing zero force.

We can turn around the definition of B and get an expression for the magnitude of the force on a moving charge:

$$F = QvB$$

This is an equation for the maximum force on a charge Q moving with velocity v through a magnetic field of strength B. As shown in Figure 18-2, the force is greatest when v is perpendicular to B. The force F is perpendicular to both v and B; as you will see later on, in this case it points downward.

Example 18-1 In a particle accelerator, a proton (its charge equals that of an electron but is positive) moves perpendicular to the magnetic field through a region where $B = 10$ N per C·m/sec. If the proton's speed is 0.1 the speed of light (3×10^8 m/sec) what is the magnitude of the force on the proton?

The speed is

$$(0.1)(3 \times 10^8 \text{ m/sec}) = 3 \times 10^7 \text{ m/sec}$$

And the force is

$$\begin{aligned} F &= QvB \\ &= (1.60 \times 10^{-19}\text{ C})(3 \times 10^7 \text{ m/sec}) \times (10 \text{ N per C}\cdot\text{m/sec}) \\ &= \boxed{4.80 \times 10^{-11}\text{ N}} \end{aligned}$$

Although this may seem to be a small force, the mass of the proton is so small that a very large acceleration is produced. The acceleration can be found by Newton's second law.

To summarize this section: The magnetic force on a moving charge is perpendicular to the direction of motion, perpendicular to the field and, for maximum force, the motion of the charge must be perpendicular to the direction of the field.

18-2 THE FORCE ON A CURRENT SEGMENT IN A MAGNETIC FIELD

Most of the moving charges with which we are familiar are the free electrons that drift through a wire carrying a current. Let's now modify the equation $F = QvB$, describing the force on a moving charge so that it describes the force on wire carrying a current. Figure 18-3 shows a short segment of wire of length Δl.* Suppose there are N free electrons, each of charge q, having a total charge of Q in this segment of wire. Also suppose that each electron has a forward drift velocity v. If the segment is placed in a magnetic field of strength B, at right angles to the wire, the total magnetic force on the N electrons is

$$\begin{aligned} F &= QvB \\ &= NqvB \end{aligned}$$

Now, how do we modify this equation to express the force in terms of current rather than the motion of electron charge? First, remember that current describes the charge that flows past a given point per unit of time (Section 14-4); that is, current = charge/time. The time for an electron to move the length of the wire Δl is $\Delta l/v$.** During this time interval all the electrons now in the wire segment will move out of it, passing through the right-hand end of the wire. (They will be replaced, of course, by a new set of electrons.) The

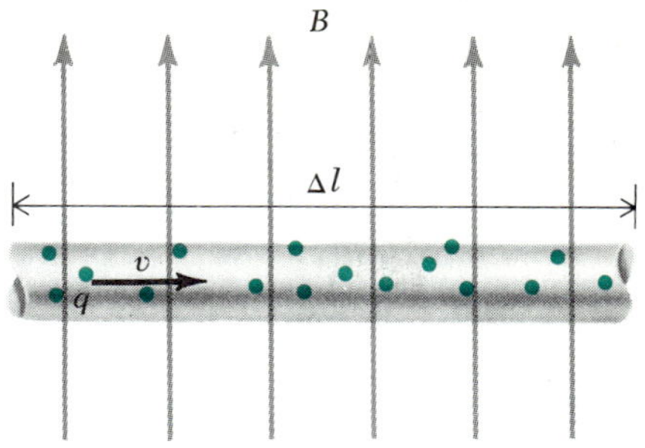

Figure 18-3 A current segment.

charge that crosses the end face of the wire in this time $\Delta l/v$ is Nq. Therefore

$$\text{current} = \text{charge/time}$$

$$\begin{aligned} I &= \frac{Nq}{\Delta l/v} \\ &= Nqv/\Delta l \\ I\,\Delta l &= Nqv \end{aligned}$$

Substituting in $F = NqvB$, we find that the maximum force on the wire is given by

$$F = I\,\Delta l\,B$$

We see that the magnetic force on a wire in a given magnetic field depends not only on the current, but also on the length of the wire. The essential factor is the product $I\,\Delta l$, which is called a *current segment.* Another definition of magnetic field strength B can be written by rearranging the last equation to get

$$B = \frac{F}{I\,\Delta l}$$

This means that we can use a short length of wire (as well as a moving charge) as a test body to explore a magnetic field. It also gives us another unit for field strength—newtons per ampere · meter. For a moving charge, the unit was newtons per coulomb · meter/second (Example 18-1). The two units are exactly equivalent, since amperes = coulombs per second. It is also important to keep the direction of each vector clearly in mind (Figure 18-4). The force F on a current segment is greatest if the current I is perpendicular to the

* In this instance Δ is used to indicate a short length of wire, not a change in the length.

** Recall that in a metal wire the current results from the motion of electrons, but because the charge of an electron is negative, we have, for convenience defined the *direction* of the current as opposite to electron motion (Section 15-7).

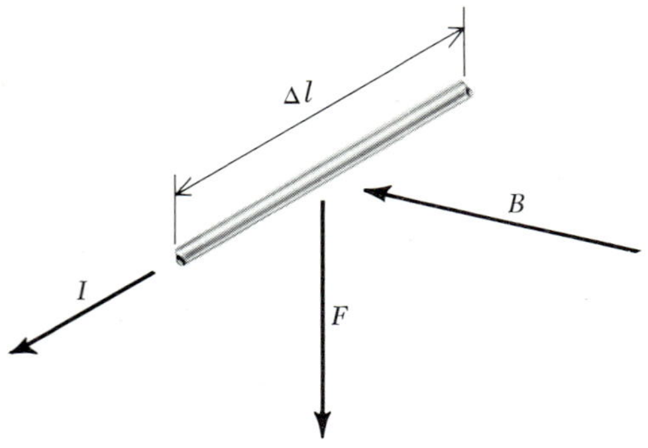

Figure 18-4

magnetic field B. In this case the force on a current, the direction of the current, and the direction of the magnetic field are all perpendicular to each other. How does Figure 18-4 compare with Figure 18-2?

Example 18-2 A wire carrying 400 A is stretched horizontally in an east-west direction between two towers 60 m apart. The earth's magnetic field at that location is to the north and downward, and the value of B at that location is 7×10^{-5} N/A·m. What magnetic force is exerted on the wire?

Since B is perpendicular to I here, we use the equation for the force on a wire segment.

$$F = I\,\Delta l\,B$$
$$= (400\text{ A})(60\text{ m})(7 \times 10^{-5}\text{ N/A·m})$$
$$= \boxed{1.68\text{ N}}$$

It may seem strange to consider a 60 m wire as a short segment. However, since the wire is straight and all parts of it are subjected to the same field, this assumption is justified. The word "short" in this case is much like the word "small" used in Section 15-5 in reference to a small volume of water. A wire is considered short if conditions do not change appreciably along its length. If the wire were curved, or if B were not constant over the length of the wire, our assumption would be incorrect.

The magnetic force on a current segment is what makes an electric motor work. The moving-coil galvanometer (which will be described in detail in Section 19-1) and its modifications, the ammeter and the voltmeter, also depend upon magnetic forces acting on current segments.

Example 18-3 A single wire in a household motor is 10 cm long and carries a current of 2 A. If the magnetic field strength B is 0.6 N/A·m and the wire is perpendicular to B, what is the force on the wire?

$$F = I\,\Delta l\,B$$
$$= (2\text{ A})(0.1\text{ m})(0.6\text{ N/A·m})$$
$$= \boxed{0.12\text{ N}}$$

Since this is about the weight of a 12-gram object, it is evident that in practical motors many current segments must be acted upon simultaneously in order to obtain a useful force.

18-3 CURRENT LOOPS

We have now seen two ways in which we might, in principle, determine the magnitude and direction of a magnetic field at any point. We can measure force per unit Qv, or force per unit $I\,\Delta l$. Both methods are based on the fact that a charge experiences a force when moving at an angle to the direction of a magnetic field, but *no* force when moving in the direction of the field. However, the methods that have been described are cumbersome and unrealistic. In particular, the idealized current segment $I\,\Delta l$ is hard to imagine, for there must be some conducting path by which the current enters and leaves the segment.

For a more practical indicator of the direction of a magnetic field, a circular or rectangular current loop can be used (Figure 18-5). Here is a complete path for the current, but the two lead-in wires need not be considered because the net magnetic force on them is zero; they are close together and carry equal currents in opposite directions. The two lead-in wires experience equal and opposite forces. Such a loop tends to set itself broadside to a magnetic field, for if it were at an angle (Figure 18-6a), the two forces F_1 and F_2—equal in magnitude—would tend to twist the loop. When in the broadside position of Figure 18-6b, the loop is in equilibrium and the net force and twisting effect are both zero, since F_1 and F_2 are equal in magnitude but opposite in direction in the same plane.* To obtain a larger effect, we can form a stiff wire into a long, narrow coil (Figure 18-7), thus adding together the effect of many current loops. When current exists in the wire, the coil tends to rotate until the axis AA' is parallel to the field B, with each loop broadside to the field. The coil points in the direction of the field, like a compass needle.

The most ancient device for indicating the direction of a magnetic field is, of course, the compass needle itself. There is a real similarity between a long, narrow coil and a magnetized iron nail hung by a thread (Figure 18-8). Electrons spin and move in orbits around the nucleus of an atom, and in iron atoms one or two of the electron orbits are oriented in approximately the same direction. These orbits will tend to rotate until they are broadside to the field. A magnetized compass needle is, therefore, a collection of current loops. The needle points northward because of the earth's magnetic force on the atomic currents.

Another way of indicating the direction of a magnetic field is to sprinkle iron filings on a piece of paper that has been placed on top of one or more kinds of magnets (Figure 18-9). The magnetic force on the current loops in the iron atoms acts to turn the slivers of iron parallel to the field. Here again, we are still making use of magnetic forces on moving charges.

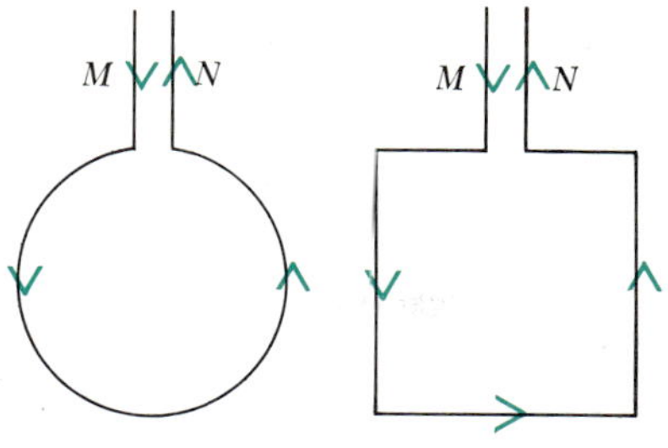

Figure 18-5 Current loops.

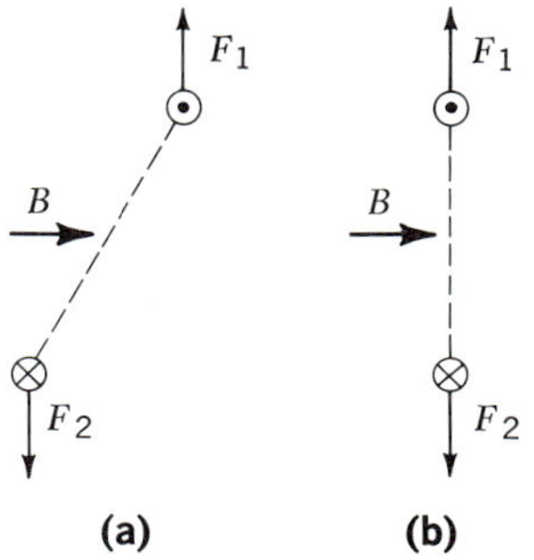

Figure 18-6 Top view of current loop. The twisting effect, or torque, is zero when the loop is broadside to the magnetic field.

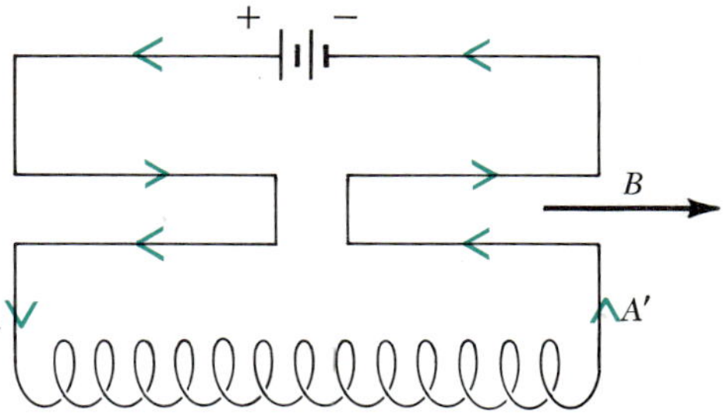

Figure 18-7

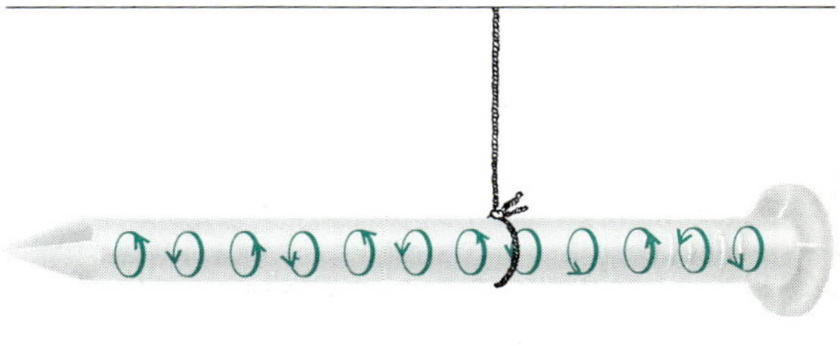

Figure 18-8

* Figure 18-6 illustrates the method that will be used for drawing vectors that are perpendicular to the plane of the paper. The current in the top wire is toward the reader, and is represented by ⊙ which may be thought of as the head of an arrow just bursting through the paper. The symbol ⊗ representing a current away from the reader can be thought of as the tail feathers of an arrow just disappearing into the paper.

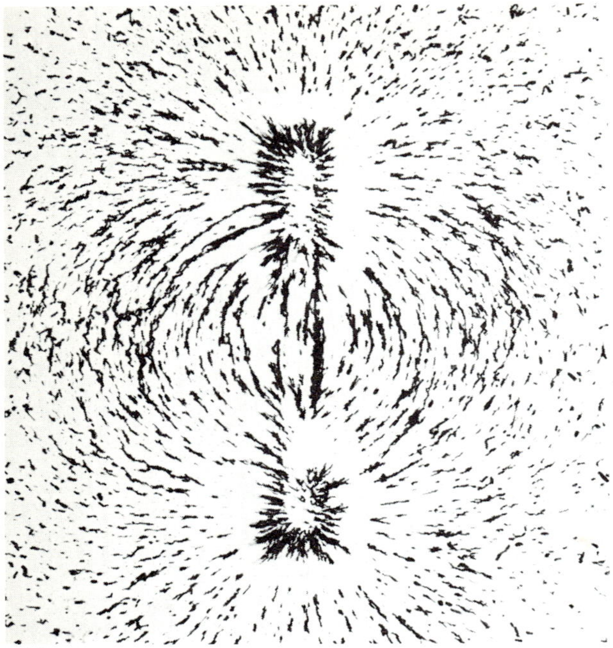

(a)

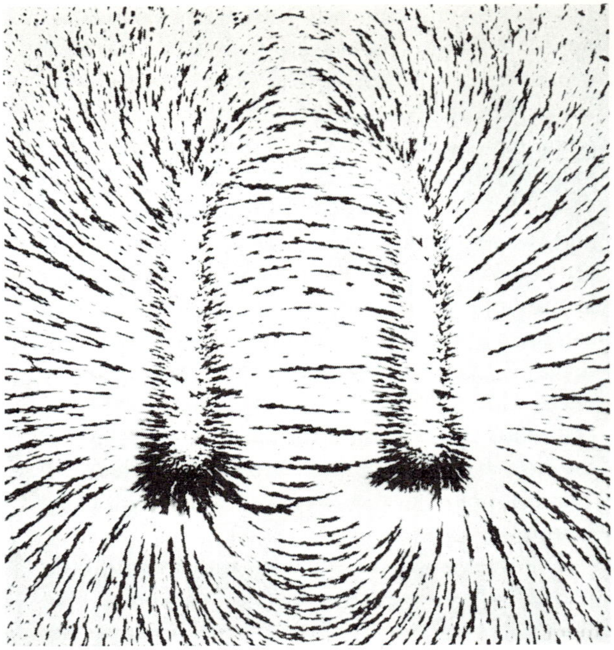

(b)

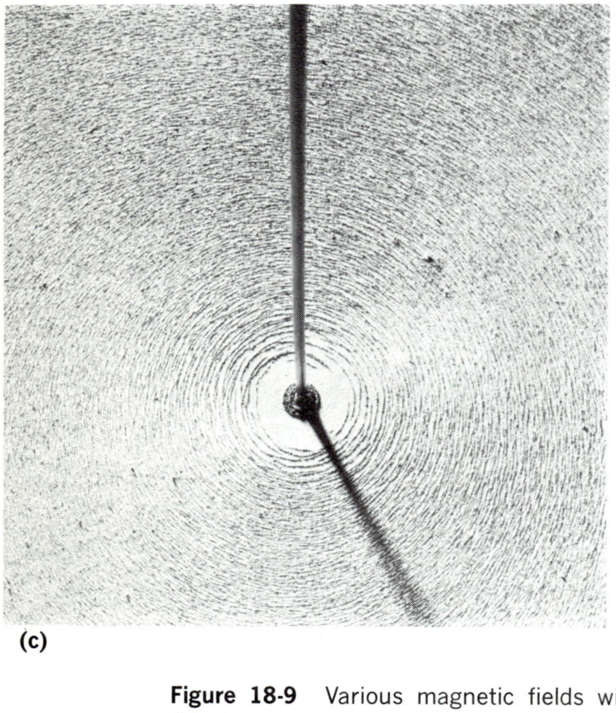

(c)

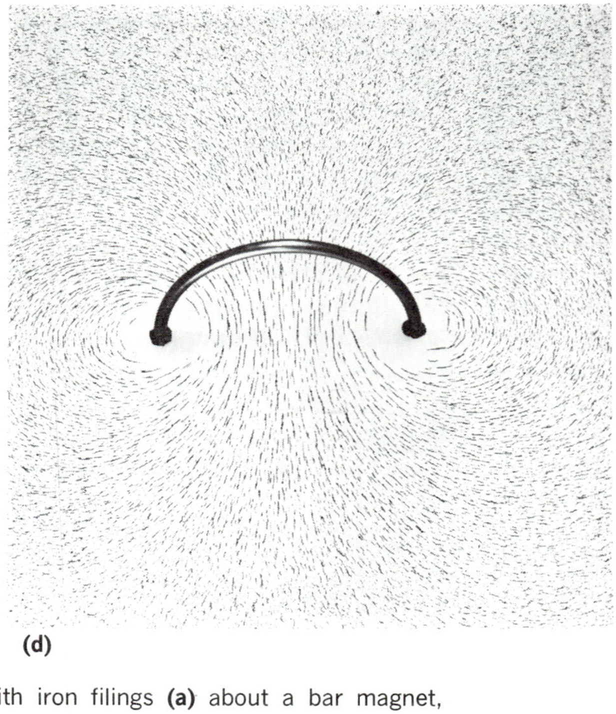

(d)

Figure 18-9 Various magnetic fields with iron filings **(a)** about a bar magnet, **(b)** about a horseshoe magnet, **(c)** around a straight wire carrying an electric current, **(d)** around a loop of wire carrying an electric current.

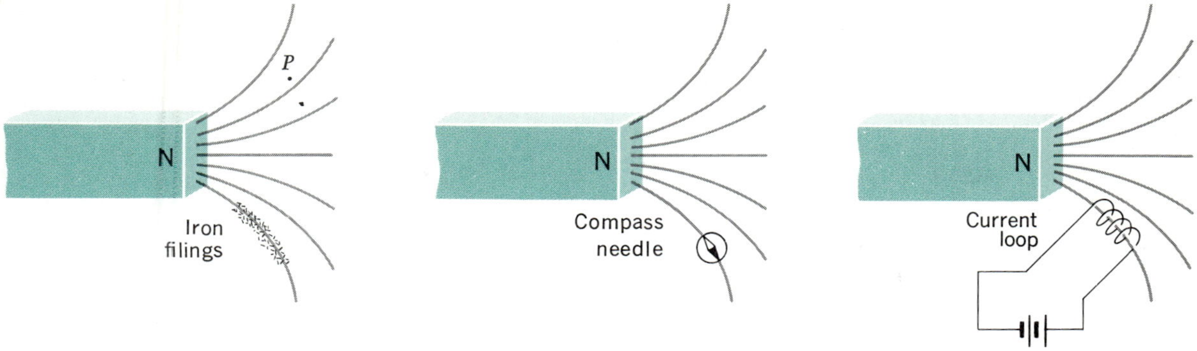

Figure 18-10 Three equivalent ways of exploring a magnetic field.

A magnetic field is considered known if the magnitude and direction of the force per unit test body are known at every point in space. Our test body may be a moving charge or a current segment. To help in visualizing the field, we draw lines of force parallel to the direction of the field. Lines of force near one end of a bar magnet are shown in Figure 18-10. The diagrams show three different ways of exploring the field.

18-4 SOURCES OF A MAGNETIC FIELD

So far we have discussed the measurement of magnetic field without saying anything about how a magnetic field is produced.

Let's first consider the magnetic field surrounding a long straight wire. The region surrounding such a wire can be explored in various ways, but perhaps it is easiest to use a small compass needle or iron filings. Figure 18-11 shows the results of such an exploration with a compass. The field can be represented by lines of force that are concentric circles around the wire. The direction of the magnetic field around a straight wire is given by a simple right-hand rule. Imagine grasping the wire with the right hand, with the thumb pointing in the direction of the current, that is, in the direction that positive charge would move. The fingers are curved around the wire and indicate the direction of the magnetic field. The north end of a compass needle placed adjacent to the fingers would point in this direction (Figure 18-12).

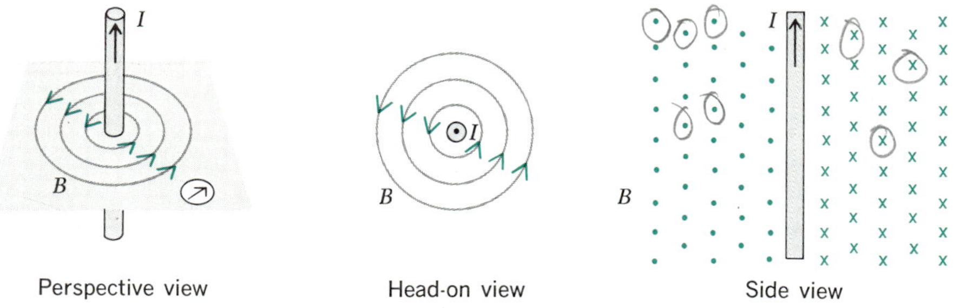

Figure 18-11 Three views of the magnetic field surrounding a long, straight wire. In each case the fingers of the right hand would represent the direction of the field if the thumb pointed in the direction of the current.

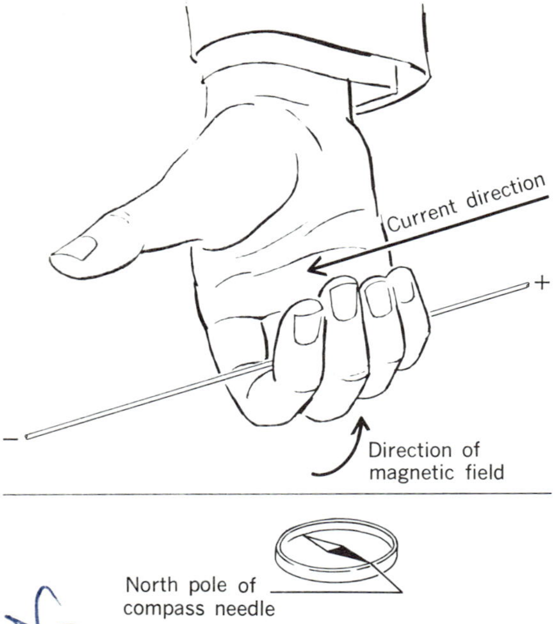

Figure 18-12 The right-hand rule.

What is the magnitude of this field at various positions relative to the wire? A quantitative investigation with a suitable test body would show that the field at a point P, a distance a from a very long straight wire (Figure 18-13) is directly proportional to the current and inversely proportional to the distance a. Thus

$$B = k'I/a$$

We are now able to define the ampere (and, therefore, the coulomb)—something that has been put off for more than four chapters. Our present discussion revolves around the constant k'. If we had already defined units of current, force, and length, then we could use $B = k'I/a$ to find k' experimentally by substituting measured values of B $(= F/I\,\Delta l)$, I, and a. You will recall that this is what was done when the constant in Coulomb's law was found (Section 14-4). However, since we have not yet defined the size of the ampere, we are free to select any numerical value for k' that pleases us. Let's choose the value of the constant k' to be 2×10^{-7} N/A^2 for historical reasons; this makes our ampere equal to the am-

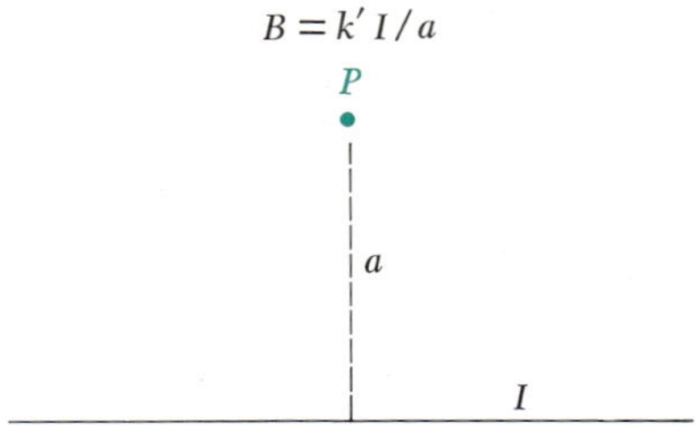

Figure 18-13

pere that had been defined in a different way by earlier workers, before mks units became popular. The equation $B = k'I/a$ now serves to define the unit of current we call the ampere.

To find out how the choice of k' leads to a definition of the ampere, let's compute the force per unit length on a long straight wire carrying $I_1 = 1$ A, which is 1 meter from a parallel wire also carrying a current $I_2 = 1$ A. The current I_1 is immersed in the magnetic field B of the current I_2, which is given by

$$B = (k'I_2/a)$$

We know that the magnitude of B is the force per unit current segment,

$$B = \frac{F}{I_1\,\Delta l_1}$$

Equating equals to equals gives

$$\frac{F}{I_1\,\Delta l_1} = k'I_2/a$$

Therefore, by multiplying both sides of this equation by I_1, we get

$$F/\Delta l_1 = k'I_1I_2/a$$

Substituting the given values gives

$$F/\Delta l_1 = (2 \times 10^{-7}\ \mathrm{N/A^2})\frac{(1\ \mathrm{A})(1\ \mathrm{A})}{1\ \mathrm{m}}$$

$$= 2 \times 10^{-7}\ \mathrm{N/m}$$

We find that the force per unit length is

$$2 \times 10^{-7}\ \mathrm{N/m}$$

Therefore, making $k' = 2 \times 10^{-7}$ N/A^2 leads to the following definition:

One ampere is the current in a long straight wire that exerts a force of exactly 2×10^{-7} newtons per meter on a neighboring long parallel wire, 1 meter distant, that carries an equal current.

The National Bureau of Standards maintains a "current balance" which is based on the definition given above, but which uses coils instead of long wires. The ammeters you use in the laboratory are (indirectly) calibrated relative to this current balance. In this way our standards of current, charge, and PD can be reproduced in terms of mechanical standards such as the kilogram, meter, and second.

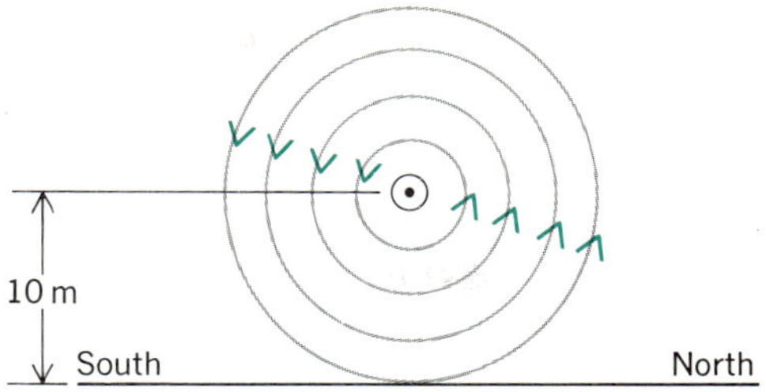

Figure 18-14

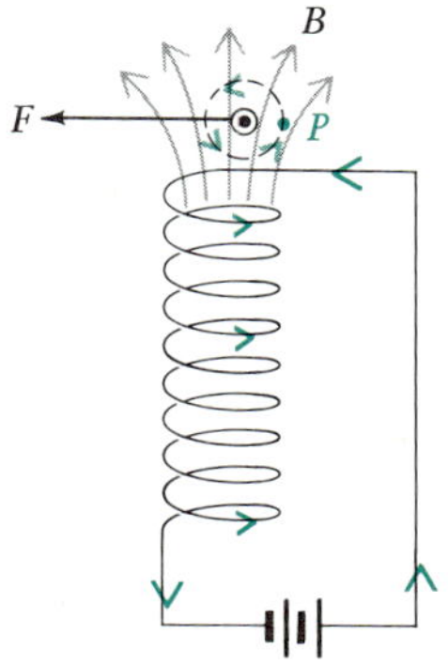

Figure 18-15 A current segment tends to move from a region of strong field into a region of weak field.

Example 18-4 At an electroplating plant, a long horizontal wire connecting two buildings carries 500 A. The direction of the current is toward the east. What are the magnitude and direction of the magnetic field at a point on the ground 10 m directly below the wire?

The direction of the field is found from the right-hand rule (Figure 18-14). The field at the ground level is toward the north. The magnitude of the field is given by

$$B = k'I/a$$
$$= (2 \times 10^{-7}\ \text{N/A}^2)(500\ \text{A})/10\ \text{m}$$
$$= \boxed{10^{-5}\ \text{N/A}\cdot\text{m}}$$

18-5 THE DIRECTION OF MAGNETIC FORCE ON A CURRENT SEGMENT

You have not yet been given a rule for finding the direction of the force on a current segment that is in a magnetic field. You have already seen, however, that the force is perpendicular to the current, and its maximum value $BI\Delta l$ is obtained when I is perpendicular to B (Section 18-1). To specify the direction completely, we need one more rule.

The magnetic force on a current segment tends to move it from a region of strong field to a region of weak field.

To illustrate this rule, consider a horizontal wire that is just above one end of a coil carrying current, as shown in Figure 18-15. If the current in the wire is toward you the wire's own magnetic field is counterclockwise (by the right-hand rule). The field due to the coil is directed upward (also by the right-hand rule). The two fields combine on the right at points such as P, but tend to cancel each other at points to the left of the wire. Hence the force is toward the left, away from the region of strong resultant field at P.

The direction of the magnetic force can also be found from various "hand rules." A convenient one is: Extend the fingers of the right hand in the direction of the magnetic field, and point the thumb in the direction of the current. The force is then in the direction you would push with the palm of your hand.

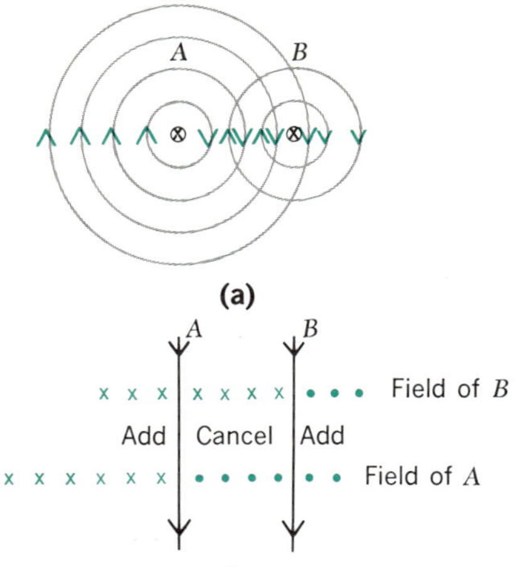

Figure 18-16 Force of attraction between two wires carrying current in same direction.

Two parallel wires that carry current in the same direction attract each other. To see why this is so, consider Figure 18-16a, which shows the two wires in cross section. Wire B is in the field caused by wire A; the resultant field is weakened on the left of B and strengthened on the right of B. Therefore the force is toward the left. Figure 18-16b shows the same situation as viewed from the side. In general, like currents attract each other, and unlike currents repel each other.

In the early days, it was thought that a lodestone or a magnetized iron rod had "poles," and that the force of attraction or repulsion between two magnets followed a force law between two kinds of poles. It cannot be denied that the north end of a bar magnet attracts the south end of another bar magnet, but we now know that these magnetic forces are essentially electrical in nature, and arise from the interactions between the moving electrons of the metal atoms.

18-6 THE EARTH'S MAGNETISM

As long ago as 1570 William Gilbert, court physician to Elizabeth I of England, constructed a permanent magnet in the form of a large lodestone sphere. Gilbert used a small compass needle to survey the magnetic field near the surface of the sphere, and found that his model successfully represented the main features of the earth's magnetic field. At a point in the Northern Hemisphere, such as in England or the United States, the field is directed downward and to the north, as shown in Figure 18-17a. These observations can be accounted for by assuming that a relatively short magnet, several hundred miles long, is buried deep inside the earth (Figure 18-17b). Since the lines of force are directed downward for an observer in the Northern Hemisphere, we must assume that the "earth's magnet" has an S pole on the end that is beneath the north magnetic pole. We know that permanent magnets are collections of current loops, and so Figure 18-17c is drawn as a modern equivalent of the outmoded permanent-magnet hypothesis. If the axis of the assumed current loop is more or less along the direction of the earth's axis of rotation, the north magnetic pole will approximately coincide with the geographical North Pole, as it actually does. The north magnetic pole is in northern Canada, which means that a compass needle does not point exactly north except at certain places. The difference between geographic north and magnetic north is called *magnetic declination;* this quantity varies from about 25° E to 20° W for different places in the United States (Figure 18-18), and also varies slowly from year to year.

The ordinary compass needle responds only to the horizontal component of the earth's magnetic field, since it is pivoted in such a way as to prevent up-and-down motion. If a magnetized needle is mounted in such a way that it can swing in a vertical north-south plane,* it is then free to point in the direction of the field, and in the Northern Hemisphere it will point downward and to the north. Such an instrument is called a *dip needle* (Figure 18-19), and the angle of dip is measured from the horizontal. Values of the horizontal and vertical components of B for the earth's field at Washington, D.C., are shown in Figure 18-19.

* Magnetic north, not geographic north.

Geographic
North Pole

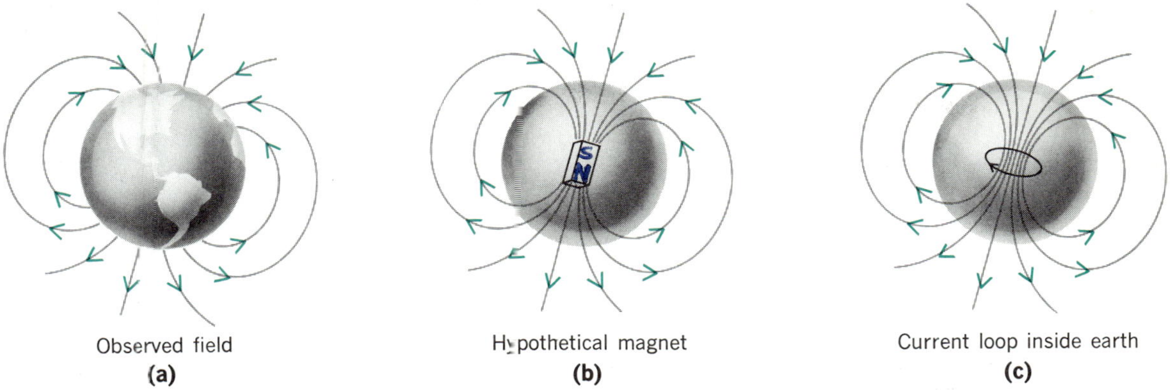

Figure 18-17 The earth's magnetism.

Note that the vertical component is about three times the horizontal component.

The ultimate explanation of the earth's magnetic field must somehow be found in circulating currents deep within the earth, or in the upper atomsphere, or both. The approximate coincidence of the magnetic poles and the geographic poles is surely significant, but the exact mechanism is still unknown. Perhaps a stream of charged particles (electrons, ions, or both) has been created by the high temperatures in the earth's core, and the motion of these charges is affected by the earth's rotation. The strength and location, or both, of the internal currents change gradually, for the angles of declination and dip change. For example, in 1580 the declination at London was measured by Gilbert to be 11° E; it was 0° in 1657, reached 25° W by 1820, and is decreasing again, having had a value of about 6° W in 1978. A second type of change is sudden and short-lived. So-called magnetic storms occur periodically, which are correlated with sun-spot outbreaks, and are a result of temporary currents of ions in the upper atmosphere. Finally, evidence points to radical changes in the magnetic axis of the earth in the geological past. During the last several million years the earth's magnetic axis is

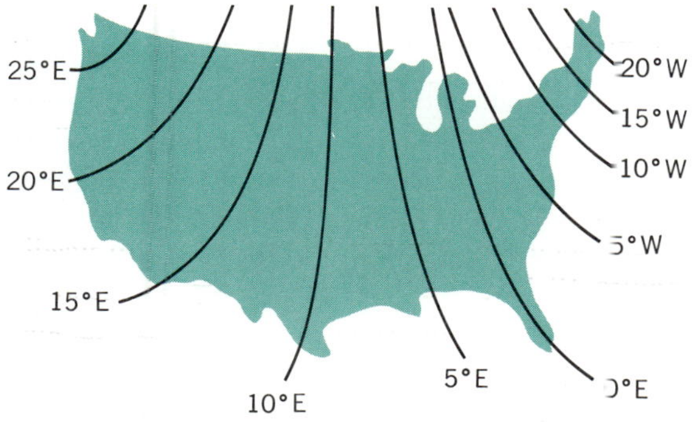

Figure 18-18 Lines of equal magnetic declination.

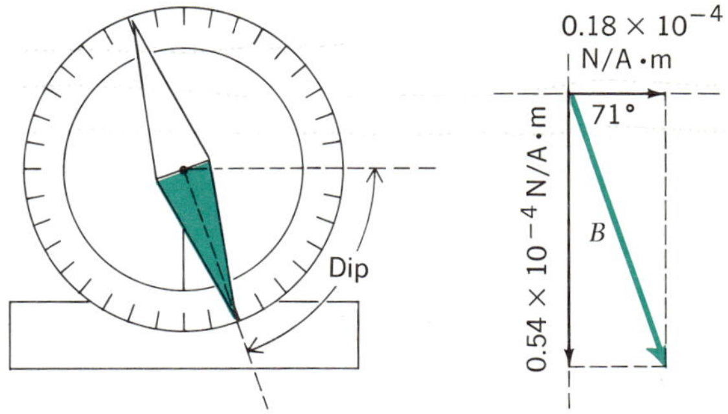

Figure 18-19 Dip needle.

known to have reversed direction several times in addition to undergoing irregular wanderings. Between 0.7 and 2.5×10^6 years ago the "north" magnetic pole was in the Southern Hemisphere. The phenomena of terrestrial magnetism are complicated and variable, and not yet well understood. So far, however, physicists are confident that the earth's field originates in the motion of charges, as do all other magnetic fields.

18-7 INDUCED EMF

You found in Section 18-1 that a charged particle moving through a magnetic field experiences a magnetic force. If, then, a conductor moves in a suitable direction relative to a magnetic field, the magnetic force acts on the free electrons in the conductor, the electrons are accelerated, and work is done as energy is transferred to them. We have here a mechanism for transforming mechanical energy into electric energy, and we say that an induced emf is generated in a conductor that is moved in a suitable direction through a magnetic field (Figure 18-20).

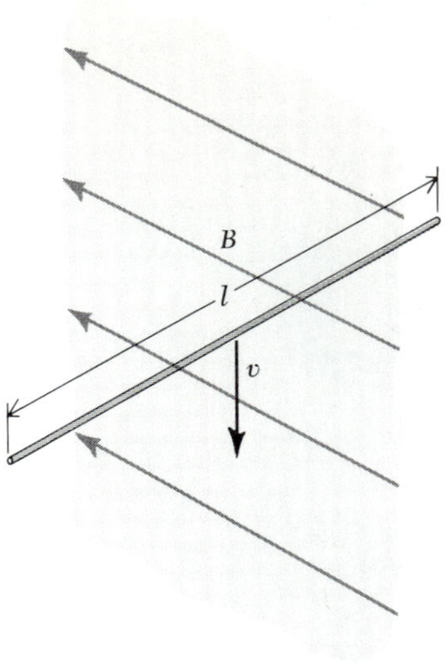

Figure 18-20

The discovery in the 1830's of this phenomenon of electromagnetic induction led to the development of the first successful electric generator. Until then, the only practical way of obtaining electric current was from cells and batteries of various types. It is interesting to consider the great advance of technology that resulted from the discovery of induced emf in about 1832 by Joseph Henry in the United States and Michael Faraday in England. The generators of the world's power stations transform mechanical energy into electrical energy with high efficiency. Now more than 99 percent of all the electric energy used in the United States is "generated" in this way, while less than 1 percent comes from other energy transformations, such as those in chemical cells.

The original investigators found that the magnitude of the maximum induced emf $\mathcal{E}$ in a wire that is moved across a magnetic field is given by

$$\mathcal{E} = Blv$$

The induced emf equals the product of the only factors that might be expected to be significant: B is the field strength, and l is the length of the wire that is moving with velocity v across the lines of force (that is, perpendicular to the direction of the field). To check units, note that

$$\begin{aligned} \mathcal{E} &= Blv \\ &= (\text{N/A}\cdot\text{m})(\text{m})(\text{m/sec}) \\ &= \text{N}\cdot\text{m/A}\cdot\text{sec} \\ &= \text{J/C} \\ &= \text{volts} \end{aligned}$$

For the maximum effect the wire should be at right angles to the field and the velocity should be at right angles to both the field and the wire. Such a motion can be described as that of a conductor "cutting" lines of force, as in Figures 18-20 and 18-21. No effect is obtained when the wire is moving parallel to the lines of force.

Example 18-5 The rotating armature of a simple generator consists of a rectangular loop $SPQR$, to which connections are made by means of sliding contacts. The armature is turned at 1200 rev/min, and

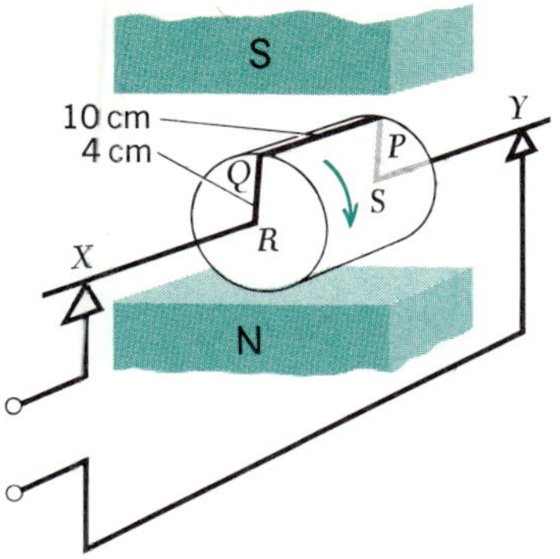

Figure 18-21

the magnetic field is 0.5 N/A·m. What is the induced emf between the terminals of the generator at the instant shown in Figure 18-21? The radius r of the armature is 4 cm and its length l is 10 cm.

The wire PQ is moving to the right with a velocity given by the circumference of its circular path times the number of revolutions per second:

$$\begin{aligned} v &= 2\pi rf \\ &= 2\pi(0.04\text{ m})\left(1200\frac{\text{rev}}{\text{min}}\right)\left(\frac{1\text{ min}}{60\text{ sec}}\right) \\ &= 5.03\text{ m/sec} \end{aligned}$$

Since the wire is cutting perpendicularly across the lines of force, the induced emf is a maximum at this particular instant.

$$\begin{aligned} \mathcal{E} &= Blv \\ &= (0.5\text{ N/A·m})(0.10\text{ m})(5.03\text{ m/sec}) \\ &= \boxed{0.25\text{ V}} \end{aligned}$$

The *direction* of an induced emf is given by Lenz's law, developed by the German physicist Henrich Lenz (1804–1865):

Lenz's law: An induced emf tends to set up a current whose action opposes the change that caused it.

Let's apply Lenz's law to the generator of the preceding example, in which a conductor moves to the right (Figure 18-22a). The best way to proceed is to assume a direction for $\mathcal{E}$, and then apply Lenz's law to see if the assumption is correct. Let's assume that if there were a complete circuit, the direction of the current would be from P to Q and on through the circuit. Using the right-hand rule to get the field due to this assumed current and then applying the force rule, we find that the force on the wire PQ would be toward the weak field, that is, to the left (Figure 18-22b). This force would indeed oppose the "change that caused it"; in this case the change is a motion to the right. Hence our assumed direction is correct. The direction of the current is from P to Q, out at X, through an external circuit, and back in at Y. The generator's negative terminal is at Y, and its positive terminal is at X at the instant shown.

Lenz's law is a deduction from the law of conservation of energy. Thus in Figure 18-22b the force F is opposite to the velocity v, and work must be done by an outside agent to move the wire against the opposing force F. Suppose the generator is connected to a lamp. Mechanical energy is put into the system, and this energy becomes transformed into Joule heat and light as induced current flows through the lamp. However, if the induced current were in the opposite direction (⊗ instead of ⊙), the wire itself would be doing mechanical work, since F and v would be in the same direction. *Also*, thermal energy

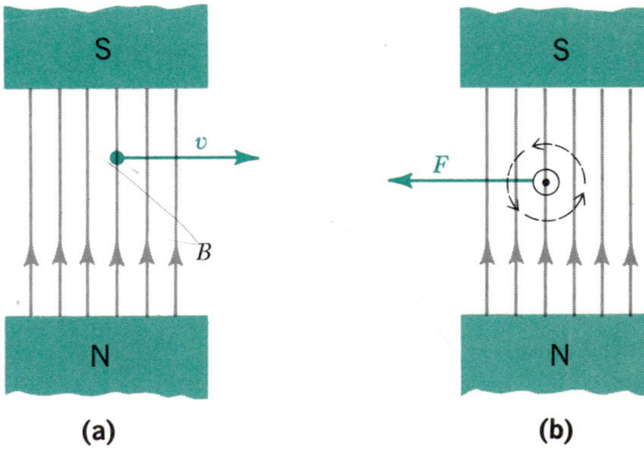

Figure 18-22 **(a)** Conductor moving to the right, cutting across a magnetic field. **(b)** Lenz's law: A conductor is moving to the right, giving rise to an induced current that in turn causes a force to the left, opposing the motion.

would be produced in the wire and in the external circuit. This could not happen according to the law of conservation of energy since there would be an expenditure of energy without any source of energy. You can see, then, that Lenz's law is a necessary consequence of the law of conservation of energy.

We have discussed one way in which induced emf's can arise—from the motion of a conductor across a magnetic field. It is also possible for an emf to be induced without *physical* motion of any conductor. An important example of such an emf is that induced in the secondary coil of a transformer, to be discussed in Chapter 19.

SUMMARY

A **magnetic field** exists in any region of space where a moving electric charge would experience an electric force other than a Coulomb force. **Magnetic forces** are caused by moving charges acting on other moving charges. We therefore speak of magnetism as a subdivision of electricity. A magnetic field can be explored by using a moving charge as a test body. The magnitude of the vector representing a magnetic field is given by the force per unit Qv product: $B = \frac{F}{Qv}$, measured in newtons per coulomb·meter per second. The magnetic force on a moving charge is always perpendicular to its motion and to the direction of the field. A moving charge experiences maximum force if its motion is perpendicular to the field. A current segment $I\,\Delta l$ can be used as a test body instead of moving charge: $B = \frac{F}{I\,\Delta l}$. If current segments are laid end to end, forming a loop or a coil, the coil tends to align its axis with the direction of magnetic field.

Permanent magnets, such as a compass needle, are essentially collections of **current loops** formed by the orbital motions and spins of the electrons in the atoms.

Magnetic fields are mapped by drawing representative lines of force. In the space surrounding a wire carrying a current, the magnetic field is represented by concentric circles. The **right-hand rule** gives the direction of these concentric lines of force.

The force on a current segment in a magnetic field equals the product of the magnetic field strength and the strength of the current segment: $F = BI\,\Delta l$. The magnetic force on a current segment tends to move it from a region of strong field to a region of weak field.

The earth's magnetic field is caused by circulating electric currents inside the earth, but the exact mechanism is not known. In the United States, the field is directed downward and to the north. Three quantities are used to describe the earth's field at any given point: the magnitude of the field, the **declination,** and the **dip.**

When a wire is moved across a magnetic field, an emf is induced in the wire. The direction of the induced emf is given by **Lenz's law** and the magnitude by $\mathcal{E} = Blv$.

QUESTIONS

1 Upon what factors does the magnetic force on a moving charge depend?

2 How would you determine the direction of magnetic field at a given point (a) using a compass and (b) using a moving test charge?

3 Give two mks units for magnetic field strength, and show that they are equivalent to each other.

4 Is there any known way in which a magnetic field can be caused other than by the motion of electric charge?

Figure 18-11, 12

5 Lightning strikes a vertical metal flagpole, and there is a momentary current up the pole. What is the direction of the magnetic field at a point in the air just east of the center of the pole?

6 Suppose that a plastic phonograph record is rubbed along its edge with fur, charging the edge with an excess of electrons. When the charged record is turning in the usual direction, in which direction is the magnetic field at a point directly above the center of the record? (out)

7 Without using iron or any other magnetic material, how could you construct a device that would serve as a compass?

8 Upon what factors does the induced emf in a conductor depend?

9 Show that the rule "like currents attract each other" applies to Figures 18-15 and 18-16.

PROBLEMS

GROUP A

1 A charge of 2×10^{-10} C is moving at 1 km/sec through a magnetic field in such a direction that the magnetic force on it is a maximum. What is the strength of the field if the maximum force on the charge is 6×10^{-8} N?

2 What is the force on an electron that moves at 1% of the speed of light horizontally through a vertical magnetic field of strength 0.12 N per C·m/sec?

3 Calculate the magnetic force on a current segment 3 cm long, placed broadside to a magnetic field of strength 1.2 N/A·m. The current in the wire is 2000 A. Could a strong man supply enough force to hold the wire in equilibrium?

4 What is the strength of the magnetic field at a point 10 cm away from a long wire in which the current is 20 A?

* 5 The current is downward in a wire that is in a magnetic field directed horizontally to the south. In what direction is the force on the wire?

6 A horizontal wire 8 cm long moves with a speed of 3 m/sec across a magnetic field of strength 0.04 N/A·m. What emf is induced in the wire?

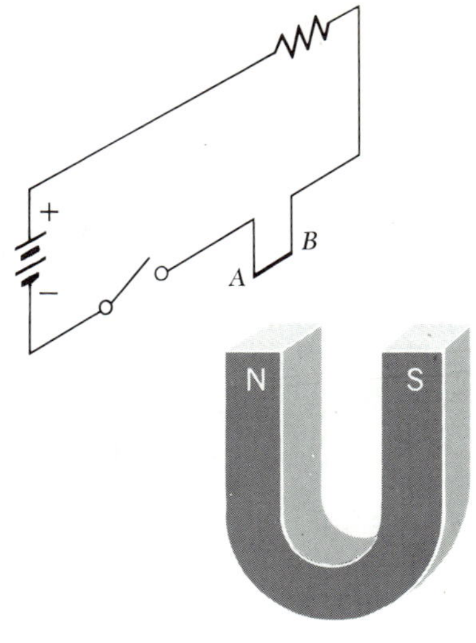

Figure 18-23

GROUP B

7 In what direction is the magnetic field between the pole pieces of the magnet of Figure 18-23?

8 In which direction (right or left) does the wire *AB* in Figure 18-23 move when the switch is closed?

9 A horizontal rod is falling through a gap between the poles of the permanent magnet in Figure 18-23. Does the induced emf cause current toward or away from you, the reader? Explain your answer using Lenz's law.

GROUP C

10 There is a current in a circular path through a horizontal coil, clockwise as viewed from above. The coil, which has 10 turns, has a radius of 3 cm, and the current is 210 milliamperes. (a) In what direction is the magnetic field at the center of the loop? (b) What is the magnitude of the field at the center? (c) Is this field stronger or weaker than the earth's magnetic field?

19 Electricity in Modern Technology

In this chapter we will discuss some of the electrical devices that are so important in everyday technology. The purpose here is to show these devices as logical applications of some of the basic principles you have already studied.

19-1 METERS

The basic electrical measuring instrument is the direct-current meter. In Section 17-5 you saw how a galvanometer can be modified by a series resistor to make a voltmeter, or by a parallel shunt to make an ammeter. What factors determine the sensitivity of a galvanometer? In other words, what factors determine the deflection caused by a given current? This is a problem in design. Let's study the galvanometer in detail, as an illustration of how physical principles are used in the design of apparatus.

The moving coil galvanometer consists of a flat coil that is supported in the magnetic field of a permanent iron-alloy magnet (Figure 19-1). When there is a current in the coil and charges move through the wires, they experience a magnetic force that causes the coil to turn against the opposing forces of the springs S and S′.

In this discussion we are concerned with the turning effect, or *torque,* of forces. This is a familiar idea: The turning effect of a force depends not only on the magnitude of the force, but also on the length of the lever arm involved. The lever arm l is the perpendicular distance from the line of application of the force to the axis of rotation (Figure 19-2). The torque is defined as force times lever arm, or

$$T = Fl$$

One torque, which might be called the "forward torque," is due to the magnetic force on the vertical segments of the coil (Figure 19-1). We

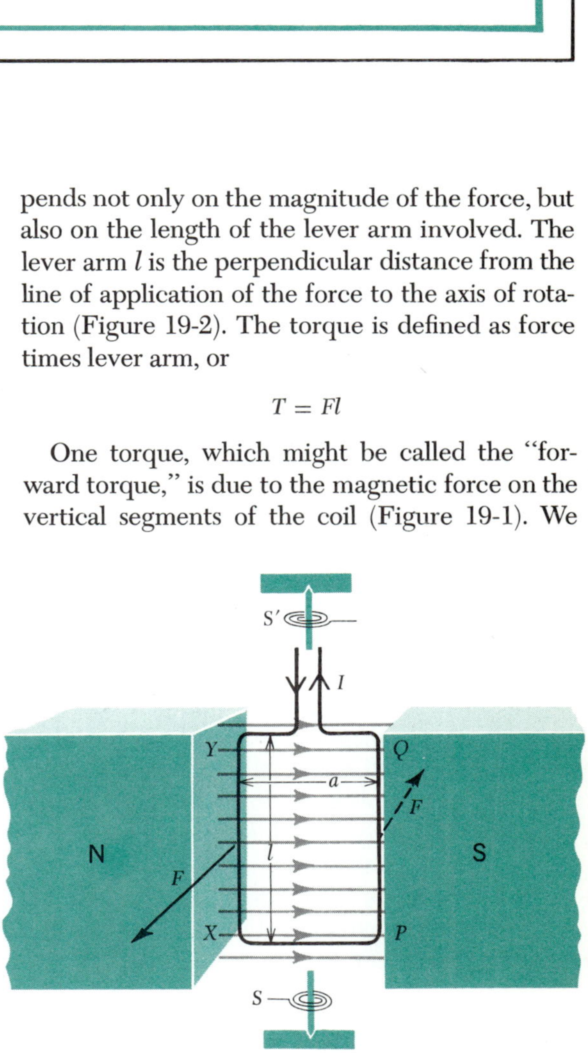

Figure 19-1 A galvanometer movement.

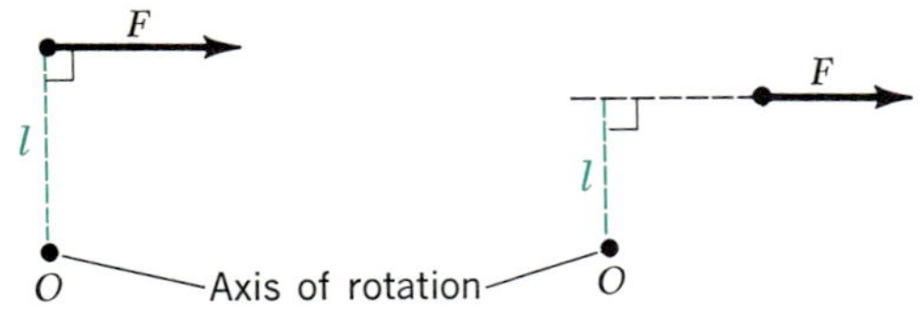

Figure 19-2 The lever arm l is drawn from the axis of rotation perpendicular to the line of application of the force.

calculate this torque T by first determining the magnitude of the force F, then multiplying by the lever arm l. In Chapter 18 it was shown that the magnetic force on a current segment is $F = I\,\Delta l\,B$. In this case the total length L of the vertical wire can be considered the length of the current segment; therefore the total magnetic force on each vertical wire is ILB. The lever arm is $a/2$ (half the distance between the two vertical wires), so the torque on each wire is

$$T = ILB(a/2)$$

The total torque acting on both wires is, therefore,

$$2 \times ILB(a/2) = ILBa$$

Finally, if the coil consists of N turns of wire, the total "forward" torque T is

$$N \times ILBa = NIBA$$

since La is the area A of the coil.*

This torque causes the coil to start to turn. As it does so, the springs wind up and give rise to a restoring torque (in opposition to the forward torque). The coil, and thus the pointer, will come to rest where the magnetic (forward) torque $NIBA$ and the restoring torque of the spring are equal in magnitude. The restoring torque is proportional to the angle of twist θ, and is also proportional to the "stiffness" of the spring, which we can symbolize by the factor k. Thus, restoring torque $= k\theta$; and since at equilibrium the restoring torque equals the magnetic torque, we see that

$$k\theta = NIBA$$

The angle θ that the coil and pointer deflect for a given current I is therefore directly proportional to the magnetic torque and inversely proportional to the stiffness of the spring, thus

$$\theta = NIBA/k$$
$$= \frac{NBA}{k} \times I$$

In making a good, sensitive meter, the design problem is to get the most deflection θ for a given current I. Our "working equation" above shows that we should strive for a strong magnetic field B and a coil of large area A having many turns N. It also shows that we should use a spring system that is easily twisted, one that has a small stiffness k.

Closer study of this simplified design of the galvanometer (Figure 19-3a) points out a defect in the arrangement. When the galvanometer coil has turned through an angle θ, the force F is no longer fully effective in producing torque. This is because the lever arm must be measured from the axis of rotation at right angles to the direction of the force. Here that length is only $a'/2$, which is clearly less than $a/2$. It follows that the torque $Fa'/2$ is increasingly less than the torque $Fa/2$. In fact if θ reached 90° the magnetic force would produce no torque at all as shown in Figure 19-3b.

To overcome this defect, the pole pieces of the magnet are curved, and a fixed, cylindrical iron core is placed inside the coil (Figure 19-3c). This helps to shape the magnetic field so that the magnetic force always acts perpendicular to a radius and keeps the torque fairly constant. The magnetic force on the current-carrying segments is the same as before, but now the lever arm is the full $a/2$, and the torque caused by the current does indeed depend only on N, B, A, and I, and not upon the coil's position (within a limited range of deflection). Figure 19-4 shows the interior of a commercially made galvanometer.

We have discussed one electrical device at some length as an example of a typical engineering procedure. A successful design can be developed from an understanding of the factors that affect

*Although these relations have been derived for a rectangular coil, the result is valid if A is the area of a flat coil of any shape.

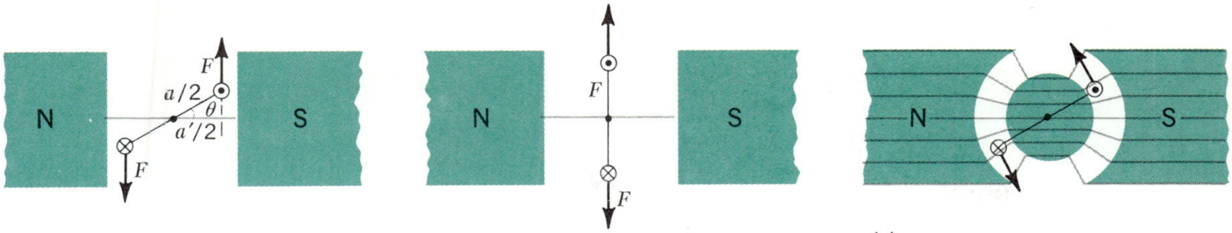

(a) (b) (c)

Figure 19-3 Top view of galvanometer coil. **(a)** Torque is reduced as coil starts to move. **(b)** Torque is zero when coil is broadside to the magnetic field. **(c)** Shaping the pole pieces and inserting an iron core restore the lever arm to its full value.

the performance of a device, as expressed in a working equation such as

$$\theta = \frac{NBA}{k} I$$

The designer uses the equation to predict the effect of changes in the design or construction. As often happens, compromises must be made. For instance, for high sensitivity a large coil (large A) is needed, but this may make the motion of the meter too sluggish for convenient use. Again, high sensitivity can be obtained by using a very delicate spring system (small k), but this would make the pointer slow to "settle down" on a reading and would also make the meter fragile and easily damaged.

Figure 19-4 Inside of commercial galvanometer. Can you recognize each essential part?

The number of turns of wire N also calls for compromise: For utmost sensitivity a coil should have many turns, but this would be physically awkward unless fine wire is used. In that event, the many turns of fine wire would have relatively high resistance, which is undesirable for a meter that is to measure small potential differences, such as the emf of a thermocouple.

In order to achieve the strong magnetic field that is desirable (large B), modern alloys such as Alnico are often used. These alloys produce strong fields with small "pole" pieces; hence modern meters are considerably smaller than those of half a century ago.

"Permanent" magnets are not permanent, strictly speaking. As a meter ages, the magnetic field B tends to weaken, and the sensitivity drops off. For this reason, in precise work a meter is calibrated at regular intervals.

Our analysis of the galvanometer movement is similar to the analysis that could be made for every instrument, machine, or device. This general procedure is typical of applied science, or engineering.

19-2 TRANSFORMERS

Power measured in joules per second, or watts, refers to the rate at which energy is transferred from one place to another or transformed from one form to another. *Electric* power refers to the rate at which electric energy is converted to other forms of energy or transferred to different locations. This rate depends upon the product of the emf and the current in the circuit. The equation $P = \mathcal{E}I$ shows, for example, that the same power can be obtained in a 12-volt headlight bulb operating at 2 amperes as could be obtained in a 120-volt house lighting circuit at a current of 0.2 ampere. For practical reasons, however, one or the other arrangement may be more suitable. The low-voltage circuit in the automobile is preferred because of the low emf of available batteries, the need for freedom from shock hazard, and for relatively light insulation. The high-voltage circuit in the home is desirable so that the current to any bulb or appliance will be relatively low; this avoids the expense of installing the heavy copper wire that would be necessary if a low-voltage, high-current circuit were used.

The transformer is a device for changing the value of the emf in a circuit. The ease with which this can be done with alternating currents (ac) has led to the almost universal present-day use of ac power for home and industry. Because the transformer is an ac device, in order to understand it we must briefly discuss some of the properties of alternating currents and how they are generated. Figure 19-5 shows schematically how a simple ac generator is constructed. (The rectangular coil of wire between the magnetic poles would actually consist of many turns wound on an iron core.) The coil is rotated about the axis by some external torque. As the wires cut across the magnetic lines of force, an emf is induced according to $\mathcal{E} = BLv$ (Section 18-7). For a given generator (B and L constant), then, the output emf is proportional to the speed with which the coil turns. Let's fix our attention on one conductor of the coil. In Figure 19-6(a) it is moving upward across the lines of force, and there is a maximum induced emf. When the coil reaches the position (b), the conductor is moving parallel to the field, and there is no induced emf. In (c) this same conductor moves down across the lines of force, and the induced current is directed opposite to that in (a). By plotting a graph of the current I in the lamp bulb vs. time t, we get an alternating current as shown in Figure 19-6(d).

The *frequency* of an alternating current is the number of times per second that the current

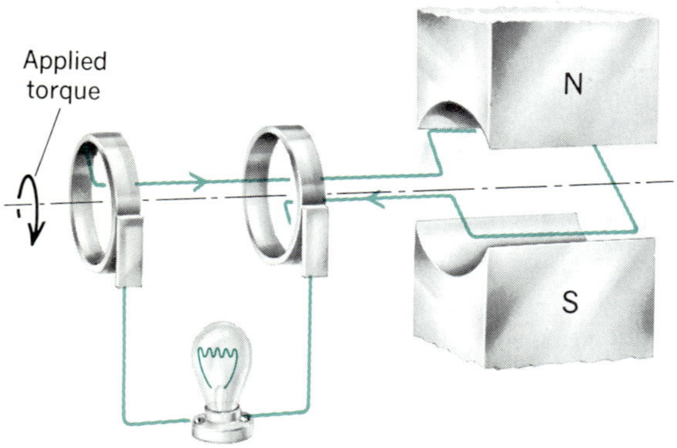

Figure 19-5 A simple generator. The coil is turned by some source of mechanical energy.

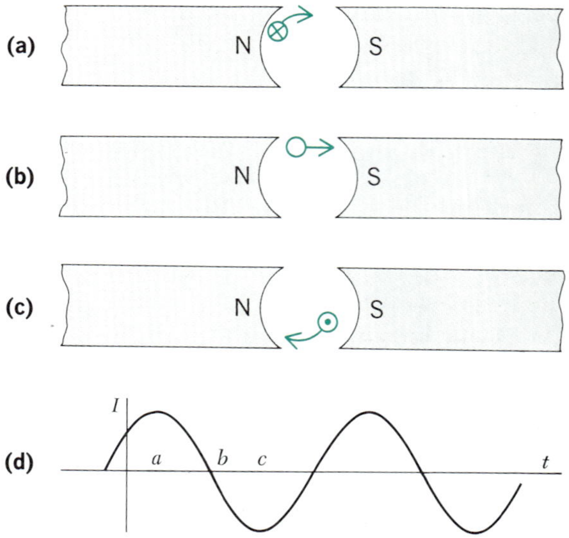

Figure 19-6

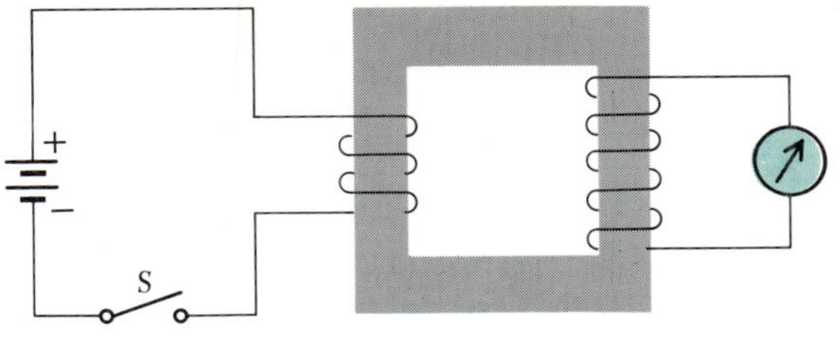

Figure 19-7

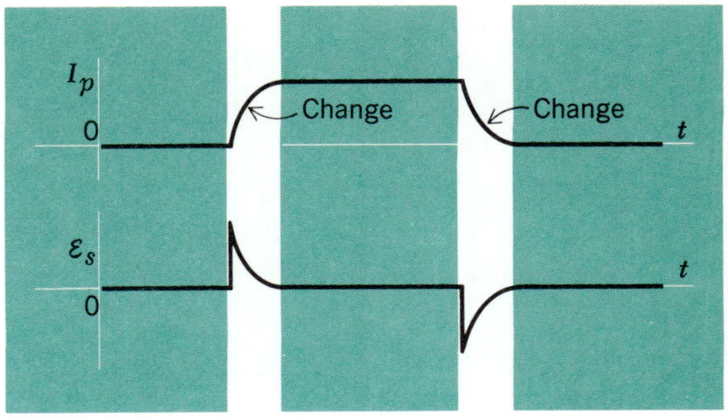

Figure 19-8 An emf $\mathcal{E}_s$ is induced in the secondary only when the primary current changes.

makes a complete alternation from one direction to the opposite one and back again. Ordinary house current in the United States has a frequency of 60 alternations per second (60 hertz).

A common type of transformer consists of an iron core upon which are wound a primary coil and a secondary coil. Let's start by imagining the primary connected to a battery through a switch S (Figure 19-7). When the switch is closed, current begins to build up in the primary circuit, causing an increasing magnetic field. The lines of force of the magnetic field cut the wires of the secondary, causing a momentary emf to be induced in the secondary coil. When the primary current reaches its steady value* (given by Ohm's law), there is no further change in the magnetic field, and the emf in the secondary becomes zero and remains zero. In other words, an emf pulse of short duration appears across the terminals of the secondary coil. If a galvanometer is connected to the secondary, there is a complete circuit. The induced emf is able to send a pulse of charge through the galvanometer, and a momentary deflection occurs. If at some later time the switch S is opened, the primary current becomes zero, the magnetic field falls back to zero, and while this is happening lines of force cut the secondary coil in a direction opposite to the cutting that took place previously when the current was increasing. A momentary pulse is observed in the opposite direction to the pulse that occurred previously (Figure 19-8). Note that a steady current in the primary produces no effect in the secondary; only a changing magnetic field induces an emf. This is similar to the induction of an emf when a wire moves through a magnetic field. It should now be clear that a continually changing current in the primary will produce a continually changing current in the secondary.

* This may take only a fraction of a second, or it may take several seconds, depending on the number of turns and amount of iron.

One fundamental relation is that the ratio of primary emf $\mathcal{E}_p$ to secondary emf $\mathcal{E}_s$ equals the ratio of the number N_p of turns in the primary coil to the number N_s of turns in the secondary coil (Figure 19-9):

$$\frac{\mathcal{E}_s}{\mathcal{E}_p} = \frac{N_s}{N_p}$$

This equation follows, in part, from the fact that if a certain emf is induced in each turn, the total emf is the sum of the emf's induced in the individual turns, and hence $\mathcal{E}_s$ is expected to be proportional to N_s. Thus, to get a large induced emf, we need many turns in the secondary coil. Trans-

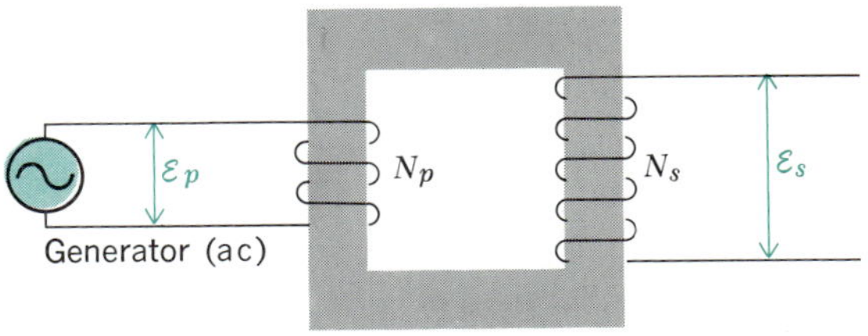

Figure 19-9

formers can be of either the "step-up" or "step-down" variety, depending on whether the secondary emf is greater or less than the primary emf.

The law of conservation of energy indicates that the power in the primary equals the power in the secondary. This statement is true for an ideal transformer of 100 percent efficiency, that is, one in which there is no dissipation of energy through Joule heating in the coils, or other processes. Even though such a transformer does not exist, modern transformers can be constructed to have efficiencies greater than 99 percent, and so our assumption of 100 percent efficiency is not far out of line.

Hence, since $P = \mathcal{E}I$,

$$\mathcal{E}_p I_p = \mathcal{E}_s I_s$$

or

$$I_s/I_p = \mathcal{E}_p/\mathcal{E}_s$$

Thus, for an ideal transformer the ratio of the currents is the reciprocal of the ratio of the voltages.

Example 19-1 A certain ideal transformer's primary has 2000 turns, and its secondary has 50 turns. The primary is connected to an ac source having an emf of 120 V; the secondary is connected to a lamp bulb having a resistance (when lit) of 0.6 Ω. Calculate (a) the emf of the secondary; (b) the current through the bulb; (c) the current in the primary; (d) the power in the primary and the secondary.

(a) This is a step-down transformer, since the secondary has fewer turns than the primary.

$$\mathcal{E}_s/\mathcal{E}_p = N_s/N_p$$

Hence,

$$\mathcal{E}_s = (50/2000)(120\text{ V}) = \boxed{3\text{ V}}$$

(b)

$$I_s = V_s/R_s = 3\text{ V}/0.6\ \Omega = \boxed{5\text{ A}}$$

(c)

$$I_s/I_p = \mathcal{E}_p/\mathcal{E}_s$$

$$5\text{ A}/I_p = 120\text{ V}/3\text{ V}$$

$$I_p = \boxed{0.125\text{ A}}$$

(d) Hence the power in the primary is

$$\mathcal{E}_p I_p = (120\text{ V})(0.125\text{ A}) = \boxed{15\text{ W}}$$

And the power in the secondary is

$$\mathcal{E}_s I_s = (3\text{ V})(5\text{ A}) = \boxed{15\text{ W}}$$

Electric energy is transmitted through long distances at high voltage in order that the current in the line will be as small as possible for a given power. Any transmission line has some resistance, and the energy dissipated as Joule heating in the line is wasted (Section 15-8). The next examples compare the energy losses when lower and higher emf is used for transmitting the same amount of electric power.

Example 19-2 A transmission line between a power station and a factory has a resistance of 0.1 Ω in each of two wires. If 200 A is delivered at 100 V, (a) what useful power is delivered to the factory? (b) What power is wasted in heating the line? (c) What total power must be supplied by the generator?

(a) The power into the factory is

$$\mathcal{E}I = (100\text{ V})(200\text{ A}) = \boxed{20{,}000\text{ W}}$$

(b) The power used in heating the line would be

$$I^2R = (200\text{ A})^2(0.2\ \Omega) = \boxed{8000\text{ W}}$$

(c) Thus the power supplied by the generator is

$$20{,}000\text{ W} + 8000\text{ W} = \boxed{28{,}000\text{ W}}$$

Example 19-3 Repeat Example 19-2 for the same line, assuming that the same useful power is delivered at a voltage of 1000 V instead of 100 V. Find (a) the current in the line, (b) the power used in heating the line, and (c) the total power supplied by the generator.

(a) Since the power into the factory is still 20,000 W, the current is now

$$I = P/\mathcal{E}$$
$$= 20{,}000\ \text{W}/1000\ \text{V}$$
$$= \boxed{20\ \text{A}}$$

(This is $\frac{1}{10}$ as much current as before.)

(b) The power used in heating the line is, then,

$$I^2R = (20\ \text{A})^2(0.2\ \Omega)$$
$$= \boxed{80\ \text{W}}$$

(c) And the power supplied by the generator is

$$20{,}000\ \text{W} + 80\ \text{W} = \boxed{20{,}080\ \text{W}}$$

By using greater emf, perhaps as a result of a different generator design, the power used in heating the wires is reduced from 8000 W to only 80 W. This energy is lost, unless it possibly helps warm some birds perched on the wires. Of course, if the power is delivered to the city at high voltage, it must be transformed to a useful (lower) emf by a step-down transformer at the place where it is to be used.

19-3 REACTANCE OF A COIL

Suppose we measure the current through a certain coil connected to a source of direct current and find it to be 6 amperes (Figure 19-10a). Using Ohm's law, we find the resistance of the coil to be

$$R = 6\ \text{V}/6\ \text{A}$$
$$= 1\ \Omega$$

However, if we repeat this procedure using a 6 V source of 60 hertz (cycle/sec) alternating

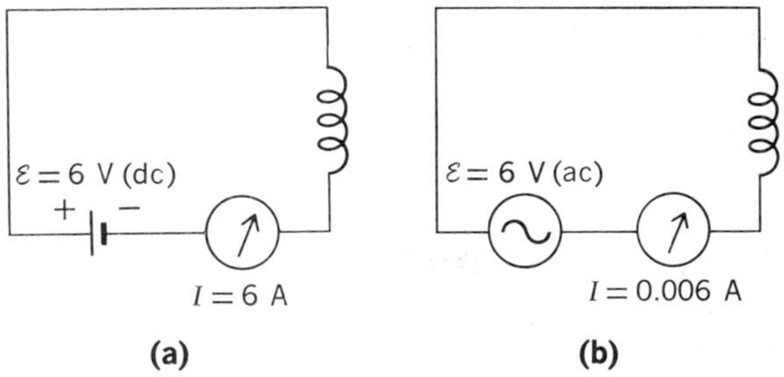

Figure 19-10

current represented by the symbol –⊚– we find a much smaller current. The "apparent resistance" in Figure 19-10b is 6V/0.006 A = 1000 Ω. This ratio is not constant for alternating current of different frequencies though. If our source had been 120-hertz, we would find that I was only 0.003 ampere, so that the ratio V/I would be 2000 ohms. The coil offers an *inductive reactance* to the alternating current. The inductive reactance X_L (also measured in ohms) can be approximately defined in the same way that resistance has been defined, but this definition holds only when the resistance is very small compared to the reactance. Any circuit or part of a circuit always has some resistance, but it can be ignored when it is small compared to the reactance. At the end of the next section you will see that another concept—the impedance—allows us to take both resistance and reactance into account.

Our approximate definition of reactance is

$$X_L = \frac{\text{alternating voltage}}{\text{alternating current}}$$
$$= V/I$$
$$\text{ohms} = \text{volts/amperes}$$

For the 60-hertz current, the coil of Figure 19-10 has a resistance of 1 ohm and a reactance of 1000 ohms. The resistance is always present, even though it can sometimes be ignored in calculations. It is important to note that a resistance always dissipates energy as heat, whereas inductive reactance refers to energy stored in the magnetic field of the coil.

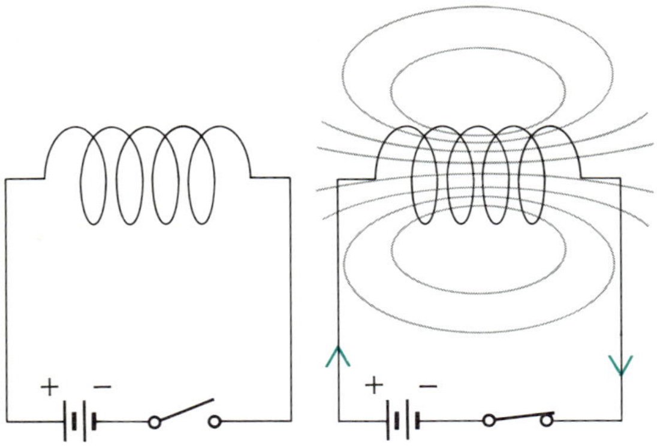

Figure 19-11

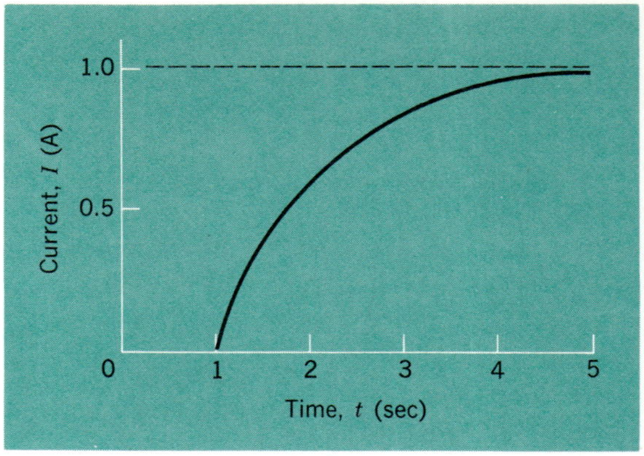

Figure 19-12 Rise of current through a coil; switch is closed at $t = 1$ sec and the current reaches 98% of its final value at $t = 5$ sec.

The reason the current in Figure 19-10b is so small is that the coil acts as a transformer, with the same winding serving as both primary and secondary. When there is a current in a coil, a magnetic field is also set up within the coil (Figure 19-11). As the current builds up lines of force cut the wires of the coil, and an emf is induced in the coil by its own changing magnetic field. This phenomenon is called *self-induction.* The induced emf is a "back emf," opposite in direction to the emf of the generator. By Lenz's law, the back emf tends to oppose the change inducing it, that is, it opposes the increase in current. Figure 19-12 shows how the back emf keeps the current in a certain large iron-cored magnet from rising instantaneously to its final value.

Any given coil can be described by a constant called its self-inductance, or simply its *inductance* L, measured in henries (H) (see Appendix, Table 14); the larger the inductance, the more the back emf for a given rate of change of current. The inductance of a coil increases with the number of turns and the cross-section area. It is related to the nature of the core (whether air or some iron alloy), and is inversely related to the coil's length. These relations are far from simple. For a given coil of inductance L, careful measurements would show that the inductive reactance of the coil is given by the formula

$$X_L = 2\pi fL$$

where f is the frequency in hertz. If the resistance of the coil is negligible, then

$$I = V/X_L$$

Example 19-4 Calculate the inductance of an iron-core coil if a 60 Hz generator, whose emf is 115 V, sends 0.1 A through the coil. Assume resistance is negligible.

$$\begin{aligned} X_L &= V/I \\ &= 115\text{ V}/0.1\text{ A} \\ &= 1150\ \Omega \end{aligned}$$

To find L, we use the reactance formula

$$X_L = 2\pi fL \qquad L = X_L/2\pi f$$

$$\begin{aligned} L &= 1150\ \Omega/(2\pi)(60\text{ Hz}) \\ &= \boxed{3.05\text{ H}} \end{aligned}$$

19-4 REACTANCE OF A CAPACITOR

The behavior of a capacitor in an ac circuit deserves a brief discussion, for there is a sharp contrast between its response to alternating and to direct currents. That is, the reactance of a capacitor is quite different from its resistance. When a capacitor is connected to a dc source of emf, such as a battery, in series with an ammeter and a lamp, there is a momentary current as the capacitor becomes charged (Section 16-4). The ammeter needle swings over for an instant (if its inertia is not too great), but it soon returns to zero and stays there. The lamp flashes momentarily.

The situation is different with an ac generator. The capacitor is charged first one way and then the other, and current rushes back and forth through the meter and the lamp (Figures 19-13a and b). No charges actually move across the gap between the plates of the capacitor (if we assume that the dielectric material is a perfect insulator), but the lamp won't know the difference.

Just as much heat and light are produced by the lamp as if charges really did flow in a complete circuit, as in Figure 19-13c. To all intents and purposes, there is a current *through* the capacitor as far as the rest of the circuit is concerned.

What limits the current through the capacitor in Figure 19-13? In other words, what factors determine the reactance of a capacitor? The larger the capacitance, the more charge flows to and fro, since more charge can be momentarily stored on the plates (according to the equation $Q = CV$). The higher the frequency of the alternating current, the greater is the rate of flow of charge, and hence the greater the current. Thus the current is directly proportional both to C and to f; and the reactance, which is V/I (ignoring resistance) is inversely proportional to C and f. Careful measurement would show that the "capacitive reactance" of a capacitor is given by the formula

$$X_C = \frac{1}{2\pi fC}$$

Example 19-5 A 110 V lamp bulb rated at 50 W is connected to the 110 V, 60 Hz line through a capacitor of 1 μF. Will the lamp light? Assume that the resistance of the capacitor can be ignored.

This problem is purposely stated in such a way that some preliminary spadework must be done. The normal current in a 110 V, 50 W bulb is given by

$$\begin{aligned} I &= P/V \\ &= 50 \text{ W}/110 \text{ V} \\ &= 0.455 \text{ A} \end{aligned}$$

If the current is much less than this value, the lamp will not light. To find the current, we next compute the reactance of the capacitor:

$$\begin{aligned} X_C &= \frac{1}{2\pi fC} \\ &= 1/(2\pi)(60 \text{ Hz})(1 \times 10^{-6} \text{ F}) \\ &= 2650\ \Omega \end{aligned}$$

Even if the bulb were replaced by a short circuit, the capacitor would limit the current to

$$\begin{aligned} I' &= 110 \text{ V}/2650\ \Omega \\ &= 0.0415 \text{ A} \end{aligned}$$

and, if the bulb is also in series with the capacitor, the current would surely be even less than 0.0415 A. Since the normal current for the bulb when fully lit is 0.455 A, we conclude that the current

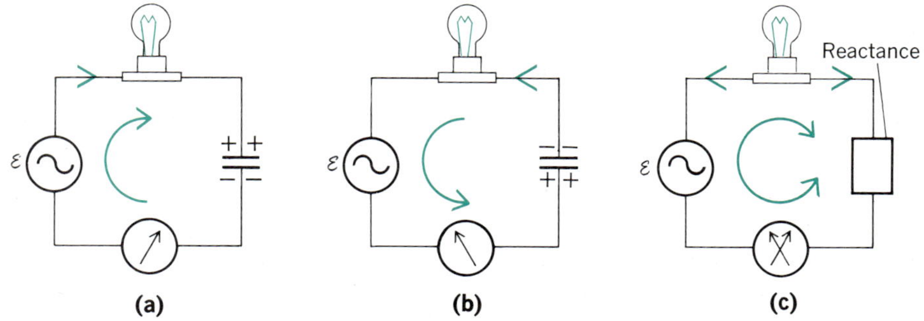

Figure 19-13 **(a)** and **(b)** Flow of charge in an alternating-current circuit containing a capacitor. **(c)** Equivalent circuit, replacing the capacitor by a reactance X_C.

through the capacitor would not be enough to cause the bulb to light even slightly.

In the discussions of both inductive reactance and capacitive reactance it has been pointed out that the definitions of reactance as the ratio of voltage to current, V/I, were only approximate. These definitions are accurate enough for practical purposes so long as the resistances can be neglected, as was illustrated by the examples worked in the text. The expressions for calculating reactances, $X_L = 2\pi fL$ and $X_C = 1/2\pi fC$, are precise and do not involve any approximations.

Now, for a coil with a fairly large resistance, the current is determined by a combination of reactance and resistance called *impedance.* The impedance Z is calculated by the expression

$$Z = \sqrt{R^2 + X_L^2}$$

Ohm's law for a coil carrying alternating current would therefore read

$$I = V/Z$$
$$= V/\sqrt{R^2 + X_L^2}$$

The resistance and capacitive reactance of a capacitor can be combined in a similar manner. The resistance of a capacitor acts as if it were a small resistance in series with the reactance.* Fortunately, this effect in most good capacitors is very small, so that the dissipation of energy in Joule heating can usually be neglected.

If we wish to combine any resistance (not part of a coil or capacitor) in series with the reactance of either a capacitor or a coil, we must be careful to use the special expressions for the impedance of such a combination. These are

$$Z = \sqrt{R^2 + X_L^2}$$
$$Z = \sqrt{R^2 + X_C^2}$$

or, in general,

$$Z = \sqrt{R^2 + (X_L - X_C)^2}$$

* The resistance of a capacitor that you would measure by applying a certain PD V and measuring the current I through the capacitor is $R = V/I$. This "leakage resistance" is usually very large and acts as if it were in parallel with the reactance. Often it is more useful to calculate an effective *series* resistance, which would be zero for an ideal capacitor.

19-5 ELECTRONICS

The vacuum tube was once considered to be the most important of all electronic devices. But this has long since ceased to be true. There are a few special applications where vacuum tubes continue to be used, but they have been almost entirely replaced by *solid-state diodes* and *transistors.* These are called "solid-state" devices because the motion of electric charge within them takes place among the atoms of a solid crystal, rather than in a vacuum. In this section, the diode and the transistor will be viewed in terms of a simple model that has proved to be very useful as a way of interpreting their behavior.

Let's examine first the properties of a solid-state diode. All diodes (and transistors) are made of materials called semiconductors. The term *semiconductor* refers to substances with electrical behavior that falls in between that of the good conductors like copper and silver, and the insulators such as glass and plastics. These semiconductors (the elements germanium and silicon, for example) have moderate numbers of charge carriers available that are free to move and thus form an electric current. In metals, there is just one charge carrier—the negatively charged free electron. Semiconductors, however, may have two types of charge carriers available: these may be free electrons, as in metals, or they may be the equivalent of positively charged ions, called *holes.* For now, let's accept the fact that a hole behaves as if it were a positive charge free to move through the crystal, and therefore can be considered a positive charge carrier. The nature of a hole will be discussed more fully in Chapter 24.

A semiconductor in which the majority of charge carriers are negative electrons is called *n-type*, and one in which the majority of carriers are positive holes is called *p-type*. In Chapter 24 you will see how the controlled addition of small amounts of impurities to a germanium or silicon crystal determines whether it will be *p*-type or *n*-type. It is the existence of two types of charge carriers in semiconductors that brings about the remarkable properties of diodes and transistors.

The device called a *semiconductor junction diode* is useful not only for its own properties but also because such junctions can also be combined to make a transistor. A model of the junction diode is shown in Figure 19-14. In this model the diode is represented by two blocks of semiconductor material (germanium, for example). In the drawing, the *n*-type region is on the left; here the electrons, the principal charge carriers, are free to move and form a current. On the right is a *p*-type region containing a large number of the positively charged "particles" called holes. The word particles is put in quotes to emphasize that, as you will see, holes are not truly particles (such as protons), but can be thought of as being like a positive ion free to move and carry positive charge.

The positive terminal of the battery in the circuit is connected to the *p*-region of the diode and the negative terminal to the *n*-region. Recall that an electric current has been defined as the motion of electric charge, and the direction of the current has been defined as the direction in which *positive* charges effectively move. This means that, when the charge carriers are positive, the current is in the direction of their motion; holes in *p*-type materials move in the direction of the current. In metals and *n*-type semiconductors, where the carriers are negative electrons, the direction of the current is opposite to the direction in which the electrons move.

To visualize the current in the circuit shown in Figure 19-14, first imagine a flow of positive charge out of the positive terminal of the battery; it moves through the connecting wire in the form of a motion of electrons in the opposite direction. This same current in the *p*-region is the motion of positive holes from right to left.

The current must somehow pass across the

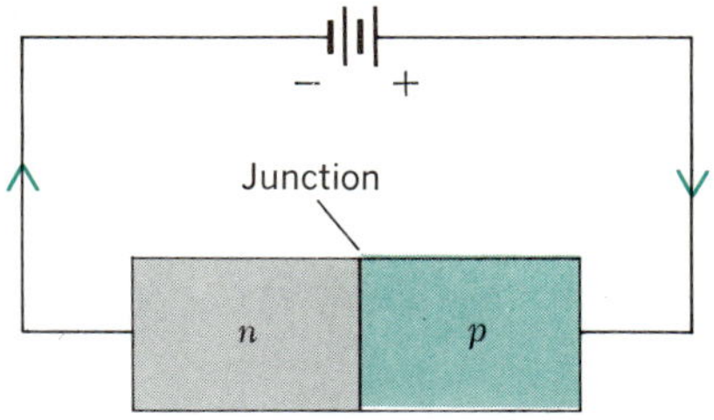

Figure 19-14

junction between the *p*-region and the *n*-region. To understand what happens at this crucially important boundary, it is necessary first to jump across and see what is happening in the *n*-region as holes are moving from right to left in the *p*-region. The direction of the current in the *n*-region must be the same (right to left), but since the charge carriers are negative electrons, their motion must be in the opposite direction—left to right (Figure 19-15).

In this situation, electrons are moving *into* the junction from the left, and holes are moving *into* the junction from the right. Do the electrons continue into the *p*-region, and the holes into the *n*-region? The answer is that they do, but only for short distances. After a short trip, the electrons and holes collide with one another, and as they collide they combine to form neutral particles. The positive and negative charges are neutralized, and the neutral particle cannot act as a charge carrier. However, this motion of charge across the junction does form a current, even though the charges are neutralized. The current can continue indefinitely because the charge carriers neutralized at the junction are continually replaced by the action of the battery.

To see why this is so, recall that electrons in the wire on the right in Figure 19-14 are moving into the battery; this represents a flow of negative charge out of the *p*-region of the diode. Each electron that leaves the *p*-region leaves an excess of one positive charge behind; this excess positive charge forms a hole. Thus the supply of holes is continually being replenished by electrons moving out of the *p*-region.

In the *n*-region electrons are continually being

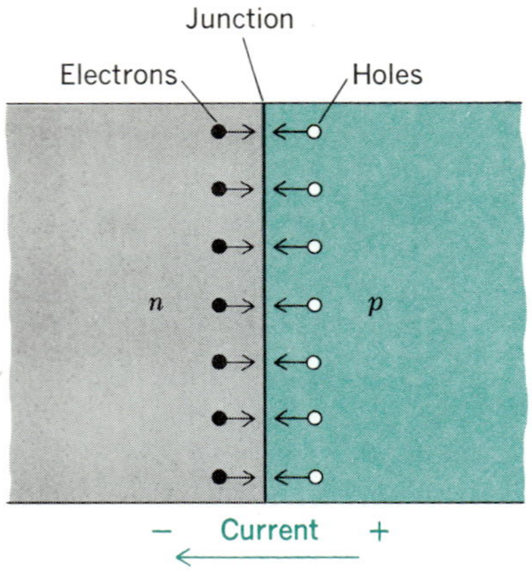

Figure 19-15

neutralized at the junction, but they too are steadily being replaced because the electrons leaving the *p*-region are driven through the battery, along the connecting wire on the left, and back into the *n*-region. As in any circuit, there is a circulation of electrons, driven by the energy (emf) of the battery; the important difference here is the presence of positive charge carriers in the *p*-region.

You may ask, why is this effect of any interest? How is this different from any circuit made with copper wires and resistors, or light bulbs? In its present form, there is essentially no difference, but can you predict what will happen when the terminals of the battery are reversed? The experimental fact, which you can check in the laboratory, is that when the positive terminal of the battery is connected to the *n*-region, and the negative terminal of the battery to the *p*-region, almost no current flows.

In terms of the model used here, there should be a current as electrons move out of the *n*-region into the positive terminal of the battery. On the other side electrons will move from the battery to the *p*-region. Holes will move to the spot where the wire joins the *p*-region, and the electrons and holes will be neutralized at this point. But, will this current continue?

The answer can be found at the junction. Is there a flow of charge across the junction to complete the circuit as there was before? The positive holes in the *p*-region are pulled *away* from the junction, and the electrons in the *n*-region are pulled *away* from the junction; there are no charge carriers left in the junction region, and no current across it. The current through the circuit rapidly stops because the supply of charge carriers is used up. But if you test this effect using a very sensitive ammeter, you will find that the current does not stop entirely. Instead it becomes very small because there are always a few electrons in the *p*-region and a few holes in the *n*-region, enough to keep a tiny current going.

Figure 19-16 shows a graph (called a characteristic curve) of the relation between the current and the voltage in a diode. To the right of the vertical axis, the voltage is positive, meaning that the positive terminal of the battery is connected to the *p*-region; the current is large and this is called the *forward* direction. To the left of the axis the voltage is considered negative because the negative terminal of the battery is connected to the *p*-region; this is the *reverse* direction. You can see that the graph is *nonlinear*, there is not a simple pro-

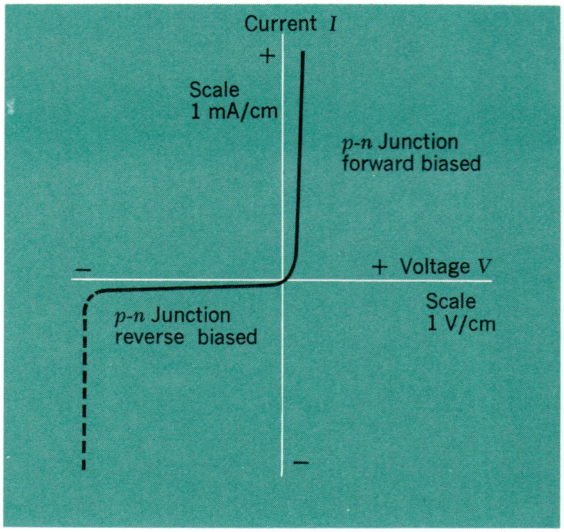

Figure 19-16 Typical voltage vs. current characteristics of a *p-n* junction.

proportionality between current and voltage, and Ohm's law does not apply. The diode is thus a one-way valve, allowing current only if the *p*-region is made positive. Diodes are often used to make direct current when an alternating-current source of potential is available, perhaps from a step-up transformer. Thus, in Figure 19-17 the secondary of the transformer *AB* is a source of alternating PD. During half of the cycle, when *A* is positive relative to *B*, there is current in the direction *AYXDB*. But during the other half of the cycle, when *A* is negative relative to *B*, the current is zero.*

The graph of *I* vs. time in Figure 19-17 shows a pulsating direct current, whose average value is given by the dashed line. If smooth, nonpulsating direct current is desired, a capacitor can be connected in parallel with the load resistor *XY*. The capacitor acts as a reservoir to smooth out variations in current. With such an arrangement, called a filter, the current can be made practically constant. The whole circuit, including the diode and the filter, is called a rectifier circuit, and the process is called *rectification*. The ac generators are simpler and more efficient than dc generators and are used increasingly today. For example, the modern automobile uses a small ac generator (or alternator) as a source of power for charging the battery. The development of high-current, compact rectifiers with solid-state diodes has made this possible.

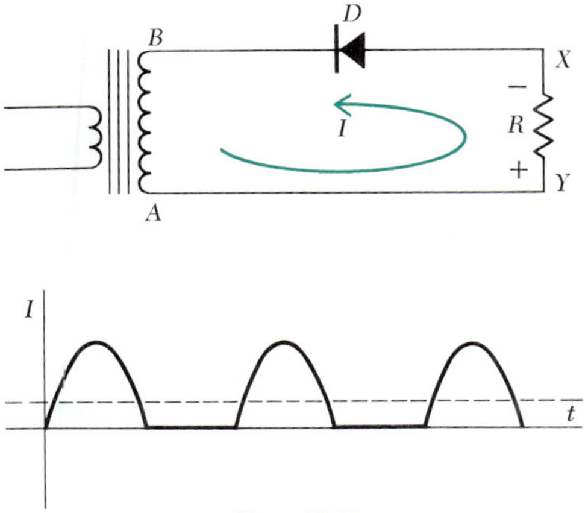

Figure 19-17

* The symbol at *D* in Figure 19-17 is the standard way to represent a diode. The direction of the "arrowhead" is the direction of the positive current.

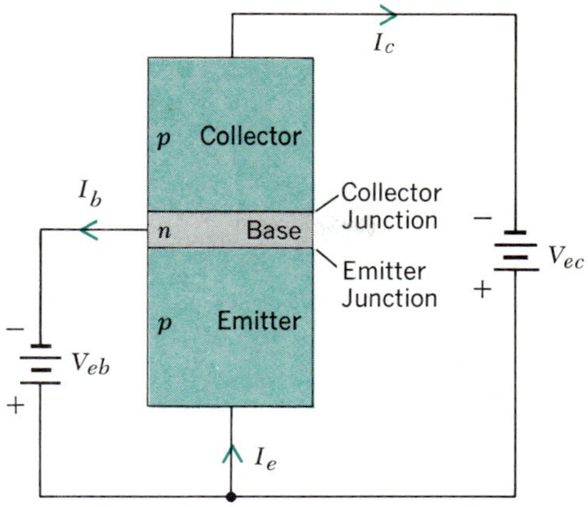

Figure 19-18

A transistor is fundamentally two junction diodes combined in a single package. The construction of a typical transistor is shown in Figure 19-18. In terms of the model of *p*-regions and *n*-regions, a transistor is a "sandwich" with *n* in the middle and *p* on the outside (a *pnp* transistor) as shown; or it can be a sandwich with *p* in the middle (an *npn* transistor). From now on, in this section, the *pnp* transistor will be used as an example, but *npn* transistors are widely used and the explanation is essentially the same.

In Figure 19-18 the batteries are connected to a *pnp* transistor. The middle of the sandwich is called the *base*, and the outer regions the *emitter* and the *collector*. The base region is the most critical one for transistor operation; it must be very thin and of highly uniform thickness. One battery is connected between base and emitter, with voltage V_{eb}; and a second battery between emitter and collector, with voltage V_{ec}.

With the batteries connected as shown, there is a current into the emitter, out of the collector, and back through the battery V_{ec}. The key point in transistor operation is that the base layer is so thin that holes moving through the emitter and across the emitter-base junction can pass right through the base region without being neutralized, and can enter the collector. The holes combine with electrons where the collector meets the wire from the negative terminal of the battery V_{ec}, and thus there is a complete circuit formed by emitter, base, collector, and battery V_{ec}.

The current into the emitter (emitter current I_e) and the current out of the collector (collector current I_c) are about equal and quite large. This is true in spite of two facts that would seem to make this current impossible: (1) there is a *pn* junction between emitter and base with the base battery V_{eb} connected in the forward direction, so that one would expect a large portion of the current entering the emitter I_e to be in the emitter-base circuit; and (2) there is an *np* junction between base and collector, with the voltage V_{ec} in the *reverse* direction, so that one would expect no current across that junction.

The fact is, however, that a large current does flow in the emitter-collector circuit, and this is because the base is so thin that holes can travel right through it and into the collector without being neutralized.

It is true, though, that some holes "emitted" by the emitter do not get across the base and are not "collected" by the collector. A few of them combine with electrons in the base and form a very small emitter-base current I_b. The current in the emitter-base circuit then, contrary to our expectations, is very small. The collector current is therefore only slightly smaller than the emitter current because just a few of the holes are drained off by recombination with electrons in the base.

The model so far has been able to explain how a current is possible in a *pnp* sandwich, but it is not clear how this arrangement is especially useful. The answer lies in the *pn* junction between emitter and base, labeled emitter junction in Figure 19-18. This is a junction diode with the voltage (from the base battery V_{eb}) in the *forward* direction. Since the voltage is forward, a large current would be expected, and this agrees with the fact that the emitter current I_e is large. You have already seen, however, that the large current of holes across this junction does *not* go out through the base but goes directly to the collector.

The operation of the transistor depends on the fact that the small voltage in the emitter-base circuit controls the very large emitter-collector current. Very small changes in base voltage V_{eb} and very small changes in base current I_b bring about large changes in collector current I_c. The transistor can therefore act as a valve in which a small amount of power (from the base battery) controls the flow of a much larger amount of power (from the collector battery).

Figure 19-19 shows a circuit that could be used to demonstrate this property of a transistor. The base current I_b can be changed by means of the variable resistor R_v, and this small change in I_b produces a corresponding large change in collector current I_c. In turn, this produces a large change in voltage across the *load resistor* R_l. This behavior, where a large output arises in response to a small input, is called *amplification*. The circuit shown is an example of an *amplifier*. The voltage output is shown as I_cR_l, because it is directly proportional to the collector current I_c through the resistor. Notice that in this drawing the transistor is represented by the standard schematic symbol; in a *pnp* transistor, the emitter arrow points inward; in an *npn* transistor, the arrow points the other way.

In a practical circuit instead of a variable resistor the input to the base would be a low-power signal (probably in the form of an alternating current), and the output would be an amplified reproduction of the signal. Usually the amplified signal would appear as an alternating voltage across the load resistor. The transistor is

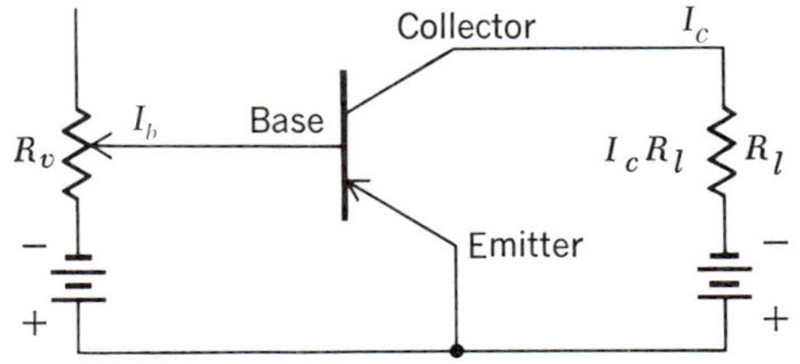

Figure 19-19

(a) (b)

Figure 19-20 A ball check valve.

used in radio and TV as an amplifier of weak message signals that enter the set.

The performance of an electronic amplifier is often described by a quantity called its *voltage gain.* This is a measure of how much the input voltage V_{in} is increased to give the output voltage V_{out}. It is commonly expressed as the ratio V_{out}/V_{in}.

Let's contrast the usefulness of a 20:1 step-up transformer and that of an amplifier whose voltage gain is 20. Both devices magnify the input voltage the same amount. The transformer, if heating losses are negligible, delivers as much power as it receives; certainly no increase of power can be obtained from a transformer. An increase of power would be contrary to the law of conservation of energy. The amplifier, on the other hand, delivers much more power than is received by the base circuit. The ultimate source of the output power, however, is the battery that supplies energy to the collector circuit; the base acts only to *control* the current and the power.

In much the same way, in the hydraulic experiments we used a very small valve to control the flow of water. More dramatically, an engineer at a hydroelectric plant can, by his own unaided muscular efforts, open or shut a valve that may control the flow of many thousands of tons of water.

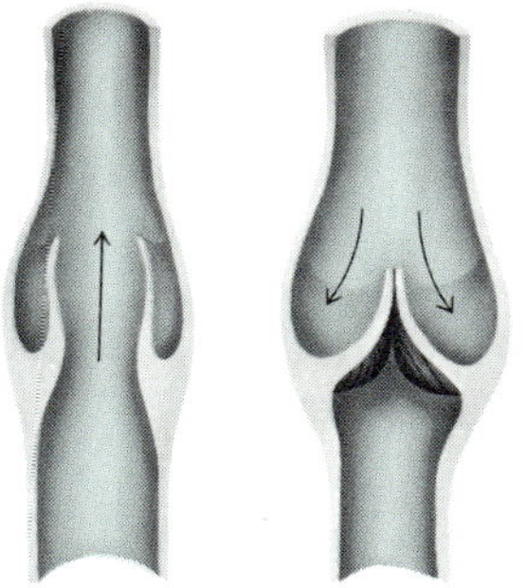

(a) (b)

Figure 19-21 Valve in a vein.

The essence of the operation of both the diode and the transistor is that they are valves. The diode is a valve in the sense that it limits the flow of charge to one direction only. It is quite comparable *in operation* to a ball-check valve in a hydraulic circuit (Figure 19-20). Similar one-way "diodes" are part of the human circulatory system (Figure 19-21). It is also worth noting that any one-way valve, whether it be a hydraulic check valve, a valve in a vein, or a solid-state silicon diode in an automobile battery charging system, is a *nonlinear* device. That is, its output is not related to its input by a simple direct proportion. It is easy to see this if we compare the output of the diode in Figure 19-17 with the graph of the input alternating current as shown in Figure 19-6.

19–6 INTEGRATED CIRCUITS

Probably the most exciting technological development of recent years is that application of solid state physics called the *integrated circuit.* This form of electronic construction, sometimes called *microelectronics,* has brought about major changes in the lives of all of us.

First, what do we mean by an integrated circuit, or IC? Of course, a complete answer would take many pages (or even many books) because making and using ICs has become a large and complex field. The techniques used are based on principles of optics, of chemistry, and of mathematics, as well as the physics of solids—and go far beyond what we will discuss here. However, the essential idea is that an IC is an extremely compact package containing a large number of electronic components.

We have studied electronic components: diodes, transistors, resistors, and capacitors. All of these are available as individual units and we could use

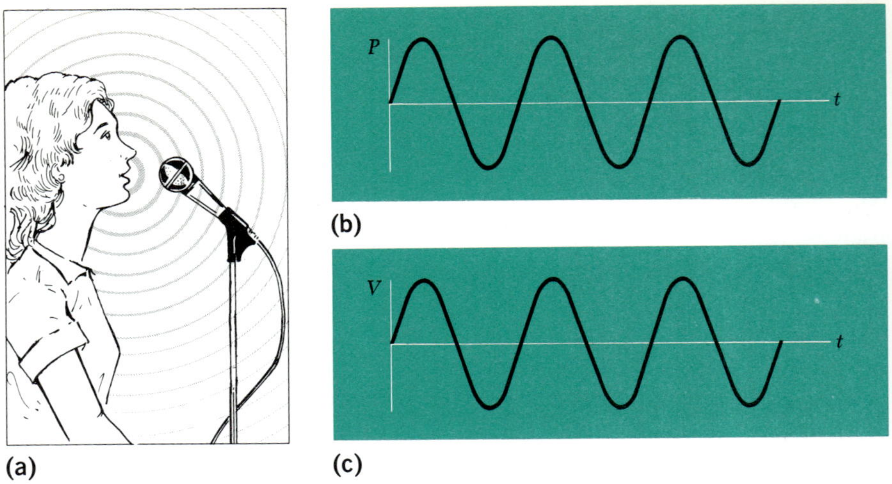

Figure 19-22 A familiar analog signal. **(a)** Singing causes pressure variations in the surrounding air—a sound wave.
(b) Wave form graph of the pressure variations at the microphone.
(c) Analog wave form graph of voltage output of microphone.

them to build amplifiers and other useful devices. An IC amplifier, however, is a "chip" of silicon, often less than a centimeter in length, containing dozens or perhaps even hundreds of thousands of electronic components in the tiny package.

The mere fact of putting so many components in a small space is remarkable enough, but there are still other benefits from this process. First is the fact that ICs can be made at very low cost. An IC for a given job costs much less than the individual components it replaces. Second, the ability to pack vast numbers of components in a small space allows us to make greatly increased use of what is called "digital" circuitry. To understand this, we must discuss briefly the two types of signals used in electronics—*analog* and *digital.*

Even though you may never have called them by that name, analog signals are familiar to us all. In radio, television, telephone, and stereo, the signals are ac voltages that we call analog signals. For example, when we are playing a record through our stereo, the signal is a varying voltage that represents the voice or music. A simple case might look just like the ac voltage shown in Fig. 19-6.

There are two important things to notice about this signal: one, it varies continuously. That is, it can take on any value positive or negative, within a certain range. Second, it is called "analog" because it can model or represent some other physical quantity. (See Section 15-2).

To illustrate what we mean, let's suppose a singer is singing a note near a microphone in a recording studio. The sound takes the form of pressure variations in the air, as described in Figure 9-13 and Section 9-5. This situation is shown in Figure 19-22. The graph in Figure 19-22b is a wave form graph of the pressure variations at the microphone. Figure 19-22c is a wave form graph of the electrical voltage output of the microphone. It is an analog signal because the voltage varies continuously and directly models or represents the pressure variations in the air.

In the studio, this voltage is weak and is therefore fed into an amplifier. This amplifier must be linear because we want the output to be directly proportional to the input (Figure 19-23).

In the past, such an amplifier would be built from a number of separate transistors, resistors, and capacitors (called discrete components). Today it is most likely that the amplifier would be an integrated circuit. On modern circuit diagrams an amplifier is often represented by a triangle; it can be thought of as an arrow pointing in the direction of signal flow (Figure 19-24).

A widely used type of IC amplifier is called an *operational amplifier* ("op amp" for short). This usually contains fifteen to thirty transistors and other components. It is customary, however, to regard an IC as a black box, and think of it only in terms of the relation between its input, V_{in}, and output, V_{out}.

Op amps have many uses: a musical instrument

amplifier or a record player might use an op amp to build up the weak voltage from the instrument or turntable pickup. Op amps are also still used for what is called "analog computing." That is, they can add, subtract, and perform other more complicated operations on voltages representing physical quantities.

Of even more interest is that type of IC known as a *digital* integrated circuit. Digital circuits (sometimes called "logic" because of their relation to logical reasoning) operate in very different ways from their analog counterparts.

First, let's contrast a digital signal with an analog one. We have seen that analog means the signal can take on any value within a given range, and that it directly models some other quantity. A digital signal, however, has only two values. The simplest form of digital device is just a switch: the familiar wall switch has just two values (or states): "on" or "off."

Digital devices (such as switches) are certainly nothing new. The telephone and electric power industries have studied their properties for many years while improving switchboard capabilities and switching systems. What is new and dramatic today is the possibility of putting enormous numbers of digital circuit units in the small IC package.

An interesting and down-to-earth illustration of the power of digital electronics is the pocket calculator. Just a few years ago, engineers used the slide rule as an aid to calculation. It was a simple, mechanical analog calculator. It was analog because positions of the moving scales were used to

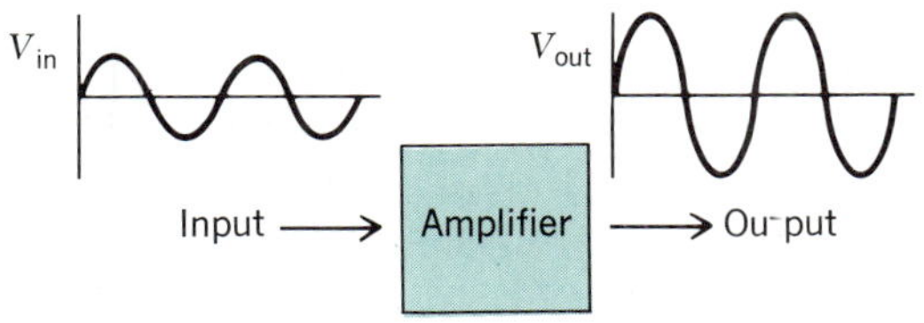

Figure 19-23 The output of a linear amplifier is directly proportional to the input. The wave form is the same, but the amplitude is greater.

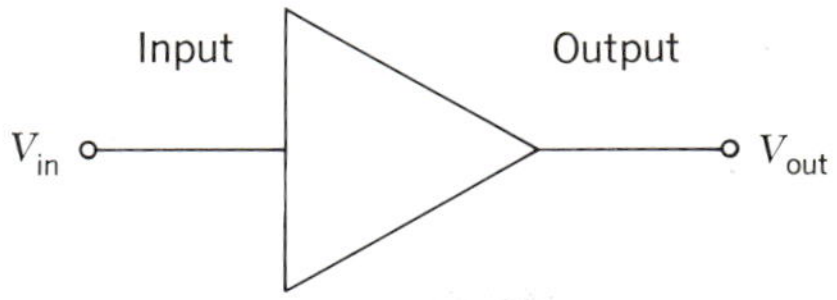

Figure 19-24 The schematic symbol for an amplifier

model or represent numbers. The slide rule could perform only a limited number of operations, was not very accurate, and required considerable skill.

The pocket digital calculator is made possible by an integrated circuit containing logic circuits permanently arranged to carry out the mathematical operations. Digital calculators have been in use for many years. However, the early mechanical ones (using wheels and gears) were large and expensive. The vacuum tube digital electronic calculators of the 1940's were also very large and extremely expensive. The IC has put all the power of these enormous machines on a miniscule chip at modest cost.

We have said that digital implies a signal with only two values—such as a switch which can be only on or off. But this seems to have only limited usefulness. How are such digital devices put together to make meaningful circuits? The basic element of digital electronics is the circuit called a "gate." It is closely allied to what is called a logical "operation." It is not our aim to present even a brief introduction to logic, but at least a nodding acquaintance with the concepts is needed to appreciate the nature of the "microelectronics revolution."

As an example, consider the basic logic operation called "AND." (The names are always written in capitals to distinguish them from the ordinary use of the word.) The concept of AND can be represented by two switches in series, as shown in Figure 19-25. These two switches form what is called an "AND gate." In words you can think of it as expressing the sentence: "The light is *on* if switch 1 AND switch 2 are *on*." AND gates are commonly available in IC form and are used as switches—perhaps to turn some circuit on and off.

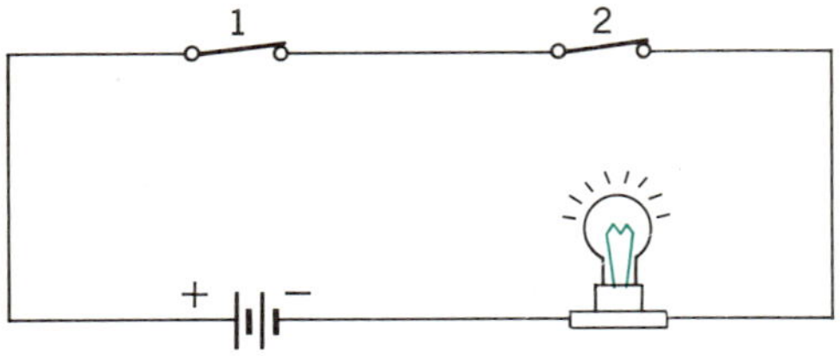

Figure 19-25 A switch model of an AND gate. The light is on if and only if switch 1 is on AND switch 2 is on.

These basic logic operations still seem very remote from a device like a calculator. Why do we discuss them? The answer is that thousands of such gates are buried deep within the black box that is a calculator IC. Although in practice we do not have to think about them, they are the "atoms and molecules" that form the IC.

Analog amplifiers and analog computers are relatively easy to understand. The physical quantities we wish to work with are represented directly by electrical quantities. For example, a speed of 10 m/sec could be directly represented by 10 volts. However, with digital devices that can have only two values, there is no such direct analogy, or modeling. Therefore, if we wish to feed a number such as ten into a digital calculator, we must code "ten" in such a way that the digital device can "understand" it.

The most straightforward code for this purpose is the binary number system. You are all familiar with the fact that our common decimal numbers are written in the base ten. That is, the columns of digits are used to express a number in powers of ten—units, tens, hundreds, thousands, etc. Since digital devices have only two states, numbers for their consumption must be written in terms of powers of *two*—the binary number system. Again, our aim is not to describe fully the system of binary notation. We wish only to give you some insight into what happens when you push the buttons and a display appears on your calculator or other digital device. A calculator, a clock, or a counter is digital not because it has a lighted display, but because its inner logic is working with numbers expressed in digital form.

Let us take a brief look at how one could code the decimal number ten in digital form. The number is, of course, written in its familiar way in decimal notation:

10

which means one ten (10^1) and no units (no 10^0).*

In terms of powers of two, decimal ten is: one eight (2^3), no fours (no 2^2), one two (2^1), and no units (no 2^0). To convert this to binary notation, keep in mind that there are only two digits in the binary system, zero and one—0 for the absence of a power of two, 1 for the presence of a power of two. Thus, decimal ten, expressed in powers of two as above, is written:

1 0 1 0

Each place signifies the absence (0) or presence (1) of a power of two. Reading from left to right; 1 for 2^3, 0 for no 2^2, 1 for 2^1, 0 for no 2^0. The number 1 0 1 0 can then be represented by four digital devices (perhaps by four switches), two turned on and two turned off.

Why are we discussing the possible ways for coding decimal ten? Because we learn something important about the behavior of the calculator keyboard: when you press the keys 1 and 0 to feed in decimal ten, the signal fed in is not 1 0, but 1 0 1 0. This number is represented in the IC by the states of four binary devices. Incidentally, each one of these 1's or 0's is called a binary digit—often shortened to "bit." As you can see, it takes four bits to represent the decimal number ten.

Perhaps you want to store the decimal number 19 in the memory of your calculator. All you need to do is press the keys "1" and "9" and the proper memory key. What is a memory, and in what form is the number stored? Basically, a memory is a circuit containing many devices—each one of which can be either on or off. In crudest form, a memory could be an array of switches, each either open or closed, and representing the binary digits 0 or 1.

* Recall that any number with an exponent of 0 is equal to 1.

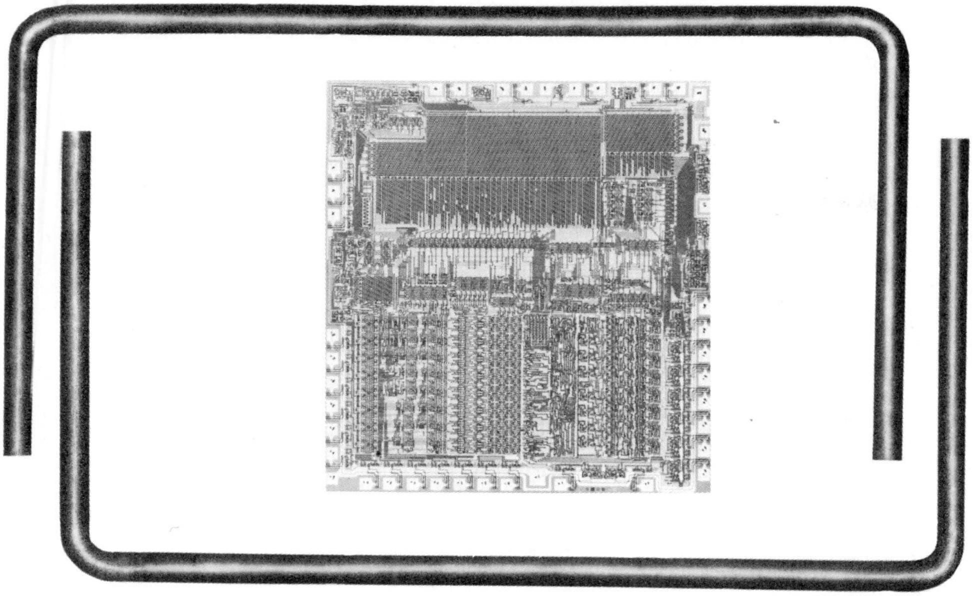

Figure 19-26 This tiny microprocessor chip contains over 7,000 transistors and is as powerful as the processing "brain" of a small computer. Its size can be seen by comparison to the two staples that surround it.

Any number, such as decimal nineteen, must first be coded into digital form. One way would be to write it as we did before: one sixteen (2^4), no eights, no fours, one two (2^1), and one unit (2^0). Written in binary digits, this becomes:

1 0 0 1 1

In practice, numbers are usually coded in a slightly different way, but all the various codes are variations of the binary number system. Digital circuits cannot "digest" any digits but zero and one.

A typical small memory (a "scratch pad" memory) that could be used in your calculator might be able to handle 64 binary digits—bits. It would probably be an IC containing 64 transistors acting as switches. There would, of course, be other parts to the circuit—some gates for example, to route the signals through the chip. By the way, a common unit of a simple memory has the graphic name of "flip-flop." It can either flip into one state—*on;* or flop into the other—*off.*

The concept of a memory as an array of digital devices for storing values of the binary digits (0 and 1) will probably be with us for the foreseeable future. Scientists and engineers are working to package ever more memory elements in a single chip. A small, inexpensive calculator may have a 64-bit memory. You can easily buy a chip for a 1000-bit memory for a dollar or so. Memories to store hundreds of thousands of bits are commonplace. Large memories are a vital part of digital electronics. It is easy to see why compact packaging and low cost are an essential part of their design.

Finally we should ask: how are calculations carried out in a calculator or in a small computer? One answer to that question lies in a complex kind of IC called a *microprocessor* chip (Figure 19-26). A microprocessor must be carefully designed to insure that all the gates and other components are arranged properly. However, for most scientists the main concern is not the gates, but the function or behavior of the device as a whole. As users, you and I can ignore the "atoms and molecules" that make up the device. We worry only about the simple electrical connections we must make to have the microprocessor carry out the operations we want.

The microprocessor (abbreviated μP) is a comparatively large (several centimeters) chip. Because of its size and complexity, this technology is called LSI for "large-scale integration." In essence, the chip provides ways of feeding numbers (data) in and taking numbers (results) out. The central part of the microprocessor (called the arithmetic logic unit) does the calculations we require.

These LSI devices provide so much computing power in such a small volume and at such low cost that the age of the personal computer has essentially arrived. Not only is the microprocessor an exciting example of applied physics, but it has opened up powerful new techniques for measurement and calculation by scientists in all fields. One example that you may encounter in the laboratory is the digital voltmeter. It performs the extremely useful function of responding to an analog quantity (voltage), converting it to digital form (analog to digital conversion), and finally displaying it on a digital readout.

SUMMARY

The **sensitivity** of a galvanometer depends upon several design factors, and the design problem is one of compromise aided by study of the working equation for the instrument. Torque is the product of force times lever arm.

Alternating currents are produced by a generator or alternator. The chief advantage of alternating currents is that **transformers** can be used to increase or decrease the emf in a circuit. No power is gained, however, and in an ideal transformer the product of emf and current is the same in the primary as in the secondary.

Coils and capacitors behave differently when in ac circuits than when they are in dc circuits. **Reactance** is approximately defined as the ratio of voltage to current for an ac circuit. The reactance of a coil is due to the back emf induced in its windings by its own changing current. The reactance of a capacitor is infinite for direct current, since there can be no steady current through a capacitor. Alternating current can flow through a circuit containing a capacitor, since the plates change in potential as they are charged and discharged at a frequency of many times per second. The reactance of a coil goes up as the frequency increases, while the reactance of a capacitor goes down as the frequency increases.

The reactance of a coil of **inductance** L at frequency f is given by

$$X_L = 2\pi f L$$

The reactance of a capacitor of capacitance C at frequency f is

$$X_C = \frac{1}{2\pi f C}$$

When resistance and reactance of two circuit components in series are combined, the result is the **impedance** Z:

$$Z = \sqrt{R^2 + X^2}$$

In modern electronics, solid state **diodes** have many applications. One important use is to change alternating current into pulsating direct current. A **filter capacitor** can then be used to smooth out the pulses. **Transistors** are used for amplification and switching, and have made possible a large number of electronic devices that are compact and require little power.

All **solid state** devices use the kind of material called a **semiconductor,** having two kinds of **charge-carrier**—negative electrons and positive **holes.** The useful rectifying action of a solid state diode occurs at the junction between a **p-type** and an **n-type** semiconductor; in the "reverse direction" there will be a shortage of carriers at the junction. A transistor is a sandwich of two junction diodes in which a small current or voltage in the base circuit controls the flow of current in the **emitter-collector** circuit.

Integrated circuits are "chips" containing large numbers of transistors and other components in a small package. **Analog** ICs are used for amplification and other linear functions. **Digital** ICs are the basis of calculators, digital voltmeters, computers, and many other everyday applications.

QUESTIONS

1 The output of a certain generator is an alternating current. Describe how this can be converted to pulsating direct current.

2 Why has alternating current been so widely adopted in place of the earlier direct-current system prevalent some 75 years ago?

3 What can you say about the construction of a transformer for which the secondary emf equals the primary emf?

4 What would happen if the primary of the transformer were supplied with 2300 V dc instead of 2300 V ac?

5 Water flows over a dam, electric energy is generated, energy is transmitted over a line and through various step-up and step-down transformers, and eventually an elevator rises in an office building. Did any *mass* flow through the transformers? Explain.

6 Can a diode be used for amplifying a weak signal?

7 Using a diagram, show why a germanium diode does not conduct a current in the "reverse" direction.

8 Modern automobile electrical systems use ac generators or alternators, rather than dc generators. Why is a large solid-state diode now an essential part of such a system?

9 Compare the behavior of inductive reactance in an ac circuit with the behavior of capacitive reactance when the ac frequency is increased.

10 What kind of input "signal" in the base circuit of a transistor controls the output, that is, the collector current?

11 Discuss the statement: "A transistor amplifier does not violate the law of conservation of energy."

12 Give two reasons why an engineer designing a computer would prefer to use integrated circuits rather than discrete components.

PROBLEMS

GROUP A

1 A certain galvanometer needle deflects 30° when the current through the coil is 6×10^{-6} A. What deflection would be produced by the same current if the magnetic field were decreased to $\frac{2}{3}$ of its original value?

2 A generator furnishes an emf of 48 V when its armature turns at 400 rev/min. All other factors remaining the same, what will the emf be if the armature turns at 500 rev/min?

3 A model-train transformer has a 600-turn primary and a 100-turn secondary. When the primary is connected to a 60 Hz, 120 V line, what is the secondary emf?

4 The transformer of Problem 3 delivers 9 A to the train's motor. What is the current from the 120 V supply circuit?

5 A coil of negligible resistance and with $X_L = 300\ \Omega$ at the ac frequency used is connected in series with a resistor of 100 Ω. What is the impedance of this combination?

6 The coil and resistor combination of Problem 5 is now connected in series with a capacitor that has $X_C = 300\ \Omega$ at the given frequency. (a) Calculate the impedance of this combination of three elements. (b) How would the impedance change if the frequency were raised? or lowered?

7 Refer to the graph in Figure 19-16 and (a) find the current through the junction diode when the forward voltage is 0.2 V. Note that the scale of the vertical axis is 1 mA/cm and of the horizontal axis 1 V/cm. (b) Calculate the resistance of the diode at this voltage. (c) From the graph estimate the current when the voltage is 2 V in the reverse direction. (d) Calculate the resistance of the diode under these conditions.

GROUP B

8 Two flat coils of wire are lying on a table, with the small one (the "secondary") completely inside the larger coil (the "primary") (Figure 19-27). (a) If the switch is closed, in which direction, clockwise or counter-

clockwise, is the current through the primary? (b) In which direction will be the momentary induced current in the secondary, just after the switch is closed?

9 If the switch in Figure 19-27 has been closed for some time and then is opened, what will be the direction of the momentary current in the secondary?

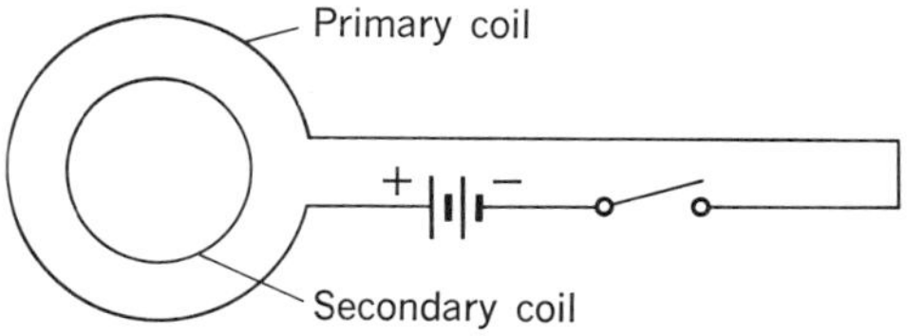

Figure 19-27

10 A transmission line between a city and a suburb consists of two wires, each of resistance 0.5 Ω; the current is 1000 A; the PD at the input end of the line is 2 kilovolts. Calculate: (a) the power input; (b) the power loss in the line; (c) the power delivered to the load; (d) the efficiency of the line.

11 Repeat Problem 10, for the same input power to the same line, but with the PD at the input end of the line 20 kilovolts instead of 2 kilovolts. (*Hint:* First find the current.)

12 An integrated circuit "op amp" is used to amplify the voltage from a microphone. The signal from the microphone is a sine wave form of amplitude 45 mV. The output of the op amp has an amplitude of 2.0 V. What is the voltage gain? The output signal has more power than the input to the amplifier. Explain the origin of this "extra" power.

13 What is the reactance of a coil of self-inductance 20 millihenries, at a frequency of 10 kHz, assuming the coil to have a negligible dc resistance?

14 Calculate the frequency at which a capacitor of value 0.02 μF has a reactance of 10,000 Ω.

15 A 110 V bulb rated at 5 W is connected to a 110 V, 60 Hz source through a 40 μF capacitor. Show that the bulb is lit to practically full brilliance. (*Hint:* Compare the resistance of the bulb and the reactance of the capacitor.)

16 Figure 19-28 shows how the collector current I_C of a certain transistor varies with the collector voltage V_C when different base currents I_B are used. (a) Assume that $I_B = 25$ μA; by how much does I_C change when V_C is changed from 0 V to 5 V? (b) Now assume that V_C is held approximately constant at 5 V and that I_B is changed from 25 to 50 μA; by how much does I_C change in response to this small change in base current? (c) Cal-

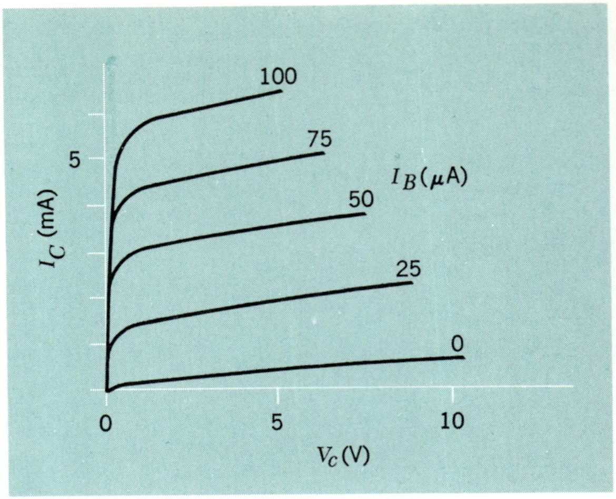

Figure 19-28

culate the current gain, that is the ratio

$$\frac{\text{change in } I_C}{\text{change in } I_B}$$

The value you obtain is typical of the current amplification of a transistor connected as in Figure 19-18. Note that a variation in I_B has a very large effect, whereas the variation in V_C (or V_{ec} in the figure) had a very small effect.

17 In the laboratory a student connects a 90 V battery to the terminals of a capacitor and measures the "leakage current" I_L through the capacitor to be 9.0×10^{-6} A. (a) What is the "leakage resistance" of the capacitor? The capacitor has a capacitance of 0.5×10^{-6} F. (b) What reactance X_C does it offer to a high audio frequency of 20 kHz? (c) Suppose now an ac voltage of 1 V is applied across the terminals of the capacitor; what alternating current would flow in the circuit at this frequency? (d) Using the relation $P = V^2/R$ calculate the power dissipated in the leakage resistance due to the PD of 1 V. (e) Calculate the effective *series* resistance of the capacitor approximately by using the relation $P = I^2R$ to calculate the resistance in series with the capacitor that would dissipate the same power you obtained in (d) if the current calculated in (c) were flowing through it.

The value obtained in (e) is extremely small and shows why the resistance of a capacitor can usually be ignored.

20 Electromagnetic Radiation

The concept of energy transfer by radiation through space was first introduced in Chapter 11 where it was stated that radiation is electromagnetic in nature, and that an electromagnetic wave transmits the energy. Electromagnetic waves arise whenever electric charges are accelerated. Unlike sound waves and water waves, electromagnetic waves require no physical medium for transmission. Table 20-1 on page 346 summarizes the properties of electromagnetic (e-m) waves over a wide range of frequencies and wave lengths.

20-1 THE RADIATION SPECTRUM

The type of electromagnetic radiation known as "light" is, of course, familiar to all of us, but the identification of light as an electromagnetic wave is a comparatively recent development. The wave nature of light has been known only since about 1800, and barely one century has elapsed since Maxwell proved (in 1864) that e-m waves are a necessary consequence of the laws of electricity and magnetism. As it was pointed out in Chapter 11, not all electromagnetic radiation is visible; the entire electromagnetic spectrum ranges from low-frequency radio waves to the high-frequency gamma rays of nuclear physics.

How shall we describe e-m radiation? At the outset, you should be warned that the description will be a strange one, full of seeming contradictions. It is known that energy can be transferred through a distance by e-m radiation—the earth is warmed by infrared radiation from the sun, a photographic film is "exposed" when it absorbs light energy, radio waves carry enough energy to control the current in a television receiver. And it is not hard to think of several other illustrations of energy transfer by e-m radiations such as ultraviolet radiation, x rays, or gamma rays. Certainly, our description (or "model") of e-m radiation must include a means by which energy can be transported from place to place.

One simple model for radiation would be a stream of particles, or corpuscles; this view was held in the earliest times. By the time of Newton people realized that energy can also be transmitted by means of waves, as with the pounding of surf upon the shore and the transmission of audible speech by sound waves. But what about light? Is it transported by means of corpuscles? or by waves? To decide between a corpuscular model and a wave model of light (and other e-m radiation), we must look for experimental results that can be described by one model but not the other.

Here we come upon one of the most baffling

Table 20-1

Electromagnetic Radiations

Type of radiation	*Usual source*	*Usual detector*	*Approximate frequency range, (Hz)*	*Approximate wavelength range, (meters)**
Radio waves	electric circuits	electric circuits		
lowest frequency used in communications			10^4	3×10^4
AM broadcast band			0.55×10^6–1.60×10^6	545–188
FM broadcast band			88×10^6–108×10^6	3.40–2.78
TV broadcast bands			54×10^6–890×10^6	5.55–0.34
microwaves and radar			10^9–10^{11}	3×10^{-1}–3×10^{-3}
highest experimental frequency			10^{12}	3×10^{-4}
Infrared radiation (heat)	hot bodies	thermocouples thermometers skin	10^{11}–3.8×10^{14}	3×10^{-3}–8×10^{-7}
Visible light	hot bodies electric arcs	eye photocell photographic plate	3.8×10^{14}–7.5×10^{14}	8×10^{-7}–4×10^{-7}
Ultraviolet radiation	electric arcs	photographic plate photocell Geiger counter	7.5×10^{14}–3×10^{17}	4×10^{-7}–10^{-9}
X rays	impact of electrons on targets	photographic plate Geiger counter ionization chamber	3×10^{17}–3×10^{19}	10^{-9}–10^{-11}
Gamma rays	radioactive nuclei	Geiger counter ionization chamber scintillation counter	$>3 \times 10^{19}$	$<10^{-11}$

Note: The boundaries between the various types of electromagnetic radiation are not sharp; for instance, radiation in the neighborhood of 3×10^{17} Hz might equally well be called extremely high-frequency ultraviolet radiation or extremely low-frequency x rays. Likewise, the range from 10^{11} to 10^{12} Hz can be produced either by electrical means (radio) or by thermal means (infrared sources).

* Wavelength = velocity/frequency. Wavelengths, when short, are often expressed in angstrom units, micrometers, or nanometers (or millimicrons): 1 angstrom (1 Å) = 10^{-10} m; 1 micrometer (1 μm) = 10^{-6} m; 1 nanometer (1 nm) = 1 millimicron (1 mμ) = 10^{-9} m.

chapters in the history of physics. Some experimental results confirm the corpuscular model, and seemingly cannot be explained by the wave model; other results are equally definite in causing us to reject the corpuscular model in favor of the wave model. In the end, we find that scientists have come to accept a new model that combines features of both. It would be meaningless to describe this dual model of radiation at this point; first you must learn about the wave model and the corpuscular model separately before attempting to fuse them into a single acceptable model.

Chapters 20 through 23 introduce the aspects of radiation that are best described by a wave model, chiefly the phenomena associated with the propagation of light, including reflection, refraction, interference, and diffraction. Chapters 24 through 26 deal with the aspects that are best described by a corpuscular model, chiefly the phenomena associated with the emission and absorption of radiation.

In speaking of a physical model, it is almost meaningless to say that the model is either "correct" or "incorrect." One must judge a model by how well it meets two standards. First, is the model based upon the best experimental evidence that we have? Do the features of the model take into account all known observations about the phenomenon? Does the model give a unified picture that correlates, or ties together, all the facts? Second, and sometimes more important, a good model should allow us to predict what *would* happen in a new situation that has not been investigated before. If the experimental results agree with the prediction of the model, then we have one piece of strong evidence for the "correctness," or usefulness, of the model. We are not going to use either the corpuscular model or the wave model unless it is "correct enough" to allow us to correlate observations and predict the results of new experiments.

20-2 ELECTRIC OSCILLATIONS

If energy is transmitted by some sort of electric wave, we need to find a type of electric oscillation that is analogous to a mechanical vibration that sends out sound waves. We have already learned a great deal by exploring the analogies between electric circuits and mechanical systems. Let's look at the characteristics of a vibrating body—say, an object of mass m on the end of a spring whose force constant is k (Figure 20-1). There is a continual interchange of energy from one form to another. Energy within the system is sometimes in the form of KE, and sometimes PE. In Section 8-2 we derived a formula for the period of oscillation T of the mass. The result was

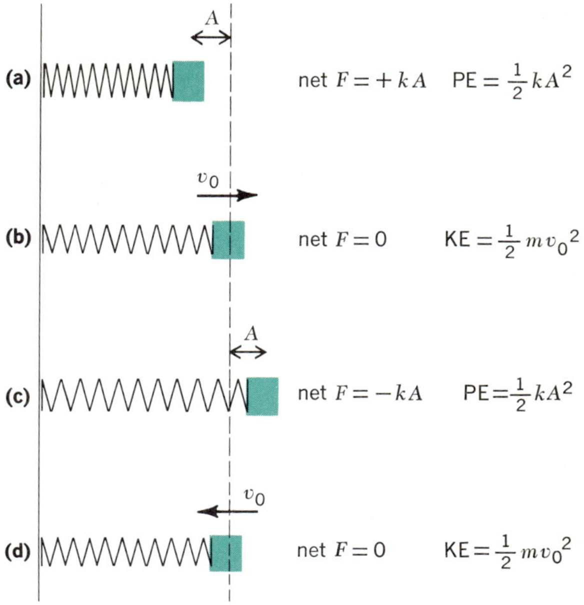

Figure 20-1

$T = 2\pi\sqrt{m/k}$. During this time, the oscillating body goes through 1 cycle of its vibration. The period depends on the stiffness of the spring (a factor related to the force constant k, and hence to the PE) and on the inertia of the body (a factor related to the mass m, and hence to the KE). Such a vibrating mass has a *natural frequency,* and can set a medium such as air or water into motion, thus causing waves to form. The frequency of the waves will be the frequency of the vibrating, or oscillating, body. Since frequency is the reciprocal of the period, we can write

$$f = \frac{1}{2\pi}\sqrt{\frac{k}{m}}$$

In searching for an electrical analogy to mechanical oscillation, we recall that electric energy in the form of PE can be stored in a capacitor. The electrical analogue of kinetic energy can be found among magnetic effects: KE is energy due to motion, and magnetism is caused by the motion of charges.

The energy in the magnetic field of a current-carrying coil can be likened to KE; the situation is similar to the energy of a moving body such as an automobile. Disregarding friction, once the car gets up to speed, no further work needs to be done, and the KE remains constant. But while it is being accelerated, there is a continual flow of energy from the fuel, and work is done. Similarly, in a current-carrying coil there is a flow of energy when the current, and thus the magnetic field, are building up. Disregarding resistance, when the current and magnetic field reach their steady value, no further work is done and a constant amount of energy remains in the magnetic field.

We are now ready to consider electric oscillations. The essence of an oscillation is the exchange of energy from KE to PE and vice versa. How this exchange takes place is shown in the circuit of Figure 20-2. Suppose the capacitor is initially charged (a), and then the switch is closed. Charge starts to flow through the coil (b), and some of the PE stored in the capacitor becomes KE as the magnetic field of the coil builds up. In (c), the capacitor is empty of charge, but there is current in the coil, and all the energy of the system is in the form of KE. The process doesn't stop, however. The charges move on through the coil and begin to accumulate (d) on the capacitor. Eventually (e) the capacitor is fully charged again (opposite to its original condition), and all the energy is once more stored as PE, since the current is momentarily zero. The process cannot go on indefinitely however. The wires, of course, have *some* small resistance, and much of the energy leaks away in Joule heating, causing the oscillation to die out eventually.

It is interesting to note, point-by-point, how similar this process is to the mechanical oscillation of a mass on a spring (Section 8-1). There, too, the oscillation dies out as work is done against internal friction in the spring and against the external air resistance. The mechanical energy is gradually converted into thermal energy, just as with the electric oscillation.

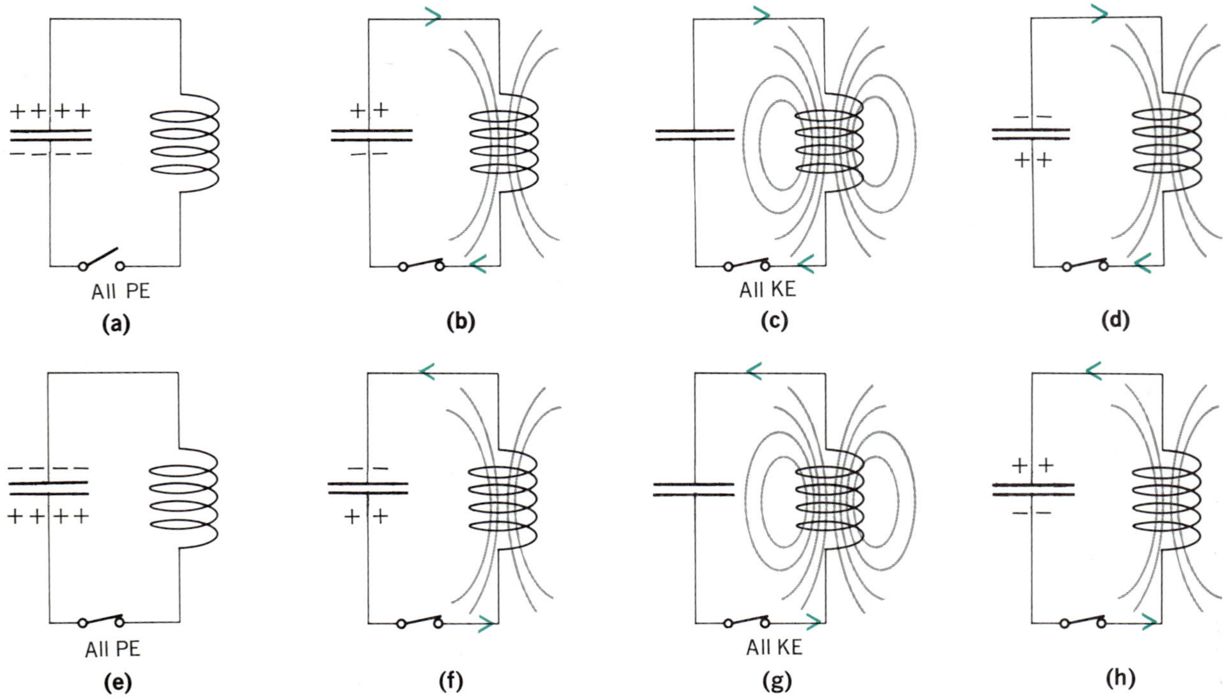

Figure 20-2 Energy transformations in an LC circuit.

We can use the analogy between mechanical and electric oscillations to help us write down a formula for the frequency of an electric oscillation. The capacitance C of a capacitor measures how "easy" it is to store charge on the plates, and thus is analogous to the *inverse* of the stiffness k of a spring. We can, therefore, replace k by $1/C$. The inductance L (Section 19-3) is directly analogous to the inertia (or mass), and therefore we can replace m by L. Hence, the formula

$$f = \frac{1}{2\pi}\sqrt{\frac{k}{m}}$$

becomes

$$f = \frac{1}{2\pi}\sqrt{\frac{1/C}{L}}$$

Simplifying, we obtain

$$f = \frac{1}{2\pi}\sqrt{\frac{1}{LC}}$$

This is the natural frequency* of the simple electric circuit of Figure 20-2 (often called an LC circuit), that is, the frequency at which it would oscillate if a certain amount of energy were quickly fed in and then the circuit was left alone.

Example 20-1 Calculate the natural frequency of a coil and capacitor (LC) circuit in which $L = 3.0$ millihenries (mH) and $C = 50$ picofarads (pF).

The natural frequency of the circuit is given by the expression

$$f = \frac{1}{2\pi}\sqrt{\frac{1}{LC}}$$

$$= \frac{1}{6.3}\sqrt{\frac{1}{3.0 \times 10^{-3}\text{ H} \times 5.0 \times 10^{-11}\text{ F}}}$$

$$= 6.2 \times 10^{5}\text{ cycles/sec}$$

$$= 6.2 \times 10^{5}\text{ Hz}$$

$$= \boxed{620\text{ kHz}}$$

* Units of cycles/sec, or Hz, are also units of $\sqrt{\frac{1}{LC}}$ (though the algebraic derivation is beyond the scope of this book).

20-3 RADIATION

Electric oscillations are similar to mechanical oscillations in an important respect: A mechanical oscillator in air, water, or other medium loses energy not only by friction but also by radiation of sound; the electric oscillator loses energy not only because of Joule heating but also by radiation of e-m waves. It loses energy in this way because charges are continually being accelerated within the coil and capacitor, and these motions of charges give rise to continually changing electric and magnetic fields. It is the close interaction between electric fields and magnetic fields that is responsible for the e-m waves (radio waves, light waves, x rays, etc.) that are broadcast by an oscillator.

We have seen that an induced emf appears when a magnetic field changes; for example, the induced emf in a generator or a transformer is caused by a changing magnetic field as magnetic lines of force "cut across" a conductor (Section 18-7). Now when an emf exists between two points in a wire, or in empty space, there must be an electric field (force per unit charge) in the region between the points. Thus, when a wire is connected to the two terminals of a dry cell, the electric field set up in the wire accelerates the free electrons in the wire and causes a current. Even if there is no wire between the terminals the electric field can be detected in the space between them. Thus an emf implies an electric field. The law of magnetic induction (Section 18-7) can be restated, then, by saying that *a changing magnetic field causes an induced electric field.* Maxwell showed that the opposite is also true: *A changing electric field causes an induced magnetic field.*

This symmetrical relationship between electric and magnetic fields provides the interaction needed to propagate a wave. When an e-m wave is passing through some point in space, both the electric field and the magnetic field at that point are changing. Maxwell showed that the electric field E and the magnetic field B fluctuate. E and B are both zero at the same time, and they reverse direction (together) twice each cycle. Another

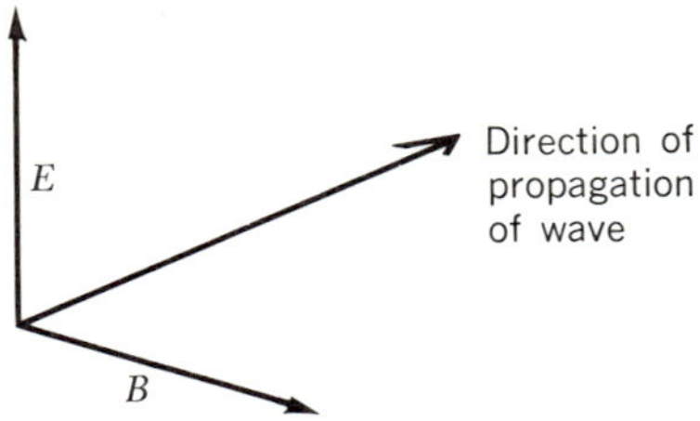

Figure 20-3 The electric field E of an e-m wave is perpendicular to the magnetic field B, and here we see that both are perpendicular to the direction of propagation of the wave.

prediction of the Maxwell theory is that E and B are perpendicular to each other and that both are perpendicular to the direction of propagation of the wave (Figure 20-3).

It is a remarkable fact that the velocity of the wave (in a vacuum) is given by

$$\sqrt{\frac{k}{k'/2}}$$

where k is the constant in Coulomb's law for charges (Section 14-4), and k' is the constant in the law of magnetic force (Section 18-4). One of the early triumphs of the Maxwell theory was predicting the numerical value of the velocity of electromagnetic waves:

$$c = \sqrt{\frac{k}{k'/2}} = \sqrt{\frac{9.0 \times 10^9 \text{ N}\cdot\text{m}^2/\text{C}^2}{10^{-7}\text{ N/A}^2}}$$
$$= \sqrt{9.0 \times 10^{16}\text{ m}^2\cdot\text{A}^2/\text{C}^2}$$
$$= 3.0 \times 10^8\text{ m/sec}$$

Thus the constant $\sqrt{\frac{k}{k'/2}}$, an electrical quantity that can be measured by electrical means, works out to be equal to c, the speed of light in a vacuum (Section 5-6). This is strong evidence for the fact that light waves are *electromagnetic* waves; the evidence becomes even stronger when we realize that all matter, including sources of light, is composed of electric charges that might be expected to oscillate and hence broadcast e-m waves. But before investigating the production of an e-m wave, let's first take a more detailed look at the description of an e-m wave.

20-4 DESCRIPTION OF AN ELECTROMAGNETIC WAVE

An e-m wave can be described in much the same way that a water wave was described in Section 9-4. We will use the technique of taking a series of "instantaneous snapshots" at equal intervals of time (imaginary snapshots, of course, since e-m waves are invisible), and our plan of attack will be to observe changing electric and magnetic fields as an e-m wave goes by. As the source of e-m waves, imagine that the antenna tower of a radio station is located at the center of town, and that a number of observers are stationed at 75-meter intervals along a street leading away from the station. The observers are equipped with devices that (in our imagination) can respond very quickly to changes in electric and magnetic fields.

To detect the electric field, each observer might have a tiny pith ball attached to the end of a light, springy fiber. The pith ball is given a charge Q, so that it will react to a force F, the direction of F being the direction of the electric field. Also, the magnitude of F is proportional to the magnitude of the field.

We start the observations at 12:00 noon, which we call $t = 0$. At that time, the observer's pith balls all indicate electric fields that are vertical in some places and zero in others. It is a matter of experimental fact that the wave emitted by a vertical antenna will, in general, have its electric field oriented vertically; this is an example of "vertical polarization." As shown in Figure 20-4a, some balls are urged upward, others are urged downward, and still others indicate the field to be zero. We have also roughly indicated movements some pith balls would make at intermediate points, and a curve has been sketched to show how the electric field might be seen to vary if there were a large number of observers.

A quarter of a microsecond later, the situation has changed (Figure 20-4b), and A's pith ball registers zero field, B's is urged upward, etc. At $t = \frac{1}{2}$ μsec (Figure 20-4c), the force on A's pith ball is downward, B's has returned to zero, etc.

Figure 20-4

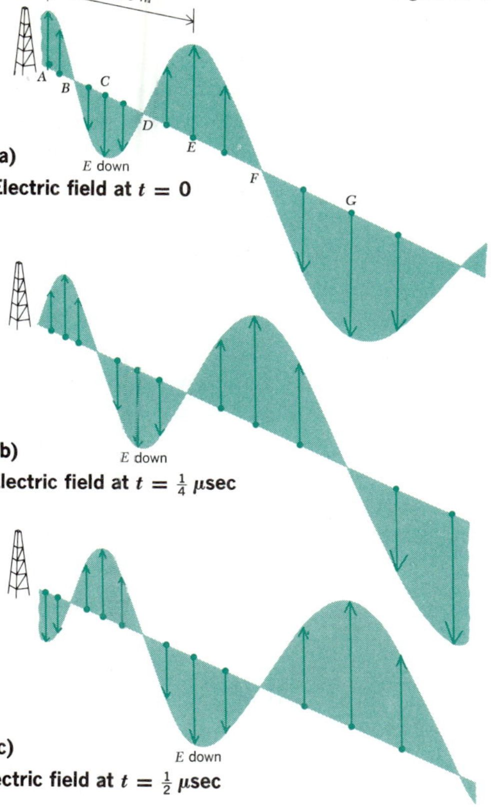

(a) Electric field at $t = 0$

(b) Electric field at $t = \frac{1}{4}$ μsec

(c) Electric field at $t = \frac{1}{2}$ μsec

Figure 20-5

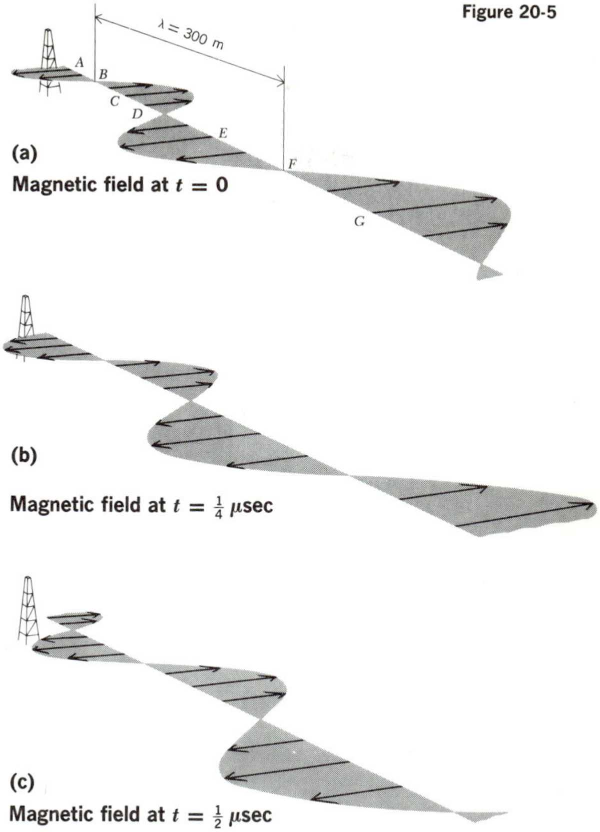

(a) Magnetic field at $t = 0$

(b) Magnetic field at $t = \frac{1}{4}$ μsec

(c) Magnetic field at $t = \frac{1}{2}$ μsec

Notice how the place of maximum downward field (E_{down}) has moved to the right. Here is all the behavior that we have learned to associate with a wave: at any given place, the electric field fluctuates; at C, for example, it is "down"—zero—"up" during the period of our observations. Thus, if we stay at one place, E is a function of t. Also, at any given time, the electric field varies along the street; thus, E is also a function of x.

The quantity above measured is an electric field, and the wave could be called an "electric" wave, but this is a bit misleading because an e-m wave has both electric and magnetic components. Recall that each observer also has a magnetic detector, and very shortly we will examine the behavior of these magnetic detectors too. It will then be possible to combine the readings of the two detectors to form the complete picture of a single electromagnetic wave. In the meantime, though, there is quite a bit to be learned by looking just at the electric part of the picture. First, the wave is a transverse wave, since the quantity that varies (the electric field E) is a vector perpendicular to the direction of propagation. This particular wave is plane polarized, although other kinds of polarization can also be obtained. The wavelength λ of our wave is 300 meters (more than three football fields). The period is 1 microsecond, which is the time for one complete cycle of changes in E at any given point, the frequency is $1/T = 10^6$ Hz. The velocity of the wave is given by

$$\begin{aligned} c &= f\lambda \\ &= (10^6 \text{ Hz})(300 \text{ m}) \\ &= 3 \times 10^8 \text{ m/sec} \end{aligned}$$

At exactly the same times that we have been observing the changes in the electric field, mag-

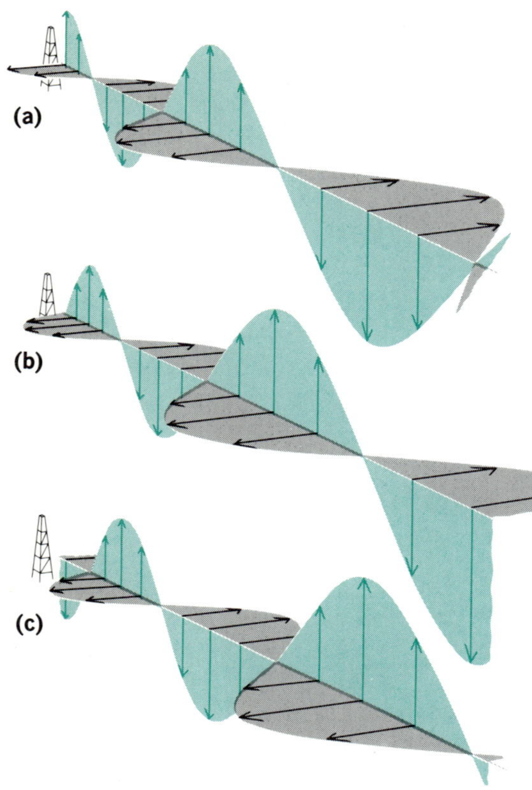

Figure 20-6

netometers at the same positions are measuring the magnetic field. Perhaps the simplest kind of magnetometer to imagine is an ideal, inertialess compass needle that would point in the direction of the magnetic field. The intensity of the magnetic field might also be measured (in imagination at least) by timing the frequency of oscillation of the needle as it "hunts" for a new position.

The results of a series of rapid-fire measurements are shown in Figure 20-5. At $t = 0$, the magnetic field at point A is directed horizontally to the east, the magnetic field at B is zero, and at C the magnetic field is directed to the west. Just as before, we interpret these measurements by means of a "magnetic" wave, which is transverse because the variable quantity (magnetic field B) is a vector perpendicular to the direction of propagation.

A radio wave is really two simultaneous waves; this is the significance of the term "electromagnetic." A purely electric wave as shown in Figure 20-4 is impossible; likewise, a purely magnetic wave as shown in Figure 20-5 is impossible. The two waves keep each other going by mutual interaction, and what *is* possible is a combined wave as shown in Figure 20-6.

Although this illustration is based on the long-wavelength e-m waves that we call radio waves, the wave model for visible light is exactly similar. Except for their short wavelengths (and high frequencies), light waves are in no way different from radio waves. They are e-m waves, and their velocity in a vacuum is given by the electrical quantity $\sqrt{\dfrac{k}{k'/2}}$, which equals 3×10^8 m/sec.

20-5 PRODUCTION OF ELECTROMAGNETIC WAVES

A simple generator of e-m waves is shown in Figure 20-7, where an antenna is formed by two metal rods connected to some sort of electric oscillator. The exact nature of the oscillator does not concern us here; an LC circuit might be used, but other circuits are also possible. The oscillator causes electrons to flow back and forth along the antenna. The system is shown at several times. In (a), electrons have been pushed to the end of the upper rod, and an excess of positive charge remains on the bottom rod. At a time that is $\frac{1}{4}$ of a cycle later (b), the charges have come together again, and the rods are momentarily uncharged. Still later, at (c), the oscillator has reversed polarity and the charges are reversed, with the top rod positively charged.

Considering the situation at any instant, (a) for example, we can see that the electric field E surrounding the antenna is similar to that caused by a pair of equal and opposite charges. At point P in (a) the electric field is upward, but half a cycle later, in (c) the field at P is downward. During the time between (a) and (c), there is a varying current in the antenna from bottom end to top end. This current produces a magnetic field, and the

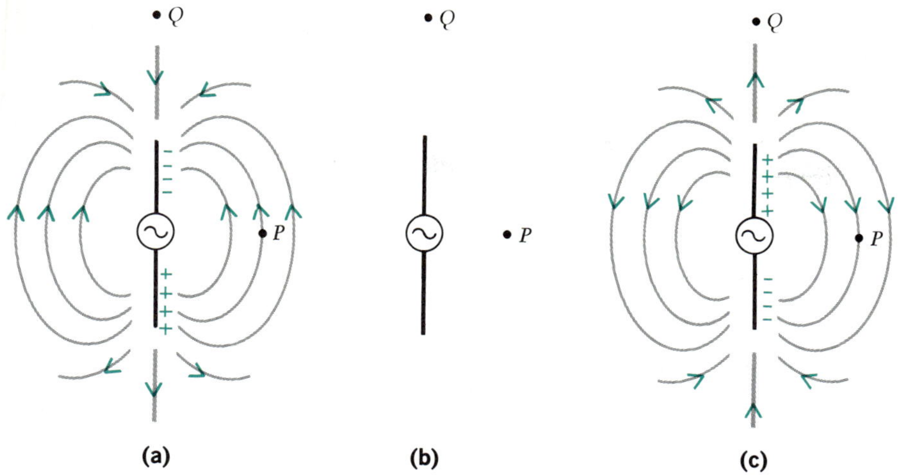

Figure 20-7 Electric field near a simple antenna.

right-hand rule (Section 18-4) tells us that the lines of force of the magnetic field are in the form of concentric circles (Figure 20-8). At point P, the magnetic field is perpendicular to the plane of the paper, directed horizontally away from us. Later, when the current is reversed, the magnetic field B reverses direction but is still horizontal.

We have had to overlook some details, but it is evident from this discussion that at point P *two* fields exist, and that the E-field and the B-field are perpendicular to each other (compare Figures 20-4, 5, and 6). At P, the wave is propagated horizontally, broadside to the rods; at a point such as Q, the magnetic field is zero (head-on view of the current; see Section 18-4), and hence no wave is propagated along any line parallel to the length of the antenna. There is a smooth variation in the intensity of the wave, from a maximum value if the antenna is viewed broadside, to a zero value if viewed head-on. If, as is usually the case, the observer is far from the antenna (compared with its length), E and B are "in phase," and both fields pass through their zero values at the same time.

At any instant, energy is stored in the fields surrounding the antenna. Figure 20-9 shows the elec-

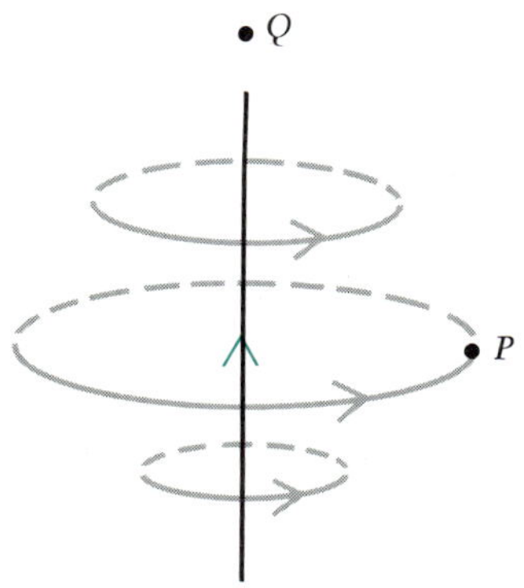

Figure 20-8 Magnetic field near a simple antenna.

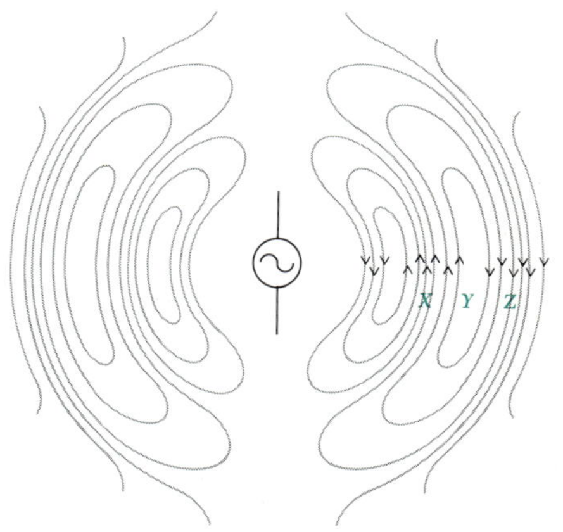

Figure 20-9

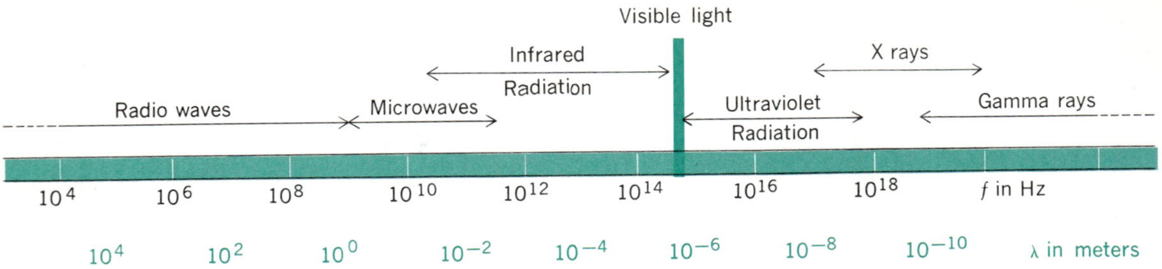

Figure 20-10 The electromagnetic spectrum.

tric field in detail (for simplicity, the accompanying magnetic field is not shown) for a given instant; the field in region X is upward, in Y is zero, and in Z is downward. Electric PE is stored wherever there is an electric field. The diagram shows that the energy is localized in regions such as X and Z, separated by regions such as Y where no energy is stored. As the wave spreads out, the energy is carried with it. This is how the wave model explains the propagation of energy by radiation. The situation is analogous to the spreading out of a water wave; gravitational PE is stored wherever there is a crest, and energy is propagated as the crest moves. One basic difference should be noted. An electric or magnetic field can exist in a vacuum, and hence no medium is needed to transmit energy by an e-m wave. In a water wave, however, gravitational PE can exist only where there is water, and hence such a wave requires a medium (the water).

20-6 SOURCES OF RADIATION

If all e-m radiation is alike, except for wavelength and frequency, it is natural that we should look for some similarity in the means by which radiations are produced. What general statement can be made that will apply to such diverse sources of radiation as radio transmitters, incandescent filaments in light bulbs, and x-ray tubes? The unifying concept here is that *electromagnetic radiation is produced when electric charge is accelerated.* You have already seen this illustrated for radio waves. As the electrons rush back and forth along the antenna, they are undergoing rather rapid accelerations, especially at the two ends. (Recall that in SHM the acceleration is greatest at the turning points.) A stream of electrons moving with a steady velocity emits no e-m wave, for then there is no acceleration of charge. True enough, a steady magnetic field surrounds a wire carrying a steady current, but an induced electric field is caused only by a *changing* magnetic field. This means that the interaction between electric and magnetic fields requires a changing current—an acceleration of charge—in order for an e-m wave to arise.

The sources of radio waves quite obviously involve acceleration of charge; free electrons, driven by an electric oscillator, drift back and forth through the metal of the wires and towers. Going down the e-m spectrum (Figure 20-10) to shorter wavelengths and higher frequencies, we next consider infrared radiation of typical wavelength 10^5 angstroms ($1\ \text{Å} = 10^{-10}$ m). Infrared radiation (often called heat waves) is emitted by the atoms of a solid or liquid that are in thermal agitation. Unless a solid is at absolute zero, the atoms are not rigidly locked in a crystal structure, but vibrate to and fro about their equilibrium positions. Since all atoms contain electric charges, it is reasonable to expect that the acceleration of these charges during thermal vibration should give rise to e-m waves.

Progressing now to still shorter wavelengths, we find visible light and ultraviolet radiation—waves often related to accelerations of orbital electrons. In a general way, we would expect the orbital electrons in atoms to vibrate with high frequency because of the small mass of an electron as compared to that of a nucleus, an atom, or a molecule. Still higher frequencies are represented by x rays. Electrons are given an extremely high speed by application of a large potential difference (say, 40,000 V). They strike a target and are made to stop. This is a negative acceleration, and an e-m wave is emitted from the target. Here, again, we find that the acceleration of charge gives rise to electromagnetic radiation.

The wave model of e-m radiation works best for low-frequency radio waves, and the model becomes progressively worse with higher and higher frequencies. As an example of the difficulty in making the model apply at higher frequencies, consider the motion of an orbital electron as it moves around the nucleus. You know that a particle in circular motion is accelerated toward the center of the circle. Why, then, doesn't the electron in a hydrogen atom, for instance, continually radiate an e-m wave? The energy of such a wave would have to come from the electric PE of the electron and its KE. The situation is analogous to the frictional loss of energy by an artificial satellite that spirals down to the earth. Calculation leads us to expect that all electrons would lose energy by radiation and spiral into the nucleus, with increasing speed. The whole universe should go up in a blinding flash of radiation—the so-called violet death. Such a disaster obviously has not taken place, and so we conclude that in this respect our wave model is inadequate. There is no doubt, however, that visible light *does* have wave properties, and so we shall continue to use the wave model whenever it is helpful to do so.

SUMMARY

A changing magnetic field gives rise to an **induced electric field,** and a changing electric field gives rise to an **induced magnetic field.** This interaction between electric and magnetic fields allows the propagation of **electromagnetic waves** in which the electric and magnetic fields are perpendicular to each other, and both are perpendicular to the direction of propagation. Electromagnetic waves are transverse and are often plane-polarized. Their velocity in a vacuum is

$$c = \sqrt{\frac{k}{k'/2}}$$
$$= 3 \times 10^8 \text{ m/sec}$$

In general, electromagnetic waves are radiated whenever electric charge is accelerated. The complete **electromagnetic spectrum** extends from radio waves to gamma rays, but only the relatively slow vibrations of radio waves are produced when charges are set into continuous oscillation by purely electrical means. Many electric **oscillators** use coil and capacitor circuits (often called *LC* circuits) whose frequency is given by

$$f = \left(\frac{1}{2\pi}\right)\sqrt{\frac{1}{LC}}$$

Infrared, visible, ultraviolet, x-ray, and **gamma-ray** radiations are also connected with acceleration of charge, but the wave model is not entirely adequate to describe the emission of high-frequency radiation.

QUESTIONS

1 Are radio waves a form of "light"?

2 Can e-m waves be propagated through a perfect vacuum?

3 What is the velocity of x rays in a vacuum?

4 A police transmitter has a vertical antenna. Should the receiving antenna on the squad car be horizontal or vertical for the most efficient reception?

5 The coil-and-capacitor oscillating circuit of a radio transmitter is often called a tank circuit. Explain why water sloshing back and forth in a tank is analogous to an LC circuit.

PROBLEMS

GROUP A

1 What is the wavelength of a radar wave of frequency 10,000 MHz?

2 What is the wavelength of a radio wave of frequency 108 MHz emitted by a transmitter in an artificial satellite?

3 Express in angstroms: (a) 4.5×10^{-5} cm; (b) 6×10^{-10} km; (c) 12 micrometers. (*Note:* 1 micrometer = 1 μm = 10^{-6} m.)

4 Express in meters: (a) 5893 Å; (b) 1.54 Å; (c) 12 μm.

5 Compute the frequency of a green light whose wavelength is 5000 Å.

6 Compute the wavelength of a standard-band AM station that broadcasts on an assigned frequency of 750 kHz.

GROUP B

7 The frequency of an oscillating circuit is determined by a fixed inductance connected in parallel with a variable capacitor. The frequency is 500 kHz when the capacitor is set at 100 picofarads (pF); what is the frequency when the capacitor is increased to 400 pF?

8 Make the following conversions: (a) 4000 Å = ? cm; (b) 0.71 Å = ? m; (c) 10^{-10} m = ? μm; (d) 5461 Å = ? nm; (e) 30 μm = ? Å.

9 Calculate the resonant frequency of a circuit for which L is 20 millihenries (mH), and C is 200 pF.

10 Calculate the wavelength of an e-m wave whose frequency is determined by a circuit with inductance 100 μH and capacitance 100 pF.

11 An amateur radio operator wishes to construct a receiver to operate in the "15-meter band" (i.e., to receive waves of wavelength in the neighborhood of 15 m). She has a variable capacitor whose average value is 50 pF, and she wishes to wind a coil to use in an LC circuit with natural frequency equal to the frequency of 15 m waves. What should be the inductance of the coil?

FOR FURTHER STUDY

Radio and TV Signals

Suppose an e-m wave moving toward the right is approaching a receiving antenna at velocity c, as shown in Figure 20-11. If the receiving antenna is placed parallel to the electric field vector E of the e-m wave, the free electrons in the metal will be urged back and forth as the field fluctuates. At any instant, the force on an electron (whose charge is Q) could be calculated from the equation $E = \frac{F}{Q}$. That is, $F = QE$. If the current in the antenna were great enough, a lamp bulb at the center could be lit to full brilliance by the charges that pulse back and forth. The effect is greatest if the antenna is parallel to the electric vector of the wave, since otherwise only a component of the force QE would be effective. An antenna whose length is equal to about half the wavelength of the radiation is much more efficient than a shorter or a longer one. This applies to both the transmitting antenna and the receiving antenna.

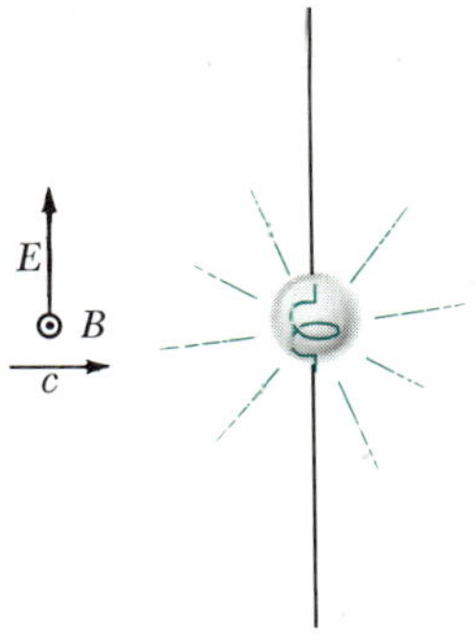

Figure 20-11

A radio or TV transmitter and receiver together constitute a system by which information can be sent from one place to another. As the term is used here, *information* can refer to any organized signal. It need not be just a voice or code. It is, in effect, the opposite of randomness. For instance, if a 1000 Hz sound is emitted by a tuning fork for a definite period of time, "information" is being transmitted to the ear of the listener. A more complex type of information is transmitted by a speaking person or by a symphony orchestra. To transmit the information of a 1000 Hz tuning fork, it would seem natural to use a system somewhat like that in Figure 20-12. A microphone would pick up the sound, producing feeble voltages that would be amplified and cause electrons to flow up and down in a vertical transmitting tower. The tower would radiate an e-m wave whose frequency is 1000 Hz, which would be received by a distant antenna. After further amplification, a pulsating flow of electrons would be caused in the loudspeaker. Unfortunately, the system of Figure 20-12 is entirely impractical because of the long wavelength of the necessary e-m wave. Since the velocity is 3×10^8 m/sec, we have

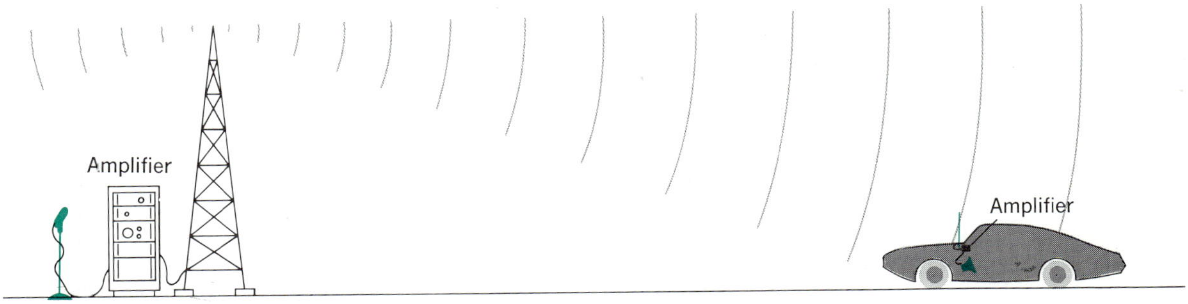

Figure 20-12 Hypothetical system of communication using unmodulated e-m waves, impractical because of long wavelengths required.

$$\lambda = \frac{\text{velocity}}{\text{frequency}}$$

$$= \frac{3 \times 10^8 \text{ m/sec}}{10^3 \text{ Hz}}$$

$$= 3 \times 10^5 \text{ m}$$

$$= 300 \text{ km}$$

The wavelength of our 1000 Hz e-m wave would be 300 km, and to obtain a useful amount of radiation both the transmitting and receiving antenna would have to be at least $\frac{1}{4}\lambda$ in height—some 75 km high! While it is true that the antennas could be stretched out horizontally, they would have to be supported many miles above the surface of the earth in order to radiate such a long wave with useful intensity. Such dimensions are fantastically large. Radio waves of frequency 10^6 Hz ($\lambda = 300$ m) are practical enough—but the frequency of 1,000,000 Hz is far above the audible range.

The problem of conveying audio or video information by means of radio waves has been solved by the technique called *modulation.* To modulate a wave, some property of the wave is varied at a slow rate. Two types of modulation are common: *amplitude modulation* (AM) and *frequency modulation* (FM). If the amplitude of a wave is periodically changed, then information can be sent at a relatively slow rate, using the rapid, easily transmitted high-frequency wave as a carrier. For instance, if the amplitude of a 1 MHz wave varies as shown in Figure 20-13, information is sent at 1000 cycles/sec. The very earliest communication by radio waves was by the use of dots and dashes (Figure 20-14), a form of amplitude modulation. An on-off switch (telegrapher's key) at the transmitter is sufficient to modulate a carrier wave in this simple fashion. This method is still widely used.

In the receiver the wave must be demodulated (or "detected") in order to recover the information. If the wave of Figure 20-15a were applied directly to a loudspeaker, nothing would be heard. The speaker has far too much inertia to

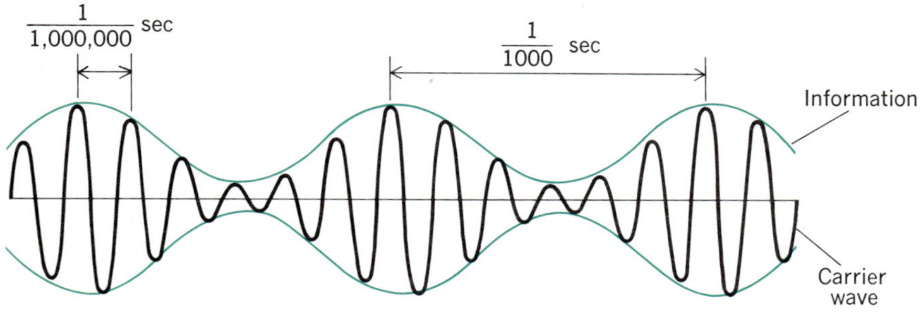

Figure 20-13 Amplitude modulation: Carrier wave of frequency 1 MHz modulated by tone of frequency 1000 Hz.

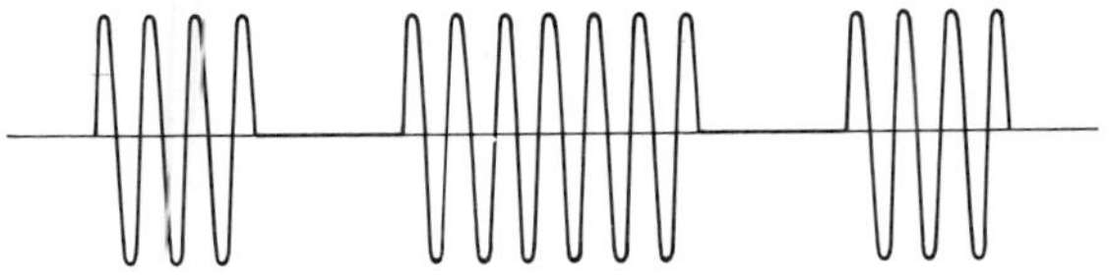

Figure 20-14 A simple form of amplitude modulation.

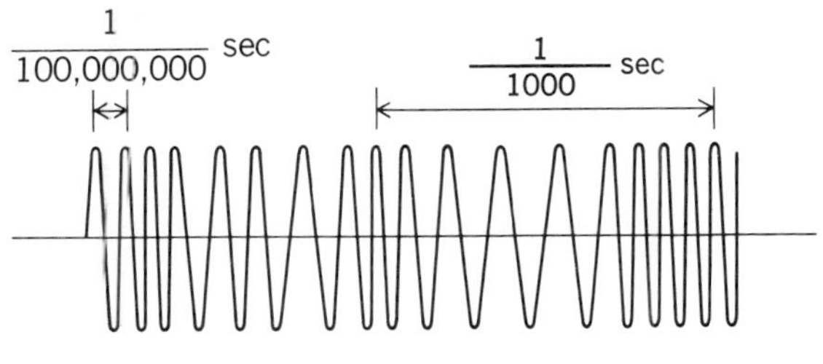

Figure 20-16 Frequency modulation: Carrier wave of frequency 100 MHz modulated by tone of frequency 1000 Hz.

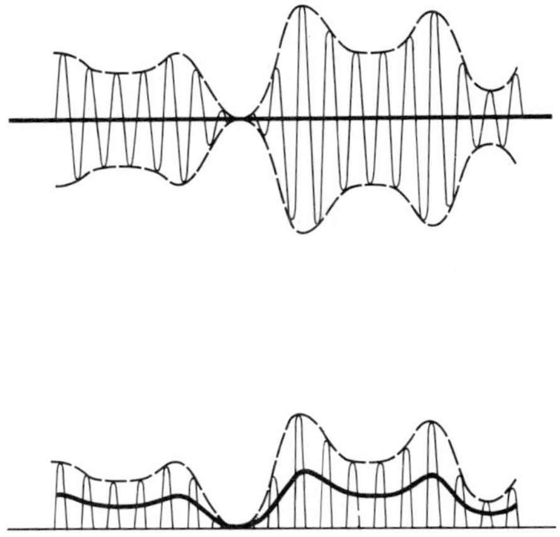

Figure 20-15 Demodulation of an AM signal.

respond to variations as rapid as 10^6 Hz, and so it responds only to the average value of the current (heavy line), which is zero. A simple way to demodulate an AM signal is to use a diode (Section 19-5), which passes current in only one direction. After the negative portions of Figure 20-15a are chopped off, the resulting wave (Figure 20-15b) has a slowly varying average value (heavy line), and the loudspeaker can follow these variations to produce a sound wave that duplicates the original sound wave in the studio at the transmitter.

In frequency modulation (FM), the amplitude of the carrier wave remains constant while the frequency changes. In Figure 20-16, which is typical of a commercial FM station whose "center frequency" is 99.9 megahertz, the frequency of the carrier wave varies from 99.8 to 100.0 megahertz, and this variation takes place 1000 times per second if a flutist is playing a tone having an audio frequency of 1000 Hz. One advantage of FM reception is that it is practically free from "static." The e-m radiation caused by thunderstorms, etc., would show up as an amplitude change, and hence the FM receiver, which responds only to frequency changes, is unaffected.

A television receiver receives two kinds of information—picture information (video) and sound information (audio). The video information is transmitted by AM, and the net result is to control the intensity of a beam of electrons. The picture is built up as the electron beam is caused to sweep back and forth, striking the face of the picture tube in a succession of spots. The brightness of the spot of light changes rapidly in proportion to the beam current, which is controlled by an electrode in the picture tube whose potential, in turn, is determined by the demodulated video signal.

The sound part of the information is received by FM; it is demodulated, amplified, and then fed to a loudspeaker. Typical carrier frequencies are 85 MHz for the center of TV channel 6, which extends from 82 to 88 MHz, and 98 MHz for the middle of the FM radio band, which extends from 88 to 108 MHz.

Working out the wavelengths, we find that λ is about 3.3 meters for the TV carrier and also for the FM-radio carrier. This has the practical advantage that it is possible to use antennas tuned for maximum efficiency—a structure of dimension $\frac{1}{2}\lambda$ is not too bulky to be placed on a rooftop or chimney. If you look carefully at TV or FM antennas, you will find that most of them have rods about a meter in length; and what is more the rods are horizontal, which shows that carrier waves for TV are polarized horizontally (that is, the electric vector is horizontal and the magnetic vector is vertical). The antennas are set broadside to the direction of propagation. The carrier waves of FM stations are also polarized horizontally, hence a TV antenna serves admirably to receive FM broadcasts, especially if it is necessary to receive distant stations.

Standard-band AM stations require tall transmitting antennas, since $\frac{1}{2}\lambda$ is of the order of several hundred meters. One way to economize is to build the tower only $\frac{1}{4}\lambda$ high. Electrons are continually rushing up and down the tower. The oscillator tends to cause electrons to flow from the top, down the tower, through the oscillator, and out into the earth. Thus, when there are positive charges at the top of the tower, there are negative charges on the earth's surface, and lines of force can terminate on these charges as shown in Figure 20-17a. Half a cycle later, the signs of all charges, both on tower and ground, are reversed. At a distance that is approximately $\frac{1}{2}\lambda$ from the tower, some of the lines of force become detached and energy is radiated. The tower and ground together are equivalent to the longer doublet of Figure 20-17b. Usually a set of heavy copper wires is buried in the ground, radiating out from the base of the tower, to facilitate the necessary to-and-fro flow of charges in the earth near the tower.

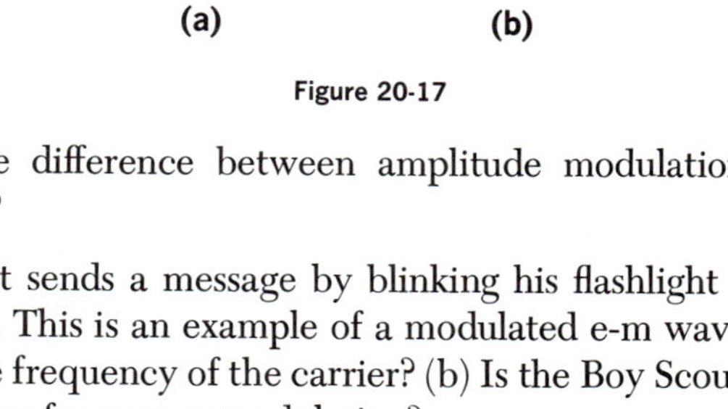

Figure 20-17

QUESTIONS

6 What is the difference between amplitude modulation and frequency modulation?

7 A Boy Scout sends a message by blinking his flashlight on and off, using Morse code. This is an example of a modulated e-m wave. (a) What is the approximate frequency of the carrier? (b) Is the Boy Scout using amplitude modulation or frequency modulation?

8 A traffic signal changes from red to green. What sort of modulation is used to convey the information that it is safe to proceed?

UNIT 5

References and Outside Reading

The purpose of the following list is to make you aware of additional reading that will extend and broaden most of the concepts you have met in this unit. These references should be easily available and are of sufficient variety to meet your individual interest.

Conductors and Semiconductors, by A. Holden, Bell Telephone Laboratories, 1964. Discusses all aspects of electricity.

"The Conservation Laws of Physics," by Feinberg and Goldhaber, in *Scientific American,* October, 1963. An excellent article that provides new insights into every aspect of physics. The treatment of the conservation of electric charge is of particular interest.

"DeForest and the Triode Detector," by R. A. Chipman, in *Scientific American,* March, 1965. A most interesting historical article about the invention of the triode vacuum tube—the beginning of modern electronics.

"The Electronic Telephone" by Peter P. Luff, in *Scientific American,* March 1978. New telephones will probably make use of integrated circuits.

"Fluid Control Devices," by S. W. Angrist, in *Scientific American,* December, 1964. The analogy between fluid circuits and electric circuits, and the description of fluid amplifiers and switches.

"High-Voltage Power Transmission," by Barthold and Pfeiffer, in *Scientific American,* May, 1964. The application of electrical theory to the solution of an important engineering problem.

"The History of the Airflow Car," by H. S. Irwin, *Scientific American,* August 1977. Ideas of fluid flow used in Chapter 15 are applied to vehicle design—streamlining to reduce air resistance.

"Intense Magnetic Fields," by Kolm and Freeman, in *Scientific American,* April, 1965. Strong fields are very useful in many areas of science; here their effects, and ways of producing them, are described.

"The Longest Electromagnetic Waves," by J. R. Heirtzler, in *Scientific American,* March, 1962. A good discussion of e-m waves in general, as well as those waves of very long wavelength.

"Microelectronics"—the entire September 1977 issue of *Scientific American* is devoted to this fascinating topic. The articles expand in detail, yet in a readable way, the concepts developed in Section 19-6. Highly recommended.

UNIT 6 OPTICS AND LIGHT

In the previous unit, you found that accelerated electric charges produce electromagnetic waves of a wide range of wavelengths. In this unit you will investigate one particular band of wavelengths to which our eye is sensitive—visible light.

First you will study the behavior of light in terms of geometrical optics. Using a ray model to analyze the path of light, you will find that light is generally propagated in straight lines but may be bent in various ways. You will investigate optical devices such as mirrors and lenses, and find out, for instance, how the convex lens shown in the photograph can form these two different images of the woman.

Next you will discover how wave concepts explain many—but not all—aspects of the behavior of light. You will apply both wave concepts and the ray model to an understanding of the spectroscope and the interferometer, experimental instruments that have been vital in the development of current physical theories.

21 Wave Fronts and Rays of Light

The study of optics is the study of a small part of the electromagnetic spectrum. Visible light waves range in wavelength from about 3800 angstroms for the violet to about 7600 angstroms for the red (3800×10^{-10} m to 7600×10^{-10} m). By making use of the relation $v = f\lambda$ connecting the speed of radiation with wavelength λ and frequency f, we find that the corresponding frequencies are high—in the neighborhood of 6×10^{14} hertz for green light of wavelength 5000 angstroms. The approximate wavelength ranges of various colors are shown in Table 21-1. In spite of the fact that visible light comprises only about one "octave" of the complete e-m spectrum, it is worthwhile to devote a considerable time to optical phenomena. For one thing, of course, much of our knowledge of the physical world comes to us through optical instruments such as the eye, the telescope, and the microscope. Then, too, our study of visible light is, to a certain extent, the study of all e-m radiation. The strange, dual, wave-particle nature of e-m radiation (Section 20-1) is perhaps best shown by visible light. Radio radiations (low frequencies) are practically "all wave" and show few or no particle properties, and nuclear gamma rays (high frequencies) are practically "all particle," with wave properties difficult to observe.

Table 21-1

Wavelengths of Visible Light

Color	*Approximate wavelengths (angstroms)*	*(nanometers)*
violet	3800–4500	380–450
blue	4500–4900	450–490
green	4900–5600	490–560
yellow	5600–5900	560–590
orange	5900–6300	590–630
red	6300–7600	630–760

Visible light occupies a position in the middle of the e-m spectrum where wave and particle aspects are about equally prominent. We shall first study the wave model that is entirely adequate for explaining how light travels in a straight line, how it is reflected by a mirror and bent as it enters a prism or a lens, and how it spreads out into a diffraction pattern when it passes through a small opening.

21-1 HUYGENS' PRINCIPLE

Christian Huygens* (1629–1695) was a versatile Dutch physicist and natural philosopher, born 14 years before Newton. His work in mechanics did much to pave the way for Newton's achieve-

* The English pronunciation is Hy′genz.

ments. And Huygens, along with Galileo, is credited with developing the pendulum clock, an invention of great importance to the science of mechanics, where precise measurement of time intervals is necessary. Huygens was a firm believer in the wave theory of light, and he used a geometric principle (without proof) to explain many properties of the transmission of light and other waves.

The concentric circles formed when a stone is dropped into a lake and water waves spread out are called wave fronts (Figure 21-1). The circle *a* represents a crest, as does the circle *b*. Between these circles is another circle *t*, which is a trough. In general, a *wave front* is defined as the locus of points of a wave having the same phase of vibration. All the points on circle *a* have the same phase, since all the molecules on this particular wave front are at their maximum displacement *above* the normal level. Similarly, *t* is a wave front on which all particles are at their maximum displacements *below* the normal level. Halfway between *t* and *b* is another wave front where the particles are (momentarily) at the normal level, but moving upward.

The word "front" is quite widely used in this way to describe the motion of some kind of disturbance or change. For example, in meteorology, a cold front is said to be passing by three cities *X*, *Y*, and *Z* when they are simultaneously experiencing a sudden drop in temperature (Figure 21-2). As a wave spreads out, the wave fronts advance, and in the example of the water wave the radius of each circular wave front increases continuously.

Huygens' principle tells us how to predict a new wave front when we know the position of an earlier one. In essence, his method is to assume that every point on a wave front is the source of a small *wavelet* that spreads out in the forward direction (Figure 21-3). To apply Huygens' construction we can draw, as in the figure, a number

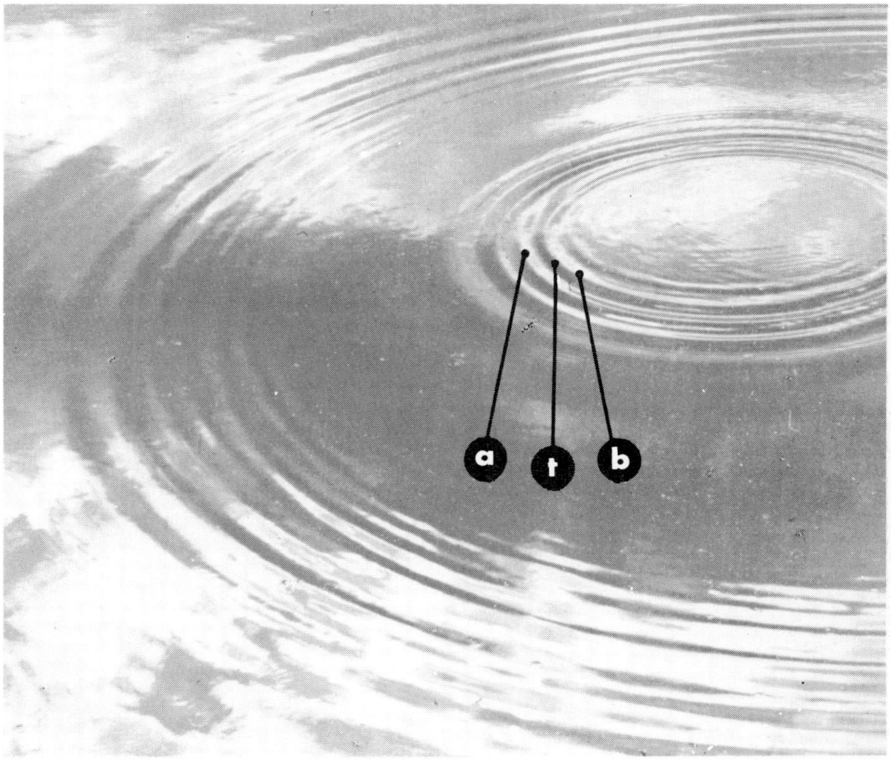

Figure 21-1 Wave fronts on the surface of water.

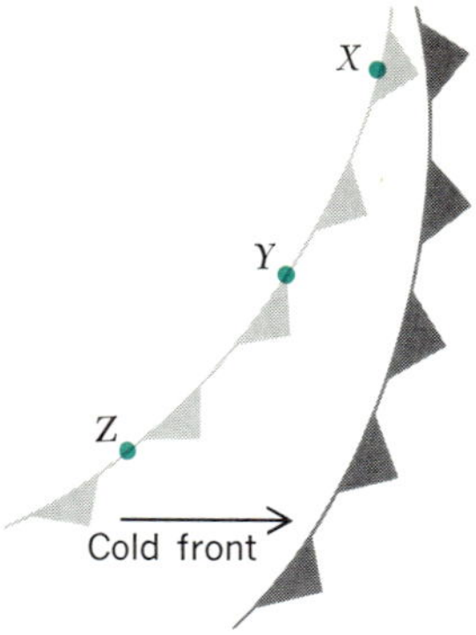

Figure 21-2 The position of the cold front is defined by a simultaneous sudden fall in the temperature at X, Y, and Z.

of arcs of circles centered on the initial wave front. The new wave front is located by the *envelope* of all the wavelets; that is, the line or surface tangent to all the little wavelets. In fact, if we imagine a huge number of wavelets, they will by themselves form a smooth line or surface in the forward direction.

There is a limit to the usefulness of Huygens' construction in this simple form. It does not help us to understand why there is no "shrinking" wave front in the "backward direction." There is nothing in this simple wavelet model to provide an answer; finding the answer would require a detailed study of how the intensity of e-m radiation varies with direction. Such a study of the intensity is beyond the scope of this book. It is enough for us to know that analysis shows that the intensity is indeed zero in the reverse direction.

If we now consider light waves spreading out in three dimensions, it is evident that the wavelets become small hemispheres, provided the speed of light is the same in all directions. This leads to the prediction, which agrees well with experimental observations, that a spherical wave front will remain spherical and that the energy of the wave will be carried away equally in all directions. Lines representing the direction of energy flow are called *rays*. The short arrows in Figure 21-3 represent rays of light that are diverging from the source S.

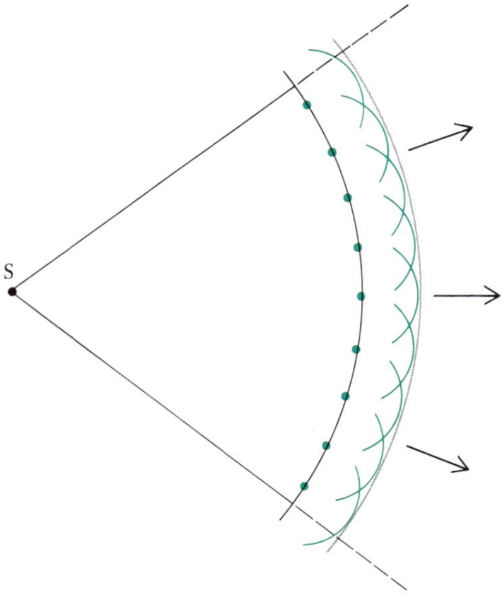

Figure 21-3

Unless specifically stated otherwise we shall assume in this chapter that light travels through a medium equally fast in all directions. Such a medium is called an *isotropic* medium.* Glass, water, air, Lucite, and most other common substances are isotropic, but certain transparent crystals have structures that give rise to "easy" and "hard" directions for the propagation of light. Such a crystal is said to be *anisotropic*. In an isotropic medium the wavelets are spherical and the rays are always perpendicular to the wave fronts (Figure 21-4).

21-2 STRAIGHT-LINE PROPAGATION OF LIGHT

Everyday experience seems to indicate that light travels in straight lines. The beam of a flashlight does not bend when we shine it through dusty air; a shadow cast by the sun has sharp edges. In fact, we use a "line of sight" as a widely acceptable standard of straightness as in Figure 21-5.

Huygens' principle describes the propagation

* From the Greek *iso*, meaning equal, and *tropic*, meaning turning or changing.

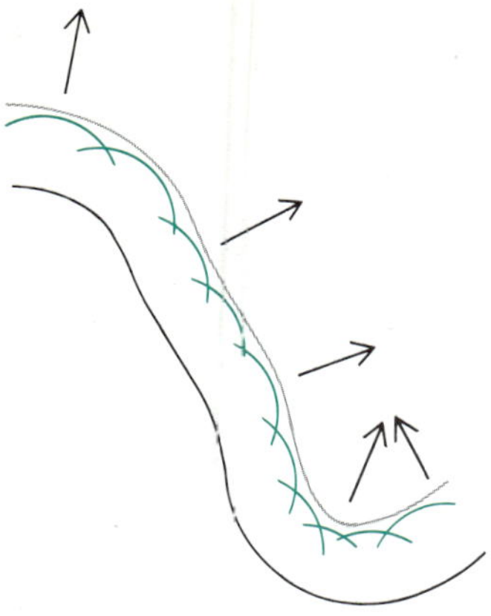

Figure 21-4

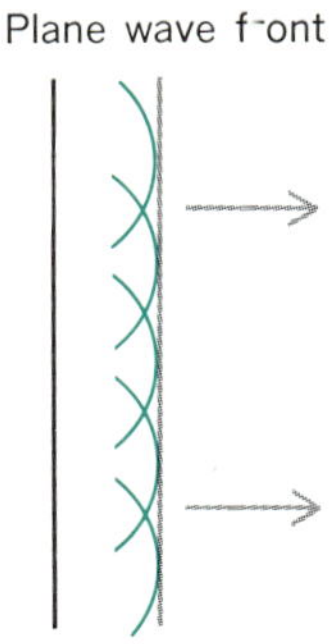

Figure 21-6

of light in a straight line, since wavelets from a plane wave front give rise to a new plane wave front (Figure 21-6), and the corresponding rays are parallel to each other. Such a plane wave front can be thought of as a section of a very large spherical wave front. For instance, sunlight strikes the earth in wave fronts of radius 149 million km. For all practical purposes, such a wave front is a "plane" and the rays are "parallel" to each other.

Newton never did accept the wave model because he felt (quite justifiably) that if light waves existed they would naturally "bend around corners" and would not travel in straight lines through openings or past obstacles, as they are observed to do. Newton did not realize that the extreme smallness of the wavelength of light compared with the size of the usual opening or obstacle could account for this behavior.

Let's use water waves to illustrate Newton's point. When straight* waves strike a gap in a long breakwater (Figure 21-7a), a disturbance spreads out in all directions in the water on the other side. And if the wave train strikes the end of the breakwater (Figure 21-7b), the waves diffuse around the corner. On the other hand, if light waves fall on an opening a millimeter or so in width or strike the end of an obstacle, the rays go straight through, with only a very slight "fuzziness" around the edges.

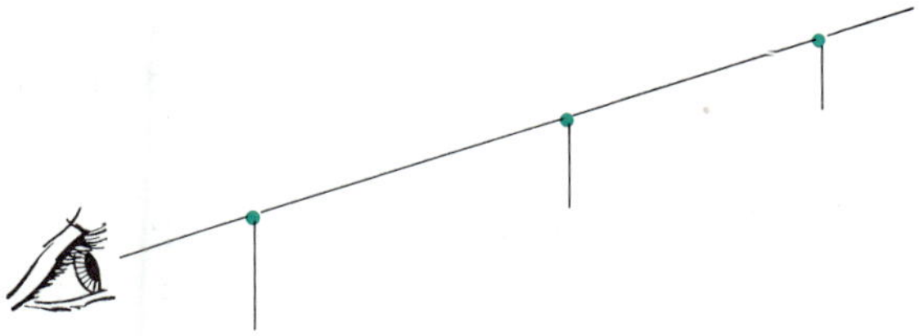

Figure 21-5 We say that three objects are "in line" if they lie on the same line of sight.

We can conclude that the straight-line propagation of a wave is a relative thing. It will be proved in Section 22-5 that the determining factor is the size of the opening or obstacle *as compared with the wavelength.* For water waves, which have wavelengths of the order of several centimeters to several meters, the "shadow" cast by a relatively small object is not sharp. But visible light waves have wavelengths of about 5×10^{-5} cm, which is small compared with the diameter of most visible

* Since water waves are on the surface of a liquid, the wave fronts form lines in two dimensions that can represent wave fronts of light waves in three dimensions.

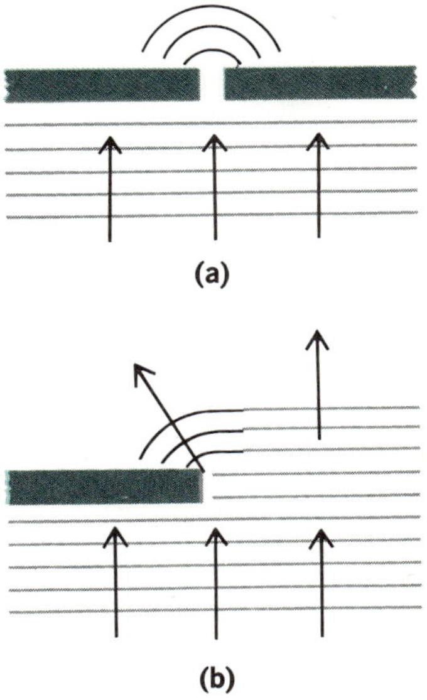

Figure 21-7

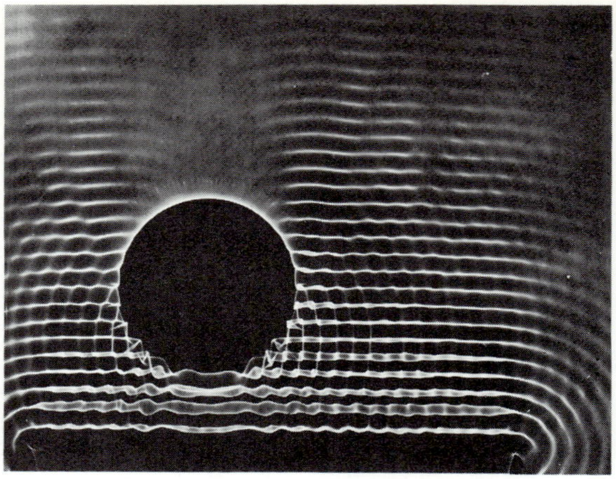

Figure 21-8 Water waves in ripple tank strike an obstacle small compared to the wavelength in **(a)** and form no sharp shadow. They strike a larger object in **(b)** and a shadow is formed.

objects or the pupil of the eye. Thus we anticipate that in the usual situations light waves are propagated in practically straight lines (see Figure 21-8).

These ideas are illustrated by sound waves. Many sounds have wavelengths of a meter or so (a 550-hertz blast from an automobile horn has a wavelength of 0.6 m), and street noises diffuse through an open window without any obvious straight-line propagation. Nevertheless, sound shadows and reflection (such as echoes) are possible; it is only necessary to have an obstacle that is large compared to a wavelength.

This can be done in two ways: an echo can be received when an ordinary-sized (audible) wave is reflected from a mountainside or the wall of a large building, or very short-wavelength sound waves of ultrasonic frequency can be reflected by obstacles of ordinary size. The latter method is used in underwater range finding (sonar) and in navigation by bats. Light and other e-m waves also show a gradual transition from diffuse nature to straight-line propagation as wavelength is decreased. Standard-band radio waves ($f = 10^6$ Hz, $\lambda = 300$ m) spread out diffusely; TV broadcasts ($f = 10^8$ Hz, $\lambda = 3$ m) are easily blocked out by buildings or by the curvature of the earth; radar waves ($f = 10^{10}$ Hz, $\lambda = 0.03$ m) are small compared with a plane and thus cast sharp shadows.

21-3 REFLECTION

When a ray of light strikes a smooth surface such as a mirror, the reflected ray leaves the mirror in a definite direction that is determined by two rules: (a) the incident ray, reflected ray, and the normal line perpendicular to the surface all lie in the same plane; (b) the angle of incidence equals the angle of reflection. These angles are measured from the normal and are denoted by i and r in Figure 21-9.

These rules can be found experimentally by means of the apparatus shown in Figure 21-10. A

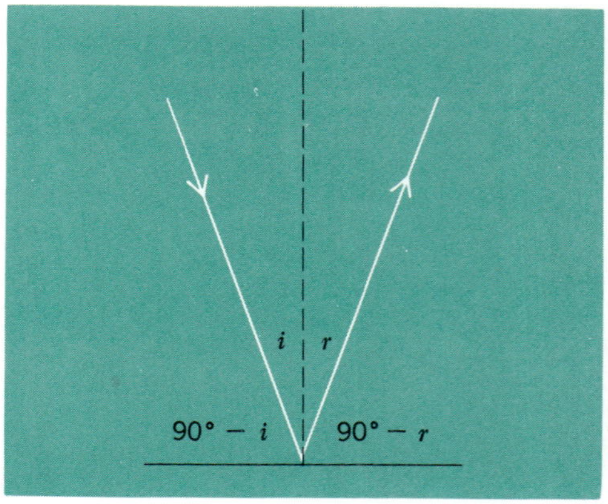

Figure 21-9

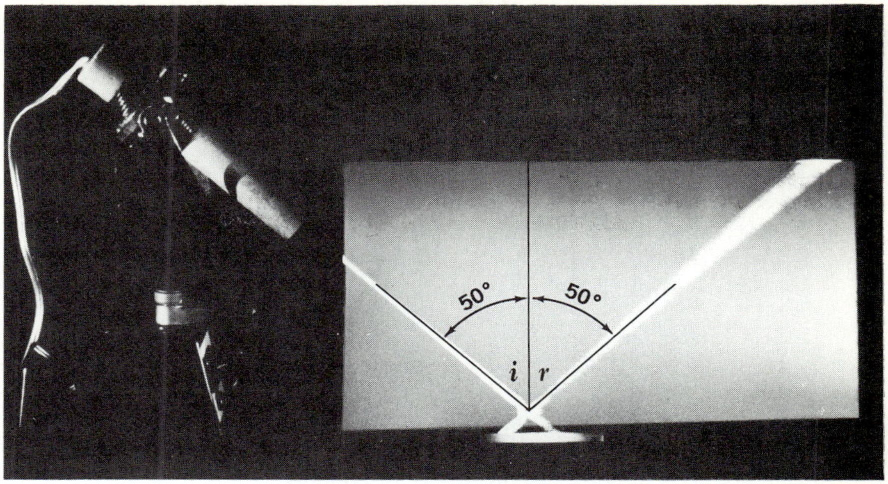

Figure 21-10 Ray of light from projector, reflected off mirror.

narrow beam of light is projected toward a mirror, and its path is traced along a white card. It is easy to see that $i = r$. And the fact that the whole path can be traced out on the card shows that both incident and reflected beams lie in the same plane. The application of Huygens' construction in this situation predicts the same result; thus we get further evidence for the validity of Huygens' model.

If a surface is not perfectly smooth, *diffuse reflection* takes place with many reflected bundles of rays coming from small flat spots. If the surface is so rough that there are no flat spots much larger than a wavelength, the light is *scattered* diffusely. We can think of scattering and also of reflection as the absorption and re-emmission of light by particles of a medium. The light of the sky reaches us after being scattered by air molecules, water droplets, or dust particles.

21-4 REFRACTION

When light passes from one medium to another, through a sharply defined boundary, or when there is a gradual change in the properties of a medium, the light is bent. This bending is called *refraction*. Just as with reflection, the angles are measured from the normal to the boundary (Figure 21-11), and, as before, the incident ray, the refracted ray, and the normal all lie in the same plane.

Refraction occurs because the speed of light in any medium (such as glass or water) is less than it is in a vacuum (or air), and the speed is different in different mediums. It is often helpful to think of refraction in terms of the following simple model. Imagine a set of plane waves traveling in air (Figure 21-12). They approach a boundary between air and glass at some oblique angle. The speed of light in glass is less than it is in air. At this angle, the right-hand edge of each wave strikes the boundary first. As each wave passes through the boundary and enters the slower medium, the right-hand edge is retarded, or slowed down, with respect to the part of the wave that is still in the air. Each wave front is therefore turned to the right as it passes through the boundary.

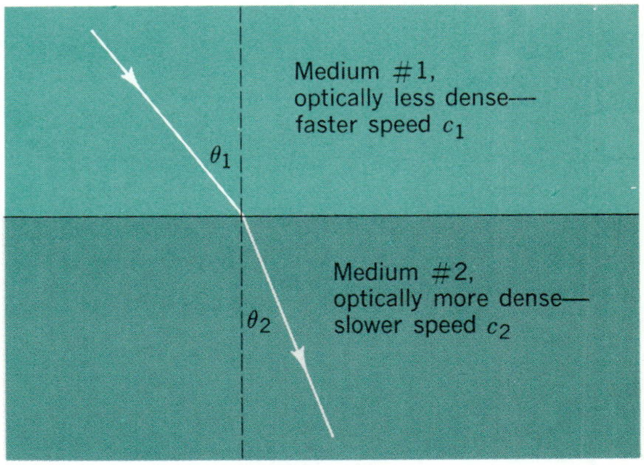

Figure 21-11

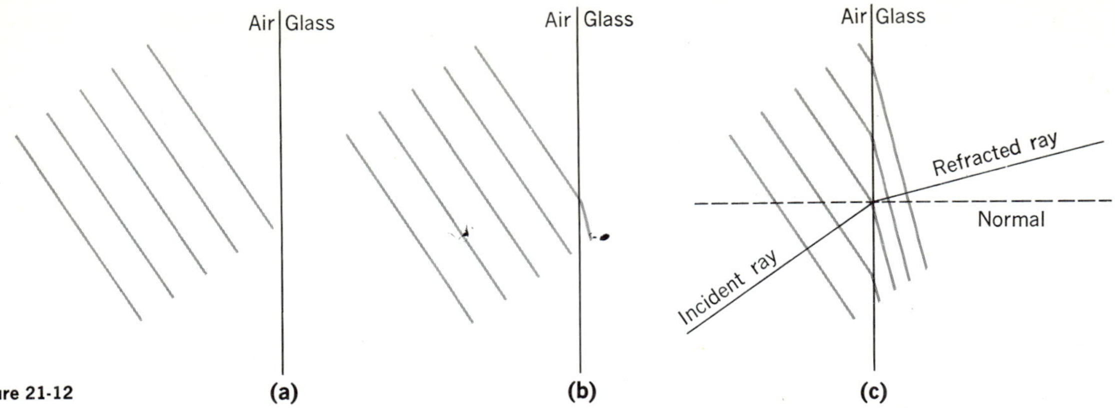

Figure 21-12

In general, if the rays are normal to the boundary (the fronts parallel to the boundary), there is no refraction because the retardation is the same for all parts of the wave. But when waves strike at an angle when going into a slower medium, the retardation always pulls the ray toward the normal.

The speed of light in a medium is denoted by c_n and the speed of light in a vacuum is denoted by $c = 3.00 \times 10^8$ m/sec. We now define a constant n, called the *index of refraction*:

$$n = \frac{\text{speed of light in vacuum}}{\text{speed of light in medium}} = \frac{c}{c_n}$$

The index of refraction is a measure of optical density. For instance, the speed of light in glass is less than in air, and so glass is said to be optically denser than air. Some values of the index of refraction are shown in Table 21-2; these are average values for visible light. The variation with color is minor. Typical values of n for window glass might be 1.51 for red light, 1.52 for green light, and 1.53 for violet light. This means that in glass, red light travels faster than violet light. In a medium, the variation of the speed of light due to color (that is, frequency) is called *dispersion.* Some of the consequences of dispersion, both useful and annoying, will be discussed more fully in Chapter 23, where we study optical instruments.

Table 21-2

Index of Refraction

Substance	n
diamond	2.42
flint glass	1.66
crown glass	1.52
benzene	1.50
water	1.333
ice	1.31
air at 0°C and 1 atm	1.00029
hydrogen at 0°C and 1 atm	1.00013

Example 21-1 Calculate the speed of light in a diamond.

Since $n = c/c_n$,

$$c_n = \frac{c}{n} = \frac{3.00 \times 10^8 \text{ m/sec}}{2.42} = \boxed{1.24 \times 10^8 \text{ m/sec}}$$

Example 21-2 The index of refraction of a certain glass is 1.50 for light whose wavelength in a vacuum is 6000 Å. What is the wavelength of this light as it passes through glass?

The frequency remains the same when light enters the glass. This frequency is

$$f = \frac{c}{\lambda} = \frac{3 \times 10^8 \text{ m/sec}}{6000 \times 10^{-10} \text{ m}} = 5 \times 10^{14} \text{ Hz}$$

The speed in this glass is given by

$$c_n = \frac{c}{n}$$
$$= \frac{3 \times 10^8 \text{ m/sec}}{1.50}$$
$$= 2 \times 10^8 \text{ m/sec}$$

Hence the wavelength in this glass is

$$\lambda_n = \frac{c_n}{f}$$
$$= \frac{2 \times 10^8 \text{ m/sec}}{5 \times 10^{14} \text{ Hz}}$$
$$= 4 \times 10^{-7} \text{ m}$$
$$= 4000 \times 10^{-10} \text{ m}$$
$$= \boxed{4000 \text{ Å}}$$

The wavelength in this glass is only $\frac{2}{3}$ as much as it is in vacuum.

We can generalize Example 21-2 by working in symbols instead of with numbers. If λ_{vac} is the wavelength in a vacuum and λ_n is the wavelength in the medium, we have

$$f = \frac{c}{\lambda_{\text{vac}}}$$

also, $$f = \frac{c_n}{\lambda_n}$$

Therefore, $$\frac{c}{\lambda_{\text{vac}}} = \frac{c/n}{\lambda_n}$$

$$\lambda_n = \frac{\lambda_{\text{vac}}}{n}$$

Thus the wavelength is reduced by a factor equal to the index of refraction.

Returning now to refraction, we can predict quantitatively how light will bend as it passes from one medium to another. The bending can be described in terms of the directions of the rays in the two mediums: θ_1 is the angle between the light ray and the normal to the boundary in one medium (Figure 21-11), and θ_2 is the angle between the ray and the normal in the second medium.

We can derive a formula relating θ_1 and θ_2 in terms of the indexes of refraction in the two mediums by using Huygens' principle. This derivation is given in *For Further Study* at the end of this chapter. We can also measure a number of angles in different mediums and try to discover experimentally the relation between them. This was the process followed by the Dutch scientist Willebrord Snell in 1621, when he discovered the law now known by his name:

Snell's law: The product of the index of refraction (for a given frequency of light) and the sine of the angle measured from the normal is a constant, the same in each medium.

As an equation, it can be expressed as follows:

$$n_1 \sin \theta_1 = n_2 \sin \theta_2$$

Thus the angle θ is less in the medium whose index of refraction is greater. That is, light bends toward the normal when it enters an optically denser medium. Conversely, light bends away from the normal as it enters an optically less dense medium.

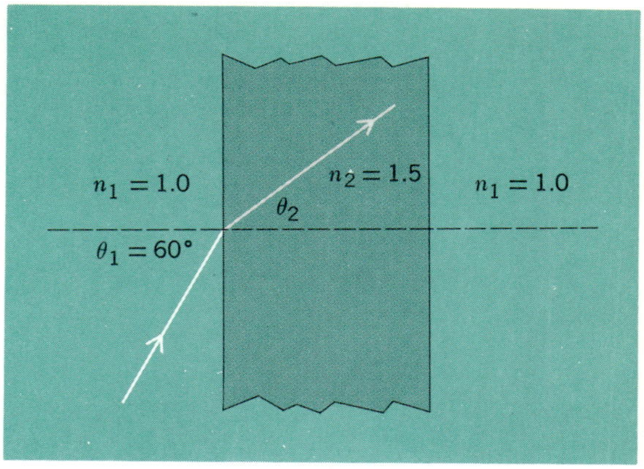

Figure 21-13

Example 21-3 A ray of light in air strikes a slab of glass, as shown in Figure 21-13. Calculate the angle of refraction in the glass, which has an index of refraction of 1.5.

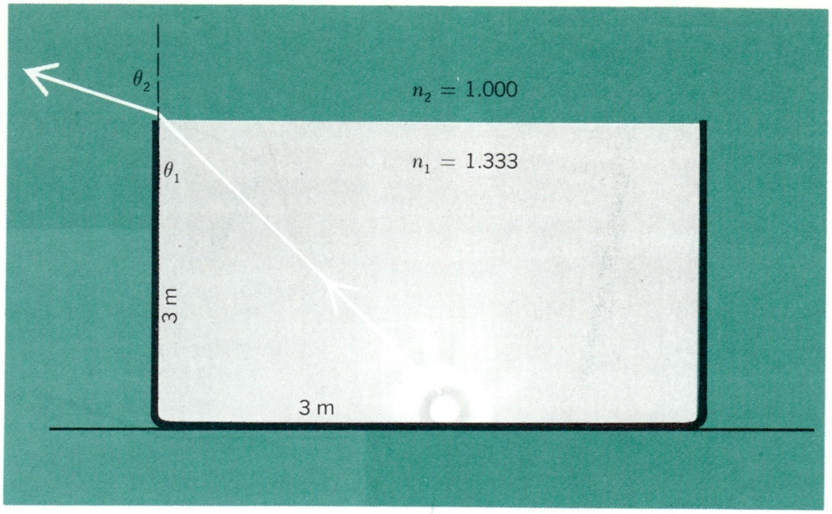

Figure 21-14

The index of refraction of air is 1.0003, and for most problems it is sufficiently accurate to call it exactly 1, the same as for a vacuum. Using Snell's law, we set $n_1 = 1$:

$$n_1 \sin \theta_1 = n_2 \sin \theta_2$$
$$(1)(\sin 60°) = (1.5)(\sin \theta_2)$$
$$\sin \theta_2 = \frac{(1)(0.866)}{1.5}$$
$$= 0.577$$

From the table of trigonometric functions in the Appendix (Table 5), we find that

$$\theta_2 = \boxed{35°}$$

Example 21-4 A lamp bulb in a watertight socket is on the bottom of a swimming pool at a point 3 m out from a wall. The pool is 3 m deep and filled to the top with water (Figure 21-14). At what angle does the light leave the water at the edge of the pool?

First we find θ_1, using simple trigonometry:

$$\tan \theta_1 = \frac{3 \text{ m}}{3 \text{ m}}$$
$$= 1.00$$
$$\theta_1 = 45°$$

Next we apply Snell's law:

$$n_1 \sin \theta_1 = n_2 \sin \theta_2$$
$$(1.333)(\sin 45°) = (1.000)(\sin \theta_2)$$
$$(1.333)(0.707) = \sin \theta_2$$
$$\sin \theta_2 = 0.942$$
$$\theta_2 = \boxed{70°}$$

As expected, θ_2 is greater than θ_1, and the ray bends away from the normal as it enters the less dense medium (air).

One common application of Snell's law is shown in Figure 21-15. Light strikes a slab of glass having parallel surfaces, such as a window pane, and the rays emerge without change of direction. Each ray is displaced slightly, parallel to itself.

21-5 TOTAL REFLECTION

You have seen how an oblique ray of light bends toward the normal as it enters a denser medium. If the ray travels exactly along the normal, it continues to do so in the second medium. But a ray in a dense medium does not always cross the boundary when it reaches a medium of lesser density. Suppose $n_1 = 2$, and $n_2 = 1$. Then, according to Snell's law, $2 \sin \theta_1 = \sin \theta_2$. Let's see what happens for various values of θ_1. When, for example, $\theta_1 = 28°$, then $\sin \theta_1 = 0.47$; therefore $\sin \theta_2 = 2 \times 0.47 = 0.94$, and $\theta_2 = 70°$. In this case, and for smaller values of θ_1, no special problems arise. But if $\theta_1 = 30°$, then $\sin \theta_1 = 0.50$, and $\sin \theta_2 = 1.0$. In this situation θ_2 is apparently 90°, which means that the refracted ray just grazes the boundary between the two mediums. What happens when θ_1 is still larger? No angle exists for which $\sin \theta$ is greater than 1; the sine of 90° is 1, and sin 91° is the same as sin 89°, which is less than 1. We then ask what happens when θ_1, n_1 and n_2 have values which would make $\sin \theta_2$, as calculated from Snell's law, greater than one.

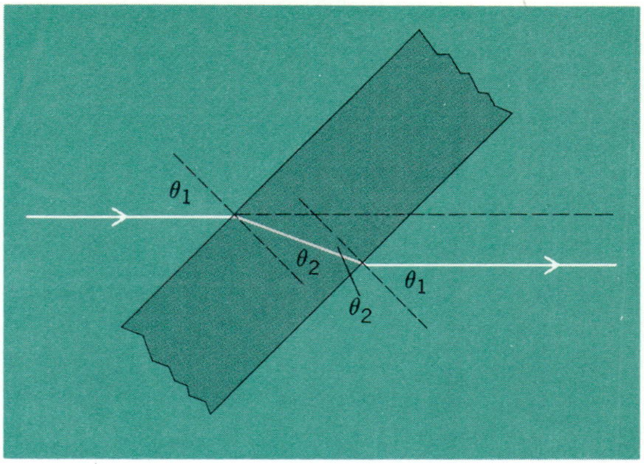

Figure 21-15

Experimenters have found that no refraction takes place; the ray is completely reflected back into the denser medium. This is called *total reflection,* and it occurs if the angle θ_1 in the denser medium is greater than a certain *critical angle.* It has also been found that *some* reflection takes place at any angle. As the angle in the denser medium is gradually increased (Figure 21-16), the reflected ray becomes stronger and the refracted ray becomes weaker, as indicated by the approximate percentages shown in the figure. In the limit, at the critical angle, the refracted ray just grazes the surface, but its intensity is close to zero. Beyond the critical angle there is no refracted ray and the reflected ray's intensity exactly equals that of the incident ray. This is one of the few examples in physics of a process that is truly 100 percent efficient; absolutely no energy is absorbed at the surface. The critical angle can be easily computed, using Snell's law. For glass surrounded by air, the angle works out to be about 42°, as calculated in the following example.

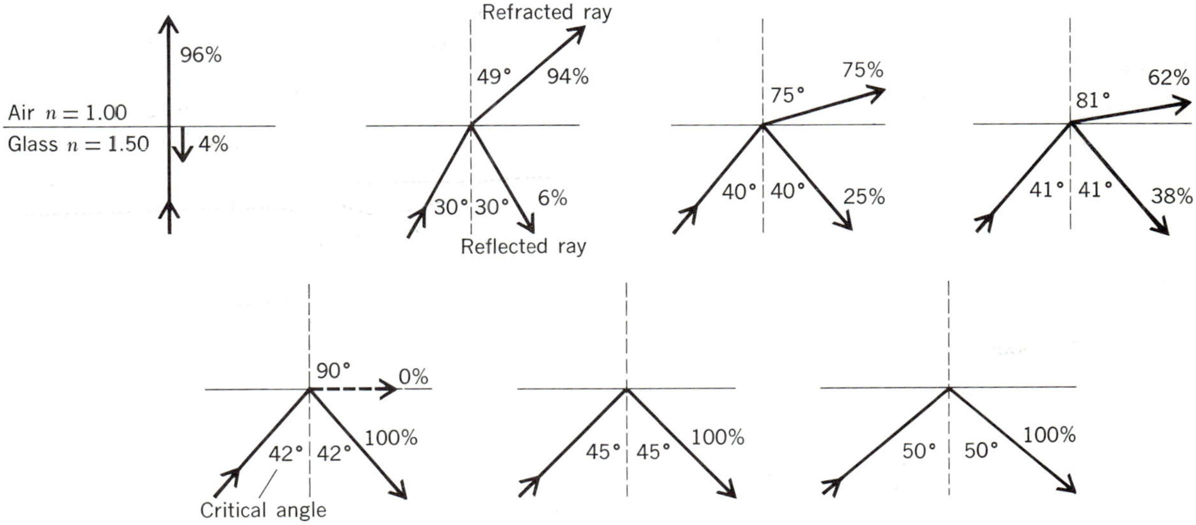

Figure 21-16

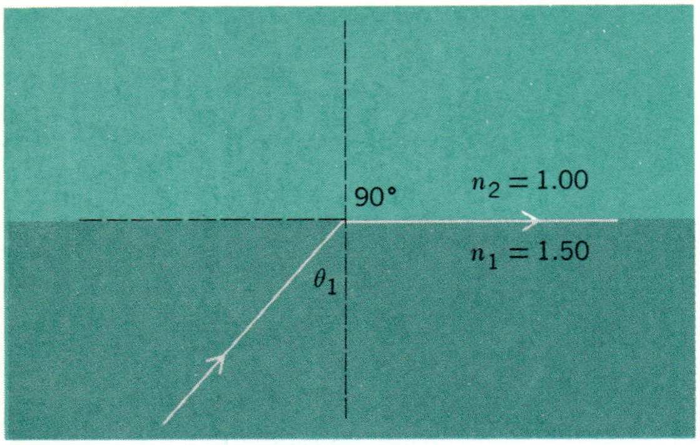

Figure 21-17

Example 21-5 Calculate the critical angle for a glass-air interface. The glass is a light crown glass with an index of refraction 1.50 (Figure 21-17).

When θ_1 just equals the critical angle, the emerging ray just grazes the surface, and $\theta_2 = 90°$.

$$n_1 \sin \theta_1 = n_2 \sin \theta_2$$
$$(1.50)(\sin \theta_1) = (1)(\sin 90°)$$
$$1.50 \sin \theta_1 = (1)(1)$$
$$\sin \theta_1 = 1/1.50$$
$$= 0.667$$
$$\theta_1 = \boxed{42°}$$

Total reflection is used in many optical instruments, including binoculars where the light rays are reflected inside prisms. Silvered mirrors might be used, but the prisms serve the same purpose and give 100 percent reflection, whereas a conventional mirror might become tarnished.

21-6 THIN LENSES

Lenses are used to concentrate or disperse light. This section deals with the ability of simple lenses to change the shape of wave fronts.

A rudimentary lens could be formed by placing two prisms back to back, as in Figure 21-18. Each of the rays is refracted toward the normal as it enters the glass, and is refracted away from the normal as it leaves the glass. Both refractions bend the ray away from the original (horizontal) direction. After passing through the "lens," the rays intersect in a zone *AA*. A much better lens is shown in Figure 21-19a, where the surfaces are curved. The various rays strike the surface at different angles, are refracted differently, and are brought to a much sharper focus *F*. If the lens surfaces are spherical (an easy shape to manufacture), the parallel rays of light converge at what is nearly a mathematical point called a *focus*. For the time being we shall consider that *thin* spherical lenses bring parallel rays of light to a perfectly sharp focus, and shall put off until Chapter 23 a discussion of the aberrations that cause fuzziness or color fringing.

In Figure 21-19a the incident rays are parallel to the *axis* of the lens, and after passing through the lens the rays converge and meet at what is called the *principal focus F*. (The axis is the line passing through the center points of both curved surfaces.) If the rays are parallel to each other but not to the axis, as in Figure 21-19b, the focal point F' is off the axis. The plane FF' is called the *focal plane*, and the distance from the lens to the focal plane is the *focal length f*. The focal length is usually measured from the center of the lens, but since the lens is assumed to be thin, the thickness is negligible compared with *f*, and the focal length can be measured from either surface without serious error.

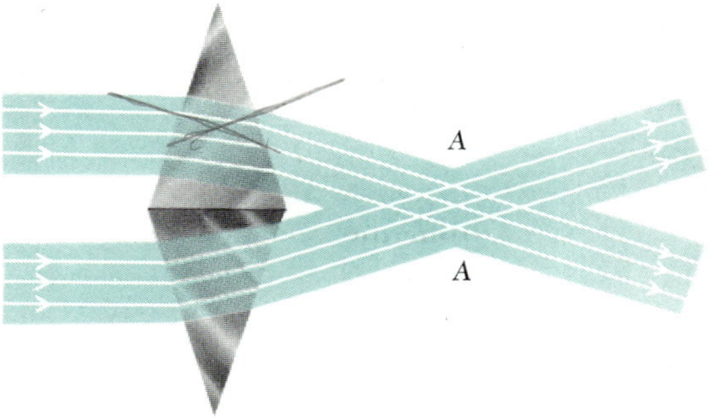

Figure 21-18

It is important to bear in mind what is meant by parallel rays. These are rays parallel to each other,

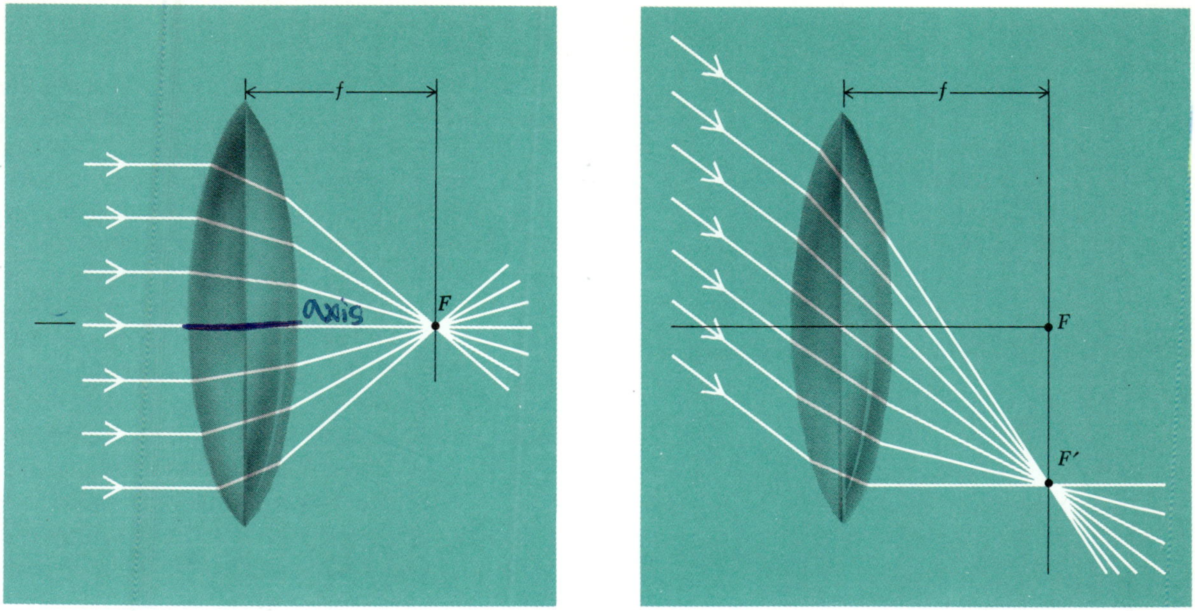

(a) (b)

Figure 21-19 Double-convex lens.

but not necessarily parallel to the axis of the lens. If you remember that rays are directions of energy flow and are perpendicular to the wave fronts, it should be clear that parallel rays must be associated with plane wave fronts of essentially zero curvature. Therefore the most common source of plane waves and parallel rays is light from a very distant source. "Very distant" need not be as remote as the sun and for many purposes the light from a lamp ten or twenty meters away may be considered parallel.

We have defined the principal focus as the point where parallel light rays *parallel to the axis* are brought to a focus. And since F is on the focal plane FF' we can assume that the focal length is the distance from the lens to the principal focus. This gives us a way of roughly measuring the focal length of a lens. Just focus the light from a distant source on any surface and measure (or estimate) the distance from lens to focal point.

Figure 21-19 shows a double-convex lens—one in which both surfaces bulge outward. Parallel rays striking such a lens (but not necessarily parallel to the axis) are focused at a point. Such a point, at which light rays intersect, is called a *real image.* This real image can be projected onto a screen, where it will appear as a bright spot.

Another useful lens is the double-concave lens of Figure 21-20. Refraction at the surfaces

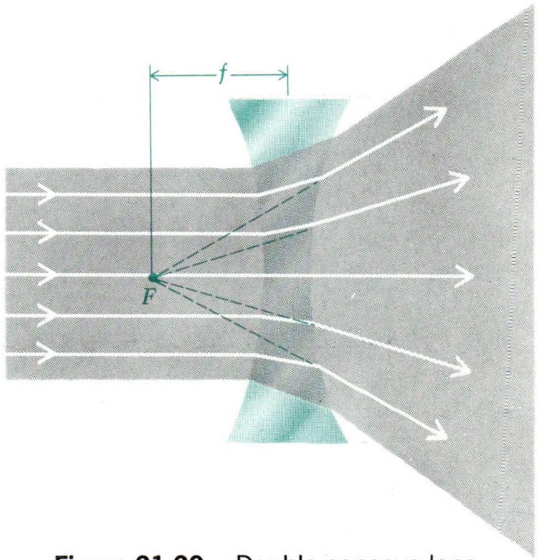

Figure 21-20 Double-concave lens.

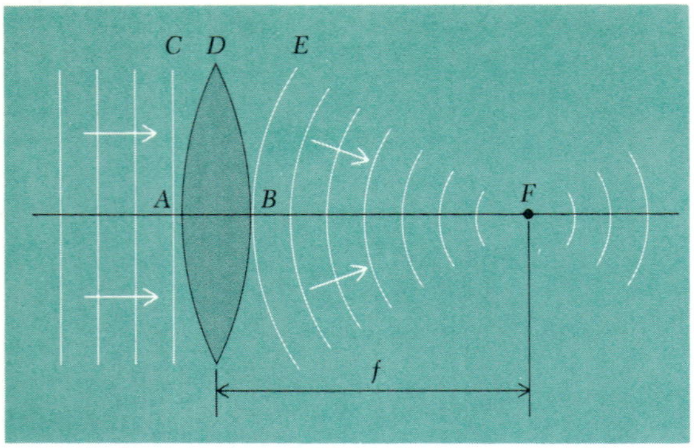

Figure 21-21

causes the parallel rays to diverge in such a way that they *appear* to come from a point F. This point is called a *virtual image.* As shown in the figure, the virtual image is located by extending the diverging rays backward. The virtual image, then, is defined as the place from which the diverging rays *seem* to come. Although a virtual image cannot be projected, as a real image can, on a screen or wall, it can be viewed directly by looking through the lens.

So that we may obtain a general idea of what happens as light passes through a lens, let's consider wave fronts instead of rays. The *curvature* of a wave front is defined as the reciprocal of its radius. Thus a sphere of radius 0.2 meter has a curvature of $\frac{1}{0.2\text{ m}}$, or 5 m^{-1}. Wave fronts reaching the earth from the sun have practically no curvature—only $\frac{1}{149{,}000{,}000\text{ km}}$. A true plane wave, which is represented by parallel rays and plane wave fronts, has zero curvature.

Spherical wave fronts are described by the behavior of the rays that are perpendicular to them. For example, a point source of light emits rays that travel outward in all directions; the wave fronts continually grow larger in radius (Figure 21-3). Such a wave front is called a *diverging wave front* and is said to have *negative curvature.* Similarly, if the rays associated with a spherical wave front are all heading for some point (called a real image), such a wave front is called a *converging wave front* and is said to have *positive curvature.*

Figure 21-19a can be redrawn, replacing the parallel rays by a plane wave front AC that strikes the double-convex lens (Figure 21-21). During a short time interval Δt while the center of the wave front is moving through glass from A to B, the outer edge of the wave front moves through air from C to E. Since the light travels faster in air than in glass, the wave front must curve around and become converging, as shown. The lens is thin, so the radius of curvature of the wave front BE is f (the focal length), but the wave front rapidly converges to a point and then spreads out. We see that the lens has had a converging effect and is therefore a *converging lens.*

The lens has acted on a wave front that had no curvature to start with, and has produced a wave front that is converging with a positive curvature equal to $1/f$. For this reason a converging lens is called a *positive lens.* Symbolically,

$$\begin{matrix}\text{original} \\ \text{curvature}\end{matrix} + \begin{matrix}\text{effect of} \\ \text{lens}\end{matrix} = \begin{matrix}\text{final} \\ \text{curvature}\end{matrix}$$

$$0 + \frac{1}{f} = \frac{1}{f}$$

The exact shape of the lens is immaterial as far as the converging effect is concerned; any lens that is thicker in the center than at the edges is a converging, or positive, lens. Of course, a lens with a

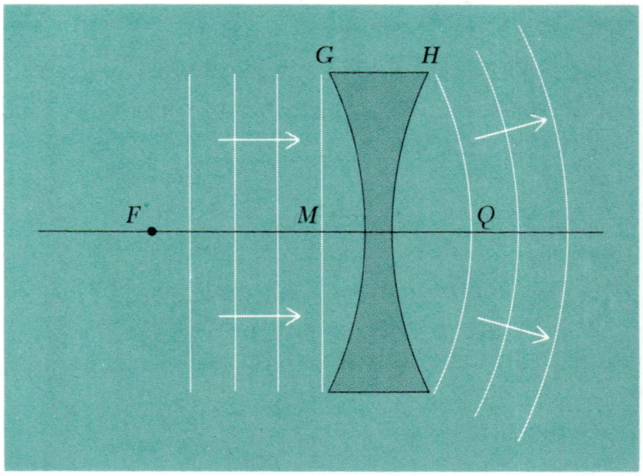

Figure 21-22

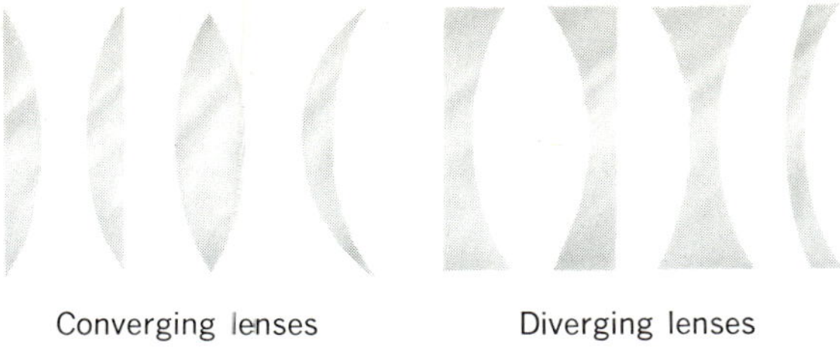

Figure 21-23

thicker bulge produces a greater effect and has a shorter focal length.

A *diverging lens* changes a plane wave front into a diverging one. Thus in Figure 21-22 the top of the wave front moves from G to H while the center of the same wave front (which has less glass to go through) moves from M to Q. After leaving the lens, the wave front continues to spread out, seeming to come from some virtual image at F. Any lens that is thinner in the middle than at the edges is a diverging, or negative, lens.

Several examples of converging and diverging lenses are shown in Figure 21-23. The effect of a lens on the curvature of a wave front is called its *power*. And since the effect of a lens on the curvature of a wave front is the reciprocal of its focal length, the power of a lens is also $\frac{1}{f}$. For any lens, the power depends upon the curvatures of the surfaces and on the index of refraction of the glass.

21-7 LOCATING IMAGES

In this section the concept of power of a lens is used to predict the location of the image of a source of light. Starting with the original curvature of a wave front (usually negative), add the power of the lens to get the final curvature of the front (which may be either positive or negative).

As with parallel wave fronts the *real image* of a point source of light is formed where rays of light are brought to a focus, or intersect. The *virtual image* of a point source of light is located where rays of light that originally started from a certain point appear to have originated after passing through the lens. Note that if the original wave front comes from a point source of light, its curvature is always negative, since light diverges from a source. Point sources may be isolated points such as a distant star or a lamp specially designed to emit its light from an extremely small area. Most commonly, however, they will be points that we imagine to be located on the surfaces of objects. That is, we "see" objects by the light reflected from them, and it is useful to imagine that the light comes from a number of points on the object. These are rather like the points on a wave front that form the basis of Huygens' principle. With very simple objects, like an arrow or a straight line, we can establish the size and position of the object by considering only two points, one at each end. If we are concerned only with the position of an object or image, we need to consider just one point at a time. The image of a complete object is the totality of the images of each of the point sources on the object. In solving lens problems, d_o and d_i are usually used to denote the distance of the object (that is, the source of light) and the image, respectively, from the lens.

Example 21-6 A point source on an object (Figure 21-24) is placed 100 cm from a lens of focal length + 20 cm. Where is the image of the source?

The lens is a converging one, since it is thicker in the center than at the edges, and so it has a positive power. The rays diverge from the object at O, and by the time the wave front reaches the lens the curvature is $\frac{1}{d_o} = -\frac{1}{100 \text{ cm}}$. Hence $d_o = -100$ cm.

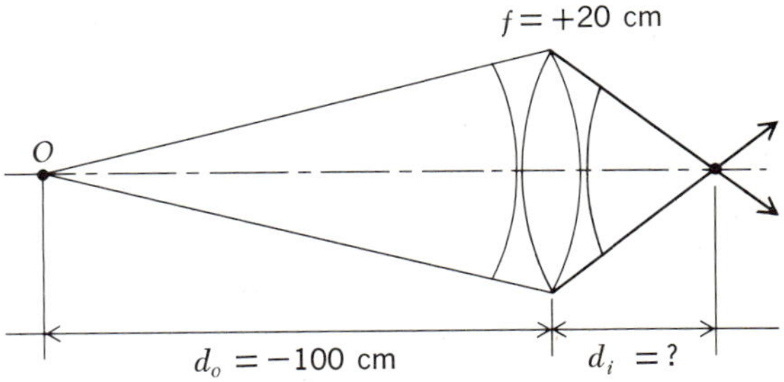

Figure 21-24

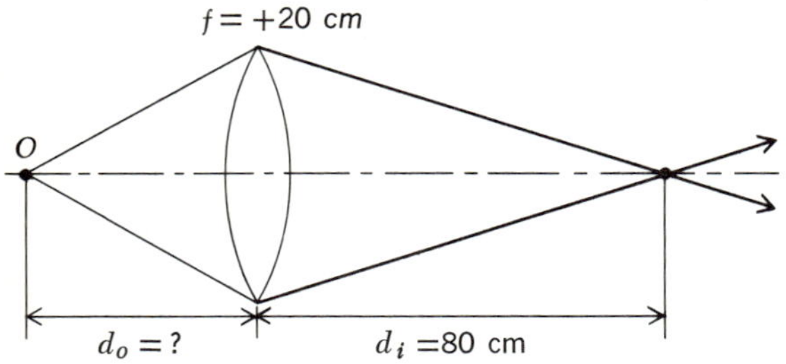

Figure 21-25

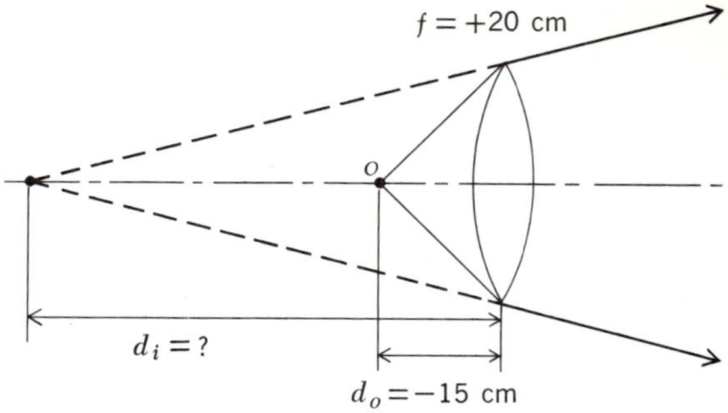

Figure 21-26

$$\text{original curvature} + \text{power of lens} = \text{final curvature}$$

$$-\frac{1}{100\text{ cm}} + \frac{1}{20\text{ cm}} = \frac{1}{d_i}$$

$$\frac{1}{d_i} = \frac{1}{20} - \frac{1}{100}$$

$$= \frac{5}{100} - \frac{1}{100}$$

$$= +\frac{4}{100}$$

$$= +\frac{1}{25}$$

Hence

$$d_i = \boxed{+25\text{ cm}}$$

Example 21-7 Here is another situation using the same lens that was described in Example 21-6. How far from this lens should an object be placed in order to form an image on a screen that is 80 cm from the lens? (See Figure 21-25.)

Since the image is formed on a screen, we are sure that it is a real image and that it must be formed by converging light. The curvature of the beam, therefore, as it leaves the lens is positive, and equal to $+1/80$.

$$\text{original curvature} + \text{power of lens} = \text{final curvature}$$

$$\frac{1}{d_o} + \frac{1}{20\text{ cm}} = \frac{1}{80\text{ cm}}$$

$$\frac{1}{d_o} = \frac{1}{80} - \frac{1}{20}$$

$$= -\frac{3}{80}$$

$$d_o = -\frac{80}{3}\text{ cm}$$

$$= \boxed{-26.7\text{ cm}}$$

The negative sign indicates that the rays are *diverging* as they strike the lens, and hence the object must be 26.7 cm to the left of the lens.

Example 21-8 An object is placed 15 cm from the same lens used in the previous two examples. Compute the position of the image (Figure 21-26).

$$\text{original curvature} + \text{power of lens} = \text{final curvature}$$

$$-\frac{1}{15\text{ cm}} + \frac{1}{20\text{ cm}} = \frac{1}{d_i}$$

$$\frac{1}{d_i} = -\frac{1}{15} + \frac{1}{20}$$

$$= -\frac{1}{60}$$

$$d_i = \boxed{-60\text{ cm}}$$

The negative sign for the image distance indicates a virtual image located to the left of the lens. Also the image lies within the focal length.

Since the simple converging lens is used so often, it is worthwhile summarizing the image-object relationships for various cases (Figure

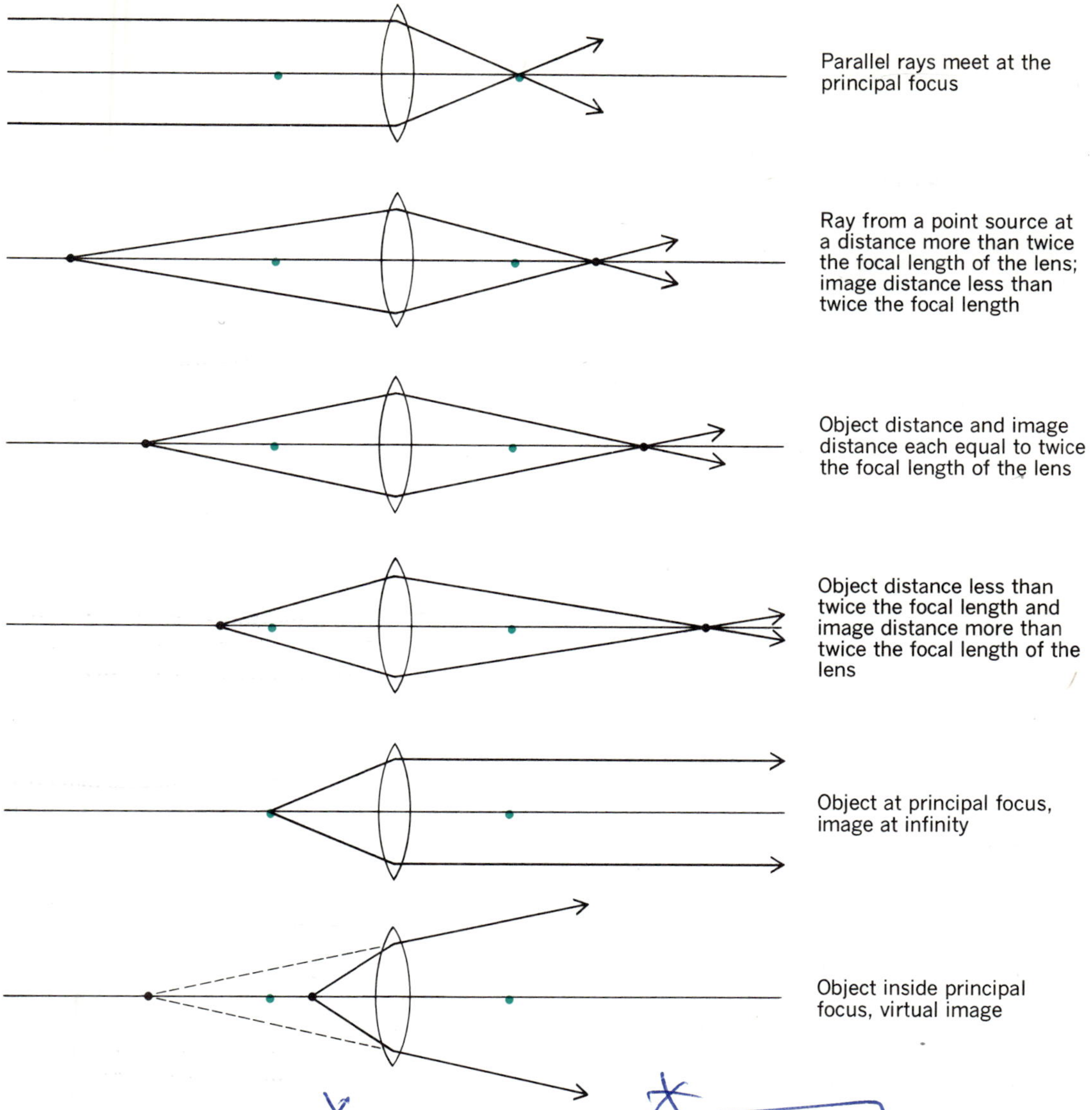

Figure 21-27 Object-image relations for a converging lens.

21-27). All the results shown in the figure can be obtained by applying the method of curvatures as in the examples. Two principal focuses are shown in color, each a distance f from the lens.

A negative lens always gives a virtual image of an object, since the original diverging wave front can only be made more diverging by the action of a negative lens.

Example 21-9 A negative lens of focal length 10 cm is placed 30 cm from an object. Where is the image? (See Figure 21-28).

$$\text{original curvature} + \text{power of lens} = \text{final curvature}$$

$$-\frac{1}{30\text{ cm}} + \frac{1}{-10\text{ cm}} = \frac{1}{d_i}$$

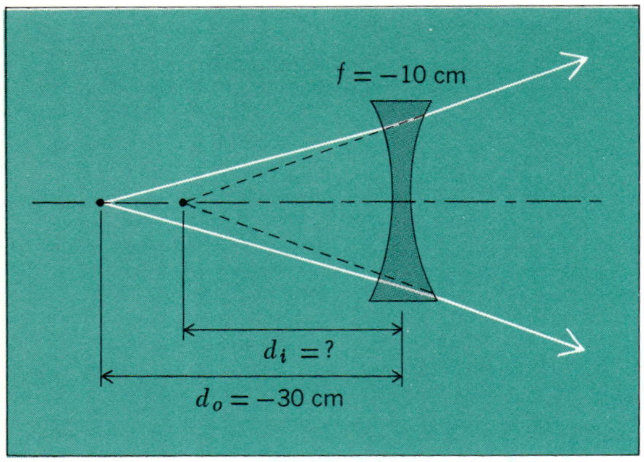

Figure 21-28

$$\frac{1}{d_i} = -\frac{1}{30} - \frac{1}{10}$$

$$= -\frac{4}{30}$$

$$d_i = \boxed{-7.5\text{ cm}}$$

The negative sign for d_i indicates a virtual image formed by diverging rays.

21-8 RAY TRACING: MAGNIFICATION

The method of curvatures is useful for finding the *location* of an image, but it is better to return to rays in order to answer detailed questions about the size and appearance of the image. An object is usually represented by a small vertical arrow to

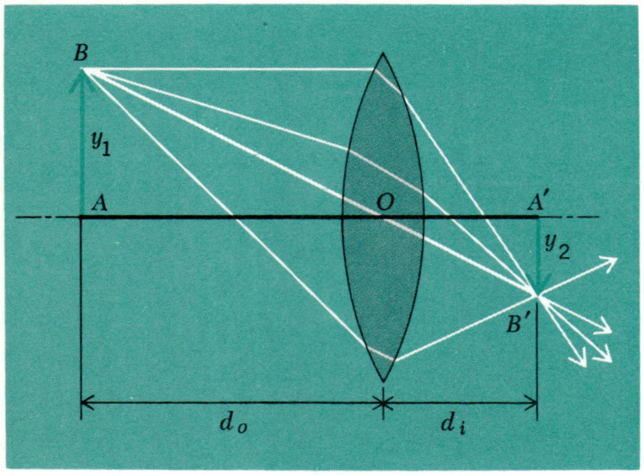

Figure 21-29

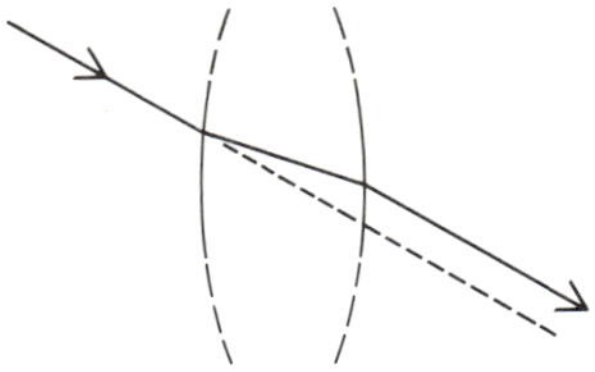

Figure 21-30

the left of the lens (Figure 21-29). The image of a point A on the tail of this arrow is A' at a distance d_i from the lens. Both these points happen to be on the lens axis. A' can be located by the method of curvatures. We assume that the image of B (the head of the arrow) is also located at the distance d_i from the lens, and we wish to compute the size of the image $A'B'$. To a good approximation, *all* the rays leaving B meet at the same image point B', as in the diagram. It is easy to trace the path of one of these rays. The ray that passes through the center of the lens does so without change of direction, since near its center the lens is like a slab of glass with parallel faces. If the lens is a thin one, the lateral displacement (shown in Figure 21-30) will be small, so we draw a straight ray from B to B', through the center of the lens. Now a simple use of similar triangles gives us the important fact that $y_2/y_1 = d_i/d_o$, or

$$\frac{\text{image size}}{\text{object size}} = \frac{\text{image distance}}{\text{object distance}}$$

This ratio (image size)/(object size) is the *magnification m* produced by the lens. The same proportion holds true for virtual images formed by converging lenses (Figure 21-31) and by diverging lenses (Figure 21-32). These diagrams also show that for any single lens, a real image is inverted, whereas a virtual image (at y_2) is erect.

Example 21-10 A slide projector has a converging lens of focal length 50 mm, and the screen is 5 m from the lens. (a) How far from the lens should the slide be placed? (b) What is the height (on the screen) of the image of a portion of the slide that is 4 mm high?

(height)(magnification)
= hm

(a) In this problem the object distance is unknown, and the image distance is 5 m.

(a) $$\frac{1}{d_o} + \frac{1}{.05\text{ m}} = \frac{1}{5\text{ m}}$$

$$\frac{1}{d_o} = \frac{1}{5} - \frac{1}{.05}$$

$$= -19.8$$

$$d_o = \boxed{-50.5\text{ mm}}$$

The slide should be placed just outside the principal focus of the lens. The negative sign is expected, since the rays leaving the object are diverging.

(b) $$\text{magnification} = \frac{\text{image distance}}{\text{object distance}}$$

$$m = \frac{5\text{ m}}{-.0505\text{ m}}$$

$$= -99$$

The image size is -99 times the object size (the negative sign indicates an inverted image). Since the object is 4 mm high, the height of the image on the screen is

$$(4\text{ mm})(99) = \boxed{396\text{ mm}}$$

The next example requires us to find the size of a virtual image.

Example 21-11 A magnifying glass of focal length 20 cm is held 16 cm from a page of fine print. How large is the image of a letter that is 1 mm high?

First locate the *position* of the image, using the method of curvatures. The object distance is 16 cm, and the image distance is unknown.

$$-\frac{1}{16\text{ cm}} + \frac{1}{20\text{ cm}} = \frac{1}{d_i}$$

$$d_i = -80\text{ cm}$$

The negative sign indicates a virtual image, which is expected, since the object is inside the principal focus of the lens. We can now compute the magnification:

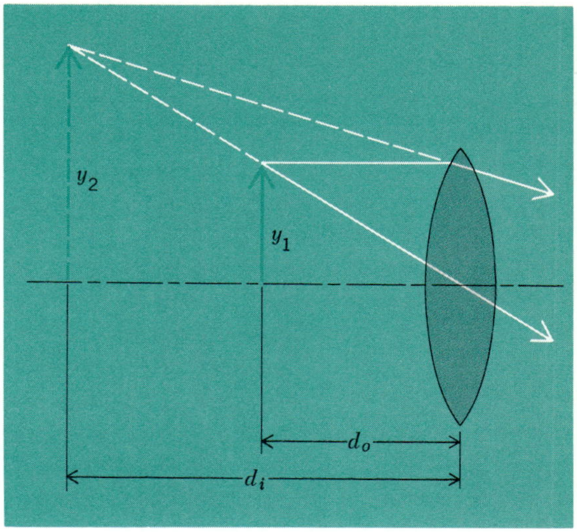

Figure 21-31

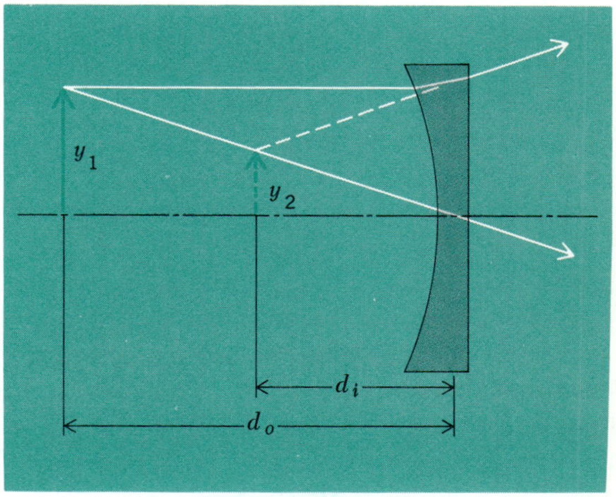

Figure 21-32

$$\text{magnification} = \frac{\text{image distance}}{\text{object distance}}$$

$$m = \frac{-80\text{ cm}}{-16\text{ cm}}$$

$$= +5$$

The positive magnification indicates an erect image (Figure 21-31). Finally, the image size is given by

$$(1\text{ mm})(5) = \boxed{5\text{ mm}}$$

21-9 MIRRORS

We have already discussed the law of reflection for plane mirrors. From a general point of view, a mirror is just another surface that can modify the curvature of a wave front. The same ideas that were used for lenses can be applied to mirrors, with the added understanding that the wave front reverses its direction of motion while being reflected. First consider the plane mirror of Figure 21-33. The wave front's curvature is not changed by reflection, and the original diverging wave front *1* becomes the wave front *2*, which is also diverging, seemingly from a virtual image *I*. Since $AB = CD$, the wave fronts *1* and *2* have the same radius of curvature. Hence the image distance equals the object distance. When you look at yourself in a plane mirror, the virtual image of your face is as far in back of the surface as you are in front of it.

When a plane wave strikes a concave mirror, it becomes converging and forms a real image at the principal focus F (Figure 21-34). While the center of the wave front moves to the right, from A to B, the top moves an equal distance to the left, from C to D. The new wave front is DBD', which is concave toward the left. By Huygens' principle, this wave front then shrinks down to a point at the principal focus and then diverges again. It can be seen that a concave mirror is a converging, or positive, mirror. A converging mirror, like a converging lens, forms a real, inverted image if the object is outside the principal focus, and it forms a virtual, erect image if the object is inside the principal focus. The six typical situations shown in Figure 21-35 are exactly equivalent to those of Figure 21-27, which were drawn for a converging lens. The focal length of a converging mirror equals half the radius of curvature of the mirror. This can be proved from Figure 21-35c, in which the object and image coincide:

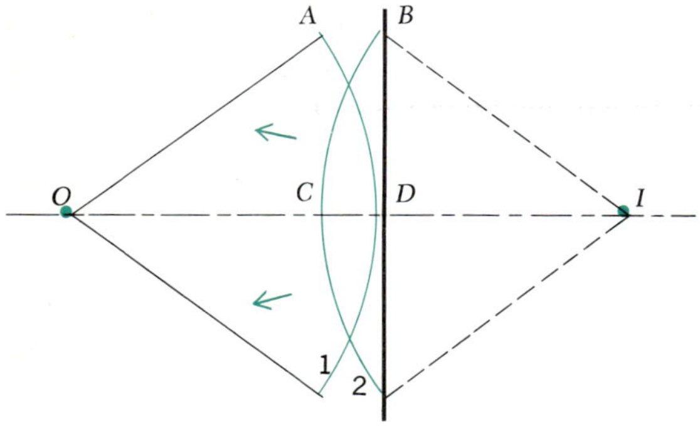

Figure 21-33

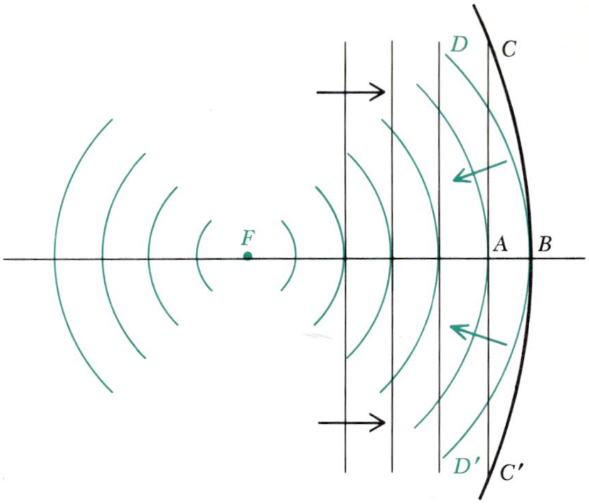

Figure 21-34

$$\text{original curvature} + \text{power of mirror} = \text{final curvature}$$

$$-\frac{1}{x} + \frac{1}{f} = +\frac{1}{x}$$

from which simple algebra gives $f = \frac{1}{2}x$. But in Figure 21-35c the rays are reflected straight back, so they must be along radii of the mirror. Therefore point C is the center of curvature of the mirror. This means that $x = R$, and hence $f = \frac{1}{2}R$.*

Example 21-12 An object of height 6 mm is placed 60 cm from a concave mirror of radius of curvature 30 cm. Where and how large is the image?

The focal length is half the radius, hence $f = 15$ cm. The mirror is concave, and therefore it is a positive mirror.

* This result is valid only for rays that make a small angle with the axis.

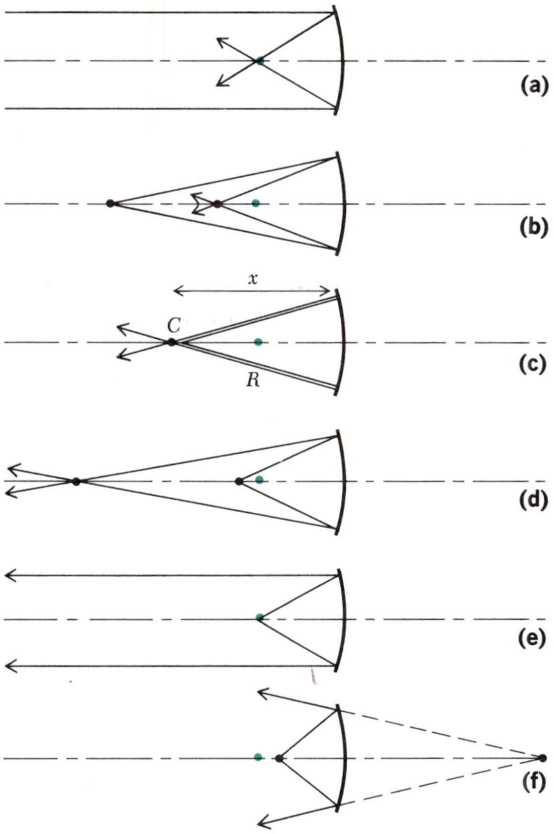

Figure 21-35 Object-image relations for a converging mirror. In each case the focal point is shown in color.

$$\text{original curvature} + \text{power of mirror} = \text{final curvature}$$

$$-\frac{1}{60\text{ cm}} + \frac{1}{15\text{ cm}} = \frac{1}{d_i}$$

$$d_i = \boxed{+20\text{ cm}}$$

Since the final curvature is positive, the image is a real one. The magnification is

$$m = \frac{d_i}{d_o} = \frac{+20\text{ cm}}{-60\text{ cm}} = -\frac{1}{3}$$

The negative sign indicates an inverted image, and its size is

$$\tfrac{1}{3} \times 6\text{ mm} = \boxed{2\text{ mm}}$$

Convex mirrors are not often used. One familiar example is the spherical ornamental mirror that may be found in a garden or a front yard. Another example is in use in department stores, where one small mirror enables an official to see all that is occurring in a large area. Can you explain how? Convex mirrors are diverging or negative. It can be shown by a method similar to that used for concave mirrors, that the focal length of a convex mirror is also equal to half its radius of curvature.

SUMMARY

Huygens' principle states that as a wave propagates, the new wave front is the envelope of wavelets that spread out from all points of the old wave front. If no surface such as a mirror, prism, or lens intervenes, a plane wave remains plane and a spherical wave spreads out spherically. These facts imply straight-line propagation of light waves, and such is actually the case for waves whose wavelength is small compared with the sizes of obstacles or openings.

When light rays strike a smooth surface, both **reflection** and **refraction** may take place; just what happens depends upon the nature of the surface. The reflected and refracted rays will be in a plane determined by the incident ray and the normal to the surface. The **angle of reflection** equals the **angle of incidence.** The **angle of refraction** can be calculated by using Snell's law, which states that $n_1 \sin \theta_1 = n_2 \sin \theta_2$. The **index of refraction** n is the ratio of the speed of light in a vacuum to the speed in the medium. The variation of the speed of light (and index of refraction) with frequency (that is, with color) is called **dispersion.** In most transparent solids and liquids dispersion is relatively slight.

The **critical angle** is the angle of incidence at which a beam of light passing from a dense medium to one that is less dense remains tangent to the surface. **Total reflection** takes place within the denser medium if this angle is exceeded.

The **curvature** of a wave front is defined as the reciprocal of its radius. A diverging wave front has negative curvature, and a converging wave front has positive curvature. Any lens that is thicker at its center than at its edges is a converging, or positive, lens. A lens that is thinner at its center than at its edges is a diverging, or negative, lens. The **power** of a lens is the reciprocal of its **focal length. Images** formed by converging rays are real, and images formed by diverging rays are virtual. Images can be located by considering changes of curvature caused by the lens; initial curvature plus power of lens equals final curvature. Image size is to object size as image distance is to object distance.

A plane mirror gives a virtual image having the same size as the object, for which image distance equals object distance. Concave mirrors are positive, and convex mirrors are negative. The focal length of a mirror equals half its radius of curvature.

QUESTIONS

1 Would you expect Huygens' principle to apply to radio waves? Explain.

2 Do any radio waves travel in straight lines?

3 A beam of white light containing all colors strikes a prism as shown in Figure 21-36, and the rays of different colors are separated because their speeds in glass are different. From a general knowledge of the relative values of index of refraction, state whether the red rays or the violet rays are bent most.

4 Can a ray of light in air be totally reflected at a glass surface if the angle of incidence (in air) and the index of refraction are just right?

5 Which of the lenses in Figure 21-37 are positive and which are negative?

6 Is it possible for a converging wave front to change into a diverging wave front without striking a surface of any kind? Explain.

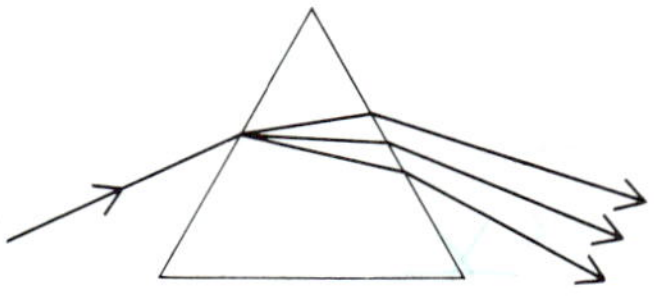

Figure 21-36

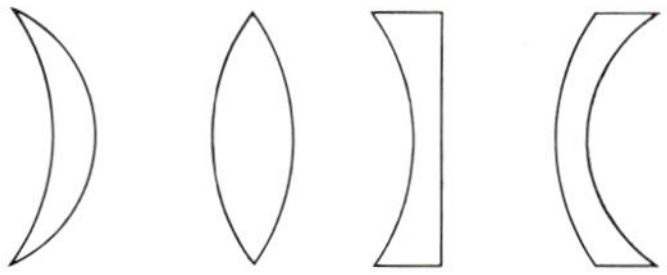

Figure 21-37

7 Is it possible for a diverging wave front to change into a converging wave front without striking a surface of any kind? Explain.

8 How is it possible for a real image to be formed by a plane mirror? What must be the nature of the initial wave front that strikes the mirror—converging or diverging?

9 Diamond has a higher index of refraction than glass. Is the critical angle for a diamond-air interface larger or smaller than for a glass-air interface? How does this affect the brilliance of a diamond gem stone?

10 Sound waves of wavelength 15 cm are obstructed by a building that is 20 m wide and 15 m tall. Is the shadow "sharp?"

11 When hot air rises above a radiator, the wall behind the air stream seems to shimmer. Explain this phenomenon.

PROBLEMS

GROUP A

1 What is the speed of light in a clear plastic material for which the index of refraction is 2.00?

2 What is the index of refraction of a liquid in which the speed of light is 2.25×10^8 m/sec?

3 What is the wavelength, in water, of orange light whose wavelength in air is 6000 Å?

4 What is the ratio of the speed of light in water to the speed of light in benzene? Use Table 21-2.

5 In Figure 21-38 a ray of light enters a liquid and is bent toward the normal as shown. Calculate the index of refraction of the liquid.

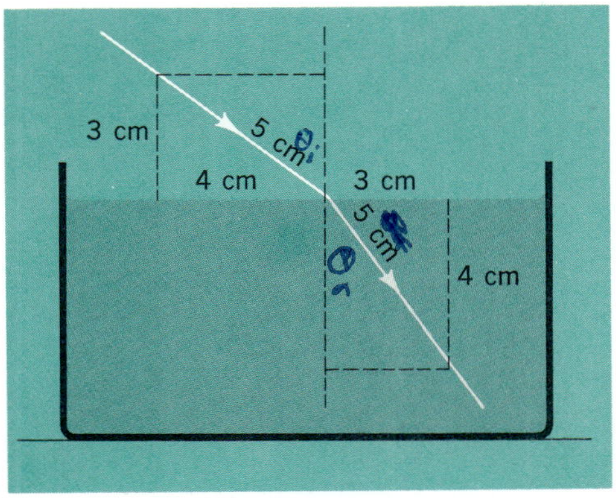

Figure 21-38

GROUP B

6 The rays of the afternoon sunlight strike the surface of a lake at an angle of 30° with the vertical. At what angle, measured from the vertical, is the refracted ray in the water?

7 A ray of light strikes a piece of glass ($n = 1.50$), making an angle of 30° with the surface. What angle does the refracted ray make with the surface?

8 A ray of light starts from a point on the bottom of a dish of liquid 3 cm deep and strikes a point on the surface that is 4 cm from the point directly above the source. What is the least index of refraction for the liquid that will allow the ray to be totally reflected?

9 A lamp bulb is placed 300 cm from a lens of focal length 5 m. (a) What is the radius of curvature of the wave front just as the light strikes the lens? (b) What is this radius as the light leaves the lens?

10 An object 6 mm high is 24 cm to the left of a converging lens of focal length 8 cm. Where is the image? Is it real or virtual? Erect or inverted? How large is it?

11 Repeat Problem 10 for the same object placed 24 cm to the left of a diverging lens of focal length 8 cm.

12 A lens of focal length 25 cm is held 20 cm from an object 1 mm high. Describe the location, nature, and size of the image.

13 A projector lens is placed 50 mm from a slide, and the real image is formed on a screen 7 m from the lens. What is the magnification? What is the focal length of the lens?

14 How far from a converging lens of focal length 30 cm must an object be placed if the image is to be a real one, three times as large as the object? (*Hint:* Let x = the unknown object distance; then the image distance is $3x$.)

15 Repeat Problem 14 for the same converging lens of focal length 30 cm, but the image is to be a virtual one, three times as large as the object.

16 A dentist holds a concave mirror of curvature radius 50 mm at a distance of 20 mm from a cavity in a tooth. What is the magnification of the image? (*Hint:* First find the image distance.)

17 A bird views himself in a convex garden mirror of radius of curvature 15 cm. If the bird is 0.6 m from the surface of the mirror, where is the image? What is the magnification?

GROUP C

18 A positive lens of focal length 20 cm forms a real image 5 times as large as the object. How far apart are the object and the image?

FOR FURTHER STUDY

Derivation of Snell's Law

We can apply Huygens' principle to derive a relation between θ_1 and θ_2. During the time interval Δt, ray *1* spreads out from *A* to *C*, while ray *4* travels from *B* to *D*. The radius *AC* of the wavelet through *C* is given by velocity $\times$ time $= c_2\,\Delta t$, whereas the distance *BD* is given by $c_1\,\Delta t$. Using plane geometry, we identify $\angle BAD$ as equal to θ_1, and $\angle ADC$ as equal to θ_2. By using the definition of the sine of an angle, we have

$$\sin\theta_1 = \frac{c_1\,\Delta t}{AD}$$

$$\sin\theta_2 = \frac{c_2\,\Delta t}{AD}$$

Dividing one equation by the other and simplifying gives

$$\frac{\sin\theta_1}{\sin\theta_2} = \frac{c_1}{c_2}$$

But since $c_1 = c/n_1$ and $c_2 = c/n_2$, we can write

$$\frac{\sin\theta_1}{\sin\theta_2} = \frac{c/n_1}{c/n_2} = \frac{n_2}{n_1}$$

and finally,

$$n_1 \sin\theta_1 = n_2 \sin\theta_2$$

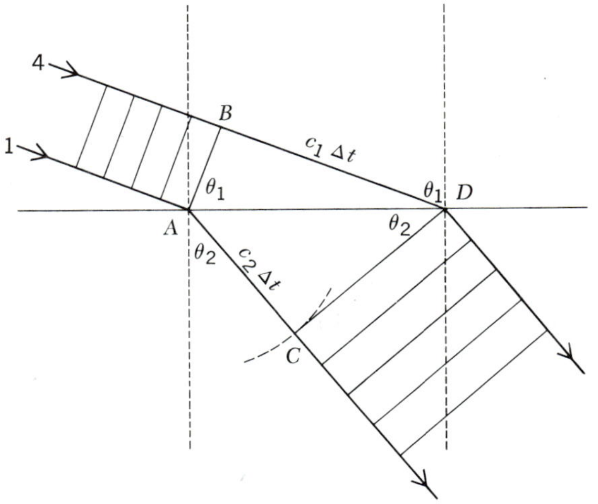

Figure 21-39 Derivation of Snell's law using Huygens' principle.

22 The Wave Nature of Light

Light is perhaps the most familiar of all physical phenomena yet it is, at the same time, one of the most mysterious and challenging. The attempt to understand the nature of light has occupied some of the greatest minds, and this has led directly to the two great theories of "modern physics"—Einstein's special theory of relativity (which was examined briefly in Chapter 5) and quantum mechanics, which will be discussed in Chapter 24.

In the previous chapter it was possible to explain many important properties of light, using a model that involved only the ideas of "wave fronts" and "rays." Rays, which were described as lines indicating the direction in which light energy was flowing, served to show the various ways in which the direction of light can be altered, as in reflection and refraction. Wave fronts and rays are a convenient kind of construction to help us solve certain types of problems.

The difficulties of model making are nowhere more evident than in the case of light. We must bear in mind that no matter what kinds of models we make for light, the essential fact is "light is light." This apparently meaningless statement is to remind you that light is unique in its own right and can only be described in terms of its properties that can be measured experimentally. It is not entirely correct to say "light is a wave" or "light is like small ripples on water." It is better to say that in certain types of experiments light has properties like those of ripples or other kinds of mechanical waves. But there are many ways in which light does not resemble water waves, and in fact there are some types of experiments in which light does not show any wave properties at all. In this chapter we are concerned with the kinds of experiments in which the wave properties are clearly shown. Later, in Chapter 24, we will examine some of the experiments in which particle properties are prominent.

In short, you should not worry about this "dual" nature of light, but rather be aware that light will reveal wave aspects in one experiment and particle aspects in another. Many times in scientific work the "answers" we get in an experiment depend on what "questions" we ask.

22-1 NEWTON'S CORPUSCULAR THEORY GIVES WAY

It was not until about 1800 that anyone was able to show clear experimental evidence for the wave nature of light. This was provided by Thomas Young (1773–1829), a British physician and experimeter. But what did people believe about light before that time? Huygens, Newton, and others recognized the *possibility* that light is a form of wave motion. In fact, the laws of reflec-

Figure 1-4 (continued from page 7) Spectrum of Krypton-86 showing the orange light used as a basis for defining the meter.

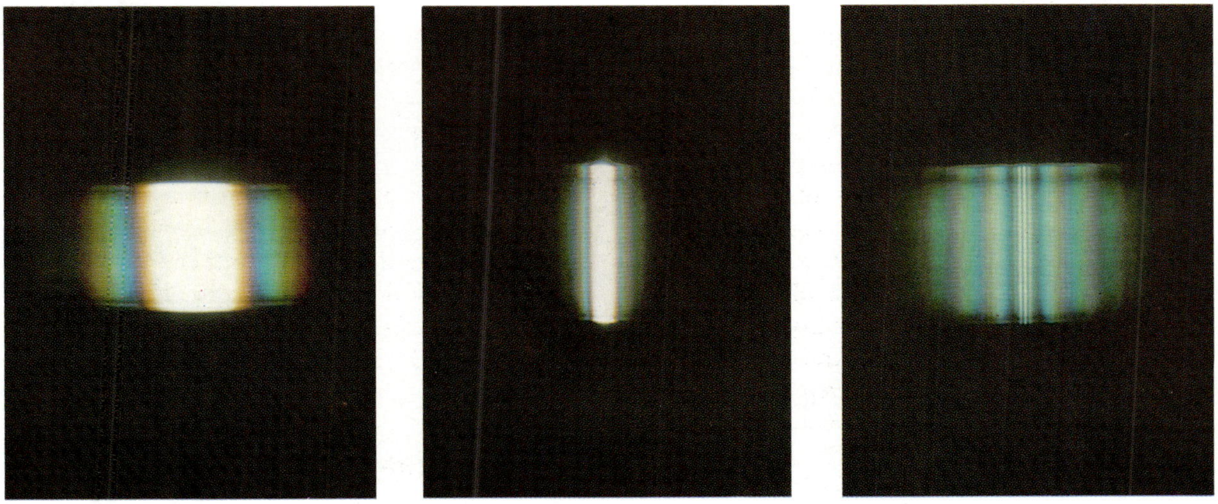

a b c

Figure 22-18 (a) Vertical lamp filament viewed through narrow slit. The central diffraction maximum is much broader than the true width of the source. **(b)** Same source as **(a),** viewed through a wider slit. The central diffraction maximum is narrower, and is a better representation of the actual source. **(c)** Vertical lamp filament viewed through two narrow slits. The closely spaced sharp fringes within the central diffraction band are the two-slit interference pattern; the over-all broad structure of colored fringes is related to the single-slit patterns of the individual slits. In each photograph, the horizontal striations are caused by the coiled structure of the vertical lamp.

TYPICAL SPECTRA

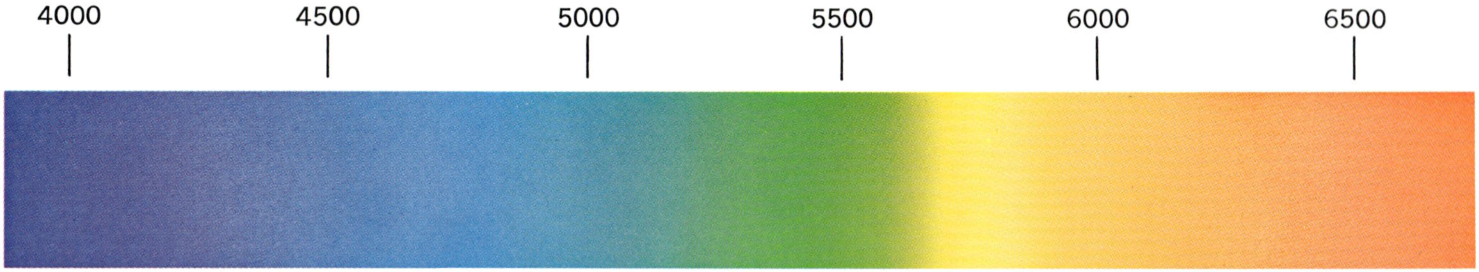

(a) Continuous spectrum of glowing solid.

(b) Absorption spectrum.

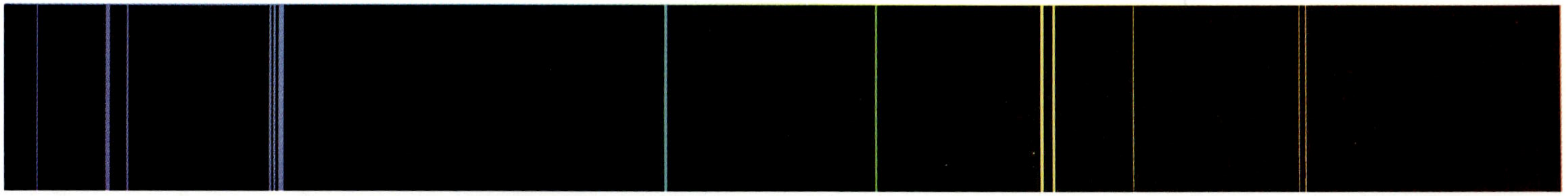

(c) Line or emission spectrum for mercury.

(d) Emission spectrum for sodium.

(e) Emission spectrum for helium.

(f) Emission spectrum for neon.

(g) Emission spectrum for atomic hydrogen (H).

(h) Emission spectrum for molecular hydrogen (H_2).

Figure 24-9 Wavelengths are in angstroms. The dark lines in **(b)** are caused by cooler vapors of various elements near the surface of the sun absorbing light of specific colors. The spectrum of molecular hydrogen **(h)** shows two of the lines of atomic hydrogen **(g)** since some of the molecules are dissociated into atoms.

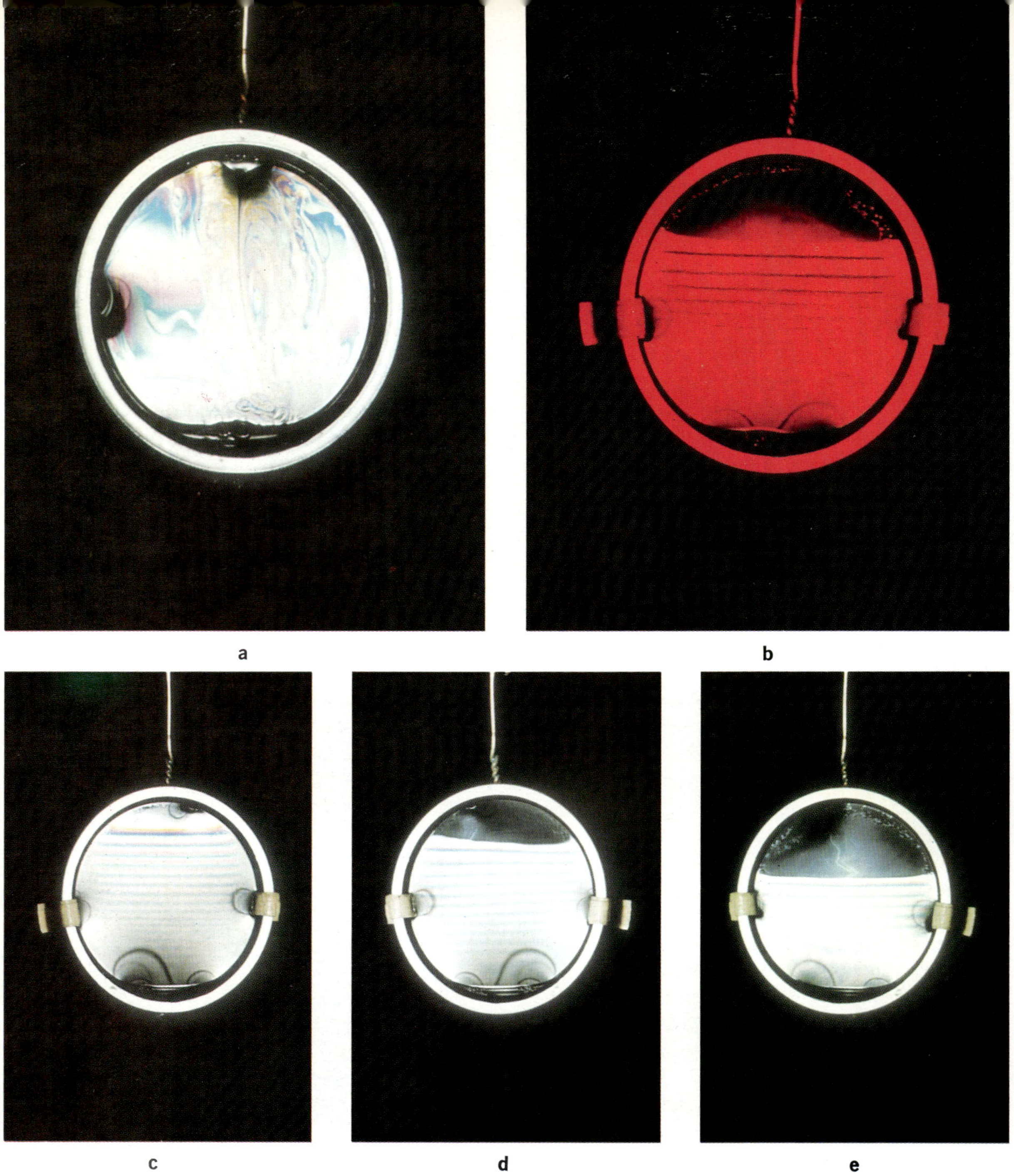

Figure 23-7 Interference of light as seen on soap films. **(a)** Pattern shown on a thick, detergent film of irregular shape. **(b)** Pattern of red light reflected off thin water-and-soap film. What does this pattern tell us about changes in the thickness of the film? **(c)** White light reflected off a thin water-and-soap film. Notice that, as most of the material settles to the bottom, the area at the top becomes invisible because it is less than $\frac{1}{4}\ \lambda$ thick for all wavelengths of visible light.

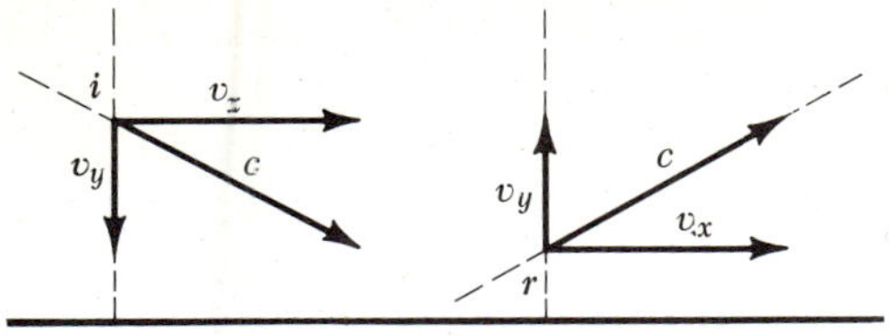

Figure 22-1

tion and refraction (geometrical optics) had been derived from Huygens' principle (Chapter 21). Yet these laws of reflection and refraction can be derived equally well from a "bullet," or "corpuscular," theory. Anyone who has played billiards or tennis knows that for an elastic ball bouncing off a hard surface, assuming there is no "spin" on the ball, the angle of incidence equals the angle of reflection.

Newton was able to show that in a perfectly elastic collision the law of reflection can be derived by applying the laws of motion. The particle, or corpuscle, approaching the surface with velocity c has horizontal and vertical components of velocity v_x and v_y (Figure 22-1). As it leaves the surface, v_x remains unchanged because this is an elastic, frictionless collision and the surface can apply no horizontal component of force to change the horizontal velocity. The vertical component v_y is reversed in direction, but its magnitude is unchanged. This is reasonable because there will, of course, be a force perpendicular to the surface. The magnitude remains the same, since this is an elastic collision in which the kinetic energy is the same before and after the collision, and so the velocity of separation is equal in magnitude to the velocity of approach.

Applying these ideas to the reflection of light, we know that the velocity of light is constant in a given medium. It can be assumed that the kinetic energy of a light corpuscle is also constant, which means that "light collisions" are elastic. Therefore the reasoning illustrated in Figure 22-1 applies. It is evident from the diagram that the angle of incidence equals the angle of reflection ($i = r$), and the law of reflection is proved.

Newton also had a corpuscular theory for refraction. He proposed that when light approaches a boundary (say, between air and glass), it is attracted toward the surface by some kind of force. He assumed, as shown in Figure 22-2, that the force of the glass on the corpuscle would be perpendicular to the surface. Therefore the perpendicular, or "normal," component v_y increases, but the horizontal component v_x is not changed. The resultant velocity in the glass c_2 makes an angle θ_2 smaller than θ_1 with the normal. Thus this theory predicts that the ray bends toward the normal, which does agree with experiment. It also predicts that the speed of light is greater in a dense medium such as glass.

Newton's corpuscular theory was, in its day, superior to the competing wave theory in certain respects: (a) it used the laws of mechanics that had been proved valid in other areas of physics, and (b) it agreed closely with the known fact that light traveled in straight lines and did not appear to bend around obstacles, as waves do.

It seemed as if an important experiment would clarify the issue—one only needed to measure the speed of light in air and in glass. According to the wave theory, light should slow down as it enters glass (Section 21-4); according to the corpuscular

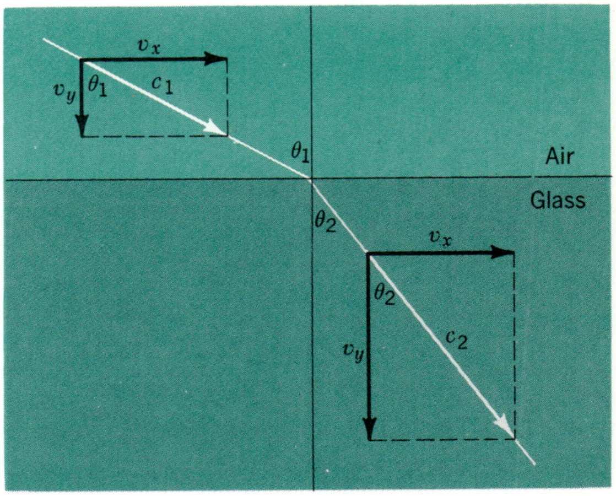

Figure 22-2 Newton's proof of Snell's law using a corpuscular model for light.

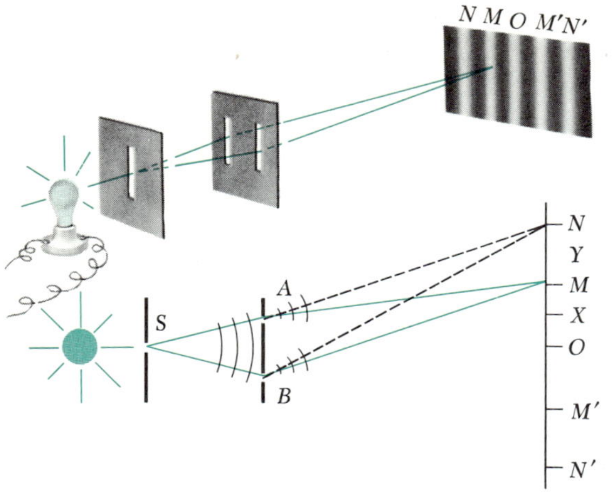

Figure 22-3 Young's experiment.

theory, it is attracted toward the surface, and should speed up as it enters glass. In Newton's time, the speed of light could not be measured in the laboratory, therefore either assumption was possible. It was not until 1850, over a century after the death of Newton, that Jean Foucault succeeded in showing that the speed of light in a dense medium is less than in air. But by that time this proof was no longer needed. In the early years of the nineteenth century the ingenious experiments of Young and others provided direct evidence for the wave behavior of light. You will find in Chapter 24 that the new corpuscular (photon) aspects of light are much more subtle than Isaac Newton's model and are based on evidence entirely unknown to Newton.

22-2 INTERFERENCE: YOUNG'S EXPERIMENT

Thomas Young was the first to design and interpret an experiment that clearly showed the *interference* of light waves. Figure 22-3 shows the important features of this experiment. A narrow vertical slit S is illuminated by a source of light. A wave front spreads out from S, falling on two narrow slits *A* and *B*. To see the effect, these slits must be very narrow and closely spaced—a small fraction of a millimeter between them. Light, in turn, spreads out from *A* and *B*, and all points on the screen should receive light from both slits. Young observed the remarkable result that the screen is not uniformly illuminated. Some points, such as those in regions *O*, *M*, and *N* are bright, and others are dark. How can we interpret this result?

Let's make use of the principle of superposition (Section 9-6), and then let's try the dangerous but often helpful practice of reasoning by analogy. We can draw upon our study of water waves (Section 9-7) to see if there is an effect that occurs with ripples which might help us to understand this behavior of the light in Young's experiment.

Figure 22-4 is a photograph of an interference pattern formed on shallow water in a ripple tank with two vibrating sources at the bottom of the picture. A light shines upward through the water. The curved surfaces of the wave crests act as converging lenses to focus bright lines of light

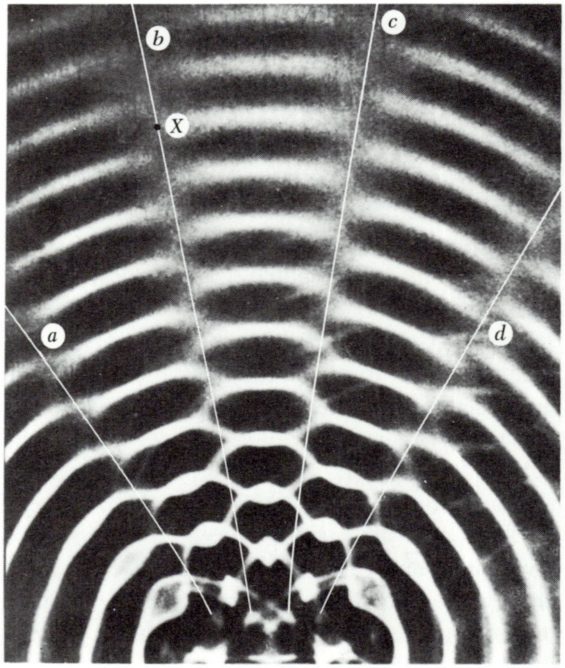

Figure 22-4 Constructive and destructive interference between water waves spreading out from two sources.

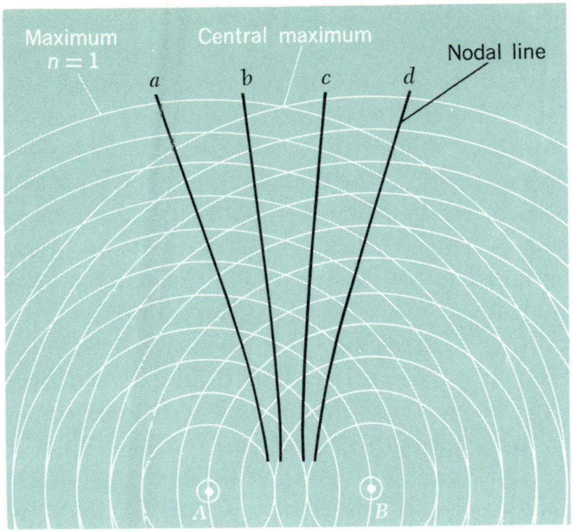

Figure 22-5

which, as you can see, stand out clearly in the photograph. To the right and the left at the bottom of the picture, are waves radiating outward, a separate set of waves coming from each source. But something unusual seems to be occurring in the region where waves from both sources are present. First, notice the lines marked *a*, *b*, *c*, and *d*. There does not seem to be any wave motion along these lines. These lines of no wave motion are called *nodal lines.* Now let's look vertically upward from the line joining the sources at the area between lines *b* and *c*. The bright lines represent crests of the ripples, and the dark spaces between mark the locations of troughs. This central region, where alternating crests and troughs are moving out through the water, is called the *central maximum.* Other regions where wave activity is similar to that in the central maximum, are also called *maximums.*

Along a nodal line, such as *b* in Figure 22-4 there is absolutely no wave activity anywhere. How can we explain this lack of disturbance of the water's surface? Look at the point on the nodal line, marked *X* in the photograph. Here, a crest from the left of the nodal line meets a trough from the right of the same line. Where they come together, the wave smears out into a region of no activity. All along the nodal line are troughs meeting crests and interfering destructively.

Figure 22-5 is a diagram showing how the pattern in Figure 22-4 was produced. At the bottom of the picture are the two sources vibrating up and down together *in phase.* Two sets of concentric circles have been drawn, radiating from source *A* and source *B*. These circles represent the crests coming out from each source; the spaces between represent troughs. Crests from *A* interfere constructively with crests from *B* in the center to form the central maximum. Constructive interference occurs in places where two crests intersect, making stronger waves.

On the left is a region where crests from *A* meet troughs from *B*, and crests from *B* meet troughs from *A*. Here we have destructive interference; the waves are *out of phase,* and a nodal line (marked *b*) is formed. Going a little further to the left, there is another region where crests from one source meet crests from the other. Again there is constructive interference and another maximum. There can, in fact be a number of nodal lines and maximums; Figure 22-4 shows three maximums and 4 nodal lines. Look at Figure 22-6; how many of each are visible there?

Figure 22-6

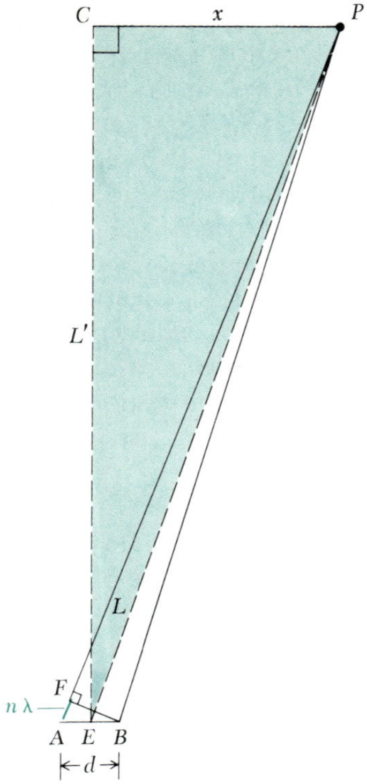

Figure 22-7

By using the ideas of superposition and interference, we have been able to move one step further in understanding this kind of interference pattern. Now we can be quantitative and try to predict just where the nodal lines and maximums will be found for various combinations of wavelengths and positions of the sources.

First, let's try to predict where points on the maximums will occur (the points where two crests intersect in Figure 22-5). Figure 22-7 shows the two sources A and B vibrating in phase. Let's consider a point P on a maximum a long way from the sources; that is, the distance from either source to P is very large compared to the separation d between the sources. In this case the lines drawn from both sources to P (AP and BP) are very nearly parallel, the dashed line L' is the perpendicular bisector of the line between the sources, and x is the perpendicular distance from the center line to the point P.

Look at the tiny triangle ABF formed by the sources and by the line BF drawn perpendicular to AP from B. For constructive interference, crests from each source must arrive at P at the same time. Therefore the path difference AF must be zero, λ, 2λ, or 3λ (in other words, a whole number of wavelengths, which is indicated on the drawing as $n\lambda$). Here n means any whole number—any integer.

Next, we must show that triangle ABF is similar to triangle ECP. The dashed line EP has been drawn from the midpoint between the sources to P, and its length is designated as L. Both ABF and ECP are right triangles because right angles have been constructed at C and F. Furthermore, the sides of the triangles are mutually perpendicular, $CE \perp AB$ and $EP \perp BF$. This means that corresponding angles of the two triangles are equal, and the triangles are similar. Therefore we know that the ratios of corresponding sides are equal, and we can say

$$x/L = n\lambda/d$$

This enables us to predict where to find a point P on a given maximum at some specified distance L from the sources, if the wavelength and the source separation are known:

$$x = L\frac{n\lambda}{d}$$

The symbol n, which stands for any integer, enables us to locate P (point on a particular maximum). By letting $n = 1$, we can select the first maximum on either side of the central maximum ($n = 0$ for the central maximum); for $n = 2$, we have the second maximum, etc. The entire extent of a maximum could be determined by calculating the positions of a number of points on it, for a range of values of L.

It is also important to examine the geometry of points on nodal lines. Figure 22-8 is almost exactly the same as Figure 22-7, but in this case P

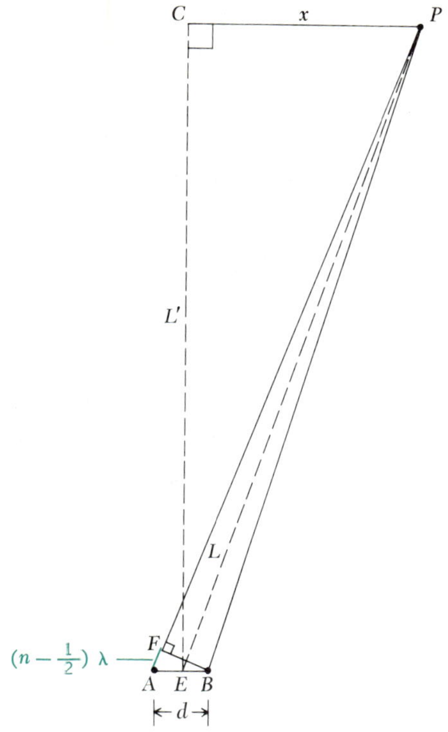

Figure 22-8

Notice in Figure 22-4 that nodal lines are fairly well defined in the ripples on the water and that the distances x and L could be measured with some accuracy. But it is hard to locate the maximums in the ripples because the center of the maximum region does not stand out clearly. It is easy though to count the value of n, which is 1 for the first nodal line to the left or right of center, 2 for the second line, etc. It is also a simple matter to measure the source separation d. We now have all we need to measure the wavelength λ in terms of the other quantities. Solving for λ, we get

$$\lambda = \frac{(x/L)d}{n - \frac{1}{2}}$$

Example 22-1 In Figure 22-9, waves on the surface of a lake strike a breakwater that has two small holes A and B in it, 2 m apart. A and B act as new sources of waves, vibrating in phase. A man in a rowboat at P, 40 m from the midpoint of a line connecting the sources, notices that he is in

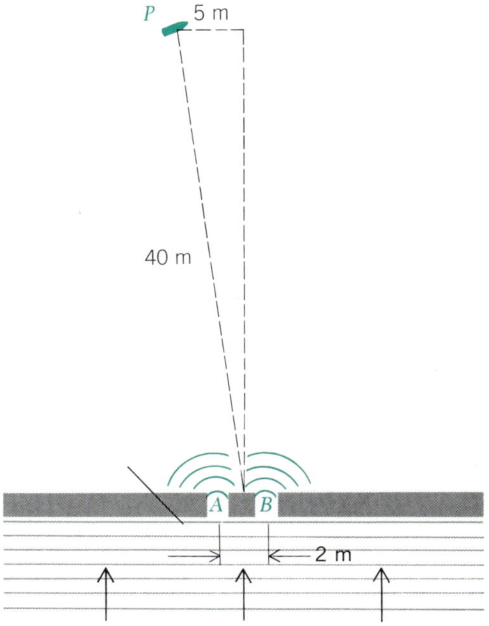

Figure 22-9

is placed on a nodal line. Since a nodal line is formed by *destructive* interference, waves from the two sources must arrive at P out of phase; the crests from one must meet the troughs from the other. This means that the path difference, $(AP - BP)$ or AF (defined by the perpendicular BF), must be equal to $\frac{1}{2}\lambda$ or $\frac{3}{2}\lambda$ or $\frac{5}{2}\lambda$, etc. In general, AF must be some *odd* number of *half-wavelengths*, and this has been indicated as $(n - \frac{1}{2})\lambda$ on the drawing. Again n means any integer. You can easily satisfy yourself that the expression $(n - \frac{1}{2})$ always gives an odd number of halves. Try it! Different values of n enable us to select different nodal lines. For example, $n = 2$ defines the second nodal line away from the central maximum. Is there a nodal line for $n = 0$?

The triangles are similar for the same reasons as before, and the ratios of corresponding sides are

$$x/L = (n - \tfrac{1}{2})\lambda/d$$

a narrow region of calm water. He also notices that there is another region of calm water between him and the central region of heavy wave activity; he is 5 m from this central region.

What is the wavelength of these waves?

How far to the left of the center line is a point Q on the first maximum, 40 m from the sources?

At P the rowboat is on a nodal line for which $n = 2$. The appropriate formula is

$$\lambda = \frac{(x/L)d}{n - \frac{1}{2}} = \frac{(5\text{ m}/40\text{ m})\ 2\text{ m}}{2 - \frac{1}{2}} = \frac{(\frac{1}{8})\ 2\text{ m}}{\frac{3}{2}} = \boxed{\tfrac{1}{6}\text{ m}}$$

For a point on the first maximum, $n = 1$ and we use the formula

$$x/L = n\lambda/d$$

$$x = n\lambda L/d = \frac{(1)(\frac{1}{6}\text{ m})(40\text{ m})}{2\text{ m}} = \boxed{3.33\text{ m}}$$

22-3 INTERFERENCE OF LIGHT WAVES

Measuring the wavelength of waves in a lake is not terribly significant in itself, but the same technique can be used to determine the wavelength of light waves as Thomas Young did. Figure 22-10 shows the diagram of Young's experiment, redrawn and lettered to correspond to Figures 22-7 and 22-8. You will recall that the two slits A and B are illuminated by light from the narrow slit S. This arrangement is very similar to that of Figure 22-9 and is one way of ensuring that the light vibrations at A and B are in phase. It enables us to say that the slits A and B are like the in-phase sources in our discussion of water waves in the previous section. With light waves, the nodal lines (often called *minimums*) lead to dark bands in the interference pattern, and the maximums lead to bright bands. Thus the bright bands at Q and P correspond to maximums with $n = 1$ and $n = 2$. The position of P (a point on the second maximum) is given by

$$x = L\frac{2\lambda}{d}$$

Since the bright bands stand out more clearly, it is easier to make measurements with them rather than with the dark bands. Therefore to determine λ we can use the formula for a maximum:

$$x/L = n\lambda/d$$

$$\lambda = \frac{(x/L)d}{n}$$

Here d refers to the spacing between the slits, L is the distance from the sources to the point on the

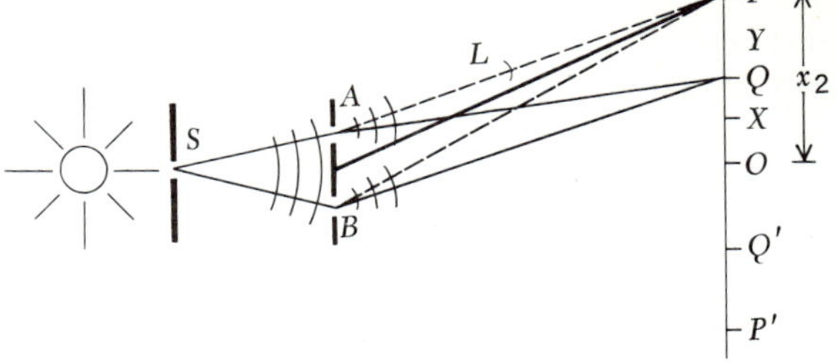

Figure 22-10 Young's experiment.

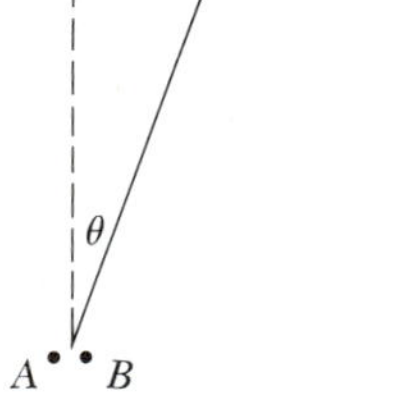

Figure 22-11

screen at the center of the bright band, and x is the distance out from the center line to the center of nth bright band.

Example 22-2 Two slits 0.03 mm apart are illuminated in phase, and the center of the fifth-order ($n = 5$) bright band is 14 cm from the central bright band when observed at a distance L, 2 m from the slits. Calculate the wavelength of the light used.

$$\lambda = \frac{(x/L)d}{n}$$

$$= \frac{(14\ \text{cm}/200\ \text{cm})(3 \times 10^{-5}\ \text{m})}{5}$$

$$= 4.2 \times 10^{-7}\ \text{m}$$

$$= \boxed{4200\ \text{Å}}$$

(One angstrom is 10^{-10} m.) This is violet light (Table 21-1, page 364).

Example 22-3 Parallel rays of light of wavelength 5000 Å fall upon a pair of slits that are 0.01 mm apart. The interference pattern is formed on a screen that is 2.30 m from the slits. Calculate the separation between the two third-order ($n = 3$) bright bands. These bands are on either side of the central bright band.

We are given the perpendicular distance L' from the sources to the screen, but not the diagonal distance to P, which has been called L (Figure 22-11). Can we make an approximation to simplify some of our calculations? Notice that if the angle θ (called the angle of deviation) is very small, then the distances L and L' are very nearly the same. We can also say that if the deviation x is small, there is very little difference between the distance L from the sources to the interference band and the distance L' measured from the sources perpendicularly to the screen. Let's make use of this "small angle approximation" and assume that L and L' are essentially the same. Therefore the formula becomes $x/L' = n\lambda/d$. Since we want the total separation between the centers of the bands,

$$2x = \frac{2n\lambda L'}{d}$$

$$= \frac{2(3)(5 \times 10^{-7}\ \text{m})(2.30\ \text{m})}{1.0 \times 10^{-5}\ \text{m}}$$

$$= \boxed{0.69\ \text{m}}$$

In many cases it is useful to calculate the angle of deviation θ. You will recall that the sine of an angle is defined in a right triangle as the ratio of the side opposite the angle to the hypotenuse of the triangle. In Figure 22-11 the side opposite angle θ is x and the hypotenuse is L. Therefore

$$\sin\theta = \frac{x}{L}$$

The formulas can now be written to include this new relation.

For a maximum:

$$\sin\theta = \frac{x}{L}$$

$$= \frac{n\lambda}{d}$$

For a minimum:

$$\sin\theta = \frac{x}{L}$$

$$= \frac{(n - \frac{1}{2})\lambda}{d}$$

Example 22-4 Calculate the angle of deviation for a third-order ($n = 3$) maximum, using the data of Example 22-3.

$$\sin\theta = \frac{3\lambda}{d}$$

$$= \frac{3(5 \times 10^{-7}\ \text{m})}{10^{-5}\ \text{m}}$$

$$= 0.15$$

From Table 5 in the Appendix, we find that

$$\theta = \boxed{8.6^\circ}$$

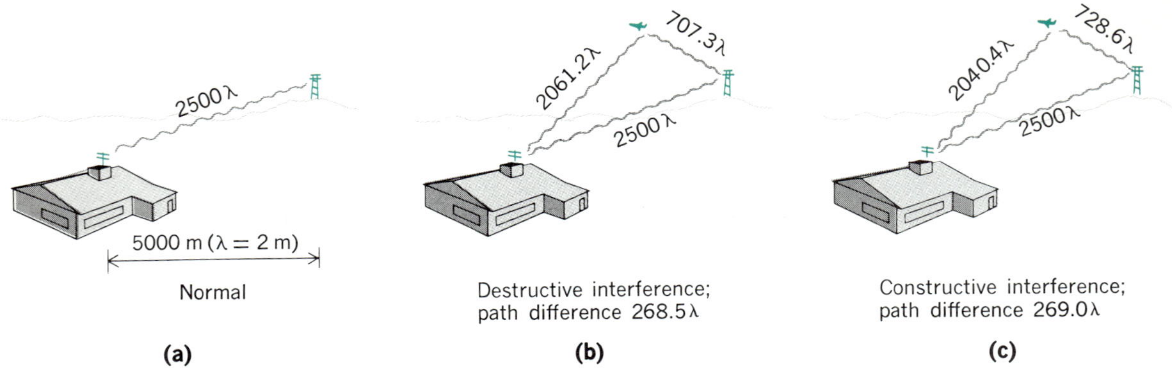

Figure 22-12 Constructive and destructive interference of TV waves.

22-4 INTERFERENCE OF TV AND RADIO WAVES

As you have already seen, the essential feature of interference is that a difference in the length of the path traveled by two wavefronts is equivalent to a difference in phase. Suppose two beams of light reach a certain point and are in phase with each other. If we stretch out the length of the path of one beam by an extra half-wavelength, this has exactly the same effect as retarding its phase at the end point of the path by one half-cycle of vibration. Thus, as you have seen, any odd number of half-wavelengths of path difference will produce destructive interference (a minimum), and an integral number of whole wavelengths will produce constructive interference (a maximum). Recall that for a nodal line, or minimum, the path difference was $(n - \frac{1}{2})\lambda$, and for a maximum, $n\lambda$.

These ideas apply to any wave motion, not just to visible light and water waves. You have probably noticed some effects of the interference of radio or TV waves, without being aware of the causes. Normally a television signal (with a wavelength of several meters) travels to the receiver by a direct line-of-sight path (Figure 22-12a). If a plane flies past, a beam reflected from the airplane may also reach the antenna. Figure 22-12b shows a situation in which the waves interfere destructively because the path difference is an odd number of half-wavelengths. In Figure 22-12c the plane has moved on, the path difference has become a whole number of wavelengths, and constructive interference takes place. The resultant signal varies in intensity while the plane is flying by, and the picture flutters. A similar effect can be caused at a slower rate by reflection from a drifting cloud or even by reflection from a trail of ionized atoms left in the atmosphere by a falling meteor.

Radio broadcasters sometimes make use of interference to beam their signals into a desired area. For instance, it would be wasteful for a station near the seacoast to broadcast signals of equal power in all horizontal directions, which is what happens if a single vertical antenna is used.

One way to solve this problem is to use a pair of antennas $\frac{1}{4}\lambda$ apart. By special electrical circuits, the current in one antenna is made to reach its peaks just $\frac{1}{4}$ cycle later than that in the other antenna (Figure 22-13). Point Y is $\frac{1}{4}\lambda$ farther from A than from B. Also A is made to radiate $\frac{1}{4}$ cycle later than B, which is equivalent to another $\frac{1}{4}\lambda$ of path difference. These two quarter-wavelengths add up to $\frac{1}{2}\lambda$, and therefore at Y there is destructive interference. No energy is radiated out to sea in the direction of Y.

Just the opposite is true at X. The tower at A is $\frac{1}{4}\lambda$ closer to X than is B, but it is fed $\frac{1}{4}$ cycle later than B, so the waves reach X in phase. There is constructive interference, and the resultant amplitude is twice as much as from either antenna

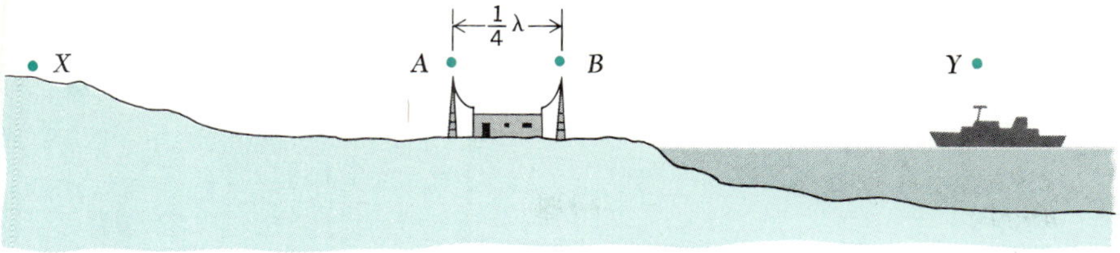

Figure 22-13 Tower A radiates ¼ cycle later than tower B. Radiation pattern is stronger at X than at Y.

separately. Can you figure what the entire radiation pattern of these antennas would be like?

Looking at this example of interference from the viewpoint of energy, we expect that the total energy radiated from the antennas must appear somewhere in the area surrounding the towers. Detailed investigation of the radiation patterns would show that this is true. All the energy is accounted for, but the *distribution* of the energy depends on the number of antennas, their spacing, and the relative phase of each. In a similar way, the total light energy falling on two slits in Young's experiment is fully accounted for, even though certain places on the screen receive no light while others are brighter than they would be if the two slits were not there.

22-5 THE SINGLE-SLIT DIFFRACTION PATTERN

Figure 22-14 and the others like it in this chapter and the next have been simplified to show only those rays that are of interest and all the other rays have been left out. In fact, Huygens' wavelets *do* diverge in all directions from each slit. But we pick a direction of observation in which we are interested, indicated by the angle θ. If someone objects by asking "How do the rays know enough to bend at an angle θ?" we know that they are not emanating at just that one angle. For the present we are ignoring the other directions and later on we can investigate the behavior of the rays in any other direction we wish.

In our discussion of Young's experiment involving the interference of light from two slits, we assumed that the slits were *narrow*. If a slit is narrow enough, it can be assumed that all the rays passing through the slit are in phase with each other. This is true enough if the slit width is small compared to a wavelength. In other words, rays *1*, *2*, and *3* in Figure 22-14 are sufficiently in phase with each other to act as a single ray; and this "super ray" is in phase with the "super ray" composed of rays *4*, *5*, and *6*, if the path difference to some distant point is a whole number of wavelengths ($n\lambda$ in the drawing). Of course, the phases of *1* and *3* do differ by a slight (but negligible) amount for any real (but narrow) slit.

With wider slits, however, phase differences between rays passing through various parts of the slit must be taken into account. Indeed, a very large number of rays are involved here, but for convenience only nine have been shown in Figure 22-15. The term *diffraction* refers to the interfer-

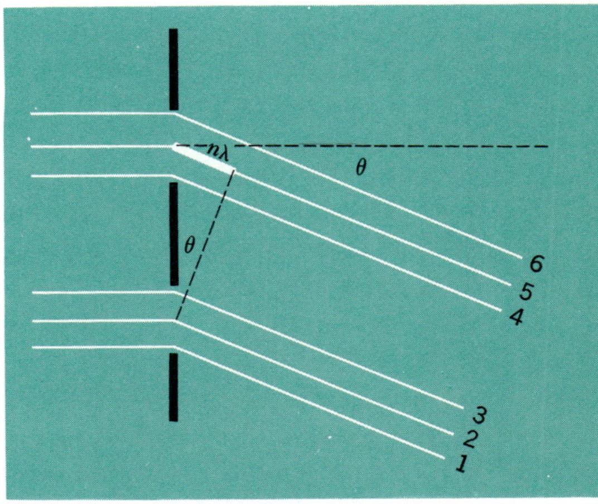

Figure 22-14

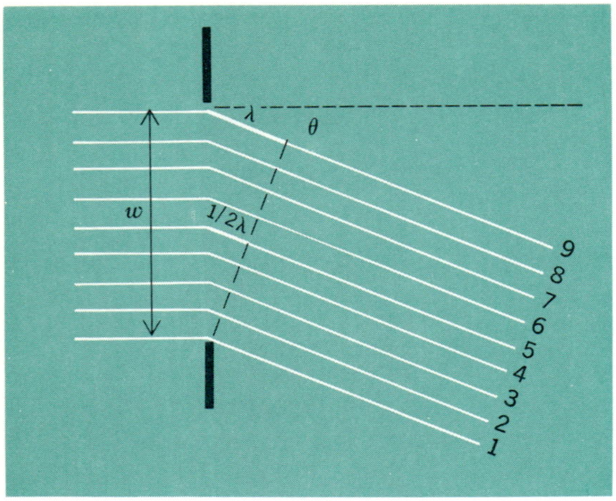

Figure 22-15 Single slit, first-order minimum.

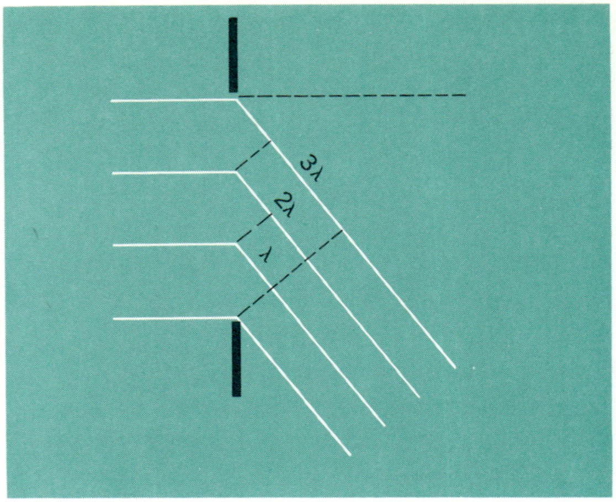

Figure 22-16 Single slit, third-order minimum.

ence between the component rays of a single broad wave front, as in this figure. Let's investigate what happens when some rays leave a slit of width w at an angle θ such that $\sin \theta = x/L = \lambda/w$. Ray *1* is one half-cycle out of phase with ray *5*. Ray *2* is one half-cycle out of phase with ray *6*, and so on for all the rays passing through the slit. This is because the path difference is $\frac{1}{2}\lambda$ for all these pairs of rays. Thus each ray in the bottom half of the slit cancels out a corresponding ray in the upper half, and the resultant amplitude of all the rays is zero for the direction given by angle θ.

The resultant amplitude from a single slit falls to zero at larger angles also. For instance, suppose the extreme path difference to a distant point (from the top and the bottom of the slit) is 3λ, as in Figure 22-16. The slit can then be divided into three equal segments, each giving zero intensity as before. We see that *minimums* in a single-slit diffraction pattern occur when the extreme path difference is a whole number of wavelengths:

$$\sin \theta = \frac{n\lambda}{w}$$

$$= \frac{x}{L}$$

Although this formula looks quite similar to the one for the *maximums* of a two-slit pattern, it is really quite different. Here w stands for the width of the single slit, and in the previous formula d represented the slit spacing.

In between the minimums, the pattern has bright maximums (Figure 22-17. See also Figure 22-18 in color section following page 388). When $x/L = 0$ (for rays that go straight through the slit), there are no phase differences, and all the rays add up to give an intense central maximum.

Now let's restate the equation for the location of the first minimum of a single-slit diffraction pattern in another way. Most of the energy of the wave lies in the central bright band. The boundary of this central band is marked by the first minimum. The location of the first minimum is measured by its deviation from the center line ($n = 1$):

$$\sin \theta = \frac{x}{L}$$

$$= \frac{\lambda}{w}$$

This leads to an important generalization:

> **The ratio of wavelength to size of opening determines to what extent light or any other wave deviates from a straight line path.**

In the same way, the ratio λ/w determines the extent that light is deviated when it passes the edge of an opaque obstacle.

The diffraction patterns of some other apertures and obstacles are reproduced in Figure 22-19. In each case, the pattern can be explained by the methods of wave optics used for the case of a single slit. The derivations are somewhat more difficult than that for a single slit, however, and are beyond the scope of this book. The shadow of a straight edge (of a razor blade, for example) is surrounded by a series of bands of uneven spacing. This is the "bending around corners" that is expected for wave motion. Such effects had actually been observed in the seventeenth century, but they had not been given a wave interpretation.

One difficulty in all experiments of this type is that shadows are usually far from sharp because of a purely geometrical reason. A sharp shadow requires a point source of light so that all the rays reaching a point on the edge of an object are traveling in the same direction. An ordinary light bulb is a broad source because of its frosted glass cover and the rays from the sun itself do differ in direction by as much as half a degree. Thus rays from different parts of the sun cast different shadows, which overlap to form a fuzzy edge. The shadow of a razor edge may be fuzzy enough to hide the diffraction effects. One must take extreme care and use a very small source.

Figure 22-19c is especially interesting. The French physicist Simeon Poisson is said to have objected to the wave theory, arguing that the theory would predict a bright spot at the center of the shadow of a perfectly circular obstacle—apparently an absurd result, since no one had ever noticed such a bright spot. When the experiment was performed by Dominique Arago in 1818, the bright spot was promptly found, and to this day it is called the Arago spot. It is hard to imagine a more dramatic piece of evidence for the wave model of light. Here is one instance where a hypothesis (the wave theory) led to a new and hitherto meaningless experiment, with results completely opposite to common sense, and yet in perfect agreement with the prediction of the theory.

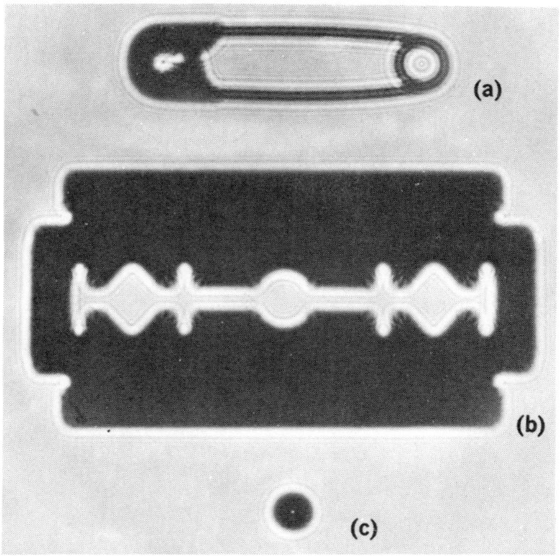

Figure 22-19 Diffraction of light. Opaque objects were mounted on a glass plate with wax and placed 4 m from a point source. A photographic plate was exposed at a position 2 m beyond the obstacles. No lenses were used. Note the diffraction bands at all edges; no shadow is sharp. The film was printed with short exposure to show only densest parts of shadows. **(a)** Shadows about a safety pin. **(b)** Shadows about the edge of a razor blade. **(c)** Shadows cast by a ball bearing. Note that the Arago bright spot is seen at the center of the circular shadow.

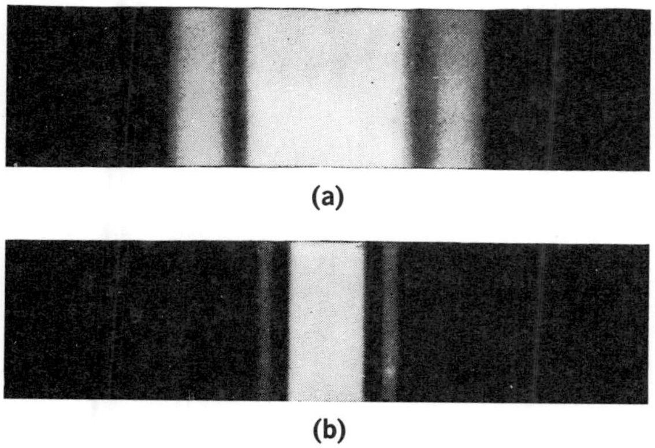

Figure 22-17 Diffraction pattern of a single slit. **(a)** Pattern of a narrow slit; **(b)** pattern of a wide slit.

SUMMARY

In trying to understand the nature of light by proposing and testing various models, you should keep in mind that light can only be described in terms of its own properties. Neither a wave model nor a particle model is complete; the essential point is that light will show some wave properties or some particle properties, depending on the nature of the experiment that is performed.

One of the earliest models was the "bullet" theory, or **corpuscular theory,** which was favored by Newton. This theory could explain reflection easily, and could explain refraction if one accepted the assumption that light travels faster in a dense medium than in a vacuum. As soon as it was possible to measure the speed of light in a medium, this assumption was found to be wrong. Newton's simple bullet theory was discredited before this, however, by Thomas Young's demonstration of interference, which can only be explained in terms of a wave model. To obtain interference, it is necessary to divide a wave front into two or more parts that come together again after traveling along paths of different lengths. Both **constructive** and **destructive interference** are possible, depending on the relative phases of the waves when they recombine. If the effective path difference between two rays is 0λ, 1λ, 2λ, etc., that is, a whole number of wavelengths ($n\lambda$), the rays interfere constructively and the resultant amplitude is a maximum. If the path difference is $\frac{1}{2}\lambda$, $\frac{3}{2}\lambda$, $\frac{5}{2}\lambda$, etc., that is, an odd number of half-wavelengths $(n - \frac{1}{2})\lambda$, the waves interfere destructively and the resultant amplitude is zero.

Using water waves in a ripple tank as the basis for reasoning, we concluded that the interference pattern of two in-phase sources could be described as a series of **maximums** (where maximum motion takes place) and **nodal lines,** or **minimums,** (where there is no motion). Maximums are spaced out from the central maximum according to the relation $\sin\theta = x/L = n\lambda/d$. In this formula, $n = 0$ for the central maximum, $n = 1$ for the first maximum on either side, etc. Nodal lines are spaced out from the center according to the relation $x/L = (n - \frac{1}{2})\lambda/d$. Here $n = 1$ for the first minimum on either side of the central maximum, $n = 2$ for the second, etc.

These same relations were found to hold true for the interference pattern formed by two in-phase sources of light. For example, they apply in the conditions of Young's experiment when two narrow, closely-spaced slits are illuminated, and the light from these two sources falls on a screen.

The **diffraction** pattern formed by waves passing through a single slit that is not extremely narrow can be explained in terms of interference between rays from different parts of the slit. Points of zero intensity in a diffraction pattern are located by the relation $\sin\theta = x/L = n\lambda/w$. The basic concepts of interference and diffraction apply to all wave motions, including electromagnetic waves, water waves, and sound waves.

Shadows are not usually perfectly sharp because sources are not usually points. Therefore, the rays from the source come from different directions and cause fuzzy shadows. When shadows are cast by point sources, the edges show diffraction bands, which again are examples of the wave nature of light. Wave effects are generally easier to observe with longer wavelengths and for smaller openings or obstacles, so that λ/w is larger.

QUESTIONS

1 Is the refraction of light evidence for the wave nature of light? What additional experimental fact makes refraction a stronger indication for the wave nature of light?

2 What must be true about the path lengths if two rays of light are to interfere constructively?

3 Why isn't there an interference pattern when the light from two closely spaced flashlight bulbs falls on a distant screen?

4 Light of a single wavelength from the source S shines on two narrow slits, as in Figure 22-20. Is the point P bright or dark? Why?

5 If white light shines on two slits, as in the previous question, what color (red or violet) appears closest to P in the first-order ($n = 1$) bright band?

6 Why didn't the early experimenters (around Newton's time) observe that light does "bend around corners?"

7 If you look at a distant linear light source (a fishtank or showcase bulb, or even a fluorescent lamp) through a narrow slit, the source appears to become broader as the slit is narrowed. Explain. This experiment can be done easily with the aid of two pencils held close together in front of your eye. Rotate or squeeze the pencils to vary the opening between them while you look at a distant lamp.

8 Hold a straight edge, such as a slide rule or a sheet of paper, about 1 m from the lamp bulb of a desk lamp. Look closely at the edge of the shadow cast on a surface held several centimeters from the obstacle. Explain why the shadow is not perfectly sharp.

9 In the ripple tank pattern of Figure 22-4, what would happen to the nodal lines if we decreased the wavelength of the ripples?

10 In this same pattern, what would happen to the nodal lines if the sources were moved closer together while the wavelength remained the same?

11 How would a small piece of cork move if placed in a nodal line of a ripple tank pattern?

12 Interference patterns formed by two slits (more commonly a grating of many slits) are used to measure the wavelength of light. How could you use an interference technique to measure the width of a small slit?

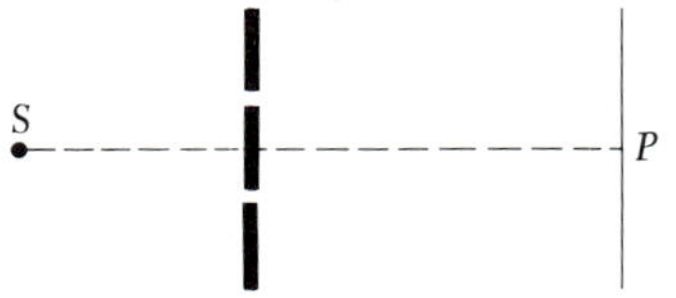

Figure 22-20

PROBLEMS

GROUP A

1 Two radio antennas are emitting waves that are in phase at the towers. An observer is 20,000 m from one tower and 22,000 m from the other tower. The wavelength is 500 m. Is reception "good" at the observer's location? Suppose the observer is listening to his automobile radio; how far should he move, and in which direction, to change the quality of reception? Explain.

2 Two slits 4.8×10^{-6} m apart, illuminated by parallel rays, form the fifth-order bright band at an angle of 30° from the center line. What is the wavelength of the light used?

3 A coarse grating designed for use with infrared radiation (so-called "heat radiation") has slits 0.5 mm apart. What is x/L of the third-order maximum for radiation whose wavelength is 60,000 Å? Remember, 1 Å = 10^{-10} m. What is the deviation when $L = 1$ m?

4 A slit 2×10^{-4} cm wide is illuminated perpendicularly with light of wavelength 5000 Å. What is $\sin \theta$ for the first dark band on either side of the central maximum? What is the deviation at a distance of 10 cm?

5 What is the deviation at a 10 cm range of the first *bright* band on either side of the central maximum for the slit of the previous problem? (*Hint:* the bright band is approximately half way between the first and the second dark bands.)

6 Make a diagram like that of Figure 22-16, showing the arrangement that produces the first bright band, as described in the previous problem. Using this diagram, try to decide whether the first bright band is as bright as the central maximum. Explain.

7 Light of wavelength 4×10^{-7} m falls on a pair of slits, and the third-order bright band falls at an angle of 30° from the main beam. How far apart are the slits?

8 When light of 5000 Å falls perpendicularly on a single slit, the first two dark bands (one on each side of the central bright band) are separated by an angle of 60°. How wide is the slit?

GROUP B

9 Prove that if the wavelength is equal to or greater than the width of a slit, light striking the slit perpendicularly passes through without forming any dark interference bands. (*Hint:* Draw a diagram like that of Figure 22-15.)

10 Light of a certain color perpendicularly strikes a slit of width 2.7×10^{-6} m. On a screen 3 m from the slit the central maximum is 1 m wide. What are the wavelength and color of the light? (*Hint:* The central maximum is bounded by the two first-order dark bands.)

11 Light of wavelength 4×10^{-7} m falls perpendicularly on a slit of width 5×10^{-7} m. What is the width of the central bright band on a screen 2 m from the slit?

12 Sound waves of frequency 2000 Hz coming from a distant outside source enter a tall, narrow, open window perpendicularly. The window is 0.7 m wide, and is in the center of a wall of a room 6.0 m on an edge. The room is padded with sound absorbing material to eliminate reflections from the walls. A number of listeners are seated in a row along the wall opposite the window. How far from the corners of the room are the two listeners who hear nothing? Assume that the speed of sound is 330 m/sec.

23 Applications of Wave Optics

The concepts of wave optics have been applied in designing many instruments and devices that have made a significant contribution to the development of modern physics. In this chapter you will see how the interference of light waves was used to make the incredibly precise measurements that contributed to the development of the theory of relativity.

You will also investigate the multiple-slit grating and the spectroscope, which played a major role in the development of atomic physics, as well as various familiar effects in nature that can be interpreted through the wave model of light.

23-1 INTERFERENCE IN THIN FILMS

Some of the effects that we now regard as evidence for the interference of light are familiar in everyday life and were known to Newton and other scientists several hundred years ago. The colors of an "oil slick," which you may have seen on a wet road, or the colors seen in a soap bubble are common examples of interference taking place in thin films.

In the laboratory, interference bands can be observed if *monochromatic light* (light of only one wavelength) strikes an air wedge, as shown in Figure 23-1. A fine wire or a sheet of tissue paper at *X* forms a thin wedge of air between two flat pieces of glass. (The slope of the wedge is exaggerated for clarity.) At any point, such as *A* or *B*, rays are reflected both at the upper surface (ray *1*) and at the lower surface (ray *2*) of the air wedge. When these rays are brought together on the retina of the eye, the resultant amplitude is large or small, depending on the path difference *PBR*. Since the path difference gradually increases as the thickness of the wedge increases, the observer sees a series of alternately bright and dark bands as she looks along the wedge. Suppose *A* is a bright band and *B* is an adjacent bright band. For any two adjacent bands, the total (round trip) path difference at *B* is just 1λ longer than the path difference at *A*.

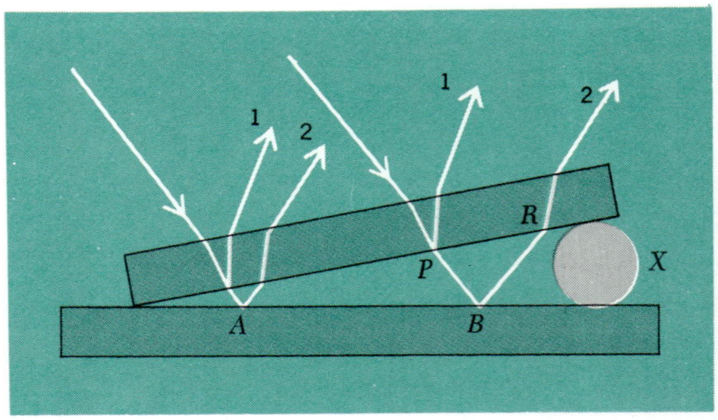

Figure 23-1

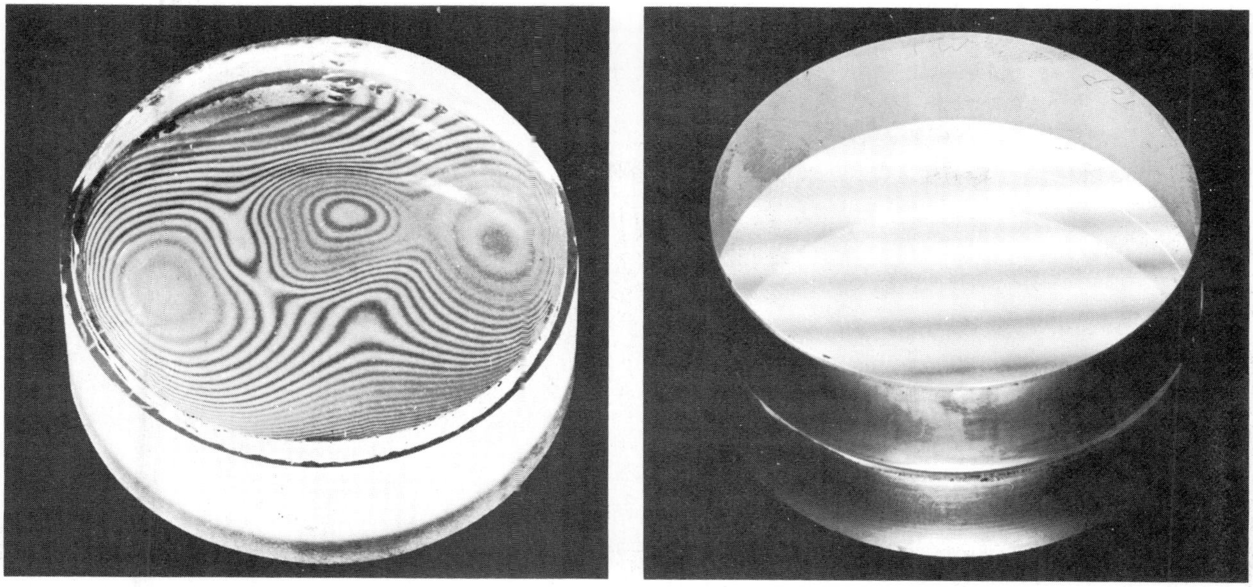

Figure 23-2 Testing an optical surface by interference. Left, initial condition of surfaces, showing three high spots; right, surfaces polished flat to within a fraction of a wavelength.

In other words, the air wedge is $\frac{1}{2}\lambda$ thicker at B than at A. The wavelength of light can be found by a simple experiment of this sort. In the manufacture of precision optical equipment, a glass surface can be tested for flatness by laying it on top of a standard surface of known flatness and wedging it slightly at one end. If the light and dark bands, or "fringes," are wavy (Figure 23-2, left photo) the surface of the glass is uneven. The high spots of the surface under test can then be polished until a parallel set of fringes is obtained (Figure 23-2, right photo).

There seems to be a strange contradiction in air-wedge experiments. Where the pieces of glass are actually in contact, a dark fringe is seen, indicating that the two reflections are out of phase and are cancelling each other. According to the wave theory, however, we would expect a bright band at such places because the path difference is zero. This is an example of a fairly common effect—the phase is changed by reflection. The effect is easier to understand in terms of waves on ropes. Suppose a wave is sent down a rope and reflected from a rigid support to which the rope is firmly tied. Let's assume that the wave travels along the rope as an upward pulse (Figure 23-3). What happens when this pulse reaches the rigid support? We know that there can be no motion of the rope where it is joined to the support. Therefore, no matter what kind of pulse moves into the support, the amplitude there must always be zero. One useful way of interpreting this is to say that the upward pulse is cancelled by an equal and opposite downward pulse. In other words, it is wiped out by a pulse of opposite phase at the rigid boundary. We can, in fact, think of the boundary exerting a force on the rope to produce such a pulse (Figure 23-3b). Most convincing is the fact that if you do send an upward pulse along a rope to a rigid support, it is followed by a reflected downward pulse traveling back the other way (Figure 23-3c).

A similar effect takes place when light waves are reflected at a boundary going into an optically "more dense" medium (Figure 23-4a). The more dense medium behaves like the rigid support in

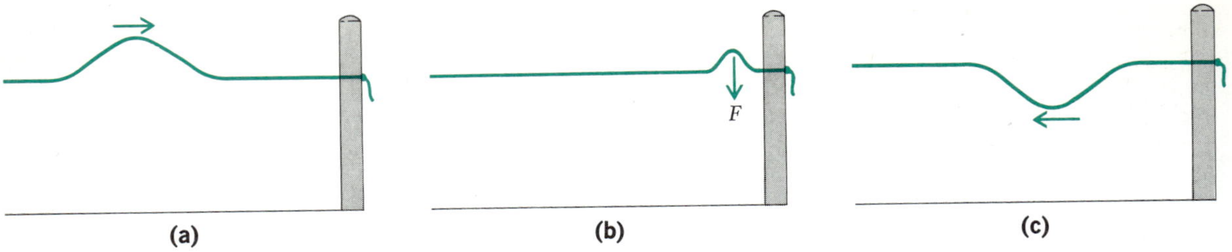

Figure 23-3

the previous paragraph. The waves of ray *B* are one half-cycle out of phase with those of ray *A* because they are reflected in air at the boundary of a glass surface—the surface of a denser medium.

However, in Figure 23-4b the waves of ray *C* are in phase with those of ray *D* because this reflection is in *glass* against the air boundary; in other words, the ray is reflected in a more dense medium against a less dense medium. Again using the analogy of a pulse on a rope, suppose you send a pulse along a rope that is completely free at one end, not tied to any support. In this case the pulse will also be reflected (it has nowhere else to go), but there is no rigid support to make an opposite pulse. The rope swings freely and the pulse is reflected back, still in the same position, in phase with the incoming pulse.

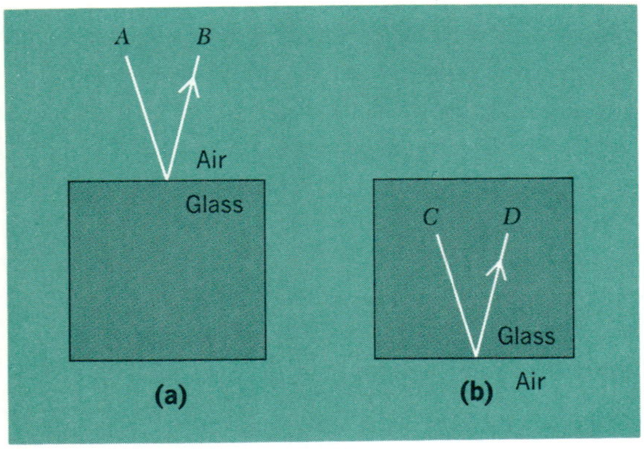

Figure 23-4 Light waves undergo a 180° phase change when reflected as at (**a**), but not when reflected as at (**b**).

The interference pattern always shows a dark band at the vertex of the air wedge, even though the path difference is zero, because one ray is changed in phase by one half-cycle when it is reflected, and the other is not. This is equivalent to an extra $\frac{1}{2}\lambda$ in the path of one of the rays, and destructive interference results.

Example 23-1 A flat glass plate 10 cm long is separated from another flat glass plate at one end by a piece of aluminum foil 0.004 cm thick. How far apart are the successive dark interference bands, viewed in light of wavelength 6000 Å? How many dark bands are there in the 10 cm length of glass?

The foil creates an air wedge, and there is a dark band at the vertex *A* (Figure 23-5), where the path difference is 0 (the slope is greatly exaggerated). The next dark band appears at *B*, where the thickness is $\frac{1}{2}\lambda$ but the total path difference is 1λ.

Hence,

$$BC = \tfrac{1}{2}\lambda$$
$$= \tfrac{1}{2}(6 \times 10^{-7}\ \text{m})$$
$$= 3 \times 10^{-7}\ \text{m}$$

The distance x can now be found by a proportion based on the corresponding sides of the similar triangles:

$$\frac{x}{10\ \text{cm}} = \frac{3 \times 10^{-7}\ \text{m}}{4 \times 10^{-5}\ \text{m}}$$
$$x = 0.075\ \text{cm}$$
$$= \boxed{0.75\ \text{mm}}$$

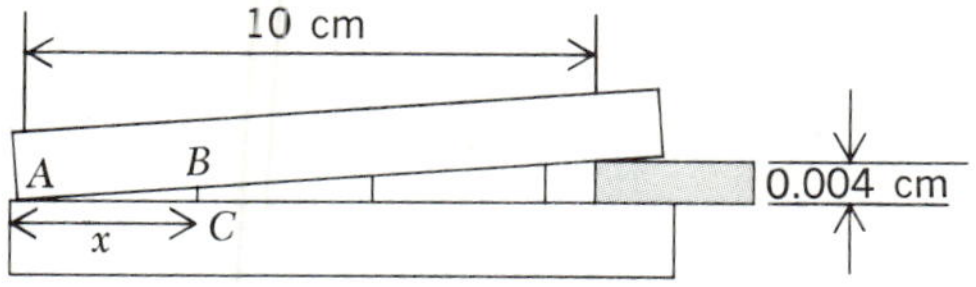

Figure 23-5

The bands are spaced 0.75 mm apart and can easily be seen with a magnifying glass. The total number of dark bands in the 10 cm length of glass is

$$\frac{10\ \text{cm}}{0.075\ \text{cm/band}} = \boxed{133\ \text{bands}}$$

One way to obtain a thin air film of variable thickness is to place a long focus lens (one with very slight curvature) on a flat glass plate. Figure 23-6 shows the arrangement, with the curvature much exaggerated for clearness. If the curved surface of the lens is almost flat, the air film varies from zero thickness (at A) to a dozen or so wavelengths at the edge. When viewed by light reflected from the glass surfaces, we see a pattern of dark rings. For any particular circular ring, such as the one at BB', the air film is of constant thickness.

It is strange that Newton himself discovered these interference rings—they are called Newton's rings—but he completely missed their significance. He stated that light particles had "fits of easy reflection and easy transmission" to explain the fact that light is partially reflected and partially transmitted at a glass-air surface. But he had no consistent theory for the rings. After two and a half centuries it is easy for us to see that Newton's simple bullet model was inadequate; our modern ideas of the particle aspects of light are much more subtle and complicated. At the time, Newton's model led nowhere and predicted no new results, unlike the great fruitfulness of the wave theory. The greatest minds (including Newton's own) were just not able to accept a model in which inconceivably small waves (it takes 2 million to make a meter) behaved like the water waves seen by the naked eye.

Other reflections are often observed. A soap bubble consists of a thin film of soap solution, and reflections at the boundaries of this film act like reflections from an air wedge. If we look at the soap film by white light reflected off its surfaces, the film appears brilliantly colored (Figure 23-7 in the color section following page 388). Suppose that for some particular color (and wavelength) the thickness is exactly $\frac{1}{2}\lambda$. An extra half-cycle phase shift is introduced at the front surface, but not at the back surface. The *effective* (round trip) path difference is $1\lambda + \frac{1}{2}\lambda = 1\frac{1}{2}\lambda$, and light of this color is not reflected, since the two rays destructively interfere. The fact that the two *reflected* rays are cancelled out means that the light of this wavelength goes right on through the film. However, other colors have longer or shorter wavelengths, and the film cannot be $\frac{1}{2}\lambda$ thick for all colors at once. Therefore the film reflects some colors more than others. As a result, a bubble appears colored, and the colors change as the thickness of the film changes while the water evaporates. Eventually, the bubble is much thinner than any visible wavelength, the path difference is 0λ for all colors, and the bubble appears black (why not white?) just before it breaks.

There is no real difference between the *interference pattern* formed by an air-wedge and the pattern formed by a soap bubble. The wedge pattern was discussed in terms of monochromatic light as the thickness of the wedge varies from point to point. However, a pattern from a thin air-wedge—Newton's rings, for example—will be tinged with color if viewed in white light. The

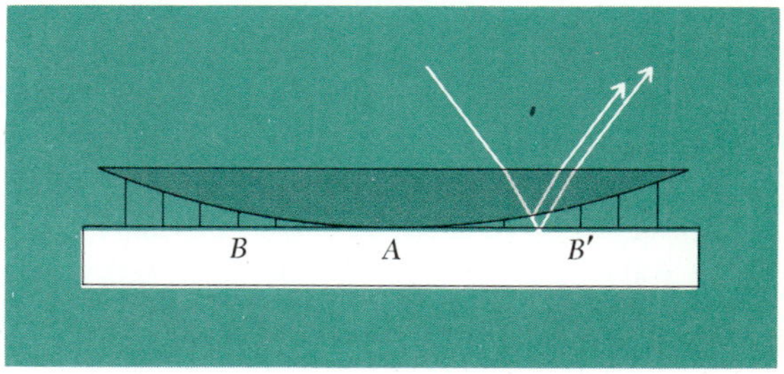

Figure 23-6

bubble reflects various components of white light at any particular time; as its thickness changes with time, the colors also change.

Example 23-2 A soap bubble 5000 Å thick is illuminated with white light. The index of refraction of the film is 1.35. What colors are *not* reflected?

The first wavelength that is not reflected is one for which 5000 Å $= \frac{1}{2}\lambda$. In this case the round trip of 10,000 Å in the film would be equal to 1λ, and the phase change at the front surface would account for another $\frac{1}{2}\lambda$, making a total of $1\frac{1}{2}\lambda$. If $\lambda = 10{,}000$ Å in the soap film, the wavelength in air would be

$$\lambda_1 = (10{,}000\ \text{Å})(1.35) = \boxed{13{,}500\ \text{Å}}$$

We multiply 10,000 Å (λ in the film) by 1.35 (index of refraction of film) because the speed of light in air is 1.35 times that in the film. Recall that the index of refraction of the film is the ratio of the speed of light in the film to the speed in air. The frequency remains constant, and by the relation $c_n = f\lambda$, or $\lambda = c_n/f$, we see that λ is proportional to the speed c_n.

13,500 Å is an infrared wavelength and is not visible.

Next, let's consider a wavelength for which the round trip is 2λ. Then 10,000 Å $= 2\lambda$, and $\lambda = 5000$ Å in film. In air, the wavelength would be

$$\lambda_2 = (5000\ \text{Å})(1.35) = \boxed{6750\ \text{Å}}$$

This is red light (Table 21-1, page 364).

The next color that is completely removed by interference is one for which 10,000 Å $= 3\lambda$; $\lambda = 3333$ Å in film. The wavelength in air, then, is

$$\lambda_3 = (3333\ \text{Å})(1.35) = \boxed{4500\ \text{Å}}$$

This is blue light. Going to still shorter wavelengths, we find that the next destructive interference is for $\lambda = 2500$ Å in film, or

$$\lambda_4 = \boxed{3375\ \text{Å}}$$

in air. This is in the ultraviolet region.

Only two visible colors, then, are totally cancelled—red light of wavelength 6750 Å and blue light of wavelength 4500 Å. Other colors are reflected to a greater or lesser extent. Since red and blue are missing, the middle of the spectrum predominates, and the bubble appears greenish.

23-2 THE MICHELSON INTERFEROMETER

One of the most significant applications of interference is the interferometer invented by the American physicist, A. A. Michelson (1852–1931). The Michelson interferometer can be used to measure distances with great accuracy, or to measure very short time intervals by comparing the phase of two light waves. It is, in effect, based on the air-wedge principle, but the reflecting surfaces are spaced many centimeters apart, instead of just a fraction of a millimeter. Figure 23-8 shows light from a broad source S (such as a frosted bulb) striking a glass plate *M*, which is "half silvered" to reflect some light and transmit the rest. Half the light is reflected to mirror *X*, and half goes on through to mirror *Y*. Beam *1* is reflected directly back from *X*, passes through *M* again, and comes out to be viewed by the eye at *E*. Beam *2* is reflected back from the mirror at *Y*, and is reflected by the half-silvered mirror at *A* to enter the eye at *E*. Interference takes place between the two beams, depending on the path difference. Notice that an extra glass plate is inserted in the path of beam *2*, so that each beam passes through the same thickness of glass.

Since *M* is a plane mirror, the eye sees a virtual image of *Y* at *Y′* (recall the discussion of virtual images, Section 21-7). In effect, the two beams *1* and *2* are coming from two mirrors at *X* and *Y′*. The great advantage of this arrangement is that the thickness of the air wedge between *X* and *Y′*

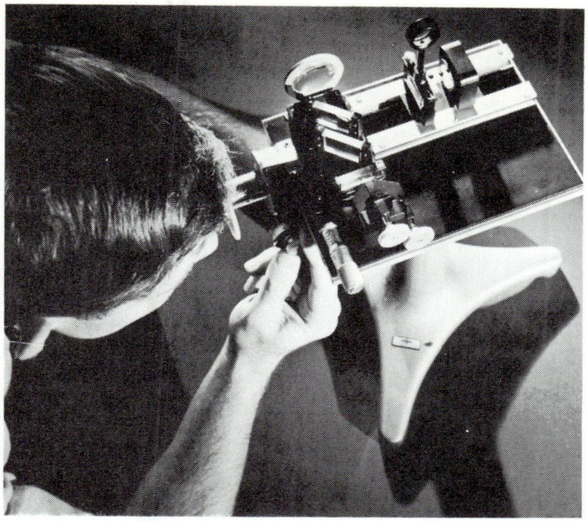

Figure 23-9 Michelson fringes.

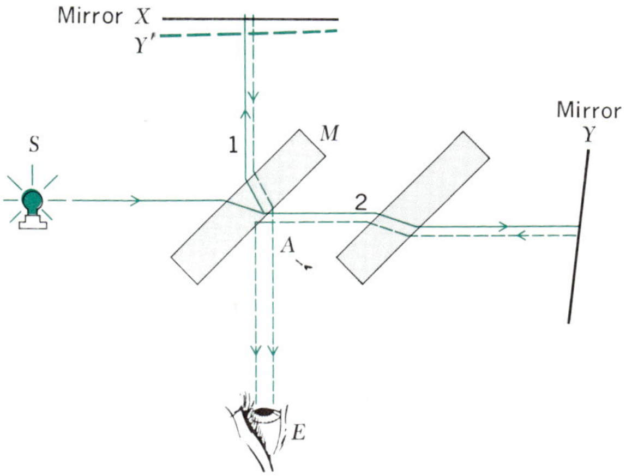

Figure 23-8 Michelson interferometer.

can be adjusted, even to zero, by moving *X* back and forth with an adjusting screw. The angle of the virtual wedge between *X* and *Y′* can be adjusted by turning screws that tilt mirror *Y*. Thus it is possible to get an interference pattern of light and dark fringes of any desired spacing (Figure 23-9). These fringes move across the field of view as the mirror *X* is made to move toward or away from the image *Y′*.

Example 23-3 Monochromatic green light is obtained by passing light from a mercury-arc lamp through a filter that removes most of the light of undesired wavelengths. A beam of this light enters a Michelson interferometer, and the mirror *X* is carefully moved through a distance of 0.273 mm. The observer watching a reference point counts 1000 fringes moving past the point as *X* is being moved. He sees a continual shifting of the pattern. Specifically, when one dark fringe is replaced by the next dark fringe at the reference point, it means that the total (round trip) path difference has changed by 1λ. Because of the "double" nature of the path, it also means that *X* has moved $\frac{1}{2}\lambda$. Calculate the wavelength of the light.

The passage of 1000 fringes means that the round trip path has changed by 1000 wavelengths. Thus

$$
\begin{aligned}
1000\,\lambda &= 2(0.273\text{ mm})\\
\lambda &= 2(0.273\text{ mm})/1000\\
&= 5.46 \times 10^{-4}\text{ mm}\\
&= 5.46 \times 10^{-7}\text{ m}\\
&= \boxed{5460\text{ Å}}
\end{aligned}
$$

Does this fall within the green range?

Michelson used his interferometer to determine the length of the standard meter in terms of the wavelength of a red line in the cadmium spectrum. He was searching for a length standard that would be more precise and more reproducible than the standard meter bar, which is made of platinum-iridium alloy and kept under carefully controlled conditions. He chose this line because it was the "purest" color known at that time and it gave the clearest and sharpest fringes in his interferometer.

Using this technique, the meter has recently (1961) been redefined as 1,650,763.73 wavelengths of an orange-red line in the spectrum of the krypton isotope of mass 86 (Figure 1-4 on page 388-1). Unlike the platinum-iridium bar, this standard cannot be lost or destroyed, and it can be reproduced virtually anywhere and at any time. With the right instrument, the meter can now be determined to an accuracy of one part in 10^8.

The Michelson interferometer was used in a famous experiment to study the nature of a mysterious "ether" in which light waves were supposed to propagate. The "ether" theory illustrates one of the disadvantages of taking a model too seriously. The familiar waves that have been used as models for the study of light waves require a mechanical medium of some kind, such as air or water. Early physicists had concluded, therefore, that light too required some kind of medium for its propagation, and the ether (which was supposed to be a very thin, fluid substance that occupied all of space) was an attempt to describe such a medium. The speed of light *relative to this medium* was supposed to be a constant c. The value of this constant was supposed to depend on the physical properties of the ether. Recall that in Chapter 9 you saw that the speed of waves along a line of balls coupled by springs depended upon the mass of the balls and the spring constant of the springs. In the same way, physicists looked for an elasticity, or stiffness, of the ether that determined the speed of light.

Referring to Figure 23-8, suppose that the distances AY and AX are equal. In this case the time for light to travel from A to Y and back again

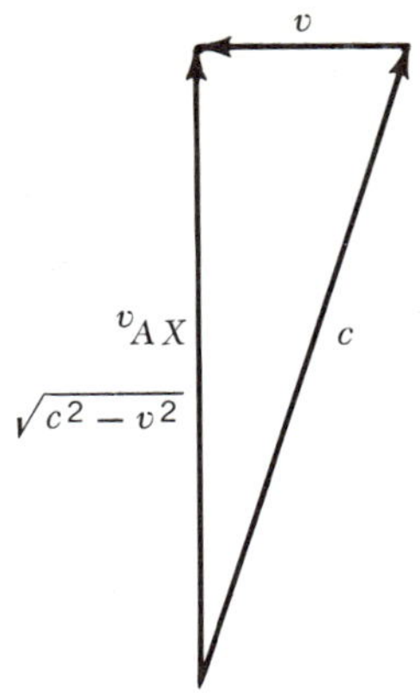

Figure 23-10

would be equal to the time required for the round trip from A to X and back again. Rays *1* and *2* would be out of phase because of the extra $\frac{1}{2}$ cycle phase shift of ray *2* at mirror M, and a dark fringe would be observed. However, the earth revolves about the sun, and so we assume that because of its orbital velocity it is moving through the ether at about 29.8 km/sec. There are quite possibly other motions with respect to the ether, due to the motion of the sun, etc. If the whole interferometer moves (with the earth) to the right with velocity v, we would notice an "ether wind" blowing to the left. The resultant velocity of light would be $c - v$ from A to Y; $c + v$ from Y to A; and $\sqrt{c^2 - v^2}$ from A to X or X to A. Figure 23-10 is a vector diagram of the velocities when the interferometer moves to the right with velocity v. The velocity of light from A to X (written as v_{AX}) is made up of the velocity of light through the ether c plus the velocity of the "ether wind" v. By the Pythagorean theorem, $v_{AX} = \sqrt{c^2 - v^2}$.

This is similar to the situation in Problem 25, Chapter 3, where it takes less time for a swimmer to cross a fast flowing river and return, than for her to swim an equal distance up the river and back. If the same reasoning is applied to the experiment with the interferometer supposedly moving through the ether, ray *1* would get ahead of ray *2*. The resulting phase difference would show up as a shift of the interference fringes.

In 1887, working at the Case School of Applied Science in Cleveland, Michelson and his coworker Edward W. Morley completed a series of experiments designed to determine the speed of the earth through the ether. The interferometer was horizontal, and could be rotated about a vertical axis so that the arm AY could be adjusted to be either parallel or broadside to the supposed ether wind. Experiments were made at different seasons of the year and at different times of day. The startling result of the experiment was that no shift of fringes was ever observed. In other words, the speed of light is the same in the directions AX and AY, hence *the speed of light is independent of the motion of the observer.* This negative result of the Michelson-Morley experiment puzzled physicists for almost twenty years. In 1905 Einstein adopted the constancy of the speed of light as one of the postulates of the relativity theory (Section 5-9), giving up the idea of an ether entirely. Strangely enough, at that time Einstein may not have been aware of the result of the Michelson-Morley experiment which has such strong implications for the theory of relativity.

23-3 THE GRATING

A series of many equally spaced slits is called a *grating*. Parallel rays falling on a grating form an interference pattern with sharper and brighter maximums than would be formed by a pair of slits, but the basic principle is the same. Figure 23-11 shows a grating of six slits forming the third-order maximum. The path length of each ray shown here differs by $\pm 3\lambda$ from the path length of the neighboring ray. Thus ray DD' is 3λ longer than ray CC', and 3λ shorter than ray EE'. The rays are all in phase at the slits, and they will all be in phase again at the wave front $F'A'$ if we choose the correct angle of deviation θ. The choice is governed by the basic relation $\sin\theta = x/L = n\lambda/d$. In this case we have

$$\sin\theta = 3\lambda/a$$

where a is used to mean the space between adjacent slits, called the grating space. As you can

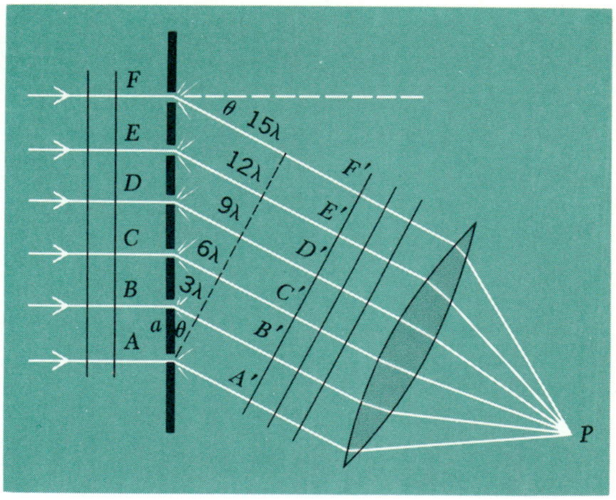

Figure 23-11 Grating with six slits. The lens might represent the eye lens, and P might be a point on the retina.

see from the similar relations involved, a pair of slits (Section 22-3) is just the simplest form of grating.

Recall what was said in Chapter 22. Rays do spread out in all directions from each slit. However, there is constructive interference only in certain directions, and only in these directions are the bright bands seen. The rays in other directions will form resultants of various amplitudes, but if the grating has a large number of slits, the intermediate amplitudes are extremely small.

The sharpness, or "narrowness," of a maximum can be determined from the following conditions. If there are 1000 slits and we are considering the first-order maximum, all the rays interfere constructively, providing θ is such that the top ray's extra path MM' is exactly 1000 λ (Figure 23-12a). If the angle is now increased very slightly to θ' (Figure 23-12b), just enough to make $MM' =$ 1001 λ, the middle ray's extra path NN' will become 500.5 λ. Hence ray *0* and ray *500* are one half-cycle out of phase and cancel each other. Likewise, ray *1* cancels ray *501*, ray *2* cancels ray *502*, etc. Every ray from a slit in the bottom half of the grating cancels one in the top half. The resultant intensity is zero at the new angle θ', which is very close to the former angle θ for which the in-

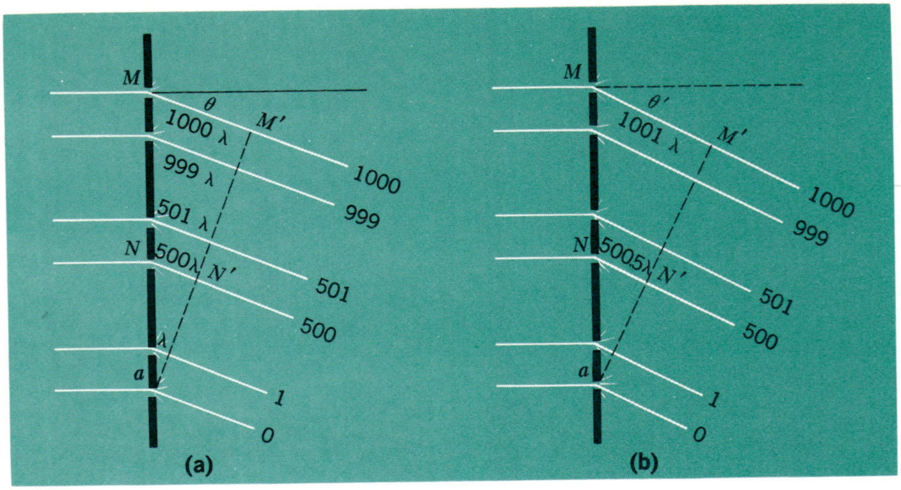

Figure 23-12

tensity is a maximum. This means that the maximum is a narrow, or sharp, one.

If there had been only two slits (say the top slit and next lower slit), a similar small change of angle would cause the top ray to be only 0.001 cycle out of phase (Figure 23-13) and it would still combine with the lower ray to give a resultant intensity almost as large as the peak. The qualitative difference between the interference pattern of a grating and that of a pair of very narrow slits is shown in Figure 23-14.

So far in our discussion of the grating, we have assumed that the light striking a grating is monochromatic, that is, it consists of a single color and has a single wavelength. If light that is composed of a mixture of colors (such as white light) falls on a grating, the equation $n\lambda = a \sin \theta$ tells us that each different color, or wavelength, has a maximum at a different θ. Thus the colors become separated, and a grating can serve the same function as a prism, though in an entirely different way. The angular separation between two colors can be quite large if the space a between the slits in the grating is small enough, as you will see in the next section (also see color sections following pages 388 and 436).

23-4 THE SPECTROSCOPE

A *spectroscope* is used to separate a beam of light into its component colors. Newton experimented with prisms while he was a student at Cambridge University, and found that white light can be broken up into the colors of the rainbow. He also found that any given spectrum color (red, for instance) is "pure" and cannot be further modified by a second prism. Light of the various spectral colors when recombined gives white again. Newton did not follow up his important discovery, which might have led to the invention of the spectroscope.

In Chapter 21 you studied Snell's law and

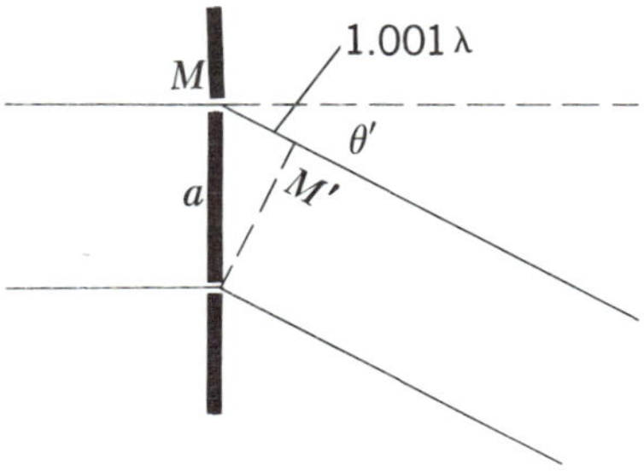

Figure 23-13

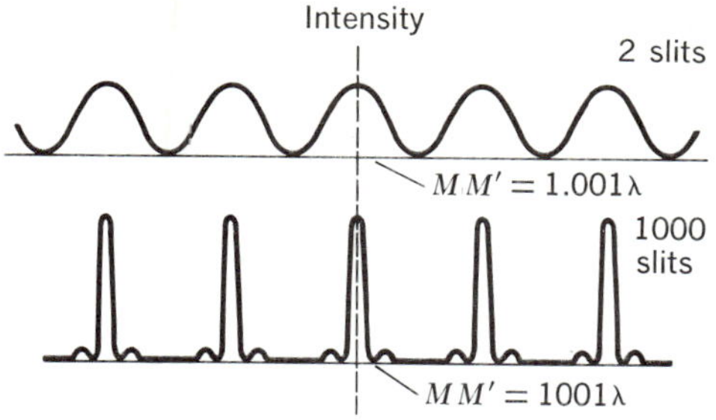

Figure 23-14 Comparison of interference pattern of two slits with that of a grating. Between the sharp maximums formed by the grating are many subsidiary maximums too weak to be drawn to scale.

learned that the index of refraction of glass is greater for violet light than for red. It is this fact that explains the operation of a prism spectroscope. In the simplest terms, the underlying idea is that different wavelengths are bent at different angles, causing the colors to become separated (Figure 23-15).

In discussing a spectrum, we often speak of spectral "lines." For example, in the mercury spectrum there is a strong line that has a wavelength of 5461 Å. The atoms of mercury vapor will emit light of this and other wavelengths when an electric current passes through the vapor, as in a fluorescent lamp. A spectral line is light of essentially a single wavelength. Light enters a spectroscope through a slit, and the width of this slit determines the width of the line as we see it for light of a single wavelength. In general, it is desirable to make the slit as narrow as possible and thus define the line as sharply as we can for greatest accuracy in measurement. However, as the slit becomes narrower, the amount of light available is reduced and the images become dimmer and harder to see.

In the previous section you saw how a grating forms a series of bright images for any given wavelength. When a beam of light shines on a grating, most of it goes straight through, but some is deviated to the side forming the interference pattern. The angle of deviation for the interference maximum of a particular color depends upon the wavelength and is given by

$$n\lambda = a \sin \theta$$

In this equation, n means the "order" of the maximum, that is, the second-order ($n = 2$) refers to the second maximum on either side of the center. The symbol a refers to the grating space, the interval between the narrow slits* that make up the grating (see Figure 23-12).

The equation above shows that bright images of different colors will appear at different angles, causing the colors to be spread out. For this reason, a grating can serve the same function as a prism. A simple grating spectroscope is shown in

* Since it is desirable for these slits to be very narrow, the widths of the slits that make up the grating can be considered negligible in these calculations.

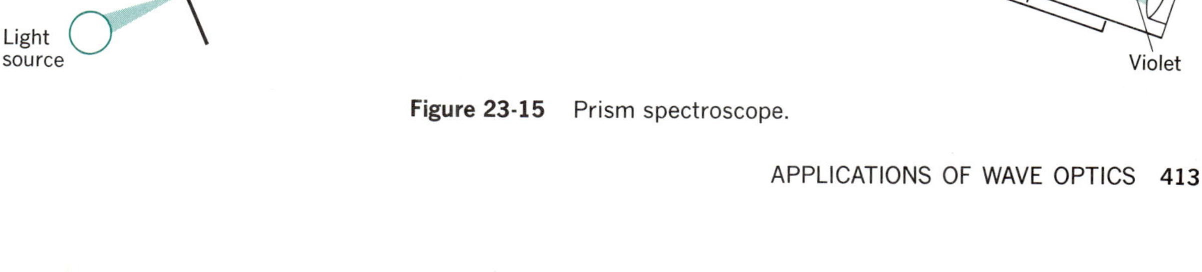

Figure 23-15 Prism spectroscope.

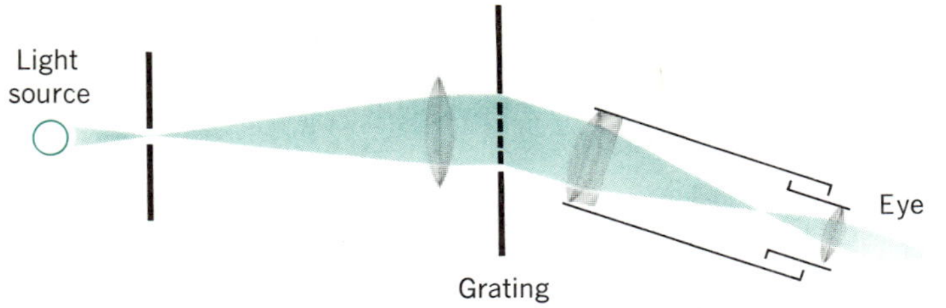

Figure 23-16 Grating spectroscope.

Figure 23-16. The grating spectroscope gives not only a single spectrum, but several complete spectra called orders. The separation between two colors can be considerable, as shown in the next example.

Example 23-4 A grating of 4000 slits/cm is used to form a spectrum of a source that emits two wavelengths, a violet line of wavelength $\lambda = 4000$ Å, and a red line of wavelength $\lambda = 7500$ Å. Calculate $\sin\theta$ for the first- and third-order images of each line. What is the angular separation between red and violet lines in the third order?

The grating space is $\frac{1}{4000}$ cm $= 25 \times 10^{-5}$ cm $= 25{,}000$ Å. We compute $\sin\theta$ for each line, using $\sin\theta = n\lambda/a$. For instance, for the red line in the third order,

$$\sin\theta = \frac{3(7500\text{ Å})}{25{,}000\text{ Å}} = \frac{22{,}500\text{ Å}}{25{,}000\text{ Å}} = 0.900$$

$$\theta = 64^\circ$$

We can tabulate the results as follows:

Line	*sin θ*	*θ*
violet, first order	0.16	9°
red, first order	0.30	17°
violet, third order	0.48	29°
red, third order	0.90	64°

The angular separation between red and violet in the third order is

$$64^\circ - 29^\circ = \boxed{35^\circ}$$

As you can see from the large differences in θ in the preceding example, the *dispersion* (the ability to spread out the colors) of a grating spectroscope can be large. One pioneer in this field was Henry A. Rowland (1848–1901) who is shown on page 436-4 of the color section following page 436. The dispersion of a grating depends only on the grating space a. The efforts of some of the best experimental physicists have been devoted to the problems of producing gratings with rulings spaced as finely and as uniformly as possible.

One way of judging the quality of a spectroscope is by how well it can separate two colors of almost the same wavelength. This is called *resolving power* and is measured by the thinness of the image of a very narrow slit. Diffraction of light (as well as loss of intensity) sets the limit to the smallness of the slit, but with a grating we can increase the sharpness of lines by increasing the total number of rulings (see the discussion at the bottom of page 409) and by using higher orders. In the second order a well constructed grating 10 centimeters long, having 5000 rulings (or slits) to the centimeter (a total of 50,000 rulings), can separate two colors that differ in wavelength by only one part in 100,000.

SUMMARY

Interference patterns are formed when light is reflected from the two sides of thin films of air, oil, water, etc. These patterns can be used as the basis of many precise measurements of length. A change of phase of one half-cycle occurs when light is reflected by an optically more dense medium, producing the equivalent of an extra $\frac{1}{2}\lambda$ of path difference. No change of phase takes place, however, when light is reflected by a less dense medium.

The **Michelson interferometer** is an important application of thin-film interference. The fringe pattern is formed by an air wedge resulting from the superposition of two images in plane mirrors. This interferometer is the basis of modern standards of length, in which the meter is defined in terms of the wavelength of light of a certain color. Also, the Michelson-Morley experiment using this device played a significant part in the interpretation of the relativity theory.

Many equally spaced slits form a **grating.** The principle of a grating is the same as that of a pair of slits, but the grating produces much more sharply defined maximums. The **spectroscope** separates light into its component colors, and a spectrometer can be used to measure the wavelengths of light. A spectral "line" is a sharp, bright image in a spectrum representing light of just one wavelength.

QUESTIONS

1 A sound wave in air strikes a hard, rigid wall and is reflected. Do you think the phase is changed upon reflection? Explain.

2 What would be the effect of covering the wall in question 1 with sponge rubber that is $\frac{1}{4}\lambda$ thick for the frequency of the sound? Assume the rubber is "more dense" than air; but "less dense" than the wall. (*Hint:* Is there a phase difference between two reflected waves?)

3 As a soap bubble evaporates, its surface appears black just before it breaks. Explain in detail why it is dark and not bright.

4 A film of oil spreads out on the surface of a wet pavement, and in some cases the thickness can be much less than the wavelength of light. Are the two reflected rays (Figure 23-17) in phase or out of phase? Does the oil slick appear bright or dark at its thinnest portion?

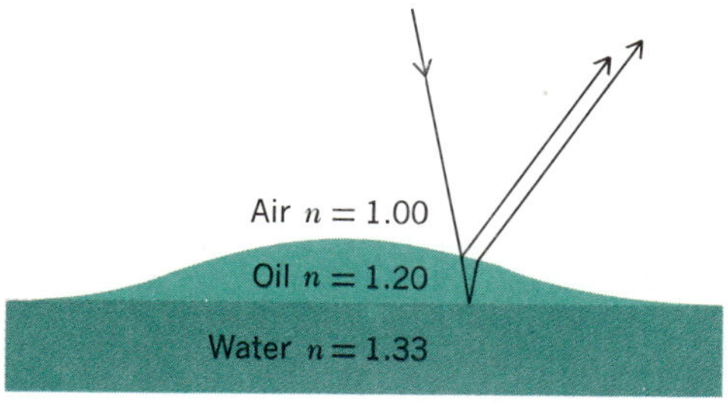

Figure 23-17

5 In Figure 23-8, which surface of the diagonal mirror M is the half-silvered one?

6 What is the difference, if any, between the interference pattern of a pair of slits separated by 0.001 cm, and the pattern of a grating having 1000 slits per cm?

7 If you look at a distant line source (such as a bulb with a straight, vertical filament) through a narrow slit, the source appears to become broader as the slit is narrowed. Explain this effect.

PROBLEMS

GROUP A

1 An air wedge is formed by placing a thin blond hair between the edges of two plane glass plates that are in contact at the other end. When viewed by reflected light of wavelength 5461 Å, 86 dark bands are observed. Calculate the diameter of the hair.

2 Two glass surfaces are separated by an air film of uniform thickness 9000 Å. If the film is illuminated perpendicularly by parallel light having a wavelength 6000 Å, explain why no light of this color is reflected by the air film.

3 Repeat Problem 2 for the same light illuminating a thin *glass* film of the same thickness as before, namely 9000 Å. The index of refraction of the glass film is 1.50. Do the reflected beams interfere constructively or destructively?

4 Light of wavelength 5461 Å is used to illuminate a Michelson interferometer. How many fringes pass the reference mark in the field of view when the movable mirror is shifted forward through a distance of 0.200 mm?

5 An observer looks at a source through a grating having 1000 slits per cm. The wavelength of the light is 6000 Å. How many images of the source can she see? (*Hint:* For what order is $\theta = 90°$?)

6 Sunlight falls perpendicularly on a grating having 1000 slits per cm. The rays leaving the grating are focused on a screen 4.00 m from the grating. How far from the central image does the green region of the second-order spectrum appear on the screen? Use an average green wavelength from table 21-1.

7 Calculate the width of the green region on the screen of the previous problem. (*Hint:* Calculate the deviation for each of the two extreme green wavelengths of Table 21-1.)

GROUP B

8 A soap bubble of thickness 2100 Å has an index of refraction of $\frac{4}{3}$. Make an analysis similar to that of Example 23-2 to show that the film appears blue when viewed perpendicularly in white light.

9 Two glass plates 20 cm long are in contact at one edge, and are separated at the other edge by a piece of aluminum foil 0.001 cm thick. When the plates are illuminated perpendicularly by light of wavelength 5000 Å, (a) what is the separation of adjacent fringes? (b) How many dark fringes are seen?

10 A "nonreflecting" coated lens has a thin film of magnesium fluoride ($n = 1.25$) that is deposited on the surface of the glass ($n = 1.56$). What should be the thickness of the film in order that the reflected rays (Figure 23-18) will be exactly out of phase for green light of wavelength (in air) 5000 Å?

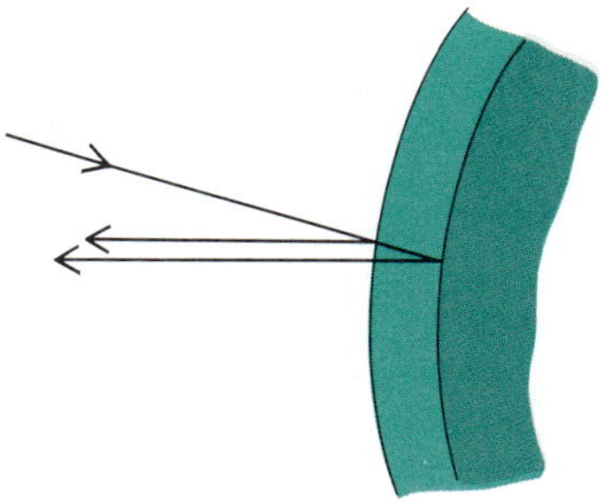

Figure 23-18

11 An experimenter wishes to determine the index of refraction of a piece of cellophane 0.0082 mm thick. She places the cellophane broadside to the beam in one arm of a Michelson interferometer, and observes a shift of 18 fringes. She is using the mercury green line, $\lambda = 5461$ Å, as a source. Calculate the index of refraction of cellophane. (*Hint:* There are more wavelengths in 0.0082 mm of cellophane than in the 0.0082 mm of air that it has replaced. For the round trip, the difference is 18 λ.)

12 Light from a flame in which sodium chloride has been placed has an average wavelength 5893 Å. What is the third-order angular deviation of a ray of this light that perpendicularly strikes a grating having 2000 slits per cm?

13 Parallel rays of red light of wavelength 6000 Å perpendicularly strike a grating having 1000 slits per cm. The rays leaving the grating are focused by the lens of an observer's eye on the retina, which is 2 cm from the eye lens. How far apart, measured on the observer's retina, are the central image and one of the second-order images?

FOR FURTHER STUDY

Polarization of Light

The electromagnetic waves described in Section 20-4 are *plane-polarized waves,* since the varying quantity (the disturbance) is an electric field or a magnetic field represented by vectors that lie in a plane. For the electric field vector in Figure 20-6, this is the vertical plane. The challenging question is not how can light be polarized, but rather why shouldn't it always be polarized? How can *un*polarized light exist when electromagnetic waves by their very nature are plane-polarized?

Light waves are emitted by molecules or atoms that are *excited* in some way, meaning that they have been given an extra burst of energy by collision with another particle. This energy does not always go to speed up their motion; instead, they sometimes are excited internally. We may imagine that an excited atom vibrates a few thousand or a few million times, sending out a relatively short burst of plane-polarized light called a *wave train.* In the usual source, such as the filament of a lamp bulb, the vibrating atom makes many collisions per second with other atoms. If the atom vibrates freely for 10^6 cycles it emits a wave train containing only 10^6 wavelengths. For visible light of wavelength 5×10^{-7} m, this wave train would be emitted during less than 10^{-8} second, and would be about half a meter long. The wave train is plane-polarized, with the plane determined by the direction of the atomic vibration.

If a beam of light from a flashlight shines on a wall 10 meters away, our model leads to a picture somewhat like Figure 23-19. Many wave trains fill the space between the flashlight and the wall. They start and stop at random, and the intensity at the wall averages out to a steady illumination. Light coming head-on from a "point" source composed of very many vibrating atoms is a combination of many plane waves, with their electric fields pointing in all possible directions. The net effect averages out, and there is no preferred direction for the combination of all these randomly oriented vibrations. Therefore, we can say that ordinary light is unpolarized, but in fact we mean that it is made up of components that are polarized in all possible transverse directions. It might seem that all these component vectors would add up to zero. But remember that we are dealing with wave trains each from a different atom, and there is no definite phase relation between them. We say that there is no *coherence* between the phases of their vibrations.

Light becomes polarized when it is reflected from the surface of an electrical insulator—for instance, the glare of sunlight reflected from a lake or a concrete pavement is partially polarized.

Another way to make a polarized beam out of an unpolarized beam is to send

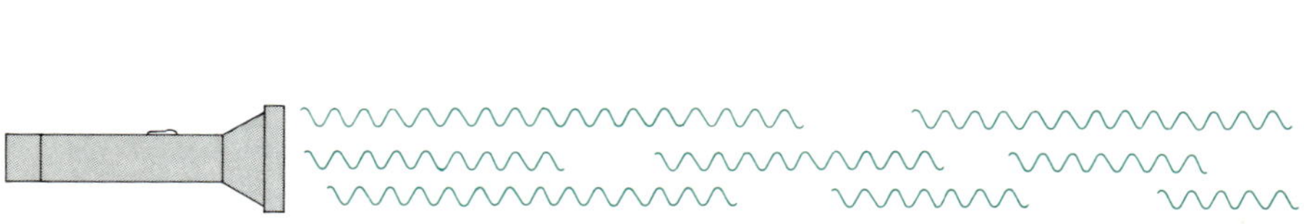

Figure 23-19

the light through a sheet of polarizing glass or film (often called Polaroid). This material consists of many tiny crystals of a quinine compound, oriented with their axes parallel to each other, and spread out on a transparent base. The crystals transmit only light whose plane of vibration is properly oriented with respect to their axes. Light vibrating in the "wrong" direction is absorbed by the crystals. A light-polarizing material can also be used to analyze the plane of vibration of a beam of light that has already been polarized by passing through one sheet of the material. This can be done by observing the source through a second polarizing sheet, to see whether the intensity changes as the second sheet is rotated.

It is worth noting that the polarization of light waves proves them to be *transverse*. Longitudinal disturbances such as sound waves show all the wave phenomena of interference and diffraction, but they cannot be polarized. We should not make the mistake, however, of thinking that polarization, as such, proves the wave nature of light. It is not difficult to imagine a polarized beam of corpuscles—for instance, a stream of flat stones being "skipped" on the surface of a lake. The orientation of the stone is all-important for the success of the operation. We could in this way separate a beam of polarized stones from among the unpolarized (randomly oriented) stones striking the water. Some nuclear particles can be polarized by reflection in this way.

Polarization, therefore, shows light to be transverse, but not necessarily a wave. Interference shows light to be a wave motion, but not necessarily a transverse wave. It takes the combined evidence of polarization and interference to establish that light is a transverse wave motion.

UNIT 6

References and Outside Reading

The purpose of the following list is to make you aware of additional reading that will extend and broaden most of the concepts you have met in this unit. These references should be easily available and are of sufficient variety to meet your individual interest.

A Source Book in Physics, by W. F. Magie, Cambridge, Mass., Harvard University Press, 1963. See Huygens' principle; the speed of light determinations by Römer, Bradley, Fizeau, and Foucault; the selections on interference and polarization from writings of Grimaldi, Newton, Young, and Malus.

"Atmospheric Halos" by David K. Lynch in *Scientific American,* April 1978. Explains some interesting optical effects in the atmosphere.

"The Bright Colors in a Soap Film Are a Lesson in Wave Interference" in the "Amateur Scientist" section of *Scientific American,* September 1978. Excellent photographs and some useful technical details.

"Experiments with Goggles," by I. Kohler, in *Scientific American,* May, 1962. Some modern experiments in human vision.

"Eye and Camera," by G. Wald, in *Scientific American,* August, 1950. A comparison of two similar optical instruments.

How to Use a Microscope, by W. G. Hartley, New York, American Museum of Natural History, 1964. More than just a "how to," it is full of interesting discussions in applied optics.

"The Interaction of Light with Light," by J. A. Giordmaine, in *Scientific American,* April, 1964. New optical effects that arise with intense beams of light from lasers.

Introduction to Photographic Principles, by L. Larmore, second edition, New York, Dover Publications, 1965. The principles of physics applied to an important area of modern technology.

Light, Color, and Vision, by Y. Le Grand, New York, John Wiley & Sons, 1958. Presents a broad discussion and thorough summary of these subjects.

"Light-Wave Communications" by W. S. Boyle, in *Scientific American,* August 1977. Describes how messages are transmitted by light from light-emitting-diodes through glass fibers.

"The Luminescence of the Moon," by Z. Kopal, in *Scientific American,* May, 1965. A study of moonlight shows that it is not all reflected sunlight.

"The Michelson–Morley Experiment," by R. S. Shankland, in *Scientific American,* November, 1964. An excellent article describing the Michelson interferometer, its relation to the science of light, and to relativity.

Michelson and the Speed of Light, by B. Jaffe, Science Study Series, Garden City, N.Y., Doubleday Anchor, 1960. An account of a great experimenter's work.

"Moiré Patterns," by Oster and Nishiyima, in *Scientific American,* May, 1963. Describes an optical effect similar to interference, and includes experiments and observations you can do.

"The Moon Illusion," by Kaufman and Rock, in *Scientific American,* July, 1962. Interesting experiments in vision, including some observations you can make yourself.

The Nature of Light and Color in the Open Air, by M. Minnaert, New York, Dover Publications, 1954. Interesting observations about mirages and other out-of-doors optical effects.

"The Optics of Long-Wavelength X Rays" by Eberhard Spiller and Ralph Feder in *Scientific American,* November 1978. Long-wavelength "soft" x rays are beginning to have many applications in the laboratory and in technology.

"The Perception of Neutral Colors," by H. Wallach, in *Scientific American,* January, 1963. Fundamental experiments in human vision.

"Photography by Laser," by Leith and Upatnieks, in *Scientific American,* June, 1965. A new photographic process with great potential.

Scientific American, September, 1968. This entire issue is devoted to light, covering many of the aspects of this important phenomenon, and its effect on other areas of science.

"Side-Looking Airborne Radar" by H. Jensen, L. C. Graham, and L. J. Porcello, in *Scientific American,* October 1977. A discussion of "seeing" at wave lengths outside the visible range.

"Three-Pigment Color Vision," by E. F. MacNichol, Jr., in *Scientific American,* December, 1964. Experiments with individual cells in the retina.

UNIT 7 ATOMS AND MATTER

Until almost the end of the nineteenth century, the nature of light was thought to be fully understood in terms of electromagnetic waves. No difficulties were found in the laws relating electricity and magnetism.

However, numerous problems were about to arise with the classical theories. The atom would be revealed as a complex structure with many component parts. The photoelectric effect—emission of electrons from a surface bombarded by light—would force drastic revision of the wave concept of light. The laws of motion and of electromagnetic radiation would have to be modified to explain the behavior of atomic particles. It was these difficulties that led to the development of the quantum concept.

Each tiny dot in the micrograph shown on the facing page represents the position of an atom in an iridium crystal. The picture was taken through a field ion microscope using techniques based on the quantum concept.

In this unit you will explore our present understanding of the outer atom and some of the applications of quantum mechanics. You will also investigate the atomic nucleus, as perhaps the most active area of physics research today.

24 The Outer Atom

Almost everyone is familiar with the concept that matter is made up of vast numbers of extremely small particles called atoms. But not so many understand why scientists believe in this atomic picture, and what is known of the underlying structure of atoms. By now you probably have a vague notion of the atom as a nucleus surrounded by a cloud of negatively charged electrons (Figure 24-1). This chapter offers you an opportunity to extend your understanding of the concept of the atom.

Probably most of us think of "atomic physics" as something very new, perhaps associated with bombs and other spectacular devices. Actually, the atomic theory of matter has a long and interesting history. As far as we know, the first definite statement of an atomic theory was made by the Greek natural philosopher Democritus around 400 B.C.

Democritus set forth many points that are still a part of the modern theory, but this early statement was not fruitful for several reasons. First, more influential philosophers, such as Plato and Aristotle, disagreed with him and promoted their own ideas more successfully. Also, Democritus' theory was based mainly on speculation, and he could offer little real evidence to back up his statements. Much more to the point, it was thousands of years before anyone could quantitatively test any of the consequences of the theory. No one can ever know whether a theory or a model is valid unless it is possible to make some definite prediction based on the theory and then experiment to see whether the prediction is a good one. Three parts of the Democritus theory are still of interest today:

(a) **From nothing comes nothing.**
(b) **The only existing things are the atoms and empty space, all else is mere opinion.**
(c) **The varieties of all things depend upon the varieties of their atoms, in number, size, and aggregation.***

It was not until the nineteenth century that chemists like John Dalton were able to build a new atomic theory based on the facts of chemistry and to perform a number of experiments that gave the atomic model a sound base. The earliest atomic theory, as well as the later theories developed by Dalton and other nineteenth-century chemists, regarded the atom as a kind of hard "billiard ball" without any definite internal structure. This kind of hard, elastic atom is quite adequate as a model for understanding such important aspects of physics as the behavior of gases, as we saw in Chapter 13. Up until the very end of

* Adapted from R. A. Millikan, *Electrons + and −*, University of Chicago Press, page 9.

the nineteenth century, it was generally quite satisfactory to regard the atom as the ultimate building block, which could not be subdivided—the very word *atom* means indivisible.

By the end of the nineteenth century, experimental evidence began to accumulate which indicated that the atom was divisible, and that it was, in fact, made up of electrically charged "subatomic" particles. Our awareness of the number and variety of these particles seems to be increasing at a rapid rate today, but in this study we shall be concerned with the very few fundamental particles that are reasonably well understood. Some early evidence indicating that the atom had structure came from experiments with x rays (discovered by Roentgen in 1895). In these experiments, x rays were found to be capable of breaking up gas atoms into two charged pieces, thus indicating that gas atoms were far from being indivisible.

We shall first consider the outer atom, which is now thought to consist of a cloud of negative electric charge, and look at the experimental evidence about electric charge and electric forces that led to our present understanding of the atom. It is worth noting that some of the experiments about to be described could not have been performed until methods were developed for producing a high vacuum. In a vacuum it is possible to study the relatively simple behavior of isolated charged particles without being distracted by the complicated effects arising from collisions with other particles. This is just one example of how science and technology benefit each other; new technological developments make new scientific investigations possible, and new scientific concepts bring about vast new areas of technology.

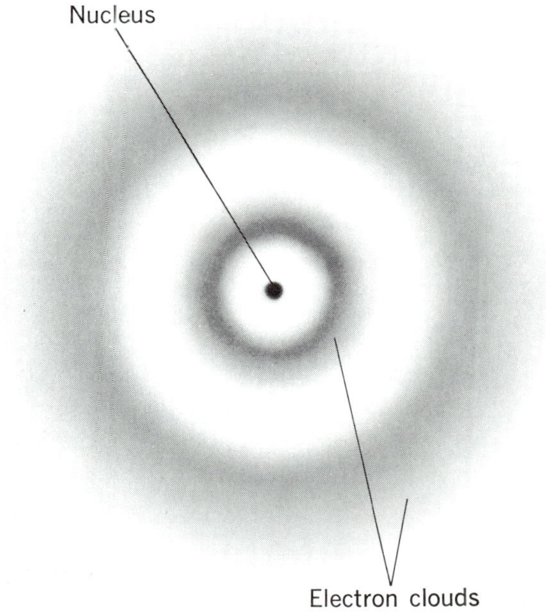

Figure 24-1 Electron cloud diagram. Electrons appear to be spread out in cloud-like shells around the nucleus. The darker areas of the clouds indicate the regions where there is the greatest probability of finding electrons.

24-1 THE ELEMENTARY UNIT OF ELECTRIC CHARGE

The facts of *electrolysis* described in Section 14-5 strongly suggest that the electric charges of ions such as Cu^{+2} are multiples of some fundamental unit of charge. It is important to note, however, that chemical experiments of this type can show only that ions have some definite *average* electric charge. They do not show anything about the charges of individual ions, nor do they show that these charges are identical to each other. We must know that the charges on all Ag^{+} ions, for instance, are identical before we can believe that there is a "natural unit" of electricity. The individual charges might differ greatly among themselves and yet always average out to the same value when many ions are considered. (Even the most delicate chemical weighing experiments deal with no less than 10^{16} ions.) You have already met with such an averaging-out in the study of the kinetic theory of gases (Chapter 13). The gas molecules have all sorts of speeds, ranging from

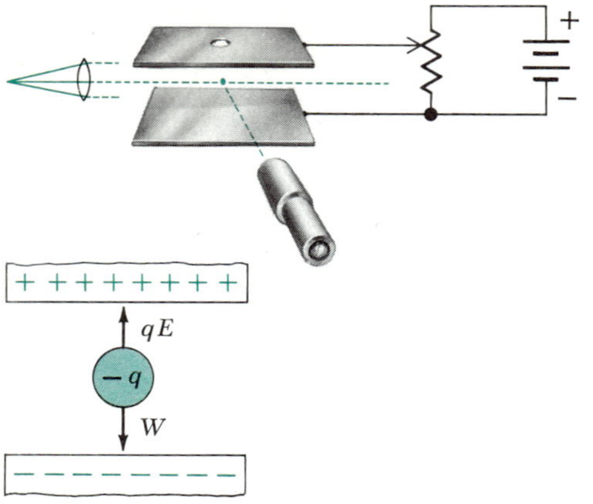

Figure 24-2 Millikan's oil-drop experiment to measure the charge of the electron.

very slow to very fast, and yet the average speed is a definite value at any given temperature.

Robert A. Millikan (1868–1953), an American physicist who worked at the University of Chicago and the California Institute of Technology, performed a series of delicate experiments that showed the effect of single elementary electric charges. These experiments, beginning around 1909, made use of an extremely sensitive kind of balance that was able to measure changes in the electric force when single units of charge were added to or removed from a small object.

Millikan's experiment has become famous as the "oil-drop experiment" because his measurements involved the behavior of tiny liquid droplets placed under gravitational and electric forces. His first experiments were done with water drops, but these were found to evaporate too fast. Later he used drops of oil, which evaporated very slowly and gave him plenty of time to watch the behavior of single drops.

In his first experiments, the water drops were part of a "fog" formed by suddenly reducing the pressure in a closed space. Millikan was able to "capture" single drops using an electric field between two charged metal plates. These drops had become electrified by picking up charged ions from the air.

In Figure 24-2, the observer has selected a negatively charged drop that he holds stationary by applying a positive voltage to the upper plate. The drop is acted on by two forces: its weight W (the downward force of gravity), and an upward force $F = qE$ due to the charge q and the electric field E. In Section 16-3 the strength of the electric field was calculated from the potential difference of the plates and the distance between them. The strength of the field E is the PD divided by the separation distance. That is,

$$E = \frac{V}{d}$$

Recall that the field strength is measured in newtons per coulomb (Section 16-1).

When the PD, and thus the electric field, are adjusted so that the drop is hanging motionless, the gravitational force and the electric force are equal but opposite in direction.

$$\text{net } F = 0$$

$$q\frac{V}{d} - W = 0$$

$$q = \frac{Wd}{V}$$

This equation gives a way of finding the charge in terms of other quantities. The quantities d and V (the plate separation and the PD) are easily measured, but finding the weight W of such an extremely small drop is far beyond the capability of any ordinary balance.

Millikan devised a simple but wonderfully sensitive method for finding the weight. A drop falling through air, like a descending parachute or a stone falling through a jar of syrup, rapidly ceases to accelerate and then falls at a constant terminal velocity. This is because the force of air resistance (or the resistance of the syrup) quickly becomes equal and opposite to the force of gravity; the net force on the drop is then zero, and the acceleration ceases. The essential relation is that the terminal velocity is proportional to the square of the radius of the drop. Therefore, a measurement of the terminal velocity makes it possible to calculate the radius, and hence the weight. Once the weight is known, the charge can be computed from the relation $q = \frac{Wd}{V}$. Recall that in this formula d refers to the separation between the plates, and V is the PD required to hold the drop motionless in the field of view.

It is possible to reproduce the essential features of this experiment in your own laboratory by making use of a modern development that was not available to Millikan (Figure 24-3). It was very laborious and time-consuming for him (and later experimenters) to determine the weight of each new drop by terminal velocity experiments. Now we can make use of very small plastic spheres, that can be produced in large quantities and closely controlled sizes. A typical sphere has a diameter of 1.8 micrometers (1.8×10^{-6} m); the density is known and it is a simple matter to calculate the mass and the weight.

The method described above allowed Millikan to measure the charge on a single droplet; it might be a small number of elementary charges, but it would probably not be just one charge. During several experiments, however, Millikan noticed that the droplet he was holding suddenly began to move up or down in the electric field. This could only mean that the droplet had either gained or lost electric charge, and was therefore experiencing a different electric force. He reasoned that the droplet was picking up single gas ions from the air and that these single ions might well carry just single units of the elementary charge. Later, instead of holding a drop fixed he used the field to give the drop a constant terminal velocity. Millikan proved that the change in speed of the drop would be directly proportional to the charge taken on or off. In his experiments, he was able to keep a "... droplet pacing its beat up and down between the plates until it could catch an atomospheric ion. ..." (These are Millikan's words from his book *Electrons + and −*.*)

The essence of Millikan's work can best be understood by looking closely at one of his early experiments using this new method. Again in his book *Electrons + and −*, he gave data (page 69 1935 ed.) reporting measurements on one oil drop that he followed first for a number of runs under the force of gravity alone. Then he followed it under the force of gravity plus the force of the electric field between the plates of his apparatus.

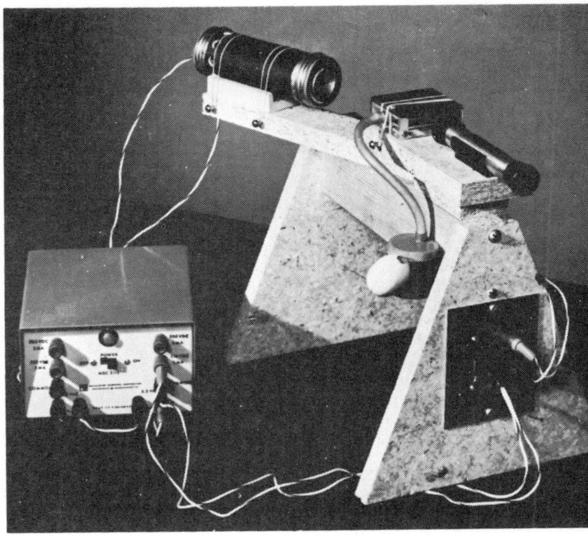

Figure 24-3 A modern version of Millikan's apparatus.

First he made seventeen measurements of the time of fall of the drop under gravity alone (he was a careful man!) These all came out closely the same and showed him, for one thing, that the drop was not evaporating (why?) Also, the average of these measurements gave him the terminal velocity of the drop, and thus its weight. He was also able to make twelve measurements of the speed of the drop when acted on by the electric force; these have been reproduced in Table 24-1. The column headed t_F gives the times taken by the drop to move between two cross-hairs in the field of view, 0.52 centimeters apart. The column headed v gives the speeds obtained by dividing the distance (0.52 cm) by the time. A bar graph of these speeds is shown in Figure 24-4.

As the drop picks up ions from the air, the charge changes, and so does the electric force. The thing to look for is how the speeds *change* because a speed change is an indication of charge coming on or off the drop. The drop in Table 24-1

* Page 65, 1935 ed.

Table 24-1

Measurement	t_F (sec)	v (cm/sec)
1	12.5	4.2×10^{-2}
2	12.4	4.2×10^{-2}
3	21.8	2.4×10^{-2}
4	34.8	1.5×10^{-2}
5	84.5	$.62 \times 10^{-2}$
6	85.5	$.61 \times 10^{-2}$
7	34.6	1.5×10^{-2}
8	34.8	1.5×10^{-2}
9	16.0	3.3×10^{-2}
10	34.8	1.5×10^{-2}
11	34.6	1.5×10^{-2}
12	21.9	2.4×10^{-2}

had a positive charge, and Millikan said that the speed change from measurement 3 to 4 represented the drop catching a unit of negative charge. Look at measurements 3 through 7. What do you notice about the changes in speed that have occurred during these intervals? How have the charges changed? What happened in the intervals just before measurements 7, 9, and 10?

The important thing to observe in these data is that the speeds always changed by the same amount or by a small multiple of the same amount: in the present case, either 0.9 cm/sec or 2×0.9 cm/sec. This means that the charge changed only in multiples of one definite unit of charge—the "elementary charge." Millikan repeated the measurements many thousands of times on hundreds of drops, and always the changes in charge came in multiples of the same unit. In our system of units this is 1.60×10^{-19} C.

The oil-drop experiments show that there is a smallest *indivisible* natural unit of charge. At least in this experiment and, by inference, in general, we are led to believe that electric charge is *quantized* (occurring as multiples of some basic quantity), rather than endlessly divisible into infinitely small fragments. The chemical experiments with electrolysis give us the same value for the size of the elementary charge, although they can give only an average value and do not indicate that charge is quantized.

24-2 THE ELECTRON

The Millikan experiment showed that electric charge was quantized, that it came in small, discrete packages of a definite size, and also showed the value of these natural units of charge. However, if the Millikan experiment were all the evidence that was available, we would have no reason to think of electric charge as being associated with any kind of material particle. That is, even though we know from the Millikan experiment that charge itself comes in packages of a definite size, we have no reason to believe that the units of charge have mass, or inertia, or other properties that are associated with particles. It is now known that negative charge is associated with a particle of measurable mass called the *electron*. Let's now explore some of the experiments that established the electron as a particle and showed its significance in the outer structure of the atom.

In the 1870's Sir William Crookes experimented with the flow of electricity through gases in tubes like that shown in Figure 24-5. In a typical case the tube contains air at low pressure. When a

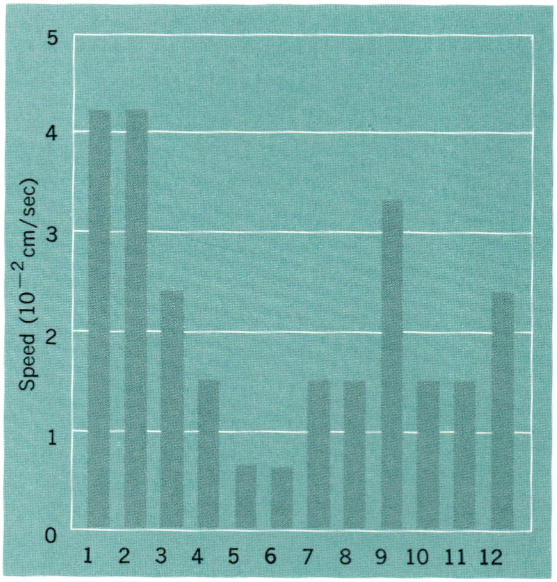

Figure 24-4

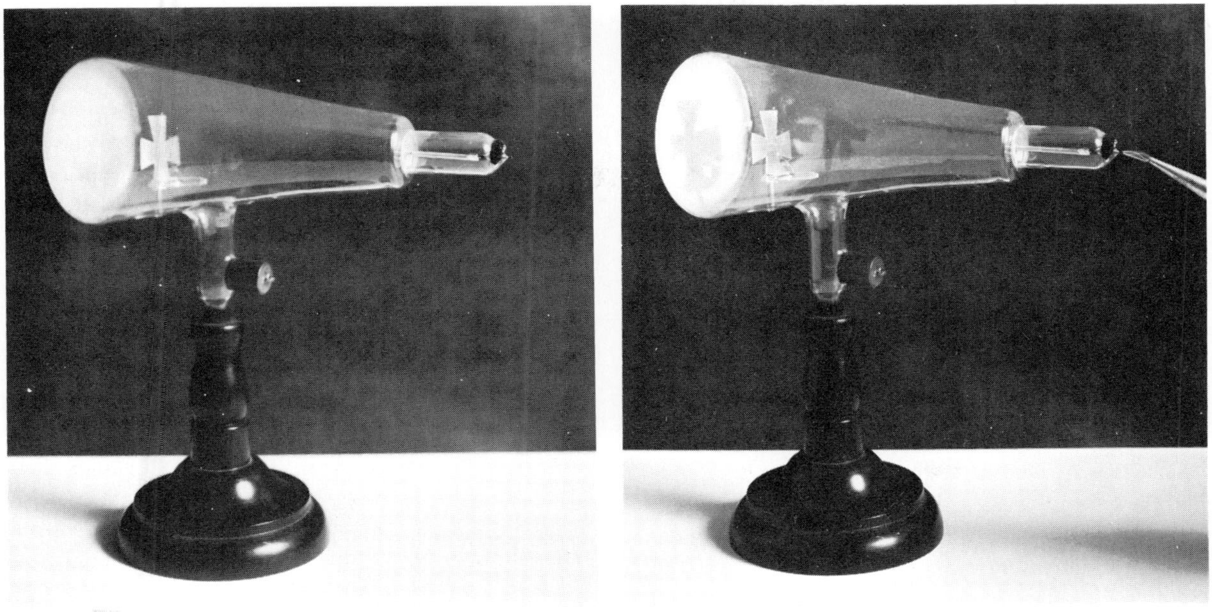

(a) Figure 24-5 (b)

high negative potential is applied (in b) by means of a *Tesla coil* or spark coil, charge flows through the tube, and a shadow is cast by the metal cross.

Using modern vocabulary and modern knowledge, we can say that charge is being carried through the tube by positive and negative ions of nitrogen and oxygen, as well as by the electrons. As Crookes showed by his experiments, a stream of particles leaves the negative terminal (the cathode), and proceeds in straight lines, perpendicular to the surface of the cathode. The presence of a stream of particles is shown by a bluish glow along their path, and the fact that the particles travel in straight lines is shown by the sharp shadow cast by the obstacle. In similar experiments, when the obstacle is a light, easily deflected vane, the particles are seen to have inertia (or mass) because they exert a measurable force on the vane. Also, the particles carry energy, for a thin metallic target can be made red-hot by the impact of the particles. The moving particles can be deflected by a magnetic field, so they must carry electric charge (Section 18-1). The direction of deflection shows them to be negative.

It is now known that these *cathode rays* are electrons that have been ejected from the cathode by the bombardment of the gas ions. The same effects, showing the inertia and energy of the particles, can be seen with the electrons given off by a heated metal. Cathode rays were studied extensively by the British physicist J. J. Thomson (1856–1940), who is generally regarded as the discoverer of the electron. Let's briefly examine the experiment that Thomson reported in 1897.

Thomson conducted his experiments with electrons in two parts. First he measured their velocities by using a magnetic field and an electric field at the same time. Figure 24-6 shows a tube like one that Thomson used. Cathode rays are emitted from the cathode *C*, are attracted to the positively charged anode *P*, and pass through a small hole in the anode, which forms them into a narrow beam. The plates *A* and *B* can be given electric charges, so there will be an electric field between them. The magnetic field is at right angles to the electric field, and is shown by the dots indicating the *B* vectors pointing out from the page.

When the magnetic field alone is turned on,

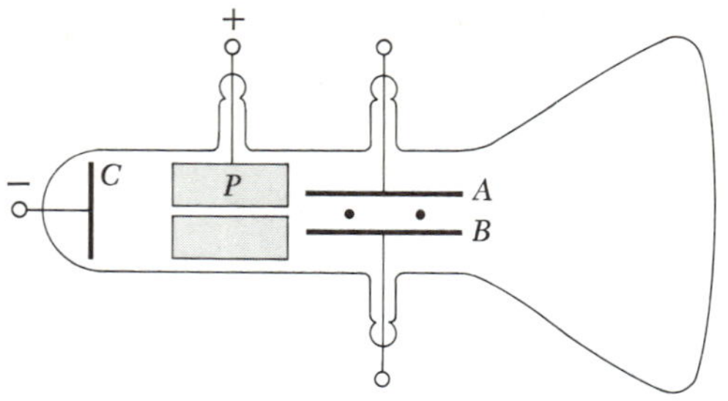

Figure 24-6 Thomson's cathode-ray tube.

there will be an upward force on the stream of particles, and the spot on the end of the tube where the particles strike moves upward. Recall from Section 18-1 that the force on a moving charged particle in a magnetic field is given by $F = Bqv$, where B is the magnetic field, q is the charge of the particle, and v is the velocity of the particle. By putting charges on A and B, we can set up an electric field opposite in direction to the magnetic force, which will tend to cancel out the force of the magnetic field. When this happens, the spot on the end of the tube is not deflected, and the electric force is equal in magnitude (but opposite in direction) to the magnetic force. That is,

$$Bqv = Eq$$

where E is the electric field between the plates. By solving this equation for v, we get

$$v = E/B$$

All we need to do, therefore, is to measure the ratio of the strength of the electric field to the strength of the magnetic field, and we have a measure of the velocity. Thomson observed speeds as high as 95,000 kilometers per second, nearly one third the speed of light.

Once the velocity of the mysterious "cathode rays" has been determined under a given set of conditions, the same apparatus can be used to find out their charge per unit mass. The charge of the electron is called e, so this ratio is e/m. If the magnetic field in the tube of Figure 24-6 is turned off, then the cathode rays are acted on only by the electric field between the plates A and B. The rays moving between these plates now act just like projectiles moving horizontally in the field of the earth's gravity. Let's call the length of the plates l, and the horizontal velocity of the particles v (Figure 24-7). Then the particles will move through the space between the plates in a time $t = l/v$. In this time, they will "fall" in the electric field for a distance $\frac{1}{2}at^2$. The acceleration a is found from Newton's law:

$$a = F/m$$
$$= Ee/m$$

where E is the electric field strength, e is the charge of the electron, and m is the mass.

We know that the downward distance d that the particle is deflected in the electric field is $d = \frac{1}{2}at^2$. By substituting Ee/m for a and l/v for t, we get

$$d = \tfrac{1}{2}(Ee/m)(l/v)^2$$

We can measure the distance d that the beam is deflected; we have measured v in the previous experiment; and finding E and l presents no problems. Therefore, e/m can be calculated by rearranging this relation as

$$e/m = 2dv^2/El^2$$

The most significant result to come out of Thomson's many measurements was that no matter what the electrodes were made of, no matter what gas was in the tube, or what velocity he used, the value of e/m always came out the same. This definitely indicated the presence of a real particle

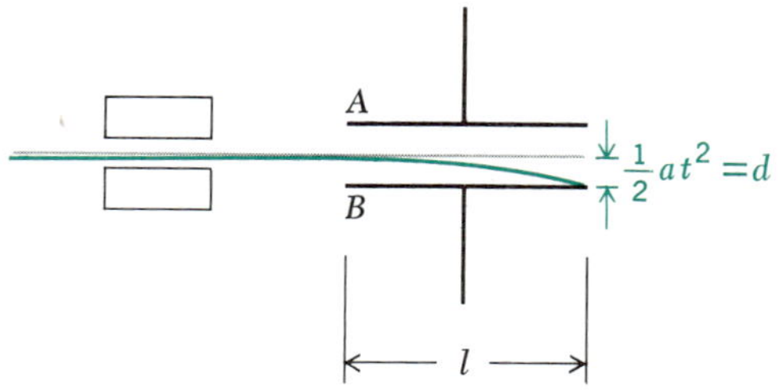

Figure 24-7

having real properties that could be measured over and over again. If the velocities are well below relativistic speeds, the value of e/m works out to be 1.759×10^{11} C/kg. From the results of Millikan's experiment, we know the value of e and can therefore calculate m, the mass of the electron. This important constant has the value 9.11×10^{-31} kg.

The combined results of the experiments described so far in this chapter show that (a) charge is quantized, and (b) free electrons exist, all of which are identical, each carrying one unit of charge always associated with the same amount of mechanical inertia. In your own laboratory, you can do an experiment to measure e/m; most modern measurements make use of a tube in which the electrons are bent into a circular path by a magnetic field.

24-3 THE PHOTOELECTRIC EFFECT

Now that we have pinned down the electron as an identifiable particle with charge and mass, we can use its behavior as a key to the mystery of the structure of the atom. To do this, however, it will be necessary to travel a path that seems to take us far from finding how electrons are arranged in atoms. We cannot really begin to understand the role of the electron until we are familiar with the relationship between light and the electron; that is, how light acts upon electrons, and how light is emitted by atoms. This will provide a start toward understanding the *quantum mechanics* that is used to describe the inner nature of the atom, rather than the classical mechanics used to describe objects of ordinary size.

Earlier in this chapter it was mentioned that the development of techniques for producing a really good vacuum made possible tremendous advances in experimental physics. These techniques had no part in Millikan's oil-drop work, but they were essential to other physicist's experiments with the photoelectric effect, which will be described next. Here you are going to see how electrons react when a beam of light shines on a metallic surface in an extremely good vacuum.

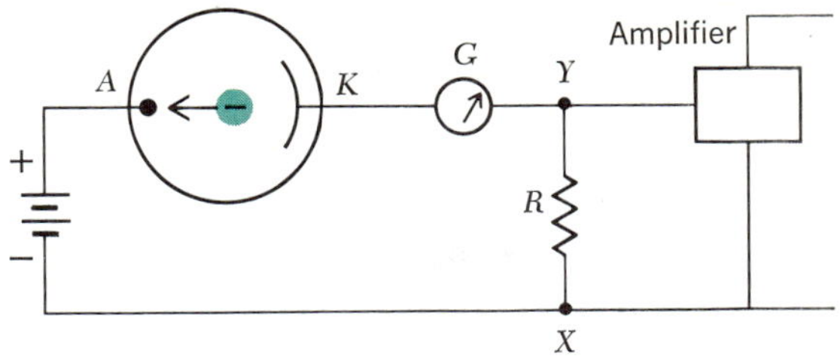

Figure 24-8(a) Photocell circuit.

For this experiment, a circuit like that in Figure 24-8a is used. The circle indicates a glass enclosure with two metal electrodes inside. The metal electrode marked K is the cathode. In most commercial photocells the cathode is curved, as shown, but this refinement is not necessary for the success of our experiment. Engineers design special cathode materials so the cells will respond to visible or even infrared light for special purposes. At the start of the experiment, the cathode is made negative by the battery.

The anode A can be made of any metal; in fact, the anode material is not critical as long as it is a good conductor. The anode is shown as a round object in the diagram because in commercial photocells they are often round in shape.

The symbol G represents a galvanometer or other sensitive current-measuring device. The purpose of the resistor is to allow for amplification of the small changes in current in the photocell circuit; a small change in current will appear as a change in voltage across R and this can be amplified by the amplifier, if desired.

For cathodes made of most metals, ultraviolet light must be used. Light from an ultraviolet source shines on the cathode; this source is usually a carbon arc, or it could be a mercury-vapor arc in a quartz bulb.

Suppose the cathode is illuminated with one ultraviolet bulb at a certain distance, and the galvanometer shows a current. If the cathode is illuminated with two such bulbs at the same distance, twice the current is produced. With three bulbs, three times the current, etc. This kind of result indicates that the current is proportional to the intensity of the light. Perhaps a reasonable interpretation of this effect is to say that somehow the beam of light knocks electrons out of the cathode, and these are attracted to the anode. The electrons leave the surface of the cathode and travel through the vacuum to the anode. The stream of electrons constitutes a current and is registered by the galvanometer.

The proportionality between current and light intensity is not surprising, but now suppose the "photocurrent" (current that arises when light shines on the cathode) is measured in relation to the *frequency* of the light. This could be done by using ultraviolet sources with various filters in front of them to pass light only in a narrow range of frequencies. With visible light, ordinary white light with filters of various colors would do the job. To obtain more precise results, a sharply defined spectral line would be best, but filters will do well enough for our purpose. Suppose the intensity of the light is kept constant, and we decrease the frequency. At a certain point the current suddenly drops to zero! Such a result indicates that for a particular cathode surface (made of a given metal, perhaps coated with a chemical compound) there is a *threshold frequency*. That is, the current remains constant until the threshold frequency is reached. Below the threshold frequency, the current is zero. In other words, below the threshold, no electrons are knocked out of the cathode, no matter how intense the light is.

This result suggests a modification of the experiment shown in Figure 24-8b. The circuit is about the same as before, except that now an ideal voltmeter *V* is connected between the anode and cathode to measure the PD between these two points. The polarity of the battery has been reversed, so that now the anode is negative and the cathode is positive. A voltage divider is connected across the battery terminals so that an adjustable negative voltage, just sufficient to bring the ejected electrons to a stop, can be applied between the anode and the cathode. In this way we are able to measure the KE of the electrons emitted from a surface. In a similar fashion, we

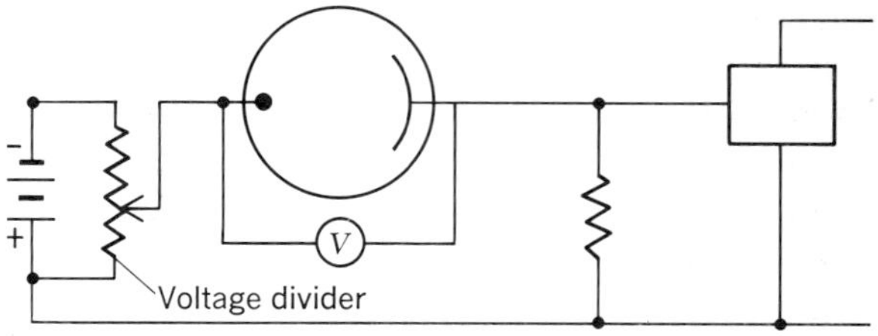

Figure 24-8(b) Circuit for measuring KE of electrons.

could measure the KE of a baseball that is thrown upward by seeing how high it can rise against the earth's gravitational field.

From this kind of measurement, we find that if the frequency is above the threshold frequency, the *photoelectrons* that are knocked out of the cathode surface have KE that depends only on the frequency of the light. A photoelectron is an electron that has been knocked out of an atom by light; otherwise it is no different from any other electron. One might expect that a bright light would eject electrons having greater KE than those ejected by a dim light, but this is not the case. The experiment shows that a bright light merely ejects *more* electrons.

To illustrate this strange situation, let's imagine a cathode surface whose threshold frequency is 6×10^{14} Hz. This corresponds to green light of wavelength 5×10^{-7} m. If the surface is illuminated by strong *red* light, not a single photoelectron is ejected, since the frequency of red light is less than the threshold frequency. On the other hand, the faint light from a distant *blue* star can eject electrons, since the frequency of blue light is greater than the threshold frequency. The KE of these ejected electrons does not depend on the star's brightness, but their KE does depend on the frequency of the light. None of the physics that we have studied so far, the so-called classical physics, is able to deal with these experimental facts.

In Chapters 20 through 23, a number of experiments and phenomena demonstrating the wave nature of light were presented. Experiments designed to show interference effects particularly emphasized the wave properties of light. However, in the introduction to Chapter 20 and again in the introduction to Chapter 22, it was pointed out that one must be very careful in thinking about models for light. Light is a phenomenon that reveals different properties depending on the kind of experiment that is done. An interference experiment will reveal wave properties. A photoelectric experiment will reveal other properties of light, *not* wave properties. There is no need to worry about any conflict because light is a complex and fundamental phenomenon, and there is no reason why it should be expected to fit purely mechanical models such as ripples or streams of bullets. In Chapter 20 the classical model was developed in terms of electric and magnetic waves, which are very different from water waves.

The wave picture does not help us to understand the photoelectric effect, and the classical wave ideas could not explain the experimental facts of radiation from heated objects, so-called blackbody radiation. (See color insert following page 436.)

In 1900 the German physicist Max Planck (1858–1947) was able to explain some of the problems of blackbody radiation by assuming that the radiation was emitted not as continuous waves, but was "quantized" in discrete packages of a definite amount of energy. Planck assumed that these packages, or *quanta*, had an amount of energy proportional to the frequency of the radiation. But he applied his quantum assumption only to the actual emission process in the atom itself.

Einstein took the next bold step and said that Planck's quantum idea applied also to the radiation as it was propagated through space and absorbed. He said that radiation is absorbed, as in the photoelectric experiment, one quantum at a time. Such a quantum of radiant energy is now called a *photon*. Since the energy is proportional to the frequency f, the energy of a photon is given by

$$E = hf$$

where the constant of proportionality h is called Planck's constant and has the value 6.63×10^{-34} J·sec.

It is not fruitful to try to answer such questions as: Is a photon a package of waves? What is a photon, *really*? In dealing with a physical phenomenon such as light, it is only possible to ask such questions *of nature* by *doing experiments*. The behavior and the properties revealed by these experiments are the answer to the question—What is light? The insight of theoretical physicists may point out great generalizations, such as the quantum ideas, that tend to indicate how many different experiments may be explained by a single underlying concept, but the concept is not likely to be something familiar. Later in this chapter (Section 24-6) the relationship between photons and e-m waves will be presented in terms of probability.

Example 24-1 What is the energy of a photon of green light?

If 5000 Å is taken as the average wavelength of green light, then

$$f = \frac{c}{\lambda} = \frac{3 \times 10^8 \text{ m/sec}}{5 \times 10^{-7} \text{ m}} = 6 \times 10^{14} \text{ sec}^{-1}$$

The energy of a photon for this light is

$$E = hf = (6.63 \times 10^{-34} \text{ J}\cdot\text{sec})(6 \times 10^{14} \text{ sec}^{-1}) = \boxed{3.98 \times 10^{-19} \text{ J}}$$

Einstein applied the quantum concept to the photoelectric effect. His photoelectric equation states that the KE of an ejected electron equals the energy of the incoming photon minus the energy w, called the *work function*, needed to remove the electron from the surface. In symbols, the law of conservation of energy gives

$$\tfrac{1}{2}mv^2 = hf - w$$

The equation tells us that no electrons can be emitted unless hf is greater than w. This is how Einstein explained the existence of a threshold frequency. The equation also tells us that the KE of an ejected electron depends only on the frequency of the light and the work function of the surface. In terms of this theory, we interpret a bright red light as a stream of many relatively weak photons; a faint violet light as a stream of a few, more energetic photons.

In the study of atomic physics, it is useful to have a unit of energy of a size suitable for atomic processes. Such a unit is the *electron volt* (eV), which is equivalent to the energy that would be acquired by 1 electron (or any particle with 1 elementary charge) moving through a PD of 1 volt. In calculating the value of this energy, remember that the PD between two points A and B is the work required to move a unit of charge from B to A and that 1 volt is equal to 1 joule per coulomb of charge.

$$W_{B\to A} = QV_{AB}$$
$$1 \text{ eV} = (1.60 \times 10^{-19} \text{ C})(1 \text{ J/C})$$
$$= 1.60 \times 10^{-19} \text{ J}$$

where $W_{B\to A}$ stands for the work done to move the unit charge from B to A or the energy equivalent of this amount of work.

It is often convenient to calculate the energy of photons in electron volts, as follows:

$$E = hf$$
$$= h\frac{c}{\lambda}$$
$$= (6.63 \times 10^{-34} \text{ J}\cdot\text{sec})\left(\frac{3 \times 10^{8} \text{ m/sec}}{\lambda \times 10^{-10} \text{ m}}\right) \times \left(\frac{1 \text{ eV}}{1.60 \times 10^{-19} \text{ J}}\right)$$

where λ represents the wavelength in angstroms. By simplifying the above equation, we get this expression:

$$E = \frac{12{,}400}{\lambda} \text{ eV}$$

The usefulness of the electron volt as a unit of energy will become quite clear to you when you study the following example.

Example 24-2 The work function for sodium metal is 2.46 eV. With what KE are photoelectrons ejected from a thin sodium surface illuminated by light of wavelength 4000 Å?

First the photon energy is calculated in electron volts:

$$E = \frac{12{,}400}{4000}$$
$$= 3.10 \text{ eV}$$

This is the energy of each photon of the light of wavelength 4000 Å. Now, using Einstein's photoelectric equation, we find

$$\text{KE} = hf - w$$
$$= 3.10 \text{ eV} - 2.46 \text{ eV}$$
$$= \boxed{0.64 \text{ eV}}$$

In practice, this is an upper limit; some photoelectrons may have less than this energy if they lose energy by collision as they make their way through the metal to the surface.

Example 24-3 What is the threshold wavelength for a sodium surface?

At the threshold, no energy is available for KE. Hence,

$$0 = hf - w$$

and the photon energy at the threshold is

$$E = hf$$
$$= w$$
$$= 2.46 \text{ eV}$$
$$\lambda = \frac{12{,}400}{E}$$
$$= \frac{12{,}400}{2.46}$$
$$= \boxed{5040 \text{ Å}}$$

The threshold is in the green region of the spectrum.

Example 24-4 Through what PD would an electron have to fall in order to acquire as much KE as a dust particle has, if the dust particle has a mass of 0.0004 g and is falling at a speed of 1 cm/sec?

The KE of the dust particle is first calculated in joules, and then converted to electron volts.

$$
\begin{aligned}
\text{KE} &= \tfrac{1}{2}mv^2 \\
&= \tfrac{1}{2}(4 \times 10^{-7}\ \text{kg})(10^{-2}\ \text{m/sec})^2 \\
&= 2 \times 10^{-11}\ \text{J} \\
&= (2 \times 10^{-11}\ \text{J}) \times \left(\frac{1\ \text{eV}}{1.6 \times 10^{-19}\ \text{J}}\right) \\
&= 1.25 \times 10^{8}\ \text{eV} \\
&= 125 \times 10^{6}\ \text{eV} \\
&= \boxed{125\ \text{MeV}}
\end{aligned}
$$

The dust particle's KE is 125 million electron volts (125 MeV), and an electron would have to fall through a PD of 125 million volts to have as much KE.

24-4 ATOMIC SPECTRA

Much of the evidence that led to our modern understanding of the outer atom came from the spectra of the elements and their compounds. The nature of these spectra was extremely difficult to interpret in terms of classical physics. Quantum physics, however, was able to provide the clue. The quantum idea, which was so successful in explaining the photoelectric effect, was applied to the problem of spectra in 1913 by the Danish physicist Niels Bohr (1885–1962). Bohr succeeded in calculating the wavelengths of the lines in the spectrum emitted by hydrogen atoms.

In a solid, liquid, or dense gas, the atoms are close to each other, and can interact to give light that is seen through the spectroscope as a *continuous spectrum*, with all frequencies (and wavelengths) represented. However, the spectrum produced by an incandescent gas is discrete; that is, there are sharp lines of definite wavelength, with no light emitted at the intermediate wavelengths. Some representative spectra are shown in Figure 24-9 (color insert following page 388). Presumably the atoms in such a gas are free enough to emit their own characteristic frequencies, much as a piano string vibrates with certain natural frequencies. Some energy must be supplied to start a string vibrating; likewise, an atom must absorb energy in order to emit radiation. There are many ways in which an atom can be given the necessary energy. For instance, the sodium spectrum can be viewed by spectroscopic examination of the light produced when solid sodium chloride (NaCl) is placed in a flame, becomes vaporized, and absorbs energy by thermal bombardment. The copper spectrum can be obtained from light produced by passing an electric arc or spark between metallic copper electrodes. The radiation comes from copper vapor between the electrodes. Gaseous substances such as helium or hydrogen can be excited electrically when enclosed in a glass tube at low pressure.

The spectrum of a *solid* or *liquid* depends only on its temperature. The light from molten iron in a furnace, and from the ceramic bricks lining the furnace, looks about the same—red-hot—and in each case produces approximately the same spectrum, which is continuous. But since the frequencies emitted by an incandescent *gas* are characteristic of the substance itself, temperature affects only the relative intensities of the lines. The line spectrum of a gas can therefore be used as a means of identification.

In most spectra the lines do not seem to be distributed according to any recognizable pattern. Hydrogen, which is the simplest atom, gives a spectrum that shows some signs of a regular pattern. For this reason, new theories about the atom have long been tested by trying to apply them to the hydrogen spectrum, and it was this spectrum that Bohr attacked. It is tempting to want to describe the frequencies in the spectrum in the same way a musical tone is described in many general science textbooks: a fundamental frequency plus several *harmonics* that have frequencies which are whole number multiples ($2f$, $3f$, etc.) of the fundamental frequency. However, the hydrogen lines do not show such a pattern of harmonics, in fact they tend to cluster near a limit of 8×10^{14} vib/sec. This behavior cannot be explained by any process known in classical physics.

There is an even greater difficulty in trying to apply classical physics to the outer atom. Suppose we imagine a model of the hydrogen atom as some kind of positive charge associated with negative charge in the form of an electron—there is much evidence to indicate that the electron is an essential part of the atom. How is the electron arranged relative to the positive charge? It can hardly be standing still, for if it were, the electrostatic force would rapidly pull it toward the positive charge and the atom would collapse. Therefore, it seems reasonable to assume that the electron is moving in some kind of orbit relative to the positive charge. But this brings up new problems. If the electron is going to move in any kind of curved orbit, then it will have a centripetal acceleration. Yet it was shown in Section 20-6 that when electric charges are accelerated, they radiate their energy in the form of electromagnetic radiation. The electron would spiral into the positive charge as it lost energy and again, if this were the case, all atoms would have collapsed long ago. Since the atoms of the universe have not collapsed, we are faced with a serious difficulty.

Bohr got out of the difficulty by saying that in the world of *atomic* physics it is an observed fact of nature that electrons do revolve in their orbits without the energy of the atom radiating away. He did *not* attempt to explain this in classical terms, but stated it as a *postulate* of a different kind of physics needed for dealing with atoms. It is also known that under certain conditions atoms can and do absorb and emit radiation. Bohr therefore postulated that atoms will absorb and emit radiation when they change from one *energy state* to another, but that there are certain energy states in which atoms can exist for long periods without radiating. It is helpful, at first, to regard these energy states as being equivalent to various circular orbits for the electron, because as the radius of an orbit is increased, there is a corresponding increase in the energy of the atom. It is important to note that the energy is a property of the complete atom (nucleus and electrons), not the electron alone. Figure 24-10 shows several circular orbits for the electron in a hydrogen atom, with the energy states indicated. This kind of diagram should be regarded as only a rough first approximation because, in fact, electrons do not follow sharply defined orbits like those of planets around a sun, but rather their position at any time is governed by the far more subtle rules of quantum mechanics. Perhaps it is best to think of all the possible positions of the electron in the atom as a "cloud" about the nucleus as in Figure 24-1, the density indicating various degrees of probability. The simple circular orbit, then, might represent a highly probable set of positions, but not the only possibility. The circular orbit idea is useful, however, and Bohr made some of his original calculations in terms of it.

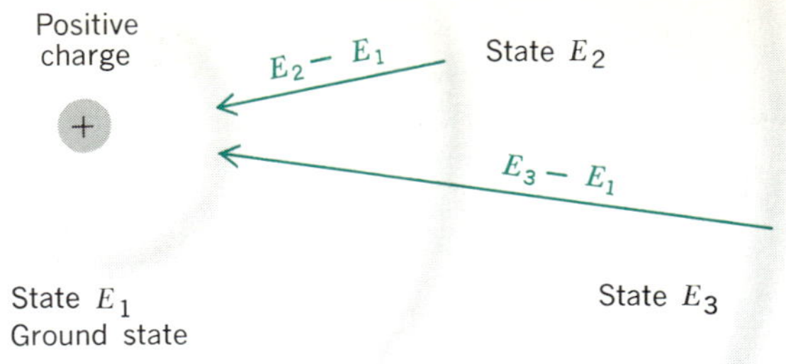

Figure 24-10

Next Bohr said that Einstein's photoelectric equation for the energy of photons applies to energy changes within the atom. In terms of the orbit model, when an electron jumps down from orbit E_2 to E_1, the atom changes from energy state E_2 to energy state E_1, and the difference in energy is $E_2 - E_1$. Bohr assumed that all this energy was used in forming a single photon, which is emitted as a result of the change in energy, and the energy of the photon is

$$hf = E_2 - E_1$$

Although the orbit diagram may help intially in visualizing energy changes in the atom, it is much more common to make use of an *energy-level diagram,* such as that in Figure 24-11, where the horizontal lines represent possible energy values. In this drawing the energies are some of those

Visual Essay

THE DEVELOPMENT OF QUANTUM PHYSICS

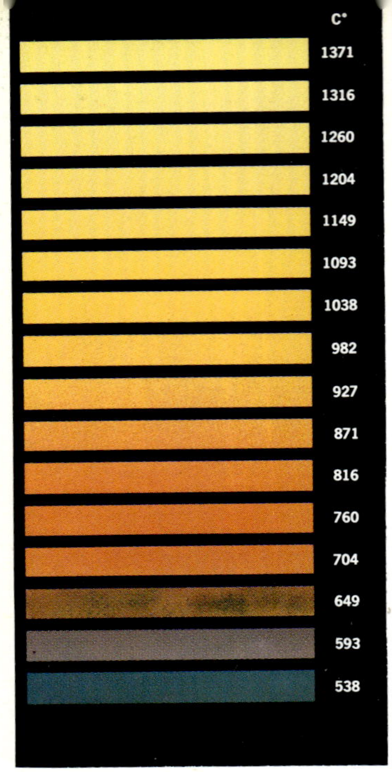

1

2

Max Planck

(1) Colors given off in an enclosed steel furnace at various temperatures.

(2) Black body radiation, from inside a steel furnace.

THE ORIGIN OF THE QUANTUM CONCEPT

When we wish to describe the behavior of very small things—atoms, molecules, or subatomic particles—we have to use the kind of physics now called *quantum mechanics.* The quantum theory was developed because the concepts of classical physics were unable to explain the observed facts of radiation. All bodies emit radiation (sometimes visible and sometimes invisible) and the spectrum of the radiation depends upon the temperature of the body. Here the word "spectrum" means how the energy is distributed as a function of wavelength.

In particular we are concerned with the radiation from a *black body.* In physics a black body is a surface that gives off radiation of many different wavelengths but reflects no light or other radiation. The interior of a furnace, or the sun are both good examples of black bodies.

An increase in temperature of a black body results in an increase in total emitted energy, and relatively more energy is emitted at shorter wavelengths. Application of the principles of classical physics (that is, the concepts of heat and the kinetic theory) predicts that the radiation from any black body at any temperature should all rapidly become concentrated into very short waves—ultraviolet radiation, x-rays, and gamma rays.

In everyday terms this would mean that the heat radiation from a furnace should rapidly become gamma rays, if classical physics applies to black body radiation. Experiment and ordinary experience clearly show that this effect does *not* occur—that visible light is still emitted.

Max Planck (1858—1947) tried to derive a relation, based on classical concepts, that would explain the experimental facts. He could not; and out of his reasoning came the quantum idea. Planck assumed that energy is radiated in packages, or quanta, where each quantum has energy directly proportional to the frequency of the radiation. This nonclassical assumption predicts a spectrum that does agree with nature.

1

2

Albert Einstein (1879–1955) extended Planck's quantum idea by applying it not only to the emission of radiation but also to the light itself as it travels through space and to the absorption of radiation by matter. The latter proposal enabled Einstein to explain the photoelectric effect, that is, how light causes the ejection of electrons from matter (Section 24-3).

3

Albert Einstein

(1) A modern application of the photoelectric effect. The sun's light moves electrons from the surface material and charges the solar batteries for mooncraft Ranger VI.

(2), (3) Detail of photocells used in various spacecraft.

(1) Continuous spectra of white light from glowing filament as seen through two different diffraction gratings; showing orders that appear immediately to the right of the source. 300 lines per millimeter; 4 orders visible.

(2) 600 lines per millimeter; 2 orders visible.

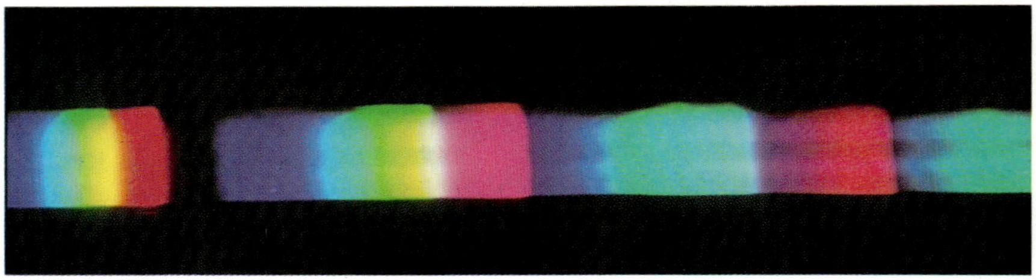

1

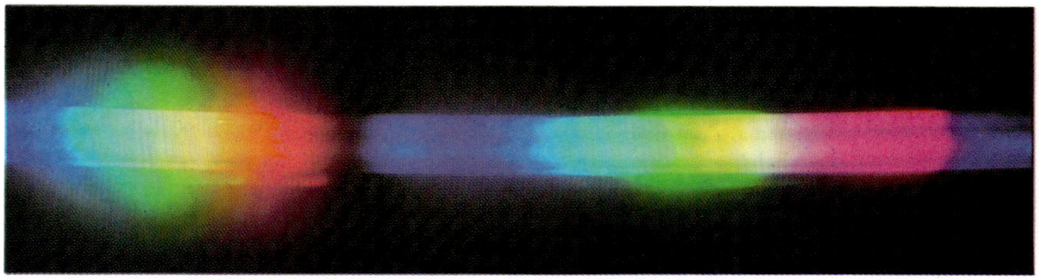

2

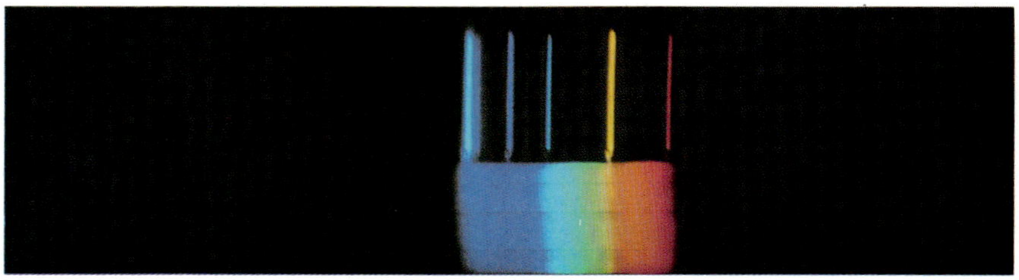

3

Henry Rowland

Henry A. Rowland (1848–1901) was one of America's greatest experimenters and made a notable contribution to modern physics in the field of spectroscopy. Rowland's diffraction gratings enabled the wavelength of the radiation from atoms to be measured with great accuracy. The pattern of this radiation was a great puzzle for many years, and was not explained until Bohr developed the concept of atomic energy levels.

Rowland developed wonderfully precise "ruling engines" in which diamond points were used to rule a pattern of many finely spaced, parallel lines to produce a diffraction grating. It is crucial that the lines be drawn cleanly and with equal spacing to preserve the sharpness of the spectrum. It is also desirable that the lines be spaced as closely as possible to obtain a large separation between different wavelengths.

(3) Low-dispersion grating spectrum of helium gas compared with continuous white-light spectrum.

(4) Diffraction gratings (reflecting studio lights); at left: 600 lines per mm, at right: 300 lines per mm.

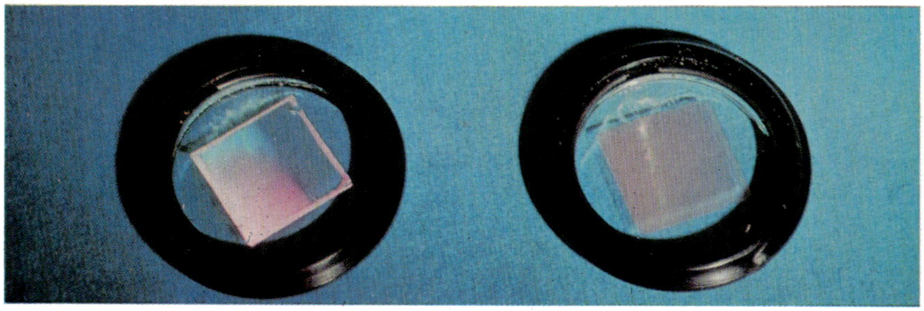

4

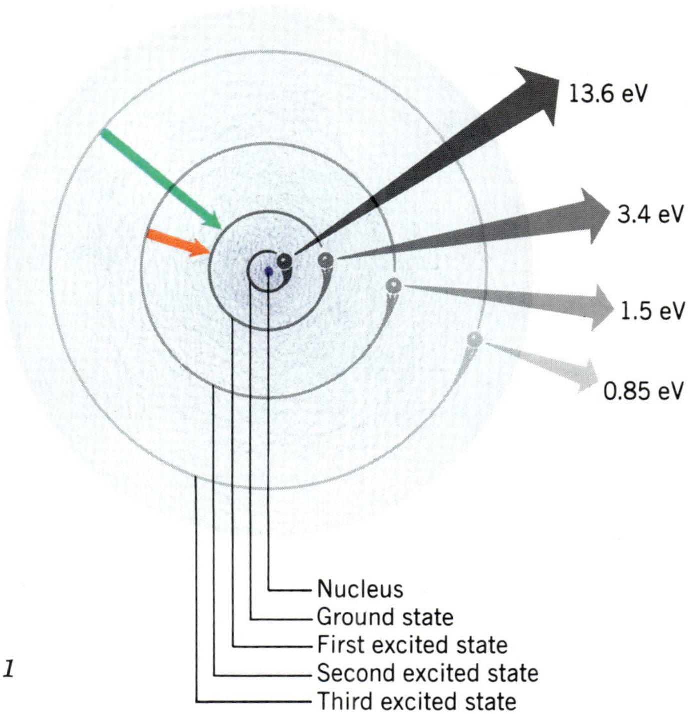

1

(1) Bohr's concept of a hydrogen atom showing various orbits, or states of excitation, and the energy required to free the electron from the atom in each of these states. Colored arrows show electron jumps from a certain outer orbit to the first excited state.

The Danish physicist Niels Bohr (1885–1962) applied the new quantum concept of Planck and Einstein to the hydrogen atom and its spectrum. He knew that if classical ideas of radiation applied to the atom, the electrons should radiate as they are accelerated in their orbits, and should rapidly spiral in toward the nucleus, as they lose energy.

Bohr gave up trying to understand the hydrogen atom in terms of the classical concepts of radiation and made use of the newer quantum ideas. His fundamental assumption was that only certain orbits are allowed, and that radiation is emitted in definite quanta of energy when an electron jumps from one orbit down to another of lower energy.

Bohr's proposals were of tremendous significance, but incomplete and primitive. They were refined and sharpened in 1925 by the theoretical work of Werner Heisenberg and Erwin Schroedinger.

Niels Bohr
Werner Heisenberg
Erwin Schroedinger

2

(2) The visible spectrum of atomic hydrogen.

Application of Quantum Physics in a Modern Device

The Transistor

(1) An early diode, a crystal set consisting of a galena crystal and wire cat's whisker.
(2) The vacuum tube (left), which has been largely replaced by the transistor (right), a solid state device.
(3) Internal view of an integrated circuit containing many circuit elements: transistors, diodes, capacitors, and resistors.

Semiconductor diodes have been used for a long time, although they were not fully understood for many years. The crystal set of the 1920's used a point contact rectifier formed by a wire *cat's whisker* touching a crystal of galena. This device was greatly improved as part of the development of radar during World War II.

Scientists at Bell Telephone Laboratories were aware for a long time of the need for better electronic amplifying devices, and a concentrated campaign to develop a solid state amplifier began in the 1930's. The effort bore fruit in 1947 when the transistor was invented by three engineers, John Bardeen, Walter Brattain, and William Shockley.

The trend of the future is toward "functional" circuit design. The engineer defines, in fundamental terms, the function or job to be done and then fabricates matter in terms of its fundamental properties to do this job, without reference to isolated circuit elements such as resistors, capacitors, or transistors.

This approach is typified by the new *microelectronics*, or *integrated circuit*, technology. In many applications such as aerospace and computers, isolated transistors are now being replaced by tiny integrated circuits containing many circuit elements (including transistors) in single, small wafers.

1

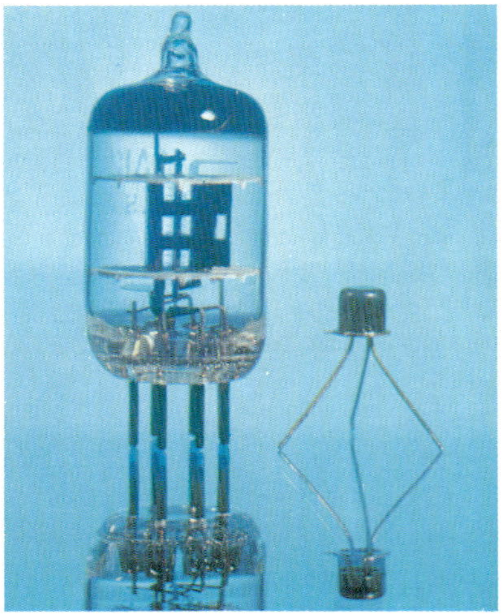

2

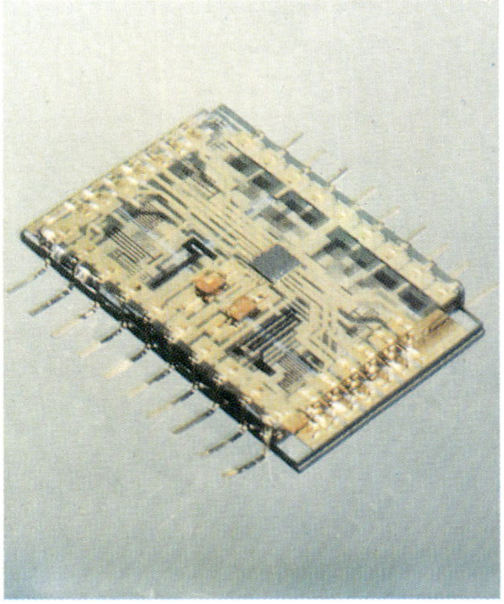

3

(4) A mounted integrated circuit wafer placed next to the perforated edge of a postage stamp.

(5) The wafer in (4) greatly enlarged.

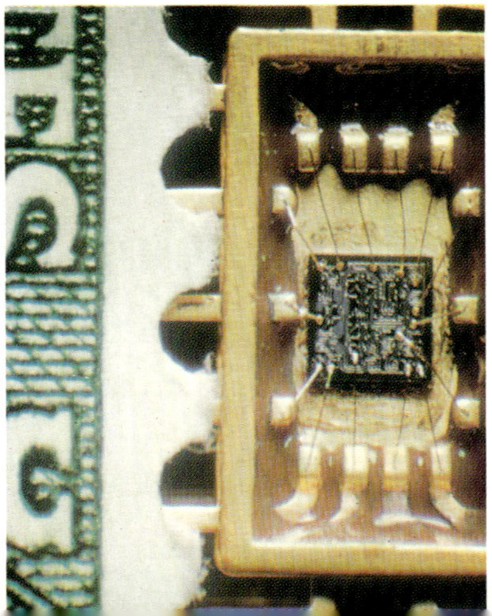

4

5

1

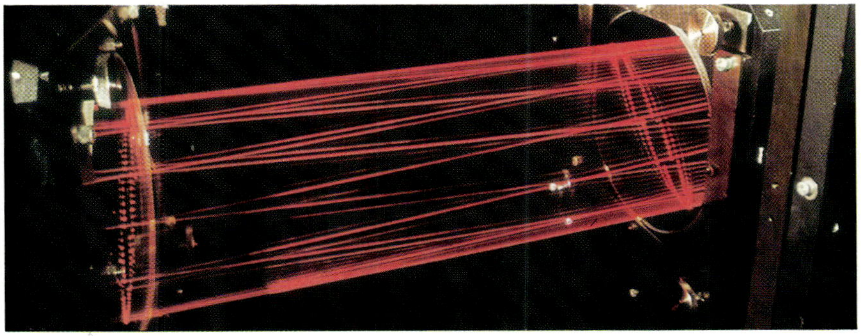
2

3

Another Modern Application of Quantum Physics

The Laser

(1) Spot-welding with a laser beam.

(2) The beam of a ruby laser bouncing back and forth between two highly polished surfaces, showing almost perfect reflective qualities.

(3) A blue beam of an argon laser is directed into a synthetic ruby (top).

The laser is an outgrowth of the original quantum mechanical oscillator, or amplifier, called the maser, invented by Charles H. Townes and others in 1954. The laser is an entirely new kind of light source, with properties very different from ordinary sources. A light source such as a light bulb radiates light energy over a wide range of wavelengths, sending out pulses of light at random in unrelated bunches of waves.

Gas discharge tubes such as a neon sign or a fluorescent tube do radiate more power at definite wavelengths. Still, their light is also incoherent in the sense that it is radiated at random by the atoms, and comes to us in unrelated bunches of waves.

Laser light is different in that the atomic oscillators are synchronized by the process called stimulated emission and radiate together. This gives a narrow, sharp beam of light of a single wavelength.

The essence of stimulated emission is that an atom in an "excited state" (that is, one at an energy level higher than its ground state) can emit a photon of light with an energy corresponding to the energy of its excited state. If it is struck by a photon of precisely this energy it will be stimulated to emit a photon, and the waves of this light will be exactly in step with the stimulating photon.

To make a practical device, there must be an excess of excited atoms, so that the emitted photons will not be absorbed by atoms in the ground state. A large population of excited atoms is maintained by "pumping" with light of a different wavelength. The whole device must be in an enclosure so the light can reflect back and forth and thus build up increasing intensity by accumulating more photons.

In the ruby laser, atoms of a ruby crystal are pumped up to excited states by the light from a flash lamp that emits a wide range of frequencies. Chromium atoms in the ruby emit coherent red light at wavelength 6943 A, built up by repeated reflection at the mirrors at each end of the crystal.

Laser Photography

(1) Reflection of light from the steps. (2) The "frozen" pattern of light waves. (3) The hologram, reproduced by passing laser light through the film.

The transistor and other semiconductor devices already have many important applications; it is reasonably clear also how such devices will develop in the future. The future of the laser is not so clear, however. Engineers foresee such applications; it is reasonably clear also how such devices will develop in the makes for efficient signaling over long distances. It is also possible that laser beams may be used for some kinds of surgery and for precision metal forming.

One use of laser light could perhaps open up a wholly new type of photography. The basic idea was proposed more than twenty years ago, before the laser made this technique useful and practical.

The laser photograph is called a hologram. The film does not record an image, but rather makes a record of the waves themselves (1). Looking at a hologram tells us nothing, for it is a complicated "frozen" interference pattern of the light waves that are reflected by the objects being photographed (2). It is a kind of optical code that records much more information than a conventional photograph.

To see the hologram reassembled, we must shine laser light through it; then the original waves themselves are reconstructed (3). So complete is the reconstruction of the scene, that there is real depth and parallax: As you turn your head from side to side you can see the objects in the foreground of the picture move back and forth in front of objects in the background.

In some ways, a hologram is like a very complex diffraction grating recorded on the film. When laser light is projected through it, a pattern of waves is formed by the grating.

1

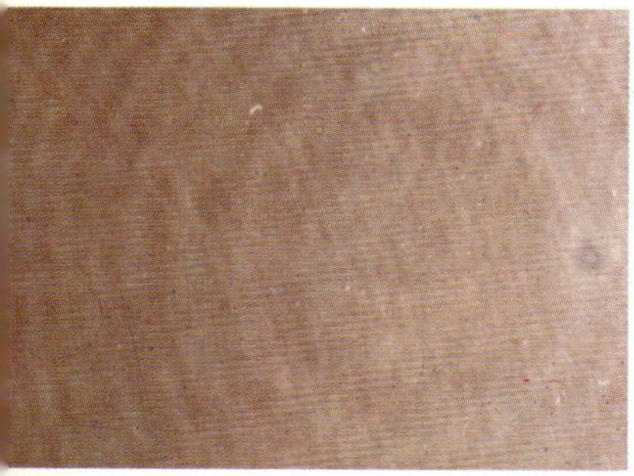

2

3

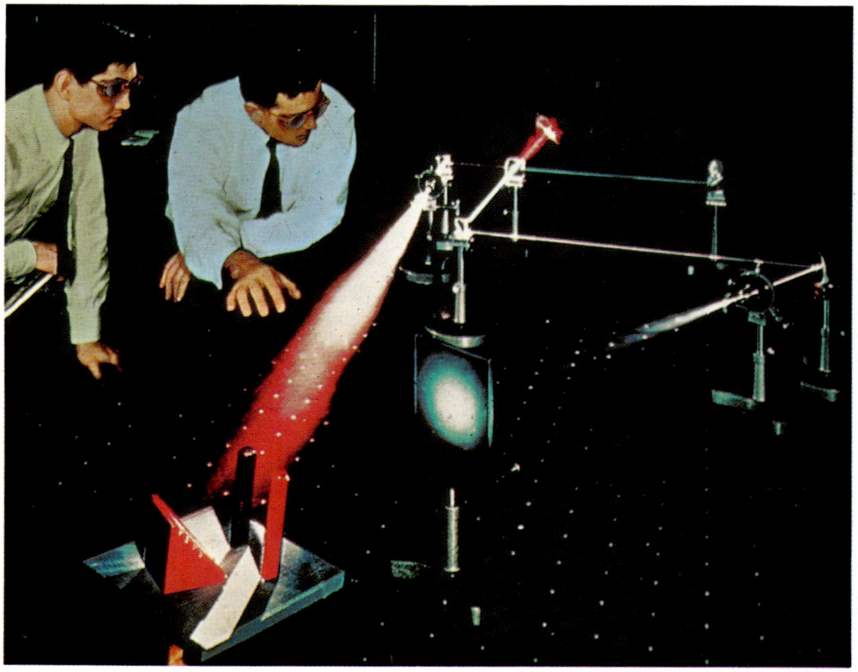

(4) Photographing several objects that are illuminated by laser light.

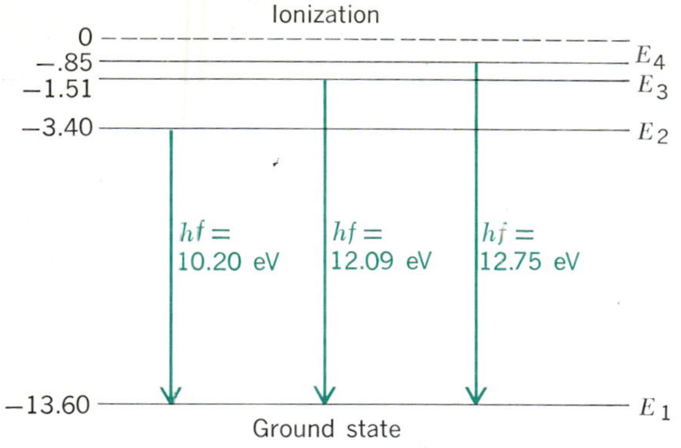

Figure 24-11 Energy levels for hydrogen.

of the Lyman series, a set of energy changes from various higher levels down to the lowest energy state, or *ground state* E_1. Bohr was able to make an approximate calculation for the radii of the possible orbits of electrons in atoms. From this he could determine the values of the energy states, or levels. Stating his conclusions in simple terms, the energies of the levels for hydrogen are given by

$$E_n = -\frac{13.60}{n^2}\ \text{eV}$$

In this formula, E_n stands for the energy of the nth energy level, n stands for a whole number representing the number of the level, or state, and the 13.60 is a constant that is characteristic of the hydrogen atom. The numerical value depends on the units chosen (in this case, electron volts), and is related to the various quantities that influence the orbit—the mass of the electron, the electron charge, the magnitude of the positive charge, the constant in Coulomb's law, and the value of Planck's constant.

Perhaps the hardest feature of this formula to understand is the minus sign on the right-hand side, which comes about because physicists have chosen to consider the energy zero when the orbiting body is removed to an infinite distance. This means that when the electron is completely pulled away from the hydrogen atom (when the atom is "ionized"), the atom is considered to have zero energy. The formula agrees with this convention, as you can see if you let n equal infinity. Note that when $n = 1$, the energy of the ground state is given as -13.60 electron volts, meaning the lowest energy is 13.60 electron volts *below* the maximum energy zero. The other energies of the diagram for the Lyman series were calculated by substituting $n = 2, 3, 4$, etc.

The arrows in the diagram represent *transitions*, or *jumps*, between energy states. The energy E of the photon emitted by the transition of a hydrogen atom from state E_2 to state E_1 may be calculated as follows:

$$\begin{aligned} E &= hf \\ &= E_2 - E_1 \\ &= \frac{-13.60}{2^2} - \frac{-13.60}{1^2} \\ &= -3.40 + 13.60 \\ &= +10.20\ \text{eV} \end{aligned}$$

The wavelength in angstroms is given by

$$\begin{aligned} &= \frac{12{,}400}{E} \\ &= \frac{12{,}400}{10.2} \\ &= 1200\ \text{Å} \end{aligned}$$

This is an invisible ultraviolet line.

The lines in the hydrogen spectrum occur in several *series*, named after the men who investigated them (Figure 24-12). The energy-level diagram and the type of calculation that was just carried out give a ready explanation for these series. For instance, the Balmer series contains lines emitted during quantum jumps $E_3 - E_2$, $E_4 - E_2$, $E_5 - E_2$, . . . , always with the second energy level ($n = 2$) as the final state. As you have just seen, for the lines of the Lyman series the level E_1 is the final state. The Lyman series lies entirely in the ultraviolet, since the quantum jumps are large (as in the calculation above, where the jump was 10.20 eV) and the photons have high frequency and short wavelength. Only one series, the Balmer series, comes within the visible spectrum (the Paschen series is in the infrared

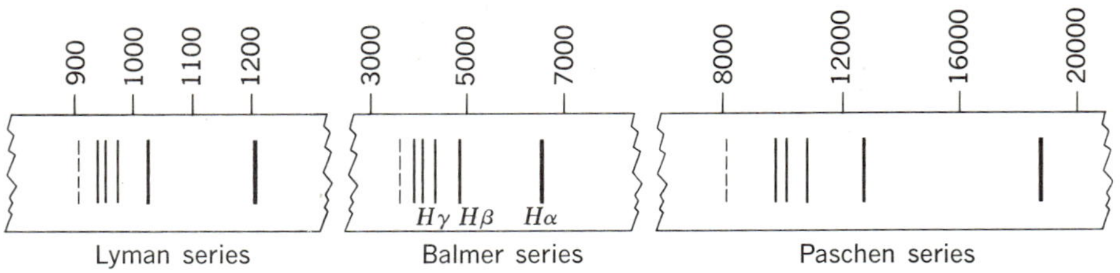

Figure 24-12 Spectral series in hydrogen. The top figures represent wavelengths in angstroms.

range). This is why the hydrogen spectrum appears so simple.

Energy-level diagrams for atoms with more than one electron are more complex than for hydrogen, and hence the spectra are also more complex. For instance, two of the mercury energy levels are very close together (E_8 and E_9 in Figure 24-13). The transitions $E_8 - E_5$ and $E_9 - E_5$ are therefore of almost equal energy. These transitions appear as a pair of yellow lines of wavelengths 5770 angstroms and 5790 angstroms in the mercury spectrum. The sodium spectrum has a strong pair of yellow-orange lines of wavelengths 5890 angstroms and 5896 angstroms, and 13 other very weak lines in the visible region. (Look again at Figure 24-9 in the color insert following page 388.)

When light is absorbed by a gas atom, a photon disappears and the atom's internal energy increases. Since the atom's energy can change only by discrete amounts, such as ΔE in Figure 24-13, only certain photons can be absorbed. In fact, it should be apparent that *an atom can absorb only those photons it is capable of emitting,* since the equation $\Delta E = hf$ applies equally well for a transition in either direction.

24-5 THE COMPTON EFFECT

The lines of the visible spectrum arise when the outer electrons of an atom gain and then lose a few electron volts of energy. Far greater photon energies are obtained by removing one of the inner electrons of an atom. This can be accomplished in an x-ray tube (Figure 24-14). Here high-speed electrons, accelerated through a PD of thousands of volts, strike a target and rip out inner electrons from the atoms of the target material. When an outer electron falls into the vacant inner energy level (or orbit), the energy change is very large because the inner level can be thousands of electron volts below the zero of energy. Such atomic energy changes result in the emission of high-energy photons called *x rays,* which, because of their high energies can penetrate solid matter much more easily than does visible light. X rays typically have energies of many thousand electron volts and wavelengths of about an angstrom—less than $\frac{1}{1000}$ of the wavelength of visible light.

The American physicist Arthur Compton was studying the scattering of x rays, and in 1922 in one experiment he noticed a peculiar small

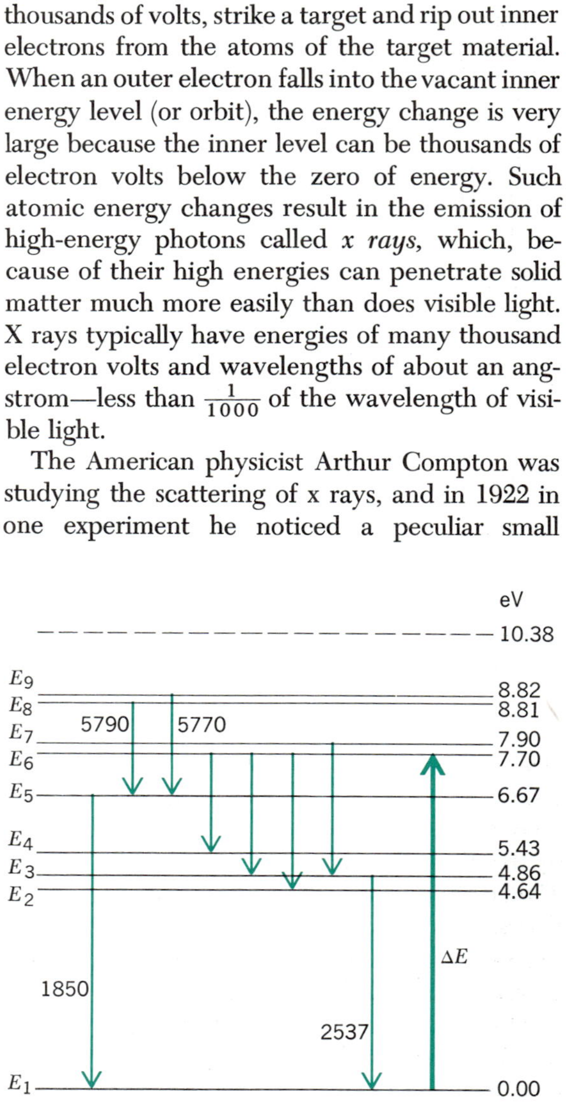

Figure 24-13 Energy levels for mercury.

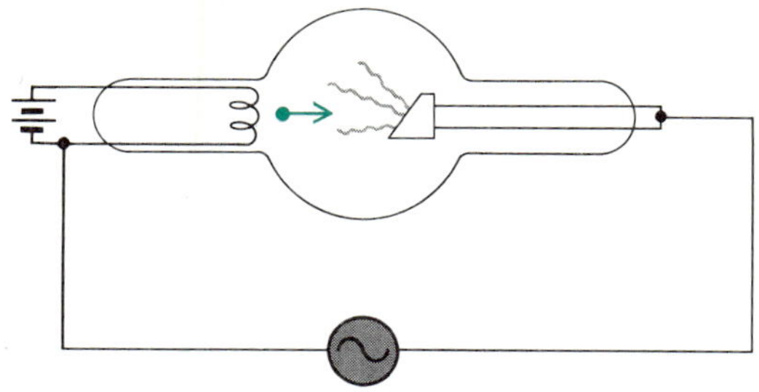

Figure 24-14 X-ray tube. Electrons, emitted from the heated filament, are accelerated during the half-cycles in which the target is positive relative to the filament.

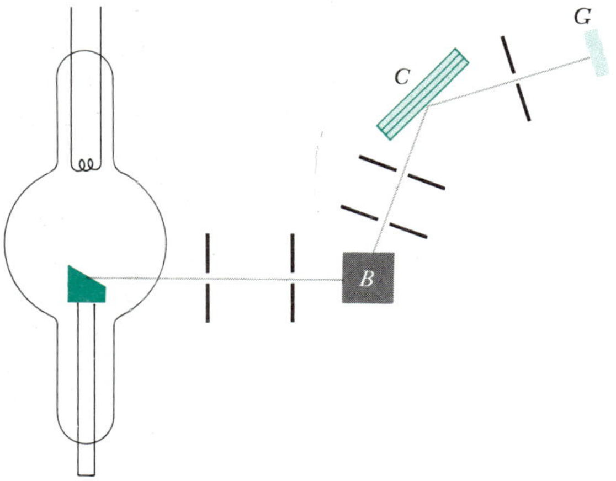

Figure 24-15 Compton effect. Experimental arrangement for studying change of wavelength of x-ray photons scattered by carbon block *B*.

change in wavelength. The experimental setup is shown in Figure 24-15. X rays from a molybdenum target strike a carbon block *B*, and the scattered rays are then analyzed by the crystal *C*. The crystal serves as the grating of an x-ray spectrograph. A manmade grating would be too coarse to give a diffraction pattern, for to be effective the grating space must be comparable with the wavelength. The x-ray wavelength is approximately the size of a single atom. Fortunately, the atoms in a crystal are about an angstrom apart, so the layers of atoms can serve as a natural grating for x-ray diffraction. (Recall diffraction gratings for visible light, Section 23-3.) The x rays are defined into a fairly sharp beam by the slits, and then detected at *G*.

Compton used this apparatus to investigate the collisions between x-ray photons and free electrons in the carbon. The energy of an x-ray photon is so great that the work function (energy needed to knock the electron out of the carbon) is negligible. The electrons are thus essentially "free," and the encounter is between photon and electron. The photon is deflected, that is, its path is changed by the collision with the electron, and the electron acquires a recoil velocity in some direction as shown in Figure 24-16. The laws of conservation of energy and conservation of momentum can be applied to these collisions, and the calculations agree well with the experimental results. The *Compton effect* is important because it shows the photon to have particle properties—to have mass—in the same sense that a bullet or billiard ball does. This goes beyond the concept of a photon that was built up to explain the photoelectric effect and the emission of spectra. So far, we have thought of the photon as a quantum, or distinct unit, of energy. Now we must think of it as also having mechanical properties such as inertia and momentum. The model of light as a stream of bullets seems inescapable—just as inescapable as the wave model discussed in preceeding chapters. We are led to the conclusion that we really must have a dual model for the nature of light.

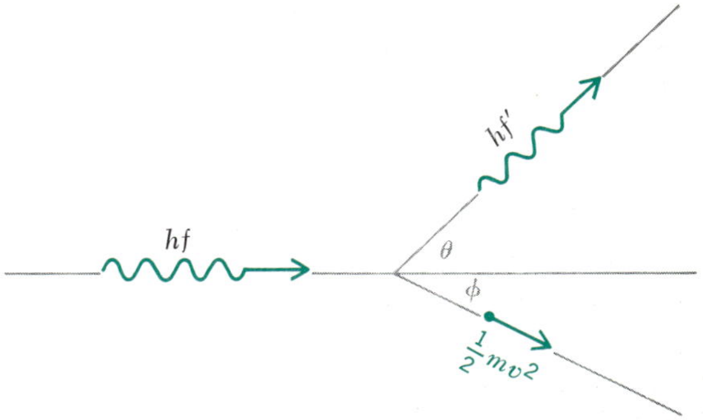

Figure 24-16 Conservation of energy during scattering of a photon by an electron.

24-6 THE DUAL NATURE OF LIGHT

The Compton effect illustrates the dual nature of light—as waves and as particles. In the very apparatus by which the bulletlike character of x-ray photons is revealed, these same x rays show their wave character when reflected by the crystal grating. In Chapter 22 we took the phenomenon of interference to be an indisputable proof for the wave nature of light and carefully built up a wave model for light (including x rays and all other e-m radiation). Now we see a corpuscular aspect of light. How can these conflicting wave and particle models be brought together?

In the first place, we should realize that the particle aspects of radiation are seen only when radiation interacts with matter. Photons are involved in the emission or absorption of light (black body radiation, lines in a spectrum, the photoelectric effect), or in the scattering of light by an electron (Compton effect). On the other hand, the wave aspects are evident when light travels from one place to another. The two aspects of light seem to be important at different times.

One way out of the difficulty is to cry "modelitis." The function of a model is to make a new phenomenon understandable in terms of familiar laws and events of our everyday, large-scale world. Recall how large-scale models were used for describing e-m waves (comparing them with water waves) and also the description of photons (comparing them with billiard balls). It may be, however, that *no* common-sense model for the small-scale phenomenon of light can exist. There is no reason why the behavior of a single photon should be the same as that of *any* large-scale object with which we are familiar. Many physicists prefer to accept light for what it is, and attempt no models. In fact, several decades of living with the strange double nature of photons has bred a certain familiarity with light; advanced workers now use photons as "understandable" models for phenomena that are even more obscure.

Insofar as a model is possible, we can visualize the e-m waves as being "guide waves" that tell the photons where to go. According to this model, e-m waves are *probability waves.* There is a large probability of finding a photon at any point where the amplitude of the wave is large. If two probability waves interfere destructively to form a node, the probability of finding a photon at the node is zero.

Here is an example that will illustrate how photons are guided by probability waves. Suppose a pair of slits and a screen are placed near a source, as in Young's experiment (Figure 22-3). The mere presence of the slits and screen sets up a pattern of probability waves, which in this case interfere constructively to form an antinode, or maximum, at *M* and a node at *Y*. The spacing of these nodes of probability depends upon the spacing of the slits, etc. If the source is now turned on, photons are emitted, and they travel to the screen. We can't trace the path of any photon from S through either slit to *M*, but we know that it is highly probable that a photon will appear at *M*. Let's assume that the screen is a photographic film with many silver bromide molecules. Each photon that is absorbed gives up its energy hf to a single grain of the film. As soon as the source is turned on, the silver bromide grains become activated,* more or less at random. After a while the picture becomes evident, and we realize that *M* is a probability antinode and *X* a probability node. Experiments of this sort have actually been performed, by recording an image on photographic film using varying numbers of photons (Figure 24-17).

In this way it is possible to compare a slit system, a diffraction grating, a lens, or a prism with a traffic light that directs the flow of cars at an intersection on the highway. The light flashes red and green even in the dead of night when no cars are on the road. The probability waves are there, ready to guide whatever car may come along. Guide waves are also involved in the photoelectric effect. Even before the Compton effect showed that photons have mass and are concen-

* The photon's energy hf is stored in the grain and used in the development process to aid the chemical reaction that produces metallic silver.

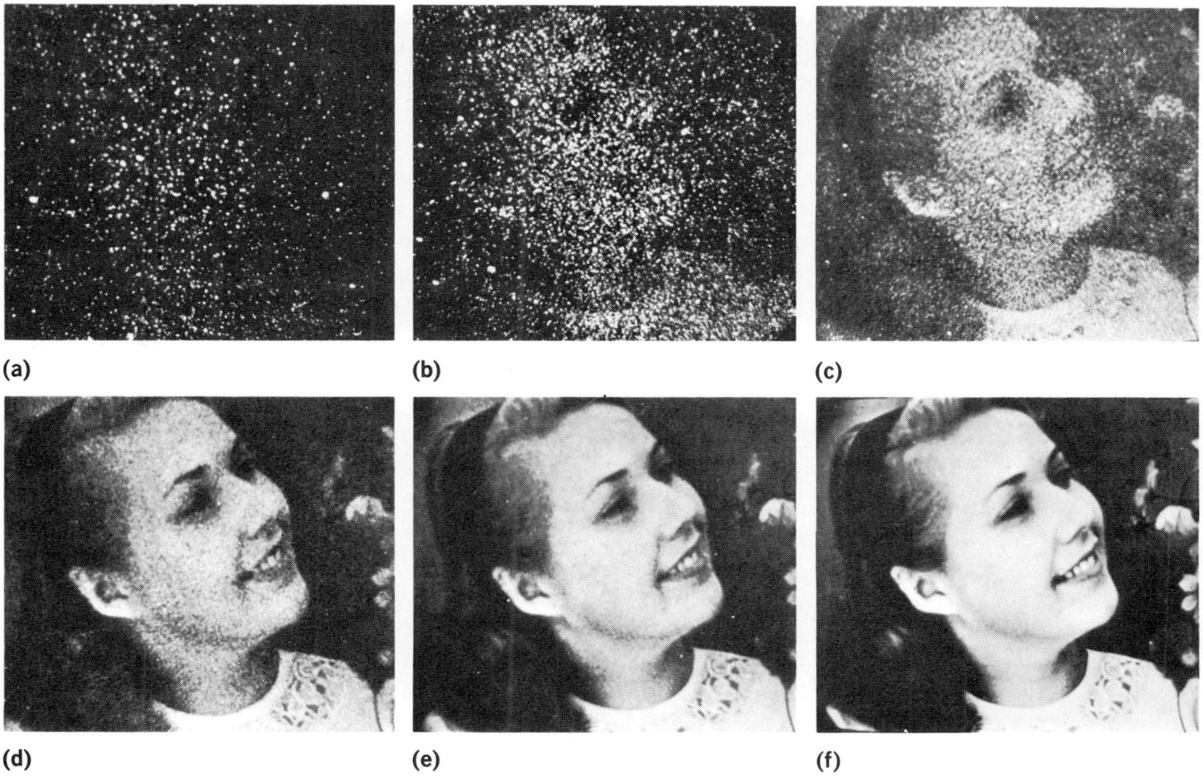

Figure 24-17 Photons of light needed to form a complete picture. The numbers given below tell how many photons were applied to form each of these reproductions of the same image. (**a**) 3×10^3 photons; (**b**) 1.2×10^4 photons; (**c**) 9.3×10^4 photons; (**d**) 7.6×10^5 photons; (**e**) 3.6×10^6 photons; (**f**) 2.8×10^7 photons.

trated bundles of energy, it was known that very weak light can eject photoelectrons if the frequency is above the threshold frequency as determined by the work function. A serious difficulty comes when the process is considered from the wave point of view. For instance, let a very weak source shine on a metal surface of area 1 square meter. About 10^{20} atoms lie in a surface of this area. To eject a photoelectron might require about 10^{-18} J ($\sim$6 eV), and if the intensity of the light is 10^{-6} J/sec · m^2, each of the 10^{20} atoms in the surface could absorb only 10^{-26} J/sec. It would require 10^8 seconds (over 3 years) for an atom to absorb enough energy to eject a photoelectron. Yet experiments show that sometimes photoelectrons are emitted within a fraction of a second after the light is turned on. Once again we are forced to assume that the energy of the incoming light is carried by small, bulletlike photons, which, in the case of a very weak source, are powerful enough, but few and far between. We visualize the e-m wave as guiding the photons in such a way that it is equally likely for a photon to be absorbed anywhere on the surface, and hence photoelectrons are emitted at random. This is the way in which a uniform beam of light is interpreted according to the guide-wave model.

If a similar dual nature were true for water waves, we would have the following strange phenomenon. A pop bottle dropped off a New York pier that is 10 meters above the water level sends out a wave front that spreads out in ever-widening circles until it reaches Lisbon, Bordeaux, and Dublin. At some unpredictable place, say Lisbon, a similar bottle floating near a pier suddenly pops up 10 meters out of the water, and simultaneously the very weak and undetectable water wave that was in the form of a large circular wave front disappears, all of its energy being given to the Lisbon bottle. If this sort of thing ever did happen, we would be tempted to talk about water waves as guide waves. The photoelectric effect for weak light is as strange as this, and equally difficult to explain by a wave model.

24-7 THE DUAL NATURE OF MATTER

According to the dual model of light, a "guide wave" is associated with every photon, and the wavelength of the guide wave can be measured by interference experiments with gratings and the like. In 1924 the French student Louis de Broglie proposed that guide waves are also associated with "ordinary" particles of matter, such as electrons and baseballs.

De Broglie assumed that a guide wave is associated with *any* moving body that has momentum. For a photon moving at the speed of light, the momentum is mc and the wavelength of the guide wave is $\lambda = h/mc$. (This agrees with $\lambda = c/f$, using $E = mc^2$ and $E = hf$.) For matter waves, de Broglie postulated that the wavelength of the guide wave is also given by Planck's constant divided by the momentum:

$$\lambda = h/mv$$

As the next example shows, the wavelengths associated with ordinary bodies are much too small to be observed.

Example 24-5 Calculate the de Broglie wavelength associated with a 20 g ping-pong ball that is moving at 400 cm/sec.

$$\lambda = \frac{h}{mv}$$

$$= \frac{6.63 \times 10^{-34}\ \text{J}\cdot\text{sec}}{(20 \times 10^{-3}\ \text{kg})(4.00\ \text{m/sec})}$$

$$= \boxed{8.29 \times 10^{-33}\ \text{m}}$$

This fantastically short wavelength is unobservable. A ping-pong ball headed for an open window proceeds in a straight line with no fuzzy "shadow." It either goes out the window or is reflected, depending on how it is aimed. Relative sizes are important. The wave nature of visible light escaped notice for so long because λ is so much smaller than the size of an ordinary opening such as a window. Only when the size of the slit or obstacle approaches λ does the diffraction of light become evident. The same argument applies with even greater force to the ball of Example 24-5. The ball's guide wave has a wavelength that is only about 10^{-26} as large as a light wave. We are quite justified in using Newton's laws of motion for the motion of a ping-pong ball.

Passing now to particles of smaller mass, we see from de Broglie's equation ($\lambda = h/mv$) that the wavelengths of the guide waves associated with moving electrons may well be observable. In 1927 C. J. Davisson and L. H. Germer, working in the United States, interpreted their experiments on the scattering of electrons by a metal surface as indicating the wave nature of the electron. The electrons were reflected (or "scattered") off a nickel crystal, and the intensity of the reflected beam was measured at various angles. The experiment showed that there were strong reflections only at certain well-defined angles depending on the electron energy. This was interpreted by saying that the atoms of the crystal acted as a kind of diffraction grating and that the definite angles were an interference effect. Independently, in England, wave effects were sought for and found in 1927 by G. P. Thomson, son of J. J. Thomson (Section 24-2). The 1937 Nobel Prize in physics was shared by Davisson and G. P. Thomson for their discovery of the *wave* nature of the electron. It is interesting to note that in 1907 J. J. Thomson had received the Nobel Prize for his experimental proof of the *corpuscular* nature of the electron. The Thomsons, father and son, brilliantly demonstrated two opposite, but complementary, aspects of the nature of matter.

The wave nature of matter has also been shown experimentally for particles such as the neutron.

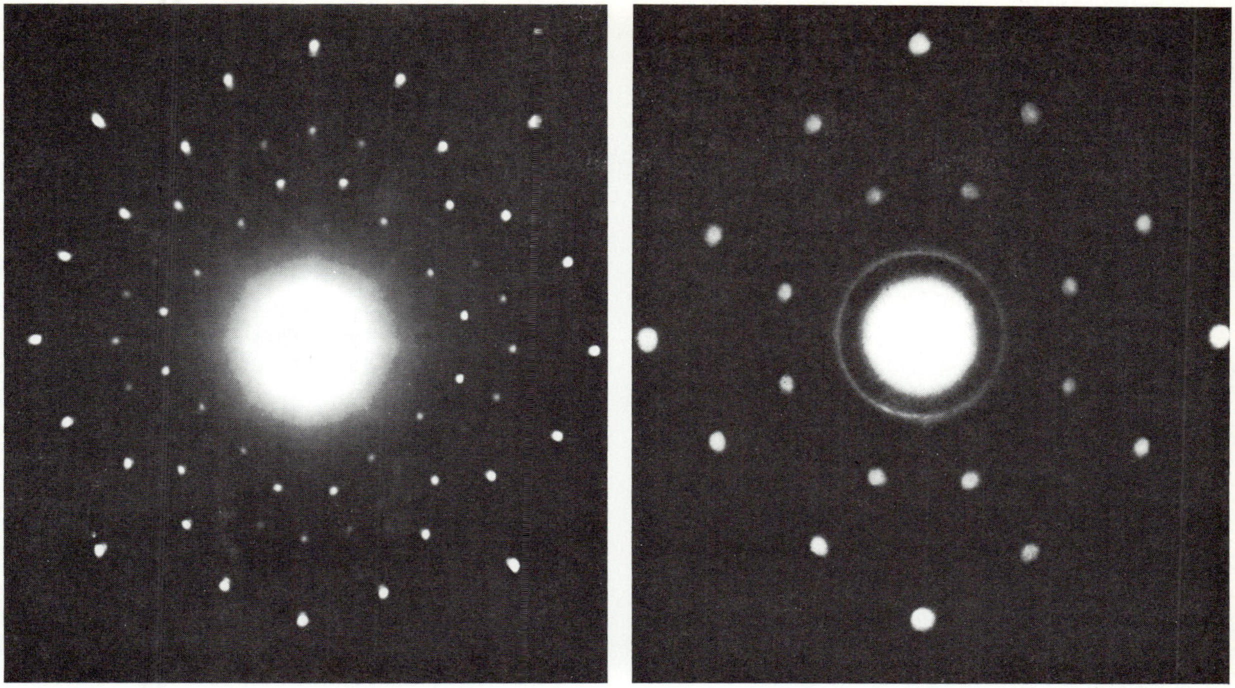

(a) (b)

Figure 24-18 Wave nature of photons and neutrons. (**a**) X-ray diffraction by a single crystal of NaCl. (**b**) Diffraction of neutrons by a single crystal of NaCl.

Example 24-6 Calculate the KE and the wavelength of a neutron whose speed is 10^3 m/sec.

The neutron mass is 1.67×10^{-27} kg (see Appendix Table 1). Hence,

$$\begin{aligned} \text{KE} &= \tfrac{1}{2}mv^2 \\ &= \tfrac{1}{2}(1.67 \times 10^{-27}\ \text{kg})(10^3\ \text{m/sec})^2 \\ &= 8.35 \times 10^{-22}\ \text{J} \\ &= (8.35 \times 10^{-22}\ \text{J})\left(\frac{1\ \text{eV}}{1.60 \times 10^{-19}\ \text{J}}\right) \\ &= \boxed{5.22 \times 10^{-3}\ \text{eV}} \end{aligned}$$

The wavelength of the guide wave is calculated by the de Broglie equation:

$$\begin{aligned} \lambda &= h/mv \\ &= \frac{6.63 \times 10^{-34}\ \text{J}\cdot\text{sec}}{(1.67 \times 10^{-27}\ \text{kg})(10^3\ \text{m/sec})} \\ &= 3.97 \times 10^{-10}\ \text{m} \\ &= \boxed{3.97\ \text{Å}} \end{aligned}$$

This wavelength is comparable with the distances between atoms in crystals, and hence a beam of such neutrons can give rise to interference and diffraction just as for x rays. Figure 24-18a shows a diffraction pattern produced by a NaCl crystal using a beam of x rays, and Figure 24-18b shows a diffraction pattern of a NaCl crystal using a beam of neutrons. The wavelengths are several angstroms in each case. The arrangement of spots is determined by the crystal structure of NaCl, and the patterns show many similarities. The bright spots are places where constructive interference of the waves occurs; the pattern of spots is an interference pattern whose size and shape depends on the wavelength and the placing of the scattering particles.

We have come a long way from the simple cause-and-effect mechanics of Newton's day. Then, it was thought that one could exactly predict the future position and momentum of any

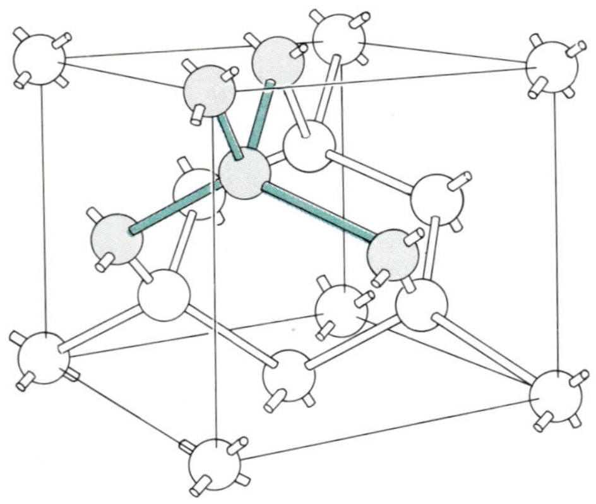

Figure 24-19 Structural unit (shown in color and gray) of diamond, silicon, and germanium crystals.

body if he knew the original position and the momentum precisely enough. It was thought that Newton's laws of motion would be sufficient to predict the future course of the universe were it not that the mathematical solution of so many simultaneous equations (one for each particle in the universe) would be too difficult. Now, we see that the future path of a particle cannot be predicted exactly because the path of a particle must be viewed in terms of the de Broglie "guide waves," which are waves of probability and do not predict the path with certainty. Newton's mechanics led to a "deterministic" kind of universe in which every event had a definite cause; that is, one could always find a definite cause-and-effect relationship. It has often been said that cause and effect have now been abandoned in physics, but this is a true statement only when applied to the motion of a single photon or subatomic particle. The predictions of modern physics are based on probability. We cannot say where any given photon will go, but the *observable* photograph of a double-slit interference pattern is quite definite, being an average effect for many billions of photons.

About 1925 the idea of probability waves was applied to develop a new science of mechanics called *quantum mechanics.* Newton's laws are a special case of quantum mechanics, valid for collections of many particles, such as a baseball or a spaceship. Quantum mechanics does not *replace* Newton's laws; it *extends* the laws of mechanics to cover situations unknown to Newton and his contemporaries.

24-8 SOLID-STATE PHYSICS

Physicists today are particularly interested in applying the quantum physics described in this chapter to understanding not just single atoms, but also the complicated collections of atoms in solid matter. The physicist has a very definite meaning for the term *solid* which may not agree completely with the everyday use of the word. For example, a window pane and a drinking glass are not considered truly solid, even though they are certainly rigid and have a definite size and shape. Here we will use the term solid to refer to a *crystal,* or crystalline substance. Therefore let's try to make clear what we mean by a crystal.

That most familiar substance, common table salt, is a good example of a crystal. A complete list of other crystalline substances would fill many pages; it would include many different minerals: chemical compounds like salt, metals such as copper and iron, and various semiconductors such as germanium and silicon. What do all these materials have in common? The outstanding feature of all crystals is that their atoms, ions, or molecules are arranged in a definite, orderly structure that is repeated over and over again to make a crystal large enough to be seen. Figure 24-19 shows one type of arrangement, or structural unit, for atoms forming a crystal. This is the diamond structure, which is also characteristic of the semiconductors silicon and germanium.

In most ordinary-sized pieces of matter, such as an iron nail or a large piece of silicon, there are in fact many small crystal grains, as shown in Figure 24-20. The study of solid-state physics involves an intimate knowledge of such crystalline structures, and has made it possible for modern metallurgists to understand how to vary the sizes and arrangements of the crystal grains to give specific properties to the metals used in industry.

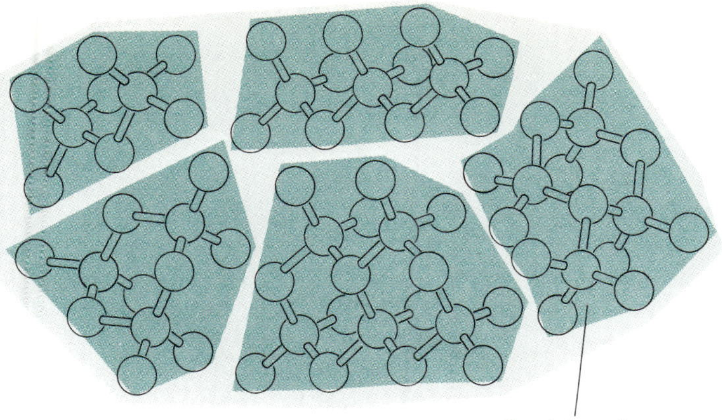

Figure 24-20

Special techniques have made it possible to grow very large single crystals of semiconductors, with one continuous structure not jumbled up into small grains (Figure 24-21). These single crystals are essential to the manufacture of the diodes and transistors described in Section 19-5.

This section will show how the basic quantum physics of the outer atom has been applied to help us understand the electrical behavior of semiconductor materials. Semiconductor devices play important roles in everyday life, and much experimental and theoretical research in physics today is being done to advance our understanding of these particular substances. Before looking closely at germanium and silicon, however, let's review some fundamental ideas about the states of matter.

In Chapter 12, the nature of gases was described. You will recall that the atoms of a gas are widely separated, with distances between them which are very large compared to their sizes. The atoms have no orderly arrangement and are moving rapidly about in all directions, occasionally colliding with one another briefly, but spending most of the time rushing around in isolation, not interacting with their neighbors. The energy levels and spectral lines described in Section 24-4 are characteristic of isolated atoms, as found in a not very dense gas.

When a gas condenses into a liquid, the separation distances are reduced to about $\frac{1}{10}$ of what they were in the gaseous state. The molecules are no longer free to fly about in all directions, but, since liquids are still able to flow, their molecules must be able to move quite freely with respect to one another. However, when a substance freezes and changes from the liquid to the solid state, very significant internal rearrangements occur. The separation distances between molecules do not change very much, but the motion of the molecules changes a great deal. Unlike liquids, solids do have a definite shape. The molecules of a liquid can slide past each other, but the molecules, atoms, or ions forming a solid crystal are "locked" in definite positions by the attractive forces between them. The atoms in a crystal are not motionless, however; they are able to vibrate

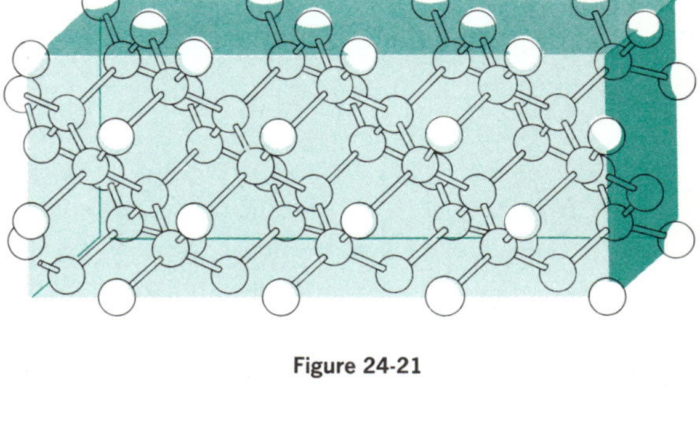

Figure 24-21

back and forth about their fixed positions in the crystal. The thermal motion of crystal atoms is vibration, instead of the completely free motion of gas atoms.

The various kinds of glass are not considered solids because they do not form a repeated structure. Glasses tend to have very long molecules that become tangled and twisted as the glass cools, forming a kind of fibrous material that becomes quite rigid at a low enough temperature. A true solid (ice, for example) has a definite melting point: At a certain temperature the atoms have enough thermal energy of vibration to overcome the attractive forces, and the crystal changes to the liquid (or, sometimes, directly to the gas). With glass, however, there is no definite melting point; rigid glass gradually softens and becomes more and more fluid as its temperature is raised.

Now let's focus attention on solids, and in particular on their electrical properties. In Section 15-7 we discussed a model for the conduction of electric current in metals. Recall that the essential electrical property of a metal was the existence of free electrons. Thinking now in terms of crystals, this means that when metal atoms come together to form the crystal structure, one or more electrons are detached from each atom and are shared by all the atoms in the crystal. As shown in Figure 24-22, a metal can be thought of as a regular structure of positive ions (atoms that have lost an electron) immersed in a "sea," or "gas," of electrons that are free to wander through the crystal. In insulators, however, the outer electrons, are not easily detached and there is no such sea of electrons able to act as carriers of charge to form an electric current.

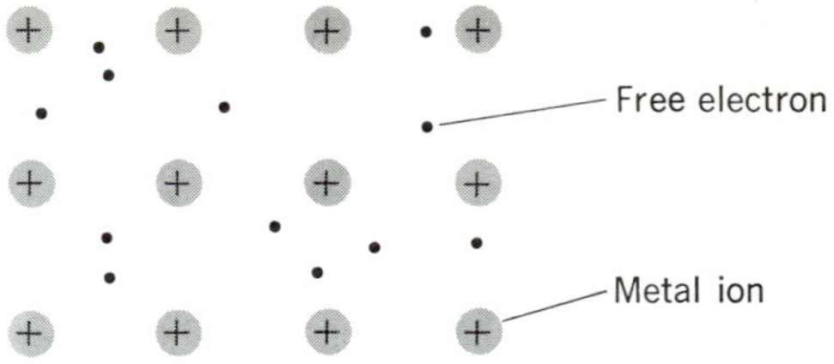

Figure 24-22

Using the concepts of quantum physics, we can think about the situation in terms of energy levels. First, consider a single isolated atom. A free electron is produced when an atom receives enough energy (usually from the thermal energy of vibration) to remove an outer electron, that is, to be raised to an energy level with an infinitely large value of n (Section 24-4). This process of tearing away an electron is called ionization. It is important, however, to be cautious in carrying over the idea of energy levels in single atoms to the more complicated situation in a crystal. The energy levels discussed in Section 24-4 apply to isolated single atoms. The atoms in a crystal are close together; they are not isolated, and the energy levels of the outer electrons are affected by their neighbors.

It may be helpful to think of the behavior of a pendulum. If you have a single, isolated pendulum, it can swing at just one frequency determined by its length alone. It is somewhat like an isolated atom, which for a given value of n, has a definite amount of energy. However, you can observe in the laboratory that when two or more pendulums of the same length joined by a spring interact with each other, new types of motion arise. There are new ways in which the pendulums can swing; Figure 24-23 shows two possible "modes" of vibration for two pendulums. It is not hard to imagine that when many pendulums are coupled together, there will be numerous ways in which they can vibrate. How does this apply to a crystal? The analogy is appropriate because the atoms of a crystal are closely coupled together by electric forces. The result of this close coupling is

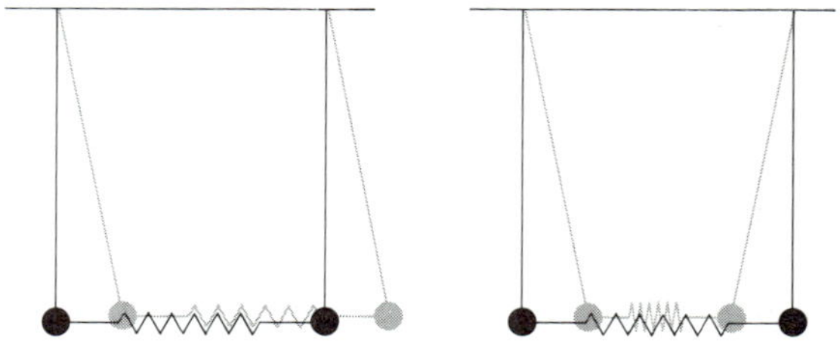
Figure 24-23

Figure 24-24 Energy bands for electrons in solids. (**a**) A conductor's valence band is only partially full. (**b**, **c**) At absolute zero no sharp distinction can be made between an insulator and a pure semiconductor, which differ only in the magnitude of the gap separating the full valence band from the conduction band. (**d**) At room temperature some electrons are in the conduction band and holes are near the top of the valence band.

that, instead of having sharply defined energy levels for outer electrons, there is a set of energies whose values are quite close together but not identical, forming an *energy band.*

For our purposes, it is useful to think of energy levels for the electrons in a solid as occurring in bands in which a range of energies is possible, separated by *energy gaps* (energies that the electrons cannot have). Usually the lowest band in which any levels are occupied is called the *valence band;* the next higher band is empty and is called the *conduction band* (Figure 24-24). Using this model, the difference between insulators and conductors becomes clear. To conduct electricity, electrons have to move; they must receive kinetic energy. A *metal* such as copper is a good conductor of electricity because the highest occupied band is only partially full. There are many empty levels to which electrons can jump by absorbing very small amounts of energy from an applied electric field. For a metal, the energy gap is of no significance because empty levels are available, close at hand, in the partially filled band. In an *insulator,* however, electrons occupy levels right up to the top of the valence band, and hence cannot receive just a small amount extra energy. No empty level is available. Such an electron would have to absorb enough energy to jump the gap, which might be from about 1 eV to more than 7 eV, or else not absorb any energy at all. Thus at absolute zero the electrons in an insulator cannot absorb energy, and so cannot conduct a current. At room temperature, a very few electrons have enough thermal energy to jump the gap, so there is a very small (usually negligible) electrical conductivity.

A pure *semiconductor* such as germanium has electrical conductivity intermediate between that of an insulator and that of a metal. In such a material, the valence band is completely filled, and the gap between the filled valence band and the conduction band is relatively small, from about 0.1 eV to about 1 eV. At absolute zero a pure semiconductor is a perfect insulator, but at room temperature some electrons near the top of the valence band can gain thermal energy and jump to the conduction band. Now, since empty levels are near-by, the electron can gain energy from an applied electric field and conduct electricity. (It is an easily observed experimental fact that the ability of a semiconductor

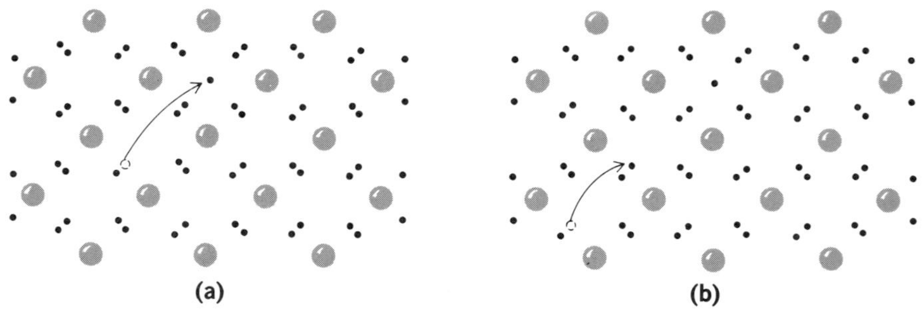

Figure 24-25 (**a**) An electron in impure germanium is given enough energy to leave its normal position; a hole is created. (**b**) A neighboring electron moves over to fill the hole, but this leaves another hole.

to conduct current increases greatly as temperature is increased.) So far, a semiconductor might seem to be similar to a metal, but one in which only a few free electrons are available. Actually the situation is very different because of the remarkable fact that when an electron is lifted into the conduction band, the effect is the same as if a physical vacancy is left behind.

These vacancies behave in such a distinctive way that they are given the special name of *holes.* We have already read in Section 19-5 how holes can be "built-in" by adding the right kind of impurity atoms, as in *p*-type material. Now we see that holes can also be created without adding any impurity at all—thermal energy is then responsible for removing an electron from an energy level in the valence band. No matter how it is formed, a hole behaves as if it were an electrically charged particle with the same magnitude of charge as the electron (but with a + sign) and with about the same mass as the electron. If you apply a PD to a semiconductor crystal (by connecting it to the terminals of a battery, for example) the hole will move through the crystal; positive charge flows, and a current is set up.

A hole is a vacancy left by the removal of an electron, therefore it moves when an electron jumps from a neighboring atom to fill the vacancy. The original hole is now filled and the vacancy has moved to the neighboring atom. Again you may say that this is merely the motion of an electron and that no useful purpose is served by the hole concept. The hole concept *is* useful, however. For one thing, the electron that jumped from one atom to the next was *not* one of the free electrons in the conduction band. It was an electron in the valence band, and it was "bound" into the crystal structure. This motion of a bound electron is a very special one, occurring in a very small region, and it cannot be described by Newton's laws of motion. It is a motion that can only be described by quantum mechanical laws involving physics far beyond the scope of this book. It is possible, however, to state the results of the quantum calculation: the motion of the vacancy through the crystal can be correctly described and correct calculations can be made if we apply familiar "classical physics" (Newton's laws and Coulomb's law) to a positive particle with a charge of $+1\ e$ and a mass equal to a quantity called *the effective mass.* The effective mass is not the mass of a completely free electron (in a vacuum) but it ranges roughly from $\frac{1}{10}$ of an electron mass m_e to about 1 m_e or somewhat more. The exact value of the effective mass depends on the nature of the crystal. Therefore, the hole is a useful and necessary concept because scientists and engineers can make useful calculations and predictions about the electrical behavior of semiconductors in terms of holes, and not in terms of the reverse motion of bound electrons.

Another way to think about a hole that may be helpful is to compare it to a bubble in a liquid. A bubble is indeed a vacancy also, in that it is a lack of liquid molecules. The bubble concept is

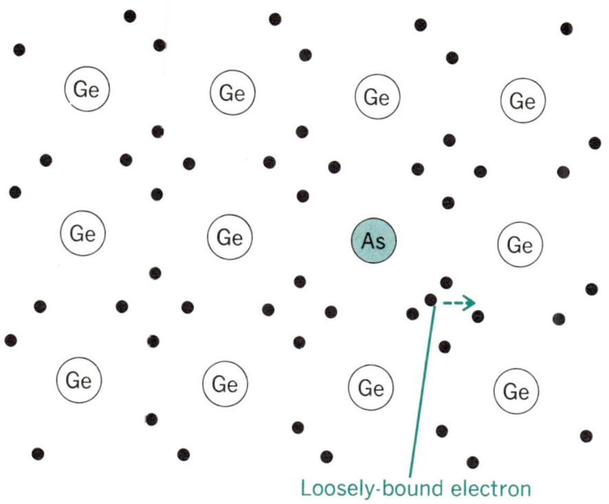

Figure 24-26

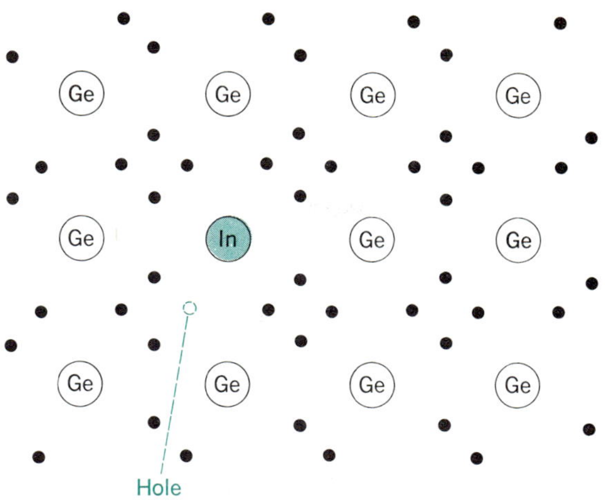

Figure 24-27

used freely without question because we can apply simple physics to describe its motion through the liquid. If we insisted on analyzing the detailed motion of all the molecules that must "jump out of the way" when a bubble goes by, the problem would be impossibly complicated.

To complete the story, we need an understanding of *p*-type and *n*-type materials. The model of a semiconductor crystal described so far applies to pure germanium or pure silicon. Each electron lifted to the conduction band has meant the formation of a corresponding hole. Such a material is neither *p*-type nor *n*-type because the positive and negative charge carriers occur in equal numbers. How are semiconductors produced that have mostly positive carriers or mostly negative carriers?

The answer lies in the marvelously precise manufacturing processes that add controlled amounts of impurity elements. The diamond structure of germanium and silicon is based on four outer electrons, as shown in Figure 24-25a. The drawing represents, in a highly simplified manner, the elevation of an electron from the valence band to the conduction band in pure germanium. A hole is formed by this process, and Figure 24-25b indicates the jumping of another electron from a neighboring atom to fill the vacancy. Figure 24-25b should not be taken literally; remember, the reason for introducing the hole concept is the unsatisfactory nature of this reverse electron motion as a means of explaining electrical conduction.

An *n*-type semiconductor is one in which the majority of charge carriers are negative electrons. How is this achieved? During the manufacturing process, when the crystal is grown, small amounts of such elements as arsenic or antimony are added, each of which have five outer electrons in their atoms. These atoms are similar enough to germanium or silicon to fit into the crystal structure, but the fifth electron is not locked in tightly. It can be lifted into the conduction band with very little energy (0.01 eV for arsenic in germanium). A free electron is added for every arsenic atom, but what about the positive charges left behind? These charges are locked in the crystal structure, they do not move like holes, they are merely fixed positive ions that do not act as charge carriers. Therefore, the charge carriers are mostly electrons and the material is *n*-type. Figure 24-26 serves to emphasize the fact that the fifth electron is an "extra" one that does not leave a structural vacancy in the crystal.

A *p*-type semiconductor is formed in a similar way. An impurity element such as indium or gallium with only three outer electrons is added. These atoms also will fit into the diamond structure, but each one will have a vacancy in the crystal associated with it, or with a neighboring atom from which an electron moved to fill up the original hole. Each of these vacancies does behave as a hole, without the addition of any free electrons to the crystal (Figure 24-27). Therefore in this type of material the majority of the charge carriers are positive holes.

There is a wealth of interesting phenomena connected with the solid state: the interactions between photons of light and a crystal structure, the interactions between mechanical vibrations (sound) and a crystal, the behavior of solids at extremely low temperatures, and many others. One of the most remarkable quantum effects is the superconductivity that occurs in some metals at very low temperatures; near 0° K, when a metal becomes a superconductor all its electrical resistance vanishes. If a current is once established in a superconducting wire, the current will continue to flow indefinitely without the need for an applied potential difference, or the addition of energy.

SUMMARY

All electrons are identical, both in charge (Millikan) and rest mass (J. J. Thomson), regardless of their origin.

Electromagnetic radiation is emitted or absorbed a **photon** at a time, with the photon energy given by $E = hf$ (Planck). The **work function** is the energy required to remove an electron from a metal surface. The KE of a **photoelectron** equals the energy of the incoming photon minus the work function (Einstein). At the threshold, the photon energy just equals the work function, and the electrons are emitted wih zero KE.

A hot solid, liquid, or dense gas emits **continuous spectrum** that depends only on temperature. A hot gas emits a **line spectrum** that depends on the nature of the gas. The **Bohr theory** of the line spectrum of hydrogen is based on three assumptions: (a) an atom can exist in certain energy states without radiating; (b) when the energy state of an atom changes by ΔE, the frequency of the photon that is emitted or absorbed is given by $\Delta E = hf$; (c) the allowed energies for hydrogen are given by a postulate that leads to the formula

$$E_n = \frac{-13.6}{n^2}\ \text{eV}$$

An atom can remain in an excited state for less than a microsecond, except in rare cases. Hence at ordinary temperatures most of the atoms in a gas are in their lowest energy state. Absorption of a photon can raise an atom to a higher state, and one or more photons can be emitted as the atom returns (in steps) to its normal state. An atom can absorb only those wavelengths it is capable of emitting.

X rays are produced when an inner electron is removed. This requires a high-speed impact (from a high-energy electron, for example) or absorption of an energetic (high-frequency) photon. X-ray photons ordinarily have energies of tens of thousands of electron volts, and wavelengths of the order of magnitude of an angstrom. Since atoms in a crystal are about an angstrom or so apart, a crystal can be used as a grating to form an x-ray spectrum. Collision experiments show that x-ray photons are concentrated bullets, having momentum as well as energy (Compton). The dual nature of light is shown by such experiments. It is not necessary to make a model for light, but if a model is desired, the interpretation of e-m waves as guide waves, or probability waves, for the photons is as good as any.

Matter as well as light has both wave and particle properties. For each, the wavelength of the guide wave is given by $\lambda = h/\text{momentum}$ (de Broglie, Davisson, G. P. Thomson). **Matter waves** have such small wavelengths that they can be ignored except when dealing with single elementary particles such as electrons, protons, or neutrons. The law of cause and effect in physics is a statistical (probabilistic) one, and the motions of individual particles or photons cannot be predicted with certainty.

Quantum mechanics is a logical development based on the dual nature of matter. Newtonian mechanics is a special case of quantum mechanics.

In **solid state physics** the concepts of quantum mechanics are applied to understanding the structure and properties of crystalline solids. A great deal of attention has been devoted to the electrical properties of semiconductor crystals, because of the many engineering applications of these effects.

QUESTIONS

1 Suppose there were two kinds of electrons: type A of charge 1.00×10^{-19} C, and type B of charge 2.20×10^{-19} C. Suppose also that the two kinds of electrons occurred at random in nature, equally plentiful in all kinds of atoms. Could this hypothesis about two kinds of electrons be tested by the oil-drop experiment? Could it be tested by electrolysis experiments? Explain.

2 In Figure 24-5, in what direction (up or down) will the shadow move if a magnetic field directed toward the reader is turned on?

3 Do all electrons have the same mass? Do all electrons have the same rest mass? (Section 5-6.)

4 Ordinary photographic film is least sensitive to red light, hence the use of red safelights in darkrooms. Explain, using the photon model of light.

5 It is harder to remove a free electron from copper than from sodium. Which metal has the greater work function? Which has the higher photoelectric threshold frequency?

6 If a proton is accelerated through a PD of 10^5 V, what is its KE in electron volts?

7 What can you say about the nature (solid or gaseous) of a source that emits a line spectrum?

8 What sort of spectrum, line or continuous, gives information about the chemical composition of a source? How can the temperature of a source be found from a study of its spectrum? Is this always possible for all sources?

9 Using Figure 24-9, on page 388-3, explain why a neon sign is red. What color do you expect a helium sign to have?

10 Why do some of the dark lines in the sun's spectrum (in the color section following page 388) coincide in position with bright lines? How do you account for the dark lines of some of the elements that do *not* match up?

11 The electric charges measured in the Millikan experiment all came in packages of a definite size. Does the charge measured in electrolysis experiments also necessarily come in packages of a single definite size?

12 Could you measure the mass of tiny dust particles floating in air by observing their motion? Describe how you would set up an experiment to make such a measurement.

13 Explain why a person descending with a parachute moves at constant speed, apparently contrary to Newton's second law.

14 Explain why it is correct to say that Millikan's experiment did not show the existence of the electron as a particle, but Thomson's experiments did.

15 Suppose you release a small bubble of air at the bottom of a swimming pool. Describe qualitatively how you would expect it to behave. Will it remain constant in size? Will it move up or down? Will it accelerate? Now try to make the same kind of description in terms of the water molecules forming the bubble surface.

16 Suppose you have three pendulums of the same length coupled together. Describe the possible motions that the three can have as a system.

17 Do you believe that some semiconductors might be able to emit light? Explain your answer, and give an approximate value for the energy of a photon emitted in this manner.

PROBLEMS

GROUP A

1 Calculate the energy of a photon of ultraviolet light whose wavelength is 1220 Å. Express your answer both in joules and in electron volts.

2 A particle has an energy of 10^{-13} J. Express this energy in million electron volts (MeV).

3 Calculate the wavelength associated with a particle of mass 10^{-20} kg that is moving at a speed of 10^{-6} m/sec.

GROUP B

4 An oil drop for which the mass density is 800 kg/m^3 is balanced between two plates 8 mm apart. The PD between the plates is 1200 V. If the drop has a charge of 5 electrons, calculate its radius.

5 What is the KE of the photoelectrons ejected from a piece of tungsten, for which the work function is 4.52 eV, by ultraviolet radiation of wavelength 2000 Å?

6 Light of wavelength 5000 Å shines on a thin metal surface, and the ejected photoelectrons have KE equal to 0.52 eV. Calculate the work function of the surface.

7 What is the threshold wavelength for a tungsten surface whose work function is 4.52 eV?

8 Calculate the wavelength of the Lyman β line (the line in the Lyman series having next-to-longest wavelength). (See Figure 24-11.)

9 Try to determine which two energy levels in the mercury atom (Figure 24-13) are involved in the emission of the green line whose wavelength is 5461 Å. (*Hint:* First find the energy of the photons in eV). Identify the line in Figure 24-9, page 388-3.

10 What is the energy of the least energetic photon that can be absorbed by a hydrogen atom at room temperature? Identify the line in Figure 24-11.

11 The fourth line in the hydrogen Balmer series (Hδ) has a wavelength of 4101 Å. (a) Identify the line in Figure 24-12. (b) Identify the line in the hydrogen emission spectrum in Figure 24-9. (c) Identify the line in the sun's absorption spectrum. (d) One of the energy levels involved in the emission or absorption of this line is shown in Figure 24-11. Use the wavelength of the line to compute the value of the other energy level that is involved.

12 Calculate the momentum of a photon of the yellow sodium light whose wavelength is 5890 Å.

13 Calculate the wavelength of the guide waves associated with a rifle bullet of mass 40 g moving at 500 m/sec. Does the bullet show any observable wave behavior?

GROUP C

14 An explorer on the moon turns on a 100 W lamp, and the light from the bulb spreads out equally in all directions. Assuming that 1% of the energy is radiated as visible light (of average wavelength 5550 Å), about how many photons per second of visible light from the bulb would strike the mirror of a large telescope on the earth? (Diameter of the mirror = 5 m; distance from the moon = 400,000 km; neglect absorption in the earth's atmosphere.)

15 The Balmer series lines appear in the absorption spectra found in hot stars, but not in absorption spectra obtained in room-temperature experiments on the earth. Explain.

25 The Atomic Nucleus

The last few years of the nineteenth century were remarkable years in the history of physics. In 1890, atoms were thought to be indivisible, and light was thought to be an electromagnetic wave that could not possibly have particle properties. However, by the year 1900 J. J. Thomson had discovered the electron (Section 24-2), x rays had been produced (which led indirectly to many other important discoveries), and a quantum process had been proposed by Planck to explain the radiation of light from a heated body. These years certainly mark the beginning of "modern physics." Today, the name modern physics refers to a kind of physics in which we can no longer describe nature in terms of common sense or everyday models that obey Newton's laws of motion. Here you will see how the ideas of modern physics along with devices of great sophistication have increased our knowledge of the atom.

25-1 EARLY DISCOVERIES ABOUT RADIOACTIVITY

In 1896 the French physicist Henri Becquerel (1852–1908) was investigating whether minerals that fluoresced when exposed to light gave off radiation similar to the newly discovered x rays. He was studying the phenomena of fluorescence and phosphorescence, in which radiation is given off by some substances when they are exposed to other radiation of suitable wavelength. Most familiar is the visible light emitted by certain minerals while they are exposed to ultraviolet radiation. The light of a fluorescent lamp is of this type. Luminous watch dials are phosphorescent; they continue to give off light for some time after being exposed to radiation.

Becquerel discovered that the compounds of the element uranium gave off highly penetrating radiation of some kind, even when they had not been exposed to light. The rays from uranium behaved much like x rays, except that they were continuously emitted, even with no apparent source of energy. At first Becquerel thought he was observing a kind of fluorescence of invisible radiation. The uranium might be releasing energy that it had stored up from sunlight, for example. To check this possibility, he used a piece of uranium ore that had been kept in a lead-shielded, light-tight box for several months. This sample was still able to expose a photographic plate (Figure 25-1) and to knock electrons out of gas molecules in the air, thus producing ionization of the air. This phenomenon was called *radioactivity,* and it presented serious problems for physicists because it was apparently an exception to the law of conservation of energy.

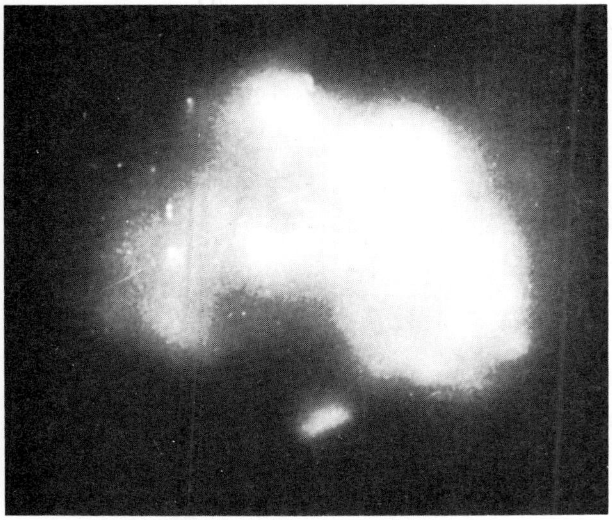

Figure 25-1 Film exposed by nuclear radiation.

Many other scientists helped to advance the study of radioactivity over the next ten years. Perhaps the most outstanding contribution was that of Pierre and Marie Curie, who processed many tons of impure uranium ore, and isolated a fraction of a gram of two previously undiscovered elements—polonium and radium. Both of these elements are more radioactive and less abundant than uranium in its most common form.

During this period, several dozen radioactive substances were discovered, and it was found that they emitted just three kinds of radiation, which were called alpha (α), beta (β), and gamma (γ) rays. The γ rays were found to penetrate matter most easily, and the α rays were found to be the least penetrating.

We now know that alpha "rays" are a stream of positively charged particles, and the direction of their deflection in a magnetic field indicates that they have a positive charge. Careful experiments have been carried out in which great numbers of alpha particles were individually counted and their total electric charge measured at the same time. The charge per particle turned out to be two elementary charges. Further experiments have been conducted in which alpha particles were shot into a thin-walled tube and collected. After a few days, the material in the tube gave the characteristic spectrum of helium when an electric spark was passed through it, showing that the alpha particles each picked up two electrons and became neutral helium atoms. An alpha particle is, therefore, a doubly charged helium ion—a helium nucleus. Alpha particles are easily stopped by a thick sheet of paper.

Beta rays were shown to have the same e/m ratio as was measured for electrons (Section 24-2), and it was concluded, therefore, that the beta rays are fast moving, negative electrons. (Their deflection by a magnetic field is opposite to that of alpha particles.) Beta rays are much more penetrating than alpha radiation; they can usually penetrate thin metal sheets, but are stopped by a few millimeters of lead. The penetrating power of the beta rays results from their higher speed. The slower alpha particles are much more likely to lose their energy by collisions with atoms. Both alpha and beta particles penetrate farther through a substance if they have greater KE to start with. In matter, a beta particle finally attaches itself to the electron cloud of some atom or ion, or becomes a free electron in a metal.

Gamma rays were found to be photons of high energy. They are not deflected by a magnetic field. The guide waves of gamma-ray photons have such short wavelength that diffraction and interference can be observed only with great difficulty. However, the wavelength of the longer gamma rays can be measured, using the atoms in a crystal to form a grating.

One of the earliest observations about radioactivity was that the amount of activity is independent of external conditions. The nature of the radiations from uranium, for instance, and the activity of a given amount of uranium were found to be completely unaffected by temperature, pressure, and other physical changes. Different chemical combinations did not affect the activity of a radioactive element. For example, the activity of one mole of UCl_3 exactly equals that of one mole of UO_2 or one mole of UF_6 gas. In the next section you will see how Rutherford's experiments with the scattering of alpha particles led to a model of the atom that explains these facts.

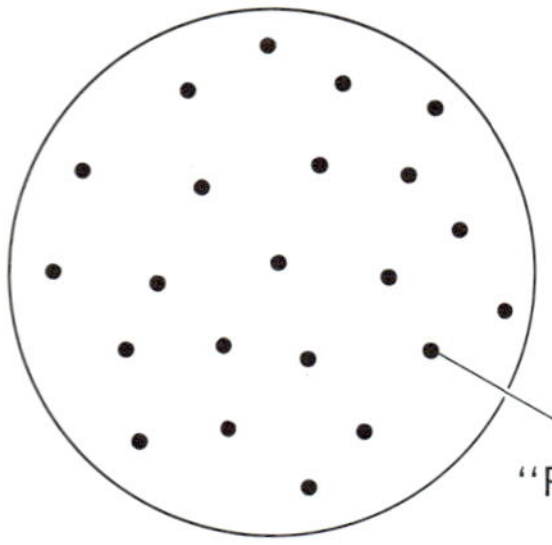

Figure 25-2 "Raisin pudding" model.

25-2 THE NUCLEAR ATOM

One of the great experimental physicists of the recent past was Ernest Rutherford (1871–1937). He was a New Zealander who worked in Canada at McGill University, later in England at Manchester University, and finally was head of the famous Cavendish Laboratory at Cambridge University in England. It was Rutherford and his co-workers who identified alpha particles as helium ions. At Manchester, he directed the work of two of his students in a crucial experiment in which they observed the *scattering*, or deflection, of alpha particles as they passed through thin metal foils.

We are so used to the planetary model of an atom, with electrons circulating around a kind of inner "sun," that it is hard for us to realize that seventy years ago the accepted picture of an atom was quite different. There was no doubt that an atom consisted of equal amounts of positive and negative charge, as shown by the existence of ions. It was also understood that electrons could be separated from atoms, as shown by the photoelectric effect, and by J. J. Thomson's measurement of e/m for electrons. But the Bohr theory of the nature of the outer electron cloud had not yet been worked out, and most physicists believed that the structure of the atom must somehow allow the electron to vibrate back and forth as it emitted e-m radiation.

One model, proposed by Thomson, pictured an atom as a ball of jellylike positive charge in which were imbedded one or more electrons. This was sometimes called the raisin-pudding model. In this model, the normal position of a single electron would be at the center of the round "pudding," and in complex atoms several electrons were supposed to exist in a stable pattern (Figure 25-2.) If an electron were somehow pulled away from its position it would oscillate back and forth in SHM and give out an e-m wave. It was very difficult with this model to explain the existence of so many spectral lines, each of a different frequency.

The Rutherford scattering experiments can be compared with trying to learn about the contents of a tightly sealed box by shooting bullets through the box. These bullets are invisible until they become embedded in a distant wall (Figure 25-3). In Rutherford's case, he did not even have a single "box" to shoot at, but rather a large number of atoms that were invisible to him because of their tiny size. With the box, there are several possible results: If the box contains nothing, the bullets will go right on through. If the box is full of a dense substance, or if it contains a sheet of some hard material, the bullets will bounce straight back. If the box is full of something resembling a pudding, with numerous very light raisins uniformly distributed throughout, the bullets will be deflected, or scattered, slightly as they pass through. However, if the box is mostly empty, but contains a few very hard and dense stones, then most of the bullets will go straight through, but a few may be deflected through large angles when they strike the stones.

Rutherford explored the structure of the atom by shooting alpha particles through a thin metallic foil (Figure 25-4a). Of course, he and his students did not deliberately "shoot" alpha particles, but rather they used a stream of alphas given off by a radioactive element. If the Thomson raisin-pudding model were correct, most of the particles would be scattered slightly out of their straight-line paths, due to the Coulomb force of repulsion between the positively charged α particle and the atomic spheres of positive charge. An α particle could hardly miss coming close to another positive charge as it passed through the metal foil, and

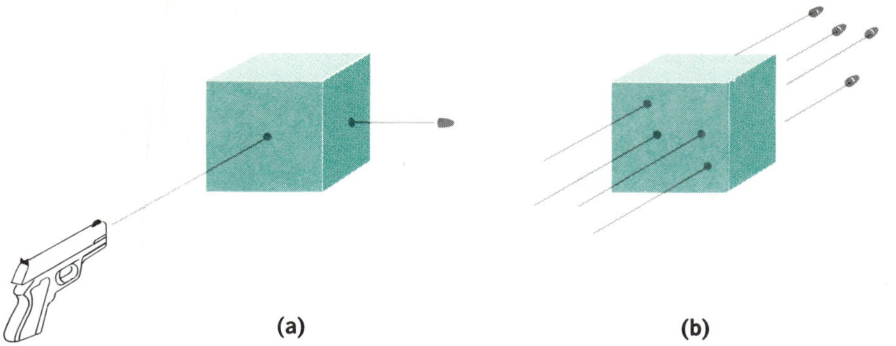

Figure 25-3 (**a**) Some dense object is inside the box. (**b**) If all the bullets go straight through, we would expect to find only light particles inside, a diffuse resisting medium, or nothing!

therefore we would expect most of the α particles to be deflected slightly, say a few degrees.

The experiment showed a surprisingly different result. Most of the α particles went right on through the foil with no change of direction, but a few were deflected through large angles (Figure 25-4b). Some even bounced straight back, a deflection of 180°. Rutherford colorfully described this situation by saying that it was as if you had "fired a 15-inch shell at a sheet of tissue paper and it came back and hit you."

To explain the large forces that could cause such large deflections, Rutherford had to assume that an α particle is itself extremely small, and that the building blocks of atoms must also be extremely small. He further had to assume that if by chance an α particle were aimed just right, it could approach to within about 10^{-14} meter of a concentrated "center of charge," which he called the *nucleus*. The atoms themselves were known to be about 10^{-10} meter apart, from the results of experiments with crystals. Rutherford concluded, therefore, that the atom consisted of a small, heavy, positively charged central nucleus, with negative electrons separated from the nucleus by much empty space.

To visualize these empty spaces, let's imagine an atom blown up to the size of a football field. The nucleus would then be a small, heavy glob of matter about the size of a pencil eraser. The rapidly moving electrons would be even smaller, but if we watched their motion for a while, there would be a reasonable probability of finding an electron at any point within the football field. To complete the picture, imagine firing another pencil eraser at the football field, and think of a strong repulsive force that only works when the two erasers are within a centimeter of each other. It's not hard to understand why so few α particles were scattered in the Rutherford experiment.

By mathematical analysis of his scattering experiments with gold, copper, and silver foils, Rutherford was able to calculate the magnitude of the central charge of the nucleus for each of these atoms. If we express them in terms of the elementary charge, then the gold nucleus had a charge of $+77.4e$, silver $+46.3e$, and copper $+29.3e$. Within experimental error, these numbers are the atomic numbers (called Z) of the elements. Later

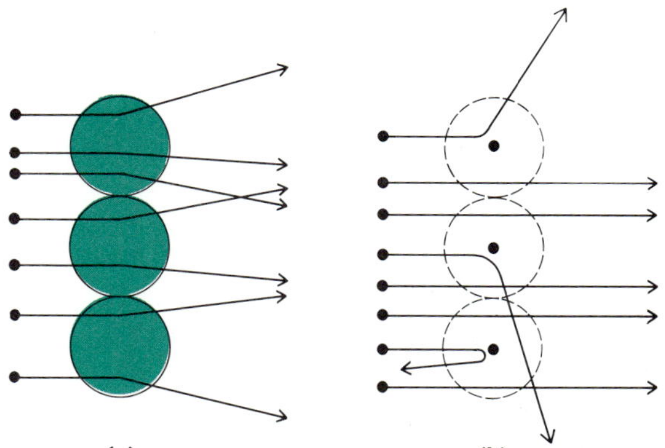

Figure 25-4 Scattering of α-particles according to two models of atomic structure. (**a**) Thomson's model; (**b**) Rutherford's nuclear model.

work of higher precision showed definitely that the atomic number in the periodic table is equal to the number of elementary charges in the nucleus. Since atoms are known to be neutral there must be Z electrons outside the nucleus, and these make up the volume of the atom. As we have seen, these electrons must be in orbits of some kind to keep from falling into the nucleus.

The Rutherford experiment is an outstanding example of how important results could (in 1911) be obtained with simple apparatus. The scattered α particles struck a fluorescent screen, which was simply a layer of small zinc sulfide crystals. The impacts produced tiny flashes of light, which were observed with a magnifying glass. Rutherford's students deserve great respect and sympathy for counting these tiny flashes for hours on end. The willing students, plus a speck of radioactive material, some slits to form a narrow beam, and the foil itself completed the apparatus. This was all that Rutherford needed (with a good measure of insight) to establish the nuclear model for the atom.

As so often happens, the experiment of 1911 raised new problems. If the electrons move in orbits outside the nucleus, then why does an electron not emit radiation continuously, since it is continuously accelerated with centripetal acceleration in its curved path? Why are atoms stable? Why don't they all collapse in a fraction of a second? Now at last, physicists were asking the right questions, and among those physicists was a young student in Rutherford's laboratory at Manchester named Niels Bohr. You saw in Section 24-4 how Bohr developed the model of the hydrogen atom in 1913, just two years after Rutherford's historic experiment.

25-3 RADIOACTIVE DECAY

The nuclear, or planetary, model of an atom explains why radioactivity is unaffected by external conditions. Chemical reactions take place when the outer parts of electron shells come close together. During collisions between atoms in a gas, the outer shells may become momentarily dis-

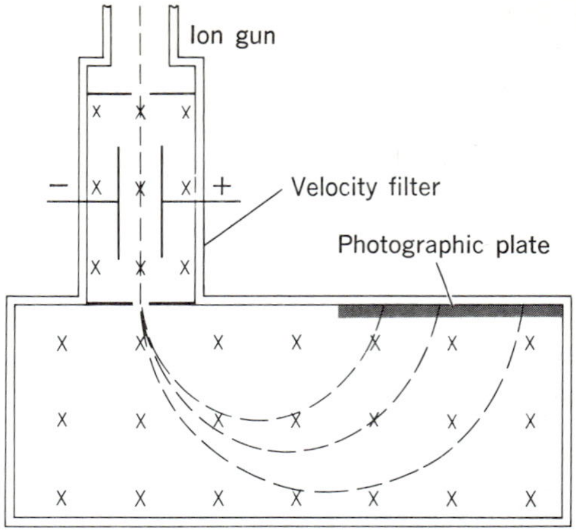

Figure 25-5 In the mass spectrometer, charged particles of different masses are separated by a magnetic field after passing through a velocity filter.

torted; however, during such physical or chemical changes the nucleus is protected by the outer electron cloud. The nuclei never get close enough to each other to react, except at high speeds corresponding to temperatures of over $10^{6\circ}$ K. Since radioactivity is practically unaffected by chemical state, temperature, pressure, and other conditions, we can conclude that it is associated with the nucleus. The α, β, and γ rays are, therefore, emitted from the nucleus. Let's now consider the composition of such a nucleus.

According to our present understanding, the nucleus contains relatively heavy particles, protons and neutrons, which are collectively called *nucleons*. Any given combination of protons and neutrons is called a *nuclide*. Several nuclides all of the same atomic number are called *isotopes* of the element in question. The symbol Z is used for the *atomic number* (the number of protons in the nucleus), and A represents the *mass number* (the total number of nucleons). Hence the number of neutrons in any nucleus is given by $A - Z$. In a shorthand notation $^{A}_{Z}E$ the atomic number is a subscript and the mass number is a superscript at the left of the elemental symbol.

Not all combinations of protons and neutrons

can stick together to form a stable nuclide. Roughly speaking, stable isotopes of light elements have about 1 neutron for every proton (for example, $^{16}_{8}$O has 8 protons and 8 neutrons). In the heavier elements, the trend is toward somewhat more neutrons than protons (for example, $^{200}_{80}$Hg has 80 protons and 120 neutrons). If a nuclide deviates too much from the normal proportion of neutrons and protons, it is certain to be unstable and radioactive. The isotopes of chromium can be used to illustrate nuclear stability and instability. Eleven isotopes of chromium are known to exist; the heaviest is $^{56}_{24}$Cr, which has 24 protons and $56 - 24 = 32$ neutrons, and the lightest is $^{46}_{24}$Cr, which has 24 protons and 22 neutrons. Ordinary chromium is a mixture of four stable isotopes.

The masses of the individual isotopes can be found by use of the *mass spectrometer*, one type of which is shown in Figure 25-5. Many types of mass spectrometer have been developed in which combinations of magnetic and electric fields focus ions on a collector. Figure 25-6a is a "photograph" of the Cr mass spectrum; each Cr ion sticks to a glass plate at a place determined by its e/m ratio. Thus isotopes of different masses are brought to different places. If the Cr^+ ions (chromium atoms that have lost an electron) strike a metallic collector, they are neutralized by free electrons, and the electron current is a measure of the abundance of the isotope (Figure 25-6b). Such experiments show four stable isotopes of Cr, with relative abundances as follows: $^{50}_{24}$Cr (4.3%), $^{52}_{24}$Cr (83.8%), $^{53}_{24}$Cr (9.5%), and $^{54}_{24}$Cr (2.4%).

The chemical atomic mass of the element chromium (52.01) is the average of these values; it turns out to be so close to 52 because the isotope of mass number 52 is by far the most abundant one in nature. Careful use of a precise mass spectrometer reveals that the atomic masses of the isotopes are not exactly integers. Thus a neutral atom of $^{50}_{24}$Cr has mass 49.9461, in terms of 12.00000 for the mass of $^{12}_{6}$Cr, a neutral carbon atom of mass number 12.

The stability of an element is measured by its *half-life*. The half-life of a substance equals the

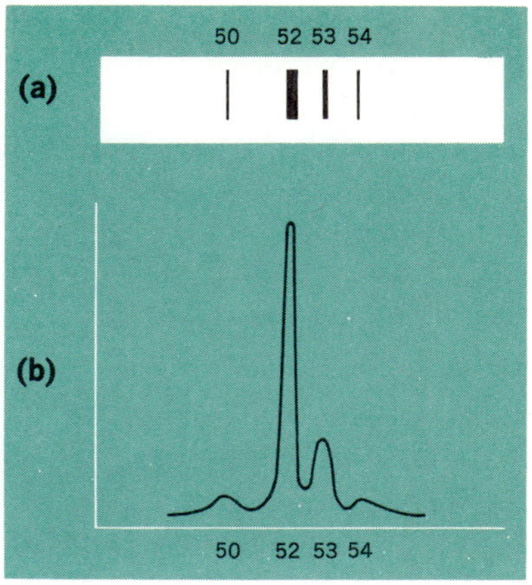

Figure 25-6 Chromium isotopes.

time required for half the nuclei to change to new nuclides. For the decay of $^{55}_{24}$Cr, experiment shows the half-life to be 3.52 minutes. This means that if, say, 8 million $^{55}_{24}$Cr nuclei are present at a certain time, 3.52 minutes later only 4 million will still be $^{55}_{24}$Cr; and after another 3.52 minutes, the number will be down to 2 million, and so on. The nuclei behave like birds sitting on a fence. Assuming that the birds take off at random, we can estimate quite closely how many will be left after, say, 3.52 minutes or 7.04 minutes, but we cannot say *which* bird will take off first, or which will be the last to go. We see that radioactivity is intimately related to probability. The half-lives of known nuclides vary from less than a microsecond (high probability of decay) to more than 10^{14} years (very low probability of decay).

There are several ways in which a nucleus can change or *disintegrate*. As one example let's look at the behavior of $^{55}_{24}$Cr. This nuclide has more than the normal number of neutrons, and there is a high probability that one of its neutrons will spontaneously change into a proton, increasing the atomic number of the nuclide by 1 unit. According to the law of conservation of charge, a negative charge would then be created to balance the newly created extra positive charge in the nucleus. This negative charge is emitted as a β particle, or negative electron, also called a

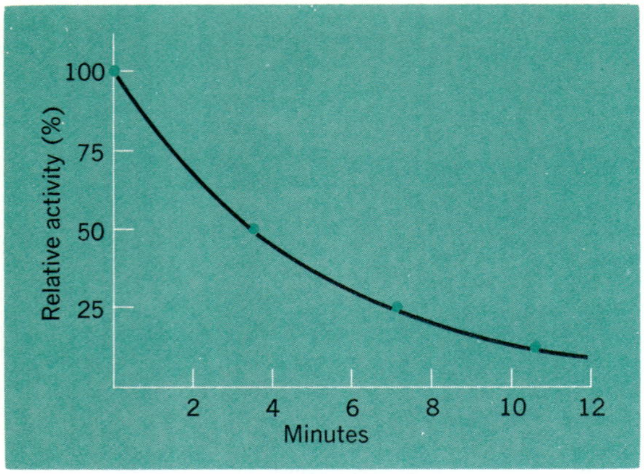

Figure 25-7

negatron ($_{-1}^{0}e$, or β^-). Since the nucleus now has an atomic number of 25, we know that it has become an isotope of manganese. The total number of nucleons has remained unaltered during the changeover of a neutron into a proton, so the mass number is still 55. Symbolically,

$$^{55}_{24}\text{Cr} \longrightarrow \, ^{55}_{25}\text{Mn} + \, _{-1}^{0}e \qquad (3.52 \text{ min})$$

Note how this equation balances in two ways; the charge balances: $24 = 25 + (-1)$, and also the total number of nucleons balances: $55 = 55 + 0$. This sort of spontaneous change is called *radioactive decay.*

Figure 25-7 shows the radioactive-decay curve for $^{55}_{24}$Cr. The activity of a sample has been plotted against the time in minutes. As you can see, after 3.52 minutes only one half of the original activity remains. This is because half of the original $^{55}_{24}$Cr nuclei are still present (the rest have disintegrated into $^{55}_{25}$Mn.) With a half life of 3.52 minutes how long would it take for a given sample of $^{55}_{24}$Cr to reduce its activity to less than one tenth of 1 percent? Will the activity ever reach zero?

25-4 PARTICLE DETECTORS

Over the years, physicists have developed many devices for detecting the particles and photons emitted by radioactive atoms. Most of these methods depend upon the fact that a moving charged particle knocks electrons out of atoms and thus ionizes any matter through which it passes.

The Geiger-Müller tube counter (now called simply a *Geiger counter*) makes use of a so-called avalanche effect (Figure 25-8). The tube contains a gas such as oxygen or argon at a pressure of about 0.1 atmosphere, and the thin central wire is about 1000 volts positive with respect to the metal wall. If an incoming particle creates just one ion pair, the negative ions are attracted to the positive wire, creating more ions by collision as they rush toward the wire. A current rapidly builds up, and the *IR* drop in the resistor *R* may amount to several volts, to be further increased by the amplifier.

The *cloud chamber* makes visible the path of a charged particle (Figure 25-9). When a charged particle passes through air, it leaves a trail of ions, and if the region is supersaturated with water vapor, the ions serve as centers upon which visible drops of water may condense (Figure 25-10). The effect is similar to the formation of "vapor trails" left in the sky by passing jet aircraft. The *bubble chamber* uses a liquid such as propane, and the passage of an ionizing particle is made evident by a trail of tiny bubbles (Figure 25-11).

Both of these devices give a great deal of information about the particle that makes a track. If the chamber is placed in a magnetic field, moving charged particles are bent in arcs of circles, and so their energies can be determined. Particles can be identified by the density of drops or bubbles along the path. The slow, heavy α particles create many more ion pairs per centimeter of path, and give a denser trail, than do the light electrons.

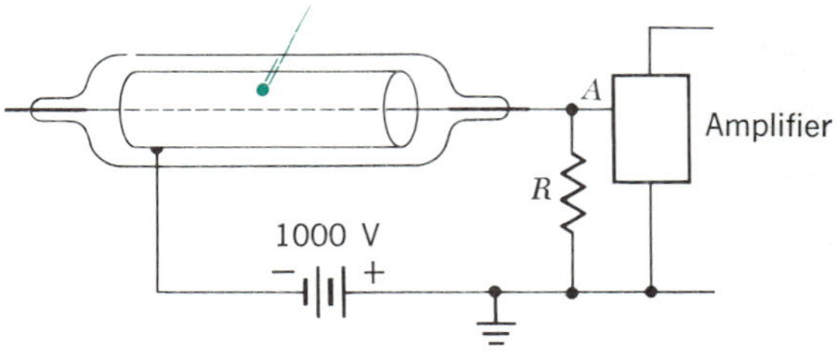

Figure 25-8 Geiger counter.

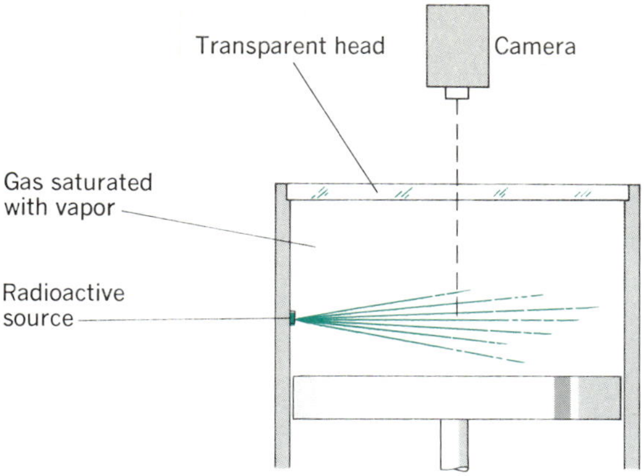

Figure 25-9 A cloud chamber.

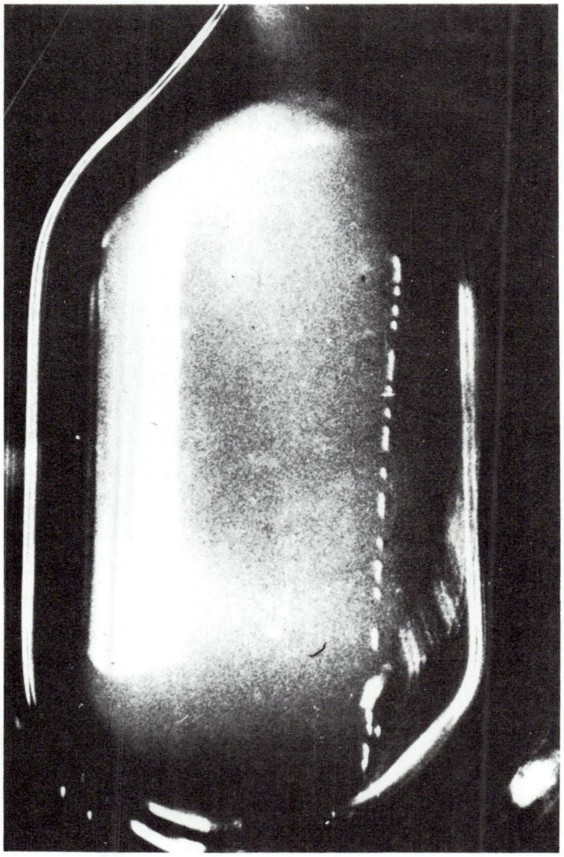

Figure 25-11 The first bubble chamber—a heavy glass bulb made by D. A. Glaser—showing a cosmic-ray particle track.

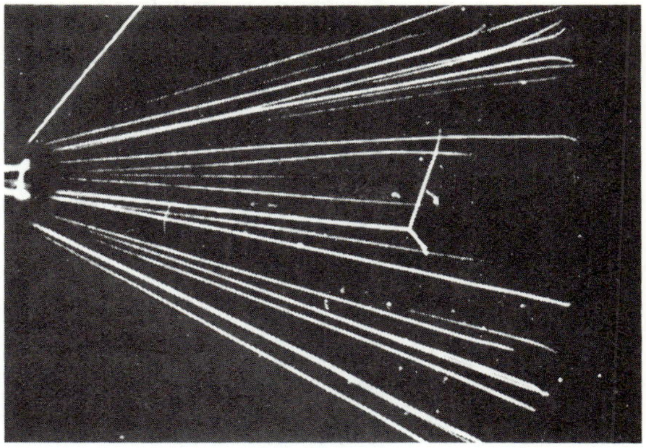

Figure 25-10 Scattering of an α particle by a heavier nucleus of fluorine. Note that all the unscattered α particles have about the same range in the gas of the cloud chamber.

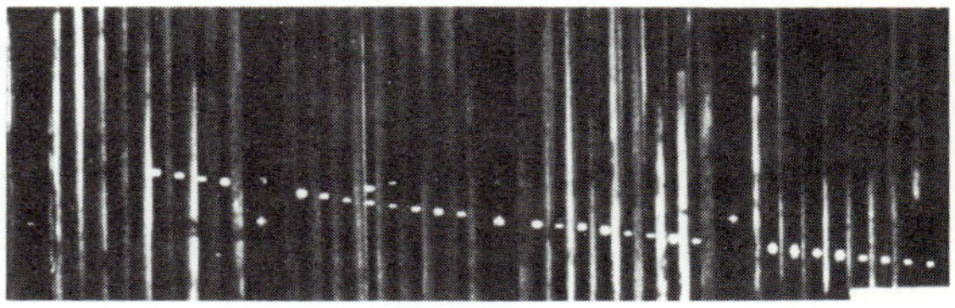

Figure 25-12 Spark tracks caused by the passage of a subatomic particle through a spark chamber.

A newer type of charged-particle detector is the *spark chamber* (Figure 25-12). It consists of many, closely spaced, parallel metal plates in a container of gas. A high PD is applied between the plates and the ionization produced by the particle causes visible sparks to jump between the plates.

Since photons are not charged, they can be detected by a Geiger counter only by the relatively few photoelectrons they eject from the wall or gas of the counter. The *scintillation detector* overcomes this difficulty. Basically an extension of the fluorescent screen used by Rutherford, it responds to photons with good efficiency because the absorption of photons is in a solid substance.

25-5 PARTICLE ACCELERATORS

Physicists use the detectors described in the previous section to study the nuclear reactions that

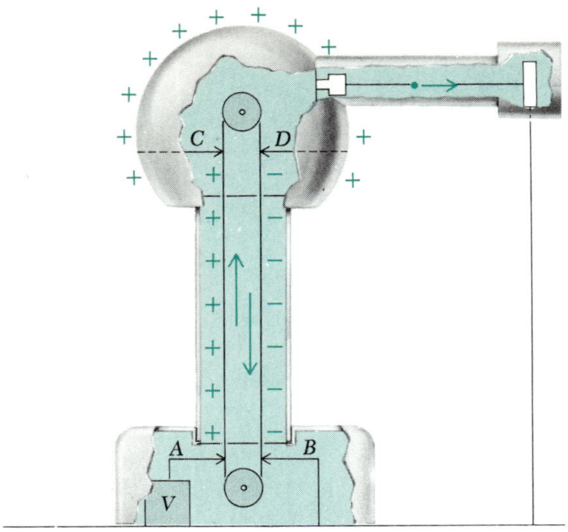

Figure 25-13 Van de Graaff generator.

take place in particle accelerators (sometimes called atom smashers) and in chain-reacting nuclear reactors.

The simplest atom smasher is the *Van de Graaff generator,* shown in Figure 25-13. The pointed electrode A is connected to a rectifier circuit V, which maintains a PD of perhaps 30,000 volts, and positive charge is sprayed onto a moving silk belt. The charge is removed by C, and replaced by negative charge at D. In an endless process, electrons are moved from the sphere to the ground, and the sphere becomes charged to a high positive potential. To use the generator as an atom smasher, a source of H^+, He^+, or He^{+2} ions is mounted inside the sphere, and the positive ions are repelled by the positively charged sphere toward a grounded target of atoms to be broken down. Large electrostatic generators of the Van de Graaff type can produce particles with energies up to about 5 million electron volts (MeV). The newest "tandem" Van de Graaff generator can accelerate particles up to more than 10 MeV.

The *cyclotron,* developed in 1930 by E. O. Lawrence and M. S. Livingston, uses a magnetic field to bend the particles into orbits (Figure 25-14). In this way a modest PD of, say, 50,000 volts can be applied over and over again to the same particle, to build up a large final energy.

The proton source injects H^+ ions at low speed, and the particles move inside two semicircular flat cavities, called dees because of their shape. The dees are enclosed in an evacuated chamber placed between the pole pieces of a powerful electromagnet, so that the particles cut across a vertical magnetic field. In Chapter 18 you saw that the magnetic force on a moving charge is perpendicular both to the field and to the motion; therefore each proton moves in a circular orbit, for which the centripetal force is supplied by the magnetic force.

The dees are connected to a rapidly varying source of alternating potential, say, $\pm 50{,}000$ volts at a frequency of 10^7 hertz. If a proton reaches A at just the right time, when D_1 is negative relative to D_2, it will be accelerated across the gap and receive an additional 50,000 eV of energy. Since it is now moving faster, the magnetic field bends it in a semicircle of larger radius. By the time the proton reaches C, the polarity of the dees has reversed so that D_2 now is negative relative to D_1, and the proton gains another 50,000 eV. In this way the proton spirals outward in a series of semicircles, and eventually strikes a target T.

The cyclotron works only because the time of each semicircular journey is the same for all particles, whether they are moving slowly (near the center) or fast (near the outside). To show that this is true, we use Newton's second law and the expression for the magnetic force on a moving charge (Section 18-1):

$$\text{net } F = ma$$

$$Bqv = m\,\frac{v^2}{r}$$

whence

$$v = \frac{Bqr}{m}$$

The time for half a revolution is given by

$$T = \frac{\text{distance}}{\text{speed}}$$

$$= \frac{\frac{1}{2}(2\pi r)}{(Bqr)/m}$$

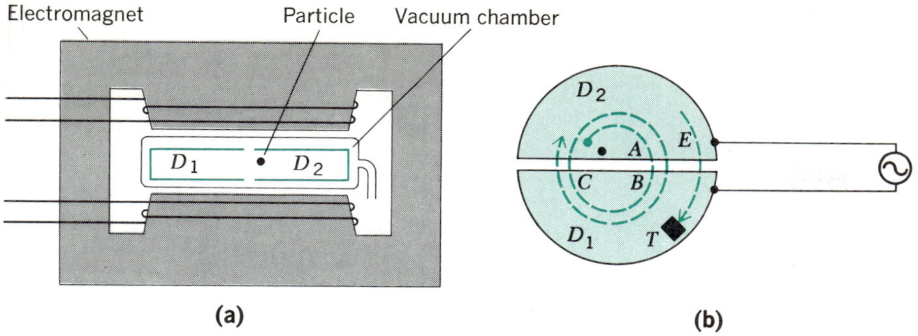

Figure 25-14 Cyclotron (schematic): (**a**) side view; (**b**) top view of dees.

or

$$T = \frac{\pi m}{Bq}$$

The radius has canceled out; therefore the time for one semicircular trip depends only on B, m, and the charge q. If these quantities are constant, then T is constant, and an oscillator of fixed frequency can accelerate *simultaneously* both the slow protons at A and the faster ones at E, which have already made many trips in their spiral path.

One might think that there is no limit to the energy that can be given to a particle by cyclotron action. To obtain a large final energy, a strong magnetic field must be used, and the diameter of the pole pieces must be large so that each particle can make many trips as it spirals outward. However, aside from cost and other practical limitations on size, there is an unavoidable limitation due to the relativistic increase of mass of the high-speed particles. The above equation shows that the time for a semicircular trip increases if mass increases, and so, as the particles gain energy—and mass—they fall behind and get out of step with the oscillating voltage applied to the dees. A 20 MeV proton moves at about $\frac{1}{5}$ the speed of light, and its mass is 2 percent greater than its rest mass. This is enough to spoil the regularity of the cyclotron action.

The difficulty can be overcome in the *synchrocyclotron* by letting protons move through the machine in short bursts, and increasing the time between alternations of the dees as the particles spiral out. Another solution is to increase the magnetic field B while the group of particles is spiraling outward, allowing T to remain fixed. By such methods, energies up to 100 MeV were obtained just after World War II.

To reach the billion-electron-volt (GeV)* range, physicists took advantage of the facts of relativity that previously were a limitation. No particle can have a speed faster than the speed of light. A 2 MeV electron is already moving at a speed of 0.98 the speed of light, and further increase of energy goes mainly into increasing its mass. Thus *all* very high-energy particles move at the same speed, about 3×10^8 m/sec, regardless of rest mass or energy (Table 25-1). The *synchrotron* takes advantage of this simplicity. If the particles

Table 25-1

Dependence of speed of various particles upon their kinetic energy†

Kinetic energy	*Electron speed*	*Proton speed*	*Deuteron speed*
0.1 MeV	.553	.0146	.0103
1 MeV	.941	.0462	.0326
10 MeV	.9988	.145	.103
100 MeV	.999987	.429	.316
1 GeV	.99999987	.876	.760
10 GeV	.9999999987	.9963	.988
100 GeV	.999999999987	.999957	.99983

† All speeds are expressed as fractions of the speed of light.

* GeV stands for giga electron volt. Giga is the newly adopted prefix meaning 10^9 (see Appendix Table 2).

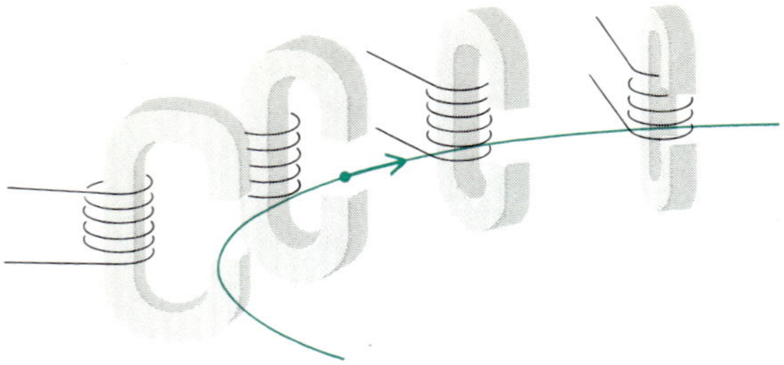

Figure 25-15 Electromagnets in a synchrotron.

are somehow brought up to nearly the speed of light, then as they gain energy their speed remains constant and the time for each revolution remains constant. The magnetic field must be gradually increased as the particles pick up energy and mass. The great advantage of the synchrotron principle is that the radius of the orbit is practically constant, and so the magnetic field need be applied only at the circumference, instead of throughout the entire area of the orbit as in the cyclotron. The beam of particles is confined to a doughnut-shaped chamber that lies between the poles of a series of C-shaped electromagnets (Figure 25-15).

Half a dozen electron synchrotrons have been constructed that yield electrons of energy greater than 1 GeV, and about a dozen proton synchrotrons in various countries are in the 3 to 70 GeV range. A giant synchrotron at the University of California is shown in Figure 25-16. The interior of the internationally operated synchrotron at Geneva, Switzerland, is shown in Figure 25-17. Suppose someone were to walk through this corridor when the synchrotron was operating? Would he be in any danger?

Figure 25-16 A 6.2 GeV proton accelerator, called the Bevatron.

Two other types of accelerators are also worthy of mention. The *betatron* uses transformer action to give energy to a stream of electrons. This was the first of the doughnut-chamber accelerators. It is practical because electrons reach essentially constant speed (equal to c) at moderate energies. In effect, the circular electron beam takes the place of the secondary of a transformer in which the primary is a large central magnet whose field changes at 60 hertz. A similar machine using protons would require an impossibly large central magnet.

A natural limitation still remaining in all accelerators, even if the relativistic increase in mass is taken care of, is the loss of energy by radiation. A proton or other charged particle moving in a circular orbit has a centripetal acceleration, and so it radiates energy. The maximum energy of a circular accelerator is reached, therefore, when the rate of e-m radiation equals the power input to the beam of particles. These radiation losses, which are inevitable as long as the particle has centripetal acceleration (v^2/r), can be reduced by using a path of large radius—another reason for the preference for largeness in the construction of atom smashers.

In the *linear accelerator,* particles move in a straight line between many properly spaced electrodes, instead of returning over and over to the same gap; no magnetic field is needed. There is no centripetal acceleration, and radiation losses occur only while the particles are crossing the narrow gaps where they gain energy. Such machines have been constructed to accelerate electrons up to 20 GeV, and are more than a kilometer long.

The foregoing descriptions have been broad outlines of the various particle accelerators in common use. Their successful operation calls for precise and imaginative design. One of the problems is focusing the particles into a narrow, stable

Figure 25-17 The interior of the cavern of a large proton synchrotron, shown when under construction at Geneva, Switzerland. Yet to be installed were magnets that supply centripetal force upon protons moving in a circle of radius approximately 100 m.

beam of high intensity. Much current research has been devoted to such questions of stability and intensity.

25-6 HIGH-ENERGY PHYSICS

Physicists seem to be preoccupied with bigger and bigger machines that give higher and higher energies to particles. Let's look at the reasons for such a trend and see what some of the results have been.

We can consider an atom-smasher as a tool with which to probe the structure of the nucleus. In order to "see" clearly, we must use a fine enough probe. More specifically, to obtain high resolving power it is necessary to use a wavelength that is much smaller than the object we are investigating, for only in this way can the effects of diffraction be made negligible. The de Broglie wavelength (Section 24-7) of any particle is given by $\lambda = h/mv$, where mv is the momentum of the particle.

As higher energies have become available, the de Broglie wavelengths of particles have become shorter, and the probing action more delicate. The newest accelerators take the projectiles far into the relativistic range, where a particle's mass is much larger than its rest mass. In a sense, relativistic mechanics is here simpler than Newtonian mechanics, since the velocity equals c and cannot increase. Under these circumstances, the de Broglie wavelength is

$$\lambda = \frac{h}{mv} = \frac{h}{mc} = \frac{hc}{mc^2} = \frac{hc}{E}$$

and so the wavelength of any *fast* particle is inversely proportional to its energy. Electrons of energy 1.3 GeV produced in a linear accelerator have energy 2.1×10^{-10} J, and their wavelength is given by

$$\lambda = \frac{hc}{E} = \frac{(6.63 \times 10^{-34}\ \text{J}\cdot\text{sec})(3.00 \times 10^{8}\ \text{m/sec})}{2.1 \times 10^{-10}\ \text{J}} = 0.95 \times 10^{-15}\ \text{m} = 0.95\ \text{F}^*$$

This is one-seventh the radius of a gold nucleus. We can expect the new high-energy machines to show the structure, if any, of protons and neutrons. Even with present energies the resolving power is sufficient to show some detail in the structure of the nuclei, as illustrated in Figure 25-18 which reveals a "skin" surrounding a gold nucleus.

The powerful atom smashers we have discussed are expensive and complicated, but fortunately we are living in a perpetual high-energy laboratory that operates day and night at no cost. The high-energy radiation that is so readily available to us is called cosmic rays, first discovered about 1900 when it was found that air is always a slight conductor of electricity. Since then, cosmic rays have been studied at high and low

*A fermi (F) equals 10^{-15} m, the same as a femtometer (fm).

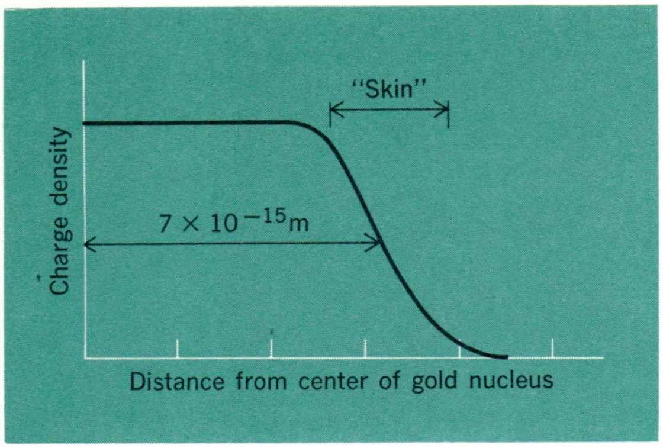

Figure 25-18 Distribution of electric charge in a gold nucleus. (Graph based on high-energy electron scattering experiments of Hofstadter.)

altitudes, with electroscopes, Geiger counters, cloud and bubble chambers, and photographic plates.

The *primary cosmic ray particles,* which strike the earth from outer space, are protons, deuterons, and other positively charged light nuclei. What is believed to be the track of a primary particle is shown in Figure 25-19. The photographic plate was exposed during a high-altitude balloon flight, and it records the stopping of an iron nucleus in the emulsion. The whole track is about 1.5 mm in length. As the iron nucleus traveled through the emulsion, orginally a bare nucleus with 26 positive charges, it picked up electrons and finally came to rest at the right as a neutral atom. The track is thicker at the start because the highly charged nucleus caused more ionization, evidenced by the many side tracks of secondary electrons. Very few of these cosmic ray primaries ever reach sea level. Close to the earth's surface we usually observe a complex mixture of *secondary particles*, which includes electrons, photons, neutrons, protons, and a variety of other particles.

25-7 NUCLEAR ENERGY

The discovery of radioactivity answered one question, but it posed another and gave a tantalizing glimpse of what we now know as nuclear energy. To the scientist of a century ago, the interior of the earth was unreasonably hot. Even near the surface this is observable; the temperature at the bottom of a mine several kilometers deep may be an uncomfortable 60° C. The earth must, therefore, be slowly cooling off as heat is conducted from the hot interior to the cool surface. How slowly? How long has this been going on?

About 1890, Lord Kelvin was able to calculate that the earth is cooling off so rapidly that it must have been a molten liquid about 10^8 years ago. However, geologists using other evidence were sure that the solid crust of the earth is at least 10^9 years old. Someone had to be wrong!

It turned out that Kelvin's work, although quite correct, was incomplete. He had rather naturally assumed that there was no source of heat within the earth; it was merely cooling off from an initial uniformly hot condition. But Becquerel's dis-

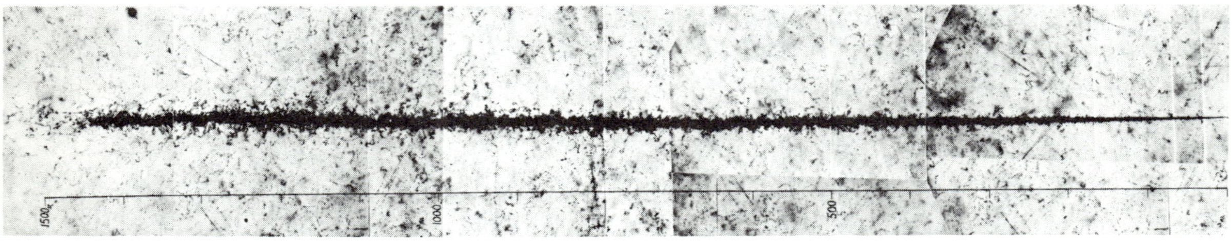

Figure 25-19 Primary cosmic-ray particle. An iron nucleus with twenty-six positive charges enters the photographic emulsion from the left, and the track tapers to a point as the ion captures electrons and finally comes to rest as a neutral iron atom. Total length of path, 1.5 mm.

covery of radioactivity changed all this. As a typical example, the uranium isotope of mass 238 units decays spontaneously by emitting an alpha particle (helium nucleus) of mass 4 units, leaving a thorium nucleus of mass 234 units. The alpha particle flies off with some kinetic energy, and the thorium nucleus also has some kinetic energy as it recoils. In symbols, we write

$$^{238}_{92}\text{U} \longrightarrow {}^{234}_{90}\text{Th} + {}^{4}_{2}\text{He} + \text{energy}$$

The radioactive minerals are therefore a source of thermal energy within the earth; this fact was unknown to Kelvin. The great debate of the late 1800's between the physicists and the geologists was thus settled (in favor of the geologists) by the unexpected discovery of this source of heat which prolonged the cooling-off period of the earth.

The discovery of radioactivity raised a serious problem for physicists. How can a radioactive nucleus supply energy to its fragments without somehow being "wound up" in advance? At first Becquerel thought that a uranium atom might store up energy from sunlight for later release during the decay process, but the rate of decay and energy release was constant even if the uranium was kept for months or years in complete darkness. There seemed to be no way to control the release of nuclear energy, and even the *existence* of nuclear energy seemed to violate the law of conservation of energy.

A piece of the puzzle fell into place when Einstein's theory of relativity appeared in 1905, a decade after the discovery of radioactivity. As we saw in Chapter 5, a body's mass increases when its speed (relative to the observer) increases. Even when a body is at rest it has a great deal of "rest-mass energy" given by Einstein's famous formula $E = mc^2$. The energy is in joules if m is in kilograms and c is the speed of light, 3.00×10^8 m/sec. According to this equation, tremendous amounts of energy are released by the annihilation of small amounts of matter.

Example 25-1 The ink used to print the word "Einstein" has mass about 1 microgram (10^{-9} kg). If its mass could be entirely converted into energy, how high would it lift a 6000-kg bus loaded with 44 students each of mass 50 kg?

$$E = mc^2$$
$$= (10^{-9}\text{ kg})(3 \times 10^8\text{ m/sec})^2 = 9 \times 10^7\text{ J}$$

The bus and students weigh $8200 \times 9.8 = 8.03 \times 10^4$ N. If the available energy is converted into PE in lifting the students, we have $W = Fs$, hence

$$s = \frac{W}{F} = \frac{9 \times 10^7\text{ J}}{8.03 \times 10^4\text{ N}}$$
$$= 1.12 \times 10^3\text{ m}$$
$$= \boxed{1.13\text{ km}}$$

The conversion of a microgram of ink could lift a bus load of students more than 1 km straight up!

The interconversion of mass and energy predicted by relativity theory has been abundantly verified by countless precise experiments in nuclear physics, first by Rutherford in 1921. Let's consider the decay of uranium in more detail. Working only to the nearest integers we saw a mass balance, since 238 = 234 + 4. Precise values of masses of these nuclei show that there has been a *decrease* in total mass:

BEFORE DECAY	
^{238}U	238.0003 u
AFTER DECAY	
^{234}Th	233.9942 u
^{4}He	4.0015
	237.9957
MASS DIFFERENCE	0.0046 u

The mass balance is not exact; 0.0046 unit of mass has been converted into the kinetic energy of the fragments. Becquerel's search for an external source of energy was fruitless; the energy was stored within the uranium nucleus as extra mass.

25-8 FISSION AND FUSION

For many years, scientists have known that a great deal of energy is "locked up" in the mass of nuclei, but they thought there could be no practical way in which man could control its release. Uranium, thorium, radium, and several other "naturally radioactive" elements were known, but even such a man as Rutherford said in 1933, ". . . we cannot control atomic energy to an extent which would be of any value commercially and I believe we are not likely ever to be able to do so." This all changed overnight when the fission of uranium was discovered in 1939.

When an isotope of uranium was bombarded with neutrons, one of the products was barium, an element much less massive than uranium. How could such a light element be formed from uranium? Physicists Lise Meitner (Figure 25-20) and Otto Frisch offered the startling explanation that a fission reaction had been produced.

Isotope ^{235}U breaks up into two large fragments when it absorbs a slow neutron of charge 0 and mass 1 unit. One of many possible reactions is

$$^{235}_{92}\text{U} + ^{1}_{0}\text{n} \longrightarrow ^{90}_{38}\text{Sr} + ^{136}_{54}\text{Xe} + 10^{1}_{0}\text{n}$$

This reaction may be a danger to our survival because the radioactive strontium isotope Sr^{90} is a component of the fallout from a fission bomb. Keeping count of the masses of the substances in the above reaction, we have

BEFORE FISSION	
1 neutron	1.0087 u
^{235}U	234.9934
	236.0021
AFTER FISSION	
^{90}Sr	89.8868 u
^{136}Xe	135.8776
10 neutrons	10.0867
	235.8511
MASS DIFFERENCE	0.1510 u

Since one mass unit is about equal to the mass of a proton or a neutron, 1.67×10^{-27} kg, we find that each separate fission event releases energy:

$$\begin{aligned} E &= mc^2 \\ &= (0.1510)(1.67 \times 10^{-27}\text{ kg})(3 \times 10^{8}\text{ m/sec})^2 \\ &= 2.27 \times 10^{-11}\text{ joule} \end{aligned}$$

Figure 25-20 Lise Meitner (1878–1968), was active in many aspects of nuclear physics. Her work with fission led to the use of nuclear energy.

This is the result of the fission of only *one* atom of ^{235}U, whose mass is $235 \times 1.67 \times 10^{-27}$ kg $= 3.9 \times 10^{-25}$ kg. In this example, the rate of energy release per kilogram of material is enormous:

$$\frac{2.27 \times 10^{-11}\text{ joule}}{3.9 \times 10^{-25}\text{ kg}} = 6 \times 10^{13}\text{ joules per kilogram}$$

As we know, the nuclear fission process can be made self-sustaining by a chain reaction in a nuclear reactor. It is the release of neutrons during fission which makes the chain reaction possible. The number of neutrons per fission is usually less than ten, for in many cases the fission products convert their excess neutrons into protons and emit electrons (beta particles) instead. Nevertheless, more than one neutron is emitted, on the average, and so the process builds up. If the multiplication factor were exactly 2, then one incoming neutron would release two neutrons, which in turn would strike two ^{235}U nuclei and release four neutrons, and so on until an explosion resulted. For controlled release of nuclear power, which is what we want from a nuclear reactor, the multiplication factor is adjusted to be only slightly greater than 1. Uranium 235 is most easily fissioned by a neutron that is moving slowly, for in this way the neutron has more time to interact with the nucleus. Therefore the uranium fuel elements are embedded in a light substance such as water or carbon, so that fast neutrons will be slowed down by collisions before striking other uranium nuclei. The other major feature of a nuclear reactor is a set of control rods made of a material such as cadmium which easily absorbs neutrons. When the rods are lowered into the reactor neutrons are removed before striking a second uranium nucleus, and the chain reaction slows down or stops.

Although nuclear reactors are now being used to generate a fraction of the electric power requirements of the United States, almost 90 percent of the energy that we use still comes to us directly or indirectly from the sun. We eat food that requires sunlight to grow; we burn coal, oil and gasoline that are fossilized remnants of biological organisms. Hydroelectric power utilizes the potential energy of rain clouds, lifted by the sun's energy only to fall into an elevated lake or on a mountainside. We now know that the sun's energy is nuclear, but the process is *fusion* (building up) rather than *fission* (breaking apart). In the sun, hydrogen is being transformed into helium. The process takes place in several steps, and in various ways, but the net result is represented by an equation in which 4 protons and 2 electrons are fused into an alpha particle:

$$^{1}_{1}H + ^{1}_{1}H + ^{1}_{1}H + ^{1}_{1}H + ^{\ 0}_{-1}e + ^{\ 0}_{-1}e \longrightarrow ^{4}_{2}He$$

BEFORE FUSION	
4 protons	4.0291 u
2 electrons	0.0011 u
	4.0302
AFTER FUSION	
1 alpha particle	4.0015 u
MASS DIFFERENCE	0.0287 u

Thus we see that for light elements such as hydrogen it is *fusion* that releases mass (and energy); for heavy elements such as uranium it is *fission* that releases mass (and energy). The sun is almost all hydrogen, and during each second over 4 million tons of the sun's mass is converted into energy as a by-product of the fusion of hydrogen into helium (Section 5-6). On earth the fusion of hydrogen has already been utilized in the H-bomb. Many scientists are working toward the goal of a controlled release of nuclear power by fusion, for which the raw material is as plentiful as the water in the sea. When this goal is achieved, an unlimited source of energy will become available for the benefit of mankind.

SUMMARY

Around 1900, Becquerel and others investigated the phenomenon of **radioactivity,** which presented a great challenge to physicists. The amounts of energy given off were great and seemed to violate the law of conservation of energy. Furthermore, the fact that radioactivity was uninfluenced by outside conditions was very difficult to explain.

Rutherford, in his famous scattering experiments, showed that the atom should be thought of as a small, very heavy, positively charged **nucleus** surrounded by electrons. Radioactivity could be explained as a nuclear phenomenon in terms of this model. The nucleus is composed of protons and neutrons, collectively called **nucleons.** Combinations of nucleons are called **nuclides,** and various nuclides of the same atomic number are called **isotopes.** Some nuclides are stable, while others are unstable and change into other nuclides by the process called radioactive decay. The **half-life** is a measure of the stability of a nuclide.

Nuclear physicists make use of various charged-particle detectors to follow the motions of charged particles and measure their energies, charges, masses, etc. Widely used detectors include the **Geiger-Müller counter,** the **cloud chamber,** the **bubble chamber,** the **spark chamber,** and the **scintillation detector.** High-energy particles that can be used for study occur naturally in **cosmic rays** or radioactivity. Physicists may also use particle accelerators, such as the **Van de Graaff generator,** the **cyclotron,** the **synchrotron,** or the **linear accelerator** as sources of high-energy particles. Particles of higher and higher energies are of great interest because their short de Broglie wavelengths enable them to probe very fine structures, not only nuclei but even elementary particles such as the proton itself.

Through their decay, radioactive minerals are a source of thermal energy within the earth. Some conversion of mass to energy occurs during **fusion** (the building-up of light elements) and **fission** (the breaking-down of heavier elements).

FINAL SUMMARY

Now let's take a backward look at the path you have followed in this course to your present knowledge of modern physics. Early in the course, you investigated the laws of mechanics, as formulated by Newton, which were concerned with how objects move and how changes of motion are related to the forces that act on bodies. You saw how the ideas of energy and momentum led to two conservation laws, which gave a powerful method for solving many problems that would otherwise be very difficult.

You studied the gravitational force as a first example of the interaction between bodies in terms of forces between them, and later saw that the concept of electric charge led to electric and magnetic forces—another way in which bodies can interact. You investigated a way of transmitting energy from place to place by radiation and waves. And the sections dealing with light gave you some background for the problems to be faced in atomic physics.

Finally, you saw how classical physics, as exemplified by the Newtonian mechanics—so useful for bodies of ordinary size moving at ordinary speeds—is inadequate when it comes to investigating the world of the atom. With the quantum ideas of Planck, Bohr, and others leading to a model of the outer atom, it took the elegantly simple experiments of Rutherford and his students to provide an understanding of the nucleus.

Some of the basic tools used by the investigator in atomic and nuclear physics were discussed, with attention focussed on how they are being used today.

As for the future, there are many areas in which some new break-through might be made. A great deal of effort and money is being spent in trying to understand nuclear forces and in trying to understand the relationships between the so-called elementary particles. Solid state physics, low temperature physics, biophysics, and the study of the nature of gravitation are all areas of strong and vital investigation. Physics today is alive and vigorous.

QUESTIONS

1 Explain why the early observations of radioactivity seemed to contradict one of the conservation laws. Is this objection still valid?

2 Compare and contrast alpha, beta, and gamma rays as regards (a) penetrating power; (b) charge; (c) deflection by a magnetic field.

3 What is the approximate ratio of the rest mass of an alpha particle to that of a beta particle?

4 The Rutherford scattering experiment may be described with the phrase "too little, too much." Explain, in terms of the angle of scattering and the numbers of scattered alpha particles.

5 Why are the chemical atomic masses of some elements not very close to integers?

6 Describe three devices for detecting the passage of a charged particle.

7 Does a gamma ray have (a) any mass? (b) any momentum? (c) any charge?

8 In Figure 25-8, when an ionizing particle passes through the counter, does the potential of point A (relative to ground) increase or decrease?

9 What is the evidence for believing that radioactivity is a nuclear process?

10 Explain why the facts of relativity limit the maximum energy that a proton can acquire in a cyclotron. Tell how this limitation can be overcome in the synchrocyclotron and in the synchrotron.

11 Why are high-energy synchrotrons constructed with a large radius?

12 Can you think of any source of useful energy in or on the earth that is not directly nor indirectly derived from solar or nuclear energy?

Note: Refer when needed to Periodic Table and to the table of masses of light nuclides (Appendix, Table 9).

PROBLEMS

GROUP A

1 Identify the elements that have the following isotopes:

$$^{118}_{50}X \qquad ^{3}_{2}X \qquad ^{50}_{22}X \qquad ^{197}_{79}X$$

2 How many neutrons are in each nucleus of Problem 1?

3 The half-life of ^{55}Fe is 3 years. What fraction of a sample of this nuclide remains after 9 years?

4 A sample of radioactive material gives 6000 counts per minute at 1:00 p.m., and gives 1500 counts per minute at 4:00 p.m. What is its half-life?

5 A reactor produces 512 g of a certain unstable nuclide, and only 32 g remains after 4 weeks. What is the half-life of the nuclide?

6 Calculate the time for one semicircular revolution of a proton in a magnetic field of 1 N/A·m, assuming that the particle is moving slowly enough so that its rest mass (Appendix, Table 1) can still be used.

7 Calculate the time for one complete revolution of a very high-energy particle in the synchrotron shown in Figure 25-15. (Radius of path = 122 m.)

8 Calculate the wavelength of the guide waves associated with a particle of energy 10 GeV that is moving almost at the speed of light.

9 A proton makes 10^7 revolutions per second in a cyclotron. What magnetic field is required? (Assume the particle's speed is much less than c, so the mass of the proton in Appendix, Table 1 can still be used.)

10 A proton in a cyclotron gains 0.05 MeV each revolution. (a) How many revolutions must it make to reach a final energy of 3 GeV? (b) About how far will it travel during this time if the average path radius is 2 m?

11 A proton in a cyclotron moves in a spiral path whose maximum radius (at point E in Figure 25-14) is 0.6 m. If the proton at this point has a speed of 2×10^7 m/sec, calculate (a) its energy, in MeV; (b) the vertical magnetic field B needed to supply the necessary centripetal force on the moving proton.

UNIT 7

References and Outside Reading

The purpose of the following list is to make you aware of additional reading that will extend and broaden most of the concepts you have met in this unit. These references should be easily available and are of sufficient variety to meet your individual interest.

Accelerators, by Wilson and Littauer, Science Study Series, Garden City, N.Y., Doubleday Anchor, 1960. A good discussion of important tools of nuclear physics.

"Advances in Optical Lasers," by A. L. Schawlow, in *Scientific American,* July, 1963. The theory of lasers and a discussion of the various types.

"The Conduction of Heat in Solids," by R. L. Sproull, in *Scientific American,* December, 1962. For further insight into the solid state; not only discusses heat conduction, but also electrical conduction.

"The Cosmic Background Radiation and the New Aether Drift" by Richard A. Muller in *Scientific American,* May 1978. Observation of the cosmic radiation left over from the "big bang" shows that our galaxy is moving at high speed.

Crystals and Crystal Growing, by Holden and Singer, Science Study Series, Garden City, N.Y., Doubleday Anchor, 1960. Not only describes the theory of crystals, but many experiments that you can do.

Electrons and Waves and *Quantum Electronics,* by J. R. Pierce, Science Study Series, Garden City, N.Y., Doubleday Anchor, 1966. These two books give an excellent treatment of the applications of modern physics.

"The Evolution of the Physicist's Picture of Nature," by P. A. M. Dirac, in *Scientific American,* May, 1963. An excellent historical and philosophical article that summarizes much of what you have studied in this book.

"Exotic Light Nuclei" by Joseph Cerny and Arthur M. Poskanzer in *Scientific American,* June 1978. Unusual combinations of neutrons and protons can be created in the laboratory. They are unstable, but some last long enough to be studied.

"Fusion Power with Particle Beams" by Gerold Yonas, in *Scientific American,* November 1978. One approach to fusion power: bombarding a pellet of fuel with beams of electrons or ions from high-voltage electric pulses.

"Heavy Leptons" by Martin L. Perl and William T. Kirk in *Scientific American,* March 1978. A new member is added to the group of particles including the electron and the muon.

"Light-Emitting Semiconductors," by F. F. Morehead, Jr., in *Scientific American,* May, 1967. Discusses the possibilities for solid-state lasers.

"Neutrino Astronomy," by P. Morrison, in *Scientific American,* August, 1962. Helpful to understanding particle physics.

The New Age in Physics, by H. Massey, New York, Harper & Row, 1960. Chapters 7 and 8 are devoted to nuclear physics.

"Nuclear Fission," by R. B. Leachman, in *Scientific American,* August, 1965. Still an important aspect of nuclear research.

"Resonance Particles," by R. D. Hill, in *Scientific American,* January, 1963. A summary of some important aspects of the physics of fundamental particles.

Restless Atom, by A. Romer, Science Study Series, Garden City, N.Y., Doubleday Anchor, 1960. A historical account of nuclear physics.

Scientific American, September, 1967. The entire issue is devoted to materials and the solid state.

"Semiconductor Particle-Detectors," by O. M. Bilaniuk, in *Scientific American,* October, 1962. A new technique of great importance to nuclear physics.

"The Structure of Crystal Surfaces," by L. H. Germer, in *Scientific American,* March, 1965. A great pioneer in research describes how this important aspect of the study of solids is carried on.

"The Two-Neutrino Experiment," by L. M. Lederman, in *Scientific American,* March, 1963. An account of a fundamental experiment in particle physics.

APPENDIX

Containing reference tables, study aids, index, and guide to key conceptual terms.

CONTENTS

mportant constants

Constant	Symbol	Value
velocity of light	c	3.00×10^8 m/sec
gravitational constant	G	6.67×10^{-11} N • m²/kg²
rest mass of electron	m_e	9.11×10^{-31} kg
rest mass of proton	m_p	1.67×10^{-27} kg
rest mass of neutron	m_n	1.67×10^{-27} kg
unified atomic mass unit	u	1.66×10^{-27} kg
energy equivalent of 1 u		931.5 MeV
Avogadro number	N_A	6.02×10^{23} mole^{-1}
Joule equivalent	J	4.185 J/cal
absolute zero, 0°K		−273.15°C
volume of 1 mole of ideal gas at STP	V_0	22.4 liters
gas constant per mole	R	8.31 J/mole • K°
charge of electron	e	1.60×10^{-19} C
Faraday constant	F	9.65×10^4 C/g.e.w.
Planck constant	h	6.63×10^{-34} J • sec
mean radius of earth		6.38×10^6 m
mean radius of earth's orbit		1.49×10^{11} m
mean radius of moon		1.74×10^6 m
mean radius of moon's orbit		3.84×10^8 m
length of year		3.16×10^7 sec

TABLE **2** **Important metric prefixes**

Prefix	Abbreviation	Meaning	Typical examples
tera	T	$\times 10^{12}$	1 terayear = 10^{12} years
giga*	G	$\times 10^9$	1 gigahertz (radar frequency) = 10^9 Hz
mega	M	$\times 10^6$	1 megaton
kilo	k	$\times 10^3$	1 kilogram = 1000 g
deci	d	$\times 10^{-1}$	1 decimeter = 0.1 m
centi	c	$\times 10^{-2}$	1 centimeter = 0.01 m
milli	m	$\times 10^{-3}$	1 milliampere = 0.001 A
micro	μ	$\times 10^{-6}$	1 microvolt = 10^{-6} V
nano*	n	$\times 10^{-9}$	1 nanosecond = 10^{-9} sec
pico*	p	$\times 10^{-12}$	1 picofarad = 10^{-12} F
femto	f	$\times 10^{-15}$	1 femtometer (approximate size of a proton) = 10^{-15} m
atto	a	$\times 10^{-18}$	

*In the older literature, certain double prefixes are used: kilomega (kM) for 10^9; millimicro (mμ) for 10^{-9}; micromicro ($\mu\mu$) for 10^{-12}.

TABLE **3** **Greek alphabet**

Α	α	alpha	Η	η	eta	Ν	ν	nu	Τ	τ	tau
Β	β	beta	Θ	θ	theta	Ξ	ξ	xi	Υ	υ	upsilon
Γ	γ	gamma	Ι	ι	iota	Ο	o	omicron	Φ	ϕ	phi
Δ	δ	delta	Κ	κ	kappa	Π	π	pi	Χ	χ	chi
Ε	ϵ	epsilon	Λ	λ	lambda	Ρ	ρ	rho	Ψ	ψ	psi
Ζ	ζ	zeta	Μ	μ	mu	Σ	σ	sigma	Ω	ω	omega

TABLE 4 Equivalents•

Length
1 angstrom (Å) = 10^{-10} m = 10^{-8} cm = 0.1 nm
1 fermi = 1 femtometer (fm) = 10^{-15} m = 10^{-13} cm
1 micrometer (μm) = 10^{-6} m = 10^{-4} cm = 10^{4} Å = 1 micron (μ)
1 nanometer (nm) = 10^{-9} m
1 light-year = 9.461×10^{12} km

Area
1 m^2 = 10^6 mm^2 = 10^4 cm^2 = 10^{-6} km^2
1 barn (b) = 10^{-28} m^2 = 10^{-24} cm^2

Volume
1 liter = 1000 cm^3 = 0.001 m^3
1 m^3 = 1000 liters = 10^6 cm^3

Velocity
1 km/hr = 16.7 m/min = $2.78 \times 10^{-}$[illegible] m/sec
36 km/hr = 10 m/sec

Mass
1 metric ton = 1000 kg = 10^6 g

Force
1 newton (N) = 10^5 dynes
An object of mass 1 kg weighs▪ 9.807 N
An object of weight▪ 1 N has mass 102 g = 0.102 kg

Pressure
1 bar = 10^6 dyn/cm^2 = 10^5 N/m^2
1 atm = 760 torr = 760 mm Hg = 76.0 cm Hg
= 1.013×10^6 dyn/cm^2 = 1.013×10^5 N/m^2 = 1.013 bar = 1013 mbar = 101.3 kPa

Work and energy
1 joule (J) = 10^7 ergs = 0.239 cal
1 cal = 4.184 J
1 kilowatt·hour (kWh) = 3.60×10^6 J
1 electron volt (eV) = 1.60×10^{-19} J

Power
1 watt (W) = 1 J/s

Specific heat
1 cal/g·C° = 4.184 J/g·C° = 4184 J/kg·C°

Latent heat
1 cal/g = 4.184 J/g = 4184 J/kg

Gas constant
R = 8.314 J/mole·C° = 1.99 cal/mole·C° = 0.0821 (atm·liter/mole·C°)

• See also Table 2 for important metric prefixes
▪ All weights are at a point on the surface of the earth where g has the standard value 9.807 m/sec^2.

TABLE 5 Natural trigonometric functions

Angle					*Angle*				
Degree	*Radian*	*Sine*	*Cosine*	*Tangent*	*Degree*	*Radian*	*Sine*	*Cosine*	*Tangent*
0°	0.000	0.000	1.000	0.000					
1°	.018	.018	1.000	.018	46°	0.803	0.719	0.695	1.036
2°	.035	.035	0.999	.035	47°	.820	.731	.682	1.072
3°	.052	.052	.999	.052	48°	.838	.743	.669	1.111
4°	.070	.070	.998	.070	49°	.855	.755	.656	1.150
5°	.087	.087	.996	.088	50°	.873	.766	.643	1.192
6°	.105	.105	.995	.105	51°	.890	.777	.629	1.235
7°	.122	.122	.993	.123	52°	.908	.788	.616	1.280
8°	.140	.139	.990	.141	53°	.925	.799	.602	1.327
9°	.157	.156	.988	.158	54°	.942	.809	.588	1.376
10°	.175	.174	.985	.176	55°	.960	.819	.574	1.428
11°	.192	.191	.982	.194	56°	.977	.829	.559	1.483
12°	.209	.208	.978	.213	57°	.995	.839	.545	1.540
13°	.227	.225	.974	.231	58°	1.012	.848	.530	1.600
14°	.244	.242	.970	.249	59°	1.030	.857	.515	1.664
15°	.262	.259	.966	.268	60°	1.047	.866	.500	1.732
16°	.279	.276	.961	.287	61°	1.065	.875	.485	1.804
17°	.297	.292	.956	.306	62°	1.082	.883	.470	1.881
18°	.314	.309	.951	.325	63°	1.100	.891	.454	1.963
19°	.332	.326	.946	.344	64°	1.117	.899	.438	2.050
20°	.349	.342	.940	.364	65°	1.134	.906	.423	2.145
21°	.367	.358	.934	.384	66°	1.152	.914	.407	2.246
22°	.384	.375	.927	.404	67°	1.169	.921	.391	2.356
23°	.401	.391	.921	.425	68°	1.187	.927	.375	2.475
24°	.419	.407	.914	.445	69°	1.204	.934	.358	2.605
25°	.436	.423	.906	.466	70°	1.222	.940	.342	2.747
26°	.454	.438	.899	.488	71°	1.239	.946	.326	2.904
27°	.471	.454	.891	.510	72°	1.257	.951	.309	3.078
28°	.489	.470	.883	.532	73°	1.274	.956	.292	3.271
29°	.506	.485	.875	.554	74°	1.292	.961	.276	3.487
30°	.524	.500	.866	.577	75°	1.309	.966	.259	3.732
31°	.541	.515	.857	.601	76°	1.326	.970	.242	4.011
32°	.559	.530	.848	.625	77°	1.344	.974	.225	4.331
33°	.576	.545	.839	.649	78°	1.361	.978	.208	4.705
34°	.593	.559	.829	.675	79°	1.379	.982	.191	5.145
35°	.611	.574	.819	.700	80°	1.396	.985	.174	5.671
36°	.628	.588	.809	.727	81°	1.414	.988	.156	6.314
37°	.646	.602	.799	.754	82°	1.431	.990	.139	7.115
38°	.663	.616	.788	.781	83°	1.449	.993	.122	8.144
39°	.681	.629	.777	.810	84°	1.466	.995	.105	9.514
40°	.698	.643	.766	.839	85°	1.484	.996	.087	11.43
41°	.716	.656	.755	.869	86°	1.501	.998	.070	14.30
42°	.733	.669	.743	.900	87°	1.518	.999	.052	19.08
43°	.751	.682	.731	.933	88°	1.536	.999	.035	28.64
44°	.768	.695	.719	.966	89°	1.553	1.000	.018	57.29
45°	.785	.707	.707	1.000	90°	1.571	1.000	.000	∞

TABLE 6 **The elements (alphabetical according to symbol)**

Symbol	Name	Atomic Number	Symbol	Name	Atomic Number	Symbol	Name	Atomic Number
Ac	Actinium	18	Ge	Germanium	32	Pr	Praseodymium	59
Ag	Silver	47	H	Hydrogen	1	Pt	Platinum	78
Al	Aluminum	13	He	Helium	2	Pu	Plutonium	94
Am	Americium	95	Hf	Hafnium	72	Ra	Radium	88
Ar	Argon•	18	Hg	Mercury	80	Rb	Rubidium	37
As	Arsenic	33	Ho	Holmium	67	Re	Rhenium	75
At	Astatine	85	I	Iodine	53	Rh	Rhodium	45
Au	Gold	79	In	Indium	49	Rn	Radon▪	86
B	Boron	5	Ir	Iridium	77	Ru	Ruthenium	44
Ba	Barium	56	K	Potassium	19	S	Sulfur	16
Be	Beryllium	4	Kr	Krypton	36	Sb	Antimony	51
Bi	Bismuth	83	La	Lanthanum	57	Sc	Scandium	21
Bk	Berkelium	97	Li	Lithium	3	Se	Selenium	34
Br	Bromine	35	Lu	Lutetium	71	Si	Silicon	14
C	Carbon	6	Lr	Lawrencium	103	Sm	Samarium	62
Ca	Calcium	20	Mg	Magnesium	12	Sn	Tin	50
Cd	Cadmium	48	Mn	Manganese	25	Sr	Strontium	38
Ce	Cerium	58	Mo	Molybdenum	42	Ta	Tantalum	73
Cf	Californium	98	Mv	Mendelevium	101	Tb	Terbium	65
Cl	Chlorine	17	N	Nitrogen	7	Tc	Technetium	43
Cm	Curium	96	Na	Sodium	11	Te	Tellurium	52
Co	Cobalt	27	Nb	Niobium▪	41	Th	Thorium	90
Cr	Chromium	24	Nd	Neodymium	60	Ti	Titanium	22
Cs	Cesium	55	Ne	Neon	10	Tl	Thallium	81
Cu	Copper	29	Ni	Nickel	28	Tm	Thulium	69
Dy	Dysprosium	66	No	Nobelium	102	U	Uranium	92
Es	Einsteinium	99	Np	Neptunium	93	V	Vanadium	23
Er	Erbium	68	O	Oxygen	8	W	Tungsten▪	74
Eu	Europium	63	Os	Osmium	76	Xe	Xenon	54
F	Fluorine	9	P	Phosphorus	15	Y	Yttrium	39
Fe	Iron	26	Pa	Protactinium	91	Yb	Ytterbium	70
Fm	Fermium	100	Pb	Lead	82	Zn	Zinc	30
Fr	Francium	87	Pd	Palladium	46	Zr	Zirconium	40
Ga	Gallium	31	Pm	Promethium	61	Rf	Rutherfordium	104
Gd	Gadolinium	64	Po	Polonium	84	Ha	Hahnium	105

• Symbol A is sometimes used instead of Ar.
▪ Niobium is also known as columbium; radon is also known as emanation (Em); tungsten is also known as wolfram.

TABLE 7 Periodic table of the elements*

1 **H** 1.0080																	2 **He** 4.0026
3 **Li** 6.941	4 **Be** 9.0122											5 **B** 10.81	6 **C** 12.011	7 **N** 14.0067	8 **O** 15.9994	9 **F** 18.9984	10 **Ne** 20.179
11 **Na** 22.9898	12 **Mg** 24.305											13 **Al** 26.9815	14 **Si** 28.086	15 **P** 30.9738	16 **S** 32.06	17 **Cl** 35.453	18 **Ar** 39.948
19 **K** 39.102	20 **Ca** 40.08	21 **Sc** 44.956	22 **Ti** 47.90	23 **V** 50.941	24 **Cr** 51.996	25 **Mn** 54.9380	26 **Fe** 55.847	27 **Co** 58.9332	28 **Ni** 58.71	29 **Cu** 63.54	30 **Zn** 65.37	31 **Ga** 69.72	32 **Ge** 72.59	33 **As** 74.9216	34 **Se** 78.96	35 **Br** 79.909	36 **Kr** 83.80
37 **Rb** 85.467	38 **Sr** 87.62	39 **Y** 88.906	40 **Zr** 91.22	41 **Nb** 92.906	42 **Mo** 95.94	43 **Tc** (99)	44 **Ru** 101.07	45 **Rh** 102.906	46 **Pd** 106.4	47 **Ag** 107.870	48 **Cd** 112.40	49 **In** 114.82	50 **Sn** 118.69	51 **Sb** 121.75	52 **Te** 127.60	53 **I** 126.9045	54 **Xe** 131.30
55 **Cs** 132.906	56 **Ba** 137.34	57 **La** 138.906	72 **Hf** 178.49	73 **Ta** 180.948	74 **W** 183.85	75 **Re** 186.2	76 **Os** 190.2	77 **Ir** 192.2	78 **Pt** 195.09	79 **Au** 196.967	80 **Hg** 200.59	81 **Tl** 204.37	82 **Pb** 207.2	83 **Bi** 208.981	84 **Po** (210)	85 **At** (210)	86 **Rn** (222)
87 **Fr** (223)	88 **Ra** (226)	89 **Ac** (227)	104 **Rf** (261)	105 **Ha** (262)													

Lanthanide Series

58 **Ce** 140.12	59 **Pr** 140.908	60 **Nd** 144.24	61 **Pm** (147)	62 **Sm** 150.4	63 **Eu** 151.96	64 **Gd** 157.25	65 **Tb** 158.925	66 **Dy** 162.50	67 **Ho** 164.930	68 **Er** 167.26	69 **Tm** 168.934	70 **Yb** 173.04	71 **Lu** 174.97

Actinide Series

90 **Th** (232)	91 **Pa** (231)	92 **U** (238)	93 **Np** (237)	94 **Pu** (242)	95 **Am** (243)	96 **Cm** (248)	97 **Bk** (249)	98 **Cf** (249)	99 **Es** (254)	100 **Fm** (257)	101 **Md** (258)	102 **No** (259)	103 **Lr** (260)

* Atomic masses of stable elements are those adopted in 1969 by the International Union of Pure and Applied Chemistry. For those elements having no stable isotope, the mass number of the "most stable" well-investigated isotope is given in parentheses.

TABLE 8 Latent heats

Substance	Melting point, °C	Latent heat of fusion Δh_f, cal/g or kcal/kg	Boiling point, °C	Latent heat of vaporization Δh_v, cal/g or kcal/kg
oxygen	−218	3.30	−183	51
ammonia	−75	108	−33	327
water	0	80	100	540
mercury	−39	2.8	357	70
lead	327	5.9	1753	206
aluminum	660	90	2450	2720
copper	1083	32	2300	1211
uranium	1133	20	3900	454
tungsten	3410	44	5900	1150

TABLE 9 **Properties of light nuclides**

Symbol	Mass, u•	Abundance or mode of decay•	Half-life•
e	0.000549		
p	1.007277		
$^{1}_{0}n$	1.008665	β^-	11.7 m
$^{1}_{1}H$	1.007825	99.985%	
$^{2}_{1}H$	2.01410	0.015%	
$^{3}_{1}H$	3.01605	β^-	12.3 y
$^{3}_{2}He$	3.01603	0.0001%	
$^{4}_{2}He$	4.002601	~100%	
$^{6}_{2}He$	6.01888	β^-	0.81 s
$^{6}_{3}Li$	6.01513	7.4%	
$^{7}_{3}Li$	7.01601	92.6%	
$^{8}_{3}Li$	8.0225	β^-	0.85 s
$^{9}_{3}Li$	9.0272	β^-	0.17 s
$^{7}_{4}Be$	7.01694	ϵ	53.6 d
$^{9}_{4}Be$	9.01219	100%	
$^{10}_{4}Be$	10.01354	β^-	2.7 My
$^{11}_{4}Be$	11.0217	β^-	13.6 s
$^{8}_{5}B$	8.0246	β^+	0.78 s
$^{10}_{5}B$	10.01294	19.6%	
$^{11}_{5}B$	11.00931	80.4%	
$^{10}_{6}C$	10.0168	β^+	19.0 s
$^{11}_{6}C$	11.01144	β^+	20.5 m
$^{12}_{6}C$	12.00000	98.89%	
$^{13}_{6}C$	13.00335	1.11%	
$^{14}_{6}C$	14.00324	β^-	5770 y
$^{15}_{6}C$	15.01064	β^-	2.25 s
$^{16}_{6}C$	16.0147	β^-	0.74 s
$^{13}_{7}N$	13.00574	β^+	10.0 m
$^{14}_{7}N$	14.00307	99.63%	
$^{15}_{7}N$	15.00011	0.37%	
$^{16}_{7}N$	16.0061	β^-	7.35 s
$^{17}_{7}N$	17.00848	β^-	4.14 s

Symbol	Mass, u	Abundance or mode of decay	Half-life
$^{14}_{8}O$	14.00862	β^+	73 s
$^{15}_{8}O$	15.00306	β^+	2.03 m
$^{16}_{8}O$	15.99491	99.76%	
$^{17}_{8}O$	16.99914	0.04%	
$^{18}_{8}O$	17.99916	0.20%	
$^{19}_{8}O$	19.00355	β^-	29.4 s
$^{20}_{8}O$	20.00402	β^-	14 s
$^{17}_{9}F$	17.00212	β^+	66 s
$^{18}_{9}F$	18.00095	β^+, ϵ	1.87 h
$^{19}_{9}F$	18.99840	100%	
$^{20}_{9}F$	19.99999	β^-	11 s
$^{21}_{9}F$	21.0001	β^-	5 s
$^{18}_{10}Ne$	18.0057	β^+	1.46 s
$^{19}_{10}Ne$	19.00189	β^+	18 s
$^{20}_{10}Ne$	19.99244	90.92%	
$^{21}_{10}Ne$	20.99395	0.26%	
$^{22}_{10}Ne$	21.99138	8.82%	
$^{23}_{10}Ne$	22.99449	β^-	38 s
$^{24}_{10}Ne$	23.99361	β^-	3.38 m
$^{20}_{11}Na$	20.0085	β^+	0.4 s
$^{21}_{11}Na$	20.99773	β^+	23 s
$^{22}_{11}Na$	21.99443	β^+, ϵ	2.58 y
$^{23}_{11}Na$	22.98977	100%	
$^{24}_{11}Na$	23.99096	β^-	15.0 h
$^{25}_{11}Na$	24.9899	β^-	60 s
$^{26}_{11}Na$	25.9917	β^-	1.0 s
$^{23}_{12}Mg$	22.99412	β^+	12 s
$^{24}_{12}Mg$	23.98504	78.7%	
$^{25}_{12}Mg$	24.98584	10.1%	
$^{26}_{12}Mg$	25.98259	11.2%	
$^{27}_{12}Mg$	26.98435	β^-	9.5 m
$^{28}_{12}Mg$	27.98391	β^-	21.3 h

• Except for the first three particles, the masses given are those of the neutral atoms.

• Abbreviations: β^-, negatron decay; β^+, positron decay; ϵ, electron capture; y, year; My, megayear; d, day; h, hour; m, minute; s, second. No nuclides are listed which have half-lives less than 0.1 s.

Density and specific gravity

Perhaps the most obvious property of matter is its density. Aluminum is a "light" metal; gold is a "heavy" metal. We define *density d* as the mass per unit volume of a substance. Thus, since each cubic centimeter of water has a mass of 1 g, its density is 1 g/cm^3. The density of this most familiar substance is expressed in the mks system as 1000 kg/m^3. In Table 10 are given the densities of some representative solids, liquids, and gases. Just as for water, the density of any substance in kg/m^3 is found by multiplying the cgs value by 1000.

For many purposes it is convenient to compare substances with one another. Pure water at 0°C (or at some other standard fixed temperature) is often used as a standard substance, and we define *specific gravity* as a ratio:

$$\text{Specific gravity} = \frac{\text{density of substance}}{\text{density of water}}$$

Any given volume of aluminum weighs 2.70 times as much as the same volume of water (both at 0°C), and its specific gravity is 2.70. Note carefully that specific gravity is a pure number, having no units, since it is a ratio of two similar quantities.• Because of its definition in terms of water, and because water has a density of 1.00 g/cm^3, the specific gravity of any substance is *numerically equal to* the density in cgs units.

•To avoid small numbers, specific gravity for gases is sometimes tabulated relative to a standard gas such as air at 0°C and 1 atmosphere pressure.

TABLE **10** **Density**

Substance	Density g/cm^3	Substance	Density g/cm^3
Solids		**Liquids**	
aluminum	2.70	water	1.00
copper	8.93	gasoline	0.66–0.69
tin	7.29	glycerin	1.26
brass	8.44	mercury	13.6
zinc	7.14	alcohol (ethyl)	0.791
magnesium	1.75	sea water	1.025
iron	7.86	carbon tetrachloride	1.595
lead	11.3	olive oil	0.918
gold	19.3		
uranium	18.7	**Gases•**	
wood	0.35–0.9		
concrete	2.7	air	1.293×10^{-3}
ice	0.917	hydrogen	0.0899×10^{-3}
limestone	2.7	oxygen	1.429×10^{-3}
bone	1.7–2.0	helium	0.1785×10^{-3}
diamond	3.51	carbon dioxide	1.977×10^{-3}
quartz	2.65	tungsten hexafluoride	12.9×10^{-3}
steel	7.8	methane	0.717×10^{-3}

•All at 0° and 1 atm pressure.

TABLE 11 Ground State Energies of the Elements*

IA	IIA	IIIB	IVB	VB	VIB	VIIB	VIII	VIII	VIII	IB	IIB	IIIA	IVA	VA	VIA	VIIA	O
1 H 13.6																	2 He 24.6
3 Li 5.4	4 Be 9.3											5 B 8.3	6 C 11.3	7 N 14.5	8 O 13.6	9 F 17.4	10 Ne 21.6
11 Na 5.1	12 Mg 7.6											13 Al 6.0	14 Si 8.1	15 P 11.0	16 S 10.4	17 Cl 13.0	18 Ar 15.8
19 K 4.3	20 Ca 6.1	21 Sc 6.6	22 Ti 6.8	23 V 6.7	24 Cr 6.8	25 Mn 7.4	26 Fe 7.9	27 Co 7.9	28 Ni 7.6	29 Cu 7.7	30 Zn 9.4	31 Ga 6.0	32 Ge 8.1	33 As 10	34 Se 9.8	35 Br 11.8	36 Kr 14.0
37 Rb 4.2	38 Sr 5.7	39 Y 6.6	40 Zr 7.0	41 Nb 6.8	42 Mo 7.2	43 Tc —	44 Ru 7.5	45 Rh 7.7	46 Pd 8.3	47 Ag 7.6	48 Cd 9.0	49 In 5.8	50 Sn 7.3	51 Sb 8.6	52 Te 9.0	53 I 10.4	54 Xe 12.1
55 Cs 3.9	56 Ba 5.2	57 La 5.6	72 Hf 5.5	73 Ta 6	74 W 8.0	75 Re 7.9	76 Os 8.7	77 Ir 9.2	78 Pt 9.0	79 Au 9.2	80 Hg 10.4	81 Tl 6.1	82 Pb 7.4	83 Bi 8	84 Po —	85 At —	86 Rn 10.7
87 Fr —	88 Ra 5.3	89 Ac —	104 — —														

58 Ce 6.9	59 Pr 5.8	60 Nd 6.3	61 Pm —	62 Sm 5.6	63 Eu 5.7	64 Gd 6.2	65 Tb 6.7	66 Dy 6.8	67 Ho —	68 Er —	69 Tm —	70 Yb 6.2	71 Lu 5.0
90 Th —	91 Pa —	92 U 4	93 Np —	94 Pu —	95 Am —	96 Cm —	97 Bk —	98 Cf —	99 Es —	100 Fm —	101 Md —	102 No —	103 Lr —

*In negative electron volts (number of electron volts below the maximum energy, zero). (See page 437.)

TABLE 12 Atomic Radii*

IA	IIA	IIIB	IVB	VB	VIB	VIIB	VIII	VIII	VIII	IB	IIB	IIIA	IVA	VA	VIA	VIIA	O
H 0.53																	He 0.93
Li 1.52	Be 1.12											B 0.80	C 0.77	N 0.74	O 0.74	F 0.72	Ne 1.12
Na 1.86	Mg 1.60											Al 1.43	Si 1.17	P 1.10	S 1.06	Cl 0.97	Ar 1.54
K 2.31	Ca 1.97	Sc 1.60	Ti 1.46	V 1.31	Cr 1.25	Mn 1.29	Fe 1.26	Co 1.25	Ni 1.24	Cu 1.28	Zn 1.33	Ga 1.22	Ge 1.22	As 1.21	Se 1.17	Br 1.14	Kr 1.69
Rb 2.44	Sr 2.15	Y 1.80	Zr 1.57	Nb 1.41	Mo 1.36	Tc 1.3	Ru 1.33	Rh 1.34	Pd 1.38	Ag 1.44	Cd 1.49	In 1.62	Sn 1.4	Sb 1.41	Te 1.37	I 1.33	Xe 1.9
Cs 2.62	Ba 2.17	La 1.88	Hf 1.57	Ta 1.43	W 1.37	Re 1.37	Os 1.34	Ir 1.35	Pt 1.38	Au 1.44	Hg 1.52	Tl 1.71	Pb 1.75	Bi 1.48	Po 1.4	At 1.4	Rn 2.2
Fr 2.7	Ra 2.20	Ac 2.0	— —														

Ce 1.65	Pr 1.65	Nd 1.64	Pm —	Sm 1.66	Eu 1.65	Gd 1.61	Tb 1.59	Dy 1.59	Ho 1.58	Er 1.57	Tm 1.56	Yb 1.70	Lu 1.56
Th 1.65	Pa —	U 1.42	Np —	Pu —	Am —	Cm —	Bk —	Cf —	Es —	Fm —	Md —	No —	Lr —

* In angstroms

Coefficients of friction

The force of friction, f, is proportional to the force, F, pressing the two sliding surfaces together. The constant fraction is called the coefficient of friction, μ, and is defined by the relation:

$$\mu = \frac{f}{F}$$

TABLE **13**

Surfaces	μ	Surfaces	μ
Wood on wood, dry	0.25 to 0.50	Metal on metal, greased	0.03 to 0.05
Wood on wood, soapy	0.20	Metal on metal, rolling	about 0.002
Metal on metal, dry	0.15 to 0.20	Rubber on concrete, dry	0.60 to 0.70

TABLE **14** **Commonly Used Abbreviations**

ampere	A	electromotive force	emf
ampere-hour	A-h	electron volts	eV
angstrom	Å	electrostatic units	esu
atmospheres	atm	farad	F
atomic mass units	amu	fermi (= 10^{-13} cm)	F
atomic weight	at. wt	frequency modulation	FM
calorie	cal	gram	g
Calorie	Cal	henry	H
Celsius	C	hertz (cycle per second)	Hz
centimeter	cm	hour	h
center of mass	c.m.	joule	J
centimeter-gram-second	cgs	kilohm	kΩ
cosecant	csc	kilojoule	kJ
cosine	cos	kiloliter	kl
cotangent	cot	kilometer	km
coulomb	C	kilovolt	kV
counts per second	counts/sec	kilovolt-ampere	kVA
cubic	cu	kilowatt	kW
cubic centimeter	cc or cm^3	kilowatt-hour	kWh
degrees Celsius (centigrade)	°C	kinetic energy	KE
degrees Kelvin (absolute)	°K	kilocalorie	kcal
direct current	dc	kiloelectron volt	keV
disintegrations per second	dis/sec	kilogram	kg
dyne	dyn	kilogram-meter	kg-m
east	E	limit	lim

cont. on p. 485

mass unit	u
maximum	max
megohm	MΩ
melting point	mp
meter	m
meter-kilogram-second	mks
microampere	μA
microcoulomb	μC
microfarad	μF
micrometer	μm
micron	μ
microsecond	μsec
milliampere	mA
milligram	mg
millihenry	mH
milliliter	ml
millimeter	mm
millimicron	mμ
million electron volts	MeV
million volts	MV
millivolts	mV
minimum	min
minute	min
nanosecond	nsec
newton	N
north	N
ohm	Ω
picofarad	pF
potential difference	PD
root mean square	rms
secant	sec
second	sec or s
south	S
standard temperature and pressure	STP
tangent	tan
ultrahigh frequency	uhf
versus	vs
volt	V
watt	W
west	W

Commonly Used Symbols

In physics as in algebra, symbols are used to represent certain unknown values, variables, or constants. An attempt has been made to establish standardized symbols for items that are frequently used, such as a for acceleration, c for the speed of light in a vacuum, λ for wavelength. However, since the subject of physics developed largely through separate areas of investigation, there has been much duplication of symbols used. Also, because of subtleties in the definition of a certain term, two or more symbols may have approximately the same meaning.

The list that follows is an attempt to show the meanings of the most commonly used symbols in this text. You will note that if the symbol is not a general one, in other words, if it is a special case, the chapters in which it is used are given in the right-hand column.

TABLE **15**

Symbol	Meaning	Chapters
A	area	general
	amplitude	8, 20
	antinode	9
	mass number	25
a	acceleration	general
	dimension, distance	13, 14, 18, 19
	space between adjacent slits	23
a (subscript)	along	7

cont. on p. 486

Symbol	Meaning	Chapters
B	magnetic field strength	general
b	base	general
C	capacitance	general
c	a dimension	13
	specific heat	11
	speed of light in vacuum	general
d	density	15
	distance, separation diameter	16, 21, 22, 24, 2
E	electric field strength	16, 20, 24
	energy	general
$\mathcal{E}$	emf	general
e	charge of an electron	general
F	force (net)	general
	principal focus	21
f	focal length	21
	force of friction	4
	frequency	general
	a certain force	7
f (subscript)	final	2, 3, 4, 6
G	gravitational constant (6.67×10^{-11} N · m²/kg²)	general
g	acceleration due to gravity at earth's surface (9.8 m/sec²)	general
h	height	general
	Planck's constant (6.63×10^{-34} J · sec)	general
h_f	heat of fusion	11
h_v	heat of vaporization	11
I	electric current	general
i	angle of incidence	21
i (subscript)	image	21
	initial	2, 3, 4, 6
K	dielectric constant	16
	Kepler's constant	7
k	a constant	general
L, l	length	general
L	inductance	19
m	magnification	21
	mass	general
	relativistic mass	5
m_o	rest mass	5
m_i	inertial mass	4
m_g	gravitational mass	4
N	node	9
	number of molecules	13
N_A	Avogadro's number (6.02×10^{23} mole⁻¹)	12
n	a number of	general
	index of refraction	21, 23
n (subscript)	in medium	21
o (subscript)	object	21

cont. on p. 487

Symbol	Meaning	Chapters
P	a point	2, 9, 10, 15
	power	15, 17
	pressure	general
P (subscript)	parallel	16, 17
Q	a point	8, 9
	heat	11, 13
Q, q	electric charge	general
R	universal gas constant	12
R, r	electrical resistance	15, 17
r	radius (distance)	general
	angle of reflection	21
	horizontal displacement (within an arc)	8
	separation	14, 15
s	displacement (distance)	general
T	absolute temperature	13
	period	2, 7
	tension	4
	torque	19
T_C	Celsius temperature	10
t	Celsius temperature	10, 11
	time	general
t (subscript)	total	6
V	PD, voltage	15, 16, 17, 24
	volume	general
v	velocity	general
	velocity after collision	6
	velocity before collision	6
W	weight	4, 8, 24
	work	5, 13, 15
w	width of slit	22
	work function, energy	24
	true weight	7
X_C	capacitive reactance	19
X_L	inductive reactance	19
x (sometimes subscript)	horizontal displacement (along x axis)	general
	displacement in SHM	8
y (sometimes subscript)	vertical displacement (along y axis)	general
Z	impedance	19
z (usually subscript)	displacement along z axis	general
α	coefficient of linear expansion	10
	nuclear particle (helium nucleus)	general
β	coefficient of volume expansion	10
	nuclear particle — electron	general
γ	nuclear radiation (photons)	general
Δ	change in	general
μ	coefficient of friction	general
λ	wavelength	general
ρ	resistivity	general
θ	an angle	general
Σ	the sum of	general

INDEX
with Guide to Key Conceptual Terms

Key conceptual terms and the summary page on which they appear are given in heavy type. Italic numbers refer to illustrations. The letter n following a page number refers to a footnote.

ARTWORK: BMA Associates, Inc.; Pencils Portfolio

Cover and Frontispiece: Jerome Kresch from Peter Arnold—Light patterns made by a swinging pendulum

Picture Acknowledgments

Key: (*t*) top; (*b*) bottom; (*c*) center; (*l*) left; (*r*) right

Unit One: *p. xii*, Jet Propulsion Laboratory, California Institute of Technology, NASA; *3*, COMSAT; *4, tl*-Wide World Photos; *tr*-University of California at Berkeley; *bl*-UPI; *c*-courtesy of Bell Laboratories; *br*-Wide World Photos; *5, tl*-Columbia University; *tr*-Tass/Sovfoto; *bl*-Cornell University Photography; *c*-Los Alamos Scientific Laboratory/University of California; *br*-Johnson Publishing Company; *7, t and b, and 8*, Bureau International des Poids et Mesures, Sèvres, France; *9*, National Bureau of Standards/U.S. Department of Commerce; *15*, HBJ Photos; *c*-Fairgate Rule Company; *17*, The Granger Collection.

Unit Two: *p. 26*, Tomas D. W. Friedman/Photo Researchers; *29, l*-from PSSC Physics, 2nd Edition, 1965, D. C. Heath and Co., Lexington, MA.; *r*-Dr. Harold E. Edgerton, Massachusetts Institute of Technology, Cambridge, MA.; *39*, HBJ Photos; *46, t*-from PSSC Physics, 2nd Edition, 1965, D. C. Heath and Co., Lexington, MA.; *58, tl*-Picture Collection, Branch Libraries, The New York Public Library; *r*-Rare Book Division, The New York Public Library, Astor, Lenox and Tilden Foundations; *61, 62*, HBJ Photos; *78*, Vannucci from DeWys; *80*, London Daily Express; *87*, © Karsh, Ottawa/Woodfin Camp and Associates; *94*, Institute of Electrical Engineers, London; *103, t*-from PSSC Physics, 2nd Edition, 1965, D. C. Heath and Co., Lexington, MA.; *108*, Dr. Harold Edgerton, Massachusetts Institute of Technology, Cambridge, MA.; *121*, Adapted from a diagram by The American Museum of Natural History.

Unit Three: *p. 134*, Space and Communications Group/Hughes Aircraft Company; *138*, Kenneth Jewell; *150*, National Ocean Survey; *151, all and 153, t*-HBJ Photos; *167*, from PSSC Physics, 2nd Edition, 1965, D. C. Heath and Co., Lexington, MA.

Unit Four: *p. 172*, Lockheed Missiles and Space Company; *175*, The Granger Collection; *176, t*-Edward Orton, Jr., Ceramic Foundation; *b*-Rhoda Galyn/Photo Researchers; *186*, Burndy Library; *187*, Sanborn C. Brown from an original reprint of "An Experimental Inquiry Concerning the Source of Heat Which is Excited by Friction," Count Rumford. Read before the Royal Society, Jan. 25, 1798; *188, and 189, t*-Burndy Library; *193*, The Westinghouse Air Brake Co.; *195*, HBJ Photos; *196*, Sears Roebuck; *202*, The Granger Collection; *205*, courtesy of The New York Public Library, Astor, Lenox and Tilden Foundations; *214*, UPI; *221*, Cooper-Bessemer Corp.

Unit Five: *p. 232*, Courtesy of RCA *241, t*-Raytheon Education Corp.; *274, tr*-James B. Minnich, Fairchild Hiller Aircraft; *275, b*-from PSSC Physics, 2nd Edition, 1965, D. C. Heath and Co., Lexington, MA.; *296, t, 305, and 310, t*-HBJ Photos; *b*-from PSSC Physics, 2nd Edition, 1965, D. C. Heath and Co., Lexington, MA.; *323, b*-HBJ Photo; *339*, courtesy of Bell Laboratories.

Unit Six: *p. 362*, HBJ Photo; *365*, Helen Faye; *368, t*-from PSSC Physics, 2nd Edition, 1965, D. C. Heath and Co., Lexington, MA.; *369, t*-from PSSC Physics, 2nd Edition, 1965, D. C. Heath and Co., Lexington, MA.; *390, b*-E. Leybold's Nachfolger, courtesy of Klinger Scientific Apparatus Corp.; *391, b*-from PSSC Physics, 2nd Edition, 1965, D. C. Heath and Co., Lexington, MA.; *399, b*-from *Fundamentals of Physical Optics* by F. A. Jenkins and H. E. White, 3rd Edition, © 1957 by McGraw-Hill Book Company, Inc. Used by permission of McGraw-Hill Book Co.; *t*-HBJ Photo; *405*, Bausch and Lomb; *409, tl*-Gaertner Scientific Corp.; *tr*-Optics Technology, Inc.

Unit Seven: *p. 422*, Dr. Erwin Mueller, Department of Physics, Pennsylvania State University; *427*, Eduquip-Macalaster Corp.; *429*, HBJ Photos; *441*, Vision: *Human and Electronic* by Dr. Albert Rose, copyright © 1974 by Plenum Press; *443, tl*-R. E. Lapp; *tr*-E. O. Wollan and C. G. Shull; *455*, D. C. Heath and Co.; *461, tr*-I. K. Bøggild, from *An Atlas of Typical Expansion Chamber Photographs*, by W. Gentner, H. Maier-Leibnitz, and W. Bothe, Pergamon Press. Copyright © 1954. Used by permission of Pergamon Press; *bl*-Dr. D. A. Glaser; *rc*-reprinted by permission of the publisher, from Derek Livesey, *Atomic and Nuclear Physics*, Waltham, MA.: Blaisdell Publishing Company, a Division of Ginn and Company, 1966; *464, bl*-Lawrence Radiation Laboratory; *465*, European Organization for Nuclear Research; *466, b*-Herman Yagoda; *468*, AIP, Niels Bohr Library.

Insert: *following p. 132*; Howard Koslow; *388-1, t*-National Bureau of Standards; *b*-HBJ Photos; *388-2 through 3*, HBJ Art Staff; *388-4*, HBJ Photos; *436-2, tl*-HBJ Art Staff; *tr*-Bethlehem Steel Corp., *bl*-Burndy Library; *436-3, tl, tr*-NASA; *bl*-AT&T; *br*-UPI; *436-4, tr, cr*-HBJ Photo; *br*-Eduquip-Macalaster Corp.; *l*-The Ferdinand Hamburger, Jr. Archives of The Johns Hopkins University; *b*-HBJ Photo; *436-5, t*-HBJ Art Staff; *r*-Danish Information Office; *b*-Bethlehem Steel Corp.; *436-6, tl, tc, b*-HBJ Photos; *tr*-Bell Telephone Laboratories; *436-7*, AT&T Photos; *436-8, tl*-HBJ Art Staff; *cl, bl, r*-Bell Telephone Laboratories.

E
F 4
G 5
H 6
I 7
J 8